Citrus

Citrus
The genus *Citrus*

Edited by

Giovanni Dugo
University of Messina, Italy

and

Angelo Di Giacomo
Stazione Sperimentale per l'Industria delle Essenze e dei Derivati degli Agrumi, Italy

CRC PRESS

Boca Raton   London   New York   Washington, D.C.

FIRST INDIAN REPRINT, 2012

This book contains information obtained from authentic and highly regarded sources. Reprinted material is quoted with permission, and sources are indicated. A wide variety of references are listed. Reasonable efforts have been made to publish reliable data and information, but the author and the publisher cannot assume responsibility for the validity of all materials or for the consequences of their use.

Neither this book nor any part may be reproduced or transmitted in any form or by any means, electronic or mechanical, including photocopying, microfilming, and recording, or by any information storage or retrieval system, without prior permission in writing from the publisher.

Direct all inquiries to CRC Press LLC, 2000 N.W. Corporate Blvd., Boca Raton, Florida 33431.

© 2002 Taylor & Francis Group, LLC CRC Press is an imprint of Taylor & Francis Group

Trademark Notice: Product or corporate names may be trademarks or registered trademarks, and are used only for identification and explanation, without intent to infringe.

Visit the CRC Press Web site at www.crcpress.com

Printed and bound in India by
Replika Press Pvt. Ltd.

ISBN 10 : 0-415-28491-0
ISBN 13 : 978-0-415-28491-2

FOR SALE IN SOUTH ASIA ONLY.

To my sweet Alice,
daughter of my first born, Paola. She joined us in this world during my initial approach to this book and since then she has renewed my enthusiasm, regenerated my energies and happily occupied my thoughts and my time.

Giovanni Dugo

To my beloved wife Carmen,
a life-time companion.

Angelo Di Giacomo

Contents

	List of contributors	ix
	Preface to the series	xi
	Preface	xiii
1	Origin and history FRANCESCO CALABRESE	1
2	Botany: taxonomy, morphology and physiology of fruits, leaves and flowers JESÚS M. ORTIZ	16
3	Soil and cultural practices FRANCESCO CALABRESE	36
4	Pests and diseases VITTORIO LO GIUDICE	49
5	Development of the citrus industry: historical note ANGELO DI GIACOMO	63
6	Flowsheet showing steps in the processing of citrus fruits ANGELO DI GIACOMO	71
7	Citrus juices technology FABIO CRUPI AND GIUSEPPE RISPOLI	77
8	Essential oil production ANGELO DI GIACOMO AND GIOVANNI DI GIACOMO	114
9	Production of bitter orange neroli and petitgrain oils LOUIS PEYRON	148
10	Production of distilled peel oils LUIS HARO-GUZMÁN	153
11	Citrus by-products GIANVINCENZO LICANDRO AND CARLOS E. ODIO	159
12	Advanced analytical techniques for the study of citrus oils LUIGI MONDELLO, GIOVANNI ZAPPIA, PAOLA DUGO AND GIOVANNI DUGO	179

13	Composition of the volatile fraction of cold-pressed citrus peel oils GIOVANNI DUGO, ANTONELLA COTRONEO, ANTONELLA VERZERA AND IVANA BONACCORSI	201
14	The oil composition of less common *Citrus* species BRIAN M. LAWRENCE	318
15	The oxygen heterocyclic compounds of citrus essential oils PAOLA DUGO AND DAVID MCHALE	355
16	Terpeneless and sesquiterpeneless oils DAVID MOYLER	391
17	Composition of distilled oils LUIS HARO-GUZMÁN	402
18	Extracts from the bitter orange flowers (*Citrus aurantium* L.): composition and adulteration LOUIS PEYRON AND IVANA BONACCORSI	413
19	Composition of petitgrain oils GIOVANNI DUGO, LUIGI MONDELLO AND IVANA BONACCORSI	425
20	The chiral compounds of citrus essential oils LUIGI MONDELLO, PAOLA DUGO AND GIOVANNI DUGO	461
21	Adulteration of citrus oils DAVID MCHALE	496
22	Contaminants in citrus essential oils GIACOMO DUGO AND GIUSEPPA DI BELLA	518
23	The market of citrus oils around the world ANGELO DI GIACOMO	532
24	Citrus oils in food and beverages: uses and analyses ENRICO COLOMBO, CLAUDIO GHIZZONI AND DIMITRI CAGNI	539
25	Citrus oils in perfumery and cosmetic products FELIX BUCCELLATO	557
26	Legislation of citrus oils FRIEDRICH GRUNDSCHOBER	568
27	Citrus species and their essential oils in traditional medicine ANTONIO IMBESI AND ANNA DE PASQUALE	577
28	The biological activity of citrus oils GIUSEPPE BISIGNANO AND ANTONELLA SAIJA	602
	Author index *Subject index*	631 638

Contributors

Dr Giuseppa Di Bella – Dipartimento di Chimica organica e biologica, Facoltà di Scienze MM.FF.NN., Università di Messina, Contrada Papardo, 98166 Messina, Italy.

Prof. Giuseppe Bisignano – Dipartimento Farmaco-biologico, Facoltà di Farmacia, Università di Messina, Viale SS. Annunziata, 98168 Messina, Italy.

Dr Ivana Bonaccorsi – Dipartimento Farmaco-chimico, Facoltà di Farmacia, Università di Messina, Viale SS. Annunziata, 98168 Messina, Italy.

Dr Felix Buccellato – Custom Essence, Inc., 53 Veronica Avenue, Somerset, NY 08873, USA.

Dr Dimitri Cagni – Fructamine, via Capitani di Mozzo, 12/16, 24030 Mozzo (Bergamo), Italy.

Prof. Francesco Calabrese – Istituto di Coltivazioni Arboree, Facoltà di Agraria, Università di Palermo, Viale della Scienze, 90168 Palermo, Italy.

Dr Enrico Colombo – Fructamine, via Capitani di Mozzo, 12/16, 24030 Mozzo (Bergamo), Italy.

Prof. Antonella Cotroneo – Dipartimento Farmaco-chimico, Facoltà di Farmacia, Università di Messina, Viale SS. Annunziata, 98168 Messina, Italy.

Dr Fabio Crupi – Consultant, Via Natoli, 79, 98123 Messina, Italy.

Prof. Giacomo Dugo – Dipartimento di Chimica organica e biologica, Facoltà di Scienze MM.FF.NN., Università di Messina, Contrada Papardo, 98166 Messina, Italy.

Dr Prof. Giovanni Dugo – Dipartimento Farmaco-chimico, Facoltà di Farmacia, Università di Messina, Viale SS. Annunziata, 98168 Messina, Italy.

Dr Paola Dugo – Dipartimento di Chimica organica e biologica, Facoltà di Scienze MM.FF.NN., Università di Messina, Contrada Papardo, 98166 Messina, Italy.

Dr Claudio Ghizzoni – Fructamine, via Capitani di Mozzo, 12/16, 24030 Mozzo (Bergamo), Italy.

Prof. Angelo Di Giacomo – Stazione Sperimentale per l'Industria delle Essenze e dei Derivati degli Agrumi, Via Gen. Tommasini, 2, 89127 Reggio Calabria, Italy.

Dr Giovanni Di Giacomo – Citrus Vita s.r.l., Z.I.R. diramazione viaria D, Pace del Mela, 98040 Giammoro (ME), Italy.

Dr Vittorio Lo Giudice – Istituto Sperimentale per l'Agrumicoltura, Corso Savoia, 190, 95024 Acireale (CT), Italy.

Dr Friedrich Grundschober – International Organisation of the Flavor Industry, 49 Square Marie Louise, B-1000 Brussels, Belgium.

Luis Haro-Guzmán – Consultant, Josè Vasconcelos, 105, Col. Jard. Vista Hermosa, 28010 Colima Col., Mexico.

Prof. Antonio Imbesi – Dipartimento Farmaco-biologico, Facoltà di Farmacia, Università di Messina, Viale SS. Annunziata, 98168 Messina, Italy.

Dr Brian M. Lawrence – R.J. Reynolds Tobacco Company, Research and Development Bowman Grey Technical Center, 950 Reynolds Boulevard, Wiston-Salem, NC 27105, U.S.A.

Dr Gianvincenzo Licandro – Citrus Vita s.r.l. Z.I.R., diramazione viaria D, Pace del Mela, 98040 Giammoro (ME), Italy.

Dr David McHale – Consultant, 19 Chantry Hurst, Epsom, Surrey KT18 7BW, UK.

Dr Luigi Mondello – Dipartimento Farmaco-chimico, Facoltà di Farmacia, Università di Messina, Viale SS. Annunziata, 98168 Messina, Italy.

Dr David Moyler – Consultant, Fivewentways Cottages, Stonestreet, Sevenoaks, Kent TN 15 OLR, UK.

Dr Carlos E. Odio – TICO FRUIT s.a., Apartado 207–1000, Pavas, San Josè, Costa Rica.

Prof. Jesús M. Ortiz – Departamento de Biologia Vegetal, Escuela Tecnica Superior de Ingenieros Agronomos, Universidad Politecnica, 28040 Madrid, Spain.

Prof. Anna De Pasquale – Dipartimento Farmaco-biologico, Facoltà di Farmacia, Università di Messina, Viale SS. Annunziata, 98168 Messina, Italy.

Dr Louis Peyron – Consultant, 14 Avenue de l'Oliveraie, F 06130 Grasse, France.

Dr Giuseppe Rispoli – Consultant, Via Luciano Manara, 8, 98123 Messina, Italy.

Prof. Antonella Saija – Dipartimento Farmaco-biologica, Facoltà di Farmacia, Università di Messina, Viale SS. Annunziata, 98168 Messina, Italy.

Prof. Antonella Verzera – Dipartimento Farmaco-chimico, Facoltà di Farmacia, Università di Messina, Viale SS. Annunziata, 98168 Messina, Italy.

Dr Giovanni Zappia – Dipartimento Farmaco-chimico, Facoltà di Farmacia, Università di Messina, Viale SS. Annunziata, 98168 Messina, Italy.

Preface to the series

There is increasing interest in industry, academia and the health sciences in medicinal and aromatic plants. In passing from plant production to the eventual product used by the public, many sciences are involved. This series brings together information which is currently scattered through an ever increasing number of journals. Each volume gives an in-depth look at one plant genus, about which an area specialist has assembled information ranging from the production of the plant to market trends and quality control.

Many industries are involved such as forestry, agriculture, chemical, food, flavour, beverage, pharmaceutical, cosmetic and fragrance. The plant raw materials are roots, rhizomes, bulbs, leaves, stems, barks, wood, flowers, fruits and seeds. These yield gums, resins, essential (volatile) oils, fixed oils, waxes, juices, extracts and spices for medicinal and aromatic purposes. All these commodities are traded worldwide. A dealer's market report for an item may say 'Drought in the country of origin has forced up prices'.

Natural products do not mean safe products and account of this has to be taken by the above industries, which are subject to regulation. For example, a number of plants which are approved for use in medicine must not be used in cosmetic products.

The assessment of safe to use starts with the harvested plant material which has to comply with an official monograph. This may require absence of, or prescribed limits of, radioactive material, heavy metals, aflatoxin, pesticide residue, as well as the required level of active principle. This analytical control is costly and tends to exclude small batches of plant material. Large scale contracted mechanised cultivation with designated seed or plantlets is now preferable.

Today, plant selection is not only for the yield of active principle, but for the plant's ability to overcome disease, climatic stress and the hazards caused by mankind. Such methods as *in vitro* fertilization, meristem cultures and somatic embryogenesis are used. The transfer of sections of DNA is giving rise to controversy in the case of some end-uses of the plant material.

Some suppliers of plant raw material are now able to certify that they are supplying organically-farmed medicinal plants, herbs and spices. The European Union directive (CVO/EU No. 2092/91) details the specifications for the *obligatory* quality controls to be carried out at all stages of production and processing of organic products.

Fascinating plant folklore and ethnopharmacology leads to medicinal potential. Examples are the muscle relaxants based on the arrow poison, curare, from species of *Chondrodendron*, and the anti-malarials derived from species of *Cinchona* and *Artemisia*. The methods of detection of pharmacological activity have become increasingly reliable and specific, frequently involving enzymes in bioassays and avoiding the use of laboratory animals. By using bioassay linked fractionation of crude plant juices or extracts, compounds can be specifically targeted which, for example, inhibit blood platelet aggregation, or have anti-tumour, or anti-viral, or any other required activity. With the assistance of robotic devices, all the members of a genus may be readily screened. However, the plant material must be *fully* authenticated by a specialist.

The medicinal traditions of ancient civilisations such as those of China and India have a large armamentaria of plants in their pharmacopoeias which are used throughout South-East Asia. A similar situation exists in Africa and South America. Thus, a very high percentage of the World's population relies on medicinal and aromatic plants for their medicine. Western medicine is also responding. Already in Germany all medical practitioners have to pass an examination in phytotherapy before being allowed to practise. It is noticeable that throughout Europe and the USA, medical, pharmacy and health related schools are increasingly offering training in phytotherapy.

Multinational pharmaceutical companies have become less enamoured of the single compound magic bullet cure. The high costs of such ventures and the endless competition from 'me too' compounds from rival companies often discourage the attempt. Independent phytomedicine companies have been very strong in Germany. However, by the end of 1995, eleven (almost all) had been acquired by the multinational pharmaceutical firms, acknowledging the lay public's growing demand for phytomedicines in the Western World.

The business of dietary supplements in the Western World has expanded from the health store to the pharmacy. Alternative medicine includes plant-based products. Appropriate measures to ensure the quality, safety and efficacy of these either already exist or are being answered by greater legislative control by such bodies as the Food and Drug Administration of the USA and the recently created European Agency for the Evaluation of Medicinal Products, based in London.

In the USA, the Dietary Supplement and Health Education Act of 1994 recognised the class of phytotherapeutic agents derived from medicinal and aromatic plants. Furthermore, under public pressure, the US Congress set up an Office of Alternative Medicine and this office in 1994 assisted the filing of several Investigational New Drug (IND) applications, required for clinical trials of some Chinese herbal preparations. The significance of these applications was that each Chinese preparation involved several plants and yet was handled as a *single* IND. A demonstration of the contribution to efficacy, of *each* ingredient of *each* plant, was not required. This was a major step forward towards more sensible regulations in regard to phytomedicines.

My thanks are due to the staffs of Harwood Academic Publishers and Taylor & Francis who have made this series possible and especially to the volume editors and their chapter contributors for the authoritative information.

Roland Hardman

Preface

The cultivation of citrus fruits probably originated at least 4000 years ago in the tropical and sub-tropical areas of the Asian continent and the Malaysian archipelago. In more recent times, citrus fruits arrived first in the Mediterranean countries, and subsequently spread to all citrus areas. Certainly citrus plants have exercised a fascination since the most ancient times, because of their evergreen leaves, the perfume of their flowers, the flavour and taste of their fruits. It is noteworthy that in Sicily the luxuriant citrus plantations are traditionally called 'gardens'.

In the past the diffusion of citrus fruits has often been linked with major historical events, such as migration, exploration, and the birth and growth of new civilizations; bearing in mind, of course, that it was possible to cultivate citrus fruits only in suitable climatic conditions, more specifically in the areas between 40° latitude North and 40° latitude South.

The world production of citrus fruits has increased enormously in the last few decades, going from an average of 48 million tons a year in the period 1970–71 to 1978–79, to about 90 million tons in the season 1999–2000. The Northern hemisphere (mainly the USA and the Mediterranean countries) contributes about 75 per cent of the total production, while the Southern hemisphere (mainly Brazil and Argentina) contributes the remaining 25 per cent. Citrus fruits constitute a considerable amount (about 30 per cent) of the world market of the principal fresh fruits, and large quantities are consumed within the areas of production. More than 30 per cent of the total production goes to industrial processing and this figure is even higher for oranges (40 per cent).

Sweet orange (*Citrus sinensis* L. Osbeck), bitter orange (*Citrus aurantium* L.), lemon (*Citrus limon* L. Burm.), mandarin (*Citrus reticulata* Blanco), bergamot (*Citrus bergamia* Risso), grapefruit (*Citrus paradisi* Mcfadyen), acid Key lime (*Citrus aurantifolia* Swingle) and acid Persian lime (*Citrus latifolia* Tanaka) are the citrus species used for industrial processing. Essential oil and juice are the principal products of such processing. Essential oils are usually cold-extracted with mechanical systems which operate both on the whole fruits and on the peel after extraction of the juice; peel Key lime essential oil is obtained mainly by distillation. Neroli oil is obtained by distillation of the flowers from some citrus species, mainly bitter orange, while the so-called petitgrain oils are obtained by distillation of citrus leaves and small branches. Distillation is also used for the recovery of secondary peel essential oils. In the next few years it is likely that new technologies, such as systems which use supercritical fluids, will be introduced for the extraction and fractionation of essential oils.

The process for the preparation of the juice can involve the following different stages: extraction, finishing, deoiling, deaeration, pasteurization, concentration, and, if necessary, drying. The use of advanced technologies is becoming more and more frequent in this field. The residues of primary citrus processing also produce numerous derivatives (pectins, flavonoids, cattle feed, brined and candied peels, wine, vinegar, etc.).

The chemistry of essential oils has made considerable progress in recent times: the introduction of advanced instrumental techniques, such as HRGC, HRGC/MS, HRGC/FTIR, HPLC, HPLC/MS, HPLC-HRGC/MS, HRGC-HRGC, HRGC-HRGC/MS, has made it possible to confirm the presence of costituents whose identity was previously unconfirmed, and to identify and define the structure of many new components, some of which are present in very small amounts, but which are nonetheless extremely important for the characterization of the oils.

The main applications of essential oils are as flavouring for alcoholic and soft drinks, food and pharmaceutical products and as fragrances for cosmetics and beauty products. They are also widely employed for their therapeutic properties (for example in aromatherapy and in antiseptics). Today the potential application of essential oils are even wider because they are 'safer' from an environmental point of view than petrochemical products, which they can sometimes replace.

The vast amount of literature on citrus bears witness to the great technical and scientific efforts which have been and are being made by researchers working in different fields, around the world (agronomists, chemists, biochemists, technologists).

All aspects regarding toxicity of essential oils are part of the scientific aims of IFRA (International Fragrance Association) and of IOFI (International Organization of the Flavour Industry), because of the lack of specific regulations. The latter organization has produced a Code of Practice that is continuously updated.

From what is reported above, citrus science is extremely wide ranging, covering numerous areas of interest. This brought us to notice how useful would have been for our work, but also for others', to have a manual, where all the different information, necessary to all who work with citrus and their derivatives, could be found. In order to accomplish such a project we have involved numerous specialists from different disciplines, to help us build this volume. We hope that we have accomplished our goal in a satisfactory manner, including up-to-date research and personal expertise on citrus and their derivatives.

1 Origin and history

Francesco Calabrese

INTRODUCTION

There is no doubt that the original genetic pool of the citrus plants originated in South-Eastern Asia. According to R.W. Scora, *Citrus maxima* (*C. grandis*), pummelo, *Citrus medica*, citron, *Citrus reticulata*, mandarin and similar, *Citrus halimii*, a recently discovered taxom, are the parent species of all the citrus known today. *Citrus maxima* was probably the first ancestor. It originated in Malaysia and Malay archipelago. *Citrus medica* originated in India and *Citrus reticulata* in China, *Citrus halimii* in Thailand and Malaya. All other species derive from cross-pollination between these biotypes:

- The common orange (*Citrus sinensis*) and the sour orange (*Citrus aurantium*) are considered hybrids of the pummelo with some mandarin.
- Limes (*Citrus aurantifolia*) and lemon (*Citrus lemon*) originated through the hybridisation of citron with some primitive Papeda (wild species from Asia of no commercial value).
- The kumquat (*Fortunella spp*) comes from mandarin.
- The grapefruit (*Citrus paradisi*) is a hybrid of pummelo.

Each species perpetuates its original characteristics by means of a biological mechanism (apogamy or nucellar embriony), which allows plants to produce embryos in seeds without fertilising. These embryos derive from cells of the mother plant and are identical to it. This is why it is possible to distinguish citrus species after millennia. There are monoembrionic (only zygotic embryos) or poliembrionic (mainly or only nucellar embryos) *Citrus* species.

Citrus fruits spread from the Asian areas to other regions following the paths of civilisation.

CITRUS WORLDWIDE DIFFUSION FROM ANTIQUITY TO THE END OF THE FIRST MILLENNIUM

China

China has been one of the main cultural centers of the world for at least three millennia.

Citrus grow in the southern areas (Hunan, Kwangsi, Kwangtung, Szechaw, Fukien, Kweichow) where there is no frost. The first traces of citrus fruit in China are reported in the *'Shih Ching'* (*'Book of odes'*), written during the Chou dynasty, which ruled China from 1027–256 BC, *'Chu'* or *'Ku'* are the names given to kumquat (*Fortunella spp*) and some other small-sized mandarins, *'Yu'* refers to pummelo and Yuzu (*Citrus junos*). Pummelo is a very popular fruit in the southern provinces even today. The more common varieties are called *'Yu'*, *'Laun'*, *'Pao'*, *'Buntan'*. Yuzu fruit is the same size as a mandarin, is yellow-skinned and of no commercial value. It is sometimes used as a rootstock.

At the end of the Chou dynasty for the first time the poet Sung Yu mentions *Poncirus* as *'a favorite tree of birds to build their nest'*. *Poncirus* is a close genus to *Citrus*, which produces an inedible, rough, seedy, yellow-skinned fruit. Its leaves are tricomposed and caducens in winter. The plant is bushy and spiny. Sung Yu calls it *'Chih'*.

During the Han dynasty (202 BC–220 AD) the flourishing of *'Fu'* literature (*'Literature of kingdom'*) reflected the splendour of the period. The works were collected in the so-called *'Imperial Conservatory'*. One of the writers was Ssu Hsiang-Ju, who died in 118 AD In his prose-poem he speaks of *'Cheng'* and *'Huang Kan'*. The first fruit is the famous sour orange (*Citrus aurantium*), nowadays used as rootstock and commercially grown for candies and marmalade. *'Huang Kan'* refers to a yellow mandarin, which is probably the common orange (*Citrus sinensis*). The linguistic confusion between orange and mandarin is due to the fact that in China there is no distinction between these two species even today.

During the kingdom of Han Wu Ti (The Martial emperor, 140–87 BC) there was a dignitary in charge of collecting the tributes of *'Kan'* due to the Emperor. *'Kan'* is the name used to indicate a big mandarin. The same term is reported by the contemporary Tung-Feng-So. It is also listed among several plants during the Han Wu Ti government. The Kan-mandarin was successfully planted in the Hunan province by the governor of Lung-Yang His.

From 265 AD to 420 AD China was ruled by the Chin (*Tsin*) dynasty. This period politically, economically and culturally corresponds to the western Middle Ages. Records regarding citrus are numerous. Citron (*Citrus medica*) is for the first time mentioned by Chi Han in his 'Nan fang ts'ao mu chuang' (304 AD) using the term *'Kuo Han'*. This species is still very important in China and Japan, where the *'Fingered citron'* is called *'the hand of Buddha'*, and considered a sort of talisman.

In the third and fourth centuries AD several herbarium books list some citrus species grown south of the Yangtze river.

The first monograph on citrus plants was written by a Chinese writer in 1178 AD, during the Sung dynasty. In this book, 27 varieties of citrus are described (mandarins, sour oranges, pummelos, limes, kumquats, *Poncirus*) as well as their propagation and cultivation techniques. The author also speaks of pests and curative effects of citrus fruit on health.

Apparently, there is no specific mention of the common orange (*Citrus sinensis*) in the ancient Chinese literature. Indeed, many citrologists agree that in some cases the term *'Kan'* corresponds to the orange. This, in fact, is a generic name for all citrus in China. Yellow Kan (Hung Kan), as we have said, refers to the common orange (Tolkowski, 1938; Cooper, 1989).

India

India is another Asian country where citrus plants have been known since ancient times. Citron (or lemon) is mentioned in the *'Wajasaney Samhita'*, a collection of sacred Brahma texts written in Sanskrit before 800 BC. *'Jambila'* and *'Jambira'* are the names used for this citrus fruit. Citrons and lemons are very popular in India, where they grow wild in the southern part of the Himalayas and Assam. A probable hybrid between citron and orange (*Citrus limonimedica*) is called '*Vijapura*' in Sanskrit. *Citrus jambhiri* ('*Jamir*', '*Jamiri*', '*Jambhiri*', '*Jambira*'), nowadays known as '*rough lemon*', is another wild Indian species closely related to lemon and citron.

There are no ancient Sanskrit references to oranges, pummelos or mandarins. Most historians believe that they were introduced later from China. In 1100 AD the orange was first mentioned (*Citrus sinensis*) in a medical book, titled '*Charaka samhita*', under the name of '*Naranga*' and '*Airvata*'. In this book citron is called '*Matulungaka*', and lemon '*Jambira*'. In the '*Amarakosha*' Sanskrit dictionary of the eighth century AD, orange is listed as '*Nagaranga*'. The word '*naranghi*' refers to both sour and common oranges in a medical text titled '*Mandanpal nighunt*', published in 1411 AD.

Zeher-ed-din Muhammed Baber, a famous king of Hindustan, wrote his memories in 1519 AD, and included information on the citrus trees under cultivation in the northern part of India. Sour orange was called '*Naranji*' (Arab term), lemon '*Limoo*', and citron '*taranj*' or '*Baleng*'. '*Kilkil*' is the term used for lime and '*Jambiri*' for the rough lemon. Baber also mentions other citrus trees such as '*Kirneh*', '*Amilbed*', '*Sadaphal*', '*Amratpal*', '*Kamilah*', '*Samterech*' (Gallesio, 1811; Tolkowski, 1938).

Japan

The first Japanese books were written during the '*Nara*' period (710–794 AD). These books are the '*Kojiki*' (Report on ancient history), published in 712, and the '*Nihon Shoki*' (Report on Japan) in 720. In the seventh century lyric poetry flourished in Japan. The best example is the '*Manyoshu*' (a collection of 100 sheets), a poetic anthology probably written by only one poet, Otomo Yakamochi. In this text we find references to a citrus tree, called '*Ch'u*', which corresponds to the '*Tachibana mandarin*'. The same term, mixed with '*kan*', means generically '*citrus-fruit*'. In '*Manyoshu*' the sour orange is also mentioned.

Around the year 1000 AD (Heian period) the writer Musaraki wrote a novel mentioning oranges and lemons.

The most popular Japanese mandarin-type fruit was introduced from China only in 1500 AD. It is the famous '*Satsuma mandarin*', which today produces about 35,00,000 tons of fruit yearly (Tolkowski, 1938).

Mesopotamia

Assyria and Babylon ruled the large region between the rivers Tigris and Euphrates until 539 BC, when they were defeated by the Persians. The Babylonians were talented farmers and created an ingenious system of canals for irrigation. We are sure that they grew some citrus trees since a few citrus seeds came to light during the excavations in the town of Nippur. The Assyrians were used to bringing back animals and plants from abroad. They left their history on terracotta tablets, including one with a list of plants known at that time. Among them there is a plant called '*iltakku*' which many historians believe to be citron.

Media and Persia

Greek writers tell Persian history from 549 BC, when Cirus the Great defeated the Medians. At that time the Persian influence extended from India to Egypt and Palestine.

When in 327 BC Alexander the Great defeated this wide area, Greek botanists found a citrus species under cultivation. This tree was the citron, called *'the fruit of Persia'* or *'the fruit of Media'*. These terms have been used for a long time in Greek and Latin literature (Tolkowski, 1938).

The Jews and Palestine

The Jews knew of citron before Christ's coming. They were slaves in Egypt until the thirteenth century BC. From the twelfth century to 597 BC they lived in Palestine. In this year, and later, in 586 and 582 BC, they were taken away to Babylon as slaves. In 539 BC Cirus the Great allowed them to return to Palestine, where they lived until Pompeus defeated them and transformed Palestine into a Roman province.

Some scholars believe the Jews knew of the citron during their stay in Egypt and, according to them, the *'Peri ets hadar'* (*'The fruit of the most beautiful tree'*) mentioned in Leviticus was indeed the citron. According to other historians the Jews discovered the citron during their captivity in Babylon.

In October of every year there is still the celebration of the Tabernacle (*'Sukkoth'*) dating back to pre-Christian times. Citron fruits with branches of palm, myrtle and willow are taken into the Temple. Evidence of citron used as a religious symbol can be seen on a coin of Simon Maccabeus (136 BC), and Joseph Flavius, a writer of Jewish tradition (37–100 AD), speaks of citrons during the rites of Tabernacle. The Judean cult of using of citrons can been found in the late chronicles (Tolkowski, 1938; Calabrese, 1998).

Greece

Hesperidium is the scientific name of the citrus fruit, containing juicy segments and covered by a peel rich in aromatic oils. The term derives from the last mythologic labour of Heracles, who was obliged to capture the 'golden fruit' from the garden of the Hesperides, the three daughters of Hesperis: Aretusa, Hyperetusa and Aegle. The legend is revealed in two Greek temples, 'the Treasure of the Athenians' at Delphi, and the temple of Zeus at Olympia. The story is also narrated in a mosaic at the Roman 'Villa del Casale' in Piazza Armerina (Sicily).

Citron was called 'Fruit of Persia' or 'Fruit of Media' by Theophrastus of Eresus (372–287 BC), who describes its spiny branches and inedible fruit. Fruit and leaves were used to perfume clothes and protect them against moths. The fruit was also an antidote to certain poisons and useful in preventing bad breath. Theophrastus also wrote about the citron propagation and planting of citron.

Dioscorides was a famous Greek physician of the first century AD who wrote a book, 'Medical Materials', translated into Italian by Matthioli in 1544. In his book citron is called *'kedròmela'*. He also says that Romans called it *'kitria'*. Citron is also referred to by the philosopher Plutarc (46–120). He was the first to say that the tender and white inside part could be eaten. Galenus (130–200) and his contemporary Nicandrus of Kalophone were Greek doctors who described the medical properties of the citron.

Atheneus of Alexandria in Egypt, who lived at the beginning of the modern era, wrote a fifteen-chapter book, the *'Deipnosophistae'* (*'wise men in conversation'*), where in there is a discussion about the origin of the citron fruit. One of the conveners quotes

'*The Beotian*' of Antiphanes (408–304 BC), where it is said that citron was introduced into Athens from the kingdom of Persia. Further information on the citron can be found in the '*Papyri Oxyrrhichus*' (third century AD), as well as in the '*Geoponica*' of Florentinus (third century AD), where four chapters explain planting care, protection, propagation and conservation of the citron fruit. He was the first to speak of special coverings for citron plants during winter.

All Greek information is only about the citron fruit, but we must be aware that there are few differences between citron, lemon and lime. The fruit of these species is yellow-skinned and acid. This is why the Greek writers did not make any distinctions between them. But we are sure that they at least knew the lemon, since it is sculptured in a '*Horn of plenty*' from the Hellenistic period and kept in the Archaeological Museum in Athens (Gallesio, 1811; Tolkowski, 1938; Calabrese, 1998).

Rome

The Romans knew of citron, lemon and lime. They called the citron fruit '*malus medica*' but later changed this term into '*citrus*' which was also used for the cedar. Indeed, '*citrus*' was used for all yellow-skinned citrus fruit. The citron was probably introduced into Italy by the Jews who lived in large communities near Rome before the Modern era. In fact the fruit is represented in some Jewish catacombs and synagogues.

The first reference to the citron fruit in Latin literature is found in Cloanzius Verus (quoted in Macrobius) (second century BC) under '*citreum*'. '*Citrea*', which is the plural form of '*citreum*', was used by the botanist Oppius one century later. The great poet Virgil (70–19 BC) gave a poetic description of the citron fruit ('*The fruit of Media*') and of its properties. A century later, in his '*Historia naturalis*' Pliny called the citron '*fruit of Assyria*' and '*fruit of Media*'. He reported all the information given by Theophrastus. At the time of Nero the citron is mentioned by Petronius in his famous '*Satyricon*', where he described the banquet held in the house of Trimalcion. Scribonius Largus, a doctor in the first century AD, suggested the citron fruit as a remedy for gout. Gargilius Martial, another third century physician, wrote a medical book indicating a citron syrup to stop a cough. In 301 AD citron ('*Citrus*') was listed in the edict of Diocletian which fixed the prices of several products. The first important culinary use of the citron fruit is described in the '*De re coquinaria*' by the celebrated gastronome Apicius Caelius (second century AD).

Rutilius Taurus Aemilianus Palladius was an agronomist from the third or fourth century AD. He wrote a book on agriculture, '*Agriculturae Opus*' with a chapter '*De citreo*' about citron, grown by the author in his garden in Naples. He described the propagation, cultivation techniques, pruning and protection of citron. He repeated concepts already known at that time.

Citron, lemon, lime fruit can clearly be seen in several mosaics, frescos, sculptures in Rome, Carthage, Sicily, Numidia (Tunisie–Algerie), Pompeii, Spain (Figures 1.1 and 1.2). The Romans probably did not know sour or common orange. As far as we know, there is only one fresco showing two round reddish fruits rather like oranges, near a lemon and a citron fruit. This fresco is in the National Museum in Rome and comes from the ruins of Tusculum (about 100 AD) (Gallesio, 1811; Risso and Poiteau, 1818; Tolkowski, 1938; Calabrese, 1998).

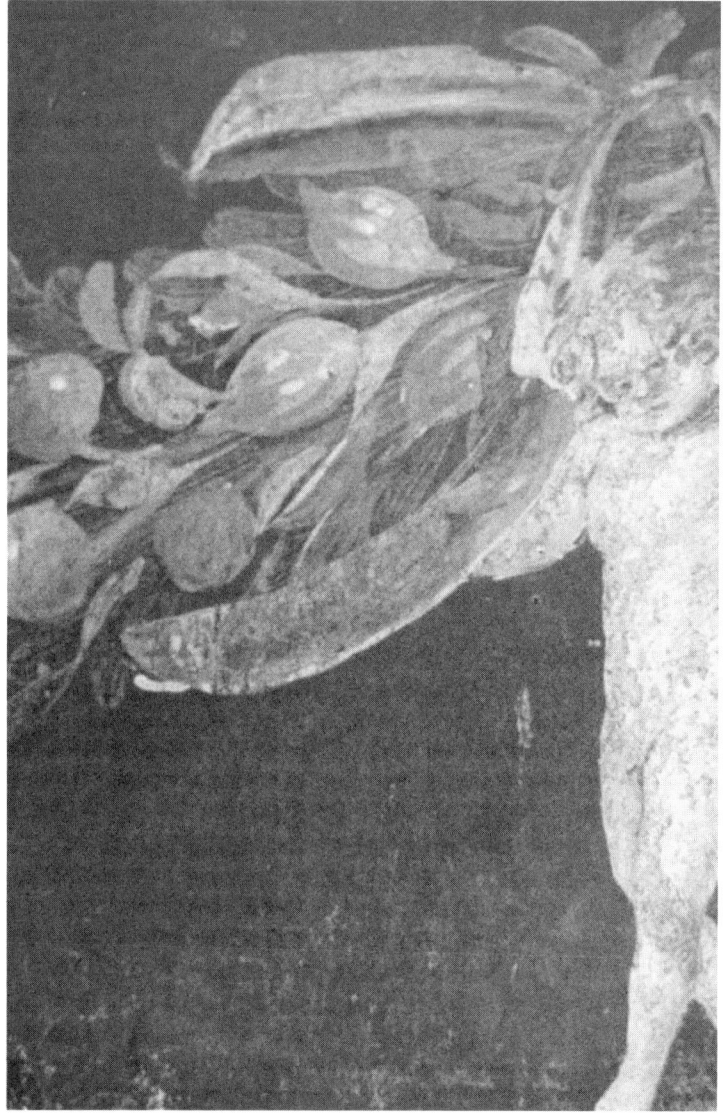

Figure 1.1 Lemons in a fresco from Pompeii (Arch. Mus. Naples).

CITRUS WORLDWIDE DIFFUSION FROM THE MIDDLE AGES TO THE NINETEENTH CENTURY

The Arabs

As far as we know, the Arabs occupied the territories of the Mediterranean sea left free by the Romans. They developed agriculture in the territories of Northern Africa and

Figure 1.2 Limes in a mosaic of the Villa del Casale (Piazza Armerina–Sicily).

Southern Europe by introducing new plants and more sophisticated cultural techniques. They adorned their gardens and mosques with the perfumed blossom of sour orange. This citrus species had an exceptional decorative value for them. They called all citrus species *'utruj'*, *'trunj'*, *'turunj'*. The terms *'limun'* and *'limuna'* were used for the lemon, *'limu'* for the lime, *'naranji'* for the sour orange. There is no mention of the common orange in ancient Arab literature (Tolkowski, 1938; Calabrese, 1998).

Ibn-el-Wahshya (904), Ibn Suleyman (tenth century), Al-Masudi (tenth century), Al-Istakhart (tenth century), Abn Ali al-Husayn Ibn Abdallah Ibn Sina (eleventh century), Abul Hassan al Muchtar Ibn Botlal (eleventh century), Ibn Jamiya (eleventh century), Muvaffaq ed-Din Abd el Latif ben Yusuf (twelfth century) all speak of the lemon, lime, citron, and sour orange in their books and treatises. A very important thirty-four-chapter treatise on agriculture was written by Ibn al-Awwâm, an Arab from Seville. Chapter seven deals with the trees in Spain, and describes the citron tree (with acid and sweet fruit), the lemon, the sour orange and the pummelo (*'bastanbon'* or *'zamboa'* or *'the fruit of Adam'*), their culture and fruit preservation. As far as we know, he was the first to speak of using the oil from the peel of the sour orange as a scent. The book declares that if a menstruating woman touches a citrus tree it will loose its leaves and fruit. Ibn el-Beithârvis (twelfth century) was another botanist and doctor who gave a clear and complete explanation on how to extract the essential oils from the peel. He describes three methods. Arab poets dedicated splendid lines to the citrus fruit. Here is an example taken from a poem by Abd Ar-Ramân (a Sicilian–Arab poet):

The oranges of the island are like flames
shining between emerald branches

*and the lemon reflects the pallor
of a lover who has spent the night in tears
for the pain of the separation*

The Late Middle Ages and the Renaissance

The Crusaders conquered Jerusalem in 1099. From then until the mid-fifteenth century they occupied part of Palestine, encountering a series of victories and defeats. Jacques de Vitry, a crusader, wrote a book on the history of Jerusalem in Latin and described the citrus species grown in Palestine. The author discusses pummelo, lemon and bitter orange. He called the bitter-orange fruit *'orenges'*.

In Europe the first mention of the common or sweet orange was found in a historical book by Hugo Falcando, who lived in Sicily from 1154 to 1169. In his historical book on the siege of Palermo, he stated that the orange fruits were full of a sweet juice. Documents that mention the presence of orange and citron trees in Italy in the twelfth, thirteenth and fourteenth centuries have been found in Tuscany, Italian Riviera and Marches. The term *'arangias'* was used by Blondus Flavius in a decription (thirteenth century) of the citrus in Amalfi and Naples.

At the beginning of the fourteenth century Piero de' Crescenzi wrote a book on agriculture which was translated into several languages. A chapter dealt with citrus trees and their cultivation and the author described the *'closed place'* to be used for growing trees under protection (a sort of green house). During the same period a medical book by Matteus Silvaticus told of the four citrus fruits known at that time: citron, sour orange, lemon, and lime. Nicolò Speciale (fifteenth century), who wrote a book on the siege of Palermo, reported that sour orange was grown in Sicily and its fruit were called *'arangias'* by the Sicilians.

At the end of the Middle Ages the citron, lemon, lime and sour orange were widely grown in Italy and in other European countries (mainly Spain). Their fruits were products for trade, as is clearly described by Francesco Balducci–Pergolotti (mid 1300s), who wrote an informative book for merchants. Several recipes using citrus fruits are also described in Italy and Spain (Volkamer, 1714; Tolkowski, 1938; Calabrese, 1998).

The introduction of the common orange into Europe

We know from the diaries of Vasco de Gama, the first man to round the Cape of Good Hope, of Gaspar Corrêa and of the Italian Andrea Corsali that Portuguese knew of the sweet or common orange in Africa, India, China. The *'orange of China'*, meaning the sweet orange, was soon changed into *'Portuguese orange'*, a term that was used in Mediterranean countries until the beginning of the last century. Valmont de Bomare was the first writer to mention that the *'orange of China'* was grown in Lisbon in 1520. The sweet orange culture spread quickly in Portugal. In 1550 they started directly exporting them to Spain and in 1610 to other European countries. The places of production were along the River Tago, at Sintra, Pavoas, Collaras and Ribeira de Barquarena.

Gallesio, Tolkoski and other scholars, on the contrary, are convinced that Genoese were the first to introduce the common orange into the Mediterranean basin. At the end of the thirteenth century the Genoeses traded widely throughout the Mediterranean. They were used to meeting caravans coming from the Far East and exchanging merchandise with them. This should have been the route into Europe for the common

Figure 1.3 'Orangerie' at Cheverny (France).

orange. In order to support Gallesio's theory there is evidence that the sweet orange was mentioned by some Italian writers, who lived in the early 1500s (Antonio Venuto, Leandro Alberti, Pietro Mattioli), as having been grown in Italy for a long time. It is easy to imagine that the common orange was present in Italy before the Portuguese introduced it. We have previously mentioned that Hugo Falcando who lived in the twelfth century told about the sweet orange grown in Sicily.

There is no doubt that at the beginning of seventeenth century the sweet orange was widely grown both in Italy and in Spain (Gallesio, 1811; Tolkowski, 1938; Calabrese, 1998).

Renaissance and Post-Renaissance

During the period from the fifteenth to the seventeenth century, Tuscany was ruled by the Medici family. Lorenzo *'The Magnificent'* was considered the most famous of the Renaissance nobility. During this period Florence became the cultural capital of Europe. Arts were flourishing. The gardens of the aristocratic villas were enlarged and enriched with botanical collections and citrus trees were given a special place among them.

Since the climate of Tuscany in winter was too cold for these exotic plants, gardeners had to grow citrus trees in terracotta pots and protect them in warm structures, forebearers of modern greenhouses. Similar houses were also built by many aristocratic families in other regions of Italy (Rome, Naples, Liguria, Garda) and abroad, in France (*'orangeries'*) (Figure 1.3), Germany, Belgium and Holland. The famous *'orangeries'* of Versailles can still be visited today.

Such interest in these beautiful plants led to the publication of the first poem on citrus written by a European poet, Jovianus Pontanus (1426–1503). The book is of low lyrical value and is of no scientific interest. At the same time Antonio Venuto, from Noto (Sicily), published his '*De agricultura opusculum*', containing the precepts for the successful growth of citrus trees. Other writers who dealt with the culture and protection of citrus trees were Porta (fifteenth century), Nicolò Pieranzoni (1510), Augustin Gallo (1560). Alonso de Herrera, a Spanish agronomist wrote '*Obra de Agricultura*' in 1513, which described citrus species such as citron, lemon, orange, pummelo and their culture. Another Spanish writer, Nicolas Monardes, was a physician in Seville. He wrote a book describing the medical properties of citrus fruits. Orange plant and fruit are dealt with in the '*Agricultura de Jardines*', published by the Spanish agronomist Gregorio de Los Rios in 1592. In the '*Théâtre d'Agriculture et ménage des champs*' (around 1600) the French agronomist Olivier de Serres described the construction of greenhouses to protect citrus plants during the cold months. Another Spanish book dealing with citrus tree culture was written by Miguel Augustin in 1617.

The importance of citrus fruit juice as a source of vitamin C was stressed at the end of the sixteenth century by the English admiral, Sir Richard Hawkins who recommended it as a remedy against scurvy. The distribution of this juice to English sailors was made compulsory in 1795.

In the Renaissance era many classical Italian artists included citrus fruits in their paintings. Among them are: Gentile da Fabriano (1360–1428), Fra' Angelico (1387–1455), Giovanni di Paolo (1403–1482), Filippo Lippi (1406–1469), Benozzo Gozzoli (1420–1497), Andrea Mantegna (1431–1506), Andrea Verrocchio (1435–1488), Botticelli (1444–1510), Luca Signorelli (1445–1523), Domenico Ghirlandaio (1449–1494), Leonardo da Vinci (1452–1519), Girolamo dai Libri (1474–1551), Raffaellino del Garbo (1476–1524), Lorenzo Lotto (1476–1555), Bernardino Luini (1480–1532), Paolo Cavazzòlo, called Morando (1486–1522), Giovanni da Udine (1487–1564), Correggio (1489–1534), Tiziano (1490–1576), Moretto (1498–1554), Parmigianino (1503–1540), Angiolo Bronzino (1503–1572), Daniele Crespi (1590–1630). Citrus fruits are also found in the sculpture by Lorenzo Ghiberti (1378–1455), Antonio Rossellino (1409–1464), the Italianised Fleming Giambologna (1529–1608) and in terracotta work by Luca Della Robbia (1400–1482). Citrus fruits are also present in the work of some Spanish painters.

From the fifteenth to the seventeenth century, an artistic movement of still life pictures flourished all over Europe. Flemish and Italian painters in particular chose citrus fruits as the subject of their paintings. Among the non-Italians we should mention Pieter Aertsen (1508–1575), Georg Flegel (1563–1638), Frans Snyders (1579–1657), Willem Claesz Heda (1593–1680) (Figure 1.4), Juan Van Der Hamen y Leon (1596–1631), Jacob Foppens Van Es (1596–1666), Pieter Claesz (1597–1661), Stoskopff Sebastian (1597–1657), Francisco de Zurbaràn (1598–1664), Jacques Linard (1600–1645), Jan Davidsz de Heem (1606–1684), Pier Dupuis (1610–1682), Maerten Boelema (1611–1654), Willem Kalf (1619–1693), Jan Jansz Van de Velde (1619–1663), Abraham Van Beyeren (1620–1690), Cerstiaen Luyckx (1623–1670), Joris Van Son (1623–1667), Joseph Heintz the younger (around 1650), Wouter Mertens (circa 1650), Juan Bautista de Espinosa (seventeenth century), Jean-Simeon Chardin (1699–1779), Luis Melendez (1716–1780). Italian painters who included citrus fruit in their works were: Giuseppe Arcimboldo (1530–1593), Vincenzo Campi (1536–1591), Jacopo da Empoli (1551–1640), Tommaso Salini (1575–1625), Fede Galizia (1578–1630), Panfilo Nuvolone (1581–1631), Bernado Strozzi (1581–1644), Francesco Gessi

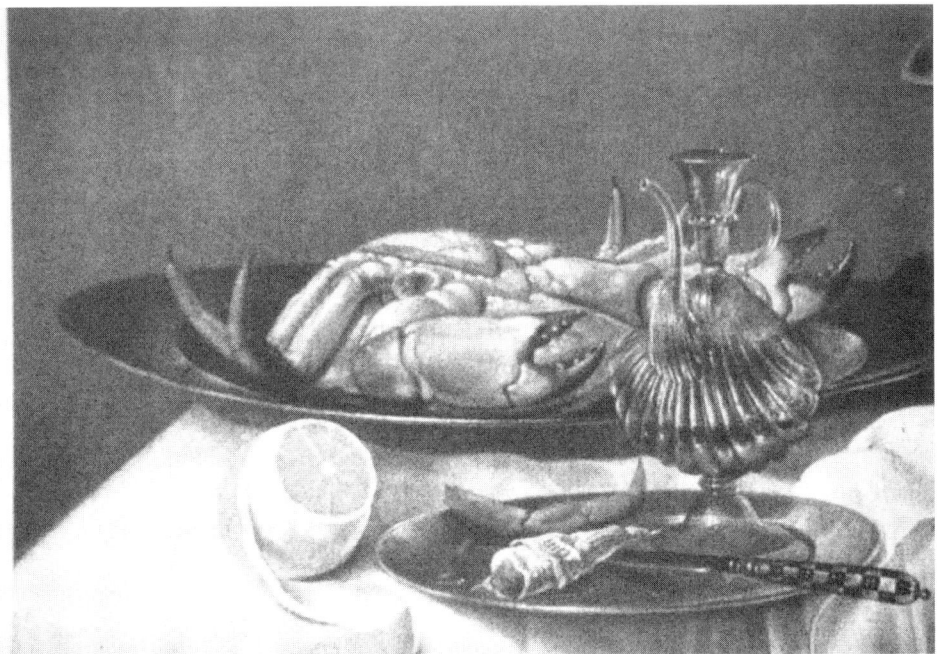

Figure 1.4 A lemon in a Willem Claesz Heda (1593–1680) painting (Ermitage, St. Petersburg).

(1588–1649), Luca Forte (1600–1670), Giovanna Garzoni (1600–1670), Pier Francesco Cittadini (1613–1681), Andrea Benedetti (1615–1649), Anton Maria Vassallo (1615), Gian Battista Ruoppolo (1629–1693), Giuseppe Recco (1634–1695), Felice Boselli (1650–1732), Giovanni Paolo Castelli (1659–1730), Cristoforo Munari (1667–1720), Tommaso Realfonso (1677–1743), Evaristo Baschenis (seventeenth century), Rinaldo Bolti (seventeenth century), Bimbi (seventeenth century) (Figure 1.5), Scacciati (seventeenth century).

In 1646 Baptista Ferrarius, a Jesuit monk from Siena, published a very relevant monography on citrus fruit. The book, entitled '*Hesperides, sive de Malorum aureorum Cultura et Usu Libri Quatuor*', was divided into four chapters. The first tells the legend of the Hesperides sisters, Aegle, Aretusa an Hesperetusa, who brought citron, lemon and orange trees to Italy. Pictures of Greek and Roman coins portraying this story are reproduced. The second chapter gives many descriptions and images of the citrus varieties, their propagation and cultivation. The third chapter deals with lemon, pummelo, and lime. In the fourth chapter Ferrarius deals with sour and sweet oranges (Figure 1.6). Seedless and navel varieties are reported, as well as blood juice cultivars.

Although the Ferrarius book is the first important scientific approach to botany and agronomy of the known citrus species, a large part is dedicated to poetry and legend. From then on, books on citrus dealt mainly with the technical and commercial aspects of culture. Economics were to become the main approach to these beautiful trees (Ferrarius, 1646; Volkamer, 1714; Gallesio, 1811; Risso and Poiteau, 1818; Tolkowski, 1938; Calabrese, 1998).

The Arabs gave remarkable importance to the extraction of oil from the peel of citrus fruit. But the Italian had extended the use of perfumed water since the XIVth century. It is first

Figure 1.5 Citrus fruit in a Bimbi painting (seventeenth century).

documented in Boccaccio's (1313–1375) Decameron. Oil was extracted from flowers and fruit and the perfumed water was said *'to exceed others in fragrance'* (Baldassare Pisanelli, 1610).

During the Renaissance another species of citrus was grown in Calabria. It was bergamot, an inedible fruit only used for its extraordinary scented oil. The essence of bergamot, with other oils, is used to prepare the famous *'eau de Cologne'*, that goes back to the XVIIth century. Paolo Feminis, who emigrated from Calabria to Germany, started the production of this scented water, which was called *'aqua admirabilis'*, sold in Giovanni Maria Farina's shop – another Italian.

'Eau de Cologne' ('Acqua di Colonia') is produced and appreciated even today.

Citrus fruit in America

Citrus trees were brought to America by the Spanish and Portuguese *Conquistadores*. Areas of penetration were Antilles, Mexico, Florida, Brazil. The presence of citrus trees in the West Indies is mentioned in the *'Natural history of Indies'* (1526), written by Gonzalo d'Oviedo. He speaks of sour and sweet oranges, lemons and citron that were considered better than the ones in Andalucia.

In 1590 Jose de Acosta published a book called *'Natural and moral history of Indies'*. He says that the most popular citrus trees of Europe such as oranges, lemons, and citrons, were grown in the West Indies and a delicious marmalade was made from their fruit.

Catholic missionaries introduced citrus plants into California and the northern part of Mexico.

Grapefruit was found in 1750 in the Barbados islands by Griffith Hughes. It was first called *'Shaddoch'* or *'Forbidden fruit'*, and only later changed its name into grapefruit, so called because trees often bear fruits resembling clusters of grape.

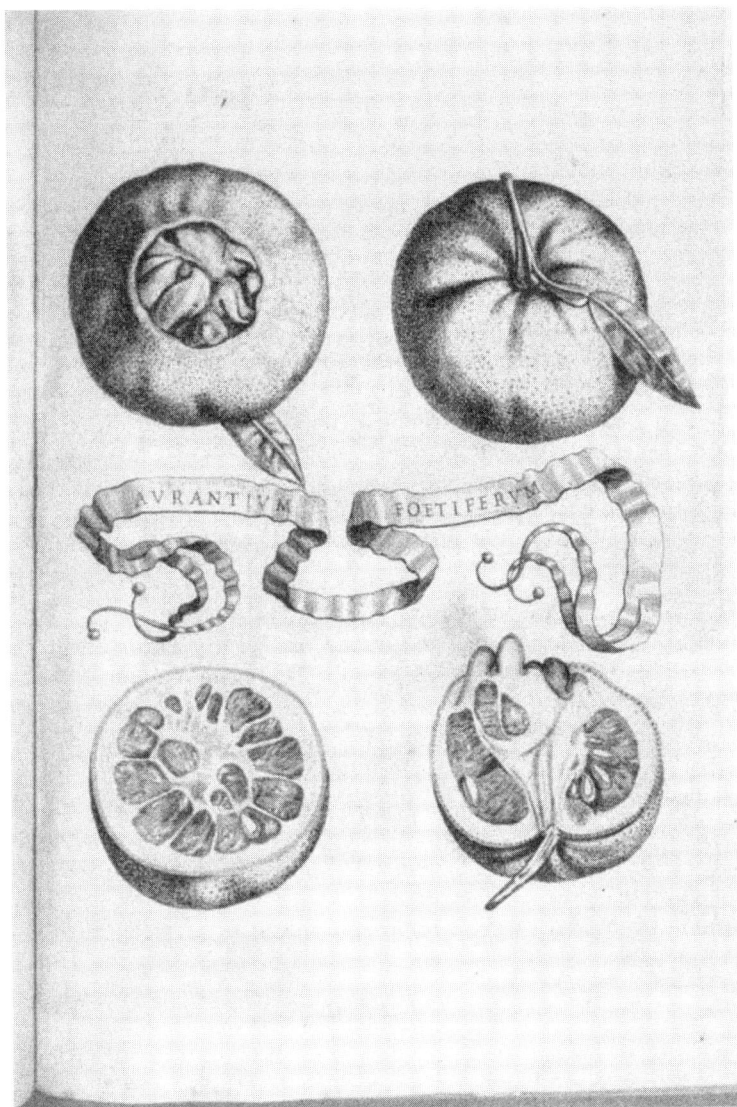

Figure 1.6 Navel oranges in the Ferrarius book (1646).

A Spanish noble, called '*don Felipe*', brought grapefruit into Florida, where the modern grapefruit was first established. Mostly world propagated varieties originated on this peninsula (Calabrese, 1998).

Citrus in Italy and Spain in the seventeenth, eighteenth and nineteenth centuries

The Italian and French Riviera were important citriculture centers in the eighteenth and nineteenth centuries, although the climate was too cold for citrus trees and frosts periodically damaged plantations (1709, 1763, 1782, 1799). At that time the commerce of citrus fruit and nursery plants spread from Liguria to the centre of Europe. The price of orange and lemon fruit were established by local decrees. Fruits were picked by public collectors. The juice was subject to licence and taxation. There were specific rules to punish transgressors.

The city of Menton had a '*Council for lemons*' made up of twenty seven members. This Council was in charge of controlling harvesting and commerce.

In February, March, April Genoese merchants used to take young lemon and orange plants as far as to the Netherlands and England.

In the seventeenth century in Liguria the sugar industry and the production of marmalade and candied fruit developed. The technique is referred to in several books (B. Scappi, G.A. D'Ezzero, Tanara).

In Italy the centre of citrus growing slowly moved from the Riviera to the South of the Peninsula. Due to its favorable climatic conditions Sicily became very important for citrus fruit production. Orchards were mostly planted around the most important cities (Palermo, Catane, Messina, Syracuse). The high income per hectare pushed growers to change culture to their farms. In 1853 there were 7,695 hectares all around the coast. Commerce increased year by year. In 1850, 22,000 metric tonne of citrus fruits (mainly lemons) were exported to the United States, Russia, Austria, Germany. In 1869, 78,345 tonne. There were specific wooden boxes for each county ('*American style boxes*', '*English boxes*', etc.). Fruits were selected according to size.

Processing quickly increased from the end of the eighteenth century, when citric acid was separated from lemon juice by Scheele (1784). Juice ('*cooked acid*' and '*raw acid*'), skin and essence were exported to foreign countries. Domenico Sestini (1870), a Tuscan monk, wrote that commerce was mainly directed to Rome, Venice, Trieste, England, the Netherlands, Denmark, Sweden, Moscow.

Due to the increasing importance of bergamot peel oil obtained in Calabria, in 1840 the Government of the Two Sicilies organised a competition for the invention of new instruments to extract essential oils. The competition was won by Nicola Barillà from Calabria.

In 1855 there were twenty four firms working on processing in Sicily. Sicily maintained its leadership in the production and commerce of the citrus fruit until the Second World War.

In Spain citrus growing first started in Oriola, Veja Baja del Segura (1700). Citriculture was transported from this area to Carcaixent, Alzira, Villareal, Almassora, Borriana. In 1873 there were 2,756 hectares, rising to 37,400 in 1908. The great increase in citrus growing started in 1852, coinciding with the agricultural crisis. Orange was preferred to other citrus species because it was the most profitable crop. Citriculture spread up along the Eastern coast, mainly around Valencia, although frosts often damaged trees (Cortes Muñoz and Zaragoza, 1992).

New citrus fruit

It has already been said that from 1700 two new citrus species entered the citrus industry in Italy and Florida: the bergamot and the grapefruit. But, while the bergamot growing has been limited to certain areas of Calabria and has only been used for its peel oil, grapefruit has become an important commercial fruit in some Western countries such as the United States, Mexico, Israel. Pummelo is preferred in Asia.

Commercial culture of the Mediterranean mandarins started in Sicily (around Palermo) at the beginning of the nineteenth century. Their growing soon spread to all the Mediterranean countries. Nowadays there are two main varieties selected in Sicily: *'Avana'* (*'Willow leaf'*) and *'Tardivo di Ciaculli'* (*'Late Avana'*).

Tangerines are mandarins of an important commercial value in the United States and in some other countries. *'Dancy'* is the prototype cultivar selected by colonel G.C. Dancy in Florida in 1867.

The clementine is probably a hybrid of sour orange and mandarin found as a seedling in Algeria by friar Clément Rodier in 1902. Many seedless cultivars are now commercially grown throughout the Mediterranean basin. It has slowly substituted the older Mediterranean mandarins, that are now only cultivated in Italy, Greece, and Egypt.

Tangelos are man-created hybrids between tangerines or mandarins and orange or pummelo. Some of them (*'Minneola'*, *'Orlando'*, *'Mapo'*) are now grown to a commercial extent.

It is important to note that the search and selection of new citrus varieties have also interested their use as rootstocks. New rootstocks tolerant to certain diseases are now propagated to substitute the sour orange. Citrange (*'Troyer'*, *'Carrizo'*, etc.), citrumelos, citremons, citrangors, citrangequats are among the most important (Calabrese, 1998).

Recent times

Citrus species and varieties spread West and reached every continent. The orange is the most extensively grown. At the beginning of the third millennium the world production is estimated to be about 59,000,000 tonne.

The citron, the oldest citrus fruit, is still a liturgical symbol in the celebration of the Tabernacle feast.

Today citrus fruits are a symbol of prosperity which bring colour and fragrance everywhere.

Citrus trees and fruit characterise the landscape of many regions in the world. They continue to be loved by artists, including famous modern painters like Manet, Fantain-Latour, Cézanne, Redon, Renoir, Rousseau, Van Gogh, Hopper, De Chirico.

We must never forget when we hold an orange or a lemon that it is possible only because a God had the courage to steal it from the heavenly Hesperides garden, and bring it to us. It is a divine present.

REFERENCES

Calabrese, F. (1998) *La Favolosa Storia degli Agrumi*, L'EPOS, Palermo, Italy.
Cooper, W.C. (1989) *Odyssey of the Orange in China*, Winter Park, Fla.
Cortes Muñoz, J.M. and Zaragoza, S. (1992) *Historia de la Naranja*, Levante, Valencia.
Ferrarius, I.B. (1646) *Hesperides sive de Malorum Aurorum Cultura et Usu Libri Quatuor*, Roma.
Gallesio, G. (1811) *Traité du Citrus*, L. Fantin libraire, Paris.
Risso, A. and Poiteau, A. (1818) *Histoire Naturelle des Orangers*, Paris.
Tolkowski, S. (1938) *Hesperides*, J. Bale, Sons and Curnow, London.
Volkamer, J.C. (1714) *Nürnbergishe Hesperides, oder Grundliche Beschreibung der edlen Citronat, Citronen und Pomerantzen–früchte*, Nürnberg.

2 Botany: taxonomy, morphology and physiology of fruits, leaves and flowers

Jesús M. Ortiz

TAXONOMY

According to the Engler system the plant kingdom has thirteen divisions; the thirteenth (Embryophyta Siphonogama) corresponds to the Spermathophyta. Lower taxonomic ranks are the following:

Subdivision: Angiospermae
   Class: Monocotyledoneae
      Order: Geraniales
         Family: Rutaceae
            Subfamily: Aurantioideae

Although other subfamilies exist within the Rutaceae, all citrus belong to the Aurantioideae subfamily, that includes two tribes: the Clauseneae, that are the very remote and remote citroid fruit trees, with five genera, and the Citreae, that are the citrus and citroid fruit trees, with three subtribes that are the Triphasiinae or minor citroid fruit trees (eight genera), the Balsamocitrinae or hard-shelled citroid fruit trees (seven genera) and the Citrinae or citrus fruit trees (thirteen genera) where the so called *true citrus genera* i.e. *Citrus, Poncirus* and *Fortunella* are included (Swingle and Reece, 1967).

Genus *Fortunella*

This genus resembles *Citrus* in the general aspect of the plant and fruits, although the fruits have small size and a low number of locules (Figure 2.1). The fruit is edible and is called kumquat. Swingle (1967) includes four species:

F. margarita (Lour.) Swing., Nagami or oval kumquat
F. japonica (Thunb.) Swing., Marumi or round kumquat
F. polyandra (Ridl.) Tan., malayan kumquat
F. hindsii (Champ.) Swing., Hong Kong wild kumquat

Hodgson (1967) adds another two species:

F. crassifolia Swing., Meiwa or big round kumquat
F. obovata Tan., Chagshou kumquat

Figure 2.1 Leaves and fruits of *Fortunella margarita* (kumquat).

Genus *Poncirus*

This is a monotypic genus because has only one species, that is *P. trifoliata* (L.) Raf. It is also called trifoliate or trifoliate orange. The tree is thorny, deciduous and has a marked cold resistance. It is mainly used as rootstocks for cultivated citrus as well as a parent in citrus breeding programmes. A botanical variety (var. *monstrosa*) is a very dwarf type of trifoliate, having curved thorns. Leaves are always with three folioles. The fruits are not edible because of the presence of a sour compound ponciridin.

Genus *Citrus*

By far the main genus within the Aurantioideae is *Citrus*, wherein practically all citrus trees grown throughout the world are included. The frequency of spontaneous mutations as well as the high occurrence of natural hybrids together to the high number of artificially obtained interspecific hybrids has originated the existence of a very large number of varieties, with more or less commercial interest, receiving local names and complicating the taxonomic identification of each of them.

Two subgenera exist within *Citrus*: *Papeda* and *Eucitrus* (=*Citrus*).

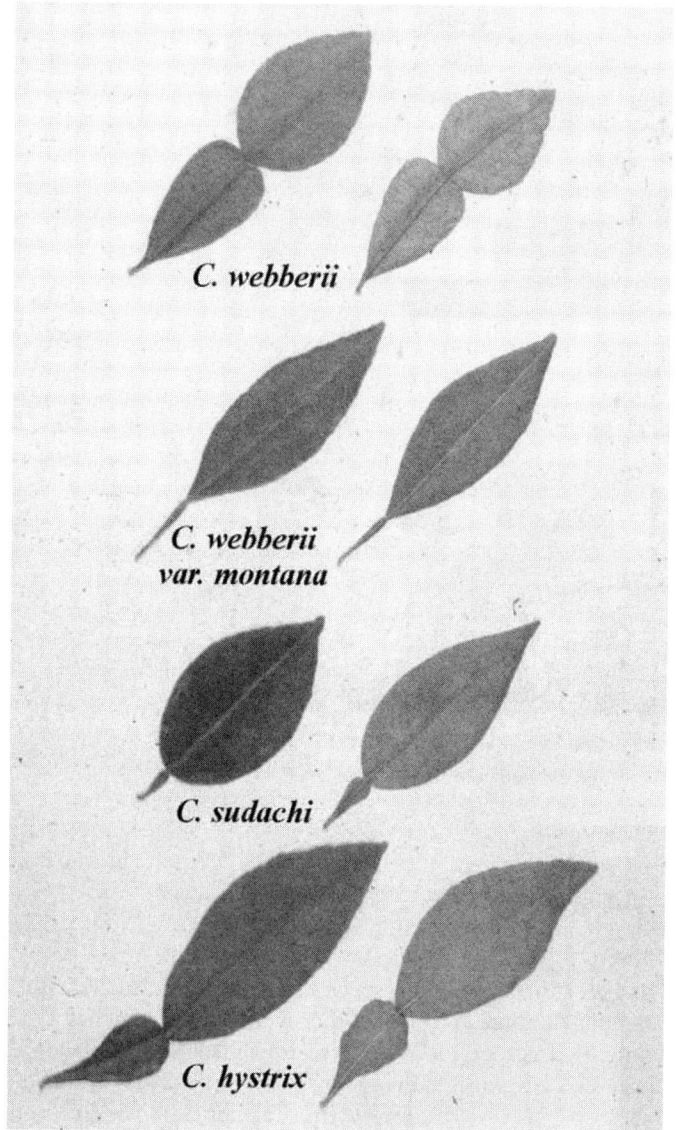

Figure 2.2 Leaves of *Citrus* species from the Papeda subgenus.

Subgenus Papeda
In contrast to *Citrus*, this subgenus has the following characteristics:

1   Bitter taste of the pulp that makes fruits inedible.
2   Wings of the petioles are markedly broad, in some cases being larger than the leaf blades (Figure 2.2).
3   Stamens are free from the base instead of fused.

4   Flowers and new growth are purple. In *Citrus* this only occur in lemons.
5   Germination is epigeous instead of hypogeous.
6   The juice vesicles, inside of the carpels, are adhered either to the external part of the membrane or to the lateral sides. In *Citrus* they are always adhered to the external part.

Subsections of the *Papeda*:

1   Subsection *Papedocitrus*: with four species that are *C. ichangensis*, *C. latipes*, *C. junos* or 'Yuzu', important rootstock in Japan, and *C. wilsonii* or 'Ichang pummelo'.
2   Subsection *Papeda*: with eight species that are *C. celebica*, *C. macrophylla* 'Alemow' or 'Kolo', used as rootstock, *C. macroptera*, *C. kerrii*, *C. combara*, *C. excelsa*, *C. hystrix* and *C. micrantha*.

Subgenus Citrus
According to Swingle (1967) this subgenus has ten species that are: *C. aurantium*, *C. sinensis*, *C. reticulata*, *C. limon*, *C. medica*, *C. aurantifolia*, *C. grandis*, *C. paradisi*, *C. indica* and *C. tachibana*.

CITRUS FRUITS

Agronomic groups

From the agronomic viewpoint, citrus can be grouped as follows:

Sweet oranges

All of them belong to the species *Citrus sinensis* (L.) Osb., being the main citrus tree grown in most of the citrus producing countries. Four groups exist within this species:

1   Navel oranges: can be distinguished by the presence of a navel at the stylar end of the fruit. They include cultivars like 'Washington', 'Thompson', 'Navelina', 'Navelate' and 'Newhall' (Figure 2.3).
2   Common oranges: also called white oranges. They include most of the older varieties grown in different countries, having local names and being difficult to distinguish. Some known cultivars are: 'Cadenera' from Spain, 'Jaffa' from Israel, 'Biondo comune' from Italy, 'Pera' from Brazil, or 'Pineapple' and 'Valencia' from USA.
3   Blood oranges: characterised by the presence of anthocyanin in the fruit, giving it a more or less intense red colour to the juice, the pulp or the rind. Cultivars of this group are 'Moro' (Figure 2.4), 'Tarocco', 'Sanguinelli' and 'Doble fina'.
4   Acidless oranges: also called sugar or sugary oranges because of the low acidity of the juice. Two cultivars are 'Sucreña' or 'Imperial' and 'Succari'.

Figure 2.3 'Newhall' navel orange.

Mandarins

It is a rather complex group, since it includes several species plus a certain number of hybrids that are considered below. The following main groups can be mentioned:

1 Satsuma group: is the species *C. unshiu* (Mak.) Marc. It is the main citrus grown in Japan, where it is called *unshû mikan*. There are cultivars with different time of ripening, from very early to late. Among cultivars include 'Owari' and 'Okitsu'.
2 Tangerine or clementine group: can be distinguished by a rather deep colour of the rind in contrast to the paler aspect of the satsuma. The clementine belongs to the

Figure 2.4 'Moro' blood orange.

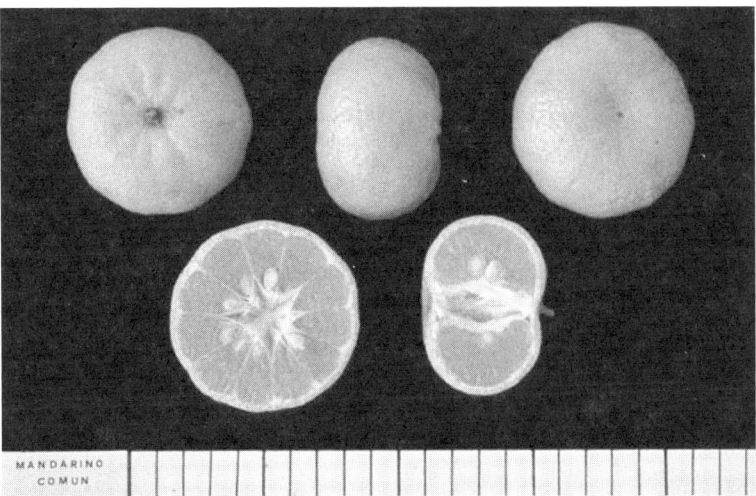

Figure 2.5 'Mediterranean' mandarin.

species *C. clementina* Hort. *ex* Tan., having several cultivars like 'Clemenules', 'Fina' and 'Monreal', while 'Dancy' is *C. tangerina* Hort. *ex* Tan.
3   Mediterranean mandarin: is the common mandarin in the Mediterranean basin. It belongs to the species *C. deliciosa* Ten. Usually with seeds. Some cultivars are 'Mediterranean' or 'Willowleaf' (Figure 2.5) and 'Avana'.
4   Other mandarins: without including the hybrids that are considered below, some other species should be mentioned: *C. reticulata* Blanco or 'Ponkan', *C. temple* Hort. *ex* Y. Tan. or 'Temple' and *C. nobilis* Lour. or 'King'.

Lemons

All of them belong to the species *C. limon* (L.) Burm. f. They have usually a high acidity, although acidless cultivars also exist. It is grown outside of the subtropical areas were limes are the substitute. Main cultivars include 'Eureka', 'Lisbon', 'Verna' (Figure 2.6) and 'Femminello'.

Grapefruits

They belong to the species *C. paradisi* Macf. They have rather large fruits, either white or pigmented. The main varieties are 'Marsh' and 'Redblush'. Among the deep pigmented varieties, 'Star Ruby' should also be mentioned (Figure 2.7).

Limes

Mostly grown in tropical and subtropical areas where there are substitutes of lemons. The Mexican lime or 'Key' lime is the *C. aurantifolia* (Christm.) Swing., while the 'Bearss' lime is *C. latifolia* Tan.

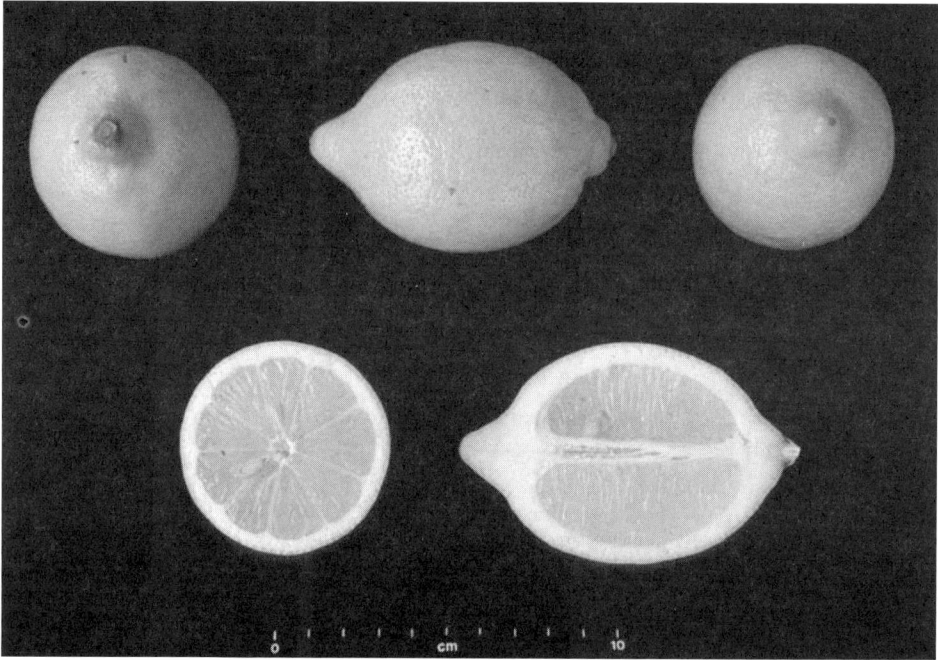

Figure 2.6 'Verna' lemon.

Sour oranges

They belong to the species *C. aurantium* L. It was the main rootstock for citrus until the appearance of the tristeza disease. As edible fruits they are grown for preparation of marmalades. They usually have a rather large number of seeds and the rind of the fruit has a deeper orange colour than most of the sweet orange varieties. Within the sour oranges the following groups exist:

1 'Standard' or 'Seville' sour orange is the main variety grown for production of fruits. Mostly grown in the South of Spain.
2 'Bouquet de Fleurs' is grown as an ornamental plant and also for production of flowers from which the 'neroli oil' is extracted.
3 'Granito' or 'Abers' is a willowleafed sour orange. Much used for extraction of essential oil from the leaves.
4 'Chinotto' used as an ornamental plant. The botanical name is *C. myrtifolia*.
5 'Bergamot' is the *C. bergamia*.

Pummelos

Also called Shaddock. These are mostly grown for fresh consumption in several countries of Asia. The botanical name is *C. grandis* (L.) Osb. The fruits are larger than grapefruits but present similarities with them.

Figure 2.7 'Star Ruby' (above) and 'Marsh' (below) grapefruits.

Citrons

Are the *C. medica* L. The rind of the fruits is used for confectionery. They are only commercially grown with this use in some countries of the mediterranean basin. The botanical variety *C. medica* var. *sarcodactylis* or fingered citron (Figure 2.8) is mainly used as an ornamental.

Citrus hybrids

The feasibility to obtain natural or artificial citrus hybrids either interspecific or even intergeneric, has originated a high number of cultivars that in some cases are difficult to identify.

A quite large number of the presently grown cultivars may proceed from natural or undocumented artificial crosses, and species names have been assessed to them by a given authority. The different existing opinions make it difficult to arrive at an international agreement since some of the considered true species can be in fact hybrids of unknown origin.

In this section only some of the known hybrids of commercial interest are included. According to the International Code of Botanical Nomenclature Code

Figure 2.8 Fingered citron.

(ICBN, 1994), parents are put within parenthesis ( ) united by the sign ×. When double crosses exist, the hybrid parent is mentioned in parenthesis ( ) while the cross is put in brackets [ ].

Hybrid mandarins

1. 'Kara' (*C. unshiu* (Mak.) Marc. × *C. nobilis* Lour.). Is a hybrid between Satsuma 'Owari' and 'King' mandarin.
2. 'Kinnow', 'Wilking', 'Encore' and 'Honey' (*C. nobilis* Lour. × *C. deliciosa* Ten.). All of them are hybrids of 'King' and 'Willowleaf' mandarins. The two first are sibling.
3. 'Fairchild', 'Lee', 'Nova', Osceola and 'Robinson' [*C. clementina* Hort. ex Tan. × (*C. paradisi* Macf. × *C. tangerina* Hort. ex Tan.)]. All of them are hybrids between Clementine mandarin and 'Orlando' Osceola tangelo. All except 'Fairchild' are siblings.
4. 'Fortune' (*C. clementina* Hort. ex Tan. × *C. tangerina* Hort. ex Tan.) is a hybrid between Clementine and 'Dancy' mandarins.
5. 'Fremont' (*C. clementina* Hort. ex Tan. × *C. reticulata* Blanco) is a hybrid between Clementine and 'Ponkan' mandarins.
6. 'Page' [(*C. paradisi* Macf. × *C. tangerina* Hort. ex Tan.) × *C. clementina* Hort. ex Tan.] is a hybrid between 'Minneola' tangelo and Clementine mandarin.

Figure 2.9 Minneola tangelo.

Tangelos

All of them are hybrids between grapefruit and mandarin. The main ones are:

1   'Minneola' (Figure 2.9), 'Orlando' and 'Seminole' (*C. paradisi* Macf. × *C. tangerina* Hort. *ex* Tan.) are hybrids of 'Duncan' grapefruit and 'Dancy' mandarin.
2   'Sampson' (*C. paradisi* Macf. × *C. tangerina* Hort. *ex* Tan.) is a hybrid of grapefruit and 'Dancy' mandarin.

Other hybrids

Only the *tangors*, hybrids between mandarin and sweet orange and the *limequats*, hybrids between lemon and kumquat have a certain commercial interest.

For a more detailed revision of the citrus taxonomy see Swingle and Reece (1967) and Ortiz (1985).

Intergeneric hybrids

Besides the hybrids with commercial interest for production of fruit, many other crosses have been carried out for plant breeding programmes. Out of them the main ones will be mentioned.

1 Bigeneric hybrids: several crosses where one of the parents is *Poncirus trifoliata* have been obtained, like the following:

>   *Poncirus trifoliata* × sweet orange = citrange ('Troyer' and 'Carrizo' among others, these two being of great interest as rootstocks)
>   *Poncirus trifoliata* × lemon = citremon
>   *Poncirus trifoliata* × mandarin = citrandarin
>   *Poncirus trifoliata* × sour orange = citradia
>   *Poncirus trifoliata* × calamondin = citraldin
>   *Poncirus trifoliata* × grapefruit = citrumelo
>   *Poncirus trifoliata* × *Fortunella sp.* = citrumquat

2 Back crosses:

>   Citrange × sweet orange = citrangor
>   Citrange × *Poncirus trifoliata* = cicitrange
>   Citrange × satsuma mandarin = citranguma

3 Trigeneric hybrids:

>   *Poncirus trifoliata* × sour orange × kumquat = citrangequat
>   *Poncirus trifoliata* × sweet orange × calamondin = citrangedin

It should be mentioned that all hybrids where *P. trifoliata* is one of the parents, produce inedible fruits because of the presence of ponciridin, and also have leaves with three folioles, since both are dominant characters.

MORPHOLOGY

In order to study the morphology of citrus, the different plant organs will be considered in detail.

Root

The root system of citrus trees is usually formed by a rootstock that frequently correspond to a different species than the variety. The main rootstocks used in Citriculture are the following:

Common name	Botanical name
Alemow or Kolo	*C. macrophylla* Wester
Carrizo citrange	*C. sinensis* × *Poncirus trifoliata*
Cleopatra mandarin	*C. reshni* Hort. *ex* Tan.
Palestine sweet lime	*C. limettioides* Tan.
Rangpur lime	*C. limonia* Osb.
Rough lemon	*C. jambhiri* Lush.
Sour orange	*C. aurantium* L.
Sweet orange	*C. sinensis* (L.) Osb.
Trifoliate	*Poncirus trifoliata* Raf.
Troyer citrange	*C. sinensis* × *Poncirus trifoliata*

Volckamer lemon   *C. volkameriana* Ten. & Pasq.
Yuzu   *C. junos* Sieb. *ex* Tan.

The characteristics of the root system depend both from the rootstock and the variety. In contrast to several fruit trees, since citrus are not deciduous, carbohydrates are not accumulated in the roots during the winter period. The two main missions of the root are the anchorage and the absorption. Deepness of the root depends on the rootstock as well as on the soil. In sandy soils, the root system can penetrate down to 5 or 6 m, while in clay soil shallow root systems prevail.

The existence of a taproot produces trees with a more erect and columnar aspect. The total length of the root system can be 30 km, that means a not much developed root system. The colour of the roots varies from yellow to olive brown.

The cross section of the horizontal roots is epitrophic (broader in the upper part, close to the soil surface) in mandarins and sweet oranges and hypotrophic (broader in the lower part, far from the soil surface) in lemons.

Trunk and branches

In most citrus, several cm above the soil surface is clearly visible the bud union (Figure 2.10), where the rootstock and the variety are fused. Three types of bud unions exist: (i) overgrowth of the stock; (ii) overgrowth of the scion; (iii) smooth union because equal growth in stock and scion. From that point on, the aspect of the trunk depends on the variety. The main trunk branches at different heights from the soil, forming the scaffold of the tree. The canopy of the tree is formed by the branching system plus the leaves. Young shoots usually have thorns, whose size and shape depends on the specific variety. Lemons tend to have vigorous thorns, while mandarins and sweet oranges do not. Thorns are modified branches, with vascular bundles, having buds at their axils.

Leaves

Out of the three citrus tree genera *Poncirus*, *Fortunella* and *Citrus*, only *Poncirus* is deciduous while the two other are evergreen. Several leaf flushes occur around the year, depending on the climate. Leaves usually stay on the trees for a period between 9–24 months or even more.

Citrus leaves (Figure 2.2) have two different parts, the leaf blade and the petiole. The union between both parts is articulated. Abscission of the leaf usually occurs at the base of the petiole, although in some cases this can occur at the articulation of the blade. The upper part of the leaf is darker than the lower one. A midrib is clearly visible and prominent at least in the lower part of the blade. The length/broad ratio is high in mandarins an lower in sweet oranges and lemons. Throughout the leaf blade, oil vesicles are clearly visible as small round dots, that are more abundant in the margin of the blade. According to the size, the juice vesicles are called primary (large), secondary (medium) or tertiary (small). Essential oils are released when leaves are crushed. The aroma is different for each species. Stomata are mostly located at the lower part of the blade; their density depends on the environment as well as on the citrus species.

The petiole is usually winged. Broadness of the wings varies with the species in the following order from broad to narrow: pummelo, sour orange, grapefruit, sweet orange,

Figure 2.10 Stock/scion union.

mandarin, citron. The Papeda subgenus, with not edible fruits has broader wings than the Eucitrus or true citrus subgenus.

Phyllotaxy in citrus is 3/8. Buds are located at the axils of the petiole being able to develop either into shoots or into flowers.

Flowers

Citrus flowers (Figure 2.11) are pentamerous and hermaphrodite. The peduncle has two abscission zones, one at the axil and the other closer to the calyx.

The *calyx* has five triangular sepals of small size, green and with oil glands. They persist in the ripen fruit.

The *corolla* has five petals white or light yellow; in the case of the lemons the outside part of the petals is purple. Oil glands are also present in the petals being clearly visible and producing a specific aroma when crushed.

The stamens placed in two verticils form the *androecium*. The total number of stamens depends on the variety. Between twenty and thirty stamen are common in citrus. The filaments are fused at least in the first third of its length and free at the upper part where yellow anthers with two visible thecas are present. Pollen grains inside of the

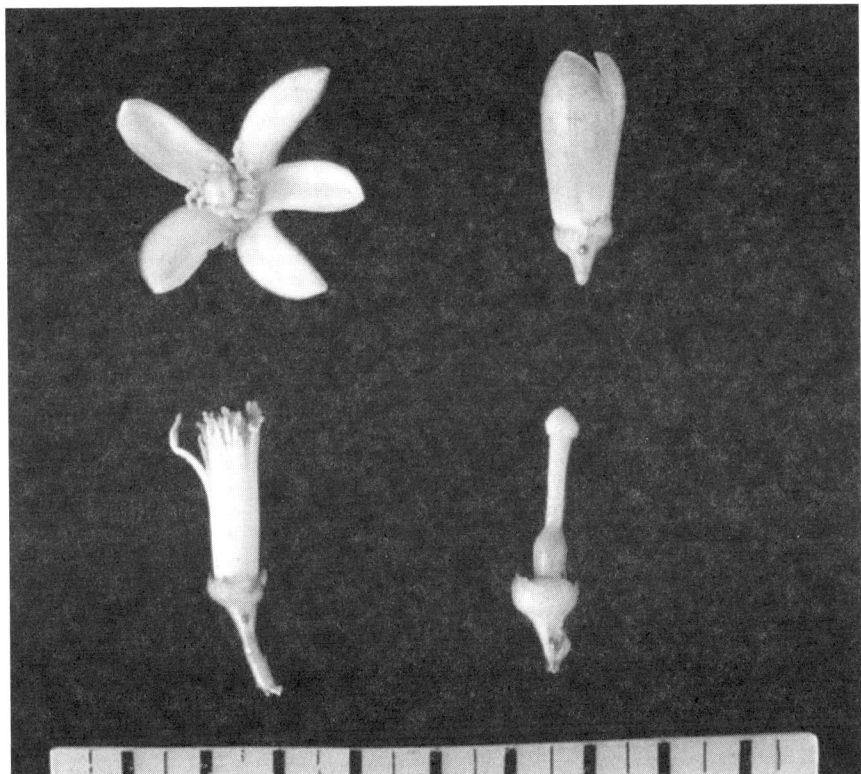

Figure 2.11 Views of lemon flowers.

thecas are entomophilous. The disc is a formation located between the androecium and the *gynoecium*. At the basis of the disc are the nectaries that produce a sugary nectar.

The ovary is superior, pluricarpelar and syncarpic. The number of carpels varies, although the most common number oscillates around eight to twelve in most citrus species. The central part of the ovary forms the placenta; placentation is central, usually with two ovules per locule. The ovules are anatropous i.e. with the micropyle looking to the placenta. The opposite part to the micropyle is the chalaza or chalazal spot and the raphe is formed between both points. The style is short and the stigma is developed at its end. Stigma is spherical and after anthesis is receptive for a short period of time, during which a secretion facilitates the adherence of the pollen grains.

Fruits

Citrus fruits are modified berries known as *hesperidia* (Figure 2.12). Characteristics of this type of fruit are the following: fruits are more or less spherical although depending on the species can be flattened like mandarins or rather ellipsoid like lemons or limes. The external part of the rind is called the flavedo because of the presence of flavonoid compounds; is rather thin, with a pigmentation that varies according to the species or variety, with colours from deep orange or reddish to light orange, yellow or greenish.

Figure 2.12 Different views of 'Madame Vinous' common orange.

Also in the flavedo are numerous essential oil glands, spherical in shape and of different sizes, that can be prominent as in pummelo, and sweet orange or grapefruit, or sunken as in sour orange or mandarin. According to their diameters they receive the names of primary, secondary or tertiary.

The internal part of the rind is the albedo, name that comes from latin (*albus* = white), has an ivory or pale yellow colour and is spongy in texture. The thickness and consistence of the albedo varies with the species. Mandarins have a very thin albedo, medium to thick in grapefruits, and very thick in shaddocks and citrons. Both flavedo and albedo form the rind, that is usually the non edible part of the fruit.

The edible part of the fruit is integrated by the segments or locules. The cross section of the fruit shows the form of each locule, frequently around 8–12 in most citrus, like corresponds to the number of carpels that were present in the ovary. Each locule is covered by a membrane, slightly coarse and covered by vascular bundles that transfer nutrients for growing of the fruit. Inside the membranes are the juice vesicles in a high number, each containing the juice within the vacuole of the cell. The central axis of the fruit is called the axis or columnella, that has a consistence and texture very similar to the albedo.

In some cases, a modification of the fruit occurs that is called the *navelisation* or formation of the navel (see Figure 2.3) that is like a secondary smaller fruit, developed at the upper part of the main fruit. When navel is present it can be visible from the outside, and can be observed in detail when the fruit is peeled.

Mature fruits have the calyx present at their base, while at the stylar end usually is only present a small scar where the style was fused. Around this scar an areole may exist, consisting in a more or less depressed and rather smooth rounded area. The shape of the fruit varies with the species, being either spheroid, ellipsoid, piriform, oblique, oblate, ovoid-oblique or ovoid. The base of the fruit may have the following shapes: necked, convex, truncate, concave, concave collared or collared with neck. The apex of the fruit can be: mammiform, angular, convex truncate or depressed (IBPGR, 1988).

Seeds

Citrus fruits are either seedy or seedless. From the commercial viewpoint citrus with very few seeds, less than 2 or 1.5 per fruit in average, are considered as seedless. Citrus seeds are different in shape depending on the variety; the main shapes are: fusiform, clavate, cuneiform, ovoid, deltoid, globose or semi-spheroid (IBPGR, 1988). A layer of mucilage and pectin, external to the seed, prevents desiccation and makes them slippery. The outer integument of the seed or *testa* is coarsy, yellowish in color and more or less ribbed. The acute angle end is the point where the seed was inserted to the fruit. The inner integument or *tegmen* is membranous and very thin. Inside is the seed, with two more or less semispherical cotyledons plus the embryo. The cotyledons are ivory or pale yellow except in most mandarins where they are green.

Citrus frequently present a modification in the formation of the seed that is called the polyembryony. Polyembrionic seeds have several pairs of cotyledons, one zygotic embryo and several nucellar embryos as well.

PHYSIOLOGY

Germination

Germination of citrus seeds is hypogeous in the subgenus *Eucitrus*, the cotyledons remaining in the soil, but epigeous in the *Papeda*. The seeds absorb moisture and swell, and the radicle emerges through the seed coats. When polyembrionic seeds germinate, several plantlets are born from a single seed; competition will make that only one of them will survive and originate the new plant. The radicle will develop into the root system and the tip apex will start the formation of the aerial part of the plant. Germination of seed will take a period of about 20 days. Viability of the seeds is rather low since they are recalcitrant seeds.

Growth

After formation of the new plantlet, a long period of growth will follow until the formed tree will arrive to the flowering period. This will occur at least after two years. During this period the main physiological occurrences will be the cell division and cell enlargement.

Flowering

Flower induction consists in the evolution of a vegetative bud into a flowering bud. Flowering buds are different in aspect, being dome shaped. In most citrus there is only one blooming period, about 2–4 weeks after the winter dormancy, while in some cases, like in limes and partially in lemons, several blooms occur throughout the year. Before opening, flowers are visible and the flower bottoms start to grow and develop while internal meiosis and formation of the gametes takes place. The time of total opening of the flowers is called the anthesis. The stigma of each flower is receptive to the pollen grains for a few days because of the production of a sticky secretion, the stigmatic fluid, where they can adhere.

Pollination, fertilisation and fruit set

Pollination is the transfer of the pollen grains from the anthers to the stigma. In citrus it is usually carried by insects, or occurs by either self or cross-pollination. Fertilisation implies the growing of the pollen tube plus the union of the spermatic nucleus with the oosphere with the formation of the zygote. At this time the petals and stamens fall, and the ovary starts to divide at a very high rate of mitosis: *fruit set* has taken place. Flowers where fruit set has not been completed, will turn brown and fall of in a very short period of time. Only the set fruits will start the following physiological period.

Pollination, growth of the pollen tube, fertilisation and the subsequent seed development seem to stimulate the production of plant hormones that prevent the ovary drop. Although in many cases fertilisation is needed for fruit production, production of seedless fruits or parthenocarpy is possible, at least in some varieties. Apparently a sufficient production of hormones in the ovarian tissues makes possible the fruit development. In some instances the stimulation of the pollen tube growth or even the fertilisation and subsequent seed abortion are needed in order to develop the fruit.

Fruits are considered as set when developed to the point that can be expected to remain until maturity, unless physiological or mechanical stress later occur.

Sexual incompatibility occurs in citrus when fertilisation is not possible although male and female components are functional. This incompatibility is due to a slow pollen tube growth, apparently caused by the presence of inhibitors in the style.

Fruit growth

Duration of this period is different according to the species or varieties, from some mandarins (f.i. early Satsuma) that will need less than five months from flowering to maturation, to some sweet oranges (f.i. Valencia late) that need more than one year. During this period both cell division and cell enlargement are the two main physiological processes.

Fruit growth includes Stages I and II of Bain. Stage I is characterised by slow growth in volume but intense cell division and cell enlargement in all tissues. Oil glands that were present at full bloom enlarge and new ones appear. Juice sacs also present at that time, continue to be formed. Most of the growth in this stage is due to the peel. Stage II is characterised by a very rapid fruit growth. Cell enlargement and differentiation predominate. The peel grows in thickness mainly due to the enlargement of the albedo cells. The spongy tissue is formed. At the end of the period, the increase of the fruit size causes a decrease on the rind thickness. The endocarp becomes the main constituent of the fruit.

Fruit maturation

Corresponds to the Stage III of Bain. It includes all changes since the fruit has the final size after the growing period until full ripeness is reached. Changes during this period can be grouped as:

External changes

Changes in the rind colour are mainly due to both degradation of the chlorophyll and increase of the carotenoids. As a consequence, the final colour of the rind becomes visible. Carotenes and xanthophylls are the main pigments responsible of the yellow, with predominant content in flavonoids in lemons or grapefruits or in carotenes in sweet oranges or mandarins. Reddish taints are usually due to the presence of anthocyanin, a pigment that accumulates with low temperatures. The beginning of the colour break in fall occurs when differences of temperature between day and night start to markedly increase.

Changes in texture are mainly due to the chemical changes in the albedo, that becomes softer and easy to separate from the segments.

Internal changes

Most of them occur within the juice vesicles, affecting the final quality of the juice. The main changes correspond to the parallel increase in sugar content and decrease of the organic acid content. The ratio between total sugars and the total acidity is called the maturity index, that is very useful for controlling the maturity in citrus, mostly in sweet oranges and mandarins.

Together with this change, pigmentation of the juice also varies. Two pigments are mainly responsible for the final colour, at least in some citrus fruits: anthocyanin, that gives red colour to the blood oranges among others, and the lycopene that causes the pink colour to grapefruits. Other pigments are also present and give the final colour of the citrus juices.

Aroma compounds, provide the final aroma of the pulp and juice. Most of them accumulate during the maturation period.

As a reference, Table 2.1 gives average values for the content of citrus fruits and citrus juices in some edible fruits.

Citrus are non climateric fruits, hence once separated from the tree the maturation process stops, consequently fruits can only be harvested and commercialised once the adequate maturation has been reached. The only point that should be mentioned on this is that external appearance of the fruit, especially the external colour of the rind can be up to a certain degree changed by external processes of curation or degreening.

Senescence

Once the fruit reaches the total maturity, senescence process starts and continues until fruit separates from the tree. During this period two main changes occur: desiccation by loss of inter or intracellular water and oxidation processes that develop into brown or brownish colours of the fruit. Citrus once mature can stay hanging on the trees for a rather long period of time until senescence, and finally the fall of the fruit will occur.

Table 2.1 Composition of citrus fruits (figures indicate amounts per 100 g)

		Moisture (%)	Calories (cal)	Proteins	Lipids	Carbohydrates		Ashes	Mineral elements (mg)					Vitamins					
						Total	Cellulose		Ca	P	Fe	Na	K	Vitamin A IU[a]	Thiamine B1 mg	Riboflavin B2 mg	Niacin Nicotinic acid mg	Ascorbic acid C mg
Limes	Fresh fruit	89.3	28	0.7	0.2	9.5	0.5	0.3	33	18	0.6	2	102	10	0.03	0.02	0.2	37
	Juice	90.3	26	0.3	0.1	9.0	tr[b]	0.3	9	11	0.2	1	104	10	0.02	0.01	0.1	32
Mandarins	Fresh fruit	87.0	46	0.8	0.2	11.6	0.5	0.4	40	18	0.4	2	126	420	0.06	0.02	0.1	31
	Juice	88.9	43	0.5	0.2	10.1	0.1	0.3	18	14	0.2	1	178	420	0.06	0.02	0.1	31
Sweet oranges	Fresh fruit	82.3	40	1.3	0.3	15.5		0.6	70	22	0.8	2	196	250	0.1	0.05	0.5	71
	Endocarp	86.0	49	1.0	0.2	12.2	0.5	0.6	41	20	0.4	1	200	200	0.1	0.04	0.4	50
	Rind	72.5		1.5	0.2	25.0		0.8	161	21	0.8	3	212	420	0.12	0.09	0.9	136
	Juice	88.3	45	0.7	0.2	10.4	0.1	0.4	11	17	0.2	1	200	200	0.09	0.03	0.4	50
Grapefruits	Endocarp	88.4	41	0.5	0.1	10.6	0.2	0.4	16	16	0.4	1	135	80	0.04	0.02	0.2	38
	Juice	90.0	39	0.5	0.1	9.2	tr	0.2	9	15	0.2	1	162	80[c]	0.04	0.02	0.2	38
Lemons	Fresh fruit	87.4	20	1.2	0.3	10.7		0.4	61	15	0.7	3	145	30	0.05	0.04	0.2	77
	Endocarp	90.1	27	1.1	0.3	8.2	0.4	0.3	26	16	0.6	2	138	20	0.04	0.02	0.1	53
	Rind	81.6		1.5	0.3	16.0		0.6	134	12	0.8	6	160	50	0.06	0.08	0.4	129
	Juice	91.0	25	0.5	0.2	8.0	tr	0.3	7	10	0.2	1	141	20	0.03	0.01	0.1	46

Notes
a IU = International Units; b tr = trace; c 440 in pink grapefruit.

REFERENCES

Hodgson, R.W. (1967) Horticultural varieties of citrus. In W. Reuther, H.J. Webber and L.D. Batchelor (eds), *The Citrus Industry*, Vol. I. University of California, Division of Agriculture and Natural Resources U.S.A., pp. 431–591.

I.B.P.G.R. (1988) *Descriptors for Citrus*. International Board for Plant Genetic Resources, Rome.

I.C.B.N. (1994) *International Code of Botanical Nomenclature (Tokyo Code)*. Koeltz Scientific Books, Königstein.

López-Fernández, J. (1995) *La naranja, composición y cualidades de sus zumos y esencias*. Generalitat Valenciana, Valencia.

Ortiz, J.M. (1986) Nomenclatura botánica de los cítricos. *Fruits*, 41(3), 199–209.

Swingle, W.T. and Reece, P.C. (1967) The botany of citrus and its wild relatives. In W. Reuther, H.J. Webber and L.D. Batchelor (eds), *The Citrus Industry*, Vol. I. University of California, Division of Agriculture and Natural Resources U.S.A., pp. 190–430.

3 Soil and cultural practices

Francesco Calabrese

CHOICE OF SOIL

Soil is the living medium where the roots of a tree grow. The first factor to take into account when choosing a soil for citriculture is texture (particle size). There are several schemes to classify soil texture. The easiest way to check if a soil is suitable for a citrus plantation is drainage. A high percentage (more than 60–65 per cent) of small-sized particles avoids percolation. Water must never flood either the surface or the root explored layer. Water accumulation discourages aeration and creates conditions for asphyxia. Rootstocks react differently to asphyxia, but even the most tolerant ones (*Poncirus* and its relatives) are severely affected by long-term flooding. The best conditions for water infiltration are found in sandy soils although they dry quickly and need frequent watering. Indeed, soil compaction does not depend only on particle size. Cementation at various depths (hardpan) or on the surface may occur for several reasons, including human and mechanical traffic.

The worst soils for citriculture are the richest in clay and lime (fine particles), which are an obstacle to water percolation. Some of these soils can be bettered to a certain extent by creating an appropriate drainage system. The so-called 'mid-compacted soils', where sand and clay-lime particles are more or less equal in quantity, are the best for citrus trees. They have sufficient colloids to avoid rapid drought, easily retain nutrient elements and allow bacterial life. The microbiological life of a soil is very important for mineral transformation. In some cases mineralized elements are inert and not easily captured by the roots.

The pH (soil reaction) and water table are other factors to be taken into account. Citrus trees can be grown in soil with a wide pH range (5.5–8.0) using appropriate rootstocks. The depth of the water table must be checked, especially where there is seasonal fluctuation. Roots must never be submerged by water.

Once orchard sites have been chosen, the soil should be checked for the amount of phosphorus, potassium and some minor elements like zinc, manganese, iron, magnesium. This is because pre-planting fertilization is the only way of correcting deficiencies by quickly adding minerals into the root-explored layers. In fact it takes many years of annual applications of phosphorus, potassium and oligoelements before roots get any benefit. However, nowadays liquid and easily soluble fertilizers reduce this inconvenience.

In tropical soils, where rains are abundant and frequent in all seasons, any pre-planting cultivation must be avoided in order to protect soil from erosion. This also applies to

Figure 3.1 6×8 m tree spacing in a lemon plantation (Tucuman, Argentina).

sandy soils, where there is no need to improve water and oxygen infiltration. In these soils the penetration of mineral and oxygen into the pore spaces is easy.

SPACING AND SOIL CARE

Spacing is chosen in relation to several factors: rootstock, climate, soil, cultural practices. Enormous differences exist from one area to another. Any tree investment must be chosen with a view to the best productive results. There are 3×4 m planted orchards (for instance, satsumas on dwarf rootstocks) as well as 6×4 m, 6×8 m (Figure 3.1), 8×8 m (Figure 3.2), 9×9 m plantations (Valencia on vigorous rootstocks in tropical areas).

When an orchard is planted, the aim of any cultivation practice is to care for the physical and biological features of the soil, to let trees produce in the most profitable way. There are different ways of managing soil during the life of the orchard. They mainly depend on climate, human resources and general economy of a farm.

A clear distinction must be made between semi-arid and rainy areas. In semi-arid areas, weed control to suppress spontaneous flora should be carried out during the dry period. This practice is not recommended under rainy conditions, where it is much better to control weed growth by periodic cutting. Where total weed control is necessary, it can be done in various ways. In many Mediterranean countries soil is cultivated in the dry season (summer) using small machinery. Superficial tillage is mainly adopted to eliminate summer and perennial weeds before they go to seed. There is no need to

Figure 3.2 8 × 8 m tree spacing in an orange orchard (São Paulo, Brazil).

control winter herbaceous flora until the end of the rainy season, when fertilizers must be applied. In these areas 2–4 tillages from February–March to September–October are satisfactory. Tillage is the least polluting technique used for keeping weeds under control, but it is difficult to apply on terraces (Figure 3.3). Negative effects of intensive mechanical cultivation are evident in soils which are rich in fine particles, where a compact '*plow sole*' below the cultivated layer will appear after a few years. Movements of equipment affect soil structure. Organic matter is quickly reduced by tillage. Because of these detrimental effects, farmers are careful to apply the '*minimum tillage*' to reduce weed growth and protect soil. Some herbaceous growth is tolerated not only in the rainy season but also in summer to an extent. In other words, weed elimination is not an automatic procedure but a choice.

An alternative technique for weed control in semi-arid and temperate regions is the use of herbicides. This technique allows the soil to remain undisturbed. Chemical control is now a world wide practice. Herbicides can be divided into three classes:

1. pre-emergence chemicals,
2. contact products,
3. root uptake herbicides.

The effectiveness of pre-emergent chemicals must be related to the capacity to controlling a specific weed. Consequently local weed species must be identified in order to choose the most appropriate herbicide. The continuous application of pre-emergence

Figure 3.3 A terraces plantation (Amalfi, Italy). In such plantation Tillage results difficult to apply.

products cleans soil. In such cases, because of their accumulation into the soil (residual effect), incorporation must be reduced, or even stopped for some time to avoid toxicity.

The contact products act as dryers of the epigeic part of a plant. Roots are not affected. They leave short-time residues in the soil. Many applications are needed for a continuous control.

The third class of herbicides are the so-called *'translocation chemicals'*. When they are applied in a solution to the foliage of weeds, they are carried to the roots, which are killed by toxic accumulation. These products must be applied in the recommended

Figure 3.4 A typical plantation in a rainy tropical country (Santo Domingo). The soil erosion that occurs in this country must be prevented by growing a cover crop.

quantities. Over dosage kills foliage and stops translocation towards the roots, which keep on living and give new stems and leaves. Chemical demolition of these materials in the soil is quick. Attention must be paid to avoid leaf wetting of the citrus trees.

In a non-tillage system roots expand near the soil surface where oxygen and minerals can be easily absorbed.

The use of herbicides is not recommended in tropical rainy regions, where soil must be protected against erosion. Under such conditions a cover crop is preferable (Figure 3.4). Annual or perennial legume crop is the best for this purpose. Leguminosae plants are compatible with the citrus root system. Besides, they are able to accumulate nitrogen in their roots. Cutting must be done periodically to reduce the leguminosae growth. Foliage is also an excellent forage. In the Mediterranean countries cover crop is useful in lemon culture to keep roots deep. It prevents the root 'mal secco' disease infection.

NUTRITION AND FERTILIZING

Citrus trees need a certain amount of minerals related to their foliar surface and productivity. The soil must supply elements to the roots in an adequate quantity and quality. So a chemical analysis of the soil is the first thing to do before planting.

Mineral intake is influenced by several factors, not least the capacity of the rootstock to capture elements from a certain chemical complex. There are two techniques to check chemical fertility: soil and leaf analysis.

Although soil analysis is the only procedure to find out the mineral composition of the root-explored layer, it has limitation, since the presence of nutrients may not correspond to their suitability for citrus trees. Some stable chemical compounds are not easily absorbed by roots. Sampling must be done in a uniform surface of soil. The sample must be taken at a depth of 10–60 cm (the main root-explored layer). Even in cases of apparent uniformity, several samples should be taken for separate analysis. Each single analysis is usually the result of mixing four soil samples taken at various points in each unit surface (one hectare or more). Analysis must be carried out for at least the main elements. The determination of the 'cation–exchange capacity' is also a good method of checking soil suitability for citrus trees. Humus-rich soils as well as clay soils have a high exchange capacity (>10 milliequivalents). Deficiencies must be corrected by an overdosage of fertilizer. Excess cannot be corrected, except in the case of soluble elements (nitrogen, chloride, etc.). Leaching is rapid in heavy rainy regions. In such cases soil permeability is very important. Water irrigation may carry salts which must be considered for the chemical soil balance.

Leaf analysis is the most efficient guide to checking elements in citrus trees. Plants accumulate nutrients in the leaves, so leaf analysis is the best tool for detecting the nutritive conditions of a citrus plant. Not every leaf is suitable for mineral analysis. Leaves must be taken from the periferic and equatorial part of mature trees scattered here and there in the orchard. They must be five–seven months old, coming from the main flush (spring cycle in the northern hemisphere) and not fruiting shoots. Leaves should be gathered early in the morning, kept in thermic bags and brought to the laboratory as soon as possible. The chemical data obtained must be compared with the standard data (Table 3.1). When mineral leaf content is satisfactory fertilization must replace the amount of nutrients lost to fruit or by leaching and vaporization.

Loss of minerals varies greatly in relation to the amount of production, soil, climate, agricultural practices. As an approximate estimation, in the Mediterranean countries a yearly loss of the main elements per hectare, in a mature orchard producing 30 tons of fruit, is as follows: $N = 180–220$ kg; $P_2O_5 = 30–40$ kg; $K_2O = 90–120$ kg. These are the amounts per hectare that must be replaced by yearly fertilizing where there is a satisfactory presence of elements in leaves. These amounts have to be increased for

Table 3.1 Percentage of different elements in 4–7 month-old orange leaves from non-fruiting terminal shoots (spring cycle) (Cohen, Israel)

Element	Unit	Deficient less than:	Low	Optimum	High	Excess more than:
N	%	2.2	2.2–2.4	2.5–2.7	2.8–3.0	3.0
P	%	0.09	0.09–0.11	0.12–0.16	0.17–0.29	0.30
K	%	0.7	0.7–1.1	1.2–1.7	1.8–2.3	2.4
Ca	%	1.5	1.5–2.9	3.0–4.5	4.6–6.0	7.0
Mg	%	0.20	0.20–0.29	0.30–0.49	0.50–0.70	0.80
S	%	0.14	0.14–0.19	0.20–0.39	0.40–0.60	0.60
B	ppm	20	20–35	36–100	101–200	260
Fe	ppm	35	35–49	50–120	130–200	250[a]
Mn	ppm	18	18–24	25–49	50–500	1000
Zn	ppm	18	18–24	25–49	50–200	200
Cu	ppm	3.6	3.7–4.9	5–12	13–19	20

Note
a approximate value.

deficiencies and high leaching. This is particularly true for nitrogen, which is easily soluble and, when applied under organic compounds, turns into gas, mainly in warm periods of the year and in certain soils.

Manuring is a common practice in tropical and subtropical countries, particularly in Asia. In the western industrialized countries where manure is rare and difficult to use, it has been replaced by chemical fertilizers. There are many commercial products. They must be chosen according to the replacing minerals and their ratio. 1 (N):0.2 (P_2O_5):0.6 (K_2O) is the ratio of the main elements lost from a citrus orchard. Fertilization must respect this ratio when the mineral content in the leaves is satisfactory. Obviously, in cases of overdosage or deficiency of one or more elements this ratio must be changed to a more appropriate one. Replacing element ratio depends on soil and climate. This is why fertilization formulas cannot be generalized. A suitable mineral ratio for a specific situation can be easily built up by using monoelement fertilizers. Complex fertilizers are more problematic, since their ratio does not always correspond to the needs of the orchard.

Several sources of elements can be used. For a cover fertilization ammoniacal nitrogen is preferred to the nitric form, especially in regions with heavy rainfall, because nitrate nitrogen is easily soluble and quickly moves to the lower layers. Nitrate compounds can be used for rapid corrective nitrogen fertilization. Nitrification speed of ammoniacal fertilizers depends on climate and soil. Urea is an organic source of nitrogen. It is transformed into an ammoniacal form in the soil. Its hydrolysis depends on microrganism activity, which is mainly related to soil temperature. Urea must not be used in water–logged soil where biuret and other toxic compounds are produced. These products have an antimitotic effect and are detrimental to trees. Foliage spray of urea is useful in correcting deficiencies. It must be used cautiously in dry warm climates where '*growth-stop*' and leaf '*burning*' may occur.

Water soluble phosphorous is the best for citrus fertilization. Phosphorous moves very slowly into the soil. As already pointed out, a pre-planting application is needed for deep penetration into the soil.

Potassium moves quicker than phosphorus, but not as quick as nitrogen. It is captured by the exchange complex of soil. Soil texture affects its mobility. In some soils potassium infiltration is very slow. Overdosage of potassium discourages magnesium absorption. Potassium sulphate is the most appropriate source for potassium fertilization, its acid reaction may not suit some acid soils. Potassium nitrate spray can be used for heavy deficiencies and to accelerate colour break-down in some citrus varieties. Heavy potassium fertilization delays acid decreasing and maturing ratio in fruit.

Calcium may be needed in some acid tropical soils. In these cases it is used as a corrective (amendment). Minor elements, such as magnesium, zinc or manganese may be needed to correct deficiencies in citrus trees. Such deficiencies are easily visible to an expert eye. Liquid foliar sprays are quickly effective, although they must be repeated periodically. Iron deficiency is frequently associated to a high calcium content in the soil or to antagonistic ions in acid soil. In these cases deficiency is qualitative, not quantitative. Chelate iron products correct this deficiency. They can be applied to soil or sprayed to foliage.

Fertilization timing depends on local conditions. In temperate and sub-arid regions fertilizers should be applied to the soil once or twice a year. Under rainy conditions and in sandy soils' various applications of nitrogen and other elements are necessary. In any case a complete fertilization must be applied to soil 30–50 days before the start of the main flush.

In cases of fertirrigation, fertilizing rates and timing must be properly scheduled. Fertirrigation is easy to apply in arid and sub-arid regions and in medium-textured or sandy soils. Under rainy conditions and in clay-soils fertirrigation is problematic. The advantage of fertirrigation is the high solubility of elements that are quickly absorbed by roots. Fertirrigation is associated to chemical weed control in mechanically undisturbed soils. Fertirrigation must be applied from the first year of plantation. Mature orchards cannot be quickly converted from a wide soil fertilizer application to localized watering and fertilizing.

IRRIGATION

Citrus trees are evergreen plants which originated in humid climates. This is why their root system is superficial. They need a periodic water supply to support the leaf, stem and fruit transpiration. Watering is usually adequate in rainy regions. In temperate and dry zones there is lack of water, at least in some periods of the year. In such areas irrigation is necessary to make the citrus industry profitable. Deficient water in soil deletes both tree growth and yield. In some cases (sub-arid climates) irrigation is indispensable to grow citrus for commercial purpose. The amount of water necessary is variable. It not only depends on climate conditions but also on the soil texture and watering system.

Water affects growth, fruit set and quality. In humid regions heavy rainfall gives excellent fruit growth, a high percentage of juice, a low sugar and acid content. If irrigation is sufficient to replace the water lost by transpiration and evaporation, the best citrus fruits are obtained in sub-arid and temperate areas. Water intake by roots is related to its presence in the explored layer and to the retaining capacity of the soil. The useful water for roots is determined by the difference between the wilting point and the field capacity. Sandy soils loose water quickly, while soils which are rich in fine particles retain water longer, sometimes too long. If there is a water shortage in the dry season, irrigation should be reduced after the fruit cell multiplication has stopped. Negative effects on fruit size are evident if a break in irrigation is scheduled in spring and early summer (when the fruit is rapidly increasing by cell division).

There are various ways of checking the amount of water in the soil. Tensiometers measure the soil moisture content. They must be used at two depths (20–25/50–60 cm). Irrigation is applied when the soil storage capacity is reduced and tensiometers show about 50 centibars. It is not convenient to keep soil moisture at the field capacity. A certain fluctuation of water is useful in getting better control over some root diseases. Tensiometers indicate the time of irrigation, not the amount of water required. In order to quantify the water needed in a certain period, a calculation of the daily evapotranspiration (ET) is necessary. This is done using an evaporimetric Class A pan tank. The amount of the evaporated water (mm) must be adjusted according to the citrus Kc factor (crop coefficient), that has been calculated by FAO (Table 3.2). It is evident that Etc (ET×Kc) for citrus trees is less than ET. In the old citriculture areas, growers build up their irrigation schedule on their experience and historical data (calendar schedule).

The replacement of water lost by evapotranspiration must be done at different times, which mainly depend on soil texture. In sandy soils, due to the rapid infiltration of water, irrigation is recommended every few days at a low volume. But, in the case of fine-particle soils, watering must be scheduled over a longer period using a higher

Table 3.2 Citrus crop coefficient (Kc) for the subtropics with autumn-winter rainfall, light to moderate wind and mature trees

	Jan	Feb	Mar	Apr	May	Jun	Jul	Aug	Sep	Oct	Nov	Dec
70% tree ground cover												
Clean cultivation	0.75	0.75	0.70	0.70	0.70	0.65	0.65	0.65	0.65	0.70	0.70	0.70
No weed control	0.90	0.90	0.85	0.85	0.85	0.85	0.85	0.85	0.85	0.85	0.85	0.85
50% tree ground cover												
Clean cultivation	0.65	0.65	0.60	0.60	0.60	0.55	0.55	0.55	0.55	0.55	0.60	0.60
No weed control	0.90	0.90	0.85	0.85	0.85	0.85	0.85	0.85	0.85	0.85	0.85	0.85
20% tree ground cover												
Clean cultivation	0.55	0.55	0.50	0.50	0.50	0.45	0.45	0.45	0.45	0.45	0.50	0.50
No weed control	1.00	1.00	0.95	0.95	0.95	0.95	0.95	0.95	0.95	0.95	0.95	0.95

volume. The effect of a correct irrigation schedule is measured by tree growth and yield, since incorrect watering affects the tree water status and its physiology.

Flood or furrow irrigation (surface irrigation) has historically been used in many citrus-growing regions. The water loss in such systems is high and its efficiency poor (about 30–35 per cent). The labour needed is also expensive. The pipe line system has improved water efficiency (more than 70 per cent) by reducing losses. Reduction of labour is consequent. Under soil and over soil pipes are now used in many citrus-growing areas.

The sprinkler system is popular in many countries where water is sufficient and its cost low. Microirrigation systems have been developed for zones where watering is problematic for some reason (shortening, cost, etc.). Microirrigation is based on the concept of wetting limited surfaces and soil depths where roots are obliged to grow. Water efficiency is high. Low pressure emitters (1.5–1.6 atm) deliver low water volumes at frequent intervals. Microsprinkler emitters release 40–80 l/h, a drip emitter only 4–8 l/h. Low-pressure lines allow the use of fertirrigation (see fertilization chapter). They need an efficient water filtration system.

Water quality must be checked before planting and samples taken in the critical period (second part of the dry season), when dissolved salts are more concentrated. The choice of rootstock depends on water quality, since, as we know, there are rootstocks which are more or less tolerant to salinity. Chloride, sodium and boron are the most frequent elements found in irrigation water. They are a limiting factor to the growth of citrus trees and their effects can be detected on leaves. The negative effects of salinity are reduced in porous soils and by delivering abundant amounts of water from time to time to leach salts in depth. Water quality can be measured by various methods: total salinity, chloride percentage, electric conducibility. FAO suggestions are reported in Table 3.3.

Table 3.3 Soil and water irrigation salinity (mmhos/cm) related to decrease in yield (%)

ECe	ECw	Decrease
1.7	1.1	0%
2.3	1.6	10%
3.3	2.2	25%
4.8	3.2	50%
8.0	–	max

PRUNING

Citrus tree manipulation is a significant cultural practice in some regions, while it is of little importance in others. It is possible to draw a geographic world map, depicting the use of pruning as a method of training trees to grow in a desirable scaffold structure and constantly produce fruit of excellent quality. Generally speaking, pruning is neglected or limited in tropical regions as well as in countries that produce fruit for processing. Scarcity of qualified labour and or its high cost is another limiting factor in some citrus growing areas, like the United States. In Mediterranean countries hand-pruning is still an important way of regulating tree growth and bearing.

As we know, deciduous fruit species need yearly cuts to promote new flush, flowering and crop. Citrus pruning is mainly devoted to thinning and opening the canopy for light penetration as well as to keeping the canopy low.

Citrus tree growth is upright in most varieties. A fruitful branch tends to bend and produce new sprouts. Some of these sprouts are too vigorous and discourage the growth of other lateral stems. During the first years, there is no need to cut all, or part of these vigorous sprouts, since research has shown that any cut to a young tree reduces root growth. This is due to the equilibrium between foliage and root system. A plant is a biological unit. Heavy cuts on a young tree, as frequently happens in some Mediterranean countries, discourage canopy growth and delay trees coming into bearing. The only pruning on young trees should be to eliminate the central leader branch. In some weak citrus varieties and species, such as limes, satsumas, etc., it is not convenient to make any cut at the beginning. After a few years, depending on the speed of tree growth, climate and cultural practices, a selective canopy thinning may be carried out. This is an annual practice in Mediterranean countries.

The choice of branches to eliminate is based on the concept that any space within the canopy must be covered by only one branch. It is not convenient to let surplus branches occupy the aerial space. Anyway, thinning must not deplete any canopy sector. Suckers should only be kept if they occupy free spaces. After a few years they bear fruit, but all interior suckers must be cut. A harmonious citrus tree grows to an almost round shape (globe). Vegetation free canopy spaces must be avoided since they reduce yield.

In some areas (United States, Brazil, etc.) pruning is either practised rarely or not at all. Unpruned trees come into bearing quickly and yield crops for many years. Growth and fruit tend to take place towards the external part of the canopy. Inner growth tends to decrease gradually. After some years crops are reduced and fruits become smaller. This is the time to make severe cuts and renew growth. In some countries (United States, Australia, etc.) it is done mechanically. Unpruned trees must be removed when yield is no longer economically profitable. The commercial life of an unpruned orchard is shorter than that of a pruned plantation.

In many countries the citrus tree canopy extends to ground level. In other areas a horizontal cut is made to detach foliage from soil (skirt) (Figure 3.5). This is made for several reasons: *Phytophtora* disease, which affects leaves and fruit, tillage, etc. It is always preferable to keep the canopy low to facilitate cultural practices and picking.

A view over a modern citrus orchard shows symmetrical separated rows. Light must reach the lower part of the trees. Thus, spacing must be in proportion to the vigour of the variety in order to avoid narrowing the free spaces between the rows. High density on the plantation causes vegetation to crowd and so makes any cultural practice

Figure 3.5 Skirt pruning in an orange orchard (California).

difficult. It must be considered an antiquated and obsolete way of growing citrus trees. The scaffold structure must be raised for cultural operations.

Lately, pneumatic and mechanical pruning have been developed and applied in many countries. Pneumatic pruning uses simple equipment consisting of an air-compressor which activates clippers, and saws placed at the tip of a hand-guided pole. Its use reduces both hand effort and pruning time to as much as 40–45 per cent. The mechanical equipment (Figure 3.6) consists of power-driven circular saws able to make vertical (hedging) (Figure 3.7) and horizontal or diagonal (topping) (Figure 3.8) cuts. The only possible pruning when using mechanical tools is indiscriminate cutting of the canopy. In other words, while hand-made pruning is guided by intelligence, mechanical pruning cuts reduce branches without selecting and thinning. This different approach to pruning is justified in the areas where labour is not economical. Mechanical pruning is extremely quick and reduces operation time by as much as 90 per cent.

Hedged and topped trees need 2 years (or more) to return to normal production. In order to avoid severe yield reduction, one side tree hedging is usually alternated to the other side or topping.

As a norm, cropping of a mechanically pruned orchard is lower than a hand pruned one, but the pruning cost is inferior. Mechanical cuts cannot be made on young trees (before 6–8 years).

Pruning comes after picking and before the main growth cycle starts. In some late varieties it is delayed until the spring flush stops (summer). In some Mediterranean countries lemon trees are usually pruned in summer to reduce mal secco–disease infection. But a new problem has arisen in recent years caused by the introduction of the

Figure 3.6 Mechanical pruning equipment (Israel).

Figure 3.7 Hedged orange trees (San Diego, California).

Figure 3.8 Topping on lime trees (Florida).

leaf-miner, *Phyllocnistis citrella*, which is unfortunately active in the warm months. Autumn and winter pruning is undesirable as stimulation of tender new flush close to the cold period must be avoided.

REFERENCES

Calabrese, F. (1991) *Frutticoltura Speciale*. REDA, Roma.
Cutuli, G., Di Martino, E., Lo Giudice, V., Pennisi, L., Raciti, G., Russo, F., Scuderi, A. and Spina, P. (1985) *Trattato di Agrumicultura*. Edagricole, Bologna, Italy.
Davies, F.S. and Albrigo, L.G. (1994) *Citrus*. Cab International, Wallingford, UK.
Reuther, W., Batchelor, L.D. and Webber, H.J. (1967, 1968) *The Citrus Industry*. Vol. **I, II**. Div. Agric. Sci. Univ., Berkeley, California (USA).
Reuther, W. (1973) *The Citrus Industry*. Vol. **III**. Div. Agric. Sci. Univ., Berkeley, California (USA).
Reuther, W., Calavan, E.C. and Carman, G.E. (1978) *The Citrus Industry*. Vol. **IV**. Div. Agric. Sci. Univ., Berkeley, California (USA).
Spina, P. and Di Martino, E. (1997) *Gli Agrumi*. Edagricole, Bologna, Italy.

4 Pests and diseases

Vittorio Lo Giudice

INTRODUCTION

Citrus fruits are grown in the world both in the Northern and Southern hemisphere, in the zones between 40° North and 40° South latitude; for this reason commercial productive zones are numerous and each with its own management problems concerning abiotic and biotic diseases. Owing to the diversity of soil, climate, variety, rootstocks, methods of cultivation and social economic situations, problems are different. Therefore information about any single pest and disease will be limited, while the methods and means of control will be illustrated as basic concepts. The choice of methods and means of management must be better taken care of because there is always a greater demand of a global quality of fruit, which must not only focus on the external and internal characteristics of the fruit but also on other parameters, among which the absence of residue treatments and the nutrition value. Market globalization involves the concentration of demand and supply and consequently, the uniformity of methods of production.

PESTS

Insects, mites and nematodes that damage the citrus plants are numerous but not all provoke outstanding economic damages. Biocenosis of the citrus orchard is particularly rich and subject to an introduction of new species, therefore the list of pests is liable to change. This chapter presents a list of some pests (Table 4.1), classified according to Couilloud (1991), and a synthetic description of the damages caused by them. Ebeling (1959), Chapot and Delucchi (1964), Talhouk (1975), Di Martino (1985), Jeppson (1989), Davis and Albrigo (1994), can be consulted for a deeper study of pests present in certain citrus areas. On citrus trees, pests may be found on the roots, trunks, branches, leaves, flowers and fruits. They spread through various methods both at short and long distances. The strengthening of the means of communication and the massive movement of people have favoured the introduction of new pests and diseases in the different citrus zones, making control difficult. Pests provoke damages to the plants, at times specific, which can generically be grouped.

Mites damage buds, they nourish themselves by damaging the fruits, provoking silvering, deformation and drop of the leaves. On the fruits mites provoke discolouration and russeting.

Table 4.1 Some representative citrus pests

Class	Group	Species	Common name
Arachnida	Mites	*Aceria sheldoni*	Citrus bud mite
		Phyllocoptruta oleivora	Citrus rust mite
		Brevipalpus phoenicis	False spider mite
		Panonychus citri	Citrus red mite
		Tetranychus urticae	Common red spider
Insecta	Whiteflies	*Aleurocanthus woglumi*	Citrus blackfly
		Aleurothrixus floccosus	Woolly whitefly
		Dialeurodes citri	Citrus whitefly
	Aphids	*Aphis citricola*	Spirea aphid
		Aphis gossypii	Cotton aphid
		Toxoptera aurantii	Black citrus aphid
		Toxoptera citricidus	Brown citrus aphid
	Soft scales	*Ceroplastes destructor*	Soft wax scale
		Ceroplastes floridensis	Florida wax scale
		Ceroplastes rubens	Pink wax scale
		Ceroplastes rusci	Fig wax scale
		Ceroplastes sinensis	Chinese wax scale
		Coccus hesperidum	Brown soft scale
		Coccus pseudomagnoliarum	Citricola scale
		Coccus viridis	Green scale
		Saissetia oleae	Black scale
	Armored scales	*Aonidiella aurantii*	California red scale
		Aonidiella citrina	Yellow scale
		Chrysomphalus aonidum	Florida red scale
		Chrysomphalus dictyospermi	Dyctiospermum scale
		Mytilococcus gloveri	Glover's scale
		Mytilococcus beckii	Purple scale
		Parlatoria pergandii	Chaff scale
		Selenaspidus articulatus	Rufous scale
		Unaspis citri	Citrus snow scale
		Unaspis yanonensis	Arrow head scale
	Margarodid scale	*Icerya purchasi*	Cottony cushion scale
	Mealybugs	*Planococcus citri*	Citrus mealybug
		Pseudococcus citriculus	Green's mealybug
	Psyllids	*Trioza erythreae*	Citrus psyllid
		Diaphorina citri	Citrus psylla
	Bugs	*Leptoglossus phyllopus*	Leaf-footed bug
		Rhynchocoris humeralis	Citrus green stink bug
	Thrips	*Frankliniella occidentalis*	Grass thrips
		Heliothrips haemorrhoidalis	Greenhouse thrips
		Scirtothrips citri	Citrus thrips
		Thrips tabaci	Onion thrips
	Beetles	*Melanauster chinensis*	Citrus trunk borer
		Otiorrhynchus cribricollis	Apple weevil
	Fruitflies	*Anastrepha fraterculus*	South American fruit fly
		Dacus dorsalis	Oriental fruit fly
		Dacus tryoni	Queensland fruit fly
		Ceratitis capitata	Mediterranean fruit fly

Ants	*Atta sexdens*	Leafcutting ant
	Iridomyrmex humilis	Argentine ant
Moths	*Phyllocnistis citrella*	Citrus leafminer
	Othreis cajeta	Indian banded owl
	Cryptoblabes gnidiella	Honeydew moth
	Ectomyelois ceratoniae	Carob moth
	Argyrotaenia citrana	Orange tortrix
	Prays citri	Citrus flower moth

Whiteflies suck the sap from leaves and from tender shoots, provoking leaf drop and producing honeydew on which saprophytic fungi develop which favour the growth of sooty mold, a blackish fungal encrustation which dirties the plants.

Aphids suck the sap from the leaves and from tender shoots. Some species provoke curling of leaves. Moreover they damage flowers, stimulating their drop and producing honeydew. They are dangerous because some of them are vectors of virus, like tristeza.

Soft scales, margarodid scale, mealybugs suck the sap from leaves and from tender shoots, producing honeydew which dirties the fruits.

Armored scales suck the sap from leaves and tender shoots without producing honeydew. The infected branches are liable to lose their leaves and twigs may dry up, sometimes in such a bad way as to cause the decline of the plant. The fruits infested are discarded. The psillid *Trioza erythreae* is the main vector of greening, a disease caused by a phloem-limited bacterium.

Beetle larvae attack roots, branches, and the trunks of trees, provoking their decline. Adults nourish themselves with leaves.

Fruitflies provoke punctures on the fruits in order to deposit their eggs from which larvae are hatched, provoking fruit rot and carpoptosis.

Ants nourish themselves on leaves and honeydew produced by other insects. They spread the attack of other insects and interfere with their natural enemies.

Moths damage flowers and fruit set. They cause miners and curling up of leaves. They make holes and miners on the fruit, sometimes stimulating fruit drop.

Among the pests there are the nematodes, microscopic vermiform organisms some of which are parasites of plants. The pathogenic nematodes of the citrus trees, living in various areas, are different (Lo Giudice, 1986), but only a few have some economical importance (Inserra and Vovlas, 1977). The most wide spread in the world is *Tylenchulus semipenetrans* (Citrus nematode). Mature female is a sedentary parasite which penetrates with the anterior part of its body into the outer cortical cells of the roots of the citrus plant, leaving on the outside posterior part of its body, where the definition semi-endoparasite derives from. The posterior part of females' body and their eggs are protected by a gelatinous mass produced by them. Males are not parasite forms. The species is bisexual, but can reproduce by parthenogenesis (Dalmasso *et al.*, 1972). They prefer well-oxygened soils but they can live on various types of soil and on species different from citrus like olive, grape and persimmon as there are different biotypes (Duncan and Cohn, 1990). The cycle from egg to egg lasts 6–8 weeks. The non-specific symptoms induced by parasitization are a reduced growth and vigour of the plant and production of under sized fruits. There are some physical, chemical and agronomic means to control this nematode, but the best control is prevention, avoiding the contamination by using, for the new plantings, the nematode free plants and using a preplanting located chemical treatment, where necessary. In case of infestation on plants in production the chemical means adopted are based on the level of infestation thought dangerous, choosing among the suitable nematocides.

Pest management techniques

The use of insecticides has evolved over time. We have passed from using hydrocyanic acid fumigation, mineral oils, tobacco, powders of pyrethrum to the organophosphorus esters, carbamates, chlorinated hydrocarbons, pyrethrin group, botanical insecticides like nicotine and rotenone, insecticides from microbial origin to follow with the development modifiers like moulting hormones, juvenile hormones and with behaviour modifiers like antifeedants and pheromones. The research continues to find other means of innovation and above all substitutes of chemical ones. Apart from insecticides, other means of control have had an evolution. For this reason we are now able to distinguish various means of control.

Biological control

Biological control involves the use of natural enemies (Table 4.2) and pathogens for the control of pests. Some insects are classified as natural enemies as they are parasitic or predatory. Parasites nourish themselves on other species, called host. Predators in various stages of their lives nourish themselves on other insects, finding them in their natural habitat. Pathogens are fungi that grow on dead or living insects or mites. Biological control is, among the various alternatives to chemical control, the most promising because it is not dangerous and often has constant effectiveness in time. Nevertheless not all the key pests can be controlled with this strategy, as for example *Ceratitis capitata*, therefore it must be integrated with chemical pesticides or other means of control (Rosen, 1986, 1990). Biological control is a strategy intended to limit or avoid the use of chemical products. In order to intensify the use of biological control the research on introduction, rearing methods and release programmes of natural enemies has been stimulated. The methodologies applied are: rearing and mass-release of natural enemies in order to have a rapid decrease of insects population; protection and increase of natural enemies, by adopting some devices (increase of alternative hosts, use of selective pesticides, use of cultural practices helping their development). In order to carry out strategies of biological control, some knowledge and experience are necessary to avoid the use of inappropriate methods.

Table 4.2 Biological control of citrus pests by natural enemies

Pest	Parasite	Predator
Dialeurodes citri	*Encarsia lahorensis*	*Clitostethus arcuatus*
Aleurothrixus floccosus	*Cales noacki*	*Synharmonia conglobata*
Parabemisia myricae	*Cales noacki* *Encarsia meritoria* *Eretmocerus debachi*	*Conwentzia psociformis* *Semidalis aleyrodiformis* –
Icerya purchasi	–	*Rodolia cardinalis*

Planococcus citri	*Leptomastix dactylopii*	*Cryptolaemus montrouzieri*
	Anagyrus pseudococci	*Scymnus includens*
	Leptomastidea abnormis	*Sympherobius pigmaeus*
Saissetia oleae	*Coccophagus lycimnia*	*Chilocorus bipustulatus*
	Metaphycus flavus	*Eublemma scitula*
	Metaphycus helvolus	*Exochomus quadripustulatus*
	Metaphycus lounsburyi	–
Ceroplastes rusci	*Coccophagus lycimnia*	*Scutellista cyanea*
Coccus hesperidum	*Coccophagus lycimnia*	–
	Encyrtus lecaniorum	–
	Metaphycus flavus	–
Parlatoria pergadii	*Encarsia citrina*	–
	Encarsia inquirenda	–
Aonidiella aurantii	*Aphytis melinus*	–
	Comperiella bifasciata	–
Aphis gossypii	*Lysiphlebus testaceipes*	*Coccinellidae*
Toxoptera aurantii	*Lysiphlebus testaceipes*	*Coccinellidae*
Panonychus citri	–	*Amblyseius stipulatus*
	–	*Amblyseius hibisci*
Tetranychus urticae	–	*Phytoseiulus persimilis*

Cultural control

It involves the use of agrotechnical practices like choice of varieties and rootstocks less susceptible, soil tillage, installing of appropriate methods of irrigation, removal of infestation sources, development of natural enemies, balanced use of chemical and organic fertilizers, quick harvest time, maintenance of canopy not in touch with soil and anything useful to keep the plant in healthy state. The use of varieties and rootstocks which are less susceptible to the attack of pests seems helpful, even though little progress has been made. This approach is compatible with the other means of control, is efficacious and cheap. Soil tillages are used to control the insects living in it. The installing of appropriate methods of irrigation avoids the irregular irrigation of plants, that can involve water surplus or water deficit. The removal of infestation sources carries out the cut through pruning of the most deteriorated parts of the plant which can shelter insects. The development of natural enemies in the agroecosystem is favoured by keeping weedy rows between the citrus trees or edge bushes. The unbalanced use of fertilizers, above all the nitrogenous ones, may increase the development of some insects. Where possible the early harvest of fruit can help avoiding the spring attacks of *C. capitata* (Bodenheimer, 1951). The keeping of canopy not in touch with soil avoids ants climbing which favours the movement of scales inside the tree and upsets the survival of natural enemies.

Mechanical and physical control

In order to discourage direct treatment on foliage some bands of various materials around the trunk are used and treated with pesticides. In some cases pesticides are not used and the band only functions as a trap. Some control devices of insects on harvested fruit have been used for a couple of years, besides the means of control that are used in field. One of these devices consists in high-pressure rinsing to remove scales and sooty mold. Another device is spreading to export citrus fruits towards countries where there are laws against the introduction of fruit attacked by *C. capitata*. Citrus fruits are sent by ship in frigo containers at temperatures, that vary according to the duration of the voyage, which are used to exterminate the unripe forms of the insect.

Autocidal control

Sterile insect technique (SIT) is based on rearing in laboratory of sterilized *C. capitata* males that are released in great quantity in order to avoid the possibility of mating between wild females and fertile males. By using this technique in Mexico and Guatemala an eradication programme of *C. capitata* has been carried out, where the intervention was integrated, including a series of activities whose last stage was mass-production and release of sterile insects (Linares and Valenzuela, 1990). This ecologically reliable method is considered expensive, even if the economic aspect should be seen into the framework of policy decision given the importance of the matter in certain environments (Cirio and Capparella, 1978). Pheromones are used either in the monitoring of insect populations or as a means of control. They are used in the mass trapping technique of *Prays citri* males. Some sticky traps are used for this aim and they are baited with synthetic female sex pheromone. This technique has managed to give a real control on *P. citri* (Sternelicht *et al.*, 1978). The use of pheromones has become widespread for the monitoring of many insects.

Chemical control

Even though at the beginning it has been the means of control which has given the most immediate results, its widespread use has led to problems in altering biological equilibrium, in increasing phenomena of plant phytotoxicity and pest resistance. In 1922 the first doubtful cases of scale resistance to hydrocyanic acid fumigation were pointed out (Quayle, 1922). The choice of the chemical product to use is based on the chemical, physical and biological characteristics. Nowadays great attention is given to the dangers of its use for both the applicator and for the consumer, taking into consideration the degree of toxicity, the persistence and the residues, while the economic aspect must not be neglected. After many years of use of chemical products, we are moving towards an age where old pesticides do not always give good results for the rising of resistance phenomena, while the search for new pesticides is increasingly reduced owing to the high costs for their production. In the chemical control of pests more and more space is given to mineral oils as they are compatible with integrated pest management (IPM) and have advantages compared with broad spectrum pesticides, as their toxicity towards vertebrate animals is low, have less detrimental effects on natural enemies, do not stimulate pest outbreaks, pests do not develop resistance phenomena and can be used with a minimum of protective clothing. In order to make mineral oils efficacious it is advisable to make a perfect spray coverage of all the tree, acting when the insect is more vulnerable. They are used with temperatures not higher than 32 °C and a relative humidity of the air not lower than 30 per cent. During summer time if they are not properly used, they may cause leaf drop, burning of fruit epidermis and carpoptosis. Moreover they may delay the ripening in the early ripening citrus varieties. On the contrary other pesticides have a disrupting effect on biological equilibrium and are toxic for humans. For this reason, it is advisable to use the most selective and the most secure ones for the environment. An example is the replacing of sulphur, which was used in the past to control mites, with more selective miticides (Rosen, 1967a,b). Even the use of poison baits, through aerial strip spray with malathion and hydrolized protein as attractant for females of *C. capitata*, has proved less disrupting for the ecosystem of the citrus orchard compared to full-coverage application of chlorinated hydrocarbons (Roessler, 1989). Some of the insect growth regulators, substances which interfere in the life processes of insects, prove to be efficacious, Darvas and Varjas (1990), but drawbacks are not to be undervalued (Viggiani, 1997).

Integrated pest management (IPM)

The current trend in pest management is to reduce the use of pesticides in order to limit the direct risks for the applicator and the environment and the indirect ones for the consumer, for the presence of residues in citrus fruits which go beyond legal amounts at the time of consuming. These are the reasons that have determined an evolution of control methods going from control on a calendar basis to a more rational one up to IPM. The traditional control on a calendar basis used pesticides in order to prevent pest attacks, without taking into consideration whether the pest was present or not, and considering only the phenological stages of trees. This method imports undesidered effects on the biological equilibrium of the ecosystem, and is expensive. Gradually IPM has developed. It uses a whole range of means in order to get a satisfactory control of pests and a greater respect for environment, without reducing the productivity of the citrus orchard. In this management a careful use of pesticides is taken into consideration, choosing among the most selective ones when necessary and avoiding broad spectrum pesticides. Biological control with natural enemies is always to be considered the main element (Rosen, 1986). Cultural, mechanical and autocidal means are used to protect the existing natural enemies and where necessary are integrated by mass-release of natural enemies. The basic components of IPM are:

1 Identification of pests and their natural enemies;
2 Monitoring of pests and their natural enemies through visual research and/or traps to evaluate the population;
3 Action threshold to know when it is time to act in order to avoid production loss;
4 Choice of management methods.

The success of IPM depends on the exact knowledge of agroecosystem of the citrus orchard.

DISEASES

Citrus orchards are subject to various diseases, but not all of them cause economically important damages. Some of them are widespread throughout the world, others have an importance restricted to certain citrus areas, others are present only in particular weather conditions. Among these diseases there are some that are made evident in field

Table 4.3 Some diseases of citrus

Causal organism	Common name
Fungal diseases	
Armillaria mellea	Armillaria root rot
Elsinoe fawcettii	Scab
Mycosphaerella citri	Greasy spot
Phoma tracheiphila	Mal secco
Phytophthora citrophthora and *P. parasitica*	Phytophthora root rot, Foot rot, Brown rot
Septoria citri	Septoria spot
Bacterial diseases	
Pseudomonas syringae	Blast and black pit
Xanthomonas campestris pv. citri	Canker

(Table 4.3) and others in post-harvest (Table 4.4). For some of these a brief description will be done.

Mal secco

It is a fungal disease restrictedly spread in the Mediterranean Basin, around the Black Sea and in Asia Minor, whose causal agent is *Phoma tracheiphila*. It has been known since 1894, even though the causal agent has been defined afterwards. It attacks mainly lemon and citron trees, but it can infect all species of Citrus, Poncirus and Fortunella. Therefore not only scions but also rootstocks like sour orange, citrange, alemow are susceptible. It is the most serious disease affecting lemon trees in Italy (Salerno and Cutuli, 1992). Infections occur mainly during autumn-winter time with high relative humidity and with a suitable temperature for the growth of fungus at 20 °C approximately. The development of infection is favoured by a vigorous growth of the tree. The infected trees are subject to a more rapid decline if the infection is in the roots and less rapid if apical. The pathogen penetration can proceed through stomata, but it is mostly through wounds deriving from winds, hail storms, frosts and other causes. Through these wounds, the fungus reaches the xylem vessels of the tree from which it spreads systematically. The infection manifests itself with vein chlorosis of leaves, which dry up and can either drop or stay on the tree according to the virulence of the infection, and with reddish or orange colouration of the xylem, that with ageing tends to brown. Quick pruning operations are used to stop apical infections, while root ones are not under control. Chemical treatments, even though repeated, have a limited effect. Cultural operations which interfere with the disease directly or indirectly are exploited as strategy of control. Among these the immediate pruning out and consequent burning of diseased parts of tree are fundamental. The only efficacious means of control against this disease is the use of resistant varieties like Monachello lemon, whose fruit quality, however, is not good.

Phytophthora-induced diseases

The best known species of Phytophthora in citrus orchard are: *Phytophthora parasitica* and *P. citrophthora*. These fungal species are widespread all over the world and they cause serious soilborne diseases in citrus. Trees are attacked in any age of their life.

In the nursery they provoke a decline of seeds in germinating stage or of seedlings and in field foot, rot and a decay of feeder roots. Fruits affected in field are subject to fruit drop and in packinghouse to brown rot. The fungus penetrates through wounds and develops provoking necrosis and abundant gum exudation of trunk and root decline. The infected trees show leaves with yellow veins. Means of spread are infested plants when they are transplanted in field, the tillage equipment, the irrigation and drainage water and the wind. The excessive soil moisture and air humidity, the low oxygenation of soil and temperatures are essential elements for mycelial growth. *P. citrophthora* has an optimum development between 24–28 °C while *P. parasitica* between 30–32 °C. Both the rootstocks and the different grafted species and varieties have a different susceptibility to infection. For instance among rootstocks, trifoliate orange and Swingle citrumelo are resistant, sour orange and Carrizo citrange are tolerant, sweet orange and grapefruit are susceptible (Castle *et al.*, 1989). The main strategy is to control the elements which favour the development of the pathogen and to strengthen the resistance of the host. Prevention is the best way to avoid the onset of this disease. Grafting beyond 15 cm high from soil, the lack of stagnant water, an efficacious airing under the tree canopy, the avoidance of every kind of injury, are very important preventive devices. The use of chemical products must be considered as a complementary measure only when the infection has reached economic thresholds.

Blast and black pit

These are diseases, whose causal agent is *Pseudomonas syringae*, a bacterium present in endemic form in citrus orchards (Salerno and Cutuli, 1992). This bacterial disease appears in cold and humid weather conditions, concerning above all the twigs of grapefruit and orange trees, and the fruits, in particular of lemon trees, when these have been damaged by wind, hail storm or other shocks that provoke wounds. Blast lesions appear on wings or petioles of leaves, as black areas. The infected leaves dry up and may remain attached to twigs or drop. Black pit lesions appear on fruit. They originate as brown spots on the peel and gradually turn black. Damaged fruit is not suitable for trade. In order to contain the damages of this disease, good cultural practices should be adopted. Orchards should be protected with windbreaks, in order to reduce wind scars, plant vigour should be controlled in autumn-winter time and, if necessary, copper fungicides should be used (De Cicco *et al.*, 1978).

Postharvest fungal diseases

Fungal attacks occur also after the fruit harvest. These infections are not neglected, either because they appear inside packinghouse, with losses for the dealer, or when they are on the market already, discrediting the product. The study of these fungal diseases is important for globalization of markets, for long distance transport needs and the necessity to reduce chemical treatments. A concise description of the symptoms induced on fruit by some of these diseases is made (Table 4.4).

Graft-transmissible pathogens

The graft-transmissible pathogens of citrus are reported in Table 4.5 and some of them will be described below.

Table 4.4 Postharvest fungal diseases

Causal organism	Common name	Symptoms
Alternaria citri	Alternaria rot	There are no external characteristic symptoms. It mostly attacks the mandarin fruits. It reveals itself with the black rot of the central axis and of the segment area near it. The pulp undertakes an unpleasant bitter taste.
Botrytis cinerea	Gray mold	At first the infected fruit shows some areas of the rind which are from yellowish brown to dark brown. Afterwards a white mold appears which later gains a characteristic gray colour.
Geotrichum candidum	Sour rot	The rot of the fruit begins with a yellow cream spot, slightly depressed with the softening of the underlying tissue. The rotting fruit gives off an acid odour.
Penicillium digitatum	Green mold	At first the infected areas are covered with a white mold which turns green (*P. digitatum*) or blue (*P. italicum*). In the advanced stage of infection the fruit decays.
Penicillium italicum	Blue mold	
Phytophthora citrophthora and *P. parasitica*	Brown rot	The deterioration of the fruit starts with a slight brown discolouration of the rind. The attacked area assumes a firm, leathery appearance. In humid condition it develops a white mycelium. The fruit gives off a characteristic penetrating acid odour.

Table 4.5 Some graft-transmissible pathogens of citrus

Diseases	Types of pathogen
Viral diseases	
Citrus variegation	Virus
Ring spot-psorosis complex	Not characterized
Tristeza	Virus
Virus-like diseases	
Cristacortis	Not characterized
Impietratura	Not characterized
Viroid diseases	
Cachexia	Viroid
Exocortis	Viroid
Systemic prokaryote diseases	
Stubborn	Spiroplasma citri
Diseases caused by unknown agent	
Blight	Not characterized

Tristeza

It is a very important pathogen of citrus for economic damages that it produces and it is liable to increasing spread in new areas. It is a virus, of which many strains have been detected with different virulence. The trees declined owing to this virus in Argentina, Australia, Brazil, Israel, Spain, South Africa, USA and other nations are millions. It infects many species and varieties of citrus. In citrus areas where sweet orange, grapefruit and mandarin were grafted on sour orange rootstock citrus growers were obliged to change this rootstock, because of its susceptibility. The infection manifests itself on sour orange with necrosis of phloem tissue below the bud union and depletion of starch levels. Moreover in the inner face of the bark there can be a thick pinholing or honeycombing. Some strains on lime, grapefruit and sweet orange provoke the stem-pitting and the seedling yellows on seedlings of lemon variety Eureka, grapefruit and sour orange. The symptoms are tree wilting, leaf chlorosis, and reduced fruit size, with final decline of the tree. Beside sour orange, other rootstocks are susceptible. The disease is transmitted by grafting and by several aphid species. *Toxoptera citricidus* is the most efficient vector among aphids. The effectiveness of transmission depends on aphid species, on strains, on the donor and on the receptor. Various screening tests are available for its identification. The typical biological test is graft-inoculation of Mexican lime seedlings, these react with specific symptoms. At the moment the enzyme-linked immunosorbent assay (ELISA) is the most widely used test because it's cheap and immediate. Spread of the disease can be prevented with quarantine, with eradication programmes, with use of tolerant rootstocks and with budwood sources certified free of tristeza.

Ringspot-psorosis complex

According to Garnsey (1999) this complex includes citrus necrotic ringspot, naturally spread psorosis, psorosis A, and psorosis B isolates. In this complex, still not well understood, are probably involved different causal agents with various symptomatologies. The transmission of all isolates occurs with infected budwoods and the control is obtained by using certified budwoods ringspot-psorosis complex free.

Exocortis

This disease, existing in many countries, is caused by a viroid. Its effects occur in citrus orchard when infected material is grafted on susceptible rootstocks like trifoliate orange, some of its hybrids and Rangpur lime (Garnsey and Barkley, 1988). The trees attacked dwarf, rarely decline and the quality of the fruit is not affected. Many species and varieties of citrus and non citrus host are susceptible to exocortis infection. The infected trees of sweet orange, grapefruit and mandarin are asymptomatic when grafted on non susceptible rootstocks. The most obvious symptom of exocortis is bark scaling on the rootstock with consequent stunting of the whole tree. Bark scaling usually begins when citrus trees are from 4 to 8 years old (Garnsey and Barkley, 1988). The disease can be spread by grafting, but it may also be transmitted mechanically by cutting graft tools and pruning equipment. It is just in this way that many trees which were immune have been infected. An indexing method for exocortis utilizes the grafting of a suspected diseased budwood on a Etrog citron clone that reacts with symptoms of epinasty. At the moment there are other detection procedures based on extracts, developed for laboratory analysis, with particular means (Garnsey and Barkley, 1988). In order to get viroid-free budwoods, shoot tip grafting can be used, a form of micrografting of the meristematic apex of a donor plant on a seedling to produce shoots immune from the pathogen. Cutting tools treated with aqueous solution containing 2 per cent sodium hydroxide plus 2 per cent formalin or freshly made 5–10 per cent solution of household bleach are used to prevent viroid mechanical transmission.

Stubborn

It is caused by *Spiroplasma citri*, a systemic prokaryote organism, smaller than common bacteria and bigger than viruses, wall free. This disease is important in hot and humid areas. Infections are rarely lethal. Among the most susceptible citrus species and hybrids there are: orange, grapefruit, mandarin and hybrids of mandarin (Garnsey and Gumpf, 1988). The causal agent can also infect non citrus plants. The trees infected show dwarfishness and branches with shortened internodes, cupped leaves and chlorosis. Fruits are frequently lopsided or acorn-shaped and small sized, with their stem ends greenish and with seeds often aborted. Stubborn is graft-transmitted from infected trees. There is also a natural transmission with some species of leafhoppers. Identification is made by graft inoculation of highly sensitive citrus varieties, such as sweet orange Madam Vinous, under warm conditions. A definitive diagnostic method is to culture the causal agent on artificial media and confirm its identity by microscopy or by serological test (Garnsey and Gumpf, 1988). Control is achieved by the use of stubborn free budwoods obtained with the shoot tip grafting technique.

Blight

It is a disease whose causal agent has not been defined, even though some hypotheses have been proposed. It is known also as Young Tree Decline and Declinio. It is widespread in Australia, Brazil, South America, Florida, Hawaii, South Africa and has widely been studied (Marais, 1990). It occurs in the tree canopy with leaves and twigs wilting, water sprouts on the trunk and zinc deficiency symptoms. It is characterized by zinc accumulation in the bark and xylem of the trunks, in whose vessels there are amorphous plugs which cause lack of water absorption. The symptoms of blight are not totally different from those of decline caused by other agents. Among the most susceptible rootstocks there are: rough lemon, volkameriana, trifoliate orange and Rangpur lime. The infection from the blight affected tree to a sound tree is root graft transmitted, and not by apical grafting. Sour orange and sweet orange rootstocks are more tolerant, but they are not taken into consideration for their susceptibility to tristeza and phytophthora. The kind of soil, cultural practices and nutritional disorders have been related to the incidence of blight. Several means of control have been used, without reducing the incidence of the disease.

CONCLUSIONS

Nowadays the defence of production, environment, and consumers' health requires a precise choice of strategy to control pests and diseases. Selective, less toxic and low persistent pesticides, are the most favoured. Control of equipment used for the distribution of pesticides is always more and more cared of, as to limit the quantity of product distributed, and possibly to concentrate treatment to trees only, without too many losses in environment. Control strategies privilege biological, agronomic, mechanical and physical means to reduce the use of chemical products to a minimum. It is only through an integrated management that production can go on in respect of environment and humans.

REFERENCES

Bodenheimer, F.S. (1951) *Citrus Entomology in the Middle East*, W. Junk, The Hague.

Castle, W.S., Tucker, D.P.H., Krezdorn, A.H. and Youtsey, C.O. (1989) *Rootstocks for Florida Citrus*, IFAS, University of Florida, Gainesville.

Chapot, H. and Delucchi, V.L. (1964) *Maladies, Troubles et Ravageurs des Agrumes au Maroc*, Institut National de la Recherche Agronomique, Rabat.

Cirio, U. and Capparella, M. (1978) Valutazioni tecniche, ecologiche ed economiche sull'impiego della tecnica di lotta dell'insetto sterile contro la Ceratitis capitata Wied. nell'isola di Procida al termine dell'esperimento pilota di controllo. *Atti XI Congresso Naz. Ital. di Entomologia*, Portici-Sorrento, 10–15 Maggio 1976.

Couilloud, R. (1991) *Insects, Araignées & Acariens*, CIRAD, Montpellier.

Dalmasso, A., Macaron, J. and Berge, J.B. (1972) Modalites de la reproduction chez Tylenchulus semipenetrans et chez Cacopaurus pestis (Nematoda-Criconematoidea). *Nematologica*, 18, 423–431.

Darvas, B. and Varjas, L. (1990) Insect Growth Regulators. In D. Rosen (ed.), *Armored Scale Insects Their Biology, Natural Enemies and Control. World Crop Pests*, 4B, Elsivier, Amsterdam, pp. 393–408.

Davies, F.S. and Albrigo, L.G. (1994) *Citrus*, CAB International, UK.

De Cicco, V., Luisi, N. and Salerno, M. (1978) Epidemiology and control of citrus blast. *Proc. Int. Soc. Citriculture*, Sidney, 1978.

Di Martino, E. (1985) I parassiti animali. In P. Spina and E. Di Martino (eds), *Trattato di Agrumicoltura*, Edagricole, Bologna, pp. 113–145.

Ducan, L.W. and Cohn, E. (1990) Nematode parasites of Citrus. In M. Luc, R.A. Sikora and J. Bridge (eds), *Plant Parasitic Nematodes in Subtropical and Tropical Agriculture*, CAB International, U.K., pp. 321–346.

Ebeling, W. (1959) *Subtropical Fruit Pests*, University of California, Division of Agricultural Sciences, Berkeley.

Garnsey, S.M. (1999) Systemic diseases. In L.W. Timmer and L.W. Duncan (eds), *Citrus Health Management*, APS, USA, pp. 95–106.

Garnsey, S.M. and Barkley, P. (1988) Exocortis. In J.O. Whiteside, S.M. Garnsey and L.W. Timmer (eds), *Compendium of Citrus Diseases*, APS Press, USA, pp. 40–41.

Garnsey, S.M. and Gumpf, D.J. (1988) Stubborn. In J.O. Whiteside, S.M. Garnsey and L.W. Timmer (eds), *Compendium of Citrus Diseases*, APS Press, USA, pp. 46–47.

Inserra, R.N. and Vovlas, N. (1977) Nematodes other than Tylenchulus semipenetrans Cobb pathogenic to citrus. *Proc. Int. Soc. Citriculture*, Florida.

Jeppson, L.R. (1989) Biology of Citrus Insects, Mites and Mollusks. In W. Reuther, E.C. Calavan and G.E. Carman (eds), *The Citrus Industry*, V, University of California, Division of Agriculture and Natural Resources, USA, pp. 1–87.

Linares, F. and Valenzuela, R. (1990) Medfly program in Guatemala and Mexico: current situation. In M. Aluja and P. Liedo (eds), *Fruit Flies Biology and Management*, Springer-Verlag, New York, pp. 425–438.

Lo Giudice, V. (1986) Citrus nematodes. In R. Cavalloro and E. Di Martino (eds), *Integrated Pest Control in Citrus-Groves*, A.A. Balkema, Rotterdam, pp. 209–216.

Longo, S., Mazzeo, G. and Siscaro, G. (1994) Applicazioni di metodologie di lotta biologica in agrumicoltura. *L'Informatore Agrario*, 50(28), 53–65.

Marais, L.J. (1990) Citrus Blight: world research review. *Citrograph*, 75(5), 119–124.

Quayle, H.J. (1922) Resistance of certain scale insects in certain localities to hydrocyanic acid fumigation. *J. Econ. Entomol.*, 15, 400–404.

Roessler, Y. (1989) Insecticidal bait and cover sprays. In A.S. Robinson and G. Hooper (eds), *Fruit Flies Their Biology, Natural Enemies and Control. World Crop Pests*, 3B, Elsevier, Amsterdam, pp. 329–336.

Rosen, D. (1967a) Biological and integrated control of citrus pests in Israel. *J. Econ. Entomol.*, 60, 1422–1427.

Rosen, D. (1967b) Effect of commercial pesticides on the fecundity and survival of Aphytis holoxanthus (Hymenoptera: Aphelinidae). *Israel J. Agric. Res.*, 17, 47–52.

Rosen, D. (1986) Methodologies and strategies for pest control in citriculture. In R. Cavalloro and E. Di Martino (eds), *Integrated Pest Control in Citrus-Groves*, A.A. Balkema, Rotterdam, pp. 521–530.

Rosen, D. (1990) IPM: Background and General Methodology. In D. Rosen (ed.), *Armored Scale Insects Their Biology, Natural Enemies and Control. World Crop Pests*, 4B, Elsevier, Amsterdam, pp. 515–517.

Salerno, M. and Cutuli, G. (1992) *Guida Illustrata di Patologia degli Agrumi*, Edagricole, Bologna.

Sternlicht, M., Goldenberg, S., Nesbitt, B.F., Hall, D.R. and Lester, R. (1978) Field evaluation of the synthetic female sex pheromone of citrus flower moth, Prays citri (Mill.) (Lepidoptera:Yponomeutidae), and related compounds. *Phytoparasitica*, 6, 101–113.

Talhouk, A.S. (1975) Citrus pests throughout the world. In E. Häfliger (ed.), *Citrus*, Ciba-Geigy, Switzerland, pp. 21–23.

Viggiani, G. (1997) *Lotta Biologica e Integrata nella Difesa Fitosanitaria*, 2, Liguori, Napoli.

5 Development of the citrus industry: historical note

Angelo Di Giacomo

ESSENTIAL OILS

As opposed to many other types of fruit whose homogeneity makes them more suitable for industrial use, the complexity and variability in the different layers of the structure of citrus fruits has been a determining factor in slowing the development of systems used in the production of their derivatives.

It is true that essential oil from citrus fruit may issue from peel when the peel is simply cut, scraped or squeezed. Consequently, its knowledge coincides with the introduction of citrus cultivation in different areas, without practical uses being found, however, for a long time. It is not until the sixteenth century that definite references to orange and lemon essential oils are found.

Conrad Gesner (1516–1565), was the first to mention distilled essential oils from oranges and lemons. He was followed by Jacques Besson (1571) and Giovanni Battista della Porta (1589). It was the latter, in his *Magiae naturalis*, who clearly described the process of distillation, starting from the triturated peel of the two fruits. The same author also described the essential oil obtained from the distillation of bitter orange blossom. In 1688 bergamot essential oil was mentioned for the first time in a pharmacist's inventory in Giessen, in Hesse, Germany (Fluckiger, 1876). Also Gaubius (1771), at the beginning of the second half of the eighteenth century, recommended the distillation process described by della Porta.

The oil of bitter orange blossom later became known as neroli from the name of the princess Anna Maria de la Tremoïlle de Noirmoutier, wife of Flavio Orsini, Duke of Bracciano and Prince of Nerola, to whom is attributed the introduction of this incomparable perfume into France and into the principal European courts. The production of neroli and of *petit-grain* developed, above all, in southern France, which still leads the field in its production. Also in Paraguay the production of *petit-grain* became important. Here the industrial exploitation of wild woodland, deriving from the plantations of the Jesuit missions, was begun by the French botanist Balansa in 1880, near Villa Rica (Guenther, 1949).

A process for the mechanical extraction of citrus peel essential oils, by breaking the tissues of the peel through the use of iron graters, was described for the first time by Geoffroy (1721) at the Académie des Science of Paris. However, the process was probably used before that date.

The extraction of essential oil by squeezing the peel dates back to 1700 (Di Giacomo, 1970). In two letters dated November 1776, published in Florence, the abbot Domenico

Sestini describes, amongst other things, the extraction of essential oil through the sponge process, with technical details practically the same as those still used in Sicily, although increasingly rarely. The manual sponge process consists of pressing the previously de-pulped peel several times against a system of natural sponges fixed on a terracotta basin, called a *concolina*, and held in place by a piece of wood resting on its rim. The pressure on the peel is accompanied by a rotatory movement of the hand. The peel, before the extraction, must be washed well in water containing lime, and left to drip-dry for a suitable period on grates or baskets. From the sponges the absorbed essential oil passes into the *concolina* together with an aliquot of liquid from the peel. At the end of the process more essential oil mixed with liquid from the peel is recovered by squeezing the sponges. Finally, by simple decanting, the essential oil is separated from the watery phase, which also contains detritus produced by the laceration of the peel tissues.

In 1909 Lo Castro patented the 'lever machine' which represented the first attempt to partially mechanize the sponge process. An interesting development of this hand-operated machine, also known as *Zona*, was produced with appreciable results during the Second World War by Mrs Z. Samish, of the Agricultural Experiment Station of Rehovoth, Israel (Braverman, 1949).

A manual process using the whole fruit, called the écuelle process, was used in Italy and southern France before 1800. It was later used for a long time in Jamaica, the island of Dominica and other islands of the West Indies. The écuelle, in the form of a shallow funnel with a diameter of about 25 cm, was initially made of wood and later of copper. The internal part of the funnel is studded with brass nails protruding for about 20 mm; in the centre a sort of hollow grip held between the knees of the operator serves to collect the essential oil that jets out together with solid matter and a watery component, as the fruit is slowly pressed and turned against the inner walls.

The most immediate development of the écuelle process is the so-called 'Calabrian Machine', built in Reggio Calabria around 1840 by Nicola Barillà specifically for the extraction of bergamot essential oil. This machine, with slight modifications, is still used in Calabria, although by an ever-decreasing number of traditional businesses. The basis of the machine is two bowls of cast zinc, called 'cups', which together form a cylindrical ring. The lower cup is fixed, and has a number of points on the bottom, and teeth on the conical walls. The upper cup, of the same diameter, contains a number of small, brass blades, which, when the machine is working, rotate around a vertical axis, dragging and pressing the fruit against the lower cup. The fruit is loaded periodically and the treatment is regulated according to the condition of the bergamot (degree of ripeness and size of fruit). The liquid that forms, made up of essential oil and a watery phase containing solid matter in suspension, is collected through holes for this purpose in the lower cup. The treated bergamots are dried with a sponge to recover any essential oil remaining on the surface of the fruit. The Calabrian machine is of low cost, and is notable for the production of an essential oil of excellent quality, with a high yield.

These manual procedures that we have briefly described above, inspired the inventors of the first extracting machines.

All the procedures that treat the entire fruit can be classified as follows, according to the way in which the surface of the fruit is treated:

1. superficial grazing, an operation generally not drastic, which causes limited damage to the tissues;
2. total abrasion or peeling of the exterior layer containing the essential oil glands.

The first machine built to extract lemon essential oil was patented in 1908 by Perroni Paladini, from Messina. He was killed in December of the same year in the devastating earthquake which struck Messina and Reggio Calabria, and was thus unable to bring his machine to perfection. If, from a mechanical point of view, this machine can be considered primitive, the technological principle on which it is based later proved its worth and was used in many subsequently produced extractors. This principle consisted in forcing the lemons into a canal fitted with graters and shaped to fit the fruit. The lemons were pulled into the machine by a belt also fitted with graters. Springs placed under the fixed canal assured a uniform pressure no matter what the dimensions of the fruit. The canal-belt system was followed by another similar, but this time fitted with sponges instead of graters, designed to soak up the essential oil left on the peel.

The principal machines came into operation immediately preceding the First World War, and their refinement has continued up to the present day.

Amongst the machines that use the superficial grazing system, it is worth noting the IFAC-Schwob extractor (Schwob, 1951), the Spanish machine 'de estilete', and the Italian 'Indelicato MK'. In the case of the latter the system may be best defined as 'pricking and vibration'.

Peeling the fruit is found in different models, amongst which: in Italy, the Lo Verde, Vinci, Cannavò, Avena and Speciale machines, all built between 1914 and 1928; in the USA, the Fraser-Brace machine, the citrus oil extractor (Hyland-Stanford Corporation), the drum extractor, and the whole citrus fruit extractor (Brown International Corporation); in Israel, the Koffler and the Jaf-Ora; and in Spain, the Calvillo.

The 'slow-folding', or *sfumatrice* machines are all based on the use of the natural pressure that the essential oil exerts inside the glands, and which forces the essential oil to jet out when the peel is pressed or squeezed.

In practice, these extractors are simple and cheap. The peel, treated with lime if necessary, is fed into the machine through a hopper, and then is forced to pass through a space, consisting of one fixed and one mobile part, which progressively narrows, and where the peel is subjected to multiple and random folding. The extracted essential oil is carried along by strong jets of water, similar to what happens in the machines which work on the whole fruit.

The Cianciolo 'slow-folding', or *sfumatrice*, machine was patented in 1927. Subsequently, other machines were built between 1930 and 1937 by Morasca, Speciale, Ramini and Avena. At present, the most commonly used 'slow-folding' machine is that produced by the firm Fratelli Indelicato of Giarre (Catania), which put their first models on the market in the 1940's.

Today, the peel is widely treated by simple pressing: a much more drastic action compared to that of 'slow-folding'. This system is particularly used in Florida (Pipkin Peel Oil Press) for orange peel, and in Sicily for lemon and mandarin peel.

In 1947, a new machine was introduced by the Food Machinery Corporation (FMC), San José, California (Guenther, 1947). This machine was completely original and operated on a different principle from all the machines that preceded it. The FMC Whole Fruit Extractor does not only extract the juice, but simultaneously squeezes the oily emulsion from the peel, without the juice and the essential oil coming into contact. This extractor, although it has kept an external appearance very similar to the original, has been produced in various models over the years, with continuous mechanical improvements. This has been a great success in all those areas of the world where citrus fruit is processed. It is for this reason that the American Society of Mechanical Engineers named this extractor 'An International Historic Mechanical Engineering Landmark' (March 24, 1983).

In the period immediately preceding the First World War, Peratoner and Scarlata from Palermo patented a method, known as the 'Peratoner process', suitable for the extraction of essential oil by reduced-pressure distillation of citrus juice directly obtained from chopped and pressed fruit.

The concentration and the deterpenation of the essential oils by fractional distillation in a vacuum was used industrially for the first time by Heinrich Haensel in Pirra in 1878, and, subsequently, also the firms Schimmel and Sachse put their deterpenated essential oils on the market.

Bennet of Messina in 1914 is considered the first to have used a centrifuge to separate essential oils from citrus juice. In 1915 the same process was patented by Peratoner and Cianciolo (Di Giacomo, 1970). After initial uncertainties on the part of the operators, the centrifuge gained in popularity to become considered the most suitable system for the separation of essential oils from oily musts.

JUICES

For a long time the industrial processing of citrus fruits was based, at least in Italy, solely on the extraction of the essential oil contained in the peel.

A partial utilization of the citrus fruit, however, can only originate from contingently anomalous conditions; but, from the point of view of an increasingly rational technological development, all the many, potential uses of the citrus fruit cannot be ignored. Its exploitation must be integrated to produce all possible derivatives.

In particular, as far as concerns the production of juice, the results of research carried out by the industries of California and Florida (Timmons, 1950; Will, 1916) since the beginning of the 1900's must be considered decisive. The advance of technology, allowing the preservation of juice over time (particularly flash-pasteurization and vacuum concentration), was an important factor for the development of juice production. The growth of this production was also made possible by the contemporary utilization of the moist or dried refuse material as cattle feed (Mead and Gilbert, 1926, 1927).

It can be affirmed that only in recent decades has there been a further noticeable increase in the development of the juice industry, thanks also to an awareness on the part of the consumer of the basic nutritional and vitamin-rich properties of fruit juice, and not simply as a drinks ingredient. There is a tendency, in consumer interest, to move away from those fermented products known since antiquity (such as wine and cider) and towards more natural products, as part of a more rationally planned diet. Naturally, this shift of consumer tastes has not only been determined by an increased awareness of the nutritional properties of fruit juice. The essential factor has been the increasing capacity of the industry to prepare high-quality products that satisfy both the organoleptic characteristics, and those that derive from the original fruit. All this with the advantage of the products reaching the consumer even when the fruit itself is out of season.

For a long time the extraction of juice from citrus fruit was carried out by hand, squeezing the half-fruit against suitably designed rosettes. The main difficulty in mechanizing this operation is that each half-fruit has to be treated separately. However, from the 1930s onwards, many attempts were made to imitate the manual method, substituting it with a mechanical procedure.

The first attempt was made by Braverman and Criss (1932), who patented a machine that cut the fruit with a circular blade, and the juice was then extracted without causing appreciable damage to the peel. Because of technical difficulties existing at the time in Palestine, the patent was never commercially used, but the inspiring principle was taken up by Brown (1935) who built a machine that was successfully used in California, later with improvements to proficiency (Brown, 1937).

In the same period a patent was granted to Segovia (1937) based on a system where each fruit is treated individually, cut into two halves which are then pushed against fixed rosettes.

All other mechanical extractors are based on completely different principles. Most of them, despite being extremely ingenious, have the defect of exerting a high pressure on the peel, causing the extraction of considerable amounts of essential oil, as well as the liquid from the *albedo*. Amongst these machines it is worth mentioning those built by FMC with the names Skinner and Polk (1937), by Rotary Juice Press (Fauld) and by Bireley's Inc. of Los Angeles (Citromat). The principle of the Citromat ('The halved fruit feed a rotating drum that presses the fruit-halves against a fixed hole; the distance between the drum and the hole is gradually reduced while the juice passes through the hole') was later taken up in Italy by the Polycitrus made by Fratelli Indelicato (Giarre).

Braverman (1949) refers to numerous other machines designed in the USA for the mechanical extraction of juice, even if almost none of them was commercially produced.

We have already referred in this chapter to the FMC-in line extractor, nowadays widespread in all the main citrus-growing areas in the world, and notable for the simultaneous extraction of juice and essential oil.

CITRIC ACID

Citric acid was isolated from lemon juice in 1784, but its composition was made clear only after almost a century. The procedure for obtaining citric acid from citrus juice (above all, lemon, lime and bergamot), also known as the Scheele process, consists of two parts: the precipitation of citric acid in the form of calcium citrate, and then the 'decomposition' of the citrate itself, which permits the production of a citric solution of sufficient purity to allow its crystallization. For a long time the Scheele process was used, above all, in England. From the eighteenth century onwards, there was the production in Sicily of *agrocotto*, a dark liquid obtained by concentrating lemon juice, containing around 40 per cent citric acid. This production was aimed at supplying the industry that was developing in England. In truth, it seems to have been a sort of *agrocotto* that was produced in Roman times for medicinal purposes, prepared from the endocarp of the citrus fruit. Later, with the aim of having available a more suitable raw material, English manufacturers encouraged the production of calcium citrate (corresponding to the first phase of the Scheele process). This production was begun in Sicily around 1860, and in the course of a few decades had replaced that of *agrocotto*. At the end of the nineteenth century, the production of *agrocotto* and calcium citrate had also gained importance in the British West Indies. In the years preceding the First World War, Italy was capable of supplying 90 per cent of world demand for calcium citrate. In 1913 in Sicily, production was equivalent to 6,000 metric tonne, in addition to around 250 metric tonne of *agrocotto*; in 1918 production was 9,087 tons. In the same period, direct production of citric acid also started in Sicily. In 1908 Italy imported citric acid (164 tons), whereas, in 1916 it exported over 1,000 tons. In the period between 1899 and 1915 citric acid was also produced in California (Braddock, 1999). Starting from 1918, processes for the production of citric acid were put into operation that were based on the fermentation of molasses from sugar cane and sugar beet. The biological product became more and more popular, so that, in practice, today citric acid is no longer produced from citrus fruit.

PECTIN

The preparation of fruit jellies was carried out long before the discovery of pectin. However, the first information on water-soluble substances having a strong gelatinizing effect, and present in fruit, was given by Vauquelin (1790). The discovery of pectin is due to Braconnot (1825), who, defining it as the gelatinous principle in fruit, described many of its properties. He derived the term 'pectin' from the Greek word πηκτος. Shortly after that, Fremy (1840) demonstrated that pectin played an important role in regulating the firmness of fruit as it ripens.

The works of Goldthwaithe (1909, 1910) gave an enormous theoretical and practical contribution to the knowledge of pectin, and constituted the basis for the development of the pectin industry in the USA. The production of pectin on an industrial scale from the waste products of lemon processing goes back to the campaign of 1923–1924, that is, a hundred years after Braconnot's discovery. The process was put into operation in the workshops of the California Fruit Growers' Exchange, under the direction of C.P. Wilson (1926) and the first factory was set up in Corona (California). The production was aided by the US government, which granted notable reductions in the price of the alcohol used on the processing. With the creation of other factories, the American industry rapidly became the supplier of the world market.

Starting in the 1930s, pectin, in the form of a concentrated liquid extract and, later, in powder, was also produced in Italy.

FLAVONOIDS

The interest in citrus flavonoids was determined by research carried out by Armentano *et al.* (1936) and Rusznyák and Szent-Györgyi (1936), who reported that 'Citrin', a crude mixture of flavonoids prepared from lemon peel, was effective in reducing capillary permeability in man, in prolonging life and mitigating tissue hemorrhaging due to scurvy in guinea pigs. The group of substances responsible for these effects was considered to be a vitamin, and was given the name 'Vitamin P'. Successive research was unable to demonstrate that flavonoids are necessary factors in the diet, and the term 'Vitamin P' fell into disuse, and was replaced by the term 'bioflavonoids' (flavonoids having a biological activity).

If flavonoids cannot be classed as vitamins, at least they have begun to be considered as pharmacological and therapeutic agents that can be used in the treatment of capillary fragility and related disorders. Among the many other clinical applications proposed for flavonoids, it is worth noting its use, although of undemonstrated effectiveness, in the treatment of the common cold.

Hesperidine has been prepared industrially since the 1940s using methods based on alkaline extraction from the peel (Higby, 1946).

BRINED CITRUS

The conservation of citrus peel using brining techniques was still considered a 'new' industry in the USA as late as the 1960s (Agricultural Research Service, 1962).

There is documentary evidence, however, that the process of brining lemons was known and used in Sicily as early as the 1700s; in particular, the abbot Domenico Sestini, during a journey through Sicily in 1776, having learned the details of the process, described it in detail in a letter sent from Catania on November 2 of that year to a friend in Florence. The process described is not greatly dissimilar from that still used today; in fact, attempts at mechanization have not met with great success.

OTHER BY-PRODUCTS

In California, between the years 1899 and 1915, many citrus by-products were prepared and put on the market. These are discussed in a report by Will (1916), which mentions, apart from citric acid, essential oils (cold-pressed, distilled, terpeneless) and pectin, the following: dried peel (cattle feed), candied peel, purees, syrup, alcohol, wine and vinegar. To this list, Chace (1922) added marmalades, jams and jellies, and also described their methods of production.

The first reports concerning the industrial production of citrus seed oil date back to 1930; during the season 1938–1939, about 90,000 lbs of grapefruit seed oil were produced in Florida. In Italy, in the years immediately preceding the Second World War, considerable quantities of lemon seed oil were produced by pressure from dried seeds. However, after the war this production was not resumed. In the USA, after a period in which production was reduced to negligible levels, there was a reawakening of interest around 1946–1947. At that time, two factories were working (one in Lakeland and the other in Plymouth) which reached a maximum production of 700,000 lbs.

A procedure for the separation of seeds from the refuse material was developed by Kirk (1967) of the Imperial Citrus By-products Corp., Lakeland, Florida.

REFERENCES

Agricultural Research Service (1962) Chemistry and technology of citrus, citrus products and by-products, *Agriculture Handbook n. 98*, United States Department of Agriculture, Washington, D.C.

Armentano, L., Bentsath, A., Béres, T., Rusznyak, S. and Szent-Györgyi (1936) *Deut. Med. Wochschr.*, **62**, 1325.

Besson, J. (1571) *L'Art et Moyen Parfait de Tirer Huiles et Eaux de tous Medicaments Simples et Oléagineux*, Paris.

Braconnot, H. (1825) Recherches sur un nouvel acide universellement répendu dans tous les végétaux, *Ann. Chim. Phys. Ser. 2*, **28**, 173–178.

Braddock, R.J. (1999) *Handbook of Citrus By-Products and Processing Technology*, John Wiley & Sons Inc., New York.

Braverman, J.B.S. and Criss, B. (1932) Palestine Patent 224 (Mar. 16) and 'An improved machine for the continuous automatic extraction of fruit juices', *Palestine Patent* 245 (June 16).
Braverman, J.B.S. (1949) *Citrus Products*, Interscience Publishers Inc., New York.
Brown, W.O. (1935) Citrus fruit juice extractor, *U.S. Patent* 2,130,610 (Dec. 2).
Brown, W.O. (1937) Juice extracting machine, *U.S. Patent* 2,199,876 (Mar. 22).
Chace, E.M. (1922) By-products of citrus fruits, *Food Technol.*, 49(9), 74, 76–77.
Di Giacomo, A. (1970) Gli oli essenziali degli agrumi, *EPPOS*, 52, 140–145.
Frémy, E. (1876) Méthode générale d'analyse du tissu des végétaux, *Compt. Rend.*, 83, 1136.
Flückiger (1876) *Dokumente zur Geschichte der Pharmazie*, Halle, 72.
Gaubii (1771) *Adversariorum Varii Argumenti Liber Unus*, Leidae, 31.
Geoffroy, Cl.J. (1721) *Mémoires de l'Académie des Sciences de Paris*, 159.
Gesner, C. (1555) *Ein köstlicher theuren Schatz EUONYMI PHILIATRI darinn enthalten sind viel heimlicher guter stuck der artzney*, 1, 238.
Goldthwaite, N.E. (1909) Contribution on the chemistry and physics of jelly-making, *Ind. Eng. Chem.*, 1, 333–340.
Goldthwaite, N.E. (1910) Studies on jellies and jelly-making, *Ind. Eng. Chem.*, 2, 457–462.
Guenther, E. (1949) *The Essential Oils*, Vol. III, D. Van Nostrand Co., New York, N.Y.
Higby, R.H. (1946) Method of manufacturing hesperidin, *U.S. Patent* 2,400,693 (May 21).
Kirk, W.A. (1967) Citrus seed separator, *U.S. Patent* 3,330,410 (July 11).
Mead, S.W. and Guilbert, H.R. (1926) The digestibility of certain fruit by-products as determined for ruminants. Part I. Dried orange pulp and raisin pulp. *Agr. Exp. Sta. Bull.* n. 409, University of California, Berkeley, CA.
Mead, S.W. and Guilbert, H.R. (1927) The digestibility of certain fruit by-products as determined for ruminants. Part II. Dried pineapple pulp, dried lemon pulp, and dried olive pulp. *Agr. Exp. Sta. Bull.* n. 439, University of California, Berkeley, CA.
Nolte, A.J. and Van Loesecke, H.W. (1940) Grapefruit seed oil: manufacture and physical properties, *Ind. Eng. Chem.*, 32, 1244–1246.
Polk, R. Sr. and Polk, R. Jr. (1937) Method of extracting fruit juices, *U.S. Patent* 2,137,414.
Della Porta, G.B. (1563) *Magiae Naturalis Libri Viginti*, Romae, 118.
Rusznyák, St. and Szent-Györgyi, A. (1936) Vitamin P: flavonols as vitamins, *Nature*, London, 138.
Segovia, C.B. (1936) Automatic fruit juice extractor, *U.S. Patent* 2,078,737 (Nov. 9).
Timmons, D.E. (1950) Citrus canning in Florida: early history and current statistics. *AE Series 50–4*, (January), Fla. Agr. Exp. Sta., University of Florida, Gainesville, FL.
Vauquelin, M. (1829) Mémoire sur l'acide pectique et la racine de la carotte, *Ann. Chim. Phys.*, 41, 46–61.
Will, R.T. (1916) Some phases of the citrus by-product industry in California, *Ind. Eng. Chem.*, 8(1), 78–86.
Wilson, C.P. (1925) Pectin manufacture, *Ind. Eng. Chem.*, 17, 1065.

6 Flowsheet showing steps in the processing of citrus fruits

Angelo Di Giacomo

PRELIMINARY TREATMENT OF THE FRUITS

The harvesting of the fruit is an operation that should always be carried out manually by specialized personnel, so as not to damage the peel, which contains the essential oil glands.

The citrus fruit then reaches the processing plant directly from the citrus groves, or via packing houses, preparation for the marketing of the fresh fruit. Transportation is usually by lorry, loaded in bulk, or in bins. As a rule the height to which the fruit is loaded is limited to avoid damage by crushing. It is certainly helpful if the lorries and bins are sterilized before use. The fruit is unloaded directly from the lorry or bin, and passes onto a level roller conveyor in order to remove the leaves, and any soil present. It is then carefully inspected to eliminate fruit which is rotten or does not reach those standards required for processing (Braverman, 1949; Agricultural Research Service, 1962). After that, a representative sample of each batch is selected automatically and is sent to the laboratory, where a series of tests are carried out to determine the yield of essential oil and juice, and the main characteristics of these two principal derivatives. In the next phase, the citrus fruit can be stored using a dry system, that is in aerated storage bins equipped with inclined shelves, where the fruit can be placed in contained layers to ensure it is kept in a healthy condition. Both the shelves and the bottom of the storage bins are constructed of seasoned wood, resistant to water or warping. Furthermore, the inclination of the shelves is sufficient to allow the unloading of the fruit without difficulty. Another system, called the 'water system', is made up of large concrete basins placed at floor level. In these basins the fruit is driven along by strong jets of water and by rotating blades. After that, the fruit, coming from one of the above mentioned systems, travels along a roller conveyor to a brush-washer machine. If collected from the storage bins, a second inspection is necessary to remove any damaged fruit, even though the bins are equipped with partitions designed to limit the crushing of the fruit. In the brush-washer, the fruit is washed by sprays of a germicidal solution (220 ppm of residual Cl_2) which completes the mechanical brush action of germ removal. A final wash using potable water is carried out before the 'Extraction' process, in order to eliminate any further contamination which may derive from the conveyor belts.

Before extraction, the fruit which is suitable for processing must be sized: first of all it passes along a roller conveyor, where it is uniformly distributed to allow the correct feeding of the sizing machine. Schematically, this consists of, on one side, a transversally inclined belt and, on the other, a rotating metal cylinder. When the fruit reaches a zone through which it can pass, the sizing machine separates it according to its diameter, thus feeding an inclined belt which, in its turn, passes the fruit on to the extractors. The belt has a number of exits, equivalent to the number of extractors being used. The different sizes are kept apart by means of dividers on the belt which guide the fruit to the extractor. Sizing is necessary for maximum efficiency of the extractor and a good quantitative and qualitative yield in essential oil and juice. After sizing, the fruit is transferred to the extractor hoppers by means of inclined belts equipped with suitable deflectors. The excess fruit is sent back to the selection phase.

PROCESSING

There are three fundamental methods for extracting the essential oil (Di Giacomo and Mincione, 1994):

1. essential oil extraction from the whole fruit, preceding juice extraction (Figure 6.1);
2. juice extraction precedes that of essential oil (Figure 6.2);
3. extraction of essential oil and juice takes place simultaneously (Figure 6.3).

In their turn, all the procedures for essential oil extraction from the whole fruit can be classified as follows, according to the way in which the surface of the fruit is treated:

- superficial rasping, generally not a drastic operation, and one which causes minimal damage to the tissues;
- total abrasion or peeling of the exterior layer containing the essential oil glands.

Furthermore, with regard to the machines that operate on the peel after the juice has been extracted, the most widespread process is that of 'slow-folding' or *sfumatura*, employing a mechanism which largely reproduces the manual sponge process. The extraction can also be carried out in a more drastic form, by means of continuously working presses.

Juice extraction can also be carried out on half-fruit using machines based on diverse principles. In some, extraction is carried out using revolving rosettes, without causing much deformation of the peel, whereas others use a more vigorous action on the half-fruit, extracting a notable amount of essential oil, as well as other liquids from the peel, above all in cases where juice extraction precedes that of oil.

During the FMC-in line process, the juice and the oily emulsions are conveyed separately; also the pulp and seeds are passed down a separate spiral channel. The finishing system for essential oils consists of a vibrating screen to separate the oily emulsion, then a centrifugal clarifier to concentrate the emulsion and after that a centrifugal polisher to completely separate the essential oil from the liquid phase. The latter is returned into circulation and re-sprayed onto the interior of the machines, washing off deposits of extracted oil to form another emulsion. From the extractors the juice goes into special collectors which pass it on to the finisher. Here the juice is separated from a certain aliquot of pulp, fiber and seeds. The system, which functions like a screw, together with the continual advance of the product through the machine, ensure that the sieves are kept clean and the juice flows quickly. A further reduction in pulp content can be obtained through a centrifugal separator.

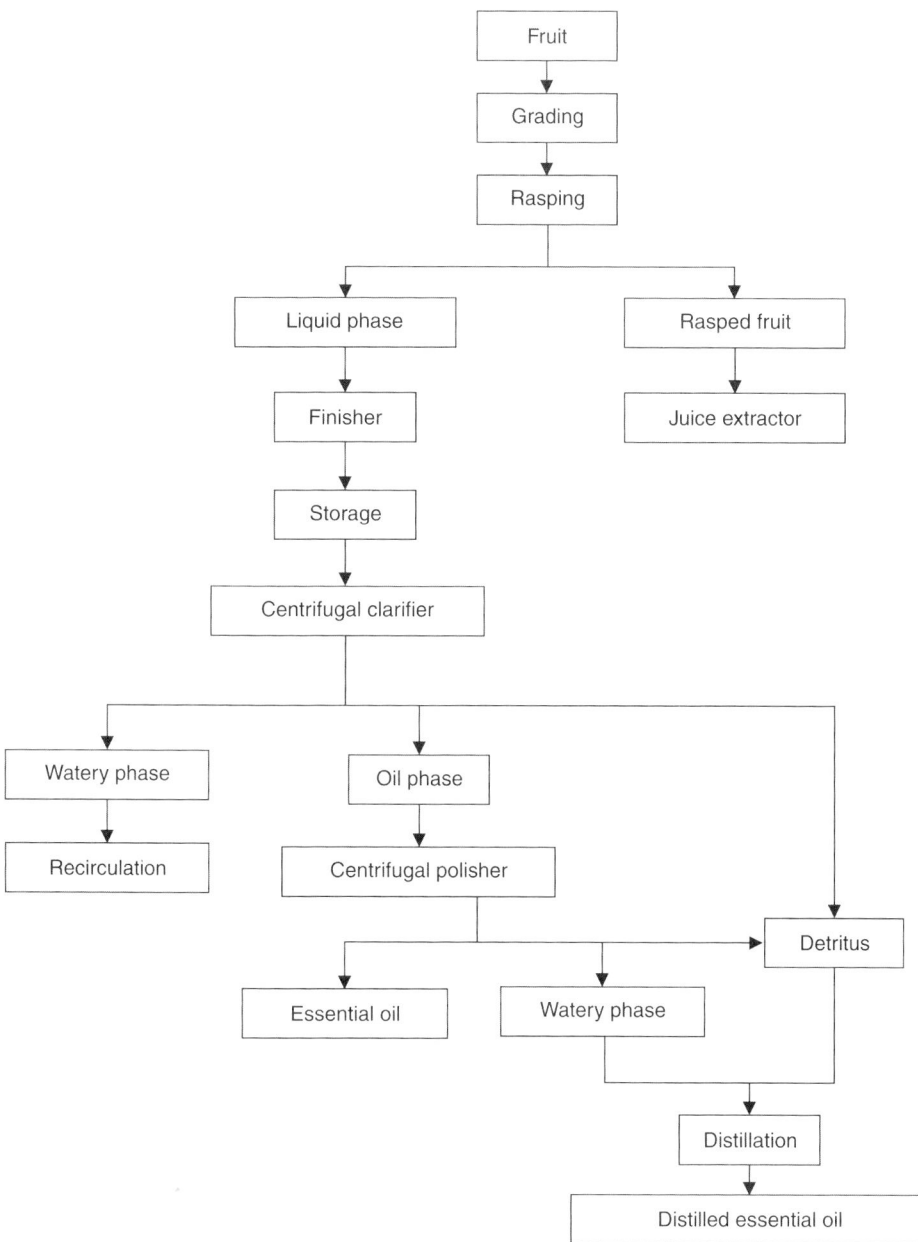

Figure 6.1 Flowsheet showing steps for citrus oils extraction (machines processing the whole fruit).

The juice obtained, especially if earmarked for immediate packaging, must be deoiled and deaerated. A high essential oil content in the juice can impair the quality, following the alteration of some of its components. Amongst other things, deaeration allows the heat exchangers to work better, and allows the final containers to be uniformly filled.

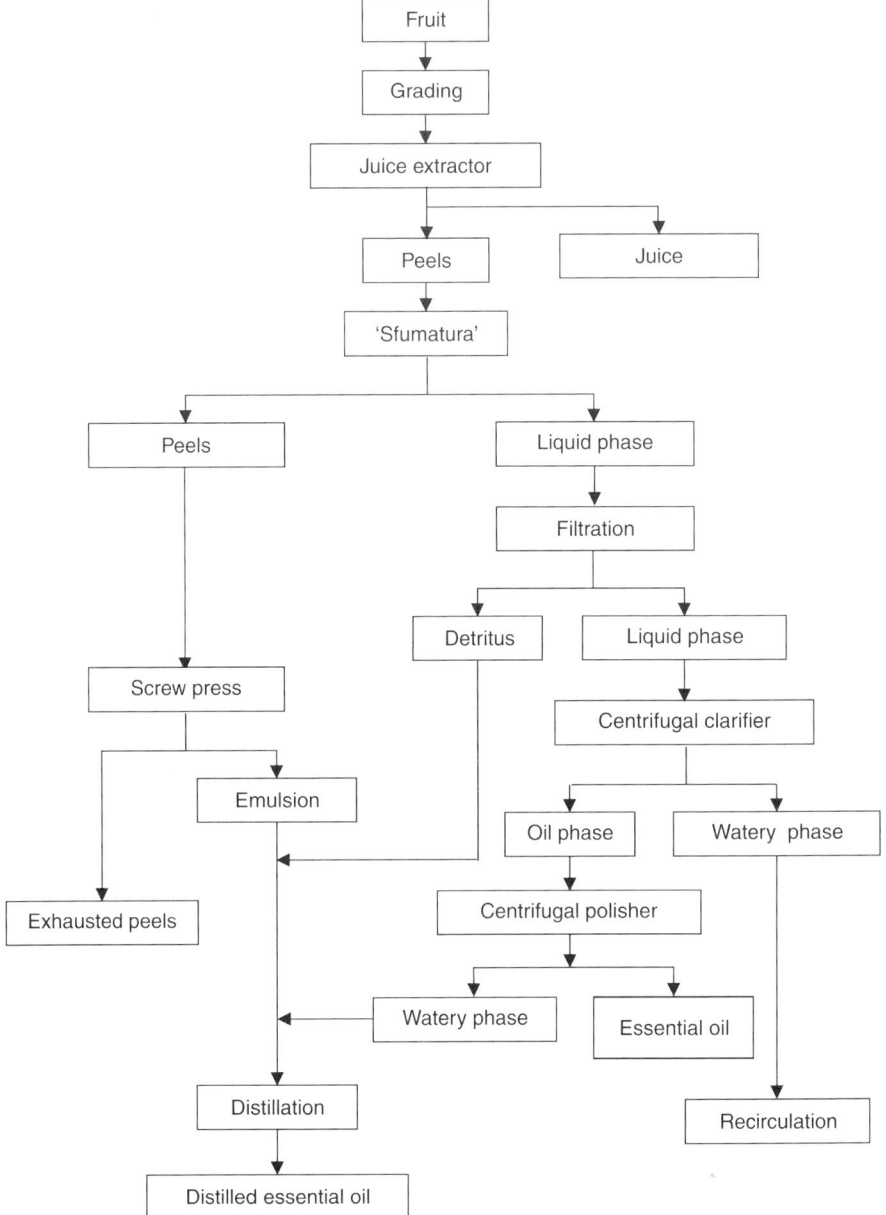

Figure 6.2 Flowsheet showing steps for citrus oils extraction (machines processing peels only).

After deaeration, the juice must be pasteurized. So that the action of the heat does not damage the organoleptic characteristics of the juice, flash-pasteurization is necessary; that is, pasteurization completed in less than a minute. Pasteurization is generally carried out using plate heat-exchangers, that work by passing the juice in

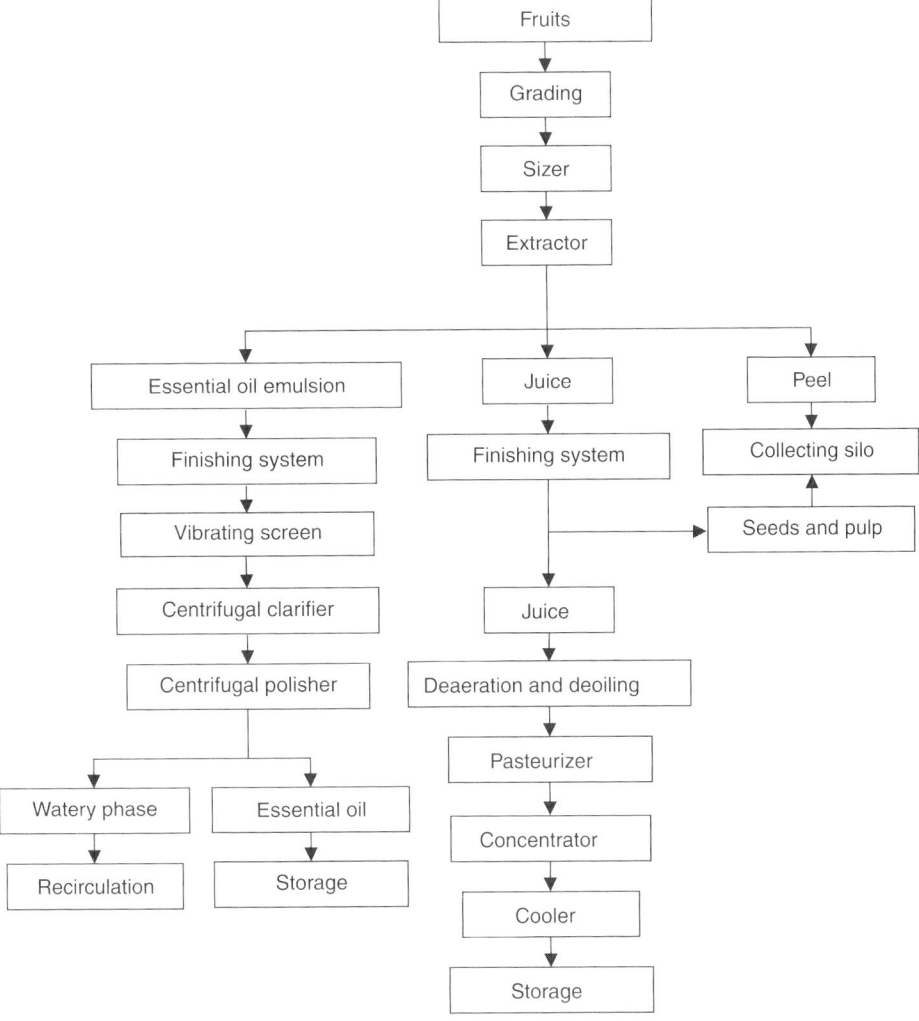

Figure 6.3 Flowsheet showing steps for a primary utilization of citrus fruits.

a thin layer between two specially designed plates. The heat exchange is effectuated using stainless steel plates heated by a fluid (hot water or low-pressure steam) that circulates in countercurrent. Apart from the liquids mentioned above, cold water is also used, to bring the juice back down to room temperature; an economy in the thermal balance is obtained by using the heat given off by the just pasteurized juice to start the heating of the juice about to be pasteurized. When the juice is earmarked for canning, the juice does not pass through the cooling phase: this allows the juice to be canned at around 90 °C. Obviously, once the lid has been clamped on, the containers are cooled rapidly.

The vacuum concentrators can be grouped in two categories:

1. concentrators working at low temperature, in which the consumption of steam and water is given relative importance;
2. rapid concentrators working at high temperature, with reduced consumption of steam and water.

In recent years, the second category of concentrators has had more widespread use. In particular, the FMC TASTE evaporator (thermally accelerated short time evaporator) is, perhaps, the only evaporator designed and built for citrus juices. The natural juice is pumped into the evaporator where, after a pre-heating cycle and an initial evaporation stage, it is pasteurized and stabilized. During evaporation the juice is nebulized inside the Tube Nest at a very high speed, expanding in the separator very quickly. Juice evaporation occurs in a single passage through a number of stages. The machine can also be equipped with an Aroma Recovery System: aromas are extracted from the juice and concentrated 150 times before pasteurization. Before leaving the evaporator, the concentrated juice passes through the flash cooler, which reduces the temperature to 10 °C.

An improvement in the quality of the juice is obtained by operation cut-back: the just produced concentrate is immediately chilled to a temperature close to freezing point, is put in tanks with a double cooling jacket and then freshly squeezed juice is added to reach 42° Brix.

Before the filling of the final containers (e.g. plastic drums, or metal drums with a plastic lining), the juice is sent to tanks where standardization is carried out if necessary, and the juice is further cooled to around −6 °C, either in the same tank or in a continual cooler of the 'Votator' type.

Finally, the concentrated juice is put in the drums by an accurately controlled system with preset weight scales.

The production of essential oil, natural and concentrated juice, and peel make up the fundamental phases of what is commonly called primary processing of the citrus fruit.

To these above mentioned phases should be added the processing that may be carried out either in the same plant, or in external plants. Above all, the complete utilization of the fruit and, particularly, of the refuse material and the peel, from which, by various procedures, other important products can be obtained (Braddock, 1999). These products will be described in a separate chapter.

REFERENCES

Agricultural Research Service (1962) Chemistry and technology of citrus, citrus products and by-products, *Agriculture Handbook n. 98*, United States Department of Agriculture, Washington, D.C.

Braddock, R.J. (1999) *Handbook of Citrus By-products and Processing Technology*, John Wiley & Sons Inc., New York. N.Y.

Braverman, J.B.S. (1949) *Citrus Products*, Interscience Publishers Inc., New York, N.Y.

Di Giacomo, A. and Mincione, B. (1994) *Gli Oli Essenziali Agrumari in Italia*, Laruffa Editore, Reggio Calabria, Italy.

7 Citrus juices technology

Fabio Crupi and Giuseppe Rispoli

INTRODUCTION

The citrus processing industry follows the demands of the market for processed products that have characteristics resembling the fresh product. On the other hand, the production of citrus destined for consumption as fresh fruit is in continuous evolution, through research of new varieties and hybrids. Besides the production of concentrated juices, (obtained by means of proceedings and devices more and more respectful of quality) the production of ready-to-serve citrus juices has become common in the last years; these juices are produced without the traditional passage through partial elimination of water and its following reconstruction at the moment of use: the single strength not-from-concentrate (NFC) juices.

This chapter deals with the principal operations concerning industrial citrus processing and associated new technologies. Production lines show some differences according to the kind of citrus fruit to be processed and to the products, concentrated or single strength, to be obtained. All of them can be traced back to a fundamental scheme with little variables.

For correct production, some strategies must be followed. Apart from the fruit quality and its preliminary treatments, citrus fruit which have been subjected to post-harvest treatments must not be processed; the removal of the sanifying agent must happen just before citrus fruit enter the extracting system, so that the fruit is protected from possible pollutions in conveyor belts and in sizing.

To avoid environmental pollution, it is necessary that squeezing plants and those of subsequent processing of juice are placed in enclosed rooms, possibly with filtrated and conditioned air; this strategy is necessary in the production of juices at mild pasteurization.

All the parts in contact with the juice must be built of stainless steel; attention should be paid for their sanification. All the plants with which the juice comes in contact must be washed and sterilized at least every eight hours, in the production of freshly squeezed juice every four hours (Attaway *et al.*, 1989). Sprays of warm water, alkaline solutions and other sanifying solution are used. For some parts of the plant, in points in which solid parts can deposit (peels, seeds, rags) it can be necessary to disassemble and manually clean all the parts.

At the end of the daily processing cycle, the plant must be left perfectly clean. After being discontinued for a lengthed time, the plant, even if already cleaned, must be treated again.

If the downtime happens during processing and is longer than two hours, it is appropriate to subject the system to the entire washing and sterilization cycle; if the downtime is included between thirty minutes and two hours, only sterilization can be sufficient. These strategies are absolutely necessary in case of fresh juice production (Beasley, 1997).

All the washing and sterilization operations can be managed automatically or manually.

The quantity of microorganisms present in the juice depends on many factors such as the condition of the processed fruit and the hygienic conditions of the plants with which the juice comes in contact.

Until the juice is inside the fruit it is sterile. The micro-flora is situated on the external surface of the citrus fruit while enzymes are contained in solid parts, juice cells and carpellary membranes.

When juice is processed, juice cells and membranes break down and permit the passage of enzymes in the juice, which, in contact with peel, but also with air and plants, enriches in microorganisms. Because of low pH, microbial flora present in citrus fruit juice is made by yeasts, bacteria and molds, whose development is connected to the quantity of free available water, and therefore with the degree of juice concentration. These are generally not pathogenic microorganisms but their presence, if in high quantity, can bring even deep alterations to juice characteristics until fermentation (Weissmann, 1991). Contaminations of fruit juices by pathogenic bacteria (Salmonella and E. coli), occasionally present in orange or apple juices, have been signalized (Parish, 2000). An accurate wash and a good sanification, both of pre-squeezed citrus fruit and of plants, lower considerably the quantity of microorganisms passing in the juice.

From the metabolism of Lactobacillus bacteria takes origin, together with other products, the diacetyl, which gives the juice a buttermilk-like flavour. The measurement of this substance is assumed in industrial practice as a sign of microbiologic pollution in the processing line. The diacetyl values in the juice are used as an alarm sign, which imposes the stop of processing and the plant sterilization.

EXTRACTION

A system of conveyor belts feeds the section in which citrus fruit is squeezed. Before squeezing, the fruit passes through sizing machines, generally roller machines that provide sizing, thus separating fruit in classes according to diameter. This operation is necessary for better rendering of extractors and for a good yield and quality of juice and essential oil.

According to citrus species to be processed, the kind of products to obtain, and in accordance with local traditions, in order to separate the juice and the essential oil from the fruit, industry uses different systems and machines. In some of these, juice and essential oil are obtained in one operation, in others this happens in different stages.

It is not suitable now to enumerate all the extracting systems, which have been used by citrus processing industry during time and across the world. These were processes needing a lot of labour and of time, nowadays incompatible with the modern conception of work.

Squeezing separates the citrus fruit in to three basic parts: essential oil, juice and peel. The systems can be subdivided in three groups:

1. Systems separating oil and juice in an only operation;
2. Systems in which oil extraction follows juice extraction;
3. Systems in which oil extraction precedes juice extraction.

Extractors separating simultaneously essential oil and juice

FMC in-line extractor

These extractors are common in citrus processing industry of the world. Built by Food Machinery Corporation of Lakeland, Florida, USA, in several models, they simultaneously separate raw juice, peel, core and essential oil. The core is constituted by two peel discs and by the central part of the fruit, included rags, pulp and seeds. Essential oil is removed under water sprays in form of an emulsion. FMC commercializes for each model two varieties of extractors characterized by a different speed, at 75 and 100 strokes per minute.

The functioning of these extractors is based on the alternate movement of a series of cups, the lower one fixed and the upper one mobile. The cups are formed by a series of finger-shaped elements, separated by a space in which the fingers of the corresponding upper cup insert when the latter goes down with a programmed movement. A proper channel system brings every fruit in the lower cup, which has a tube cutter in the inferior point, constituting the upper part of a strainer. The upper cup, going down, gradually presses the fruit, like when two hands close down on an object with intersected fingers. The pressure exerted on the fruit against the tube cutter of the lower cup brings away a peel disc and forces the internal part of the citrus fruit, juice, pulp, rags and seeds to pass inside the strainer through the circular hole made by the lower cutter. In the meantime, another tube cutter, placed in the upper cup, cuts a peel disc from the top of the fruit. A perforated piston, the restrictor with orifice tube, entering inside the strainer, pushes out the juice separating it from the rest of the core material. The compression between the two cups, while squeezing the citrus fruit, bursts the oil glands containing essential oil; this is removed in the form of emulsion under water sprays and separately recovered. The peel is discharged through an annular opening near to the upper cutter.

From the basic principle characterizing the in-line FMC extractors other constructors have drawn their inspiration; for example Bertuzzi of Brugherio, Milano (Citrostar MU1), Fomexa, Speciale.

Extractors in which juice extraction precedes essential oil extraction

A lot of models of Brown extractors built by Automatic Machinery and Electronics, Inc., Winter Haven, Florida are available. They all have the characteristic of operating juice extraction on half-cut fruit.

Brown 400

The essential part of the machine is made up by two revolving flat discs slightly leaned, at whose periphery are put some plastics serrated reamers, also revolving. The two halves of the fruit are picked up by synthetic rubber cups and positioned in connection with the reamers (half a fruit for each reamer). Because the disc bringing the reamers is slightly leaned, the gradual juice squeezing is completed in one disc revolving. The raw juice is collected in a small basin to be sent to finishing, while the peel is expelled by means of a deviator. From the peel, essential oil is recovered subsequently.

The working capacity of Brown 400 is of about 350 fruit per minute. This type of extractor is usually employed for grapefruits and oranges of big diameter.

Brown 700

The squeezing principle is analogous to that of Brown 400. The revolving reamers are installed on two vertical discs, not leaned, and the cups are mounted on two tracks, one for each disc. The tracks are disposed so as to determine a deeper penetration of the reamers in the beginning stage of the squeezing and a gradual pressure in the rest of the path. A compressed air mechanism regulates the penetration of the reamer in the half fruit, depending on the peel thickness, so as to obtain the complete extraction of the juice without excessive pressure on the albedo.

The potentiality of Brown 700 is exactly of 700 fruit per minute.

Brown 1100

The extractor is constituted by three pairs of revolving circular synthetic rubber tapered discs inside which a fixed knife is inserted, together with a double fixed wedge-shaped perforated grid, so that the distance between each revolving disc and the corresponding part of the grid reduces more and more. The distance is regulated according of the peel thickness.

The citrus fruits fall down, one by one, in each couple of discs that pushes them against the knife by means of which they are cut in two halves. Continuing the turning, every disc drags the half fruit and presses it against the perforated grid, causing the gradual extraction of the juice. The squeezing area is connected with two collecting basins; in the first one arrives the rate of juice extracted in the milder pressure area, which gives a better juice, with less essential oil, less pulp and no bitter taste. The remaining juice is collected in the second basin.

Essential oil is recovered from the spent peel.

In lemon processing by means of Brown 1100, the essential oil extraction precedes juice extraction. In this case, before arriving to the 1100, the lemons pass in a BOE (Brown oil extractor), which is a scarifying equipment with running shafts, each having thousands of sharp points of stainless steel, cutting the fruit peel. The essential oil is removed in form of aqueous emulsion and recovered by centrifugation.

The fruits, scarified and still entire, before being directed to squeezing, pass on a roller conveyor which has first a washing section with water sprays and then a drainage section (Carter, 1992).

Birillatrici

These are machines in which juice is extracted by means of revolving reamers by citrus fruit previously cut in halves and laid down on plastics cups. The principle is ancient but still valid and it is used in Italy and in other Mediterranean countries above all in lemon processing, where essential oil has a pre-eminent value. By spent peel, essential oil is recovered by means of 'sfumatrici' or screw presses. Even if continuously updated, they remain suitable for special processing, not for mass-production.

Systems in which oil extraction precedes juice extraction

The use of revolving drums, which press the half fruit against a grid, is a system utilized for a long time in industrial extraction of citrus juice. The extractors of this class are preceded by systems for the extraction and recovering of essential oil; they are generally paired to machines rasping fruit peel.

On this principle are based the juice extractors 'Indelicato' (Giarre, Catania) and 'Speciale' (Giarre, Catania), built in Italy and diffused above all in the Mediterranean area.

Briefly, they are made up by two drums with converging rotation and two fixed perforated grids with a curving, which slightly differs from that of the drums. A fixed knife cuts the fruit in two halves, which fall down between drum and grid, oriented so as to present the external part toward the drum and the internal part toward the grid. Because the distance between the drum and the respective grid is making smaller and smaller, the pressure on the half fruit gradually increases according to the drum speed. To a higher speed corresponds a higher working capacity per hour of the extractor, but also not optimal extraction conditions.

If the productive scheme includes the use of peel for the production of candied fruit, these extractors permit a variation: at the grids ends are located knives separating the flavedo from the spent peel. When the peel is destined to the production of candied fruit, the fruit does not pass through the rasping equipments.

The Rotary Juice press, used above all in grapefruit processing, is a further variation: it is also made up by two drums with converging rotation, but in these are shaped some series of semi-cups which coincide at the point of maximum approaching. The fruits, laid in the spherical space thus obtained, are cut in halves by a fixed knife. The juice is extracted by the half fruits by means of reamers, which penetrate in them with an alternate revolving movement in order to reduce air absorption.

Finishing operations

The raw juice produced by extractors contains a notable quantity of pulp, little seeds and pieces of membranes, which have to be removed.

For this reason finishers are used; they are screening systems linked by each producer to the extractors of his own production.

The finishers usually consist of cylindrical screens through which the juice is forced by means of a helical worm screw. The juice enters through an end and while the liquid, together with the finest pulp, is filtrated through the perforated wall, the roughest parts are pushed forward and discharged by the other end. Because the space between the perforated wall and the worm screw decreases with the advancing of the pulp parts, the pressure deriving by the volume variation between the initial coil and the final one of the finisher is determined on them. A pneumatic- or mechanical-operated closing system further regulates pressure. In some cases, the space between the perforated wall and the endless screw does not vary and the pressure on the pulp is exerted only by the closing system.

Another system resumes, modifying and softening it, the screening principle used for years in the tomato processing industry: the 'passatrici'.

In these cases the finisher is made up by a horizontal perforated cylinder, or by a slightly leaned one, in which a shaft rotates, bringing a series of sheet metal paddles. These push the juice against the perforated wall and determine its screening. The roughest parts are discharged by the other end.

A third finishing system uses screens moved by a double movement, rotating and vibrating. They are constituted by three horizontal screens laid upon one another, with different meshes. The raw juice is separated in coarse particles, cells and juice with fine pulp.

After the screening the juice still contains a quantity of pulp, which is too high for normal concentration. In this case, the pulp undergoes a further reduction, up to 3–5 per cent by means of discs centrifuges or decanters. If one wants to collect juice cells to separately freeze them, the juice is freed by heavier particles, basically constituted by pieces of seeds and by embryonic seeds, making it pass through hydrocyclons located before screening. In this case, in order to avoid excessive breakdowns of the cells, softer screening systems are used.

Citrus fruit processors have a wide choice range for what concerns squeezing machines and juice finishing. A conditioning exists, that is the possibility of relying on specialized maintenance workshops and of supplying oneself rapidly with the necessary spare parts. Efficiency and a correct management of the machines is a basic condition for respecting product quality and yield.

The good sizing of fruit is always a useful operation but it becomes necessary when extractors with fruit holders are used. If during squeezing the fruit or the half fruit is not held well by the rigid walls of the cup, because this is too big or too small, the operation becomes anomalous, with loss both of juice and of essential oil.

When FMC in-lines are used, the small fruit, if positioned in a big cup, 'explodes' for lack of support, while the big fruit, if positioned in a small cup, breaks.

The extractors feeding must be adjusted so that the quantity of citrus fruit arriving to squeezing is not significantly lower than the potentiality of the installed machines. But it does not have to be excessive either. The surplus of fruits must return to the washing and sterilization systems. In order to minimize the return, it is necessary to install optical sensors, which count the fruits and regulate their access to extractors. A level sensor at the squeezed juice tank regulates the speed of the conveyor belt to the extractors.

Even for extractors based on drums action the fruit sizing is convenient. In fact, the distance between grid and drum, which determines the pressure on the open half fruit, must be well regulated in order to avoid the excessive squeezing of the peel (passage of substances from the peel to the juice) or a too mild pressure (low juice yield). The fruit feeding must be regular to avoid superimposition inside the extractors or lower feed.

It is important also to take care that in the sizing machines there is no fruit superimposition, which could invalidate the success of the operation.

Moreover, washing of extractors and finishers must be done often, possibly in countercurrent and automatically, with alkaline solutions, in order to avoid incrustations of hesperidin and pulp particles. The occlusion of the holes of the perforated screens in which the separation of juice from pulp occurs can lead to loss of juice in the pulp. The finishers regulation control is led by means of the 'Quick Fiber' test.

DE-AERATION AND DE-OILING

During squeezing, the essential oil coming from epicarp passes in the juice. If the oil quantity is limited, no more than 0.15 per mil, it participates to the bouquet formation, of the typical flavour which characterizes the species; a higher content is a negative factor also for juices with short shelf life stored at low temperature.

In addition to the enzymatic degradation occurring in juices with mild or no thermal processing, the acid pH causes chemical alterations, most of all at the expense of monocyclic and bicyclic terpenes (Dugo et al., 1990; Verzera et al., 1990). The single strength lemon juice taste, even if stored at low temperature, can show significant degradations depending upon its essential oil content.

The oxygen absorbed in the juice during technological operations is another cause of decay: ascorbic acid is oxidized and d-limonene is attacked with formation of terebenthic taste.

The recent trend of industry is to reduce to a minimum these inclusions with the use of adequate equipments, like extractors modified in order to avoid oil excess; for what concerns oxygen, it is recommended to reduce the close contact between juice and air.

The essential oil excess can be removed centrifuging the juice with de-oiler centrifuges or through partial distillation in vacuum: a limited juice rate, between 5 and 6 per cent, is transformed in vapour at about 50 °C; in these conditions, the 80 per cent of essential oil is removed together with the air present in the juice. The non-terpenic aromatic fractions, condensed again, are reintroduced in the juice after the oily phase has been separated by means of decantation or centrifugation.

When the juice must not be confectioned but concentrated, the operations of de-oiling and de-aeration occur inside the system of concentration in vacuum, in a different way depending on the constructing scheme of the used evaporator. Several systems are used in the case of freeze-concentration and of osmotic concentration, depending on the productive cycle.

The contemporary reduction of essential oil and oxygen from citrus fruit juice is obtainable also with the use of a spinning cone column, constituted by a vertical stainless steel cylinder in vacuum, which has a series of inverted cones inside; alternatively, a cone is soldered to the cylinder wall and is therefore fixed, while the following one is linked to a rotating shaft. At the top of the column the juice to be treated enters, flowing because of gravity on the upper surface of the first fixed cone, and from here it drops on the following cone connected to the rotating shaft. Because of centrifuge force, the juice ascends toward the brim of the rotating cone and flows on the brim of the third cone, which is fixed, and so on until the juice, de-oiled and de-aerated, reaches the bottom of the column. Due to rotation, alternate with fixed cones, a thin turbulent film is generated. From the bottom of the column a gas current rises, generally water vapour, which helps the stripping of volatile components; they go out from the head of the column and are condensed. Operating at 27 °C, more than 90 per cent of oil and more than 95 per cent of oxygen present in the juice are removed (Schofield, 1994; Schofield et al., 1998).

The oxygen present in citrus fruit juices can be strongly reduced by means of glucose-oxidase which catalyses the oxidation of glucose to gluconic acid.

PASTEURIZATION PROCESSES

In industrial citrus processing, pasteurization has a double function: on one hand it reduces microbiologic charge to minimum values and on the other it inactivates the enzymes present in the juice.

In any case, the goal is to increase the shelf life of the juice itself, that is the time that passes between production and use. In this time interval the product has to maintain practically intact its chemical and physical characteristics and its sensory attributes, provided that the expected storage conditions have been respected (Rysstad et al., 1993).

For what concerns the complex of enzymes, except polyphenoloxidase, the most dangerous presence is that of pectic enzymes, such as pectin-methyl-esterase which, hydrolysing pectin, produce juice clarification and gelification. When pectin has been destroyed, the juice appears in two completely separate layers, the pulpy lower one and the clear upper one. On the contrary, if the juice has been opportunely processed and if pectic enzymes have been inactivated, the upper phase maintains its cloudy aspect and the same colour of the sediment because of suspended pulp particles. On the whole, the product is more appreciated by the consumer because it has an aspect similar to that of freshly squeezed juice.

Pectin, whose basic structure is that of a polygalacturonic acid, bonds metallic ions, particularly those of calcium and magnesium present in the juice. The quantity and strength of these bonds is related to the number of COOH free because of demethylation due to pectin-methyl-esterase (Rouse et al., 1952; Versteig et al., 1980).

Of this enzyme two different forms, a thermolabile one and a thermostable one, have been isolated in Marsh withe grapefruit pulp.

Averagely, the complex of pectin-methyl-esterase is constituted for about 90 per cent by the first form and for about 10 per cent by the second form. The thermostable form differentiates from the thermolabile one for the unlike quantity of attached carbohydrate. The partial hydrolysis of carbohydrate produces moiety of enzyme thermostability (Seymour et al., 1991).

In Navel orange three isozymes of pectin-esterase have been isolated. The first two are thermolabile and are rapidly inactivated at 70 °C, while the third one is inactivated only at 90 °C (Versteig et al., 1980). The remaining activity of pectin-esterase in Valencia orange pulp rapidly decreases until 4 per cent because of heating at 80 °C for 19 seconds and decreases very slightly because of subsequent heating at 80 °C for 180 seconds. The D values (that is, the time necessary to inactivate the 90 per cent of the enzyme) at 90 °C for the thermolabile fraction and for the thermostabile one are respectively of 0.225 and 32 seconds (Wicker et al., 1988).

The kinetics of thermal inactivation between 70° and 90 °C of the pectin-methyl-esterase coming from orange and grapefruit peel shows the presence of thermolabile fractions and of thermostabile fractions. The latter are found above all in the albedo both of orange and of grapefruit (Cemeroglu, 1998).

In lemon fruits two major pectin-esterase have been isolated, one exclusively present in the peel and the other in the endocarp. In order to reach the optimum activity both require the presence of cations, in higher concentration those of the peel. They have almost similar molecular weighs, 35 KDa and 33 KDa respectively. They are completely inactivated at temperatures higher than 88 °C and do not attack the cloud of lemon juice, while they destabilize that of orange juice (MacDonald et al., 1993).

In lemon fruits two minor pectin-esterase have been separated and purified, one stabile at low pH with same initial pectin-esterase activity at 88 °C as at 25 °C. This pectin-esterase, still active also after having been processed at 85 °C for 9 minutes and destabilizing the cloud of lemon juice, could not be a real isozyme but it could derive from the association of a major enzyme with pectin (MacDonald et al., 1994).

Probably most phenomena observed in citrus juices during storage, such as cloud loss, are caused by insufficient pasteurization processes to inactivate thermostabile forms of pectin-methyl-esterase.

Generally producers and confectioners of citrus fruit juices are pushed to drastic pasteurization processes in order to assure to their productions an extended shelf life.

Excessively higher temperatures and times are thus used, also in case of productions in which, like in chilled orange juice and in frozen concentrate orange juice, less excessive conditions are sufficient (Fellers et al., 1993).

Nowadays the market is oriented towards juices, which give the sensation of 'naturalness'. Citrus processing industry tends then to reduce to the strictly necessary the use of those means, such as heat prolonged in time, which cause the sensation of an over industrialized product, excessively denatured.

On the other hand, citrus fruit juice, to be marketable, does not need absolute sterility because of its low pH. Moreover, the shelf life for a correct marketing is different according to juice typology. The diffusion of distribution systems at refrigeration temperature permits to minimize pasteurization processes.

Processors and confectioners have generally focused their attention on the state of processed fruit and on the state of processing and confectioning plants with which juice comes in contact. Accurately removing unsuitable fruit for breaks and crushes reduces the microbiological population present in the finished product.

Traditional thermal pasteurization

For the thermal pasteurization of citrus fruit juices citrus processing industry uses heat exchangers, generally plate-type or tubular ones.

Pasteurization, as the stage preceding concentration, occurs in the initial parts of the evaporator, in order to obtain the total exploitation of given calories. If, on the contrary, it is an operation apart, then plate-type exchangers or exchangers with tubes divided in four areas are used:

1. Recovery area in which juice to be pasteurized is heated in counter-current with warm pasteurized juice;
2. Pasteurization area in which pre-heated juice reaches pasteurization temperature at the expense of the heating medium (normally warm water or low-pressure steam);
3. Pause area in which juice maintains pasteurization temperature for the provided time;
4. Cooling area in which juice, after having exchanged part of its heat with the entering juice, completes its cooling against water and/or another cooling liquid.

In plate-type exchangers, the plates are shaped so to exalt juice turbulence and are linked each other with gaskets; the number of the group of plates for each area is calculated according to the foreseen surface and depending on viscosity and therefore on juice temperature. Inside the so formed chambers, juice flows with a speed up to 2 meters per second. The resulting advantage is that pasteurization can occur with a high thermal difference at the wall, without thermal alteration in the juice. Because of the limited chamber thickness and of plates shape, this kind of exchanger is not usually indicated for very pulpy juices (above all when large cells are present) or for pulp. Plate-type exchangers specifically studied for these productions are now available.

Cleaning and internal sanification of the exchanger is a necessary operation at least at the end of every working cycle. It is opportune to remove possible crusts formed by deposits of polysaccharides, pectin and pulp, that can sediment in points at which flowing speed is lower.

In plate-type heat exchangers the operation is facilitated by the fact that the interior of the exchanger is easily examinable by opening the packets of plates, but the gaskets can cause microbial and enzymatic contaminations, above all when they are worn out.

The tubular heat exchangers (with the multi-tube variation in every shell) are constituted by pieces of a tube (usually 6 meters long), linked each other with elbow curves. The almost total absence of gaskets, except the connection rings to the curves, is a first advantage. Moreover, they can be built for medium and high pressures. They can be used also for pulpy juices and for cells made pumpable with a small liquid quantity. The negative aspect is constituted by the fact that the juice inside the tube flows at different speeds, more slowly along the wall and faster in the centre, with the risk of overheating for a part of the juice and underheating for another part.

Some expedients can be used to decrease this risk:

- Insertion of static mixers in every tube so to increase turbulence and improve thermal exchange;
- Use of corrugated tubes, spiral-shaped, to increase turbulence;
- Reduction of thermal wall difference with subsequent increase of the necessary surface;
- Instead of the internal tube in which juice flows, a set of parallel tubes with small diameter can be inserted in a shell. The exchange surface thus increases, together with juice speed along the wall and thermal exchange.

From the sanitary point of view, when tubular exchangers are used, the recovery section of heat (in which the entering juice heats at the expense of warm pasteurized juice) is not easily reachable for sanification. In these cases, therefore, it is preferable to divide the shell in two sections and use as thermal transportation liquid the water in a closed circuit; water warms up in a section at the expense of warm pasteurized juice and gives its heat to the entering juice in the other.

For enzymatic inactivation and destruction of microbial charge through heat action, industry uses different processing temperatures and times depending on the kind of juice to be processed and on its pH and according to the product to obtain. For orange, mandarin and tangerine juices, temperatures of 90/98 °C are used for about 60 seconds, for grapefruit 85/90 °C for 30/40 seconds and for lemon 75/85 °C for 30 seconds. In this last case, high pasteurization temperatures connected with low pH cause loss of cloud for acid-thermal degradation of pectin. These values, purely indicative, are generally used in the production of cloudy and pulpy concentrated juices; they are lower when clear juices are pasteurized and they change more and more for the confectioning of juices re-diluted from concentrate. They depend a lot on the kind of 'life' that the product has had and on the assigned destination.

NOT-FROM-CONCENTRATE JUICES

The not-from-concentrate citrus fruit juice (NFC) is one of the most ancient productions of citrus processing industry. Citrus juices confectioned without passing through concentration and subsequent dilution have been present on the market for years in several confections such as bottles, tins, and so on.

In order to give the product a long shelf life at room temperature, it is necessary to subject juice to drastic thermal processes tending to the destruction of microbial population and to enzymes inactivation. The thermal damage reveals itself with the alteration of natural flavour compounds present in the juice and with the formation of an undesirable 'cooked' flavour.

In these last years beside traditional NFC juice, whose presence in the market is in continuous decline, products with sensory attributes closer to those of domestic juice squeezed from fresh oranges have become more and more diffused. In fact, the request of the market is now addressed toward alimentary products presenting less 'technological' characteristics than those of the past.

To the request of higher 'naturalness', 'healthiness', 'genuineness', citrus industry has answered with ready-to-serve juices with short and medium shelf life. These productions do not derive from specific technologic innovations, but from a different approach to productive and distributional problems and from a more careful management of pre-existing systems, with some plant improvement.

Different directions have been followed: on one hand a rigid observation of hygienic norms in the choice of raw materials and in plant conduction in order to minimize from the beginning the quantity of micro-organisms present in the squeezed juice, and on the other a careful maintenance of suitable temperatures in order to slow the development of the even limited microbial population of the juice. Also some interventions on the plants have been made, to obtain a higher respect of the characteristics of the product processed by means of technologies reduced to the indispensable minimum.

With the exception of traditional productions, there are two kinds of ready-to-serve NFC citrus fruit juice on the market: the freshly squeezed juice with short shelf life and the NFC juice with medium shelf life. In the strict sense, in the first productive line no thermal process is admitted, either of pasteurization or of freezing, while the second one is characterized by a mild pasteurizing process. In both cases, storage and distribution must occur at refrigerating temperature. In the production of freshly squeezed juice no pasteurizing process is foreseen, therefore very careful surveillance is essential to avoid phenomena of microbiologic pollution during processing or subsequent phases.

Some incidents of contamination by Salmonella and *Escherichia coli* have been noticed in non-pasteurized fruit juices. Subsequently, it has been proposed to make pasteurization compulsory for every kind of fruit juices. Alternatively, it could be suitable to provide packages with warning labels, in order to alert consumers about the possible presence of bacteria. It is also recommended the use of hazard analysis critical control point (HACCP) protocols.

The word 'fresh' is exclusively dedicated to the first product, which reproduces, even on an industrial scale, the quality of home-squeezed juice. Citrus fruit juices included in the two above-mentioned categories possess freshness, aroma and flavour of the fresh fruit of good quality. The pre-eminent bouquet is typical of endocarp. The typical taste of essential oil coming from the peel, even if present, must not prevail.

The presence of notable quantities of entire juice sac particles is appreciated both because it causes in the consumer the visual and taste characteristics of the fresh fruit, and because it improves the 'body'.

The production of juice with short and medium shelf life begins with the choice of the citrus fruit to be processed, which must be more accurate than in normal processing.

The fruits must be carried in bins and loose on lorries. At the arrival in the processing plant they have to undergo a first operation of selection, washing and sterilization and they can be placed in silos for a short stop after having been dried, but the direct discharge is preferable. The storage for some days in the refrigerator at 0 °C with relative humidity of 80/85 per cent is useful also because it permits an even limited standardization mixing fruits of different origin.

At the moment of drawing, the citrus fruits undergo a second selection. The rate of rejected citrus fruits for this production is usually at least double than that of concentrate production. Small fruits must in any case be rejected so that essential oil in the squeezed juice is less likely.

Also the washing and sterilization operations of fruit and equipments must be more accurate than in usual processing in order to minimize microbiologic pollutions. Discharging and conveyance equipments must be built in a way, which does not provoke injuries or other types of damages in the fruit. Both equipments and fruit must be treated with cleaner and sanitizer sprays at the moment of unloading.

In the brushwasher the first set of nozzles sprays on the citrus fruits phosphoric acid anionic cleaner. If the state of fruit recommends it, a second set of nozzles sprays the fruit with the same cleaning and sanitizing solution. Follow a rinse with water and the process with chlorine solution or iodophor compounds. The remaining halogen is removed with a wash at the moment of the entrance of fruits in the extractors (Beasley, 1997).

For what concerns the extracting and screening operations, it was necessary to modify normal procedures because of particularities required for these juices: small quantity of essential oil, absence of bitter taste, much pulp, presence of floating juice cells.

FMC has built a new series of in-line extractors, the premium juice extractors (PJE) with small diameter cutters and a softer system of extraction, operating also on peel expulsion, on cups and strainer tubes configuration.

Also finishers have been modified. Screens of different kind, according to desired characteristics, can be used. Finishers can be divided in two parts, with large holes in the first sections and smaller holes in the second one; juice with large cells with sizes included between 4 and 10 millimetres can be obtained.

All the defects, that is embryonic seeds, broken seeds and membrane pieces are separated through centrifuge force by means of a hydrocyclon.

Through other extractors, operating according to principles different from those of In-Line, juices with large cells can be obtained without modifying the machines. It is clear, anyway, that in order to obtain first quality juice, yield must be sacrificed.

Soon after finishing, juice is rapidly cooled at the temperature of 0 °C/–1.5 °C. Confectioned and put on the market at these temperatures, it constitutes the freshly squeezed juice. In these conditions, the shelf life reaches 21 days (but commercial life is considered of 17 days as a precaution), while at 4 °C a limit of 10 days is considered (Gray, 1989).

The shelf life can be notably increased if in the productive line a variation is introduced: ultra-filtration. After all squeezing and refining, juice passes through a plant with VPDF membranes, obtaining about 65 per cent permeate free from microorganisms and enzymes and about 35 per cent of pulpy retentate, which is pasteurized at 85 °C and soon cooled. Mixing the two fractions, juice is reconstituted and maintains its characteristics for at least 90 days if kept at the temperature of 2 °C (Gherardi *et al.*, 1992).

If grapefruit juice is added with carbon anhydride, saturating the headspace, the shelf life increases from 21 to 31 days when the storage temperature is of 4 °C. The same result is obtained if the juice is treated at 56 °C for 12 seconds (Soffer *et al.*, 1996). Saturating with CO_2 the headspace of confectioned not pasteurized orange juice, the shelf life passes from 21 to 25 days if the juice is maintained at 4 °C and from 5 to 10 days if temperature is of 10 °C (Shomer *et al.*, 1994).

The producing technique of above-described fresh juice in any case originates some conditionings. Production can occur only in the period in which the interested citrus fruits, above all oranges, but also mandarins and grapefruit, come to ripening and it is not possible to balance the peculiarities of the different varieties mixing raw materials, which have different harvesting times. This is a very heavy limitation in countries, such as Italy, which have a short producing window, which is not sufficiently extendable by fruits refrigeration. In any case, the finished juice, which cannot be standardized, has big sensory attributes differences even for what regards the *ratio*.

In order to be in condition of supplying the market for the whole year with juices at limited variability, without penalizing the finished product characteristics, a producing technique has been got ready: it inserts as an intermediate phase the rapid freezing together with final mild pasteurization.

The juice, soon after squeezing and refining, is rapidly cooled at temperatures near to the freezing point and then let in plate refrigerators.

These are made up by a series of plates empty at the interior, where refrigerating fluid passes, generally ammonia. The plates are shaped so that, when they are pressed against each other by a piston, they form chambers of desired dimensions, for example 1070×550×75mm. The chamber are filled up with cold juice at 1 °C, and already standardized in refrigerated containers. The freezing of the cooled juice occurs in a short time, according to the chambers thickness, because of the direct expansion of the ammonia inside the plates. When the juice has reached the temperature of −18 °C, an automated system widens the plates and puts the sheets of frozen juice on pallets, that are wrapped in plastics and stored in the refrigerator.

As a variant, the freezing of the cooled juice can be started by means of refrigerators of the Votator kind in which at least 15 per cent of water contained in the juice changes into ice. The completion of the state passage and the cooling at −18 °C occurs in metallic drums placed in tunnels with air cooling at −40 °C, or placed in a basin bathing in glycol at −40 °C.

Granulators, which can be of a different type than Votators, can be a very dangerous source of microbiologic pollution. They must thus be built in a way that easily permits disassembly and sterilization.

The frozen juice, in sheets or in barrels, is stored in refrigerators equipped for this purpose. At the moment of confectioning, the frozen juice is taken from cells according to predetermined mixing plans. The juice is grinded, rapidly de-frozen at 2 °C and, through sterile equipment, is introduced in the final containers after mild pasteurization at 70 °C for a few seconds. With storage temperature of 2 °C the shelf life exceeds 40 days.

For what concerns the pigmented oranges, typically Italian, of the varieties Moro, Tarocco and Sanguinello, the production of NFC juice without any standardization is not possible. Together with other characteristics, the anthocyanins content, the compounds that give the juice its peculiar red colour, varies, not only among the different varieties, but also in the range of the same variety, in such a marked way to bewilder the consumer. For this production the passage through the storage stage of the frozen juice is unavoidable.

As already said, the presence of large cells is a really appreciated characteristic not only because it remembers the home-squeezed juice, but also because it improves the body of the product. This kind of pulp can be separated from juice after the screening by means of vibrating strainers or of small holes finishers, which operate in a not drastic way. The cells are then pasteurized at 90 °C for 60/70 seconds in tubular heat exchangers with corrugated tubes and, after the cooling at 5 °C, are placed for freezing in 20-kilo cartons or in 180-kilo barrels with polyethylene bags (Kolb, 1993). The juice cells are added to the juice when the final confectioning occurs.

The NFC juice can be produced according to a different scheme: soon after the finishing the juice, after having been cooled at 0 °C, is introduced in blending tanks. It is then warmed at 50 °C, through a tubular heat exchanger and de-aerated in vacuum. Volatile aromas, recovered by a surface condensing system, are re-introduced in the juice, which is subsequently pasteurized at 93/95 °C, cooled at 2 °C and aseptically introduced in big refrigerated tanks under nitrogen atmosphere or in aseptic bag-in-bins of about 1,000 litres.

The tanks are built in polished stainless steel: for those of bigger dimension, with capacity of even 3,800 cubic meters, it is preferable to use carbon steel with epoxy liners. The accurate system sterilization is an essential condition for the good maintenance of the juice.

The structure that constitutes the bags must guarantee the maintenance of juice characteristics until the moment of its final confectioning. Various kinds of bags exist, either for the maintenance of juices in well-controlled temperature situation, or for juices, which have to be shipped, and for which it is impossible to know whether maintenance conditions are constantly optimal. In any case, these are multi-layer structures; generally inner layers are made of polyethylene and outer ones of nylon or metallized polyester.

If storage temperature is maintained between 2 and 4 °C and the material constituting bags has the characteristics of a good barrier for oxygen and aromas, the juice remains intact for a year (McFarlin, 1997).

SINGLE STRENGTH JUICE FROM CONCENTRATE

This production still represents a great part of the market of ready-to-serve citrus fruit juice. From the qualitative standpoint, it is a less valuable product than NFC juice even because it undergoes two pasteurizations, one at the moment in which the juice is introduced in the concentrator and one at the moment of confectioning after dilution. The first pasteurization is drastic because all the forms of pectin-esterase have to be inactivated; the second is milder, around 80 °C for 15/30 seconds, if the juice is distributed under refrigeration, while it is more drastic, beyond 90 °C, if the confectioned juice is put on the market at room temperature with long shelf life.

CONCENTRATION

The main component in citrus fruit juice is water. In industrial practice the trend has always been the decrease of juice weight and volume, reducing water presence, thus concentrating it, altering as little as possible the sensory attributes and nutritional characteristics of the starting product. The water reduction implies some evident advantages: it limits the package, storage and transportation costs; the activity of the disposable water, and thus the degree of enzymatic and microbiologic activity, decreases.

The systems to remove a considerable quantity of water are substantially three:

1   Through evaporation, in which the undesired quantity of water is removed under steam form (thermally in vacuum concentration);
2   Through freezing, in which water is removed under ice form (freeze concentration);
3   Through osmosis by means of membranes which permit the passage of water and not of dissolved substances (membrane concentration).

Concentration through evaporation

For many years in citrus processing, industries have used evaporators operating at high vacuum for long times. The juice was pasteurised in heat exchangers external from the evaporator and introduced in concentrators operating at low temperature. In this way, they attempted to reduce thermal damages but, because of high vacuum, volatile aromas got lost. For some juices, such as lemon, a product with good cloud stability was obtained because the acid-thermal attack to pectin was practically inexistent. The steam consume was very high, being the evaporators at a single effect.

Later, the principle has been changed: evaporators at several effects have been used, bringing the juice at a higher temperature for a shorter time; the recovery of volatile aromas occurs by a system of partial condensation. With the term 'effects' the phases of the vapour path and with 'stages' those of the product path are meant.

An evaporator is schematically constituted by:

- A heat exchanger giving the juice both the sensitive heat necessary to bring it to boiling temperature (depending on the applied vacuum) and the latent heat of vaporization;
- A juice/vapour separator, generally a cyclone one;
- A condenser to eliminate evaporated vapour.

Citrus juices are very sensitive to heat; their physical and chemical characteristics (viscosity and content of pectin, pulp, hesperidin, and so on) contribute in different ways to the formation of fouling and salting. The choice of the concentrator is then depending on several requirements such as:

- The characteristics of the starting juice and of the product to be obtained;
- The evaluation of the thermal damage which depends on the contact time (above all at high temperature) and of the thermal wall difference between the concentrating juice and the warming media;
- The recovery of volatile aromas, as a constituting part of the evaporator system;
- The plant cost depending on the potentiality and on annual working hours;
- The operating economy.

Many types of thermal evaporators are available for citrus processing, because concentration is common to other industrial activities. Some plants strategies are then necessary because of the peculiarities of citrus fruit juice, such as heat sensitivity and viscosity.

Thermal evaporators more commonly used in citrus processing, as considered more suitable to the whole of industrial necessities, are HTST (high temperature short time) with falling film, both with tube bundle and plates. Other kinds of evaporators (heat pump, agitated film, at forced circulation, through centrifugation) are used for particular requirements.

Tube bundle evaporators with falling film

Among the different HTST concentrators available for citrus processing industry, a very diffused one is the TASTE, thermally accelerated short time evaporator, continuous evaporator, single pass, long vertical tube falling, film type, multiple effect, multiple stage, high vacuum, no vapour recompression, high temperature, short residence time (Chen *et al.*, 1992). It is made by FMC (Food Machinery Corp.), Lakeland, Florida, USA.

Substantially, every stage is constituted by a distribution cone, by a long tube bundle and by a cyclone centrifuge separator. The distribution cone is connected at the beginning, through a pipe suitably reduced by a Venturi, to the feeding pump of the warm juice coming from the pasteurization system or from a preceding effect. At the basis the distribution cone is connected with a tube bundle, which is in vacuum. The vacuum degree of every tube bundle depends on its position in the product path, less vacuum and then higher temperature at the beginning of concentration, more vacuum and then lower temperature when the juice becomes more and more viscous. Inside the distribution cone the pressure difference causes the evaporation of a part of the water from the juice. A turbulent mix juice/vapour then forms, and its volume is about six times bigger than that entering the cone. The volume increase permits a good distribution in the different tubes of the bundle and improves the 'wetting'.

In the external part of the bundle containing the tubes of every effect, that is in the shell, circulates the warming fluid, constituted in the first effect by live steam coming from the boiler and in the other effects by the steam coming from the preceding effect. At the top of the shell a pipe conveys not condensable vapours and juice volatile aromas to the aroma recoverer and to the barometric condenser.

The long tube bundle implies some advantages, diminishing the number of tubes for every effect and, maintaining the warming surface, on one hand a good distribution at the head of the tube bundle is assured, and on the other the turbulence inside the tubes and thus a better 'wetting' increases.

The separator is a cyclone in which the juice/vapour mass enters tangentially: the so originated centrifuge force helps the separation of the two phases.

The choice of the effects number is a function of an economic calculation deriving from the comparison of the energetic consumption and the cost of the plant, including maintenance expenses (Guatelli, 1998). Normally there are from four to seven effects. When juices very sensitive to heat are used, like lemon juice, damages caused by residence time are limited lowering entry temperature, even if this implies the reduction of hourly evaporative capacity of the concentrator.

The reachable concentration degree depends on final product viscosity. Viscosity can be lowered by about 20 per cent if from one of the effects the juice is extracted when it reaches about 50° Brix (in case of orange juice) and, after having been homogenized at 1,000 bar, reintroduced in the concentrator. Homogenization brings also the reduction of sedimentable pulp (Crandal *et al.*, 1988, 1991; Grant, 1990). The reduction of the viscosity can also be obtained by adding pectolytic enzymes with low content of pectin-methyl-esterase. If these enzymes are added directly to citrus juices before finishing, this increases the yield (Baker *et al.*, 1995).

The concentrate juice goes out from the last effect at 35/40 °C, according to temperature and available quantity of cooling water at the barometric condenser. The juice is soon brought to 10 °C by means of a flash cooler. The concentrate temperature is further on lowered, between −5 °C and −10 °C by means of a cool glycol exchanger. After passing in refrigerated tanks for the standardization, the juice is usually pumped in great tanks under nitrogen atmosphere placed in refrigerators at −10 °C. Alternatively, metallic drums with plastic bags can be used and stored in refrigerators at temperatures lower than −18 °C.

Evaporating, the water brings away most of volatile substances contained in the juice and that are constituted by aromas coming from endocarp and by essential oil, which passed in the liquid phase from epicarp during squeezing. The whole constitutes the juice bouquet, which would get lost during concentration. It is recovered, to add it to the concentrate at the moment of use, in aroma recovery, with as less thermal damage as possible, in two different phases: the oil phase, essentially constituted by terpenes and the water phase, which contains principally hydrosoluble oxygenated compounds, which characterize the aroma.

The aroma recoverer is basically constituted by a series of condensers in which a part of the evaporated in the juice, the richer part in aromatic compounds, concentrates and condenses, exchanging its heat first with the entering single strength juice, then with water and subsequently with cooling mix. About 90 per cent of aromatic richness of the origin juice is thus recovered.

When the juice to be concentrated enters the first stage, a part of its water evaporates and the vapours pass in the shell of the subsequent stage, giving their heat to the juice inside the pipes. A limited part of these vapours, the part containing the more volatile compounds, is sucked by the head of the shell in the recovery system where it is concentrated, condensed and refrigerated. A decantation system separates the oil phase from the water phase.

Single strength juice to be concentrated is usually introduced in one of the final effects (for example in the third effect of a four effects evaporator) where temperature is not very high because vacuum is high. Consequently, a part of the most volatile aromatic compounds remained in the uncondensable vapours and could not be recovered. Nowadays the principle has been modified: it is preferable to introduce the single strength juice in the first effect, where higher temperature and less vacuum facilitate the recovery of low-boiling aromas.

Falling tubes evaporators with multiple effects have some disadvantages that are both common to other multi-effects and specific. The principal negativities are:

- Rapid fouling in tubes and necessity to proceed to frequent stops for the cleaning, above all when early season orange juice is used, with higher acidity and higher hesperidin concentration;
- Longer residence times with higher stages number.

Tube bundle evaporators with thermal recompression

In order to decrease both the plant cost and fouling, the Gea Wiegand of Ettlingen, Germany, proposes an evaporation system with five effects with thermal recompression in which a part of the evaporate of the fourth effect is recovered (Wiegand, 1993).

In these conditions, the quantity of live vapour necessary to evaporate a kilo of water is the same necessary for a standard seven effects USDA. The vapours coming from the first effect go to the aroma recoverer.

The residence time of the juice in the evaporator is lower than that of the seven effects, both totally and in the part in which the temperature is higher.

For what concerns fouling, the particular design of the parts in contact with the juice permits to hesperidin to deposit in rather thick layers before detaching. A longer operative time would pass before intervening with the cleaning.

The first effect condense, deriving from boiler steam mixed with the vapour from the juice, cannot be introduced in the boiler. To recover its calories, an exchanger can be used against the water entering the boiler or against the juice.

Plates evaporators

Various kinds are available: with falling and rising film or only falling and also with thermal or mechanical compression like those made by APV Baker Ltd., by Schmidt Bretten GmbH (Sigmaster concentrator), by Alfa Laval (ACE concentrator by Alfa Laval cassette evaporator).

Schematically, they are all made of thin plates distanced by gaskets so to form adjacent little chambers: alternatively, in one of them passes the juice to be concentrated, in the subsequent one passes the warming steam, like in plates pasteurizers. The plates are shaped so to assure fluid turbulence necessary to permit thermal exchange. It is thus possible to have higher viscosity concentrates, up to 400 cP against the value of 200 cP normally obtainable with tubular evaporators with falling film. Because of the limited thickness between the plates, determined by gaskets, the pulp content and, above all, its dimensions must be carefully controlled.

The limited residence time permits the concentration of thermally sensitive juices, such as lemon juice, without great thermal damages. The gaskets represent the delicate point of the system, both because microbiologic pollution points can occur, and because an incorrect operating economy can lead to even fast consumption. Internal inspection is an easy operation, it is sufficient to loosen the plates and carry on a further manual cleaning after the usual alkaline cleaning. Normally, because of juice turbulence in the plates, fouling and salting phenomena occur even after 20 working hours; it is then necessary to clean with alkaline detergent solutions, eventually operating with automatic systems such as cleaning in place.

During the years the system has been modified; instead of short plates with alternate flow, rising and falling, longer plates with only falling juice flow are now used (FFPE, falling film plate evaporator by APV Baker Ltd., West Sussex, England). Juice distribution on the plates surface has been improved. The plates can be vertically divided in two parts; there is then a double stage in which juice circulates in series or in parallel and a better 'wetting' is obtained.

Subsequent modifications have brought to FFLE system (falling film long plate evaporator) in which plates are 2.5 metres high, 50 per cent more than before. This causes a further improvement in juice distribution on the plates and thus on the wetting, it reduces stages number and decreases residence time in the plant. It is thus possible to increase the number of effects exploiting a higher total thermal difference as much as possible. The presence of the thermal re-compressor has positive influences on the plant cost, even if, as a negative factor, the problems caused by the above-mentioned gaskets remain, together with the unsuccessful recovery of condensate for the boiler. Evaporators of notable capacity are available, of about 80,000 pounds per hour on six effects.

The capacity, proper of evaporators of this kind, of reaching higher viscosity in the leaving concentrate, permitted the construction of mixed evaporators (with aroma recovery), in which the first two effects are with tube bundle with falling film and, subsequently, four plate effects of the FFLE type, working at lower temperatures.

Cryogenic evaporators with heat pump

In this kind of evaporators, built by several producers such as Mojonnier Bros., Blaw-Knox, APV Co., Gulf Machinery, Mazzoni SpA, it is exploited the property of some refrigerating substances, such as ammonia, which has steam tension much higher than that of water, of notably reducing its volume when these substances are highly compressed. Therefore, during compression, warm ammonia gases give their warmth to the juice passing in the tube bundle as falling film. Steam coming from concentrating juice is condensed by cold ammonia gas, which derives from the expansion of liquid ammonia, originated by compression and condensation of expanded ammonia. This last state passage occurs at the expenses of the cooling water in a separate condenser.

Equipments with two effects and multiple stages have been built exploiting the thermal interval of juice vaporization in working conditions between 27 °C and 39 °C.

The evident advantage of this evaporation system is the low temperature at which it operates. There are some disadvantages:

- The high plant cost;
- The onerous operating economy above all in countries with high cost of electrical energy;
- The strong loss of aromatic substances because of high vacuum in which the operations take place;
- The permanence of the product in the concentrator at a temperature which is favourable for the microbial growth and the enzymatic action for generally longer times than those of other evaporators; from here the necessity of making a good pasteurization first.

Because of particular operative conditions, this concentration system has been used in the citrus processing field for some particularly delicate productions such as lemon processing and, in Sicily, the concentration of pigmented orange juice, characterized by the presence of red pigments, the anthocyanins, very sensitive to heat.

Other types of evaporators

For particular requirements, citrus processing industry can use thermal evaporators of different kinds, such as:

- Centrifugal evaporators (Centritherm by Alfa Laval);
- Agitated film evaporators (Luwa) for high viscosity concentrates, generally used as last stage in combination with other evaporation systems. Concentrates with 20,000 cP can be obtained (Ramteke *et al.*, 1993);

- Forced circulation evaporator, taken from other agro-alimentary processes and used in citrus processing industry almost only in the production of citrus molasses with high Brix.

In the field of thermal evaporation it has also been proposed to subject the single strength juice to cross-flow filtration and to introduce in the evaporator only clear permeate, free from any colloid. In these conditions it is easy to reach high concentration degrees, even beyond 80° Brix, without any particular fouling problems. At the desired moment, the concentrate is mixed with pulpy retentate, which has to be kept under opportune refrigeration. In this case, membranes retaining in the retentate as much aromatic substances as possible must be chosen.

Freeze-concentration

The principle of freeze-concentration consists in cooling the juice below its freezing point, so that part of the water changes into ice, while dissolved substances concentrate in the water left at the liquid state with the lower freezing point. The two phases, ice and concentrated juice, are subsequently separated. The separation is conditioned by several factors such as viscosity of the slurry and crystals diameter.

During concentration loss of soluble solids can occur, both because the concentrate sticks to the ice crystals and because a part of the concentrate can remain into the crystals when their development is very fast. In order to reduce losses it is necessary to form big spherical crystals and proceed to the washing of the crystals with cold water after separation.

The state passage of the part of the water to be removed is obtainable by means of several processes (Muller, 1967; Tyissen, 1975; Van Pelt, 1981; Desphande *et al.*, 1982).

It is usually preferable to operate on two separate sections: a scraped surface heat exchanger, in which the juice remains for a very short time and where nuclei are formed, while the crystals growth occurs in ripening vessels. It is in this second section that the smaller crystals melt and crystallize again on the surface of the bigger crystals already contained in the crystallizer. Crystals with dimensions between 0.2 and 3 millimetres are thus formed.

The separation of the concentrated juice from the ice can be obtained with different systems: through centrifugation (Struthers, Gasquet, Union Carbide, Votator systems) or through filterpress (Daubron system), or also through column with filtrating walls (Philips system) or through washcolumn. In the Grenco system (Grenco Process Technology B.V., s-Hertogenbosh, The Netherlands) are used both washcolumn and, for 'Superbrix' process, the centrifugation with decanter.

The washcolumn is schematically constituted by a vertical cylinder with a movable piston with perforated surface in the bottom part and an ice scraper at the top. When the washcolumn is full of the slurry made by the ice mixed with the concentrate, the ice is pushed upwards by the piston, from whose holes the concentrated juice goes out. In the higher part of the washcolumn the crystals are washed with icy water to recover the concentrate on the ice surface. The scraper removes the washed ice from the head of the column. Since the washcolumn operates in a closed space, the aroma characteristics of the juice remains unmodified.

The above-described system refers to an only stage process.

The main difficulties in the separation of the concentrate from ice depend on viscosity. When orange juice is processed, it is necessary to reduce strongly the pulp percentage under 3–5 per cent (van Weelden, 1994), preferably under 2 per cent in volume.

A more advanced system is the three-stage counter current process, in which the water that must be removed under solid form leaves the machine at the starting stage, that in which the single strength juice enters under slurry form. The three exchangers, feeding the respective ripening vessels, operate at different temperatures depending on the Brix that the juice must acquire at that stage. The liquid phase increases its concentration passing from a stage to the following one and leaves the system from the wash-column connected to the last stage. The ice follows the path in the opposite direction. In these conditions the losses of soluble solid substances are negligible. For orange juice, the reachable concentration degree is about 45° Brix, higher than that obtainable with an only stage process.

With the 'Superbrix' higher concentration values can be reached, up to 55° Brix for orange juice (van Nisterooij, 1991). Applying the centrifuge force with the decanter, the detachment of the concentrate from the crystals is easier, even because, operating with the counter current multistage, it is not necessary to free ice from the juice totally.

If single strength juice, before being introduced in the 'Superbrix', is centrifuged or passed through ultra-filtration, the pulp separated in this phase can be added to the concentrate to reconstruct the sensory characteristics of the starting juice. The influence of the pulp content on the taste is well known. Submitting the clear permeate coming from cross-filtration to freeze concentration, the separation from ice is facilitated. Juice pasteurisation is fundamental with every used process.

The qualities of the concentrate obtained through freeze-concentration (or cryo-concentration) are better than those of the juice processed by means of thermal evaporator.

Even if cryo-concentration process, as a technical principle, has been known for many years, its industrial application is limited to some special production because of the high plants cost, of their limited potentiality, of operating economy difficulties and because the obtainable concentration degree is lower than that reachable with multiple effects evaporators.

In conclusion, even if freeze-concentration and also osmotic concentration which we will examine later on, give a very good citrus concentrated juice, almost completely respecting sensory characteristics of the starting juice, the most common concentration system remains thermal evaporation. These evaporators, which have been improved during the last years, are still preferred for the plant cost, for their operating economy facility, for their reliability when they are used well. The processors have the possibility of choosing among the different kinds the one more suitable to their work.

CLEAR CITRUS JUICES

In the field of citrus processing the clear juices have a minor relevance. The bigger part is represented by lemon juices and lime juices, often used as nectars and syrups acidifying and in kitchen products. Juices of other species, as orange juice, are used in cordials and squashes above all in the northern markets.

For many years, clear citrus juices production has been based on the action of pectolytic enzymes naturally present in the juice. It was placed in tanks and left there until the end of the destructive action of pectinase. When the upper phase had become clear, it was taken away by means of decanting, filtrated on kieselgur and concentrated. Because the attack time of pectic enzymes naturally present in the juice is very long, even of several weeks, and depends not only on their concentration but also on pH and temperature, it was necessary to add also some sulphur dioxide to prevent juice fermentation. During concentration, a part of the SO_2 was removed; but a part remained in the concentrate, above all that chemically linked with the monosaccharides of the juice.

Subsequently, in order to accelerate times and eliminate the use of preservative substances, added industrial enzymes are employed. The juice to be transformed in clear is pasteurized, cooled, added with industrial enzymes and placed in tanks. Proceeding to the inactivation of 'native' enzymes causes the increase of the quantity of necessary industrial enzymes, but the standardization is better both for the enzymatic concentration in the juice, and for what concerns operative conditions.

Opportunely dosing the enzymatic concentration, the clarification of the juice occurs in much shorter time than that necessary with traditional technology. The test with alcohol (or, better, with acetone) is useful to verify if pectin breakdown has occurred.

The upper opalescent phase is taken away, added with clarifying such as silica sol, and filtrated on kieselgur. If the juice has been de-pulped before being treated with enzymes, the lower pulpy phase is not quantitatively relevant. It is passed through centrifuge, added with clarifying substances and filtrated on kieselgur.

A simplified technology, used for lemon juice, excludes pasteurization. After having added industrial pectinase, the juice is left at rest for 6–8 hours; it is then added with silica sol and placed for decantation. The upper phase, practically clear, is filtrated with kieselgur on pressure filter while the sedimentation is filtrated, with kieselgur too, on rotary filter in vacuum. For the whole process no more than 24 hours at room temperature are necessary.

Complete pectin breakdown can occur only when all the pectolitic enzymes that act on the pectic molecule are present in a correct ratio also enzymes attacking the 'hairy region' (Urlaub, 1999).

The silica sol acts as inhibitor towards pectolitic enzymes; it is thus necessary to avoid its addition in the juice before the pectic enzymes have completed their work.

During the storage opalescence phenomena can occur, because of flavonoids precipitation, when they are present in a very high quantity in the juice. For this problem the solution is strongly cooling the juice so that its crystallization is accelerated, and facilitated by hesperidin crystals (Fisher-Ayloff-Cook *et al.*, 1991).

The most recent technology implies the use of crossflow filtration, in which pectolytic enzymes are employed to decrease juice viscosity and improve permeate flow. The complete pectin breakdown implies an increase of galacturonic acid content and of its oligomers in the clear juice. Also in clear juices obtained with crossflow filtration it is possible that the cold stored concentrate becomes opalescent. The permeate treatment with absorbing resins before concentration avoids the formation of opalescences and increases juice stability (Lenggenhager, 1998).

The membranes used can be of the polymeric kind in polyvinylidene fluoride (PVDF) or in polysulphone (PS), with plain, tubular, or hollow fiber configuration, or of the ceramic kind.

When the citrus fruit juice passes through a membrane system, the permeate is free from insoluble solids, practically free from pectin and enzymes. Sugars, acids, vitamins and minerals do not undergo any change, except the decrease of phosphorous content, clearly due to the rate of this element contained in the juice under organic form.

The slight lowering of the Brix is related to the quantity of pectin removed (Capannelli et al., 1990).

The permeate does not possess pectin-esterasic activity, even when the membrane cut-off is higher than the molecular weight of the enzyme. The same happens with pectin. This phenomenon can be due to the fact that pectin in juice is associated with cellulosic material, while pectin-methyl-esterase is associated with suspended solids. The gel layer formed on the membrane wall can reduce its pore size (Hernandez et al., 1992).

Comparing polymeric membranes of different kinds (PS and PVDF) and different configurations, fed with orange and lemon juice with constant composition, it has been noticed that the permeate fluxes are largely independent from the kind of membrane and from the nominal molecular weigh cut-off. Cross sections of the membranes at the end of the process, observed using a scanning electron microscopy, show a fibrous deposit on the surface of the membrane, which appears as a dynamic membrane which influences the system behaviour (Capannelli et al., 1992).

In cross filtration systems using polymeric and ceramic membranes with tubular configuration, the permeate fluxes strongly depend on the tangential feed velocity at the membrane and are almost independent from the driving force at average pressure over a membrane greater than 0.2 MPa.

Because of its different superficial structure and roughness, the pectin-pulp deposit formed on the membranes has different properties in the two cases. The ceramic membranes show higher fluxes at a lower Reynold number than the polymeric membranes (Capannelli et al., 1994).

In the industrial production of clear citrus juices two systems of crossflow filtration are used: batch and continuous processing.

In the first case the retentate returns to the feeding tank and joins the new juice to be treated, until the mix viscosity in the feeding tank is so high (and the permeate flux so low) as to break the operation and proceed to cleaning. Therefore, the system works with increasing viscosity and pulposity.

The continuous processing consists in a series of elements of ultra-filtration or microfiltration in more stages. The retentate going out from the first stage feeds the second and so on.

Because the juice circulating in the crossflow filtration system becomes warmer by friction, exchangers for cooling must be installed in the plants.

At the end of each cycle the membranes must be washed to remove the substances left on the membranes surfaces and to restore flux capacities. Alkaline solutions are used, followed by hypochlorite solution or by nitric acid and then by warm water. To help cleaning, pectolytic enzymes can be used: hydrolizing pectin, they help their removal (Jansen, 1997).

The lime processing shows some peculiarities, and therefore European Community had to include in its rules on fruit juices a derogation allowing the production of lime juice from the whole fruit, while for other citrus fruit the juice must derive only from endocarp.

The older system of lime processing in Mexico is based on fruit pressing, after removal of unsuitable ones and after washing, in screw-press. The juice-essential oil emulsion thus obtained is steam distilled and essential oil is separated from condensate by decantation.

The drastic thermal treatment in presence of an acid medium such as lime juice brings deep changes in oil essential components. Lime distilled oil has thus chemical, physical and sensory characteristics very different from essential oil extracted without thermal treatment. But this flavour is appreciated for some uses. Moreover, having undergone a drastic acid-thermal action, its flavour does not change when it is used in acid liquids (lemonades, cordials, and so on), even for long commercial times.

Because of the drastic treatment, the juice has a bad quality, while pectin can be produced from the peel. To obtain a good quality juice and thermally degraded essential oil, the greatest part of primary emulsion juice-oil produced by screw-presses is refined in finishers and sent to disc centrifuges, where an oil richer emulsion is separated. The juice is cleared, filtrated and concentrated while rich emulsion is mixed to the rest of primary emulsion (corresponding to about 10–15 per cent of the whole quantity of processed lime) and steam distilled. The acidity given by this limited juice quantity (which goes lost) is enough to give the pH necessary to the desired degrading of oil aromatic compounds.

The juice, placed in settling tanks, is added with pectolytic enzymes and left at rest. When native enzymes together with added ones have completed their work, the upper layer is decanted, added with bentonite and silica, filtrated on kieselgur and concentrated. Thus, sacrificing a limited juice quantity, the principal two products required by the marked are obtained: human alimentation juice and distilled oil. From the pulp left in tanks further distilled oil is produced.

Because of higher acidity of lime juice, the time necessary for enzymatic clarification is very long, even of more than four weeks, and the addition of preservatives is necessary. The operation becomes much faster using membrane filtration.

Using suitable rasping equipments followed by juice extractors (or with screw press and centrifugation systems for essential oil) not thermally degraded juice and oil can be obtained. With normal processes cloudy concentrate juice can be produced.

DEHYDRATED CITRUS JUICES

The complete removal of water, which is the most important component in weight, brings to almost total elimination of many alteration phenomena of citrus fruit juices. In fact, the condition permitting the development of micro organisms and the action of enzymes and chemical agents is water activity. Moreover, the weigh reduction determines the reduction of package, transport, handling and storage costs. The storage, not needing low temperatures, influences less the final cost of the product.

Many technologies have been proposed for the production of dehydrated citrus fruit juices; the main inconvenient which industry has always had to face is the hygroscopicity of the finished product. This characteristic involves on one hand great difficulties to bring the juice at a residual humidity lower than 3 per cent, and on the other the necessity of confectioning the dehydrated juice in limited dimension containers, in a dry atmosphere and in presence of dehydrating agents.

To improve dehydrated juice stability, before proceeding to drying, more or less relevant quantities of corn syrup or maltodextrin can be added. The so obtained product cannot be legally declared as 'juice' and is used for particular necessities.

In any case, the preparation of dehydrated citrus fruit juices is an activity which interests a limited part of the market, also because of the rather high processing cost.

From the standpoint of quality, the best dehydrated juice is that obtained through freeze-drying, in which water removal occurs by sublimation: the juice is frozen and the water removed in high vacuum without passage through the liquid state.

In order to reach easily the eutectic point, the orange juice must be introduced in the equipment at a not very high concentration, preferably not higher than 30° Brix, with notable increase of producing costs.

The drying of citrus fruit juices can be conducted on a stainless steel belt moving into an in vacuum room, at about 2 mm Hg (Vacuum 'puff' Drying) or in the form of foam in presence of foaming agents (Foam-Mat Drying). In Spray Drying system, the product to be dehydrated is introduced atomized in form of little drops in the upper part of a cyclone in which it is transported by a warm air current at 150 °C. Because of the process rapidity, the temperature inside the juice drops does not reach high values since a steam layer wraps the granule while it forms. In Spray Drying the presence of notable percentages of corn syrup or maltodextrin is necessary. In order to obtain a good drying, the concentration of the mix to be dried up must not be high.

INNOVATING PROCESSES

Some technologies proposed in recent years for the process of citrus fruit juices will be now treated. Some of them are already used in several plants, even if they are still substantially being modified for a better setting-up; other are still in a phase of advanced study and we can reasonably think that their application can be usefully employed in citrus juices production. In particular, the technologies related to non-traditional pasteurisation systems, debittering and osmotic concentration will be treated.

Non-traditional pasteurization

In substitution of the above-described pasteurization processes, alternative systems are available for citrus processing industry. Some of them use heat to inactivate microbiologic population and enzymes, but with different technologies, others operate with methods which do not foresee significant thermal variations.

Non-traditional thermal processes

To reach in a very short time the pasteurization temperature and avoid the overheating risks present in traditional technique, several heating technologies are available, such as direct injection of steam and microwave use.

In the first case, food-grade steam is introduced in the juice to be pasteurized. When ready-to-serve juice must be confectioned from concentrate, the condensed steam can remain in the juice as part of the water necessary for dilution. The cooling is obtained injecting sterile-filtred-deaerated water (Muller *et al.*, 1988). If the purpose of the thermal treatment is the production of NFC juice, a quantity of water corresponding to the introduced steam must be removed subjecting the juice to flash evaporation. Thus the removal of air but also that of part of the aroma is obtained (Knorr, 1994). But if the introduced steam comes from meteoric water, the characteristic isotopic rates of the so treated juice can undergo some variations because a corresponding quantity of 'total' water is removed from the juice, not only the meteoric water introduced under steam form. The isotopic rates would not undergo notable variations if the steam needed for pasteurization is obtained from condensate of juice separated from the same kind of juice during thermal concentration and opportunely purified. In the use of condensate the necessary cautions must be taken to avoid pollutions. The steam evaporated from the juice is not constituted by simple distilled water; juice sprays are present and are transported in the steam phase; the organic substances, even if diluted, permit the growth of a significant microbial population. Moreover, if thermal energy present in the condensate is lowered in the refrigerating tower, the pollution possibilities increase. Contaminations from thermo-resistant spores of acidophilic microorganisms of Alicyclobacillus genus can occur.

The pasteurization technique by means of microwave energy has been studied above all at the Florida Department of Citrus, Scientific Research Department, Lake Alfred, Florida, USA, in the production of orange juice. Using as thermal source microwave energy at 2,450 MHz, the juice reaches more rapidly than the traditional conduction/convection method the temperature of 96 °C. Maintaining these conditions for 15–25 seconds, the 99.9 per cent of pectin-methyl-esterase is inactivated and the total microbial charge is reduced to values lower than 20 Colony Forming Units per millilitre (Nikdel et al., 1993). Operating at 70 °C for a time comprised between 12 and 26 seconds, the pectin-esterase inactivation is less marked while micro-organisms are practically destroyed also in these conditions. Since using microwave energy as heat source, the heat forms inside the heated body itself, the risks of off-flavours due to localized overheating are avoided.

From the point of view of thermal balance, it was noticed that at least 90 per cent of given heat is absorbed by the juice, with limited heat losses in the environment (Fox, 1994).

The electromagnetic radiations in the field of radio frequency can be used in a different way: when electromagnetic waves are passed through the juice, this behaves as a dielectric component of a capacitor (with microwave energy the heating occurs through irradiation). The electric field established between the two flat electrodes generates a disordered movement among the bipolar molecules subject to the action of inverted polarity million times a second. A rapid and even increase of temperature then occurs (Fox, 1994; Di Giacomo, 1995).

For pasteurization of alimentary products with particulates up to 24 mm 'Ohmic heating' can be used, with alternating current in the low frequency (50–60 Hz). During the process, both solid parts and the liquid phase are heated in the same way (Ladwig, 1988).

It is not thus necessary to overheat the liquid phase to ensure the reach of the desired temperature inside even large particulates, as it is required with traditional heat transfer equipment.

Non-thermally processes

The inactivation of the enzymes present in citrus fruit juice and the destruction of the microbiologic population are, as it has been already said, necessary operations to increase the shelf life of the finished product. The use of heat implies always a more or less marked alteration of the characteristics of the treated juice. Citrus processing industry tends to minimize these alterations reducing processing times and temperatures and using new heat transmission systems. In any case, to reach the purpose it is fundamental that used citrus fruits are as much as possible intact and the plants are built and managed in a suitable way for an accurate sanitation.

The destruction of microbial charge and the enzymatic inactivation (or, at least their reduction) is obtainable with technologies, which do not foresee heat use but only alternative systems. Above all when the purpose is not to reach a long shelf life but to ensure to the produced juice a commercial life sufficient to the normal supplying to the consumer, under predetermined cooling conditions.

High hydrostatic pressure (pascalization)

It is the non-thermally pasteurization system more used in citrus processing in alternative to the traditional one. It is based on the principle that irreversible modifications of microorganisms morphology can be caused by the application of high pressures. Generally, the microbial vegetative cells are destroyed, while the spores, if present, appear to be stronger.

The Bacillus stearothermophilus spores are destroyed when they are subjected to six cycles of oscillatory pressurization at 70 °C and 600 MPa (5 min/cycle). The destruction is not complete with a continuous pressurization system for the time of 60 minutes at the same conditions (Hayachawa *et al.*, 1994).

The destruction speed depends on several factors such as product concentration, sugars concentration, pH and temperature, as well as applied pressure and processing time. At equal conditions, enzymes and micro-organisms in the concentrate orange juice show a higher resistance than single strength juice (Knorr, 1994).

The application of high hydrostatic pressure implies irreversible alteration to the protein structure of enzymes too. These modifications also depend on several factors such as pH, quantity of dissolved substances, temperature, and kind of enzyme. The inactivation of pectin-methyl-esterase in orange and grapefruit juices is obtainable with pressures of 500–900 MPa.

Pressures higher than 600 MPa inactivate thermolabile but not thermostabile pectin-esterase (Goodner *et al.*, 1998). If the pH juice is lowered through CO_2, the inactivation of thermolabile pectin-esterase can be obtained with pressures lower than 100 MPa (Balaban *et al.*, 1995).

Aromatic substances do not undergo substantial modifications, as well as vitamins, colourings and aminoacids.

High-pressure technology has used at industrial level from the 90s. In Japan, Wakayama Prefectural Agricultural Processing has developed a system used for the production of tangerine juice. The treatment occurs at 4,000–6,000 bar and permits the production of single strength juice without original taste modifications. Also Wakayama Nokyo Food Industry has perfected a process for the production of pasteurized citrus fruit juices by means of high pressure, while Pokka Corp. obtains not bitter grapefruit juice using a technology realized in cooperation with Mitsubishi Heavy Industry (Di Giacomo, 1995).

The industrial plants using high hydrostatic pressure to increase the shelf life of citrus fruit juices have a different structure according to the state in which the product to be worked is. The Alstom ACB, Nantes, France, provides two different equipments, one suitable to the treatment of the juice already confectioned in the final container and another for bulk processing.

In the first case the high-pressure chamber is filled with containers. These must be built with flexible materials for the transmission of pressure variations and have suitable form for the best room exploitation. Moreover, the headspace inside the container must be reduced to minimize oxygen and optimise effects. When the high-pressure chamber has been charged and closed, empty spaces are filled with water or with water containing small oil quantities, carefully removing all the air. Instead of water, another pressure transmitting medium can be used. High pressure is generated through direct compression and uniformly transmitted until predetermined values.

For bulk processing a semi-continuous system can be used with three dephased pressure cells, one for the filling, one for the treatment and one for discharge. Because the treatment is conduced on the not confectioned juice, in the upper of each cell a floating piston separates the juice from the pressure medium (Bignon, 1996).

Freshly squeezed orange juice, treated at 3,500 bar for one minute at the temperature of 30 °C acquires, if maintained in conditions of opportune refrigeration, a shelf life of 60 days keeping the organoleptic characteristics of the fresh product (Donsì *et al.*, 1996).

Treatments of pressurization/depressurization between 50 and 400 MPa for 15 minutes with mild thermal processes between 20 and 60 °C on freshly squeezed orange juice can bring significant reductions of the activity of peroxidase and of pectin-methyl-esterase. At a temperature of 35 °C these activities reduce of 50 per cent (Cano *et al.*, 1997).

The fresh orange juice subjected to high pressure treatments maintains similar characteristics to those of referring juice for 30 days at 4 °C (Pelletier, 2000).

Ultra high pressure technology

In order to reduce the micro-biologic charge in fruit juices, the Coca Cola Company has patented in 1993 a process using ultra high pressure, about 15,000 psi, obtained with a homogenizator APV Model 30CD cell disrupter. The high-pressure difference between the inlet in the homogenizer and the outlet implies also pulp smashing.

Treated with this technique, the freshly squeezed orange juice has a shelf life of 40 days at 4 °C, while the corresponding not treated juice shows a shelf life not longer than 10 days. Flavour and other chemical, physical and organoleptic characteristics do not undergo any variations (Fox, 1994).

High electric field pulses technology

When microorganisms are subjected to strong electric fields, they undergo such alterations to cause their death. Because the cellular membrane is not conductive, when an electric field is applied to a cell, a *trans*-membrane potential is generated. If the critical point is overcome, the phenomenon becomes irreversible with the formation of spores on the membrane because of the repulsion between the molecules and thus causing the cell death. With the application of stronger forces, also the spores can be inactivated.

This technology, applied to freshly squeezed orange juice, according to proofs conducted by FMC and Krupp, brings a strong reduction of the total microbial charge but it does not inactivate pectin-esterase. The organoleptic characteristics, as well as vitamins, colour and flavour do not undergo any alterations (Knorr, 1994).

The Pure Pulse Technologies Inc., San Diego, California, USA, in order to not thermally pasteurize juices, proposes a system based on the application of multiple, short-duration, high-intensity electric field pulses. The product to be treated passes between the two electrochemically inert electrodes and is subjected to electrical forces for a time of 1–10 microseconds with the destruction of microorganisms and without damages for the treated juice (Lander, 1996).

Debittering

The bitter taste sometimes present in citrus fruit juices is fundamentally due to the excessive presence of two substances: naringin and limonin. As it is well known, limonin is present in orange juice, limonin and naringin in grapefruit juice, but also in other citrus fruit juices.

When the concentration of the former overcomes 500 mg/L and that of the latter 5 mg/L, the bitter taste becomes perceptible and, at a higher concentration, the juice acquires a clearly bitter taste (Maier *et al.*, 1977).

The high content of bitter substances can be linked to several factors: the variety (for example, Navel oranges also in the USA and Biondo Comune oranges in Italy), climatic factors such as for Temporona oranges (Hofsommer *et al.*, 1991). Also the rootstock, independently from variety, can concur to determine a high content of limonin in oranges (Di Giacomo *et al.*, 1977).

Because, as it is known, limonin and naringin, but also other flavonoids, are contained in the maximum part in the peel and in membrane materials, their concentration in the juice depends also on the processing system. A hard squeezing in juice extractors and in finishers increases the juice yield but also increases the quantity of bitter compounds passing from solid parts into liquid.

Many technologies have been proposed to reduce bitterness in citric products and this abundance of attempts shows the importance that this problem has for processors and for the market.

At present, two absorbing substances are generally employed for this purpose by citric industry: activated carbon and non-ionic resins.

The use of activated carbon in citrus processing is usually linked to the production of clear juices. In these cases, the treatment occurs in bulk; after the enzymes action, both naturally present in the juice and added, necessary for pectin demolition, the juice is added with flocculation agents and, if necessary, with activated carbon in powder form. For carbon removal the juice is filtrated on kieselguhr.

When cloudy citric products must be debittered with activated carbon, this must be used in granular form. Carbons with absorbing inner surface bigger than 1,000 square meter per gram and with granules having an average diameter higher than 1 mm must be chosen. The pore volume must be higher than 0.8 ml per gram.

The debittering operation is conduced at 'fixed bed', passing the juice on a carbon layer high at least a meter and placed in stainless steel vessels. Every treatment unit includes two dephased vessels, working alternatively, one in debittering phase and one in washing and regenerating phase, so as to realize a semi-continuous process. The employed technique is basically equal to that used for absorbing resins. The pulp must not be brought through centrifugation at values higher than 1 per cent, but it is not necessary to pay much attention to the content of essential oil present in the liquid to be treated because the absorbent capacity of the carbon is not damaged by it.

Since carbon has a specific gravity, which is double of the absorbing resins used in citrus industry (2.0–2.2 instead of slightly more than 1), the removal of the pulp stuck on the carbon granules is not a difficult operation. It is just necessary to have enough water pressure in the backwashing; carbon losses for transportation together with the pulp are limited. In this stage it is important to care that every carbon particle, eventually formed for granules breakdown, is removed.

To have a correct debittering of citric products and to exploit completely the capacities of absorbing substances in a vessel system with fixed bed, it is important that the product to be treated is uniformly distributed on the entire mass surface.

The advancing front of the liquid inside the mass must be as much as possible parallel to the absorbing surface. If privileged advancing directions form, a part of the absorbing contained in the vessel will not carry out its function.

During the backwashing the absorbing must be well mixed. If pockets not reached by water (and subsequently by alkali) are formed, areas of microbiologic contamination will occur.

For regeneration, solutions of alkaline hydroxides are used; potassium hydroxide is preferred to sodium hydroxide even if it is more expensive. Generally, after one or two campaigns it is opportune to proceed to high-temperature reactivation, which permits to restore the initial absorbing potentiality.

The carbon action is not only due to the absorbing activity of Van der Waals forces. The complexes carbon-oxygen, which make the surface of the carbon slightly polar, have at the same time an oxidative action polyphenols and anthocyanins. In fact, carbon has also a decolouring action.

The debittering with activated carbon is used both in juice treatment and in core-wash and citrus extracts treatments. These products, which are employed in the processing industry of drinks and squashes, are characterized by high viscosity and a higher cloud than the juice. Cloud stability is the most appreciated factor in these productions.

The alternative to carbon to debitter and upgrade citrus products is constituted by resins. Styrene-divynil-benzene hydrophilic absorbents are used. They are macro-porous resins, FDA approved for the treatment of alimentary products, with surface area greater than 700 square meters per gram (on dried resin). In the United States the use of absorbing resins for the debittering of Californian Navel oranges and Florida grapefruit juices is legal (Lenggenhager et al., 1997), for the type for manufacturing without obligation of declaration on label. Since 1997 also in Italy the debittering with absorbing resins is permitted as physical treatment of fruit juices.

Many studies have been carried out on the activity of this kind of resins as limonoids and flavonoids absorbents from citrus products. All the researches confirm that the so treated juice loses the excessive bitter and maintains its characteristics improving the total quality. Chemical and physical data substantially remain the same, except the content of limonin, hesperidin, naringin and of other flavonoids. Not only orange and grapefruit juices have been tested in relation to debittering on absorbent resin, but also bergamot and sour orange juices. In this case the taste can be improved reducing acidity by means of the contemporary use of absorbing resins and ions-exchanging resins. Also the debittering of pulp-wash and by-products is possible, even if with some more difficulties (Grohmann et al., 1999).

The absorbing mechanism is linked to Van der Waals forces between the resin surface and the phenolic compounds. This link is broken by alkaline solutions.

To realize industrially the debittering of citrus products by means of resins, it is necessary first to remove or at least strongly reduce the pulp. For this operation two different ways can be followed: centrifugation or ultra-filtration. The removal of the greatest part of the pulp through centrifugation reduces the quantity of solid particles retained inside the resin layer, but it does not eliminate the difficulties for their total removal in the backwashing phase. Since the not high relative density of the used resins, big losses of resin spheres can occur together with the pulp. On the contrary, it can happen that the smaller particles of pulp and cloud are not removed with loss of absorbing activity. To avoid these risks, in some plants the removal of the resin bed from the vessel is carried out at the end of each cycle and also the treatment in a scrubbing system (Lyndon, 1996).

The essential oil content in the juice must be very low because the resins absorb the oil, which is not removed by alkaline solutions. Moreover, the specific gravity of the resin is lowered, and it can float. Therefore, after pasteurization, the juice is de-oiled. For this purpose the first stage of the evaporator can be used.

After debittering, the juice is mixed with the pulp separated during centrifugation to restore the characteristics of the initial juice.

In the other alternative, the crossflow filtration, the clear permeate passes on the resin bed (Akin *et al.*, 1991). The absence of solid particles and terpenes, which constitute the main part of essential oils, makes the treatment easier, without the risks linked to the pulp presence.

The clear debittered juice is thus mixed with the retentate; the pulp retained by ultra-filtration membranes maintains all its starting limonin and naringin and thus its rate of bitter taste. If the bitterness of the final mix is still excessive, it can further be lowered subjecting the retentate to dialysis. Instead of meteoric deionised water it is better to use condensate obtained during thermal concentration of the juice opportunely purified and, if necessary, sterilised. The liquid at low Brix coming from dialysis is used as final phase of debittering.

Membrane concentration

The removal of water from juices using reverse osmosis (RO) has the advantage of not needing state passages; it is not necessary to consume energy to transform the water rate that one wants to remove in steam (thermal evaporation) or in ice (freeze-concentration). It is necessary just to apply on the juice a pressure higher than the osmotic one while it passes on suitable semi-permeable membranes. Unfortunately, with regard to this, citrus juices, as well as having a rather high content of dissolved solids, to be commercially valid as concentrates, must have a concentrate degree of 4–6 times. To reach these values very high pressures must be applied; orange juice at 60° Brix has an osmotic pressure of about 200 bar, not easily maintainable in industrial equipments also for membranes collapse.

Therefore, RO concentration is usually applied when much lower concentrations must be reached, 20–25° Brix for orange juice.

This technology is usually applied in particular cases, like when one must retrograde an orange juice 65° Brix, thermally concentrated, to 42° Brix as cutback. This kind of frozen concentrated is usually prepared adding to the thermally obtained concentrate single strength juice to improve its taste and aroma. In consequence of the industrial introduction of reverse osmosis, several producers use pre-concentrated juice at about 24° Brix, osmotically obtained with pressures of about 60 bar, instead of single strength juice. The favourite configuration, dealing with pulpy juice, is the tubular one. The proposals of using RO as a pre-concentration stage for semi-processed product, to be completed with the other above-described technologies, seem not to have been followed.

Higher concentrations are reachable recurring to multi-stages systems, combining a first stage of high rejection membranes with a second of low rejection membranes. In this last stage a part of the soluto crosses the membrane increasing the osmotic pressure of the permeate (Gostoli et al., 1995). The permeate, containing a not negligible quantity of soluble solids, is reintroduced in the first stage at high rejection.

The system studied by Watanabe et al. (1990) is based on three stages, of which the first and the second equipped with high rejection membranes while the third brings low rejection membranes. A spiral module is used. The applied pressures are: 7.5 MPa at the first stage and 9.5 at the second and third stages. The concentration degree reachable with clear juice is of about 45° Brix.

On the same principle is based the 'FreshNote', set up by SeparaSystem with FMC and Du Pont. The squeezed juice is soon cooled at 10 °C and placed under nitrogen; these conditions are maintained for the whole process.

As first step, the juice is completely de-pulped in an ultra-filtration system with membranes at flat configuration, with plate and frame modules. Two phases are obtained, the clear permeate passing to RO and the pulpy retentate constituting about 1/10–1/20 of the starting volume and containing the macro-molecules and all the insoluble solids, enzymes, pectins, micro-organisms. The retentate is pasteurised and cooled to be subsequently mixed to the RO concentrate.

The clear permeate, which contains sugars, acids, salts, amino acids, vitamins and aromatic oxygenated substances, goes to the RO concentrating section, constituted by a series of hollow fine fiber permeators. The membranes, made of aromatic polyamide, have an inner diameter of 42 micron and an outer one of 93 micron, not to collapse under the high pressure they are subjected to. The first concentration section contains high rejection membranes, followed by a second one with low rejection; this last permeate, which contains a quantity of dissolved solid substances, returns at the beginning of the first section together with the UF permeate to be concentrated. In order to recover the small quantities of dissolved solids and aromas, which can escape together with the water going out from the first section of the concentrating system, there is the possibility of inserting a polishing section.

The number of the ultra-filtration stages is function of the plant potentiality and of the concentration ratio of insoluble solids. The number of the permeators is function of the plant dimensions and of the concentration ratio of insoluble solids.

The reverse osmosis systems works at a pressure included between 105 and 140 bar. Adding the pasteurized UF retentate to the RO concentrate, the final juice at 45° Brix is obtained, when orange juice with aromatic richness corresponding to 80 per cent of the initial juice is processed (Goettsch et al., 1991).

Alternatively to reverse osmosis, the Cogia SA of Palaiseau, France, proposes the Osmotic Evaporation, which consists in removing water from the juice with a brine solution through microporous hydrophobic membranes. High concentration degrees can be reached (Anonymous, 1995).

Water is extracted from the juice by means of an extractor before the hydrophobic porous membrane. The passage of the water from the juice to the extractor occurs through the membrane pores under steam form without any penetration of juice or brine solution in the membrane, whose pores remain full of gaseous phase.

STORAGE AND SHIPMENT

Only a limited percentage of citrus juice production is confectioned and commercialized as finished product by citrus processing industries. Generally, these industries produce half-processed products, that is not destined to direct use, which are sold to other factories. These confection their finished products through further processing operations, and these products have different formulations and characteristics according to the market typologies to which they are destined.

For the storage of juices that citrus processing industries produce, the most suitable preservation media is without any doubt low temperature.

In some cases, chemical preservatives such as benzoic acid and its derivates and sulfur-dioxide are used, substances which were once the principal media to ensure fruit juices preservation. Chemical preservatives still find some use in the production of juices destined to the production of some products in which juice is only a limited percentage component, for example refreshing drinks (orangeades and lemonades, and so on) and the squashes, provided that the laws ruling these productions allow it. In the European Community the presence of sulphur dioxide is permitted (until 350 mg/l) in lemon juice destined to kitchen use. When this is allowed, sulphur dioxide is preferred to benzoic acid because it also has an anti-oxidizing action and because it can be removed, even if not completely, through vacuum distillation.

The conditions for the low temperature storage depend on several factors, such as the kind of juice, its concentration degree and use destination; for NFC juice they have already been indicated. For what concerns concentrate juices, the most common options are the storage in drums, tanks and bag-in-bins.

The drums are metallic containers with a high density polyethylene double bag. They are filled with juice previously cooled at $-5\,°C$ and placed in a refrigerating cell at a temperature not higher than $-18\,°C$. In these conditions, the juice remains unaltered for at least a year.

The storage in big tanks is the most common system. The tanks are placed in a refrigerating cell or are singularly provided with refrigerating devices with circulation of a cooling media. The maintenance temperature is chosen according to the concentration degree of the juice and to its viscosity, in order to avoid that the content freezes inside the container or that it becomes uneasily extractible by means of a pump. For the orange juice at $60/65°$ Brix temperatures of about $-10\,°C$ are used. In order to avoid that juice undergoes oxidative variations during time and to prolong its life, nitrogen atmosphere is created in the tank.

The very low storage temperature maintains unaltered the chemical and physical characteristics of the juice preventing the formation of off-flavour and colour alteration. But changes in the appearance, which do not mean a product decay, can occur. In concentrate orange juice, crystalline agglomerates of monopotassium citrate with small amounts of sugars, hesperidin and pulp can be generated; their formation is favoured by acidity and high concentration. Under the same conditions of the other parameters, the lowest temperature accelerates the phenomenon (Kimball, 1991; Filomena *et al.*, 1998).

The tanks destined to citrus fruit storage are built in stainless steel; the alloy composition is function of the kind of juice to be stored. The presence of molybdenum, as well as of nickel and chromium, makes the metal resistant to the attack of reducing acids as SO_2, while titanium increases the welding stability.

The resistance to corrosion also depends on the conditions of the superficial metallic layer. Besides the surface passivation with nitric acid, the electro polishing permits the elimination of the small pores in which micro organisms can hide.

The tanks must be carefully washed with cleaning agents, to which steam sterilization must follow. The steam condensation on the containers walls is avoided by introducing warm sterile air.

The tanks dimensions vary according to the productive and stocking structures linked to them; those of pre-storage and standardization have more limited dimensions, while those destined to real stocking above all in clearing centres, have a capacity up to 1,000 cubic meters. Pipes and valves systems, coordinated by a computer in the biggest structures or manually operated in smaller factories, link the tanks each other and permit charge and discharge operations. Pipes and valves are built so to be easily disassembled for cleaning.

Before being introduced in the big tanks, the juices must be standardized. This operation cannot be carried out in big dimension tanks, in which mixing is not easy.

The transportation from the producing industries to the markets occurs in tanks or cisterns. From Brazil, which is the greatest producer of concentrate orange juice, the transportation occurs in suitable supply-ships too. The biggest European importer for this kind of juice is Holland. Some European ports, for instance Rotterdam, Amsterdam, Gent, Antwers are equipped with a suitable systems for pumping directly from ships to refrigerated tanks for the storage.

Citrus fruit juice, concentrated or single strength, can be stored and transported in bag-in-bins. The filling is sterile. The juice, after de-aeration, pasteurisation and cooling, is sent to the collecting tank under nitrogen and then to the aseptic filling machine.

Sterilized bags, made of food grade polyethylene and metallic film, are used as a barrier for oxygen. The bags are supported by drums of 55 USA gallons or by bins of 300 USA gallons.

Before starting work, all the line, included the filling machine, must be carefully sterilized with culinary steam followed by a treatment with warm sterile air. For all the working time, the filling head area is under the action of sterilizing agents. In such a way, the sterilization of the interested outer part of the bags before the cap removal is assured. Subsequently, inside the sterile chamber, the cap is automatically removed and the filling head is introduced into the bag spout. When the filling is over, the cap is put again, always in a sterile way.

REFERENCES

Akin, D., Milnes, B., Gomes Jòia, M., Rocha Neto, A.J. and Graça Maset, R. (1991) Citrus juice upgrading using the combined technologies of crossflow filtration and adsorptive resins. *Report of XI Intern. Congr. Fruit Juice*, Sao Paulo, Brazil, 415–419.

Anonymus (1995) Osmotic evaporation. *Fruit Processing*, 5, 33.

Attaway, S., Carter, R.D. and Fellers, P.J. (1989) Die hestellung und beandlung von frish gepresstem, nicht pasteurisiertem, orangensaft. *Fluessiges Obst.*, 56, 606–612.

Baker, R.A. and Grohmann, K. (1995) Enzyme application in citrus processing. *Fruit Processing*, 5, 332–335.

Balaban, M.O., Marshall, M.R. and Wicker, L. (1995) Inactivation of enzymes in food with pressurizes CO_2. *U.S. Patent*, 5, 393, 547.

Beasley, L.R.M. (1997) Recommendations for fresh juice production with the FMC extractor. *Fruit Processing*, 7, 296–298.

Bignon, J. (1996) Cold pasteurizers hyperbar for the stabilization of fresh fruit juices. *Fruit Processing*, 6, 46–48.

Cano, M.P., Hernandez, A. and De Ancos, B. (1997) High pressure and temperature effects on enzyme inactivation in stawberry and orange products. *J. Food Sci.*, 62, 85–88.

Capannelli, G., Lister, D.G., Maschio, G., Bottino, A., Munari, S., Ballarino, G., Mirzaian, H. and Rispoli, G. (1990) Clarification of orange and lemon juice using membrane technology. *International Congress on Membranes and Membrane Processes*, Chicago, USA, Vol. I, p. 286.

Capannelli, G., Bottino, A., Munari, S., Ballarino, G., Mirzaian, H., Rispoli, G., Lister, D.G. and Maschio, G. (1992) Ultrafiltration of fresh orange and lemon juices. *Food Sci. Technol.*, 25, 518–522.

Capannelli, G., Bottino, A., Munari, S., Lister, D.G., Maschio, G. and Becchi, I. (1994) The use of membrane processes in the clarification of orange and lemon juices. *J. Food Engineering*, 21, 473–483.

Carter, B.A. (1992) Lemon and lime juice. In S. Nagy, C.S. Chen and P.E. Shaw (eds) *Fruit Juice Processing Technology*, Agscience Inc., Auburndale, Florida, pp. 228–230.

Cemeroglu, B. (1998) Heat inactivation kinetics of pectin methylesterase from orange and grapefruit peels. Peroxidase as an indicator of peel blanching. *Fruit Processing*, 8, 158–161.

Chen and C.S., Shaw, P.E. and Parish, M.E. (1992) Orange and tangerine juices. In S. Nagy, C.S. Chen and P.E. Shaw (eds) *Fruit Juice Processing Technology*, Agscience Inc., Auburndale, Florida, 126.

Crandal, P.G., Davis, K.C., Carter, R.D. and Sadler, G.D. (1988) Viscosity reduction by homogenization of orange juice concentrate in a pilot plant taste evaporator. *J. Food Sci.*, 53, 1477–1481.

Crandal, P.G. and Davis, K.C. (1991) Viscosity reduction and reformation of structure in orange concentrate as affect by homogenization within commercial taste evaporators. *J. Food Sci.*, 56, 1360–1364.

Deshpande, S.S., Bolin, H.R. and Salunkhe, D.K. (1982) Freeze-concentration of fruit juices. *Food Technol.*, 36, 68–82.

Di Giacomo, A. (1995) Tecnologias inovativas en la trasformaciòn industrial de los agrios. *Essenz. Deriv. Agrum.*, 65, 327–351.

Di Giacomo, A., Calvarano, M. and Tribulato, E. (1977) Sul contenuto di limonina del succo di arancia. Nota IV. Ruolo del portinnesto sulle cultivar 'Valencia Late' e Moro. *Essenz. Deriv. Agrum.*, 47, 156–166.

Donsì, G., Ferrari, G. and Di Matteo, M. (1996) Stabilizzazione del succo d'arancia a mezzo di alta pressione: valutazione dell'effetto delle condizioni di processo. *Ital. J. Food Sci.*, 8, 99.

Dugo, G., Del Duce, R., Verzera, A., Stagno D'Alcontres, I. and Daghetta, A. (1990) Trasformazioni dei componenti degli aromi solubili agrumari in ambiente acido acquoso. Monoterpeni aciclici e monociclici. *Industria Bevande*, 21, 217–222.

Fellers, P.J. and Carter, R.D. (1993) Effect of thermal processing and storage of chilled orange juice on flavor quality. *Fruit Processing*, 3, 436–441.

Filomena, M., Valim, C.F.A. and Menezes, H.C. (1998) Formation of crystalline agglomerates during storage of concentrated orange juice. *Fruit Processing* 8, 232–236.

Fisher-Ayloff-Cook, K.P. and Hofsommer, H.J. (1991) Neue technologische aspekte (IV). Verfahreustecnische moeglichkeiten zue hestellung von citrusspezialprodukten. *Fluessiges Obst.*, 58, 596–601.

Fox, K. (1994) Innovations in citrus processing. *Fruit Processing*, 4, 338–348.

Gherardi, S., Vicini, E., Trifirò, A., Barbieri, G. and Decio, P.L. (1992) Uso dell'ultrafiltrazione per la produzione di succo d'arancia di elevata qualità. *Industria Conserve*, 67, 307–311.

Goettsch, H.B.G. and Kannami, M. (1991) FreshNote premium juice concentration systems for premium orange juice production. *Report XI Intern. Congress. Fruit Juice*, S. Paulo, Brazil, 399–401.

Goodner, J.K., Braddock, R.J. and Parish, M.E. (1998) Inactivation of pectinesterase in orange and grapefruit juices by high-pressure. *J. Agr. Food Chem.*, 46, 1997–2000.

Gostoli, C., Bandini, S., Di Francesca, R. and Zardi, G. (1995) Concentrating fruit juices by reverse osmosis. The low retention-high retention method. *Fruit Processing*, 5, 183–187.

Gray, L.E. (1989) Produktion von frischem orangensaft. *Fluessiges Obst.*, 56, 15–21.

Grant, P. (1990) Homogenizing concentrate in a juice evaporator. *Citrus Engin. Confer., Florida Section ASME*, Lakeland, Florida, 55–71.

Grohmann, K., Manthey, J.A., Cameron, R.G. and Busling, B.S. (1999) Purification of citrus peel juice and molasses, *J. Agric. Food Chem.*, 47, 4859–4867.

Guatelli, G. (1998) Energy saving by T.A.S.T.E. evaporator. *Essenz. Deriv. Agrum.*, 68, 373–383.

Hayakawa, I., Kanno, T., Tomita, M. and Fuyio, Y., (1994) Application of high pressure for spore inactivation and protein denaturation. *J. Food Sci.*, 59, 159–163.

Hernandez, E., Chen, C.S., Shaw, P.E., Carter, R.D. and Borros, S. (1992) Ultrafiltration of orange juice: effect on soluble solids, suspended solids and aroma. *J. Agric. Food Chem.*, 40, 986–988.

Jansen, E. (1997) Enzyme applications for tropical fruits and citrus. *Fruit Processing*, 7, 388–393.

Hofsommer, H.J., Fisher-Hayloff-Cook, K.P. and Radke, H.J. (1991) Neue technologische aspekte–zur enterbitterung von citrussaeften. *Fluessiges Obst.*, 58, 62–64.

Kimball, D.A. (1991) *Citrus Processing*, AVI Book, Van Nostrand Reinhold, New York, NY, USA, 272–278.

Knorr, D. (1994) Novel processes for the production of fruit and vegetable juices. *Fruit Processing*, 4, 294–296.

Kolb, E. (1993) Citrus finisher pulp processing and its use in fruit juices. *Fruit Processing*, 3, 6–8.

Ladwig, H. (1988). Ohmic heating a continuous sterilization method for media with particulate products. *Confructa*, 33, 178–185.

Lander, D. (1996) Microbial kill with pulsed light and electricity. Fruitful possibilities. *Fruit Processing*, 6, 50–51.

Lenggenhager, T. and Lyndon, R. (1997) Profit-generating benefits of ultrafiltration and adsorber technology. *Fruit Processing*, 7, 250–256.

Lenggenhager, T. (1998) Ultrafiltration and adsorber technologies combined produce top-grade concentrates and juices. *Fruit Processing*, 8, 136–142.

Lyndon, R. (1996) Commercialisation of adsorber technology in the fruit juice industry. *Fruit Processing*, 6, 130–134.

Macdonald, H.M., Evans, R. and Spencer, W.J. (1993) Purification and properties of the major pectinesterase in lemon fruits (*Citrus limon*). *J. Sci. Food Agric.*, 62(2), 162–168.

Macdonald, H.M., Evans, R. and Spencer, W.J. (1994) The use of continuous-flow electroforesis to remove pectin in the purification of the minor pectinesterase in lemon fruits (*Citrus limon*). *J. Sci. Food Agric.*, 64(1), 129–134.

Maier, V.P., Bennet, R.D. and Hasegava, S. (1977) Limonin and other limonoids. In S. Nagy, P.E. Shaw and M.K. Veldhuis (eds) *Citrus Science and Technology*, AVI Publishing Co., Westport, Connecticut, USA, Vol. I, pp. 355–396.

Mc Farlin, G.P. (1997) Storage and shipment of NFC-juice in aseptic bag-in-bin conteiners. *Fruit Processing*, 7, 217–221.

Muller, H., Moya Donoso, J. and List, D. (1988) Direct sterilization/pasteurization and its effects on the quality factor of fruit and vegetables juices. *Confructa*, 33, 170–178.

Muller, J.G. (1967) Freeze-concentration of food liquids: theory, practice and economics. *Food Technol.*, 21, 49–61.

Nikdel, S., Chen, C.S., Parish, M.E., MacKellar, D.G. and Friedrich, L.M. (1993) Pasteurization of citrus juice with microwave energy in a continuous-flow unity. *J. Agr. Food Chem.*, 41, 2116–2119.

Parish, M. (2000) The relevancy of *Salmonella* and pathogenic *E. coli* to fruit juices, *IFU-Workshop Microbiology*, April 12, Cologne, Germany.

Pelletier, B. (2000) High-pressure treatment for fruit juice. *Fruit Processing*, 10, 202–204.

Ramteke, R.S., Singh, N.I., Rekha, M.N. and Eipeson, W.E. (1993) Methods for concentration of fruit juices: a critical evaluation. *J. Food Sci.*, 30, 391–402.

Rouse, A.H. and Atkins, C.D. (1952) Heat inactivation of pectinesterase in citrus juices. *Food Technol.*, 6, 291–294.

Rysstad, G., Fredhammer, T. and Osmundsen, J.I. (1993) Preservation of fruit juice quality. *Fruit Processing*, 3, 393–399.

Schofield, T.F. (1994) Aroma improvement by means of the spinning cone column. *Fruit Processing*, 4, 144–147.

Schofield, T.F. and Riley, P. (1998) Developments with the spinning cone column to extract natural concentrated aromas. *Fruit Processing*, 8, 52–55.

Shomer, R., Cogan, U. and Mannheim, C.H. (1994) Thermal death parameters of orange juice and effect of minimal heat treatment and carbon dioxide on shelf life. *J. Food Processing and Preservation*, 18, 305–315.

Seymour, T.H., Preston, I.F., Wicker, L., Lindsay, J.A., Cheng, I.W. and Marshal, M.R. (1991) Stability of pectinesterase of Marsh white grapefruit pulp. *J. Agric. Food Chem.*, 39, 1075–1079.

Soffer, T. and Mannheim, C.H. (1996) Effect of minimal heat treatment and carbon dioxide on shelf life of grapefruit juice. *Fruit Processing*, 6, 99–101.

Tijssen, H.A.C. (1975) Current developments in the freeze-concentration of liquid foods. In S.A. Goldblith, L. Ray and W.W. Rothmayr (eds) *Freeze-Drying and Avanced Food Technology*. Academic Press, London, pp. 149–155.

Urlaub, R. (1999) Enzymes from genetically modified microorganisms and their use in the beverage industry. *Fruit Processing*, 9, 158–163.

Van Nisterooij, M. (1991) Superbrix: the continuing development of freeze-concentration technology. *Report XI Intern. Congress Fruit Juice*, S. Paulo, Brazil, 409–414.

Van Pelt, W.H.J.M. (1984) Economics of multistage freeze-concentration process. *Confructa*, 28, 225–239.

Van Weelden, G. (1994) Freeze-concentration: the alternative for single strength juices. *Fruit Processing*, 4, 140–143.

Versteig, C., Rombouts, F.M., Spaansen, C.H. and Pilnik, W. (1980) Thermostability and orange juice destabilizing properties of pectinesterase. *J. Food Sci.*, 45, 969.

Verzera, A., Del Duce, R., Donato, M.G., Trozzi, A. and Gigliotti, G. (1990) Trasformazioni dei componenti degli aromi solubili agrumari in ambiente acido acquoso. Monoterpeni biciclici. *Industria Bevande*, 21, 379–386.

Watanabe, A., Nabetani, H., Nakajama, M., Ohmori, T., Yamada, Y. and Isiguro, Y. (1990) Development of multistage RO combined system (MRC) for high concentration of apple juice. *International Congress on Membrane*, Chicago, USA, 282.

Weissmann, S. (1991) Microbiologische aspekt from fructhprodukten und deren verarbeitung. *Fluessiges Obst.*, 58, 655–658.

Wicker, L. and Temelli, F. (1988) Heat inactivation of pectinesterase in orange juice pulp. *J. Food Sci.*, 53, 162–164.

Wiegand, B. (1993) Production of orange juice in Brazil example: evaporation. *Fruit Processing Special Report*, VIa, 13–17.

8 Essential oil production

Angelo Di Giacomo and Giovanni Di Giacomo

DISTRIBUTION AND LOCATION OF THE ESSENTIAL OILS IN THE FRUIT

When viewed in cross section, moving from the outside inwards, a citrus fruit has the following layers (Figure 8.1):

- a thick cuticle belonging to the *epidermal* cells, containing the stomata;
- the *epicarp* or *flavedo*, made up of parenchymatous tissue which is rich in pigments (chloroplasts and chromoplasts);
- the *albedo*, parenchymatous tissue composed of irregular-shaped spongy cells which are white in colour and separated by large air pockets;

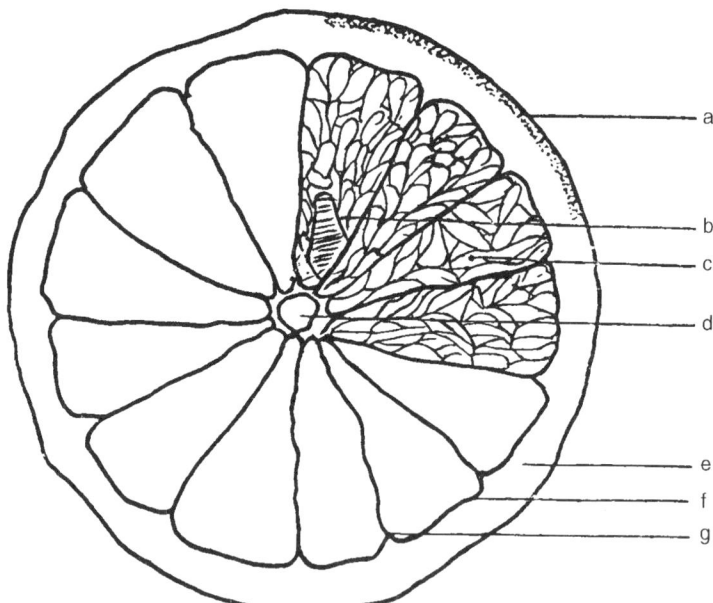

Figure 8.1 Cross section of a citrus fruit, showing: (a) flavedo with oil glands; (b) seeds; (c) juice vesicles; (d) central axis; (e) albedo; (f) segment; (g) segment membrane.

- the *endocarp*, which is the edible portion of the fruit, made up of segments containing the juice vesicles which have ultra-thin walls. The segments are covered by a thin carpellary membrane and are distributed around a central axis which has the same composition as the albedo;
- the seeds, found around the central axis embedded in the endocarp.

In the epicarp, and more precisely in the region immediately below the epidermis of the fruit, the *oil glands* containing the essential oil are found. These glands are between 0.4 and 0.6 mm in diameter and have no walls as such, but are enclosed instead by the remains of decayed cell matter, with no openings or excretory outlets. The oil glands are surrounded by cells containing an aqueous solution which is rich in salts, sugars and colloids. When the peel comes into contact with water, the higher osmotic pressure inside these cells causes them to absorb water and become turgid, thus putting the oil glands under pressure from all directions. Under these conditions, if the peel is subjected to a mechanical process such as pressing or grinding, the oil spurts out with considerable force when the oil glands are ruptured. Turgor of the peel can be attained more readily in freshly-picked fruit. The degree of maturity of the fruit is also important. On the other hand, as already mentioned, the tissues around the oil glands are spongy and can absorb liquid rapidly, and so they will tend to soak up any essential oil in their vicinity. Essential oil extraction methods take these factors into account and so the mechanical process is always performed under strong jets of water, which wash the oil away before it can be reabsorbed by the fruit.

DETERMINATION OF THE ESSENTIAL OIL CONTENT OF THE FRUIT

None of the mechanical or manual processes used to extract essential oils from fruit or peel is able to recover the total quantity of these important derivatives. The maximum attainable yield of oil from the raw materials depends on a number of factors, some of which relate to the nature of the fruit and others to the nature of the machinery used to extract the oil. These factors will be discussed in some depth later on, but it is in any case essential to have a method at our disposal with which we can determine the total essential oil content of the fruit. Such information provides us with an indispensable point of reference when we need to make decisions about the quality of the fruit (from the point of view of oil yields) and about the efficiency of the machinery employed for the extraction and recovery of the oils.

In general, distillation is the best method to determine the essential oil content of the fruit. In order to achieve reliable results, the fruit must be representative of the batch under investigation and there must be no oil loss during the grinding process. A convenient procedure is therefore to take a sharp knife and remove the thin outer layer containing the oil glands from weighed fruit (the quantity should be sufficient to provide not more than 2 mL of essential oil). The peel should be soaked for at least 12 hours in an aqueous sodium chloride solution (ca 20 per cent). The softened peel is ground in a standard meat grinder, care being taken to sluice thoroughly, and about 2 L of slurry are collected in a 3 L, narrow-neck, round-bottom flask. A Clevenger (1928)

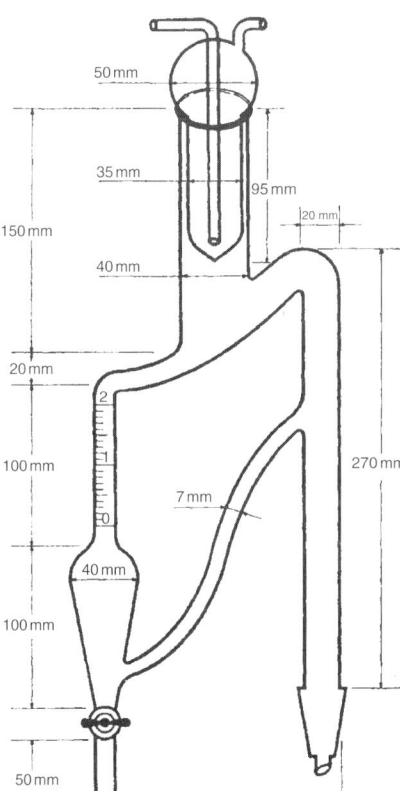

Figure 8.2 Apparatus for the determination of total essential oil content of citrus fruit.

oil separator (Figure 8.2), filled with water up to the graduated part, and a condenser are fitted to the flask and the mixture is brought to boiling point. The mixture is refluxed for about an hour at the rate of 50 drops per minute. The distillation is then interrupted, the separator is removed from the flask and water is drained off through the tap provided until the essential oil reaches the graduated part. The quantity of essential oil is measured from the base of the lower meniscus to the highest point of the upper meniscus. To calculate the weight in grams, the measured volume is multiplied by the mean specific gravity of the distilled oil. Duplicate determinations must not give rise to differences greater than 0.1 mL.

COLD-PRESS EXTRACTION OF ESSENTIAL OILS

Rodanò (1930) identified three distinct types of phenomena that occur during the cold-press extraction of essential oils from citrus fruit:

1   rupture of the epidermis and of the oil glands that contain the oil;
2   creation in the peel, of compressed areas, surrounded by areas which are under less pressure, through which the oil can be expressed;
3   abrasion of the peel, producing small pieces of debris.

Essential oil can be extracted either from the whole fruit or from the peel. When the whole fruit is processed, all three of the above-mentioned phenomena occur, while only the first two take place when the peel alone is processed. In addition, one of the extraction procedures in widespread use today (the FMC recovery system) involves the simultaneous extraction from the whole fruit, of both juice and essential oil.

The readiness with which the oil spurts from the oil glands is affected by the freshness and ripeness of the fruit. Essential oil issues more violently from unripe fruit.

In the case of bergamots, however, it is often helpful to store the fruit for a time prior to processing; it is easier in this way to separate the oil from the emulsion.

Opinions differ greatly over the relative merits of extracting the essential oil by processing the whole fruit or by processing the de-pulped peel. Processing of the whole fruit is better suited to continuous production (partly because an easier and fuller separation and clarification of the essential oil is achieved) and therefore to the industrial processing of large quantities of fruit. Processing of the peel, on the other hand, involves a discontinuous production cycle and requires a lot of space because the peel has to be treated with lime and left to stand for a period prior to extraction. However, the essential oil obtained from peel (especially that extracted by the *sfumatura* (slow-folding) method) is of higher quality than that obtained from the whole fruit. As a result, the method chosen will depend on the circumstances in each case. Large plants processing hundreds of metric tons of fruit every day will choose to work on the whole fruit with highly automated production systems, paying a price in terms of essential oil quality. Conversely, small companies processing limited quantities (tens of metric tons) of fruit per day are better off working on the peels using the *sfumatura* method, which produces higher quality oils.

In order to provide a full picture of the advantages and disadvantages of using the two different oil extraction processes discussed so far, the main implications of these methods for the production of other important and related citrus derivatives, namely juice and peel, are summarised below:

- if the essential oil is extracted before the juice, the quality of the juice is less likely to be compromised by the presence of high levels of oil. However, it is always good practice in any case to de-oil the juice prior to packaging;
- whole fruit which has undergone drastic surface treatment for oil extraction is more likely to release bitter substances from the albedo into the juice. Moreover, the recycled water used to wash the oil away from the surface of the fruit may contaminate the juice with micro-organisms. These two drawbacks can be minimised by washing the fruit again with water after the oil has been extracted. It is also a good practice to add an antiseptic to the water cycled through the extraction machinery and the system used to separate the emulsion;
- pulpless peel can only really be used satisfactorily to produce brined peel, since the lime used prior to the extraction of the essential oil is detrimental to pectin production (Di Giacomo and Lo Presti, 1956).

To conclude this brief discussion, we should point out that for many years, a number of Sicilian citrus processing plants have used the *sfumatura* method to extract oil from the peel without the pre-treatment with lime, which was considered indispensable in the past (Di Giacomo and Signorino, 1971). With improvements in technology, this last approach is certain to become ever more widespread and will lead to better yields and higher quality oils, and therefore the disadvantages of oil extraction from the peel will no longer apply. With this technique, a continuous production process can be used to extract the oil, since it is no longer necessary to let the peel stand prior to extraction. Furthermore, after extraction, the peel will still be suitable for pectin production.

Machinery for the treatment of whole fruit

The procedures for essential oil extraction from whole fruit can be classified as follows, according to the way the surface of the fruit is treated:

- superficial *grazing*, generally an operation which is not too drastic and which causes minimal damage to the tissues of the fruit;
- total *abrasion* of the surface layer containing the essential oil glands.

Both procedures are mechanised versions of manual prototypes:

- the *écuelle method*, which originated in the area around Nice (France) and was used in the West Indies;
- the *circular grater process*, which in the past was used in Spain.

The manual extraction procedures and the famous 'Calabrian Machine' (still in limited use today) have already been discussed in our earlier chapter on the historical development of the citrus industry.

Brown oil extractor (BOE) (Brown International Corporation, Covina, CA, USA)

The BOE (Figure 8.3) is installed in the fruit stream after the washing stage and before the juice extractors. Citrus fruit is delivered to the BOE by a metering elevator which ensures a steady flow. An interruptor gate is incorporated into the inlet chute feeding the BOE. This brief pause in fruit flow allows additional time for the fruit to settle in the machine for maximum performance. The BOE extracts essential oil from whole citrus fruit. Oil removal is achieved by lightly puncturing the entire surface of the fruit with sharp stainless steel spiked rollers, rotating in a pool of water. An adjustable speed differential between adjacent rollers controls the extent to which the fruit is punctured. For complete coverage of an elongated fruit such as a lemon, the rollers not only rotate, but oscillate horizontally in opposite directions, causing the entire peel surface to be treated, releasing the oil. The puncturing takes place beneath the surface of a shallow pool of water to ensure that there is no loss of oil to the atmosphere. Drying rollers remove any water and oil clinging to the fruit surface after its discharge from the machine. This liquid is added to the oil-water stream leaving the BOE. The oil-water mixture then passes through a fine screen in preparation for centrifugation. A first centrifuge produces a stream of rich oil and the aqueous phase from this centrifuge is recycled back to the BOE to eliminate losses. The rich oil stream passes through a second centrifuge which continuously produces pure polished oil.

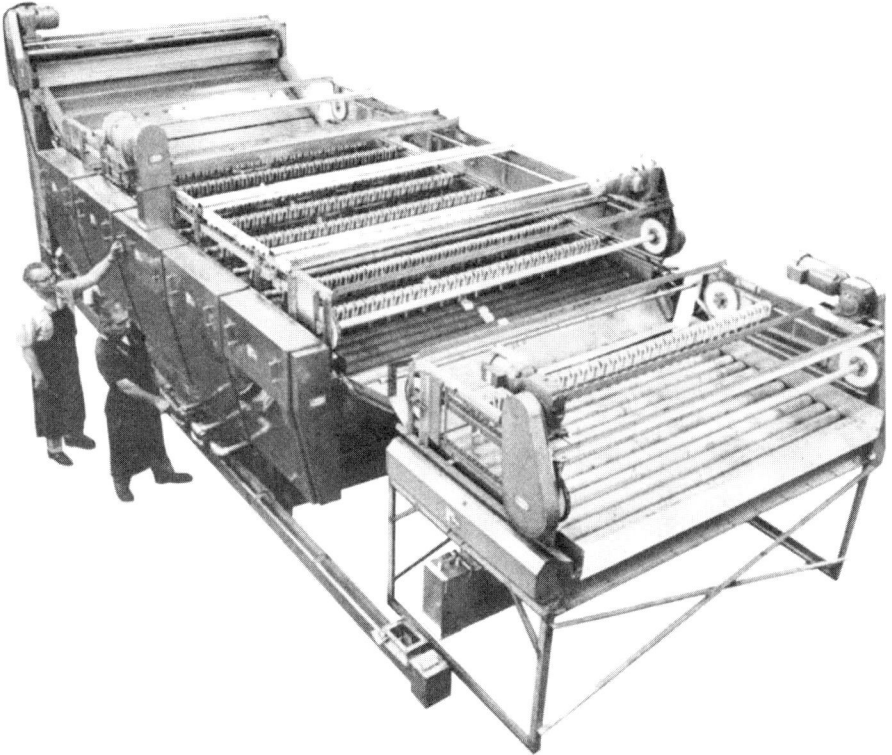

Figure 8.3 Brown oil extractor (BOE).

This process is described in detail by Waters (1993), who reports that it is possible to recover more than 50 per cent of the oil from Valencia oranges.

Polycitrus extractor (F.lli Indelicato, Giarre, Catania, Italy)

The Polycitrus extractor (Figure 8.4) is equipped with an automatic selecting device to expel rotten fruit through a screw conveyor, so that only sound fruit is squeezed. It can process simultaneously fruit of any size. The machine is composed of two units arranged in a cascade: one for the extraction of essential oil by rasping the surface of whole fruit, the other for the extraction of juice. All parts in contact with the products, including the covers, are made of AISI 304 stainless steel. The fruit arrives on the hopper where an automatic device regulates the flow to maintain a constant level. A batcher draws the fruit into the machine, where it is pushed forward by paddles between pairs of rotating rasping cylinders. The speed of the paddles can be regulated, so that the fruit remains on the cylinders for the time required for a complete rasping. This varies according to fruit variety and ripeness. The essential oil is collected by jets of water and the mixture flows into a finisher, where the solid particles are eliminated. The clarified mixture can then be conveyed to the centrifugal separators to recover the essential oil.

Figure 8.4 Polycitrus (F.lli Indelicato, Giarre, Italy).

After oil extraction, the fruit falls into the juice extracting section, where every piece of fruit is cut in half and each half is squeezed against a stainless steel screen.

Different models of this machine are available. For these the maximum working capacity with oranges are the following:

Type ZX2–M6   8,000 kg/h
Type ZX2–M10 12,000 kg/h
Type ZX2–M15 19,800 kg/h

With some modifications, the Polycitrus extractor can be operated without using water to remove and carry away the oil extracted from the fruit. The raspings obtained by the rotating action of the graters are collected in the appropriately-shaped lower section of the machine and from there are channelled out of the machine by two screw conveyers. The raspings are then pressed to extract the essential oil. The yields are fairly modest, but the oil obtained from oranges has an intense colour due to its high carotenoid content. This characteristic is much appreciated by producers of soft drinks, especially in those countries where the use of artificial colourings is not permitted.

Pelatrice speciale (F.lli Speciale, Giarre, Catania, Italy)

The 'Pelatrice Speciale' (Figure 8.5) is completely automatic and does not require any manual input while in operation. It can process oranges, lemons, bergamots, mandarins, limes and grapefruits, all in optimal yields. The machine can process roughly-washed fruit of any shape, size or degree of ripeness (including very ripe). As a result, no additional machinery is required for sorting, washing or brushing of the fruit prior to extraction.

The extractor consists of a series of abrasive rollers and a screw conveyor mechanism with an abrasive surface. The fruit is fed automatically from a drawer in the bottom of

Essential oil production 121

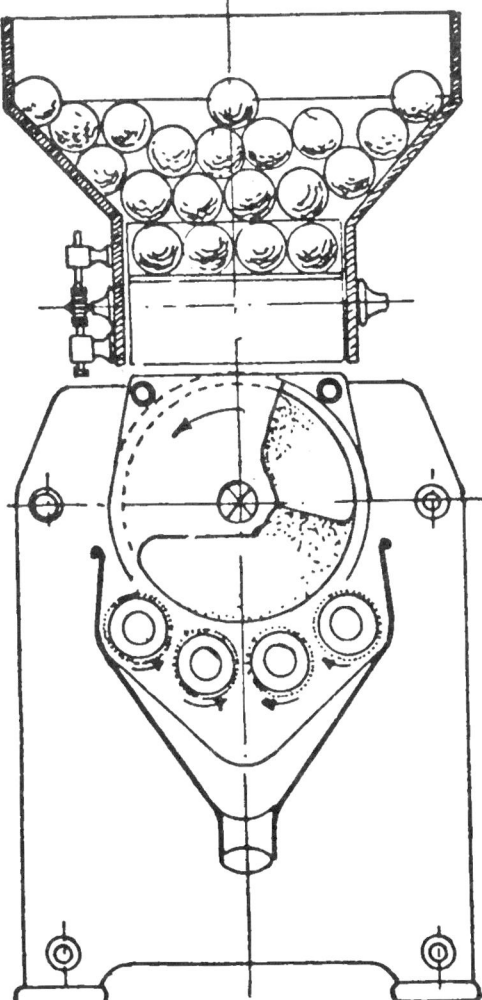

Figure 8.5 'Pelatrice Speciale'.

the hopper into the processing boxes defined by the abrasive rollers and the screw conveyor. The extraction takes place under a constant mist of water from spray nozzles placed along the axis of the machine. The water is recycled and a special device, operated by an external winch, keeps the spray nozzle holes clear while the machine is in operation. The extractor is equipped with three variable speed drives which control: the rotation of the abrasive rollers (according to the hardness of the fruit); the screw conveyor (according to the kind of product desired); the feeding roller. All parts of the machine which come into contact with the oil-water mixture are made of stainless steel. The external surfaces of the rollers and the screw conveyor are covered with stainless steel sheets with hard-wearing abrasive spikes. The machine is usually supplied together with a filtering and pressing machine, to filter and press the solid

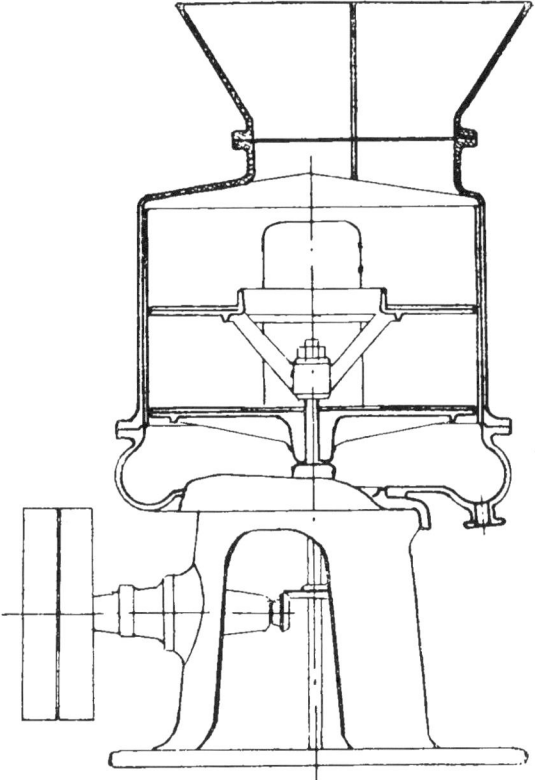

Figure 8.6 Avena extractor.

residues produced by the abrasion of the epidermis of the fruit, and with a water-recycling pump.

Avena extractor (Giuseppe and Placido Avena, Pistunina, Messina, Italy)

In his famous book on essential oils, Guenther (1949) wrote that the Avena extractor (Figure 8.6) 'today represents a most efficient and attractive machine of general application'.

The extractor, which exploits the centrifugal principle, is very flexible and is able to handle different varieties of citrus at different degrees of ripeness (Braverman, 1949). As a result, it was widely used for many years, especially in Italy, Spain, Brazil and Argentina.

It resembles a large sugar centrifuge, about 1.2 m in diameter, in which the rotating portion is not the basket but two horizontal circular plates, the upper of which has a large central opening; the surrounding walls are stationary. The upper surfaces of the plates are fitted with stainless steel segments that have replaceable, pyramidal, abrasive spikes covering the peripheral area. The fruit is loaded by means of a hopper which is divided into two compartments, each of which has a discharge opening at the bottom.

Each compartment contains one plate-load (ca 25 kg). At intervals, the shutters at the bottom open and fruit is discharged onto the two plates. Because of the centrifugal force, the fruit is slammed to a greater or lesser extent, according to the speed setting used, the length of the treatment and the length of the spikes on the plates. Inside the machine, water is sprayed to carry away the essential oil and the debris produced. When the pre-set abrasion time has elapsed, the abraded fruit is expelled and a new cycle starts.

The average capacity of the most common model is about 1 metric ton of fruit per hour, but higher capacity models have also been made. Speed regulators allow the operator to vary the speed of rotation of the plates and the length of the treatment, according to the quality of the fruit and the kind of product desired. Good quality essential oil can be produced with this machine if care is taken to ensure that the length of the abrasion treatment never exceeds 90 seconds, with a plate rotation speed of about 70 rpm. Under these conditions, 'normal' yields are obtained. The quantity of oil extracted can be increased by applying a more vigorous treatment, but the quality of the oil is inevitably lower in this case.

The Avena extractor is better suited to processing hard-skinned citrus fruit. It can also be used on thinner-skinned and more delicate fruit, providing that the abrasive spikes are changed to avoid causing excessively deep incisions.

The use of the Avena extractor has been steadily decreasing in recent years and, in fact, it is hardly ever employed today. The main reason for this decline is undoubtedly its non-continuous mode of operation, which makes it difficult to incorporate this machine into continuous production cycles.

Pelatrice Moscato (CMA, Pellaro, Reggio Calabria, Italy)

It works on the same principle as the Avena extractor.

This machine, currently used for the bergamot essential oil extraction, consists of three rotary parts that will force the fruit to change direction of rotation, from that at the entrance of the extractor to that at its end.

Fraser-Brace excoriator (Fraser-Brace Engineering Co., Tampa, FL, US)

The Fraser-Brace extractor consists of a corridor formed by four horizontal, abrasive, carborundum-covered rolls revolving at high speed. As fruit passes through the extractor it is rapidly turned over and over on the abrasive rolls. A fine spray of water plays upon the fruit and rolls to carry away the oil, wax and grated peel. The oil and wax emulsion with suspended solids is pumped to a screen to remove the coarse particles and then it flows into settling tanks, where it is held from 3 to 12 hours. It is then decanted and passed into a centrifuge to separate the oil. The machine is completely enclosed and very little loss of oil occurs.

According to Guenther (1949), orange oil recovered by the Fraser-Brace extractor exhibits a dark red colour and a high evaporation residue, specific gravity, refractive index and ester content because of the high wax content. The optical rotation is low. The oil is low in aldehydes because of the very large amount of spray and carrying water required to remove the emulsion and cell detritus (100 parts of water per part of oil) from the fruit. Moreover, settling of the oil and cell detritus mixture for a period ranging from 3 to 12 hours results in major changes in the composition of the oil, due to enzyme action, etc.

AMC scarifier (American Machinery Corporation, Orlando, FL, USA)

The scarifier consists of a tilted frame on which 22 revolving cylinders are mounted. The cylinders are covered with stainless steel sheets which have been pierced, causing sharp points of the metal to protrude, similar to the points in a grater. These points puncture the oil glands of the peel. The punched holes are of such size and spacing to prevent clogging with peel while, at the same time, ensuring excellent piercing action. The cylinders are mounted at right angles to the flow of the fruit. A variable speed drive controls the speed of rotation. Inside the machine, a mist of water coming from fog-type spray nozzles washes the released oil from the fruit and saturates the flavedo with moisture to minimise reabsorption of the oil by the peel.

A similar method of extraction is also used by the Israeli *Koffler* machine (Taglith Ltd, Tel Aviv, Israel) and by the Italian *Citrorap* (Bertuzzi, Brugherio, Milan, Italy), models 700, 1400 and CK series (Figure 8.7).

Indelicato MK extractor (F.lli Indelicato, Giarre, Catania, Italy)

The Indelicato MK extractor (Figure 8.8) consists of two vibrating platforms, one above the other, made up of fine-pointed spikes, which prick the peel of the fruit whilst leaving the surface apparently intact. The flow of fruit onto the vibrating platforms is regulated by a hopper at the back of the machine. The fruit is pushed through the machine by paddles made of stainless steel and plastic. The operation of the paddles is controlled by a variable speed drive which can be adjusted according to the citrus variety and degree of ripeness of the fruit being processed. Processing time can therefore vary from 50 to 200 seconds. Because of the vibration, the fruit bounces and turns around in all directions, allowing the spikes to prick and burst the oil glands. The expressed oil is washed away by sprays of water and passed through a screen to eliminate any detritus. This allows a fairly clean emulsion to be sent to the centrifuge.

The working capacity of this extractor is 4–4.5 and 1.8–2.5 metric tons per hour of oranges and lemons respectively.

Machinery for the treatment of de-pulped peel

The ideal term of comparison for all the machines that extract essential oils from peel is the manual technique of the sponge (*sfumatura*). In fact even the best machinery is not able to produce oil of a quality which even remotely approaches that produced manually. At best, as Cultrera (1954) put it, we descend 'from the noble to the merely good'. It seems therefore appropriate briefly to describe this method.

Peel which has been manually de-pulped with a special knife (*rastrello*), which is spoon-shaped with cutting edges, is thoroughly washed in limewater. It is then left to drip dry on woven mats or on special baskets. According to the degree of ripeness of the fruit, this takes from 3–4 hours (in hot or humid weather) to 24 hours (in cold weather).

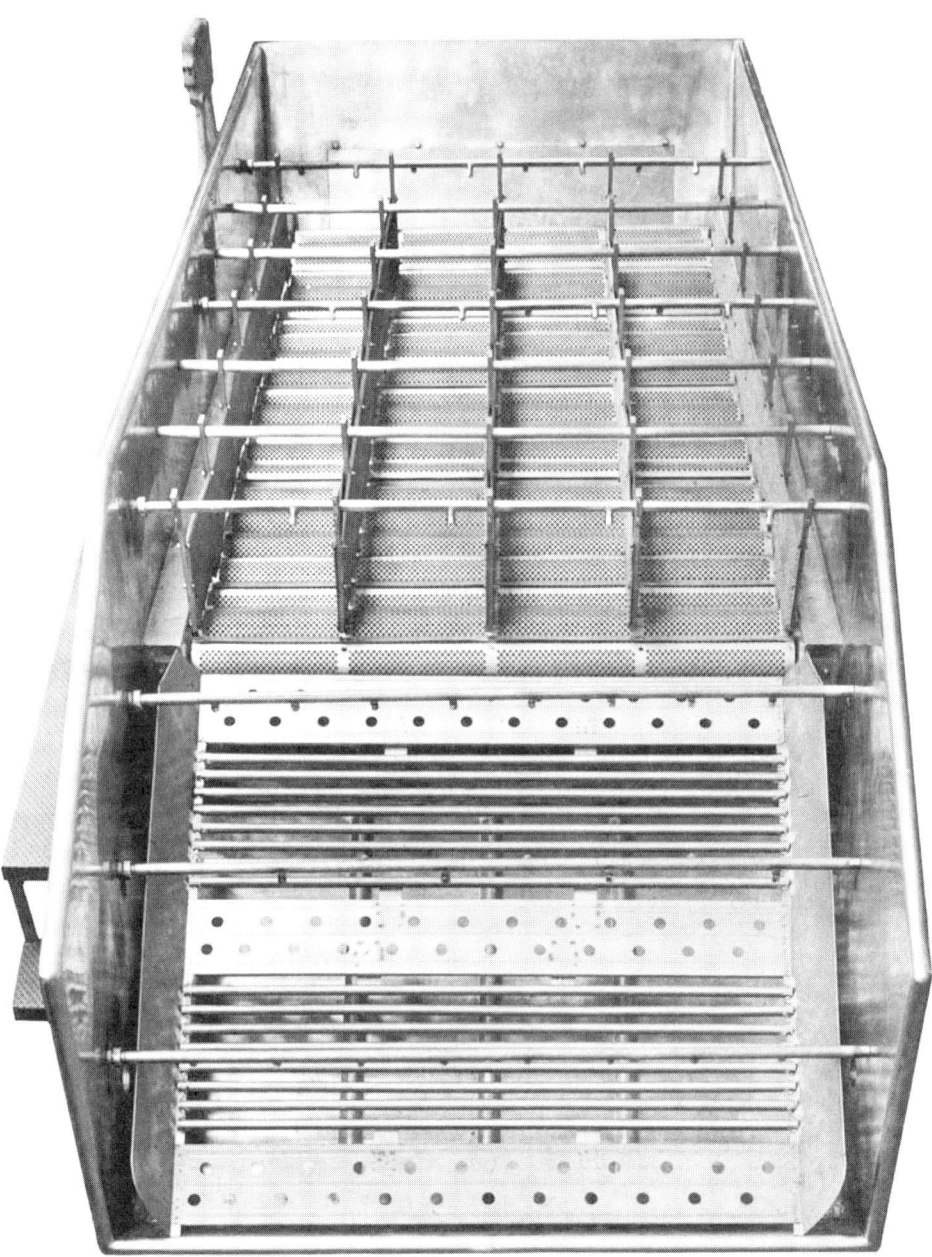

Figure 8.7 Citrorap machine.

This technique produces a hardening of the tissues which makes the oil spurt out of the oil glands more readily, particularly if the *sfumatura* (slow-folding) process is to be carried out by hand. Furthermore, liming has the effect of neutralising the remaining acidity of the peel (which is especially useful if it has been obtained by *birillatura*).

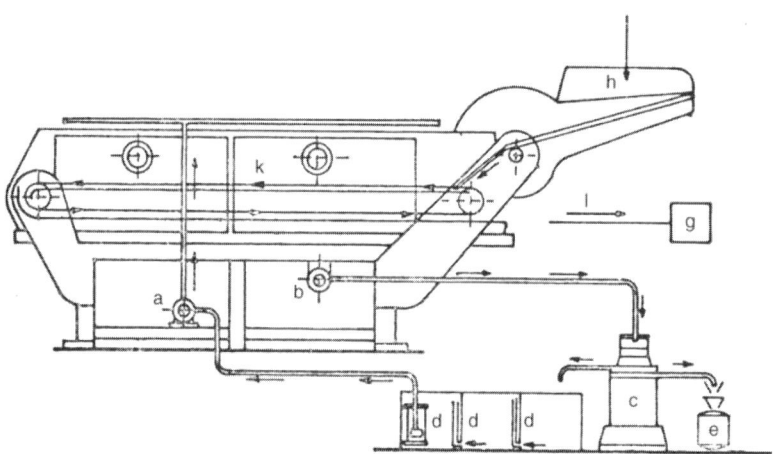

Figure 8.8 Indelicato MK extractor. Diagram showing: (a) pump; (b) oil slurry outlet; (c) centrifuge; (d) decantation tanks; (e) essential oil collector; (f) juice extraction line; (g) fruit loading hopper; (h) fruit de-oiling pathway; (i) de-oiled fruit discharge outlet.

The process involves folding and pressing the peel several times with a circular motion against a series of natural sponges arranged around a terracotta dish (*concolina*). The sponges are held in place by means of a bar across the top of the dish. The essential oil and other liquid from the peel pass through the sponges into the dish. Additional oil is recovered from the impregnated sponges by pressure and torsion. The essential oil is then simply decanted off from the other liquids and from the detritus produced during the breaking of the tissues.

Sfumatrici (slow-folding) machines

The *sfumatrici*, or slow-folding, machines originated in Sicily, where they are still built and used, predominantly for the extraction of lemon essential oil. The mechanism of action of all these machines exploits the natural pressure exerted by the essential oil inside the oil glands, which causes it to spurt out when the peel is folded and compressed. This action has to be applied gently to minimise the seepage of other liquid from the peel. The *sfumatrice* machine can be judged to be working efficiently when the percentage of broken peel discharged from the machine is low. Obviously, this must be accompanied by a low residual oil content in the treated peel.

Virtually all *sfumatrici* machines are of fairly simple design and are relatively cheap, despite the fact that they operate in a continuous mode. The peel, which may or may not have been previously treated with lime, is fed by a loading hopper into the machine, where it is forced to pass through a confined space formed by a fixed part and a moving part. Here it undergoes multiple, random folding to express the essential oil. The moving part usually consists of a conveyor belt with specially-shaped protuberances facing the fixed surface on which there are similar protuberances. The distance between the conveyor belt and the fixed surface becomes progressively narrower as the peel passes through the machine, until it is finally discharged.

The extraction method described so far is a fairly faithful reproduction of the manual technique (see chapter: 'Development of the Citrus Industry: Historical Note'). However, instead of being absorbed by a sponge, in the mechanised process the extracted oil is sluiced out by strong jets of water, just as it is in the rasping machines.

Birillatura and sfumatura unit AZF204 (F.lli Indelicato, Giarre, Catania, Italy)

This is a combined unit for the extraction of juice and essential oil from citrus fruit. The fruit is loaded by means of an oscillating hopper with four channels and adjustable stroke and tilt. The feeding system consists of rotating discs which are suitably shaped to pick up four pieces of fruit at a time from the hopper and send them to the fixed knives which cut them in half. The two halves end up in plastic cups which carry them towards rotating rosettes (*birilli*). The juice is then squeezed out of the fruit halves by the action of these rotating rosettes: a process which is known as *birillatura*.

The juiceless halves are conveyed down chutes to the oil extraction stage, where they undergo repeated folding between a fixed and a mobile specially-shaped plane. The curling and deformation of the peel cause rupture of the oil glands. The essential oil which has been extracted is then washed away by jets of water from spray nozzles. The emulsion formed is collected below in a tank and then sent to the next stage to be finished and separated by centrifugation (Figures 8.9 and 8.10).

Brown peel shaver (Brown International Corporation, Covina, CA, USA)

Nowadays, this machine has been almost entirely replaced by the Brown oil extractor. Nevertheless, it is still used in some specific instances, such as the extraction of oil from tangerine peel.

Peel from juice extractors is fed into the machine, where it is split by a wide knife blade into flat slices of albedo and flavedo. The knives can be set to provide slices of

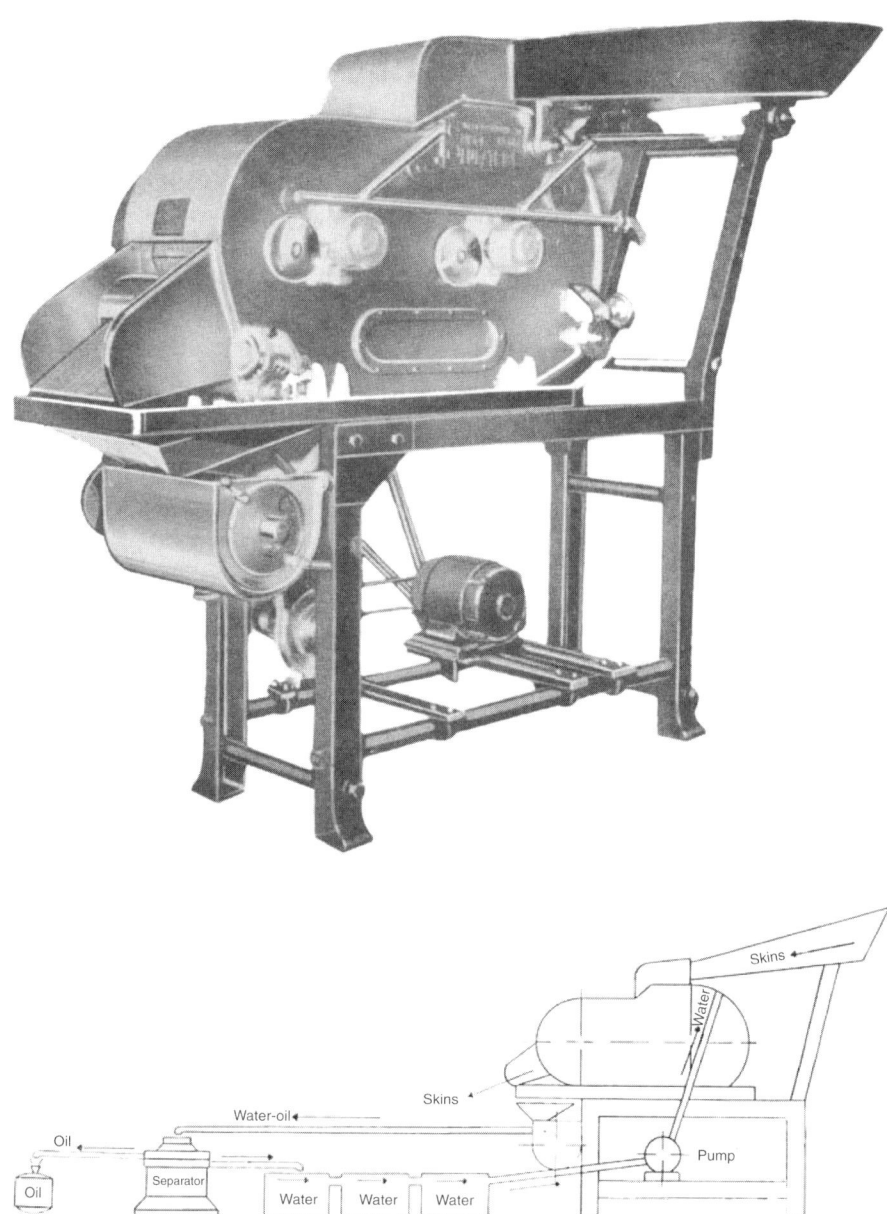

Figure 8.9 Indelicato *sfumatrice* and relative diagram.

different thickness, from paper-thin flavedo slices to thicker ones which include some of the albedo layer. The albedo slices are discharged from the machine, while the flavedo slices are pressed between knurled rolls to extract the oil under a spray of water. The oil-water emulsion is separated from the flavedo pieces with a finisher.

Figure 8.10 Indelicato *sfumatrice*: detail showing the processed peel discharge outlet.

Press method of extraction

The continuous screw presses (Figure 8.11) used in Italy nowadays for the extraction of essential oil generally consist of two screws which turn in opposite directions, forcing the peel, which is fed in randomly, to advance through the whole length of the press under great pressure. The screws operate inside a filtering sieve or cage with calibrated slits.

The presses in common use can hold quantities of peel corresponding to roughly 1.8 metric tons of fruit per hour. The yield of oil is directly proportional to the surface area of peel coming into contact with the screen. Consequently, five 1.8-ton presses will give a greater yield of oil than one 9-ton press. The pressing is usually carried out in two stages: in the first, the press applies a mild degree of pressure, with a mode of action which resembles that of the *sfumatrici* (slow-folding) machines; in the second stage a more powerful pressure is applied. The liquid obtained during the first stage is filtered through a screen and then centrifuged to recover the oil. The denser slurry from the second stage, diluted if necessary with the aqueous phase separated during centrifugation of the first-stage liquid, is distilled to recover a further quantity of oil, of lower-quality. For the two stages described above, the press is set up in a different way. In the first stage the screws have rounded edges and a conical shape; in the second stage they have plain edges which extend right to the walls of the surrounding cage. Jets of water can be applied inside the press via the hollow axis of the screw.

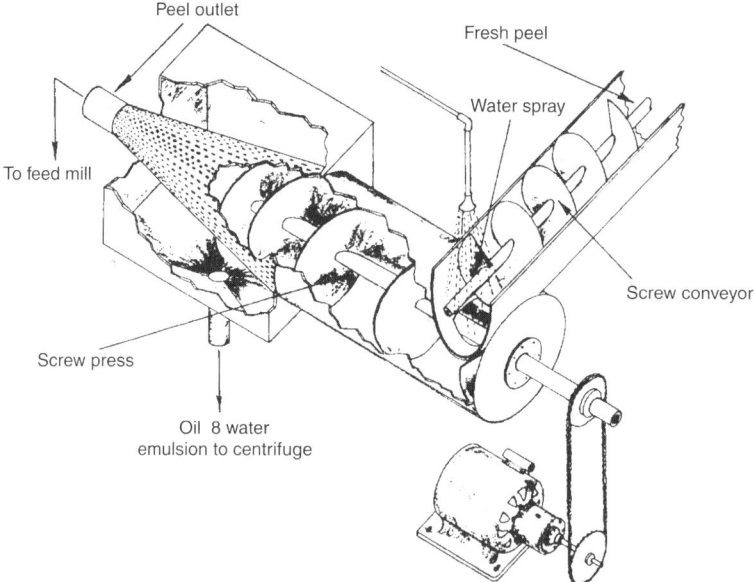

Figure 8.11 Screw press.

The pressing method just described is in widespread use in Sicily (Italy), particularly in the province of Palermo. Screw presses are also used in the USA, especially in Florida. These have a working capacity of about 3 metric tons per hour. They consist of a conical screw enclosed by a perforated jacket with 1/4 inch holes. During pressing, the liquid obtained flows out through the holes, while the pressed peel is expelled from the front end of the jacket. Because of the conical shape of the machine, the pressure on the peel progressively increases and reaches its maximum in the vicinity of the discharge outlet.

The operation can be carried out with the screws in either a vertical or a horizontal position. Water may or may not be used in the pressing operation.

Screw presses can also be used on whole fruit (especially mandarins), and in this case the oil and the juice are separated and recovered by centrifugation.

FMC peel oil recovery system (FMC Food Machinery, San José, CA, USA)

The recovery of essential oil by the FMC system (Figure 8.12) occurs during juice extraction. This simultaneous extraction of juice and oil is considered by many essential oil producers to be economically advantageous.

The FMC citrus juice extractor employs the *whole fruit extraction* principle. This principle relies on the unique design of the juicing components. The components interact during the extraction cycle in such a way that various parts of the fruit are separated simultaneously.

During the extraction process, the peel deflects, causing the oil glands to burst and release their oil. At the same time, water is introduced through a specially designed

Essential oil production   131

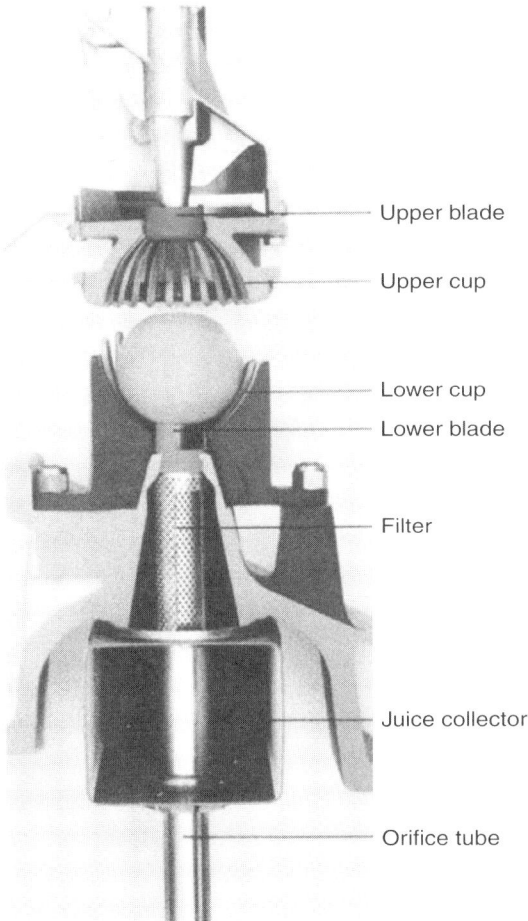

Figure 8.12 FMC in-line extractor: detail showing water application to extractor cup during extraction.

spray ring, located at the upper cup adapter, to capture the oil and small pieces of peel in an emulsion called oil slurry. It is important that the oil released from the oil glands be washed away to prevent the oil from being reabsorbed by the peel.

Each of the extractors in the line must be set up with juicing components that match the size of the fruit being processed.

To optimise recovery of the oil during extraction, water is piped to the extractor via a line, usually located in the tilted feed belt. The water is delivered during the extraction cycle through a stainless steel tube assembly, rubber hose and spray rings. Each of the five cup positions in the extractor must be equipped with a water spray ring. The spray ring is made of stainless steel and contains six holes of 1/16 inch evenly distributed around the ring. The spray ring is located in the upper cup adapter, just above the upper cup. Careful installation, maintenance and cleaning of the spray rings are crucial for efficient oil recovery. The amount of water used is very important to the oil recovery process. This quantity depends not only on the amount of oil contained in the fruit, but also on the amount of peel surface that is fed to the extractor per unit of time. Therefore, the amount of water depends on the type of fruit that is being processed, its size and the extractor that is being used. In order to maximise oil recovery, the concentration of oil in the emulsion should be between 0.9 and 1.1 per cent. To achieve this, it is necessary to use from 28 to 45 L/min/extractor, depending on the variables mentioned above. Adequate water pressure should be maintained and if necessary a pump should be installed in the water supply line. Another important consideration is the quality and temperature of the water. In order to ensure the proper functioning of the spray ring, the water used must be free of impurities, such as sand or insoluble solids, which might plug the holes in the spray ring and reduce the efficiency of the system. To maximise oil recovery, the temperature of the water used during extraction should be approximately 15 °C.

FMC has developed a water recycling system to minimise water consumption and waste discharge. In this recycling system, the aqueous discharge effluent from the desludger is sent to a decant/surge water recovery device. It is then passed through an automatic filtering system before being recycled back to the extractors. This process permits a citrus processing plant to reduce its water consumption and waste discharge considerably, without affecting oil yield and quality.

METHODS OF SEPARATION

The recovery of essential oil from the emulsion obtained during the extraction phase is a crucial step in the processing cycle.

When the extraction is performed manually, either by the sponge process or the écuelle method, water is not used and the tissues of the peel are only subjected to superficial grazing. In this case the liquid extracted separates rapidly after standing, into three layers: the upper layer is practically pure essential oil which merely needs to be filtered through paper; the intermediate layer contains emulsified oil and includes colloidal particles in suspension; the lower aqueous layer contains no oil and cannot be used. In order to recover oil from the intermediate layer a suitably absorbent material (sponge or wool) is impregnated with the emulsion and squeezed in a press. In this way the emulsion is easily broken, allowing the recovery of the essential oil.

When mechanised extraction processes are adopted, varying amounts of water are used to wash away the essential oil from the surface of the fruit or the peel to prevent it being reabsorbed and no spontaneous separation of the essential oil occurs. In this case the liquid obtained tends to separate very slowly into two layers. The upper layer, containing essential oil, is highly viscous, rich in mucilage and detritus and tends to form foam on its surface. The lower layer is coloured but contains no oil and, being of no value, can be eliminated. The emulsion is collected in a series of Florentine vessels. These are kept filled to the same level and linked up by connecting the lower part of each to the upper part of the next one, thereby allowing the accumulation of the emulsion in the upper part of the vessels. The aqueous phase from the last Florentine oil-collecting vessel is sent back into the extractor.

The absorption method of oil recovery, using sponge or wool followed by pressing and separation via hydraulic press, was used for a long time. This system has been

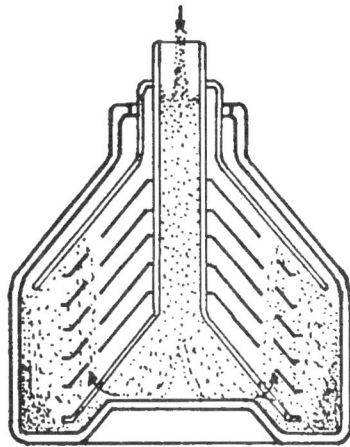

Figure 8.13 Drum of disk centrifuge.

practically abandoned because as well as being lengthy and costly, the quality of the oil may be severely compromised by its prolonged contact with the aqueous phase.

Today, the essential oil is separated using centrifuges (Figure 8.13) for the following reasons: higher yields, speed of separation and thus reduced contact of the essential oil with the aqueous phase, and the possibility of utilising continuous operating cycles.

Centrifugation is preceded by the elimination or drastic reduction of the coarsest waste particles. The crude oil emulsion is passed through a finisher (screw or paddle type), having sieves with 0.5–1 mm openings, or over a shaker screen (20–80 mesh). Excessive finisher pressure should be avoided, otherwise the emulsion becomes enriched with pectic substances and suspended solids, causing increased viscosity which makes breaking the emulsion in the centrifuge more difficult. To this end it is convenient to apply the technique known as the Bennet process (1931). This consists of adding 2 per cent sodium bicarbonate with sodium sulphate to the water in circulation. The use of this additive offers undoubted advantages which facilitate separation in the centrifuge: firstly, to increase the difference between the specific gravity and the surface tension of the oil and the aqueous phase; secondly, sodium bicarbonate gives rise to the formation of sodium pectate, which is a less powerful emulsifier than pectin, and also neutralises the acidity present which has a negative effect on the quality of the essential oil; lastly, sodium sulphate tends to neutralise the electrical charge of the colloidal micelles.

A number of commercial enzyme preparations, added to the circulating water, can also have positive effects in the process of separation of the essential oil (reduction of viscosity and breaking the colloidal oil emulsion). The benefits obtained are: improved centrifugal performance; higher oil yields; lower consumption of fresh water and reduction in the quantity of waste water produced. The increase in oil is obtained indirectly, in the sense that the losses of oil during processing are reduced by using the selected enzyme preparation. The amount of enzyme used will depend on such factors as: variety, type and maturity of the fruit; amount of water recycled in the system; the length of time the emulsion remains in the system; pH and temperature of the emulsion.

A screen or filter at the point of discharge of the oil emulsion into the emulsion-holding tank is a good safeguard against contaminating an entire tank in the event that a finisher screen is ruptured. The recovered oil emulsion will normally contain a small but significant amount of sand. This can cause undue wear on the centrifugal equipment unless it is removed. However, its removal can be accomplished quite readily and satisfactorily by using a liquid cyclone in the line to the desludger feed tank from the small surge tank next to the oil finisher.

The finished emulsion contains from 1 to 2.5 per cent oil, with suspended solids not exceeding 4 per cent.

The centrifugal process includes a number of different stages which can vary according to the size of the processing plant and the prevailing practice in the different citrus production areas.

In the USA, especially in Florida (Kesterson and Braddock, 1976), the finished emulsion is fed to a desludger centrifuge to produce a concentrated oil emulsion or 'cream' (70–80 per cent or higher oil content). The aqueous discharge effluent should not contain more than 0.1–0.25 per cent oil under optimum operating conditions. The oil-rich emulsion is fed to a self-thinking type polisher centrifuge (16,000–18,000 rpm) without adding water. The feed rate to the polisher should not exceed 1–1.5 gallons per minute, depending on the capacity of the polisher and the oil concentration in the feed. The aqueous discharge from a self-thinking type polisher should contain no more than 5–7 per cent oil. If the polisher does not have the self-thinking feature, water should be added to the polisher feed at a rate of between 3 and 20 parts to 1 part of oil-rich emulsion from the desludger. This water is not needed to break the emulsion, but is used primarily to keep the polisher clean.

In Mediterranean countries the recovery of essential oil from emulsion is carried out by two-stage centrifugation: a primary separation, which removes 50–70 per cent of the aqueous phase from the emulsion, and a second centrifugation to achieve complete separation of the oil phase from the residual water.

It has been demonstrated (Di Giacomo *et al.*, 1999) that the temperature increases during the second centrifugation, and that some alteration of the components most liable to deterioration occurs, because of the close contact of the essential oil with air and the acidity of the emulsion. This problem can be avoided by using the following devices:

- a pressure tank for storage of the emulsion under nitrogen;
- a cooling apparatus to refrigerate the emulsion before centrifugation;
- centrifugal separation equipment operating in hermetic conditions.

All of these contribute to prevent or significantly reduce the molecular transformation phenomena occurring in the standard process for separating the rich emulsion which has been used in industry to date.

PRESERVATION AND STORAGE

In order to avoid deterioration of citrus essential oils during storage, it is necessary to protect them from the deleterious effects of atmospheric oxygen, light, heat, residual traces of moisture and acidity, and to avoid contact with metallic surfaces which may act as oxidation catalysts, etc.

Although cold-pressed oils contain natural antioxidants, they also contain many unsaturated components, such as terpenes, which polymerise easily and form resinous materials, thus provoking alteration of the aroma and fragrance.

The oxygen in the atmosphere has a direct oxidising effect on the stored oils, light affects the colour and the fragrance, while heat accelerates these processes (Braverman, 1949). The development of undesirable flavours, as well as the deterioration of citrus essential oils, is related to limonene auto-oxidation (Kesterson *et al.*, 1971). The formation of some compounds (e.g.: *p*-cymene in lemon oil) can be considered characteristic indicator of deterioration and the level of peroxides is a further index of essential oil ageing. Normal essential oils should not have a peroxide content higher than 0.32 per cent.

Given that most of the factors which give rise to alteration of the composition of essential oils are known, it is possible to select the most appropriate conditions for storage.

First of all, citrus essential oils must be stored in well-filled, stainless steel, aluminium, tin-dipped or polymer-coated steel drums or containers. It should be considered, however, that some essential oils, as lemon and lime, may be stored and shipped in tin-dipped drums, which adds expense to the value of these oils (Braddock, 1999). The polymer-coated drums may be satisfactory for a short time but not for long-term storage. In fact, many polymers are incompatible with terpenes.

The essential oil should not contain residual moisture: any traces of water left after the centrifugal process should be eliminated by treatment with anhydrous sodium sulphate. If present, water will induce slow hydrolysis and formation of α-terpineol. Acidity from the juice is detrimental to the stability of the oil and it is therefore good practice to neutralise it by adding sodium bicarbonate.

In Sicily for many years essential oils were stored in containers made of copper coated with tin (*ramieroni*), with a capacity of about 1000 litres. These have now been replaced by stainless steel tanks.

During storage, a waxy residue is deposited on the bottom of the essential oil tanks. This process is also favoured by the cool storage temperature. This waxy material can be removed either by an additional centrifugation or by using tall storage tanks with drainage taps at different levels. The essential oil is drained off from the highest tap first and then progressively from the other taps. In the end, the waxy layer and any residual water present are evacuated from the lowest tap which is fitted to the conical base of the tank.

While the storage tanks are full, the effect of the atmospheric oxygen present is negligible, but later on, once batches of essential oil have been removed, flushing with nitrogen or CO_2 is necessary to minimise oxidation. Guenther (1949) suggests using a very simple method by which a small amount of 'dry ice' (solid CO_2) is placed in the oil. The solid carbon dioxide rapidly evaporates, drives out the air, and thereby ensures the presence of a neutral gas in the container before it is tightly sealed.

Essential oils should be stored in fairly cool conditions, at temperatures no higher than 15–20 °C.

Prior to storage, many essential oil plants use the winterisation technique, which consists of a partial depletion of the non-volatile fraction of the oil. The components of this fraction are the ones that make the oil turn cloudy or produce precipitates on cooling or during extended storage. The winterisation procedure to reduce the wax content, which is considered a defect in cold-pressed oils, is carried out in large conical-bottomed tanks equipped with a double jacket in which a cryogenic fluid circulates. The oil is kept at $-20\,°C$ in these containers for at least seven days and then stored after any sediment formed has been centrifuged off. Winterisation, carried out as just described, does not produce any appreciable variation in the physicochemical properties of the essential oil. The analytical parameters remain within the limits established by ISO and comply with the main Pharmacopoeia standards (Di Giacomo et al., 1970). De-waxing of citrus essential oil is not always carried out efficiently because of the presence of colloids which may act as protecting agents, preventing wax separation. However, these agents can be inactivated by rinsing the oil with an aqueous solution containing an enzyme preparation (esterase).

The addition of anti-oxidising agents to essential oils generally makes them more stable. However, their use has to comply with any restrictions and limitations in force in the country where the oil is going to be commercially employed. Furthermore, when selecting an antioxidant, the criteria listed by Riemenschneider and Ault (1944) should be taken into account, so that, in addition to its inhibitory properties, the agent employed must:

- have no harmful physiological effects, even in quantities considerably larger than those likely to be used and even when ingested over a long period of time;
- be sufficiently soluble in the product in question to facilitate its use; the greater the solubility the more advantageous its use;
- impart no objectionable odour, colour, or flavour, even after storage;
- be stable under whatever processing conditions it must be subjected to;
- be relatively inexpensive and available in large quantities.

Kesterson and McDuff (1949) evaluated the performance of 10 anti-oxidising agents added to cold-pressed Valencia oil and found that α-tocopherol was the most effective at a concentration of 0.1 per cent. According to Kenyon and Proctor (1951), the best results for orange oil are obtained with α-tocopherol or nordihydroguaiaretic acid (NDGA), at a concentration of 0.05 per cent. Propyl gallate, ethyl gallate, butylated hydroxytoluene (BHT) and butylated hydroxyanisole (BHA) also give satisfactory results, either separately or in combination, at an overall concentration in the range of 100–400 ppm. Storage for 12 months under an inert gas such as nitrogen or in full containers with added butylated hydroxytoluene (50–100 ppm) has been found to preserve the original quality of the limonene in essential oils (Newhall and Kesterson, 1961). The efficacy of each antioxidant can be measured in a number of ways: by an organoleptic test; by the Warburg method of following the rate of oxygen uptake in a respirometer (Kenyon and Proctor, 1951); or by determining the amount of peroxides formed (Flores and Morse, 1952).

FACTORS WHICH INFLUENCE THE QUALITY OF ESSENTIAL OILS

The quality of citrus essential oils obviously depends to a large extent on factors deriving from the nature of the fruit itself (provenance, type of soil, climate, citrus variety), but the processing of the fruit also has a significant effect.

Variety, maturity and storage of fruit

The physical properties of essential oils can be dependent at a certain stage, on the fruit varieties used, on their maturity and on the storage time that proceeds the industrial transformation.

Some information, regarding the variation that can occur in essential oils for these factors will be found in the chapters of this volume where the composition of the volatile fraction and the oxygen heterocyclic fraction of cold-pressed oils are described.

Treatment of the peel

In those factories where the traditional *sfumatura* process is still used, after the juice extraction the peels are soaked for several hours in a diluted solution of milk of lime and left to stand for a while in suitable containers (draining trolleys, baskets, etc.) in order to drain off the excess liquid.

This treatment can be counterproductive if it is allowed to go on for too long. The appropriate length varies from a few hours (in the hot season) up to 24 hours (in winter). When the weather is extremely humid, the peel rapidly loses turgor and oil yields are lower. Furthermore, some of the volatile components of the oil may undergo transformations: for instance, sabinene easily hydrates to form terpinene-4-ol (Calvarano et al., 1971).

Extraction technique

The extraction method used has an effect upon the physical properties of citrus oils.

The qualitative characteristics of an essential oil are almost always closely related to the yield obtained. An exhaustive extraction procedure produces a larger quantity of high-boiling components, with high molecular weights. As a result, the oil has high specific gravity, non-volatile residue and refractive index values, while the optical rotation value is lower because of the lower relative percentage of d-limonene present.

In ideal conditions, the extraction of essential oil would not involve the use of water. However, almost all the industrial procedures employed today require the use of water to wash away the oil from the surface of the whole fruit or peel. The relative ease of separation of the type of emulsion which is obtained during processing depends on the extraction method employed.

Water acts as a solvent of oxygenated compounds which play a fundamental role in the formation of the bouquet. This behaviour is due to the presence of hydrophilic moieties in aldehydes, alcohols and esters. Terpene hydrocarbons on the contrary are strongly hydrophobic. It is likely that the effect of the aqueous phase is not due to a simple partition effect, but somehow involves the components of the peel and the pulp which have specific affinities towards carbonyl compounds. It is not clear whether only an absorption process is involved, or whether the formation of an intermediate complex is required, or if enzymes also play a role (Stanley and Vannier, 1959). Possible chemical interactions should not be underestimated, especially for the esters (such as linalyl acetate in bergamot oil) which can be hydrolysed by the action of enzymes or acids present in the aqueous layer.

Since the fragrance of an essential oil is directly related to the content of aldehydes and esters, the amount of water which circulates during processing is very important.

Kesterson and McDuff (1948) carried out experiments on oranges in an industrial plant in Florida. While keeping all other factors constant, they varied the amount of water used from extremely large quantities down to a 100:1 ratio of water to oil extracted. The respective aldehyde content increased from 1.08 to 1.64 per cent. In order to produce orange oil with a high aldehyde content, the amount of water used should be reduced to the minimum necessary to wash away the essential oil.

Research has been carried out on the relative merits of open vs closed circuit operating cycles: that is, whether it is useful to recycle the water separated by the centrifuge and use it to respray lemon peel during the *sfumatura* (slow-folding) process (Calvarano *et al.*, 1971). Whilst the re-use of the water, saturated in components containing hydrophilic moieties, results in an oil which is richer in the constituents fundamental to the bouquet, two problems arise:

1   increase in colloidal substances which favour emulsification of the oil and therefore hinder separation in the centrifuge;
2   the presence of large quantities of juice which can cause fermentation and consequently alter the essential oil composition.

In the industrial cold-pressing of lemon peel by *sfumatura*, the first difficulty can be overcome by using an automated two-step centrifugal procedure. A first centrifugation separates the aqueous phase to be recycled from an oil-rich emulsion (80 per cent essential oil) which is then recentrifuged to eliminate the colloidal and waxy components from the oil. As far as the second problem is concerned, it has been shown that no qualitative deterioration of the oil takes place when the water is recycled for a period of up to eight hours.

Biological formation of α-terpineol

Citrus oil emulsions provide an ideal medium for bacterial growth, leading to the biological formation of α-terpineol from *d*-limonene (Murdock and Hunter, 1958). It has been demonstrated that bacteria grow very rapidly in dilute oil solutions, especially when adequate standards of hygiene have not been observed. The increase in α-terpineol concentration produces a corresponding decrease in *d*-limonene, which is the most abundant component present in citrus oils. The highest concentration of α-terpineol was found in lemon oil emulsions containing less than 1 per cent oil.

Bacterial growth in other citrus oil emulsions (orange, grapefruit, tangerine) also causes undesirable changes in the composition of the oils. It is therefore essential to employ high standards of hygiene in the production line from the first stage of the citrus oil recovery process. Obviously, holding of slurries for any period of time prior to dislodging should be avoided.

RECOVERY BY DISTILLATION

Dalton's law applies when a heterogeneous system made up of two unmixable liquids, such as water and essential oil, is heated. The vapours coming from the two liquids do not reciprocally influence each other, and, therefore, the total vapour pressure is equal to the sum of the two. The mixture boils when this sum is equal to atmospheric pressure. As a result, essential oil, despite having a high boiling point, boils at a considerably lower temperature under reduced pressure.

Furthermore, if distillation is carried out under reduced pressure, the consequent lowering of the boiling point avoids, or significantly reduces, reactions harmful to the integrity of the original composition of the essential oil and allows recovery of components which are not distilled under ordinary pressure.

The distillation apparatus is built to different specifications according to the needs of the producers.

A very simple still is described in the chapter of this volume on production of distilled peel oils.

Distilled essential oil should be almost colourless. A pronounced yellow colour indicates that the distillation has been carried out either too quickly or too slowly. If distillation is interrupted too soon, a low quality essential oil may be produced, too rich in hydrocarbons and deficient in oxygenated compounds.

Distilled essential oil is notably different from cold-pressed oil, both in appearance and odour, as well as in its chemical composition. The first is colourless or faintly yellow and has a pronounced terpenic odour, the second is green-brown and smells of fruit.

It is undeniable that the conditions of distillation, at a pH lower than 3, determine important molecular transformations in many particularly sensitive components. Slater and Watkins (1964) have shown that in distilled oils d-limonene yields α-terpineol, terpinen-4-ol, 1,4-cineol, 1,8-cineol, α- and β-terpinene.

In the USA distilled lime oils are obtained from peel which is first processed to obtain cold-pressed oil and subsequently steam-distilled to produce the distilled oil. The combination of heat and acid has a marked effect on the aldehyde content of the oil.

In Florida and California (Kesterson *et al.*, 1971), distilled orange oils are recovered from the de-oilers installed in the cannery. The juice is sprayed into the de-oiler, which is essentially a small concentrator that works under reduced pressure (11–26 inches) and at a temperature that oscillates between 50 and 90 °C. The mixed vapours of water and essential oil are condensed and the essential oil recovered by centrifugation or distillation, whereas the water is returned to the juice (during the operation between 3 and 6 per cent of the juice is evaporated). This treatment removes about three quarters of the essential oil present and thus allows the juice to comply with the USA standard, which, for grade A, does not permit orange juice to contain a percentage of recoverable essential oil higher than 0.03 per cent. These distilled oils have an aldehyde content about 24 per cent higher and an ester content about 10 per cent lower than the corresponding values for the best quality cold-pressed oils. Similarly, distilled grapefruit oils from juice will generally run 90 per cent higher in aldehydes and 60 per cent lower in esters than the expressed oils.

Other sources of distilled essential oil production are the aqueous discharges from the centrifugation of the cold-pressed essential oils, which contain on average 0.1 per cent of essential oil.

An essential oil of decidedly inferior quality is stripper oil, a by-product of citrus molasses. This is obtained from the liquid extracted from minced and pressed limed peel before it is dried (to make animal food). The peel oil present in this liquid is recovered by heating at 110–115 °C and subsequent flashing at atmospheric pressure. Usually stripper oil is a mixture of various citrus essential oils, since the pressed liquid is indiscriminately obtained from orange and grapefruit peel.

Distilled essential oil is also recovered during production of the volatile aromas from the juice. In this case, the essential oil is separated by centrifugation from the aromatic condensate.

ALCOHOL SOLUBLE CITRUS FLAVOUR

Alcohol soluble citrus flavour, or alcoholate, has been used for a long time in the perfume industry and above all in the food industry to give flavour to syrups and drinks. Its main properties can be summarised as follows:

- highly soluble in aqueous phase;
- marked resistance to alteration processes if compared with the essential oils from which it derives;
- high concentration of the oxygenated fraction compared with the hydrocarbon fraction (terpenes and sesquiterpenes).

The cold preparation of alcoholates (also called soluble essential oils) relies on the ability of ethyl alcohol of suitable strength to dissolve the oxygenated compounds and thus separate them from the hydrocarbon fraction, which is only slightly soluble. The oxygenated compounds are responsible for the characteristic flavour and fragrance of a citrus essential oil.

The alcoholates are generally put on the market with an indication of their flavouring strength in relation to the final syrup (1:100, 1:200, 1:400, 1:800).

A general outline for the production of citrus alcoholates is as follows:

Essential oil	10 kg
Ethyl alcohol (60°/70°/80°)	10 kg

Mix well for at least one hour. Decant and separate the two phases.

The oily fraction, largely containing hydrocarbons, is re-treated with a second 10 kg aliquot of alcohol of the same strength. The mixture is stirred and decanted, and the alcohol layer is combined with the one previously separated to give 20 kg of alcoholate.

The oily fraction can be further treated with alcohol for one or more washes. The aromatic alcohol obtained requires additional processing.

The separation of the hydrocarbon and oxygenated fractions becomes more selective as alcohol strength is lowered. Overall, the distribution of the oxygenated components between the two phases depends on the following main factors: alcohol strength, number and length of extractions, and temperature. The processing techniques may vary considerably and are not usually disclosed by the producers.

The apparatus used is very simple (Fenaroli, 1963): it essentially consists of a vessel with a conical bottom contained within another which functions as a thermostatic bath. The capacity of the vessel varies according to the production required. A mechanical stirrer fitted with two sets of differently angled blades is inserted into the vessel. The vessel has a lid and a drain in the bottom fitted with two taps, between which is a small glass window. Essential oil and then alcohol are pumped into the apparatus, after which the stirring process is started. Using cold water circulating in the jacket the temperature is brought to around 13 °C and maintained at that level for the whole operation. The stirrer is left to work for the required length of time, and then the mixture is left to stand. The hydrocarbon phase settles at the top, while the alcoholate is drained off and filtered.

Alternatively, in a heat-based procedure the alcoholates can be obtained directly from the flavedo previously separated from the fruit or, more commonly, from the whole peel (Di Giacomo, 1974). To obtain the flavedo the two poles of the fruit are fixed in a machine designed for this purpose, while a special instrument pares off the part containing the essential oil glands in the form of a long, thin, continuous strip. The whole peel usually comes from a standard *birillatrice*. Both the flavedo and the peel, with a sufficient quantity of water added to make them fluid, are finely chopped in a grinding mill with 3 mm diameter holes. More water is added to this chopped material together with 95° alcohol, so that the final alcohol concentration of the slurry is 4°. After a certain period of maceration, the mixture is continuously pumped into a plate distillation column and after a preheating phase it reaches the high part of the plates in the column. The rectified vapours are condensed (alcohol strength: 50°–60°, according to the requirements of the end-user) and then separated from the terpenes and sesquiterpenes using Florentine oil-collecting vessels. During this heat-based procedure to obtain alcoholates, hydrolysis, oxidation, isomerisation, hydration, and dehydration reactions may take place which induce changes, particularly in the ratios between the various oxygenated components (Dugo *et al.*, 1988). From an olfactory point of view, these alcoholates have a distinctive 'floral' fragrance.

All alcoholates, like the corresponding essential oils, have individual characteristics which indicate their origin. Sweet orange alcoholate is characterised by the presence of δ-3-carene and valencene, which are absent, or present only in traces, in alcoholates derived from other varieties of citrus fruit. Mandarin alcoholate is characterised by the presence of α-sinensal, methyl N-methyl anthranilate and thymol. Bergamot alcoholate is characterised by a high linalool and linalyl acetate content.

Lemon alcoholates have characteristic α-thujene/sabinene, α-thujene/β-pinene, α-pinene/β-pinene and β-pinene/α-terpinene ratios. The citron alcoholates present carachteristic composition that allow to distinguish these from other citrus alcoholates (Dugo *et al.*, 1988)

In order to make a sensory test of the 'strength' and the quality of the alcoholate, a drink is prepared as follows: citric acid monohydrate 2 g; sucrose 80 g; water up to 1000 mL; flavour with a quantity of alcoholate between 200 and 800 µL per litre of drink and inject CO_2 at 3 atm at room temperature.

SPECIAL TECHNOLOGIES

Effects of enzyme treatment

In the emulsion formed after essential oil is extracted, significant quantities of colloidal substances are present, especially pectin. This increases viscosity and adversely affects the centrifugal yield and, as a consequence, recovery of the essential oil.

In order to improve the conditions under which oil is separated, enzymes can be usefully employed in all the extraction processes used in industry. The use of enzymes increases essential oil yield, reduces water consumption and consequently decreases the quantity of waste effluent.

The requisite characteristics of the enzyme preparation (Janser, 1995) are as follows:

- hemicellulytic-pectolytic activities;
- high level of polygalacturonase (PG) activity;
- low level of pectinesterase (PE) activity;
- high PG:PE ratio;
- active in a very wide pH range (2.5–6.0), excellent at pH 3.0–5.5;
- excellent levels of activity at room temperature (15–40 °C);
- highly suitable for oil-water separation.

The quantity of enzyme preparation necessary depends on a number of parameters: citrus type (orange, lemon, etc.), fruit variety, maturity of fruit, extraction method, quantity of water used in the cycle, mean life of the emulsion between extraction and separation, time necessary to recycle the aqueous phase.

Citrozym CEO, a food grade enzyme preparation produced by *Aspergillus niger*, is an enzyme preparation tailor-made for this application.

When optimised, this enzyme technology allows the aqueous phase to be recycled back to the extractors after the first centrifugation. Thus the consumption of fresh water and production of waste water are drastically reduced, thereby minimising oil loss. The increase in the yield of essential oils is in the range of 5–15 per cent. The overall process is improved and becomes more economical.

Extraction processes using supercritical solvents

Processes based on the use of supercritical solvents enable temperature labile compounds to be extracted in relatively mild conditions. They potentially offer some undoubted advantages over traditional processing methods in the case of some particular classes of compounds.

In particular carbon dioxide, which above the critical point (31 °C and 74 bar) assumes precise supercritical characteristics, and is preferable to other fluids for a number of reasons: it is odourless, tasteless, and colourless; it is chemically inert; it presents no danger of explosion; it is an apolar solvent; it is easy to remove.

In the essential oil sector, it has been found that the use of supercritical CO_2 at low temperatures in a system equipped with columns operating in countercurrent allows problems relating to quality to be addressed that traditional processes, such as distillation, are unable to resolve. Traditional methods are inadequate at maintaining the stability of some elements of the bouquet, avoiding loss of volatile components and at achieving a sufficient degree of selectivity towards various components.

It should be remembered that the main oxygenated components of essential oils (esters, alcohols, aldehydes and ketones) of low and medium molecular weight are soluble in liquid CO_2 and in supercritical CO_2. Apolar compounds of low molecular weight (hydrocarbons) are similarly soluble. The solubility of these compounds, within the same family, generally decreases as molecular weight increases. Since chlorophyll, caretonoids, sugars, aminoacids and the majority of inorganic salts are insoluble, it is evident that this property can be exploited in the extraction process to obtain considerable selectivity which discriminates against compounds that contribute nothing to the aroma and whose presence in the aromatic extract may be counterproductive.

Interesting applications forecast in the citrus area are: in the extraction of essential oils from flowers and from leaves; the enrichment of certain classes of compounds (deterpenated and desesquiterpenated oils); and the elimination of undesirable components (for example furanocoumarin from bergamot essential oil).

The prospect of employing supercritical solvents in the extraction of essential oil directly from the peel, however, is of little practical interest given that cold-press systems currently used in industry already enable high quality cold-pressed essential oil to be prepared at a reasonable cost.

Ultrafiltration and reverse osmosis

The citrus industry has an interest in limonene recovery techniques and in the application of membranes.

Commercial ultrafiltration (UF) and reverse osmosis (RO) membranes are used to concentrate limonene present in coldpressed oil centrifuge effluent and molasses evaporator condensate (Braddock, 1982). UF membrane rejection rates are 78–97 per cent for mixtures with initial limonene concentrations from 0.04–0.6 per cent v/v. RO membrane rejection rates of limonene range from 87–99 per cent for feed streams containing 0.06–0.23 per cent limonene. Initial membrane flux rates for centrifuge effluents are in the range of 10–100 kg/m^2/h. Evaporator condensate fluxes are higher, 25–400, while pure water rates range from 25 (RO) to 1,000 kg/m^2/h (UF).

Contact with limonene adversely affects membrane flux rates in decreasing order of severity as follows: polysulfone > cellulose acetate > teflon-type.

Coloured orange essential oil

In countries in which artificial colourings in drinks are prohibited there is a special demand for a type of sweet orange essential oil with an elevated content of carotenoids (ca 1,000 ppm in terms of β-carotene), while, usually, significantly reduced levels are found in the essential oil obtained industrially using the most common extraction processes, even from late season fruit.

For some years technological strategies have been studied to enable essential oil to be enriched with carotenoids to reach the levels expressly demanded by the market (Rispoli and Di Giacomo, 1968).

One procedure adopted is as follows: fresh flavedo from ripe oranges is cut into fine strips using a Citromat-type machine, then finely ground without adding water in a grinding mill with 1.5 mm sieve openings. The product obtained in the form of a paste is rendered fluid with the necessary quantity of an essential oil obtained using one of the traditional methods and stirred for several hours. The quantity of essential oil may vary depending on the carotenoid content of the flavedo. Some operators use orange pulp in place of flavedo. The mass obtained is pressed and the liquid obtained is centrifuged to separate impurities which might be present.

Using this method there is a significant increase in the waxy fraction–the vehicle through which the carotenoids are transferred to the essential oil. The finished essential oil when compared to common essential oil, has significantly higher refractive index, specific weight and evaporation residue values; at the same time the optical rotation value is lower.

Furanocoumarins free bergamot essential oil

The possibility of producing bergamot essential oil bergaptene free is a major issue for the citrus industry in Calabria which must face the danger that this essential oil may be used less and less because of problems associated with the phototoxic properties of this component.

Three conditions must be fulfilled so that furanocoumarins free essential oils can be used in the same way as ordinary oils:

- the properties of the fragrance must be maintained;
- the storage life must not be affected;
- the physicochemical parameters, especially those detectable by instrumental analysis, should not be altered, with the exception of those which obviously derive from the reduction of the furocoumarin content.

The ideal process should aim at the specific elimination of bergaptene, without modifying the content of all coumarin and psoralene compounds. The overall composition of the non-volatile fraction certainly should not be compromised.

The process currently used in Italy is based on the treatment of the essential oil with an aqueous alkaline solution. The bergaptene level can be substantially reduced without compromising the quality of the essential oil by appropriate control of the following variables: the concentration of the alkaline solution, the type of stirring apparatus, the contact time between the two phases, and the temperature.

DRIED FLAVOURINGS

The production of *dried flavourings* is mainly based on the technique known since many years: the microincapsulation by *spray dry*. Shortly that method can be described as follows (Di Giacomo and Mincione, 1994):

- formation of an emulsion or dispersion, enough stable and with small particles, formed between the essential oil and the material the forms the external coating of the microcapsules;
- successive spraying for the formation of the microencapsulated material.

Among the materials commonly used are different starch types, gums, organic polymers, and introduced more recently, also some modified gels. The choice of these materials is limited by the instrumentation available to spray and by the specifications necessary for the good dispersion, the viscosity and a commercially profitable concentration. Moreover, if the product is used in the food industry, the encapsulating material must be accepted as food additive.

Usually the amount of dry material in the emulsion can vary between 25 and 50 per cent.

The dispersion can be obtained above or below room temperature depending on the essential oil character and on the material used for the dispersion: the product obtained is usually water soluble. The water resistant capsules can be obtained by using some modified starches.

The spray drying procedure is usually performed by atomisation of the dispersion, or of the emulsion, in a large drying chamber; initially is formed a fine dust made of essential oil particles contained in a film of the solvated coating material. The solvent is rapidly eliminated from the liquid film, leaving the solid dry coating of the capsule that contains the essential oil.

In the plant the emulsion is transferred to the spray chamber by a pump, and successively to the dryer chamber. The air used during the process, pre-filtered and pre-heated, is introduced by a flow regulator in the same chamber (temperature 200–280 °C), and mixed with the fluid drops.

The time of permanence in the chamber can be adjusted depending on the system of ventilation.

Most of the water evaporates almost instantaneously. The humid air is introduced in a twister (90–120 °C) where a further amount of powder is carried and recovered by a venting system.

The dry powder is collected on the bottom of the main chamber, cooled with cold air and added with the recycled powder, then sent to the last twister.

The final product is sieved and stocked in containers. The amount of water in the powder is between 3–5 per cent. The size of the particles can vary between 5–2000 µm, but generally ranges from 5–400 µm.

The encapsulation can also be obtained by cryo-drying.

Numerous are the proposed applications for the microencapsulated products in the food and flavour industries. The most important could be the essential oil micro-encapsulation for bakery products. The micro-encapsulation of essential oils with a coating the is resistant to the high temperatures of the oven during the initial stage of cooking, allows the reduction of the amount of aroma used, without loss of the flavour.

The advantages offered by using encapsulated products as flavours are summarised below:

- the strength and the quality of the aroma are totally protected against evaporation, oxidation and can be maintained at high temperatures and for periods of time higher than 12 months;
- the microencapsulated product can be completely introduced in powder mixtures by using simple mixing procedures, obtaining a fine and uniform dispersion;
- the powders, that are bacteria free, usually present excellent hygienic properties. They contain a reduced amount of water, and can be easily mixed to different dry food products.

REFERENCES

Bennet, H. (1931) Process for the recovery of essential oil from lemons and other fruits of the genus Citrus, *U.S. Patent* n. 1,814,888 (July 14).

Braddock, R.J. (1999) *Handbook of Citrus By-products and Processing Technology*, John Wiley & Sons, New York.

Braddock, R.J. (1982) Ultrafiltration and reverse osmosis recovery of limonene from citrus processing waste streams. *J. Food Sci.*, 47, 946–948.

Braverman, J.B.S. (1949) *Citrus Products*, Interscience Publishers Inc., New York.

Calvarano, M., Bovalo, F. and Di Giacomo, A. (1971) La 'sfumatura' meccanica delle scorze di limone. Nota II. Esami analitici. *Essenz. Deriv. Agrum.*, 41, 147–159.

Clevenger, J.F. (1928) Clevenger method. In J.B. Redd, Ch.M. Hendrix and D.L. Hendrix (eds) *Quality Control Manual*, Intercit Inc., Safety Harbor, Fl, 1986, pp. 40–41.

Cultrera, R. (1954) Breve e patetica storia del commercio delle essenze agrumarie. *Industria Conserve*, 3, 171–172.

De Villiers, F.J. (1930) Citrus by-products research: orange oil. *Farming in South Africa*, 4, 515–516.

Di Giacomo, A. (1974) *Gli Oli Essenziali degli Agrumi*. EPPOS, Milan, Italy.

Di Giacomo, A., Bovalo, F. and Postorino, E. (1970) Sulle caratteristiche analitiche dell'olio essenziale di limone decerato. *Essenz. Deriv. Agrum.*, 40, 143–150.

Di Giacomo, A. and Calvarano, M. (1978) Il contenuto di bergaptene nell'essenza di bergamotto estratta a freddo. *Essenz. Deriv. Agrum.*, 48, 51–83.

Di Giacomo, A. and Lo Presti, V. (1956) Il trattamento con calce delle scorze di limone pregiudica il rendimento in pectina. *Essenz. Deriv. Agrum.*, 26, 122–125.

Di Giacomo, A. and Mincione, B. (1994) *Gli Olii Essenziali Agrumari in Italia*, Laruffa Editore, Reggio Calabria, Italy.

Di Giacomo, A. and Signorino, G. (1971) La 'sfumatura' meccanica delle scorze di limone. Nota I. Aggiornamenti per una tecnologia più razionale. *Essenz. Deriv. Agrum.*, 41, 137–146.

Di Giacomo, G., Crupi, F., Gionfriddo, F., Rispoli, G., Signorino, G. and Di Giacomo, A. (1999) Improvement of the recovery of essential oils from emulsion. Improvement of the quality of the recovered essential oils. *Essenz. Deriv. Agrum.*, 69, 27–31.

Dugo, G., Cotroneo, A., De Filippo, V. and Daghetta, A. (1988) Sulla genuinità delle essenze agrumarie. Nota XXI: Gli alcoolati di cedro. *Industria Bevande*, 17, 17–31.

Fenaroli, G. (1963) *Sostanze Aromatiche Naturali*, Vol. I, Ulrico Hoepli, Milan, Italy.

Flores, H. and Morse, R.E. (1952) Deterioration of orange oil. *Food Technol.*, 6(1), 6–8.

Gionfriddo, F., Mazza, B., Postorino, E. and Di Giacomo, A. (1981) Gli acciai inossidabili nella conservazione delle essenze agrumarie. *Essenz. Deriv. Agrum.*, 51, 317–324.

Guenther, E. (1949) *The Essential Oils*, Vol. III, D. Van Nostrand Co., New York, N.Y.

Guenther, E. (1941–1942) Sketches from French Guinea and French Guinea sweet orange oil. *Amer. Perf. Essent. Oil Rev.*, Sept.–Dec. 1941 and Jan. 1942.

Janser, E. (1995) Improvement of citrus oil recovery and waste water treatment. *Fruit Processing*, 5, 328–331.

Kenyon, E.M. and Proctor, B.E. (1951) Effect of antioxidants on orange oil. *Food Res.*, 16, 365–371.

Kesterson, J.W. and Braddock, R.J. (1976) By-products and specialty products of Florida citrus. *FL. Agr. Exp. Sta. Tech. Bull.*, n. 784, University of Florida, Gainesville, FL.

Kesterson, J.W. and McDuff, O.R. (1948) Florida citrus oils. Commercial production methods and properties. *FL. Agr. Exp. Sta. Tech. Bull.*, n. 452, University of Florida, Gainesville, FL.

Kesterson, J.W., Hendrickson, R. and Braddock, R.J. (1971) Florida citrus oils. *FL. Agr. Exp. Sta. Tech. Bull.*, n. 749, University of Florida, Gainesville, FL.

Kesterson, J.W. and McDuff, O.R. (1949) Anti-oxidant studies. *Amer. Perf. Essent. Oil Rev.*, 54(10), 285–287.

Moyler, D.A. (1993) Extraction of essential oils with carbon dioxide. *Flav. Fragr. J.*, 8, 235–247.

Murdock, D.J. and Hunter, G.L.K. (1968) Bacteriological contamination of some citrus oils during processing. *Fla. State Hort. Soc.*, 81, 242–252.

Newhall, W.F. and Kesterson, J.W. (1961) Factors affecting the auto-oxidation of d-limonene during storage. *Proc. Fla. State Hort. Soc.*, 74, 239–243.

Riemenschneider, R.W. and Ault, W.C. (1944) How to evaluate and improve the stability of fatty foods. *Food Ind.*, 16, 892–894, 935, 939.

Rispoli, G. and Di Giacomo, A. (1968) Le essenze di arancia dolce ad alto tenore carotinico. *Essenz. Deriv. Agrum.*, 38, 138–156.

Rispoli, G., Di Giacomo, A. and Crupi, F. (1962) Il *p*-cimene nell'essenza di limone alterata. *EPPOS.*, **47**, 118–122.

Rodanò, C. (1930) *Industria e Commercio dei Derivati Agrumari*, Hoepli, Milan, Italy.

Safina, G. (1984) *I Derivati Agrumari*. Stazione Sperimentale per l'Industria delle Conserve Alimentari, Parma, Italy.

Slater, C.A. and Watkins, W.T. (1964) Citrus essential oils. IV. Chemical transformations of lime oil. *J. Sci. Food Agric.*, **15**, 657–664.

Stanley, W.L. and Vannier, S.H. (1959) Effects of environmental and processing factors in the citral content of lemon oil. *Food Technol.*, **13**, 96–99.

Waters, R. (1993) Cold-pressed citrus oil recovery. *Trans. Citrus Eng. Conf.*, **39**, 28–48.

9 Production of bitter orange neroli and petitgrain oils

Louis Peyron

INTRODUCTION

Neroli oil is obtained from the flowers of bitter orange (*Citrus aurantium* L.); the origin of its name is the consequence of the predilection for the orange flowers water by the princess Anna Maria de la Tremoïlle de Noirmutier, wife of Flavio Orsini duke of Bracciano and prince of Nerola (La Face, 1942; Igolen, 1946).

The term petitgrain was first used to indicate the small green fruits (orangettes) of bitter orange, and used to name the oil obtained from them; later this term indicated the essential oil obtained by distillation from the buds, and leaves. In the nineteenth century production was limited to the Provence area, and later extended to Italy, Spain, North Africa, Haiti and Brazil. After 1880 it was also largely produced in Paraguay.

Bitter orange, probably originated from the South-East Asiatic, first propagated in India, and Persia, then in the tenth and eleventh centuries the Arabians divulged it among the Mediterranean countries (Syria, Palestine, North Africa, Sicily, Sardinia, Liguria, Provence and Spain). Following the discovery of the New World, bitter orange was introduced to the Caribbean islands, and to the rest of the American continent. Its cultivation greatly improved during the eighteenth century, but since the sixteenth century, in Italy, was practised the distillation of its flowers and in the seventeenth century in Grasse (France) were prepared the bitter orange flowers perfumes (Igolen, 1946; Guenther, 1949; Gildmeister and Hoffman, 1959; Peyron, 1989; Rolet *et al.*, 1998).

The main regions were bitter orange production developed were in part the same as today: Tunisia (Cape Ban peninsula, Nabemb, Hammamet), Morocco (Khemiset, Sidi Slimone), Algeria (plane of Mitidja), France (Maritime Alps), Italy (Sicily and Calabria), Spain (Andalusia), Guinea, Comore islands, Haiti, Argentina and Paraguay (Guenther, 1949; Gildmeister and Hoffman, 1959).

The varieties of bitter orange are numerous. Following are reported the main morphological characters of some of them found mostly in Provence (Rolet *et al.*, 1998):

- wild bitter orange, obtained directly by the seeds, with long shaped leaves, small flowers, light and not abundant, but intensively scented;
- bitter orange with small spine and medium size flowers. Big leaves, numerous flowers and high content of essential oil. Largely diffuse;
- bitter orange with big spine and double flower. The spines also persist on the old branches. Large leaves, and large flowers with thick petals; poor content of essential oil;
- bitter orange with medium size flowers, arranged as clusters at the tip of the branches;

- bitter orange with small flowers light with thin petals; dark leaves, wings at the petioles;
- big calyx bitter orange, very large candid white flowers;
- 'bouquetier' from Nice, a very interesting hybrid with flowers rich in essential oil.

The different clones cultivated today are reproduced by grafting on wild bitter orange, spontaneously grown from the seeds.

Bitter orange from the Mediterranean area grow poorly in Paraguay, while different hybrids of 'bitter sweet orange' such as the 'Apepu-Jhai' are easily cultivated (Urbieta-Rehnfeld and Jennings, 1974).

Bitter orange cultivation and the consequent production of neroli and petitgrain oils has been subject to a considerable contraction in Europe during the last century; in fact, in Provence the cultivated area in 1914 was of about 320 hectares, and in 1997 was less than 10 hectares; the flowers production in the same region was about 2,500 tons at the beginning of the twentieth century and was reduced to only 12 tons in 1998. The neroli oil production can be estimated, today, of about 1 ton yearly, and the main productive countries are Tunisia and Morocco. Today, the bitter orange petitgrain oil is scantly produced in the Mediterranean area (Spain, Italy, Greece, Tunisia and Morocco), while its production is considerable in Paraguay, where, from about 100 tons produced in 1978, the total production in 1997 was about 250 tons (Igolen, 1946; Guenther, 1949; Gildmeister and Hoffmann, 1959; Cadillat, 1969).

The yield and the sensory characters of neroli and of petitgrain oils are related to numerous factors, such as the raw material conditions at the harvest time, the transport and storage conditions used, and the extraction procedure (Guenther, 1949; Gildmeister and Hoffmann, 1959; Rolet et al., 1998).

FLOWERS HARVEST, TRANSPORT AND STORAGE

The flower production start to be considerable when the tree is about 10 years old, and reaches its maximum, when the tree is about 20–30 years old (5–25 kg of flowers per tree per year). The harvest of the flowers requires a considerable amount of work, is time consuming and therefore expensive. Generally one worker collects daily from 8 to 12 kg of small and light flowers, and about 20 kg of 'heavy' flowers (large and semi large). The flowers are picked at the time of their blooming with warm and sunny weather conditions. Those harvested in the early morning give oil yields higher than those harvested in the afternoon. The closed blossoms give a lower yield, and give to the oil a green note. The collection of the flowers should be selective, avoiding small leaves and petioles. These would contribute to the oil with a petitgrain character. The oil yields are low if the flowers are harvested when the weather is cloudy, foggy or rainy. They increase during the productive season, ranging from 0.7 per cent at the season beginning, to the 1.2 per cent at the end of it. The flowers harvested in Autumn are sent to the herborist industry (Rolet et al., 1998).

The transport and storage conditions prior to the extraction (time, temperature, layers thickness, aeration) considerably influence the yield, the storage state, and therefore oil quality. Overheating phenomena (occassionally the temperature of the sacks in which the flowers are transported can reach 50 °C) and the humidity can cause enzymatic reactions that determine variation of the natural components of the oil in the flowers. Dipping the flowers in water, prior to the distillation, increases the oil yield. This is higher for the fresh flowers than that obtained from the distillation of the flowers kept for seven hours in storage. After twenty four hours storage of the flowers the yield increases of about 0.2–0.3 per cent of that obtained from the fresh flowers (Igolen, 1946; Guenther, 1949; Gildmeister and Hoffmann, 1959; Rolet, 1989).

FLOWERS DISTILLATION

In nineteenth century the flowers producers distilled the flowers directly in small alembic with straight flame. Larger distilling flasks later were used with indirect heating (water jacket, coils, etc.), or by external steam injection. The most common procedure in the Mediterranean countries was the hydro distillation, without the distilled or cohobated water recycle, at low pressure inside the alembic, for three hours. The average yield of oil, by this procedure was about 1‰ (Igolen, 1946; Guenther, 1949; Gildmeister and Hoffmann, 1959).

Recently, it was evaluated how different technological parameters (e.g. the flowers load in relation to the alembic volume; the presence or absence of water contact with the flowers; the temperature and pressure in the alembic; the total time of the extraction process) would influence the yield, the chemical composition and the sensory characters of the oil. For example, the use of saturated steam and an internal pressure higher than 0.8 bar, improves about 0.2‰ the oil yield, reducing, at the same time, the length of the process, and therefore production cost. This also limited some chemical changes such as the hydrolysis of linalyl acetate, and the formation of alcohols such as α-terpineol (Rolet *et al.*, 1998).

From the fresh distilled waters after the neroli oil separation, is prepared the bitter orange flower water absolute, by counter current extraction by hexane. The yields from the waters is of about 0.25–0.35‰. This product is particularly rich in methyl anthranilate (Pecout, 1973).

FLOWERS EXTRACTION BY VOLATILE SOLVENTS

The bitter orange flowers concrete is obtained by extraction with hexane or light petroleum with a yield that ranges between 0.2–0.3 per cent for the flowers harvested in spring. The extractors used are the classical static type, with a load of 1000–2000 L. The yields from the flowers obtained in Autumn are higher; up to the 0.4 per cent. These are slightly higher for the flowers harvested in the morning hours, than for those harvested in the afternoon, and for the flowers extracted immediately after the harvest than those stored for several hours. Moreover, dipping the flowers in the solvent at different times gives a higher yield than one single extraction, even if prolonged. The chemical composition of concretes, as that of the essential oil, is influenced by the period of blooming, by the physical state of the flowers, by time between harvest and extraction of the flowers.

From the concrete it is possible to obtain the bitter orange flower absolute by using conventional methods for the absolutes (extraction with alcohols, winterization, filtration and concentration). The yields is of about 48–52 per cent. The bitter orange flower absolute presents a persistent odour that reseamble the fresh flowers note, more than the neroli oil does (Naves, 1970).

LEAVES HARVEST

The bitter orange varieties used for the distillation of their leaves, are the same used for the flower extraction; to these must be added the Paraguay bitter orange (bitter sweet orange). For the petitgrain extraction in Italy, France and Tunisia, is used the foliage obtained from the annual pruning of the trees (leaves and canes without raw fruits). The pruning is usually practised from the end of June–October in Provence (France), and in April in Sicily and Calabria (Italy). Generally are obtained 3–5 kg of vegetable material by pruning each tree. The foliage obtained from sunny and aerated field is preferred, for the quality of the oil obtained, than that obtained from thick cultivated fields. The foliage collected must be clean, prior to the distillation, of all woody material. The yields of oil are higher from the young leaves that from the aged ones, and from the fresh one than those faded. The trees cultivated on the plain give lower yields than those cultivated uphill and the yields between January and June are higher than for the pruning performed in the period August and October. The distillation of raw material must be carried out immediately after the harvest, in order to avoid possible enzymatic phenomena that would modify the chemical composition of the oil. Storing of the foliage for different days, however, can determine a linear increase of the yields, from about 0.4 per cent to 0.6 per cent, but with a subsequent rapid decrease to less than 0.2 per cent (Igolen, 1946; Guenther, 1949; Gildmeister and Hoffmann, 1959; Rolet *et al.*, 1998).

LEAVES STEAM DISTILLATION

In the Mediterranean region the distillation is usually performed in alembics of 1500–2000 L and more, with steam injections of between two and two and a half hours, placing inside the alembic an amount of water equal to 400 L for 500 kg of foliage. For one kg of foliage used for distillation, are collected about 0.8 kg of water. The oil yield varies from 2‰ to 2.8‰ (Peyron, 1965).

In Paraguay the distillation was traditionally carried out by rudimental methods, by direct steam without the presence of water in the alembic. Improvements were obtained only after 1980. The oil yield varies between 2‰ to 2.5‰. To avoid the esters hydrolysis it is better to distil rapidly with vivacious steam and with under slightly high pressure. The use of pierced double bottom stills and of large and efficient condensers render more rapid the distillation (Igolen, 1946; Guenther, 1949; Gildmeister and Hoffmann, 1959; Mollan 1962; Pecout, 1973; Urbieta-Rehnfeld and Jennings, 1974; Baker, 1980).

The leaves extraction is also performed by using volatile solvents to prepare the concrete (Naves, 1970).

REFERENCES

Baker, D.M. (1980) Petitgrain–Paraguay. Distillation technology. *PAFAI J.*, 3(2), 137–141.
Cadillat, R.M. (1969) Observation on essential oils of citrus. *Fruits*, 24, 389–405.
Gildmeister, E. and Hoffmann, F. (1959) *Die Ätherischen Öle*, Vol. 5, 627–639. Akademie Verlag, Berlin.
Guenther, E. (1949) *The Essential Oils*. Vol. 3 D. Van Nostrand Co., New York, NY.

Igolen, M. (1946) Oranger Bigaradier. *Ind. Perfumerie,* (1), 3–12.

La Face, R. (1942) Le essenze di petitgrain. *Bollettino Ufficiale Stazione Sperimentale Essenze Derivati Agrumari,* 20(4,5,6), 27–35.

Mollan, T.R.M. (1962) Oil of petitgrain Paraguay. *Perfumery Essent. Oil Record,* 53, 13–14.

Naves, J.R. (1970) *Technologie et Chimie des Parfums Naturel,* 190–194. Masson, Paris.

Pecout, W. (1973) Huile essentielle de petitgrain in Paraguay. *Fruits,* 28, 301–303.

Peyron, L. (1965) Petitgrain. Production zone mediteraneenne. *France et ses Parfumes,* 299–314.

Peyron, L. (1989) Un siècle de production de plantes à parfum dans le Sud-Est de la France. *Parfums Cosmétiques Arômes,* (89), 97–113.

Rolet, A., Peyron, L. and Laporte, J.F. (1998) *Plantes à Parfums et Plantes Aromatiques.* C.M.E., Paris.

Urbieta-Rehnfeld, J.C. and Jennings, W.C. (1974) Bitter orange of Paraguay. *Chemie Mikrobiologie Tecnologie der Lebensmittel,* 3(2), 36–38.

10 Production of distilled peel oils

Luis Haro-Guzmán

INTRODUCTION

History

The lime fruit and juice industry originated in the West Indies, largely in the British islands of Antigua and Barbados. In the early nineteenth century the lime tree was cultivated in Dominica on a small scale. Around 1850 a more intense cultivation took place in Montserrat which at that time became the headquarters of the lime industry remaining so until superseded by Dominica (Guenther, 1949).

In the nineteenth century distilled lime oil was obtained in the British West Indies as a by-product of the production of lime juice. At that time vitamins were unknown but painful experience had shown that lime juice prevented scurvy. This disease was rampant among the sailors in the sixteenth, seventeenth and eighteenth centuries. In the days of sailing ships, British ships always carried ample supplies of lime juice. It was found that where ships carried a supply of lime or lemon juice the disease did not occur. The juices were supplied unsweetened and 'fortified' with 15 per cent of rum to act as a preservative.

Of the two juices, lime was preferred because its keeping qualities prior to the addition of alcohol were much better than those of lemon juice, due to the fact that it contains hardly any sugar, whereas lemon juice contains enough to cause fermentation unless it is preserved when absolutely fresh (Rose, 1965).

In 1867, a patent was granted to L. Rose for the preservation of juice with sulphur dioxide, thus the first fruit drink (Lime Juice Cordial), became a practical proposition.

At that time the fruit was pressed in sugar cane roller mills and the resulting juice sent to decant in wooden vats for 2–4 weeks. After this time the intermediary layer of more or less clear juice was drawn off, filtered, preserved with sulphur dioxide and filled into wooden casks. The upper and lower layers were distilled in copper or wooden stills. Some heated with direct fire but the majority by steam, either directly or indirectly or both.

In Mexico the lime tree was known in semiwild state on the Atlantic coast before it was regularly planted near Acapulco on the Pacific coast around 1870 (Guenther, 1949). The oldest existing factory goes back to 1933 but distilled lime oil production in Mexico must have started much earlier.

The first factory in Peru was installed in 1968. The number has grown significantly since then.

Distilled lime oils around the world

Distilled lime oils are produced from two varieties of the acid lime tree. The majority of the distilled oil is produced from the West Indian, Key or Mexican lime (*Citrus aurantifolia* Swingle). Some amounts are produced from the Persian, Tahitian or seedless lime (*Citrus latifolia* Tanaka). The countries where distilled lime is produced are listed below in order of decreasing importance: Mexico, Peru, Ivory Coast, Cuba, Guatemala, Haiti, Jamaica, Dominican Republic, Trinidad and Tobago, Dominica, Bahamas, USA, India, Tanzania, Swaziland, South Africa, New Zealand.

Peratoner oils

With the exception of lime oils, the production of distilled citrus peel oils, commonly named Peratoner is considered complementary and dedicated exclusively to the recovery of those amounts of oil contained in stable emulsions very difficult to separate (Di Giacomo, 1974).

The production of these distilled oils, forbidden in Italy until recently, is now permitted and regulated given that the modern procedures permit the production of oils with acceptable quality.

TECHNOLOGY

Distillation of key and Persian limes

Process used in Mexico and Peru

The main source of raw material for the industry is represented by the packing house culls. This fruit has been discarded because it is over- or undersized, not homogeneous in colour, damaged, odd-shaped, etc. The volume of the culls is regulated by the balance between production and market requirements for the fresh fruit.

The limes are stored in wooden bins prior to processing. This operation is necessary given that the delivery of fruit to the factory happens in a span of 6 or 7 hours during the afternoon and the processing takes place 24 hours a day. The first step of processing involves washing of the fruit to remove dirt and other impurities. This operation is normally carried out by passing the limes over revolving cylindrical brushes under a water spray. A different way to wash the fruit is to put them in a tank filled with water which is agitated by some mechanical means. Normally pure water is used. It is unusual to add any kind of detergent.

The distilled oil is obtained by pressing the whole fruit by means of a screw press and by distilling the resulting juice. The screw press consists of a stainless steel cone provided with holes of 4–6 mm in diameter, in the axis of which a screw forces the raw material, compressing it against the inner side of the cone. The oil cells burst expelling the oil that is carried out by the juice. The pulpy liquid obtained goes through the holes and the skin fragments are expulsed through the narrowest section of the cone.

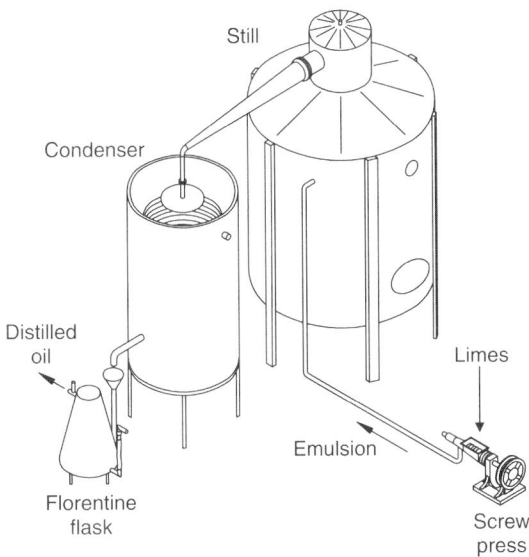

Figure 10.1 Diagram of the equipment for the production of distilled lime oil.

A distillation system consists of three main sections: a still, a condenser and a continuous separator for essential oil and water. The still and the condenser are joined by a 'gooseneck' (Figures 10.1 and 10.2).

The juice obtained by pressing the whole fruit in a screw press is sent to a tank directly feeding the stills. These are usually very simple. They generally have a cylindrical shape and their size depends on the desired working capacity. The most usual capacities go from 7–10 cubic meters. Their width is normally larger than their height in order to make the separation of the essential oil easier and to avoid returns. In the old days they were made of tinned copper or wood. Presently all are made of stainless steel. Heating the juice mass is done by steam, either by means of a coil or jacket or, much more commonly, by direct injection. This last procedure is preferred by most of the processors because it stirs the juice and facilitates the separation of vapours. Every producer has his own preferences and works according to his personal experience. Steam is produced in boilers fed with bunker C or diesel fuel. Distillers who prefer the direct injection method adjust the pressure in order to obtain the appropriate agitation. In any case, if the indirect heating system is used, the direct steam injection method is also used to maintain the level and to compensate for any loss of water due to the distillation, as distilled water is not cohobated. Steam injection is effected by means of a perforated annular tube and/or by two perforated tubes placed crosswise.

The stills are usually provided with thermal insulation to avoid any premature condensation on the inner sides.

The stills, usually containing several thousand litres are filled with juice up to 3/4 of their height. The level is controlled through a peep hole. The most advisable form for the top part of a still has always been a subject of issue. The gooseneck must not offer any resistance to the vapours path.

Figure 10.2 Still, condenser and Florentine flask used in the distillation of key lime oil.

The condenser usually comprises a coil with a 5 cm inner diameter pipe and the necessary length to reach the desired exchange surface. This length is usually around 100 m and is immersed in a tank in which cold water is circulating. The head of the coil often has a discoidal widening where vapours are partially condensed by sudden expansion. In the past the piping was of copper but nowadays only stainless steel is used. Multi-tubular condensers have been tried several times but their use was discontinued due to the fact that they get plugged often and are difficult to clean.

The dimensions of the condenser are very loosely calculated. In practice it has a surface up to five times larger than the one calculated. It is believed that the condenser is more efficient when the cooling water leaves at a temperature of 80 °C and when the oil is collected at 36–38 °C. If the surface is relatively small, the distillation speed is lowered and the heating of the still is consequently reduced. This results in an oil of abnormal composition, hence a poor quality.

The essential oil and water leaving the condenser are separated continuously in a Florentine flask. During the first hours of the operation, the oil and water are easily separated. Subsequently, compounds with higher boiling points–some of them being more soluble in water–may form a light emulsion. Separation can be better obtained if the agitation of the fluid is reduced to a minimum. The capacity of the Florentine flask should be at least equivalent to the hourly capacity of the still. If necessary, several Florentine flasks can be placed in series, each one placed at a lower level than the preceding one. However, this practice is seldom used.

The distillation time depends on the volume on the still, its form and the efficiency of the condensing system, as well as on the availability of cooling water and the consistency of the juice. A thick juice requires more time. This is why very ripe yellow limes give a more viscous juice, which takes more time to be distilled. Distillation must be effected very carefully, gently and slowly at the beginning, with the steam at a low pressure and carried on with more pressure towards the end of the operation. Live steam must boil gently through the liquid mass with a sufficient supply to maintain it agitated so as to promote a good contact between the oil droplets and the hot juice to make the necessary chemical reactions to take place to the necessary extent to produce a good quality oil. This is the practice generally in use to obtain the quality of oil commercially accepted.

Expected performance can reach 4.3 Kg of distilled oil per ton of fruit. However, 5 Kg per ton can be obtained when green fruits are being processed.

The oil must be pale yellow. A bright yellow colour indicates an exaggerated steam pressure or, on the contrary, a too slow distillation. A premature interruption of the distillation results in a colourless, low quality oil mainly composed of terpene hydrocarbons and lacking oxygenated and sesquiterpene compounds.

Processes used in other countries

A similar process is in use in Ivory Coast the only difference being the use of cohobation, which is the return of the water from the decanter to the still in order to minimize the loss of water soluble components.

Other variations are used in different countries. McHale (1980) describes several: in some places the fruit is treated in a roller mill instead of a screw press. The traditional process used in the West Indies involves the treatment of the limes in a screw press or roller mill. The resulting juice-oil emulsion is left to settle in vats and after two to three weeks the upper layer of top pulp juice with oil is separated from the middle layer of clear or 'racked' juice and distilled. An alternative process incorporates a limited settling period after which the juice-oil emulsion is filtered. The majority of the oil is retained by the filter cake and is subsequently recovered by steam distillation. The composition of the oils obtained varies according to the method used (McHale, 1980).

Peratoner distillation

Peratoner and Scarlata in 1910 were the first to apply vacuum distillation for the recovery of lemon oil. This method is described in detail by Romeo (1934) and later by Di Giacomo (1974). This method, known as Peratoner, originally consisted in the crushing and pressing of the fruit, followed by steam distillation, under reduced pressure at the lowest possible temperature, of the liquid obtained.

Obviously, higher quality essential oil was obtained from freshly produced unfermented liquid, but the distilled oil had in any case a distinctive burnt odour and, because of the chemical transformations that take place during distillation, was of a decidedly inferior quality compared to that obtained from the various cold processes.

In the standard process, lemons yielded an *industrial juice* which could only be used for the production of citric acid. In consequence, the Peratoner method has been modified over the years, following new trends in the citrus industry. This method had developed as part of a fruit processing sector that was mainly geared towards the production of citric acid. However, *industrial juice* has not been extracted on a large scale for some time, the production of juice for human consumption being much more profitable. The Peratoner method has thus been limited to the treatment of peel, depleted of majority of the essential oil by common slow-folding, or sfumatrice machines or by pressing. The new treatment includes a further pressing of the peel in a special continuous press, that maintain a high pressure, avoiding a possible re-absorption of the essential oil by the albedo. This operation is assisted by moderate jets of water.

Other liquors that are also distilled are the discharges of the centrifuges and the water that is recycled to the oil extractor. The production of these oils is normally made by batch distillation and, sometimes when convenient because of the amounts processed, by continuous distillation.

Recently, there have been some lemon and mandarin distillates of good quality on the market. However, the absence of a non-volatile residue (including effective antioxidants) notably reduces the length of time the quality of the oil may be preserved.

Improvement in the quality of distilled oils is also connected to the continual perfecting of the distillation technique and also the use of apparatus of reduced capacity which allow the processing to be completed in short time.

REFERENCES

Di Giacomo, A. (1974) *Gli Oli Essenziali degli Agrumi*, EPPOS, Milan, Italy.
Guenther, E. (1949) *The Essential Oils*. Vol. III. D. Van Nostrand Co., New York, N.Y.
McHale, D. (1980) Effect of processing conditions on composition of distilled oil of lime. Paper No. 67, *VIIIth International congress of essential oils, Cannes*.
Romeo, G. (1934) *Un Nuovo Metodo di Estrazione Diretta di Acido Citrico dal Succo di Limone*. Industrie Riunite Editoriali, Palermo, Italy.
Rose, L. (1965) *History of L. Rose and Company*, Unpublished work.

11 Citrus by-products

Gianvincenzo Licandro and Carlos E. Odio

INTRODUCTION

The products resulting from the processing of citrus are the juices, the essential oils and the peel. The latter, together with the pulp and seeds, are the industrial residue and account for 40–60 per cent of the weight of the raw material. The utilization of this residue is a fundamental requirement of the fruit processing industry, not only for economical reasons, but also to reduce the severe environmental impact that this could produce.

The main by-products produced from this residue, are: dry peel, citrus molasses, d-limonene, pectins, clouding extracts, soluble solids, flavoniods, limonoids, peel fibre, seed oils, ethyl alcohol, peels in brine, candied peel, fermentation products, etc. Other products may be produced from the pulp separated from the raw juice, such as pulp wash, juice pulp, whole juice cells and carotene extracts.

The essential oils result in interesting by-products such as concentrated oils, terpeneless oils and alcohol formulations. It should be noted that the American industry defines the orange essential oil as by-products and considers the juice as the sole primary product from this fruit (Di Giacomo *et al.*, 1992; Kesterson and Braddock, 1976). Other products that cannot be defined as by-products are the comminuted and fruit sections.

Many of the above mentioned by-products are only sporadically produced; therefore, this chapter will be focused only on those by-products that are commonly produced by the major processors and, therefore, have higher commercial value.

DRY PEEL, MOLASSES, d-LIMONENE

From a quantitative point of view, the main component of the residue is water (80–85 per cent). The dry matter is fundamentally made up of soluble sugars (glucose, fructose and sucrose, together with a modest amount of pentose), insoluble carbohydrates (cellulose, emicellulose, protopectin), organic acids (citric, malic, isocitric, oxalic) and a considerable amount of flavonoids (hesperidin in the majority of citrus species, naringin in grapefruit, bergamot and sour orange). Inorganic constituents differ from those in the juice mainly for their higher calcium content.

The lime, bergamot and lemon peels are mostly used for the extraction of pectin while the peel obtained from other citrus fruits are almost totally used as animal feeds.

Because of the high moisture content and the high tendency to ferment, it is not practical to transport the residue for long distances or store it for long periods of time. Therefore, the most practical treatment is drying which is done in the part of the processing plant known as the feed mill.

Feed mill

The feed mill is of fundamental importance because of the following reasons:

- it is the collection and treatment centre for all solid residue as well as the more highly polluting liquids that present more than 1.5–2 per cent soluble solids and/or more than 0.1 per cent essential oils;
- it is the plant's main water and air pollution control centre;
- it is here where the most profitable by-products are produced, namely: animal feed, d-limonene, molasses and in some cases, pectin and ethanol.

It is important to note that the feed mill is the largest energy consumer in the plant (Odio, 1993). Figure 11.1 shows the schematic diagram and mass balance of a typical feed mill. By far, the major cost of the entire feed mill operation is related to the energy required to remove the water from the peel. To reduce this cost, the peel, after milling and liming, is mechanically pressed to reduce its moisture. The resulting press liquor together with the most polluting liquid streams from the processing plant are concentrated in a multiple effect waste heat evaporator (WHE) that uses, as its sole source of energy for evaporation, the dryer exhaust vapours. Therefore, that portion of the water removal is accomplished at zero energy cost. The pressing conditions are important, because the higher the pressure (the lower the resulting moisture in the press cake) the larger the load that goes to the WHE. The concentrate produced in the WHE (above 50° Brix), referred to as molasses, is mixed with the press cake and both are dried together. Occasionally, the molasses are commercialized separately in concentrations that range from 50° to 72° Brix for various uses, including ethanol production.

Milling

The hammermill must produce as uniform a particle as possible (0.6–2.0 cm) for a better reaction with the lime. Smaller particles (fines) are difficult to separate from the gas stream in the dryer, therefore, some are recirculated to the furnace, if the dryer is so equipped, where they will be incinerated and others will end up in the waste heat evaporator black water. Larger pieces will not dry uniformly and could result in mould formation ultimately leading to spontaneous combustion. Grinding of the peel is generally done in hammermills.

Lime treatment

The purpose is to remove the slimy texture of the peel thus making easier to release water: the divalent cations, Ca^{+2}, from the lime probably combine with the demethoxylated pectin and the end result is syneresis, or the natural expression of fluid (Kimball, 1991). The treatment can be done 'dry' or 'wet.' In the 'dry' treatment the peel is mixed with the calcium oxide, for approximately 15 minutes, in larger reaction screw

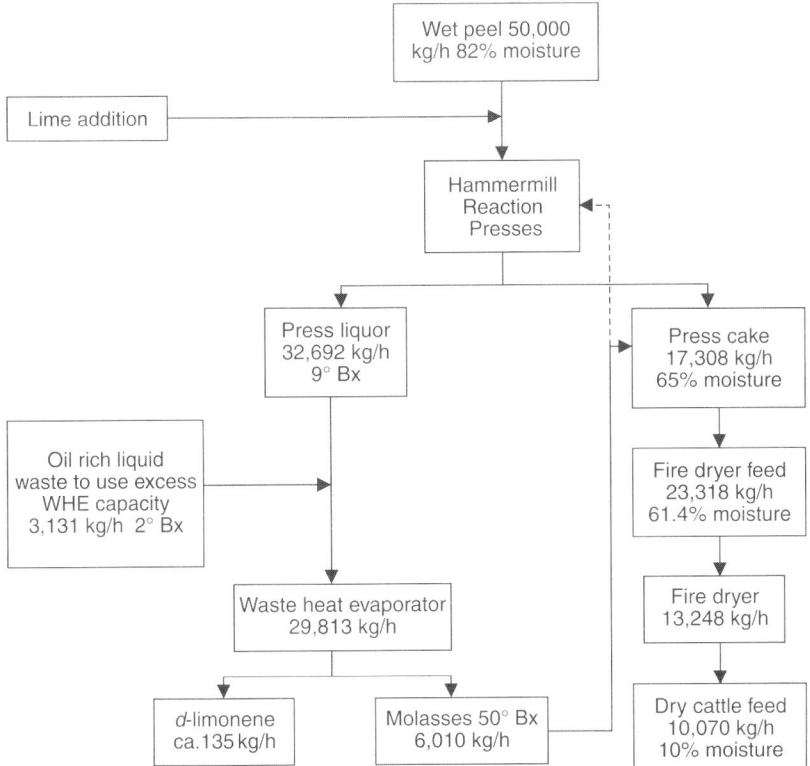

Figure 11.1 Flow and mass balance of a typical feed mill.

and mixers. In the 'wet,' the most recently used treatment, the peel reacts with the lime for a few minutes in a tank or in pipe mixed with receded liquor to make it flow through the pump. After the reaction, the peel and liquor are separated. The peel goes to the presses and the liquor is returned to the ground peel tank where it will once again mix with the peel. All of this translates into less equipment, energy consumption, breakdowns, maintenance, reaction time and WHE scaling. The pH can be raised to 5.8–6.0 instead of the 6.5–7.0 required in the 'dry' system (Odio, 1993; Marques, 1995).

Pressing

From an economical stand point, this is the most important operation in the feed mill because it is here where the flows that are going to the dryer and to the WHE are determined. A strong pressing results in a dryer press cake, and consequently a larger press liquor volume that is concentrated into 40–50° Brix molasses and mixed with the press cake thus further reducing the product moisture at the dryer inlet. This moisture content, which varies between 60–75 per cent, depends on the efficiency on the reaction with lime, on the particle size, on the type of press used, on the configuration (single or double pressing), on the WHE capacity, on the molasses' degree of concentration and the point where these molasses are mixed with the press cake. Additionally, increasing the molasses/peel ratio and non re-circulating the reaction liquor, reduces the essential oil content in the peel at the dryer inlet and this reduces VOC (volatile organic compound) emission. The most commonly used presses are single screw slow rotating presses or double screw presses; these latter reduce moisture significantly in one pass and require very small particle size for proper operation. Double pressing with vertical presses may yield the same results. Horizontal presses yield the highest moisture content and should not be used when plenty of WHE capacity is available and low VOC emissions are desired.

Drying

Generally, single pass rotary drum dryers are used. Occasionally, triple pass are also used. The majority are direct fire with the furnace and drum placed parallel to one another thus forcing the combustion gases to travel through two 90° bends to make sure that the peel does not see the flame. The gases of combustion, which are pulled by fans in the suctions box, push the peel through the dryer drum. In the drum, the peel is continually remixed and moved from the periphery of the drum, to the centre and vice versa. The goal is to attain a uniformly dried peel with the least amount of energy consumed, at the maximum possible yield and to exhaust gases-of-combustion/water-vapour mixture at the highest possible moisture content, to increase the WHE efficiency. Generally this is accomplished by recycling 50 per cent of the exhaust vapours back to the discharge of the combustion chamber. This results in a very low oxygen content in the dryer gases (less risk of fire) and allows the flame's radiant energy to heat the gases that transfer this energy to the peel at much lower temperatures. Moreover, by lengthening the section immediately after the combustion chamber and before the first 90° bend, VOC's can be reduced as a result of the recycled gases' greater turbulence, higher temperature and longer treatment timer (Odio, 1993). Maximum operating efficiency (uniform product quality at low cost) of the system requires constant feed flow (moisture and weight) and constant drying capacity.

Waste heat evaporator

This is the most important piece of equipment in the feed mill because:

- it drastically reduces energy consumption by eliminating most of the water from the peel using, as its sole source of energy for evaporation, the dryer discharge gases;
- it is the feed mill's pollution control centre because it is the dryer gas scrubber, and by having a capacity in excess to that required to concentrate the press liquor, it can be economically used to concentrate the most highly polluting liquid streams from the processing plant;
- it recovers d-limonene, thus converting a source of VOC emissions into a source of revenue.

The WHE uses tube bundles that operate under the principle of 'descending turbulent mist'. Allowances must be made however to allow for large gas volumes on the vapour side and to be able to concentrate liquids which are subject to fermentation with the resulting large volumes of CO_2. Additionally, the outside of the tubes must be continually washed to avoid scaling and to scrub the dryer gases prior to the release into the atmosphere.

Pelletizing

The peel comes out of the dryer with a moisture content of 12–15 per cent and an apparent density of 0.4 kg/lt. Through hot extrusion the peel is transformed into approximately 1.5 cm long pellets with twice the apparent density. The composition of the product depends not only on the process employed but, also, on the raw materials used. The table below shows the nutritional average composition of citrus pellets and their relative digestibility. The values are reported as nutrient composition and related digestibility of dried citrus pulp adapted by Kesterson and Braddock, 1976; Sinclair, 1984.

	% Composition	% Digestibility
moisture	10	–
protein (N×6.25)	6	50
fat	3	85
fibre	12	68
nitrogen free extract	64	89
ash	6	–

The average total digestible nutrient (TDN) is about 75. The calcium content varies depending on the amount of lime added to the product prior to pressing.

In the drying process described above, which is certainly the most commonly used, 65 per cent of the cost goes to the energy required for evaporation. Today's fuel cost and its future tendencies forces the feed mill operators to consider the dry citrus peel not only as animal feed but also as a possible fuel with an energy value of 4.137 kcal/kg (Kesterson et al., 1979). Moreover, the search for less polluting technologies draws attention to alternative systems which were previously considered of little interest.

Drying with superheated steam

The possibility of drying orange peel with pressurized superheated steam, using a process similar to the one used in the sugar industry to dry the pulp left from the beets after the extraction of the sugar, has been analysed (Jensen, 1990). The product to be dried is sent into a cellular fluid bed under pressure where the fluid bed is driven by a flow of superheated steam, which is recycled through a heat exchanger. All this is contained in one pressure vessel. The steam produced by the evaporation is discharged from the pressure vessel and by using this in, e.g. evaporators, nearly 100 per cent energy recovery is achieved. The main advantages are: environmentally friendly due to absence of emissions, product improvement because of the greater digestibility of the dry pulp, and fuel economy. This process, however, requires that the orange peel be thoroughly washed prior to drying to remove the soluble sugars to avoid caramel formation in the dryer. Therefore, the evaluation of the process must take into consideration the resulting wash waters and their necessary subsequent treatment.

d-Limonene production

d-Limonene production is certainly related to the subject dealt in this chapter. Commercially, by this name we do not refer to the pure compound but, instead, to the essential oil that is steam distilled in the WHE and decanted from the condensate.

Its composition is therefore quite similar to that of orange cold pressed essential oil, but differs from it for the absence of non volatile compounds and, depending on the evaporation parameters, also part of the volatile components, mainly the high boilers, can be lost. Through the use of adequate treatment and using highly efficient distillation systems a > 99.5 per cent pure product can be obtained (Caratozzolo, 1998). Among the multiple uses of *d*-limonene its applications in the production of synthetic resins, its use as a solvent in substitution of mineral solvents and as a base for the synthesis of aromatic compounds, are the most important. When orange cold pressed essential oil is concentrated through vacuum distillation, the distilled product is similar to *d*-limonene but organoleptically superior.

SOLUBLE SOLIDS, CLOUDING EXTRACTS

The citrus processing industry, after extracting the juice, is economically interested in recovering other soluble solids that can be present in the finisher pulp or other parts of the fruit. This results in liquids with physical-chemical and organoleptic characteristics which, depending of the method of extraction used, can approach or differ from those of the primary juice; based on their origin and constitution, they can be added to the juice or be used as bases for soft drinks. It is important to note that this secondary recovery of soluble solids is done almost exclusively in the processing of oranges.

Pulp wash–wesos

The most common secondary soluble solids recovery is pulp wash or wesos (water extracted soluble orange solids). The process consists of washing the pulp and other particles discharged by the finishers, it is practically a soluble solids extraction through water diffusion. As such, the process' behaviour is a function of the physical phenomena of solubility and diffusion and as such it is influenced by operating variables such as temperature, diffusion coefficient, solubility index, differential concentration of the solutes, solvent/raw material ratio. Other factors, that play a part in the final solids recovery, are particle size and contact time. The operation is carried as a counter current multiple stage in which each stage consists of a water/pulp mixing phase in a mixing screw or vessels with agitators, followed by a liquid/pulp separation phase in finishers similar to the ones used in the primary finishing of the juice. Generally, 3–4 stages are used and the water/pulp ratio falls in the 1.5–2.0 range resulting in a recovery of 85–90 per cent of the available soluble solids, which could result in a total yield increase on the order of 6–8 per cent. It is not recommended to operate with a larger number of stages because the product, thus recovered, is of inferior quality (Morales, 1998). The pulp wash is treated like the juice, i.e. it is centrifuged and concentrated in the same evaporator to up to 65° Brix. Analytically, the composition of these liquids is similar, by some parameters, to the primary juice (Nonino, 1996), but, generally, showing a higher water soluble pectin and flavonoids content. Noticeable increase of limonin, can be observed when the process is carried out with the presence of high quantity of rag and under extreme washing and extraction conditions. Therefore, a pectolitic enzyme treatment is almost always necessary to reach concentrations above 40° Brix. The main quality parameters are, therefore, the limonin content, cloudiness, suspended pulp, diacetyl index and viscosity (Ballantine and Ferguson, 1985). Generally, extraction is done using normal potable water or condensate from the evaporation of the juice. In this case, care must be exercised to avoid contamination with thermophile bacteria that can contribute to the development of bad off-flavours during storage of the concentrate (Braddock, 1999).

Concerning the legislation for the production of wesos in the USA, this is regulated by the Food and Drug Administration (FDA, 1993), which permits adding the wesos to the juice prior to concentration, provided both products are made on-line from the same fruit. This differs from the wesos produced in a different line; NFC (not from concentrated), for example, cannot be added or blended with other juice but must be sold separately as 'pulp wash' and can be legally used in the production of soft drinks. Under Florida legislation the pulp wash must be packaged in easily identifiable containers and, as a tracer, sodium benzoate must be added between 50 and 100 ppm. This amount is well below the value necessary to guarantee preservation of the product and below the harmful limits when the product is used under the law requirements, in the preparation of drinks (Florida Department of Citrus, 1998). In Brazil, the world's largest producer of FCOJ (frozen concentrated orange juice), the legislation is similar to the USA's. The EU's legislation, on the other hand, forbids the use of diffusion processes in the production of concentrated juices (EC, 1993).

Core-wash

The FMC extraction system separates the centre of the fruit (core), the two peel plugs, seeds and the membrane from the pulpy juice, and allows for the recovery of soluble solids from these materials that represent 15–18 per cent of the entire fruit. The extraction process is similar to that employed for pulp wash but, because of the nature of the material, fewer washing stages are used and less agitation. Even then, the resulting product contains more pectin, is more viscous and cloudier, and has a limonin content on the order of 3–5 times that of the juice (Ballantine and Ferguson, 1985).

Second pressure extracts

In the Mediterranean area many processing plants recover soluble solids from all juice extraction wastes. This is done in screw presses or through diffusion with water. These means of solids recovery are used primarily when the juice extraction process leaves a certain amount of juice in the waste. The resulting product, known commercially as 2nd pressure extracts, are characterized by its clouding and colouring properties which prove to be quite desirable in the preparation of bases for the production of low juice beverages and soft drinks (Postorino et al., 1998). The main steps in the process are extraction and centrifuging. Generally, the product must be characterized by a high and stable cloud and by a very low pulp content. The extraction methods are numerous, among them the most common are: screw presses, fine chopping (>1 cm) or water diffusion after medium to coarse chopping (1–3 cm). The critical parameters in the process are the temperature and the enzymatic treatment. In fact, the temperature, not only improves the separation between the solid and liquid phases, but also improves the migration to the liquid phase of the substances responsible for the cloud and the colour. The treatment with pectolitic enzymes must reduce the viscosity of the extract without compromising the pectin's colloid property. Each producer varies the operating parameters (type and quantity of enzymes, reaction time and temperature) as a function of the raw material being used and the desired product. There are many theories concerning the composition of the cloud, generally these can be traced back to the complexes of lipids, coagulated proteins, flavonoids, very fine pulp particles with pectic substances (Shomer, 1991; Shomer et al., 1991; Klavons et al., 1991). Since the peel and membrane are very rich in bitter substances, all products that use these raw materials in their formulation must be debittered. The installation and technology utilized for debittering are similar to those used in debittering the juice (Grohmann et al., 1999; Adami et al., 2000).

The chemical composition of pulp-wash, core-wash and 2nd pressure extracts is different from that of the juice. Extensive research has been conducted to study the composition of these extracts and many analytical methods have been proposed to identify their eventual presence in the juice. Several methods, based primarily on the difference in concentration of some components in pulp and peel with respect to juice, have been proposed (Hammond, 1996). Since the phlorin (3,5-dihydroxyphenil beta-D-glucopyranoside) content in the albedo is considerably higher than that of the juice, high levels of phlorin in orange juice should indicate adulteration with peel (Johnson et al., 1995; Louche et al., 1998). It is possible (Leuzzi and Licandro, 1997) to distinguish, with ionic chromatography, orange juice from 2nd pressure extract. The presence of pulp-wash and peel extracts in the juices can be determined through flavanone glucosides and polymethoxylated flavones profiles (Grandi et al., 1994; Pupin et al., 1998; Ooghe and Detavinier, 1999; Leuzzi et al., 2000).

PECTIN

The name pectin covers a group of substances formed by complex carbohydrates made out of long anhydro-galacturonic chains connected through the 1,4 glycosidic linkages. Such polymers can be formed by hundreds, even thousands of monomers and have a MW of 50,000–150,000. The carboxyl groups of the galacturonic acid are partially esterified (methyl) or neutralized with one or more cations. In citrus the pectin is present in the form of insoluble protopectin (pectin tied to the cellulose) and of soluble pectin in different degrees of polymerization and esterification. One of pectin's characteristics is its capacity to gel sugar solutions acidified with organic acids. Such property is due particularly to the present methoxyl groups.

Commercial pectins are divided according to their degree of methoxylation (DM). The degree of gelation, jelly grade (JG), is expressed as the weight of sugar that can be jellified by a given weight of pectin. From a productive standpoint, the jelly units (JU) are very important. These are obtained by multiplying the per cent yield of pectin (dry base) by its gelation strength, thus indicating both the quantity and quality of the pectin obtained in the industrial process.

Production

Pectin production has been controlled by a few large processors, which, with regard to some phases of the process, have adopted some confidential procedures. Pectin is obtained from citrus peel and from press residue from the apples. The one from the citrus is more convenient because it has a higher JG and because it is less difficult to purify. The JG decreases with fruit maturity and, for oranges, it is acceptable only from immature fruit. Generally, the JG obtained from lemon or lime peel is 300–350°, from grapefruit it is 250–300° and from oranges 150–200°. The JU decreases in the following order: lime > lemon > grapefruit > bergamot > orange. It is important to point out that peel must be treated as soon as possible and never more than 12 hours after juice extraction (Di Giacomo, 1987). This is necessary to avoid fermentation and to limit the deterioration of the pectin that the pectolitic enzymes, present in the peel, can cause. In practice, it is always convenient to have the peel, ground and washed, kept in suspension in water and pasteurized at 95–98 °C for approximately 10 minutes to eliminate the enzyme activity (Swisher and Swisher, 1977). The excess peel is dried after washing and stored to be used for pectin extraction during the off-season. Following we touch briefly on the main process points.

Grinding

Grinding must be as uniform as possible ($2 \, cm^2$ pieces) to optimise the washing but, above all, the drying and the acid hydrolysis.

Washing

The aim is to get rid of the soluble substances to improve the purity of the product. Washing is generally done counter currently, in mixing screw conveyors and the conditions used result from the compromise between the need to use large volumes of water, to accomplish a thorough washing, and the amount of water available. When the peel comes from very ripe fruit, modest amounts of aluminum sulphate are added to the wash water to block the non-degraded water-soluble pectin.

Extraction

After washing, the pectin is almost totally present as protopectin. Solution is generally obtained by hot hydrolysis with mineral acids taking the precautions to avoid drastic conditions that could break the pectin molecule or too soft that would jeopardize yield. Taking into consideration fruit maturity, the parameters to control are temperature (70–100 °C), the pH (1.2–3.0) and reaction time (30 min to 4 hours). Since these parameters are inter dependent, shorter time requires higher temperature and acidity. The ratio water/peel must be such as to reach a pectin concentration in the liquid, after extraction, on the order of 1 per cent.

Filtering

This can be done in two stages: coarse filtering to eliminate the waste peel and a fine one to clarify the liquor. Generally, rotary vacuum or under pressure filters are used with cellulose or diatomee as filter aids.

Precipitation, purification, drying and grinding

Pectin is separated from the clean pectin liquid through various methods, of which there are two categories:

Alcohol precipitation

To reduce the volume of alcohol required, the pectin solution is concentrated in vacuum evaporators. Concentration systems using membranes have not given positive results due to irreversible blockings occurring in the medium-long period. The liquid, concentrated up to 4 per cent pectin content, is mixed with a quantity of alcohol (methanol, ethanol, isopropanol) that will result in a final alcohol content of 55°–60° Trl. The pectin precipitates as a foamy fibrous mass and is separated using filtering conveyors, filter presses, rotative presses. After washing with an increasingly richer alcohol mixture, the precipitate is pressed and dried with hot air at 40 °C. The dryers used allow the recovery of the vapours thus permitting the recovery of the alcohol. The precipitate is then ground (60–80 mesh). This process has the disadvantage of large alcohol losses that have a negative impact on the cost.

Precipitation with aluminum salts

An aluminum chloride or sulphate solution is added to the pectic solution, then sodium carbonate or ammonium hydroxide is added until a pH 3.8–4.2 is reached. Most batches exhibit optimum precipitation near pH 4.0. After a first treatment with acid alcohol, the operations that follow (washing, drying and grinding) are similar to those performed with alcohol precipitated product. Alcohol losses, compared to the previous procedure, are noticeably reduced.

Standardization

The powder produced with either precipitation method, which can have a JG significantly superior to 300°, is standardized to the desired grade by mixing with other batches or through the addition of sugar (sucrose, dextrose) so as to have batches with a uniform JG. The finished product, eventually added with one or more buffer salts (citrate, tartrate, or phosphate), is appropriately packed and stored in cool areas to avoid possible degradation. Generally citrus pectin is standardized to 150°.

Drying

As mentioned before, the peel is either immediately taken to extraction or, alternatively, dried and stored for future extraction. In such a case, after pressing without the addition of lime, the peel is dried at temperatures that, preferably, must not exceed 110 °C, to avoid possible browning. Rotary or direct fire dryers are used, avoiding direct contact between the flame and peel. Also, steam tube dryers may be used. After cooling, the dry peel must not have a moisture content above 10–12 per cent to reduce the risk of fermentation and spontaneous combustion. The drying and storage process results in a pectin loss compared to the fresh peel extraction, but permit storage and long-distance transportation of the product.

Commercial pectins are separated into two groups: high methoxyl (HM) pectin, in which more than 50 per cent of the carboxyl groups are methoxylated, and low methoxyl (LM) pectin, in which methoxylated carboxyl groups are lower than 50 per cent.

The HM pectin gels only in the presence of high sugar concentrations (near 65 per cent) and a low pH that must stay within the limits of 3.3 and 3.5. Additionally, the HM pectins can be differentiated as rapid-set (RS) and slow-set (SS). The RS pectin which, practically, are those produced with the process described above, form, at high temperatures, under specified conditions (concentrations of soluble solids and pH), an irreversible gel. These are used in the preparation of small volume jams. The SS pectins, on the other hand, are generally used in the production of large volume jams. The gelation takes place at temperatures below 80 °C and is thermally reversible. The SS pectins can be obtained either optimizing temperature and pH during the extraction process, either slightly modifying the process final steps: the pectin is left in alcohol solution with strong acids at controlled temperature. This treatment probably results in a slight reduction in the methoxylation degree and a reduction in the molecular weight. The LM pectin forms gel without the addition of sugars or organic acids; in fact, all that is needed is the presence of polyvalent cations, mainly calcium, which form bridge bonds among the free carboxyl groups, resulting in a stable reticle. The pH has little influence allowing operations between 2.5 and 6.0. Their use is also facilitated by the gel low melting point and its thermal reversibility capacity; moreover, the gel is translucent and syneresis rarely occurs. Therefore, the LM pectin is used in the production of puddings in which the calcium, contained in the milk, is enough to allow formation of gel, or for the production of low sugar gelatin (dietetic products). The LM pectins are produced either using pectin obtained from the last purifying step or using finely powered pectin. These are divided into two groups: conventional LM pectins, in which demethoxylation is done with acid hydrolysis, and amidic pectins, in which demethoxilation is done, in alkaline hydrolysis, with ammonia in an alcohol environment (min 65° Trl). The residence time regulates the final methoxyl content (Rispoli and Di Giacomo, 1962; Sinatra, 1999).

Fibre pectins

The fibre pectin, known in the past as crude pectin, is a particular pectin obtained through variations to the classical production method. The peel, after grinding and crushing, is pressed and mixed, in a 1:2 ratio, with acidic ethyl alcohol solution (96° Trl). After 22 hours residence at 60 °C the solution is filtered and the solid phase is washed with increasingly concentrated alcohol solutions and, in the final step, with acetone. This process avoids that the pectin, liberated by hydrolysis from the protopectin, go into solution. The steps, that follow in the production process, are the same used in the production of classical pectin. The fibre pectin, produced only from lime peel, is therefore a product rich in fibre and has a flavour that is not totally neutral. It can be used for those products in which purity is not necessary and in which the residual raw material flavour is compatible with that of the finished product (Askar and Triptow, 1997).

FLAVONOIDS

Citrus fruits contain numerous flavonoids with a typical pattern for each species, resulting good markers of the relative products and by-products. Most flavonoids are in the fruit albedo and flavedo and they can be grouped in flavanones, flavones and anthocyanins (Horowitz and Gentili, 1977). The most numerous are the flavanones which are, almost invariably, in the form of glycosides belonging to two groups: rutinosides and neohesperidosides. The most common aglycones are naringenin and hesperetin; the combination of β-neohesperidose with the above aglycones can result in naringin and neohesperidin, respectively; the same aglycones and β-rutinose can form narirutin and hesperidin, respectively. Rutinosides are mostly tasteless and normally found in sweet oranges and lemons while neoespheridosides are bitter and can be found in bitter oranges, bergamots and grapefruits. The flavone glycosides and aglycones are more numerous than the flavanones, but they occur in low concentration and are generally tasteless. The flavone aglycones are characteristic constituents of citrus fruits. They are essentially found in the fruit flavedo and generally are highly substituted compounds with many methoxyl and few, if any, hydroxyl groups. Remarkably, the fully methoxylated compounds are absent in *citrus lemon*. Flavanones glycosides and polymethoxylated flavones present typical pattern in different species or part of fruits (Reminiac *et al.*, 1989; Ooghe and Detavernier, 1997; Robards *et al.*, 1997; Mouly *et al.*, 1998; Postorino and Gionfriddo, 1999). The anthocyanins are glycosidic derivates of the 3-hydroxyflavylium cation; they characterize the blood orange variety and are found in the endocarp (Maccarone *et al.*, 1985). Flavonoid-correlated compounds, also contained in citrus fruits, are esters and glycosides of trans-hydroxycinnamic acids (Fernandez de Simon *et al.*, 1992), which have been studied in blood orange juices (Mouly *et al.*, 1997; Rapisarda *et al.*, 1998). Flavonoids have been extensively studied for antioxidant, anticancer, antiviral, and anti-inflammatory activities, effects on capillary fragility, and an observed inhibition of human platelet aggregation (Attaway, 1994; Benavente-Garcia *et al.*, 1977; Miyake *et al.*, 1997; Bocco *et al.*, 1998; Chen *et al.*, 1997; Chen and Ho, 1977; Kawai *et al.*, 1999; Rapisarda *et al.*, 1999). An additional exploitation could be their use as food natural antioxidant agents.

Commercially valuable flavanoid-based industrial citrus products are hesperidin, naringin and their by-products.

Hesperidin production

The raw material mostly applied is the orange peel, but unripe mandarin peel has also been used (Di Giacomo and Lo Presti, 1959). The production process normally consists of peel grinding into homogeneous pieces, and washing under water to remove most of the soluble solid; this is followed by a 1.0:1.5 blending with water and the addition of NaOH solution at pH 11.0–11.5. During this phase pH should not exceed the established one to avoid any breaking in the glycosidic bond. After about one hour at room temperature, insoluble solids will be separated and the liquid phase filtered. The filtered product pH is raised to 4.2–4.5 by means of mineral acids, heated at 40–45 °C and left for about 12–24 hours. Any hesperidin crystal is collected and dried. This procedure allows a flavonoidic complex with 60–70 per cent hesperidin content. Hesperidin having a >95 per cent purity rate can be obtained from repeated crystallization in alkaline solutions (Rispoli, 2000). A purification procedure of the extraction liquid using adsorbent styrene-divinylbenzene, non-ionic, resin has also been suggested; hesperidin is firstly fixed on resin, then eluted with 0.5 N NaOH solution containing (10 per cent) ethanol e finally acidified at pH 3.5. The final product purity rate is 86–98 per cent (Di Mauro *et al.*, 1999). To increase the scant solubility in water, hesperidin can be transformed into the corresponding methylchalcone or dihydrochalcone by means of alkaline treatment and subsequent methylation or hydrogenation.

Naringin production

The same procedure as hesperidin production applies but the greater solubility of naringin requires a lower pH (9–10) during the extraction phase. Grapefruit limed peel showed a high recovery rate (49 per cent) but the purity of product was the lowest (72 per cent) and its residue could not be reused. Hot water with yeast treatment gives the following advantages: higher purity of product (93 per cent); faster precipitation (in 2–3 days); higher recovery rate (33 per cent); and finally the possibility to recover further pectin from the residue (Wu *et al.*, 1991a). Using bergamot peel with a naringin content of 2.9 g/kg, extraction in hot water, followed by clearing and adsorption on polystirenic resin, produced a high purity product with a yield close to 90 per cent after under vacuum concentration and crystallization. The post-extraction peel is suitable for pectine extraction with no pre-treatment. (Calvarano *et al.*, 1996).

Naringin and neohesperidin dihydrochalcones have an edulcorating power exceeding by 300 and 1100 times, respectively, that of sucrose. Due to scarce availability of neohesperidin, the relative dihydrochalcon is derived from naringin; this latter is transformed, by alkaline degradation, into floroacetofenone-4'-β-neohesperidoside which, in turn, is transformed into neohesperidin by Claisen-Schimdt condensation with isovanillin. Pilot plant tests (Wu *et al.*, 1991b) of naringin transformation into neohesperidin showed a 16 per cent yield and a 93 per cent purity rate, while the corresponding dihydrochalcon was obtained with a 90 per cent yield and a 94 per cent purity rate. Due to the rising demand for natural colouring, the possibility to apply similar procedures for blood orange anthocyanin extraction has been considered (Timberlake and Henry, 1988). Recovery from processing residues was also proposed (Calvarano *et al.*, 1995). However, the lower anthocyanin content of blood oranges does not make such processing commercially valuable.

BRINED AND CANDIED PEEL

The citrus peel, most commonly used for candying, comes from sweet and bitter oranges, lemon, citron and grapefruit. Peel is selected in relation to its thick, possibly turgid albedo which should easily separate from the carpellar membranes. Separation used to be hand made but, nowadays, it is done mechanically during juice extraction. Depending on the targeted finished product, the peel is cut in cubes or strips. Candying operation consists of the absorption on the peel (previously washed and softened) of an increasingly concentrated solution of sucrose and glucose. Candied peel should appear translucent, turgid, doughy, no mushy and homogeneous, with dry, not-cracked surface, a regular colouring, a sweet, not off-flavours or cooked flavour; all these features should remain constant keeping the product at room temperature. The best procedure implies using fresh fruit. However practical reasons such as fruit seasoning, inconstant quality of raw material, candying duration, and the eventual candying of other products as pumpkins, watermelons, cherries, peaches, flowers, etc., account for the frequent use of previously preserved peel. Brined preservation of citrus peel is one of the most widespread. The peel is plunged in a 5–7 per cent sodium chloride solution for 4–5 days. The solution is then drained and peel placed in appropriate vessels, packed in layers with salt and kept in fresh brine for about two months. The brine is then replaced with a new more diluted solution and, most of the times with the addition of a 500–1000 ppm sulphur dioxide solution. During the first phases of the brine processing, a lactic fermentation occurs which will proceed throughout the following phases. The fermentation soften the cellulose parts which enables osmotic changes between the peel and the brine. This treatment allows a translucent, soft, no mushy peel with a good turgidity that can be further improved by adding a modest quantity of calcium hydrate. A further system, which avoids the fermentation, implies that the peel is repeatedly treated with calcium hydrate and with increasingly concentrated sodium chloride solution (2, 4, 6, 8, 10 per cent), each treatment lasting for about one hour at 45 °C. This procedure allows a better keeping of the fruit albedo after desalination (Crupi, 2000). The raw material obtained from one of the above procedures is sent to candied fruit processing industries where more or less rapid systems are used all based on peel desalination in cold and/or hot water. This phase, called cooking (softening), is one of the most delicate of the entire processing. Its purpose is to enable osmotic exchanges during syrup preparation, while extending the tissues and causing fruit softening and oxidase inactivation. Slow candying process, not as common anymore, envisage peel washing and cooking followed by the arrangement in pools or small baskets and plunging in syrup which flows countercurrent from one pool to the other. During this sequence the syrup concentration will decrease, until when, in the last section the syrup is drawn, re-concentrated and recycled. The whole processing cycle will take 10–15 days. More rapid methods are performed under vacuum conditions and at lower than boiling temperatures. This allows a shorter time for candying processing (18–24 hours) as the syrup concentration is maintained and the sugar penetration in tissues is made easier by the interstice dilation. When the vacuum is interrupted the pressure increase will push a greater quantity of syrup in the peel. The above procedures are repeated and the syrups are gradually raised from 50° to 75° Bx either by adding extra solutions or, by widening the vacuum, through concentration in the same candying vessel. All the procedures of washing desalination, candying and finishing can be carried out in the same device. This generally consists of baskets set in controlled under vacuum autoclaves. The product is placed in the baskets and the syrup, which is heated by an external heat exchanger, flows on it. After candying, the product is drained and centrifuged (Decio and Lodigiani, 1990). A further system (Swisher and Swisher, 1977) envisages about one hour boiling of softened and dried peel in sugar solutions (sucrose), at increasingly higher concentration up to 50° Bx, and with a slight use of citric acid. After 20 hours standing, the product is boiled and concentrated until the temperature of 104 °C is reached. Candied peel is hot drained, dried and covered with maize starch or powered sugar. The remaining humidity can be eliminated drying the product in hot air.

Nowadays, candied peel production has been, mostly, taken over by bakery and ice-cream industries.

COMMINUTED

The term comminuted refers to the product obtained by grinding and homogenizing the whole citrus fruit or blends of different parts of the fruits components (albedo, flavedo, pulp, single strength or concentrated juice, essential oil), in proportions that depend on the targeted final product.

Comminuted is almost exclusively produced from oranges and lemons and it is characterized by a suitable colloidal density (thick and stable cloud), a brilliant colour due to flavedo pigments, a good flavouring supplied by the fairly good content of essential oil and, if appropriately kept, a long shelf life guaranteed by the high content in ascorbic acid and other antioxidant agents present in the peel which mainly consist of polyphenols and tocopherols. The product is therefore particularly suitable for soft drinks as it avoids additive addition. Comminuted is almost exclusively sold in some European countries where its use in citrus drinks and squashes is allowed. It can be either single strength or concentrated up to three times. Production procedure requires careful fruit selection to discard even slightly defective fruit; after fruit washing, juice and a part of the essential oil are extracted, this latter depending on the targeted final content. Separated peels are ground in a mill, using the previously extracted and refined juice as a vehicle; especially in the case of lemon comminuted, seeds or parts of them should be discarded prior to grinding, as they determine a bitter taste and would be easily noticeable in the final product. The remaining formula components are then added to the ground peels and the blending is homogenized by a carborundum mole colloidal mill until a dough whose suspended particles will not exceed a few micron in diameter is obtained. De-aeration and pasteurization can be performed between 80–98 °C for 45–90 seconds, depending on the product pH. To obtain a more homogeneous product and to disperse and emulsify the eventually added essential oil, an additional homogenization might be necessary. The product is canned or packed in aseptic drums, otherwise added with preservatives such as sulphur dioxide and/or benzoic acid or its derivatives. Storage is normally at room or lower temperatures between 0 and 5 °C ; in the case of orange comminuted it has been remarked (Saénz et al., 1982) that, after four months, the lower temperature allows a better keeping of vitamin C and carotenes, while reducing off flavours. A 3:1 lemon comminuted kept at 10 °C showed good colour and fluidity, while the product kept at +27 °C was dark and jelly; moreover aromatic fraction showed more remarkable variations in the oxygenating components of the product kept at higher temperature (Calvarano and Di Giacomo, 1984). The atomization-drying process of orange comminuted showed the influence of two variables–temperature of inlet air and product feeding–on some quality parameters as humidity, carotenoids and vitamin C content. Higher quality with lower humidity (1 per cent) could be obtained if the temperature of the air outgoing from the dryer ranged between 100 and 110 °C (Welti and Lafuente, 1983).

JUICE CELLS AND STANDARD PULP

The presence of floating, big pulp in citrus juices is greatly appreciated by consumers. Such a presence gives a sensation of 'naturalness' improving the product flavour and providing useful fibres. The usual technological processes are, however, incompatible with consistent insoluble particles. It is therefore necessary to draw large cells from the processing juice and to re-add them during product packaging. The various production systems are conceived so as to reduce the juice extractor breaking of large cells naturally present in fruit. Extractors operating with rotating reamers, as a few Brown models and so-called 'birillatrici', do not require any modification. When FMC extractors are used, on the contrary, standard streiner tubes are appropriately replaced with others having larger holes (0.062″–0.090″) or vertical, window-shaped slots. Refining, anyway, will require the use of wide-holed paddle finishers to allow the passage of juice, cells and fine pulps. Two hydrociclons in series will discharge the defects (broken or embryonal seeds); a narrow holed paddle finisher (0.015″–0.040″) will separate the juice from the large cells to be stabilized in tubular heat exchanger with corrugated tubes or scraped surface, at 90 °C for 60–70 sec. Large cells are then freed from the excessive juice by means of a narrow holed finisher and packed into 20 kg cartons or 180 kg barrels destined to refrigeration, in case of aseptic packaging, or to freezing. In this latter case and especially with 180 kg drums packaging, longer freezing and thawing times are required, which may result in the development of off colours, off flavours, and/or spoilage (Johnson, 1987; Kimball, 1991; Kolb, 1993). The importance of heat on cell stability should be stressed. Carbohydrates, pectin and crude fibres are the major constituents of washed juice cells. The hemicellulose and cellulose portions of crude fibre provide structural support to the tissues which make up individual cells. Heat affects the complex carbohydrates that bind these supporting elements, and if the temperature is too high, they can be disorganized resulting in a soft pulp mass which can be easily broken up into small pieces (Johnson, 1987).

Using standard juice extractors and finishers and applying the same procedure as juice cells production, juice pulp will be obtained. This is smaller and softer than cells and provides 'body' to the product, although the produced sediments might result in an unwanted appearance.

CITRUS SECTIONS

The procedure for citrus section production envisages fruit peeling, fraction separation and carpellary membrane removal. Using traditional methods, steam, boiling water and lye solution are applied (Braddock, 1999). After separation from the peel, membranes can be removed with a chemical treatment in a hydrochloridic acid diluted solution (0.6–0.7 per cent) for 40 seconds at 30–35 °C, followed by quick washing under water, passage in a sodium hydroxide diluted solution (0.4–0.5 per cent) for 15 seconds at 35 °C and a final rinsing. This required labour-intensive process and is a drawback. To reduce production costs and facilitate membrane and peel removal, alternative systems, based on pectolitic enzymes, have been proposed (Janser, 1996). First, citrus fruits undergo a piercing treatment which allows dipping into the superficial flavedo and penetration of the enzymatic solution. The pectolitic treatment is carried out under vacuum conditions, at determined pH, temperature and duration, depending on the enzymatic activity; peel is manually removed under a water jet. The following step is the separation from the carpellary membranes. This is obtained with a treatment similar to the previous one, and the resulting sections are cleaned in water. Sections can be packed and commercialized as fresh, frozen, canned or dry products (without a covering liquid). When refrigerated at 1–2 °C, the fractions keep their organoleptical features for about three weeks (Baker and Gromann, 1995). Considerable liquid could be discharged during shelf-life. The drawback can be avoided by covering the segments with a wax micro-emulsion (Baker and Hagenmaier, 1997).

FERMENTATION PRODUCTS

Due to their composition, citrus processing discharges and waste water are usable in fermentation processes. One of the most exploited industrial usage is the alcohol production from pressing liquid (molasses) obtained from peel drying process. Press liquor concentrated at 72° Bx was used in the past while, nowadays, more diluted products are chosen, despite their higher content in essential oils which normally inhibits the yeast growth (Bradoock, 1999). Beer with about 5 per cent alcohol content is obtained from fermentation. The beer is first distilled and then rectified so as to obtain alcohol at 96° destined to the liquor industry. The quantities produced are limited however since, nowadays, most of the molasses are added to the drying peel to improve the nutritional value of this latter (Merrit and Fox, 1996). A weaker rectification will produce a 93° product, which is employed as transport fuel in countries where this is legal. This possibility is a hedge against the downfall of the citrus pellets on the international market (Schlottfeldt, 1991). Citric acid can also be obtained from press liquor, through fermentation with *Aspergillus niger* with yields comparable to those obtained from other raw materials such as beet or cane molasses (Aravantinos-Zafaris *et al.*, 1994). The use of orange peel to produce single cell proteins (SCP) suitable as a feedstuff has also been described. A promising process is to submit the peel to a mild acid pre-treatment to solubilise hemicellulose and pectin and carry out fermentation exclusively on the liquid. By using G*eotrichum candidum*, yields are about 14 per cent of SCP and 50 per cent of solid residue to be used as feedstuff for ruminants or to obtain hesperidin in larger quantities and at lower costs than using traditional methods (Vaccarino *et al.*, 1989; Lo Curto *et al.*, 1992).

Production of biogas

Biogas production from citrus processing waste water has already been performed in a number of plants for some time (Szendrey, 1989; Nonino and Ribeiro, 1991). The possibility to apply the same procedure to orange peels was evaluated (Lane, 1984). However, this technology was considered not feasible for the necessity to reduce drastically–up to 96 per cent–the content of essential oil, which strongly inactivates the fermentation bacteria, and depends on fruit seasoning. Nowadays, new peel-storage systems in out of season periods combined with appropriate distillation techniques and an optimal combination of biochemical parameters, make the peel-derived biogas production profitable, especially, when the biogas is used to generate valuable energy (Ramnensee and Agiorgitis, 1999).

REFERENCES

Adami, A., Carlini, D., Di Giacomo, G., Mentuccia, L. and Postorino, E. (2000) Deamarizzazione dei succhi di arancia mediante resina adsorbente. *Essenz. Deriv. Agrum.*, 70, 101–104.

Aravantinos-Zafaris, G., Tzia, C., Oreopoulou, V. and Thomopoulos, C. (1994) Fermentation of orange processing wastes citric acid production. *J. Sci. Food Agric.*, 65, 117–120.

Askar, A. and Treptow, H. (1997) Tropical fruit processing waste management. Part I: waste reduction and utilization. *Fruit Process.*, 7, 354–359.

Attaway, J.A. (1994) Citrus juice flavonoids with anticarcinogenic and antitumor properties. In M.T. Huang, T. Osawa, C.T. Ho and R.T. Rosen (eds), *Food Phytochemicals for Cancer Prevention I*, ACS Symp. Ser. 546, Washington, USA, pp. 240–248.

Baker, R.A. and Gromann, K. (1995) Enzyme applications in citrus processing. *Fruit Process.*, 5, 332–335.

Baker, R.A. and Hagenmaier, R.D. (1997) Reduction of fluid loss from grapefruit segments whith wax microemulsion coatings. *J. Food Sci.*, 62, 789–792.

Ballantine, P.L. and Ferguson, R.R. (1985) Water extracted soluble orange solids: pulp wash. In FSHN Dept., Univ. Florida Gainesville (eds), *The 25th Annual Short Course for the Food Industry*, FL, USA, pp. 233–247.

Benavente-Garcia, O., Castillo, J., Marin, F.R., Ortuno, A. and Del Rio, J.A. (1977) Uses and properties of citrus flavonoids. *J. Agric. Food Chem.*, 45, 4505–4515.

Bocco, A., Cuvelier, M.-E., Richard, H. and Berset, C. (1998) Antioxidant activity and phenolic composition of citrus peel and seed extracts. *J. Agric. Food Chem.*, 46, 2123–2129.

Braddock, R.J. (1999) *Handboock of Citrus By-products and Processing Technology*, John Wiley & Sons, Inc., New York, USA.

Calvarano, M. and Di Giacomo, G. (1984) El 'comminuted' (tridurado de citrus): un producto discutibile. *Essenz. Deriv. Agrum.*, 54, 200–219.

Calvarano, M., Calabrò, G., Postorino, E., Gionfriddo, F., Calvarano, I. and Bovalo, F. (1996) Naringin extraction from exhausted bergamot peels. *Essenz. Deriv. Agrum.*, 66, 126–135.

Calvarano, M., Postorino, E., Calvarano, I. and Gionfriddo, F. (1995) Recupero di antocianine dai residui delle lavorazioni della arance pigmentate. *Essenz. Deriv. Agrum.*, 65, 557–566.

Caratozzolo, C. (1998). Personal communication.

Chen, J., Montanari, A.M. and Widmer, W.W. (1997) Two new polymethoxilated flavones, a class of compounds with potential anticancer activity, isolated from cold-pressed Dancy tangerine peel oil solids. *J. Agric. Food Chem.*, 45, 364–368.

Chen, J.H. and Ho, C.-T. (1977) Antioxidant activities of caffeic acid and its related hydroxycinnamic acids compounds. *J. Agric. Food Chem.*, 45, 2374–2378.

Crupi, F. (2000) Personal communication.

Decio, P. and Lodigiani, G. (1991) Tecnologia di canditura. *Tecnologie Alimentari*, 6, 82–88.

Di Giacomo, A. (1987) *Sulla Trasformazione Industriale degli Agrumi*. Stazione Sperimentale Industria Essenze Derivati Agrumari, Reggio Calabria, Italy.

Di Giacomo, A. and Lo Presti, V. (1959) Ricerca per le condizioni più favorevoli per l'estrazione dell'esperidina dai mandarini. *Essenz. Deriv. Agrum.*, 29, 57–59.

Di Giacomo, A., Rapisarda, P. and Safina G. (1992) *L'industria dei Derivati Agrumari.*, Stazione Sperimentale Industria Essenze Derivati Agrumari, Reggio Calabria, Italy.

Di Mauro, A., Fallico, B., Passerini A., Rapisarda, P. and Maccarone, E. (1999) Recovery of hesperidin from orange peel by concentration of extracts on styrene-divinylbenzene resin. *J. Agric. Food Chem.*, 47, 4391–4397.

EC (1993) Concil Directive 93/77 ECC of 21 September 1993 relating to fruit juices and certain similar products. *Official Journal of the European Communities* 30.9.93, No L244/23-31.

FDA (1993) U.S. Food and Drug Administration, CFR 21 § 146.146.

Fernandez de Simon, B., Perez-Ilzarbe, J., Hernandez, T., Gomez-Cordoves, C. and Estrella, I. (1992) Importance of phenolic compounds for the characterization of fruit juices. *J. Agric. Food Chem.*, 40, 1531–1535.

Florida Department of Citrus (1998) Water extracted soluble fruit solids. *Official Rules Affecting the Florida Citrus Industry*, Ch. 20–64.021, March 19, 19–21.

Grandi, R., Trifiro, A., Gherardi, S., Calza, M. and Saccani, G. (1994) Characterization of lemon juice on the basis of flavonoid content, *Fruit Process.*, 4, 335–359.

Grohmann, K., Manthhey, A., Cameron, R.G. and Busling, B.S. (1999) Purification of citrus peel juice and molasses, *J. Agric. Food Chem.*, 47, 4859–4867.

Hammond, D.A. (1996) Authenticity of fruit juices, jams and preserves. In P.R. Ashurst and M.J. Dennis (eds), *Food Authentication*, Chapman & Hall, London, UK, pp. 32–36.

Horowitz, R.M. and Gentili, B. (1977) Flavonoids constituen of citrus. In S. Nagy, P.E. Shaw and M.K. Velduis (eds), *Citrus Science and Technology*, AVI Publ. Comp., Inc., Westport, Co, USA, Vol. I, pp. 397–426.

Janser, E. (1996) Enzymatic peeling of fruit. *Fruit Process.*, 6, 92–95.

Jensen, A.S. (1990) Neue Erfahrungen mit dem DDS-Verdampfungstrockner. *Zuckerindustrie*, 115, 827–833.

Johnson, T.M. (1987) Citrus pulp recovery. *Citrus Industry Magazine*, 68, 36–39.

Johnson, R.L., Htoon, A.K. and Shaw, K.J. (1995) Detection of orange peel extract in orange juice. *Food Technology Australia*, 47, 426–432.

Johnston, R.B. (1998) Pumped peel–five years later. *Trans. Citrus Eng. Conf.*, 44, 79–90.

Kawai, S., Tomono, Y., Katase, E., Ogawa, K. and Yano, M. (1999) HL-60 differentianting activity and flavonoid content of readily extractable fraction prepared from citrus juices. *J. Agric. Food Chem.*, 47, 128–135.

Kesterson, J.W. and Braddock, R.J. (1976) Dried citrus pulp. In *By-products and Speciality Products of Florida Citrus*, Fla. Agr. Exp. Sta. Tech. Bull. No. 784, University of Florida, Gainesville, FL, USA, pp. 4–25.

Kesterson, J.W., Crandall, P.G. and Braddock, R.J. (1979) The heat of combustion of dried citrus pulp. *J. Food Process Eng.*, 3, 1–5.

Kimball, D.A. (1991) *Citrus Processing Quality Control and Technology*, Van Nostrand Reinhold, New York.

Klavons, J.A., Bennet, R.D. and Vannier, S.H. (1991) Nature of protein constituent of commercial orange juice cloud. *J. Agric. Food Chem.*, 39, 1545–1548.

Kolb, E. (1993) Citrus juice finisher pulp processing and its use in fruit juices. *Fruit Process*, 3, 6–8.

Lane, A.G. (1984) Anaerobic digestion of orange peel, *Food Technology Australia*, 36, 125–127.

Leuzzi, U. and Licandro, G. (1997) Ion chromatography as a rapid method to distinguish orange juices from second pressure concentrates. *Ital. J. Food Sci.*, 9, 313–322.

Leuzzi, U., Caristi, C., Panzera, V. and Licandro, G. (2000) Flavonoids in pigmented orange juice and second-pressure extracts. *J. Agric. Food Chem.*, 48, 5501–5506.

Lo Curto, R., Tripodo, M.M., Leuzzi, U., Giuffrè, D. and Vaccarino, C. (1992) Flavonoids recovery and SCP production from orange peel. *Bioressource Technology*, 42, 83–87.

Louche, L.M.-M., Gaydou, E.M. and Lesage, J.C. (1998) Determination of phlorin as peel marker in orange (*Citrus sinensis*) fruits and juices. *J. Agric. Food Chem.*, 46, 4193–4187.

Maccarone, E., Maccarone, A. and Rapisarda, P. (1985) Acylated anthocyanins from oranges. *Annali di Chimica*, 75, 79–86.

Marques, D.S. (1995) Short time reaction system for improved citrus peel processing. *Trans. Citrus Eng. Conf.*, 41, 92–105.

Merrit, G. and Fox, K. (1996) Technologies for adding profits to citrus processors. *Fruit Process.*, 6, 96–98.

Miyake, Y., Yamamoto, K., Morimitsu, Y. and Osawa, T. (1997) Isolation of C-glucosylflavone from lemon peel and antioxidative activity of flavonoid compounds in lemon fruit. *J. Agric. Food Chem.*, 45, 4619–4623.

Morales, A. (1998) Water extraction operation in citrus processing. *Fruit process.*, 8, 146–151.

Mouly, P., Gaydou, E.M., Faure, R. and Estienne, J.M. (1997) Blood orange juice authentication using cinnamic acid derivates. Variety differentiations associated with flavanone glycosides content. *J. Agric. Food Chem.*, 45, 373–377.

Mouly, P., Gaydou, E.M. and Auffray, A. (1998) Simultaneus separation of flavanone glycosides and polymethoxylated flavones in citrus juices using liquid chromatography. *J. Chromatogr. A*, 800, 171–179.

Nonino, E.A. and Ribeiro, R.E. (1991) Enviromental control in the Brazilian citrus industry. *Report of XI International Congress of Fruit Juice*, Sào Paulo, 17–21 November 1991.

Nonino, E.A. (1996) Orange juice quality upgrading. *Diffusion Process for Citrus Juice*, Interlaken, 20 May 1996. Bruxelles.

Odio, C.E. (1993) Tank reaction system for citrus peel. *Trans. Citrus Eng. Conf.*, 39, 1–14.

Ooghe, W.C. and Detavernier, C.M. (1997) Detection of the addition of *Citrus reticulata* and hybrids to *Citrus sinensis* by flavonoids. *J. Agric. Food Chem.*, 45, 1633–1637.

Ooghe, W.C. and Detavernier, C.M. (1999) Flavonoids as authenticity markers for *Citrus sinensis* juice. *Fruit Process.*, 8, 308–313.

Postorino, E., Gionfriddo, F. and Di Giacomo, A. (1998) Il succo di arancia dolce di seconda pressione. *Essenz. Deriv. Agrum.*, 68, 12–37.

Postorino, E. and Gionfriddo, F. (1999) I flavanoni glucosidici dei succhi di arancia italiani. *Essenz. Deriv. Agrum.*, **69**, 149–158.

Pupin, A.M., Dennis, M.J. and Toledo, M.C.F. (1998) Polymethoxylated flavones in Brazilian orange juice. *Food Chem.*, **63**, 513–518.

Ramnensee, W. and Agiorgitis, G. (1999) Verfahren zur Energiegewinnung aus Citrus Abfaellen. *Deutsches Patent* De10020832.0.

Rapisarda, P., Carollo, G., Fallico, B., Tomaselli, F. and Maccarone, E. (1998) Hydroxycinnamic acids as markers of italian blood orange juices. *J. Agric. Food Chem.*, **46**, 464–470.

Rapisarda, P., Tomaino, A., Lo Cascio, R., Bonina, F., De Pasquale, A. and Saija, A. (1999) Antioxidant effectiveness as influenced by phenolic content of fresh orange juices. *J. Agric. Food Chem.*, **47**, 4718–4723.

Reminiac, C.C., Bourrier, M.J. and Goiffon, J.P. (1989) Différenciation des produits de l'orange et de l'orange amère par analyse chromatographique des hétèrosides de flavanones. *Ann. Fals. Exp. Chim.*, **82**, 471–479.

Rispoli, G. (2000) Personal communication.

Rispoli, G. and Di Giacomo, A. (1962) Le pectine a basso tenore metossilico. Nota seconda: caratteristiche principali della gelatinizzazione, *Boll. Lab. Chim. Prov.*, **13**, 122–128.

Robards, K., Li, X., Antolovch, M. and Boyd, S. (1997) Characterisation of citrus by chromatographic analysis of flavonoids. *J. Sci. Food Agric.*, **75**, 87–101.

Sàenz, C., Gasque, F., Nadal, M.I. and Lafuente, B. (1982) Tridurados de naranja- III. Estabilidad de los tridurados de naranja ('comminuted') durante el almacenamiento. *Rev. Agroquìm. Tecnol. Aliment.*, **22**, 403–412.

Schlottfeldt, G. (1991) Energy by-products: energy generation from sugar cane bagasse; alcohol production from citrus press liquor. *Report of XI International Congress of Fruit Juice*, São Paulo, 17–21 November 1991, IFU, Paris, pp. 253–260.

Shomer, I. (1991) Protein coagulation cloud in citrus fruit extract 1 Formation of coagulates and their bound pectin and neutarl sugars. *J. Agric. Food Chem.*, **39**, 2263–2266.

Shomer, I., Vasiliver, R. and Salomon, R. (1991) Protein coagulation cloud in citrus fruit extract 2. Structural characterization of coagulates. *J. Agric. Food Chem.*, **39**, 2267–2273.

Sinatra, I. (1999) Personal communication.

Sinclair, W.B. (1984) Dried citrus pulp. In *The Biochemistry and Physiology of the Lemon and Others Citrus Fruits*, University of California press, Oakland, CA, USA, pp. 764–772.

Swisher, H.E. and Swisher, L.H. (1977) Specialty citrus products. In S. Nagy, P.E. Shaw and M.K. Velduis (eds), *Citrus Science and Technology*, AVI Publ. Comp., Inc., Westport, Co, USA, Vol. II, pp. 290–345.

Szendrey, L.M. (1990) The anaerobic treatment of food and citrus processing wastewaters. *Trans. Citrus Eng. Conf.*, **36**, 40–54.

Timberlake, C.F. and Henry, B.S. (1988). Anthocyanins as natural food colorants. In V. Cody, E. Middleton, J.B. Harborne and A. Beretz (eds), *Plant Flavonoids in Biology and Medicine II Biochemical, Cellular and Medicinal Properties*, Alan, R. Liss, Inc., New York, NY pp. 107–121.

Vaccarino, C., Lo Curto, R., Tripodo, R.R., Patanè, R., Laganà, G. and Ragno, A. (1989) SCP from orange peel by fermentation with fungi acid treated peel. *Biol. Wastes*, **30**, 1–10.

Welti, J.S. and Lafuente, B. (1983) Secado por atomizaciòn de disgregados (-comminuted-) de naranja. I. Effecto de la temperatura del aire y del caudal de alimentaciòn del producto sobre su calidad. *Rev. Agroquìm. Tecnol. Aliment.*, **23**, 97–106.

Wu, H., Calvarano, M. and Di Giacomo, A. (1991a) Improvementes of extracting naringin from grapefruit peel. *Essenz. Deriv. Agrum.*, **65**, 187–191.

Wu, H., Calvarano, M. and Di Giacomo, A. (1991b) Sulla preparazione della naringina diidrocalcone e della neoesperidina diidrocalcone in un impianto pilota. *Essenz. Deriv. Agrum.*, **65**, 56–60.

12 Advanced analytical techniques for the study of citrus oils

Luigi Mondello, Giovanni Zappia, Paola Dugo and Giovanni Dugo

INTRODUCTION

Citrus essential oils obtained from the peel of fruit are used in the food and perfume industries. They are mixtures of more than 200 components (Shaw, 1979), that can be grouped essentially into two fractions:

- a volatile fraction, that constitutes 85–99 per cent of the whole oil, and contains the monoterpenes and sesquiterpenes hydrocarbons and their oxygenated derivatives along with aliphatic aldehydes, alcohols and esters;
- a non volatile residue, that ranges from 1 per cent to 15 per cent of the whole oil, and contains hydrocarbons, fatty acids, sterols, carotenoids, waxes, coumarins, psoralens and flavanoids.

The analysis of the components of the volatile fraction of an essential oil is characterized by the complexity of the separation of the components which belong to different classes of compounds and are present in a wide range of concentrations. Many components cannot be resolved in a single GC analysis and the best approach is to fractionate the essential oil before the gas chromatographic analysis. High resolution GC, with conventional stationary phases, is the technique which best helps the analyst for the characterization of the citrus oils (Dugo *et al.*, 1994a). The information obtained with the GC analysis of the volatile fraction of the oils can be sufficient to determine whether the product is genuine or not, and sometimes, when the product is adulterated, the kind and the level of adulteration can be detected. Sophisticated adulterations make the oil very similar to a genuine one. In such cases, the addition of natural and/or synthetic products (of lower price) are used to get an economic profit while attempting to maintain the qualitative or even the quantitative composition of the natural oil, making the detection of the adulteration very difficult. Since the volatile fraction of essential oils are complex mixture, quality control analysis is usually carried out on very long columns with 0.32 or 0.25 mm i.d. and slow temperature program (Dugo *et al.*, 1994a, 1999a).

Sometimes, to improve our knowledge of the composition of essential oils, the analysis of some components of the non-volatile residue can be also carried out. Oxygen heterocyclic compounds, that are coumarins, psoralens and polymethoxylated flavones, are present in all the cold-pressed citrus oils. Because of their structural diversity and diverse occurrence in different citrus peel oils, oxygen heterocyclic compounds may have an important role in identification of the various oils and also in their quality and authenticity control (Dugo *et al.*, 1997). Obviously, distilled oils, that have a considerably lower value than cold-pressed oils, do not contain oxygen heterocyclic compounds. The analysis of these compounds is usually carried out by normal or reversed phase HPLC.

This chapter reports the use of different multidimensional techniques that are HPLC-HRGC/MS, HRGC-HRGC and some innovative techniques such as HRGC/MS with Linear Retention Indices, and Fast GC. These techniques were successfully applied to the analysis of citrus essential oils.

ON-LINE HPLC-HRGC AND HPLC-HRGC/MS

High performance liquid chromatography (HPLC) coupled to high resolution gas chromatography (HRGC) is one of the most current powerful analytical techniques because of its selectivity and sensitivity in the analysis of complex mixtures. On-line coupling (Munari *et al.*, 1990; Dugo *et al.*, 1994b,c; Mondello *et al.*, 1994a,b,c, 1995a, 1996a,b, 1998a) permits the separation and identification of compounds of the same polarity in mixtures of compounds of different polarity even when the concentrations of the various classes of compounds are considerably different. Moreover, there is no sample pre- or post-treatment as the separation analysis is fully automated. In comparison with off-line methods, on-line high performance liquid chromatography-high resolution gas chromatography (HPLC-HRGC) offers some advantages: the amount of sample required is less, no sample work-up is needed, and very complex sample pre-treatment is possible in a fully automated way. In on-line HPLC-HRGC, the sample is first separated by HPLC using a single column or a combination of columns to isolate the components of interest and then to directly transfer them to a capillary column where a further separation is carried out. Using an automated HPLC-HRGC system the analysis with the so-called retention gap transfer technique is possible. This technique is based on the mechanical stabilisation of the liquid using an uncoated pre-column at the entrance of the GC system until all the solvent is evaporated. The retention gap method (Grob, 1987a) represents the best approach in the case of qualitative and quantitative analysis of samples containing highly volatile compounds. In fact, retention gap allows analysis of substances eluting immediately after the solvent peak, due to the reconcentration of those components by the so-called *solvent effects* (primarily *solvent trapping*) (Grob, 1983). On the other hand, this method is restricted to fractions of only modest volumes, and the use of a long uncoated precolumn. Working under conditions which still produce a zone flooded by the eluent (providing solvent trapping), but which cause a large amount of eluent to evaporate during its introduction, we are able to work with a short uncoated pre-column or to transfer larger fraction volumes. This method is the so-called *partially concurrent evaporation* (Grob, 1987b): part of the eluent is evaporated concurrently, that is, during its introduction into GC. The introduction of an early-vapour exit greatly improves partially concurrent evaporation and protects the GC detector. Figure 12.1 shows the on-column type interface, that allows the partially concurrent eluent evaporation described above.

The coupling of a mass spectrometer to an HPLC-HRGC instrument greatly enhances the detection capabilities of the system and allows components to be reliably identified. The HPLC pre-separation into classes of compounds eliminates interferences from co-eluted peaks and produces better resolved gas chromatograms. The mass spectra obtained with this system can be more easily interpreted than those obtained with a simple HRGC/MS system.

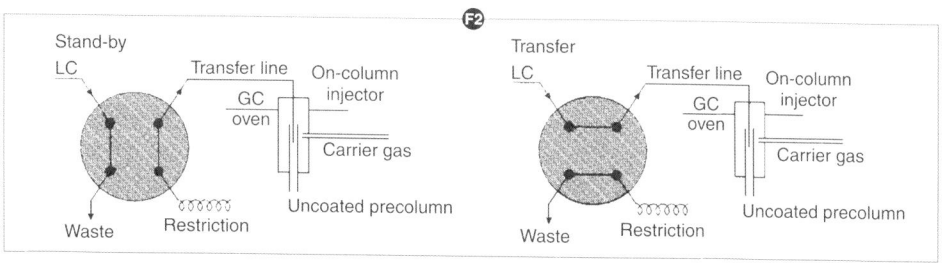

Figure 12.1 On-column interface. The eluent leaving the HPLC detector enters the valve and flows to waste when in the stand-by position. When the valve is switched on, the eluent is pumped through the transfer line into the inlet of the on-column injector. At the end of the transfer the valve is switched off, and the eluent again flows to waste.

Applications of HPLC-HRGC and HPLC-HRGC-MS techniques to the analysis of citrus oils have been developed (Dugo *et al.*, 1994b,c; Mondello *et al.*, 1994a,b,c, 1995a, 1996a,b, 1998a,g). Tables 12.1 and 12.2 report HPLC and GC conditions, respectively, of HPLC-HRGC analyses carried out on various citrus oils using a fully automated HPLC-HRGC instrument (DualChrom 3000 series, Fisons).

As can be seen, the technique has been used to analyze only selected components of the oil, such as aldehydes in sweet orange oil (Mondello *et al.*, 1994a). In fact, their identification and quantitative calculation is difficult because of the possible overlapping of peaks. In sweet orange oil, the aliphatic aldehydes content may furnish useful information concerning the origin of the oil and the ripeness of the fruit at harvest time. The HPLC-HRGC analysis offers a method for quantitative analysis of aldehydes free from interferences, even for those components present in trace amounts. Other applications include the fractionation of citrus oils into two fraction: F1, containing mono- and sesquiterpene hydrocarbons, and F2, containing oxygenated compounds. The two fractions obtained, are much simpler than the whole oil, and also chemically more omogeneous. These fractions give chromatograms in which all the peaks are better separated and thus easier to identify by retention times or MS. As an example, Figure 12.2 shows the total ion chromatogram (TIC) of the LC fractions of bitter orange petitgrain oil, together with on-column GC-chromatogram obtained with the same column system. The chromatogram of the whole oil shows overlap between the following pairs of peaks: 6-methyl-5-hepten-2-one and myrcene; 1,8 cineole and limonene; *trans*-linalool oxide and terpinolene; δ-elemene and an unknown oxygenated compound; α-copaene and citronellyl acetate; δ-elemene and geranyl acetate. HPLC pre-separation allowed the separation of all detectable components present in the oil.

On-line HPLC-HRGC has been also applied to the determination of the enantiomeric distribution of compounds present in small amount in citrus oils. Altogether, monoterpene alcohols represent about 0.5 per cent of the volatile fraction of lemon, mandarin, sweet and bitter orange oils, the determination of their enantiomeric distribution can give useful information about the authenticity, quality and origin of the essential oils. As can be seen from results reported in Table 12.3, the enantiomeric distribution of terpinen-4-ol permits to differentiate cold-pressed and distilled mandarin and lemon oils. Moreover, the very different ratio between (−) and (+) linalool in sweet and bitter

Table 12.1 HPLC conditions of HPLC-HRGC analysis (Mondello et al., 1996a)

	Time interval (min)	Eluents	Flow (μL/min)	Fraction	Transfer time (min)	Solvent vapour exit closed at (min)
Aldehydes of sweet orange oil	0–2	pentane	180			
	3–12	pentane:diethyl ether, 97:3		F_1^a	9.5–10.5	10.9
	13–17	pentane:diethyl ether, 90:10		F_2^a	13.0–14.3	15.0
	18–23	pentane:diethyl ether, 85:15		F_3^a	16.5–17.8	18.5
	back-flush with diethyl ether (1 mL)					
Monoterpene and sesquiterpene hydrocarbons of citrus oils	0–5	pentane	180		1.6–3.6	
	back-flush with diethyl ether (1 mL)					
Bergamot oil	0–3	pentane	180	F_1^b	1.0–3.0	3.8
	4–13	pentane:diethyl ether, 97:3		F_2^b	12.0–15.0	16.5
	14–19	pentane:diethyl ether, 90:10		F_3^b	17.0–19.0	19.9
	20–30	pentane:diethyl ether, 80:20		F_4^b	21.0–23.5	24.9
	back-flush with diethyl ether (1 mL)					
Neroli oil	0–5	pentane	180	F_1^a	1.5–3.5	4.0
				F_2^c	7.5–9.5	10.1
	back-flush with diethyl ether (1 mL)					
Enantiomeric distribution of linalool and terpinen-4-ol*	0–3	pentane	220	F_1^d	17.0–19.0	
	4–15	pentane:tert-butylmethyl ether, 95:5				
	16–30	pentane:tert-butylmethyl ether, 90:10				
	31–50	pentane:tert-butylmethyl ether, 70:30				
Petitgrain oils	0–5	pentane	180	F_1^c	1.5–3.0	
				F_2^c	5.5–7.0	
	back-flush with diethyl ether (1 mL)					

Notes

F_1^a, aliphatic aldehydes; F_2^a, monoterpene aldehydes; F_3^a, sesquiterpene aldehydes; F_1^b, mono- and sesquiterpene hydrocarbons; F_2^b, aliphatic aldehydes and esters; F_3^b and F_4^b, alcohols; F_1^c, mono- and sesquiterpene hydrocarbons; F_2^c, oxygenated compounds; F_1^d, linalool and terpinen-4-ol.

Table 12.2 GC conditions of HPLC-HRGC analysis (Mondello et al., 1996a)

	Separation column	Carrier	Column temperature	Eluent evaporation rate (μL/min)
Aldehydes of sweet orange oil	25 m × 0.32 mm i.d. fused silica, SE-52, 0.4–0.45 μm	He, 120 kPa	45 °C (6 min) to 220 °C at 3 °C/min	151
Monoterpene and sesquiterpene hydrocarbons of citrus oils	21 m × 0.25 mm i.d. fused silica, DB-5, 0.25 μm	He, 140 kPa 32 cm/s	45 °C (6 min) to 240 °C at 3 °C/min	138
Bergamot oil	21 m × 0.32 mm i.d. fused silica, SE-52, 0.4–0.45 μm	He, 120 kPa	45 °C (6 min) to 220 °C at 3 °C/min	145
Neroli oil	26 m × 0.32 mm i.d. fused silica, SE-52, 0.4–0.45 μm	He, 120 kPa 33 cm/s	45 °C (6 min) to 250 °C at 3 °C/min	150 (pentane) 139 (diethyl ether)
Enantiomeric distribution of linalool and terpinen-4-ol	25 m × 0.32 mm i.d. fused silica, 2, 3-dimethyl-6-pentyl-β-cyclodextrin 30% and OV-1701 70%, 0.25 μm	He, 120 kPa	45 °C (3 min) to 150 °C at 1 °C/min	155

Table 12.3 Enantiomeric distribution of linalool and terpinen-4-ol in citrus oils (Mondello et al., 1996a)

	(−) Linalool	(+) Linalool	(−) Terpinen-4-ol	(+) Terpinen-4-ol
Lemon oil (c.p.)	54	46	80	20
Lemon oil (distilled)	53	47	78	22
Mandarin oil (c.p.)	17	83	87	13
Mandarin oil (distilled)	17	83	73	27
Bitter orange oils (Italy)	83	17		
Sweet orange oils (Italy)	7	93		
Bitter orange oils (Spain)	82	18		
Bitter orange oils (Brazil)	68	32		
Bitter orange oils (Ivory Coast)	67	33		
Italian sweet orange oils				
Blonde oranges				
Biondo commune	8	92		
Navelina	9	91		
Washington navel	8	92		
Valencia late	9	91		
Ovale	5	95		
Red oranges				
Moro	6	94		
Tarocco	5	95		
Sanguinello	5	95		
Mandarin oils				
Mandarin oils (Italy)	16	84		
Mandarin oils (Uruguay)				
Malvasio	12	88		
Ellendale	14	86		
Ortanique	16	84		
Comun	21	79		
Malaquina	30	70		

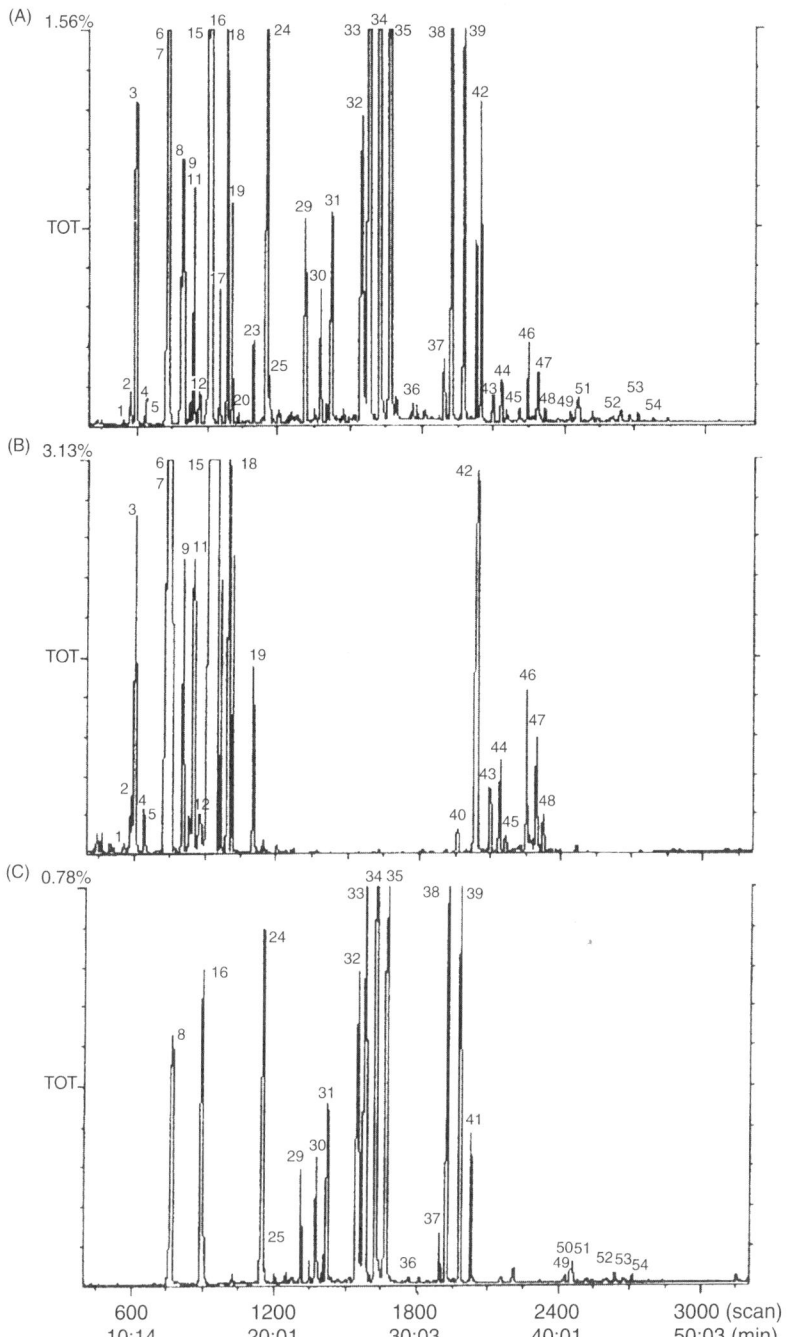

Figure 12.2 Total ion chromatogram of a bitter orange petitgrain oil (a) and of the F_1 (hydrocarbons) (b) and F_2 (oxygenated compounds) (c) fractions from its LC separation. Peak identification: (1) tricyclene; (2) α-thujene; (3) α-pinene; (4) camphene; (5) sabinene; (6) β-pinene; (7) 6-methyl-5-hepten-2-one; (8) myrcene; (9) α-phellandrene; (10) δ-3-carene; (11) α-terpinene;

orange oils allows the detection of contamination by as little as 5 per cent of sweet orange oils in the more valuable bitter orange oil. The enantiomeric ratio of (−)/(+)-linalool may also varies with the geographical origin of the oil (see values reported for genuine Italian and Spanish bitter orange oils, compared to those of genuine bitter orange oils from Brazil and Ivory Coast) and with the cultivar of the fruits (see values reported for Italian sweet orange oils and for Uruguayan mandarin oils of different cultivars).

The examples here reported show that on-line coupled HPLC-HRGC is an excellent method for the analysis of complex mixtures, such as citrus essential oils. The further coupling with a mass spectrometer enhances the power of the system, and permits a more reliable identification of components than a GC/MS system.

MULTIDIMENSIONAL GAS CHROMATOGRAPHY (MDGC)

The MDGC system permits the transfer of fractions containing one or more compounds of the same volatility belonging to different chemical classes. Such system permits several fractions to be transferred from the pre-column to the analytical column and analysed during the same GC run. This makes the GC-GC system more versatile than HPLC-HRGC system, which allows the transfer of compounds of the same polarity from the LC to the GC, even of more than one LC fraction, but each of them have to be analysed by GC only after the end of the previous GC run.

MDGC have been used for many years in many different ways. In particular, the transfer of one or more selected groups of compounds eluted from a GC column onto a second column is usually referred as 'heart cutting'. The main application of heart-cutting is the optimization of chromatographic resolution, in which unresolved sample components eluted from the first column are selectively diverted to a second column of higher efficiency or better selectivity. At the same time, it is possible to collect information on the retention time in two different stationary phases, which facilitates the identification of the compounds. Applications of MDGC to citrus essential oils mainly refer to the determination of the enantiomeric distribution of some components of the oils. Gas chromatography with chiral columns coated with derivatized cyclodextrins is the analytical technique most frequently employed for the determination of the enantiomeric ratio of volatile compounds. Depending on the complexity of the sample to be analyzed, preseparation procedures are sometimes needed to achieve a reliable stereodifferentiation especially of those components present in small amount. In particular, enantioselective MDGC with the combination of a non chiral precolumn and a chiral main column has been demonstrated to be a powerful method for the direct stereoanalysis of chiral volatiles without any further clean-up or derivatization procedures.

GC-GC heart-cutting has been used since mid-1950s (Simmons and Snyder, 1958; Deans *et al.*, 1971; Kaiser, 1971; Schomburg *et al.*, 1973; Deans, 1981; Martin and

---

(12) *p*-cymene; (13) limonene; (14) 1,8 cineole; (15) (Z)-β-ocimene; (16) (E)-β-ocimene; (17) γ-terpinene; (18) *cis*-linalool oxide; (19) terpinolene; (20) *trans*-linalool oxide; (21) linalool; (22) citronellol; (23) terpinen-4-ol; (24) α-terpineol; (25) nerol; (26) neral; (27) linalyl acetate; (28) geranial; (29) δ-elemene; (30) α-cubebene; (31) α-terpinyl acetate; (32) citronellyl acetate; (33) α-copaene; (34) neryl acetate; (35) geranyl acetate; (36) β-elemene; (37) *N*-methyl-methyl anthranilate; (38) (E)-caryophyllene.

Winters, 1963) either with two packed columns, a packed column and a capillary column, or two capillary columns.

In the early years, MDGC system regularly employed a rotary switching valve. Installation and operation of a mechanical valve seemed easy, but dead volume and absorption effects, limited maximum operating temperature, potential gas leakage or flow path plugging, and limited flexibility were the disadvantages of these systems. The use of off-line solenoid valves to alternate the line pressure at various times to direct or change the carrier gas flow direction overcame the drawbacks of multiport valves.

In 1968, Deans reported a valveless switching system that allowed no valves or moving parts to be in either the sample flow path or the higher temperature zone. This technique was based on a pressure balance between the two columns (Deans, 1981; Deans and Scott, 1973; Rijks et al., 1974; Bertsch et al., 1977; Jennings et al., 1977; Anderson et al., 1979; Brotell et al., 1982), which is made possible by in-line restrictors and the use of additional make-up gas. An improvement of this system was presented by Shomburg et al. (1982), in which the original valveless column connection of Deans's system was replaced by a 'live' switching system containing a special coupling unit. The two capillary columns were inserted over a thin platinum capillary, which was the central component of the coupling piece. Supplementary carrier gas was added trough two control lines and adjusted with needle valves. A decade after the introduction of the switching technique proposed by Deans (1968), a commercial instrument enabling solvent flushing, heart cutting, and backflushing was developed by Siemens. Many papers reporting the use of this instrument for the analysis of citrus oils can be found in the literature (Mosandl, 1995; Juchelka et al., 1996; Juchelka and Mosandl, 1996; Mosandl and Juchelka, 1997; Casabianca et al., 1995). A schematic diagram of this system is reported in Figure 12.3.

Mechanical valve-switching was described again by Jennings (1984) and reviewed by Gordon et al. (1985). Technological progress of valve design rendered miniaturized connectors available for the assembly of MDGC system, eliminating previously observed problems, such as adequate thermal stability and memory effects. In 1998, an MDGC system for multitransfer purposes based on a high temperature valve to heart-cut fractions from the first capillary column to a second capillary column with a hot transfer line and a system to maintain a constant flow during the transfer has been

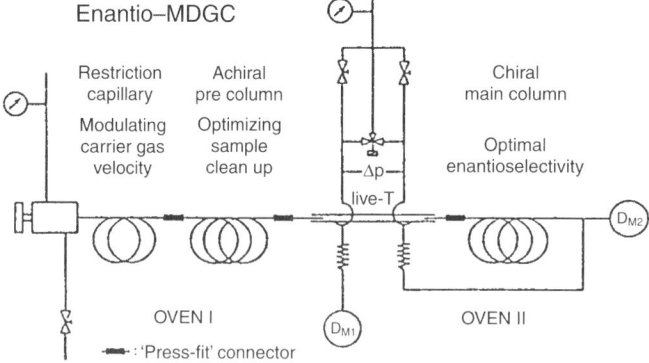

Figure 12.3 Schematic diagram of enantioselective multidimensional gas chromatography.

developed (Mondello *et al.*, 1998b). This system and its subsequent versions (Mondello *et al.*, 1998c,d,e,f,g, 1999), shown in Figure 12.4, allows the multitransfer of different fractions during the same GC analysis (cut position), and the use of the two GC independently when the multitransfer option is not used (stand-by position). Pneumatic and electronic circuits permit to maintain constant retention times in the pre-column even for those components eluted after numerous transfers. This system has been used to determine the enantiomeric distribution of β-pinene, sabinene, limonene, linalool, terpinen-4-ol and α-terpineol in citrus oils. Linalyl acetate was also determined in bergamot oils (Mondello *et al.*, 1997, 1998b,c,d,e,f, 1999; Dugo *et al.*, 2001). The same MDGC system has been used in combination with a mass spectrometer detector (MDGC-MS) (Mondello *et al.*, 1998c). The analysis of a bergamot oil using a nonpolar precolumn (SE-52) and a polar main column (carbowax 20M) has been carried out. Figure 12.5 shows the chromatogram of the oil on the SE-52 column. The zone of the chromatogram that has been transferred onto the Carbowax column is shown (around 23 min). The chromatogram of this fraction on the Carbowax 20M column, acquired with the MS detector, is shown in the upper part of the figure. As can be seen, after the

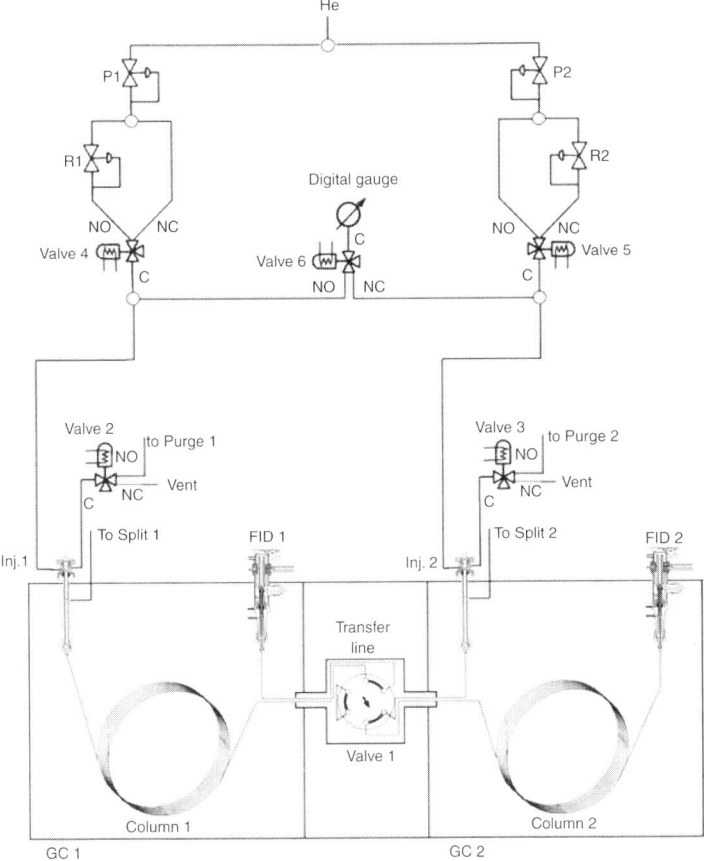

Figure 12.4 Pneumatic and electronic scheme of the GC-GC system, in the standby position.

transfer, the fraction showed seven peaks, four of which have been identified on the basis of the MS data.

STABLE ISOTOPE RATIO MASS SPECTROMETRY

Although enantiomeric distribution provides a useful parameter for the evaluation of authenticity and quality of the oils, there are a lot of non-chiral but sensorily important essential oil compounds. Recently, the determination of the natural abundance of stable isotope by means of stable isotope ratio mass spectrometry (IRMS) has been demonstrated to be useful in the origin assessment and authenticity control of flavours and fragrances. Conventional IRMS requires relatively large sample volume in a purified gaseous form. Recently, an 'on-line' capillary GC-IRMS system has been developed, which combines the high purification effect of GC with the high precision of IRMS. A schematic of this system is shown in Figure 12.6. The substances eluting from the GC column are converted into CO_2 in a combustion oven and then directly analysed in the isotope mass spectrometer, adjusted for the simultaneous recording of masses 44 ($^{12}C^{16}O_2$); 45 ($^{13}C^{16}O_2$, $^{12}C^{16}O^{17}O$) and 46 ($^{12}C^{16}O^{18}O$) in the nanomole range and with high precision (<0.3‰). Isotope ratios are calculated relative to standard values (δ-values, given as ‰), where δ-values is the relative difference in the isotope ratio of the sample to that of the international standard PDB (belemnitella Americana from the crustaceous Pedee formation, South California, USA).

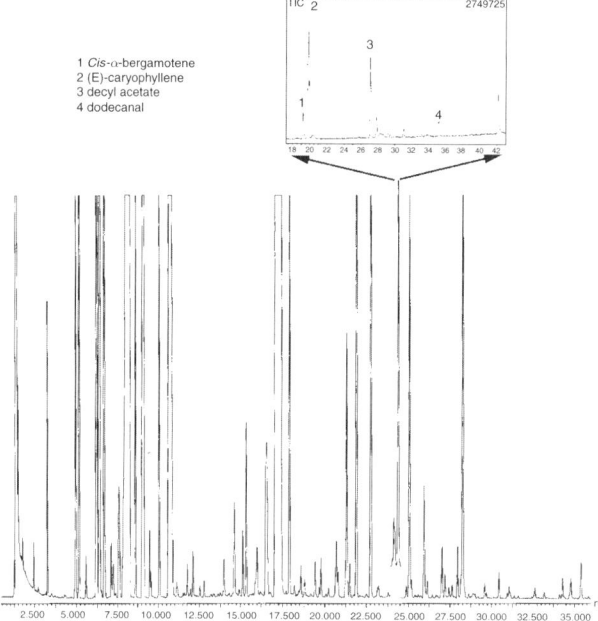

Figure 12.5 Lower: HRGC7-HRGC-FID chromatogram on a SE-52 column of a bergamot essential oil with the transfer heart-cut. Upper: HRGC-HRGC/MS chromatogram of a Carbowax 20 M column of the heart-cut.

Faulhabert *et al.* (1997a,b) reported $\delta^{13}C_{PDB}$ values of characteristic flavour components of mandarin oil, obtained by GC/IRMS analysis. They analyzed genuine Italian cold-pressed oils, commercial oils of different geographical origin and also distilled mandarin and cold-pressed sweet orange oils in order to assess blends of these less valuable oils with cold-pressed mandarin oils. In order to get reliable isotopic data from GC/IRMS measurements, the components of interest must be separated well by GC. Since monoterpene hydrocarbons are present in high amount in the oil, $\delta^{13}C_{PDB}$ values for these components have been measured by direct GC/IRMS analysis. The determination of α-sinensal, methyl-*N*-methyl anthranilate, linalool and octanal was carried out after PLC (preparative layer chromatography) pre-separation of the oxygenated fraction. The sample clean-up procedure must be carried out without any isotope fractionation. This can be proved by experiments with reference substances. Table 12.4 reports the isotopic values for the oils analyzed. As can be seen, differences in ^{13}C content of the components analyzed among oils of different origin are observed. Isotope effects caused by growing conditions influence the $\delta^{13}C_{PDB}$ values of plant materials. To overcome this limitation, the use of a suitable internal isotopic standard (i-IST) can be used. In this case, myrcene has been used as i-IST, because it has the necessary requirements. Figures 12.7A and B show authentic cold pressed mandarin oils from Italy (shaded ranges) compared to commercial samples of different provenance, using $\delta^{13}C_{PDB}$ values (A) and $\delta^{13}C_{myrcene}$ values (B). Comparing the two profiles, it can be seen that introducing the i-IST method individual influences of provenance are eliminated. Samples from Greece and Brazil show a profile similar to the authenticity profile, thereby the genuineness of the oils can be confirmed. Argentine oils show small differences.

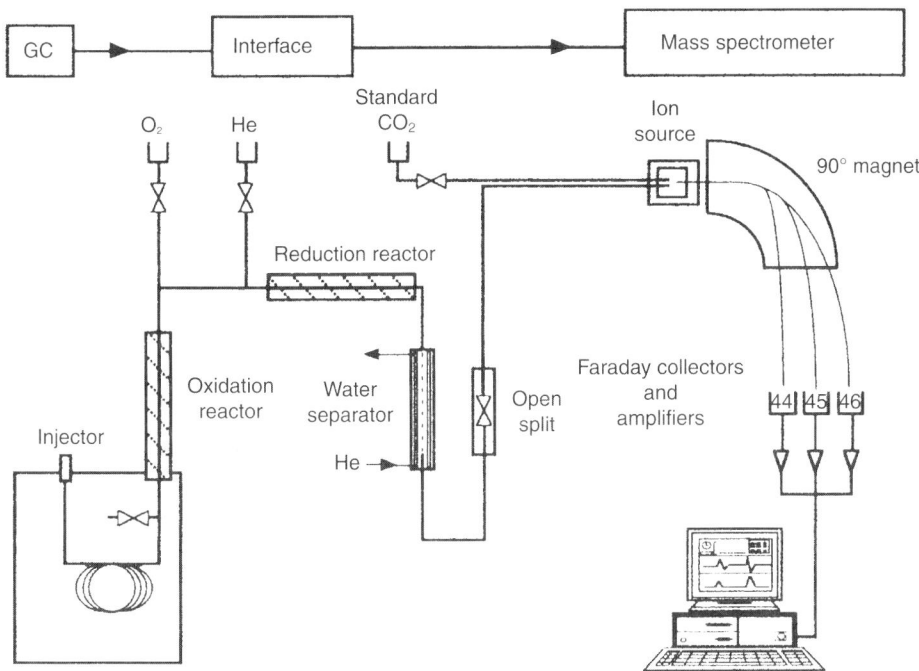

Figure 12.6 Schematic diagram: cGC-IRMS.

Table 12.4 $\delta^{13}C_{PDB}$ values of the investigated citrus peel oils (Faulhaber et al., 1997)

Samples	α-sinensal	limonene	γ-terpinene	α-thujene	β-pinene/sabinene	myrcene	terpinolene	methyl N-methyl anthranilate	linalool	octanal
Authentic mandarin oils from Italy Au 1–10*										
MIN	−28.53	−31.01	−30.16	−28.60	−30.08	−27.49	−30.37	−31.07	−28.60	−30.90
MAX	−26.41	−29.40	−28.83	−27.18	−28.76	−25.83	−28.63	−29.54	−26.39	−28.30
MW	−27.95	−30.25	−29.60	−27.81	−29.32	−26.56	−29.17	−30.24	−27.79	−29.68
[σ]	[0.77]	[0.52]	[0.46]	[0.43]	[0.44]	[0.47]	[0.56]	[0.50]	[0.71]	[1.03]
Commercial mandarin oils from Italy										
Co 1	−27.30	−30.27	−29.86	−28.55	−29.95	−27.51	−29.52	−32.42	−27.62	−28.48
Co 2	−27.99	−30.92	−30.49	−29.29	−30.56	−27.94	−30.56	−31.42	−28.39	−29.81
Co 3	−26.90	−30.10	−29.66	−27.37	−28.87	−26.32	−29.06	−29.97	−27.00	−29.67
Commercial mandarin oils from Greece										
Co 4	−26.05	−29.41	−28.34	−27.06	−28.20	−25.35	−28.54	−29.11	−26.42	−28.25
Co 5	−26.81	−29.36	−29.10	−27.35	−29.03	−26.48	−29.08	−30.37	−27.10	−29.31
Commercial mandarin oil from Brazil										
Co 6	−30.35	−32.36	−32.00	−30.53	−31.30	−28.55	−31.29	−32.66	−29.68	−31.01
Commercial mandarin oil from Argentina										
Co 7	−30.22	−31.91	−30.95	−28.94	−30.62	−27.95	−31.47	−32.66	−29.40	−30.22
Commercial mandarin oils unknown origin										
Co 8	−27.02	−30.14	−29.50	−28.16	−29.84	−27.05	−29.21	−30.22	−27.30	–
Co 9	−27.46	−29.94	−28.97	−27.41	−29.11	−25.30	−28.38	−30.42	−27.41	−28.59
Co 10	−29.92	−30.50	−31.52	−30.01	−31.16	−28.82	−29.19	−32.05	−28.81	−31.31
Distilled mandarin peel oils from Italy										
Co 11	–	−30.32	−30.50	−29.37	−30.04	−26.93	−29.75	−32.48	−28.03	−30.47
Co 12	–	−31.57	−29.66	−29.18	−30.05	−27.36	−29.97	−33.42	−28.29	−30.89
Au 13*	−28.97	−30.45	−30.82	−28.58	−29.73	−26.92	−29.25	−30.30	−28.60	−31.76
Orange peel oils from Italy										
Au 15	–	−29.48	–	–	–	−25.78	–	–	−27.07	−27.01
Au 16	–	−28.86	–	–	–	−25.45	–	–	−26.02	−25.60
Au 17	–	−28.68	–	–	–	−26.29	–	–	−26.30	−26.54
Au 18*	–	−29.61	–	–	–	−26.25	–	–	−27.50	−28.00

Notes
$\delta^{13}C_{PDB}$ values [‰] average of three measurements; standard deviation [σ]; authentic samples Au 1–10, Au 13, Au 15–18 (Au 7–10* and Au 18* are laboratory prepared by solvent extraction; Au 13* are laboratory prepared by distillation); commercial samples Co 1–7 and Co 8–12; (–) not available in sufficient amounts or not contained.

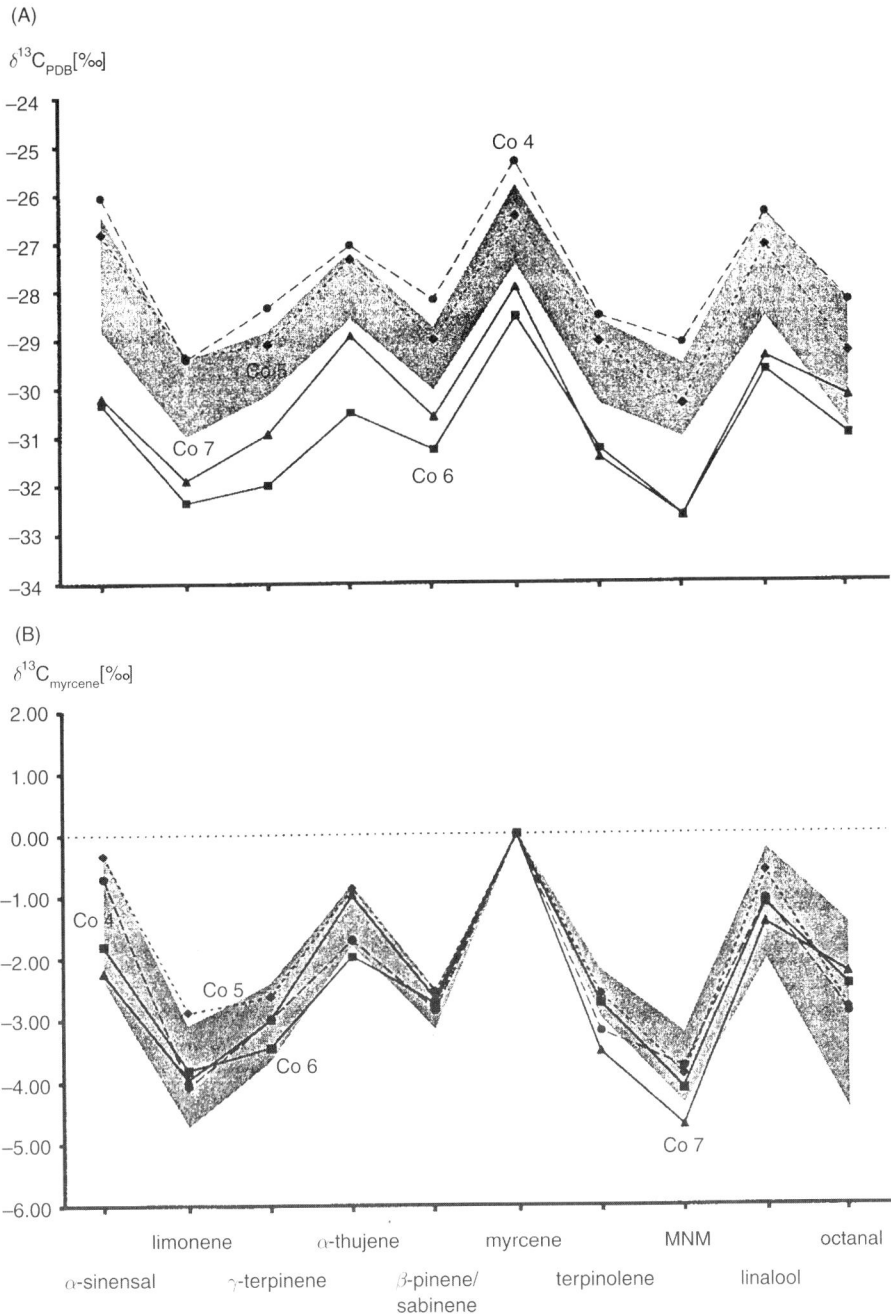

Figure 12.7 Authentic cold-pressed mandarin oils from Italy (shaded ranges) compared with commercial samples of different provenance (Greece Co 4, 5; Brazil Co 6; and Argentina Co 7): (A) authenticity range (shaded range, including minimum and maximum $\delta^{13}C_{PDB}$ values of samples Au 1–10); (B) authenticity profile (shaded range, calculated for myrcene as I-IST). MNM, methyl *N*-methylanthranilate.

FAST GC AND FAST GC-MS

Often, when dealing with complex matrices, the time required for complete gas chromatographic separation of the components of interest can be very long. The use of fast GC techniques can drastically reduce analysis times, maintaining the same or even improving resolution.

Fast GC (David *et al.*, 1999; Mondello *et al.*, 2000) and Fast GC-MS (Mondello *et al.*, 2000) have been applied to the analysis of citrus essential oils. Narrow bore columns have been used for fast GC determination of volatile and also semi-volatile components of some citrus oils.

Narrow bore columns, characterized by a reduced internal diameter and reduced film thickness, allow fast separation as indicated by the Golay equation (Sandra *et al.*, 1987). In fact, as the optimum carrier gas velocity is higher and the H-u plots are flatter for narrow-bore columns, it is possible to work with higher average linear velocities without loss of efficiency. Using narrow-bore columns and hydrogen as the carrier gas, high linear velocity can be applied allowing fast separation with excellent resolution. In order to perform fast GC, several requirements are necessary:

- the instrumentation must allow high inlet pressure and highly controlled split flows (because narrow-bore columns have limited sample capacity, split injections are usually used)
- fast linear heating rates
- fast electronics for detection and data collection.

Fast GC techniques also requires that sample-preparation time is short enough to justify the technique itself. For citrus essential oils this requirement is largely satisfied, since the oils were analysed after only a dilution.

Figure 12.8 shows a comparison between a conventional GC analysis and a fast GC analysis of a lime essential oil. As can be seen, in the fast analysis the same components have been separated with a speed gain of 4.6.

Simply by varying the experimental conditions slightly it is possible to obtain information on the volatile and non-volatile fractions during the same fast GC run, as shown in Figure 12.9. Here, a fast GC chromatogram of a cold-pressed lemon oil is shown. The peaks of herniarin, isopimpinellin and δ-3-carene are separated; the possible contaminations with sweet orange (δ-3-carene) and lime oils (herniarin and isopimpinellin) are easily detectable.

An FID detector is not an information-rich detector in GC analysis. Sometimes, the information required for peak identification is not enough and GC-MS analysis is require. As for the GC-FID analysis of citrus essential oils with conventional columns, GC-MS is often very long. The coupling of a fast mass spectrometer (6750 amu/sec) has been performed. Figure 12.10 shows a TIC fast chromatogram obtained for a lime essential oil. Fifty-five components have been identified. Table 12.5 reports the list of peaks identified and comparison between quantitative results obtained by conventional and fast GC analyses. As can be seen, results are in good agreement. This analysis permits very reliable mass spectra to be obtained. An example is shown in Figure 12.11, which reports the mass spectrum of α-pinene, together with library MS spectrum for comparison. As can be seen, spectrum obtained with the fast GC-MS analysis was similar to that from the MS library.

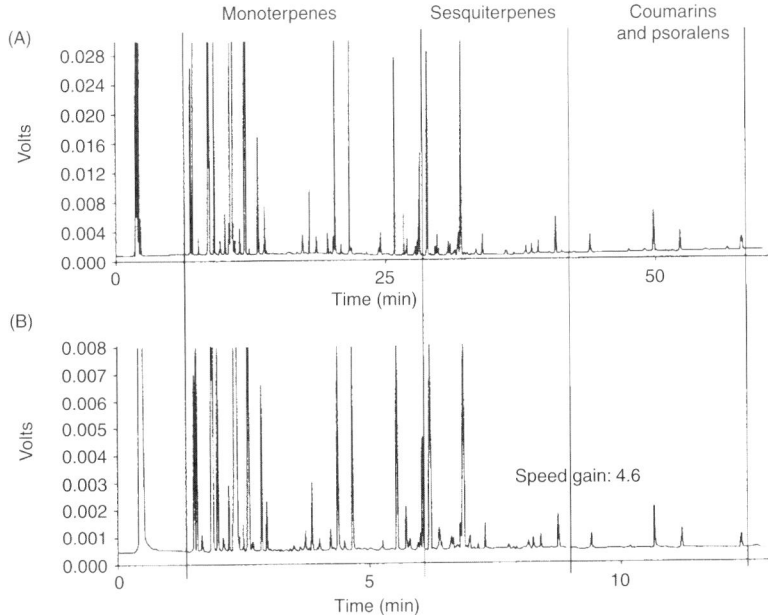

Figure 12.8 (A) GC chromatogram of a lime essential oil obtained using an RTX-5 MS column 30 m × 0.25 mm i.d., 0.25 μm film. Temperature: 50–350 °C at 3 °C/min. Pressure: 54–84 kPa at 0.3 kPa/min. Carrier H_2; u: 36 cm/s; detection FID H_2: 60 kPa; air: 50 kPa; make-up: 80 kPa (He); sampling frequency: 10 Hz. (B) Fast GC chromatogram of the same oil obtained using an RTX-5 column 10 m × 0.1 mm i.d., 0.1 μm film. Temperature: 50–350 °C at 14 °C/min. Pressure: 186–294 kPa at 5 kPa/min. Carrier H_2; u: 57 cm/s; detection FID H_2: 60 kPa; air: 50 kPa; make-up: 80 kPa (He); sampling frequency: 50 Hz.

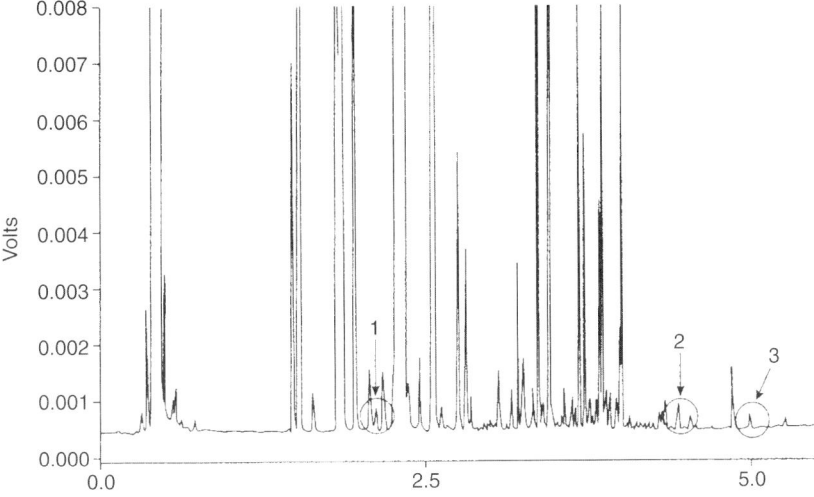

Figure 12.9 Fast GC chromatogram of a lemon essential oil adulterated with sweet orange and lime oils. Peaks: 1 = δ-3-carene; 2 = herniarin; 3 = isopimpinellin.

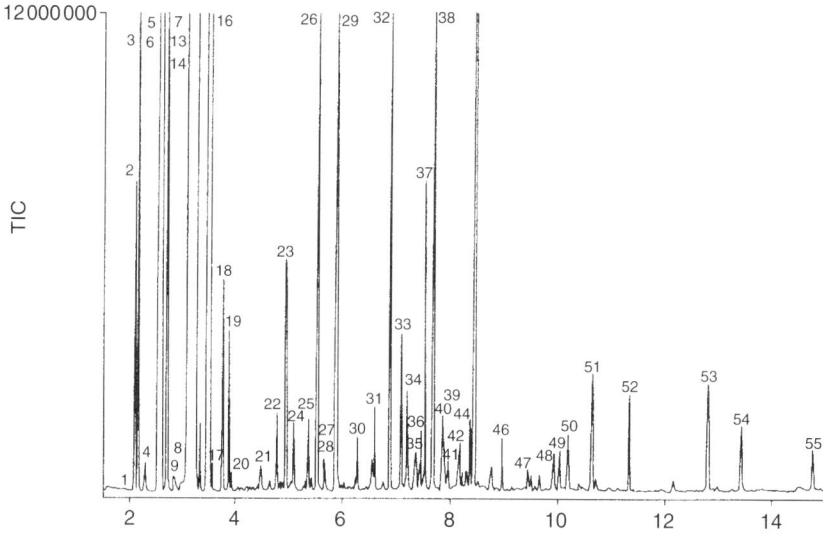

Figure 12.10 TIC fast chromatogram of a lime essential oil.

Table 12.5 Peak identification and comparison (conventional and fast GC) of the quantitative results of lime oil (Mondello et al., 2000)

Compound	Conv.	Fast
1 tricyclene	0.005	0.006
2 α-thujene	0.601	0.585
3 α-pinene	2.308	2.132
4 camphene	0.064	0.059
5 sabinene, 6 β-pinene	12.038	12.128
7 myrcene	1.662	1.546
8 octanal, 9 α-phellandrene	0.058	0.061
10 δ-3-carene	0.006	0.007
11 α-terpinene	0.249	0.265
12 p-cymene	0.149	
13 limonene	54.502	56.546[a]
14 (Z)-β-ocimene	0.059	
15 (E)-β-ocimene	0.107	0.097
16 γ-terpinene	14.210	13.255
17 cis-sabinene hydrate	0.035	0.034
18 terpinolene	0.376	0.402
19 linalool	0.218	0.203
20 nonanal	0.004	0.006
21 citronellal	0.024	0.027
22 terpinen-4-ol	0.106	0.092
23 α-terpineol	0.309	0.299
24 decanal	0.096	0.094
25 nerol	0.115	0.113
26 neral	1.215	1.195
27 geraniol, 28 piperitone	0.061	0.061

29	geranial	1.984	1.884
30	undecanal	0.046	0.044
31	δ-elemene	0.095	0.102
32	neryl acetate	0.992	0.903
33	geranyl acetate	0.212	0.199
34	β-elemene	0.085	0.089
35	dodecanal	0.044	0.038
36	cis-α-bergamotene	0.077	0.071
37	β-caryophyllene	0.544	0.498
38	trans-α-bergamotene	1.132	1.143
39	α-humulene	0.048	0.047
40	(E)-β-farnesene	0.106	0.099
41	β-santalene	0.041	0.042
42	germacrene D	0.072	0.078
43	α-selinene	0.030	0.034
44	(Z)-α-bisabolene	0.127	0.130
45	(E,E)-α-farnesene + β-bisabolene	1.862	1.843
46	germacrene B	0.107	0.111
47	tetradecanal	0.028	0.030
48	2,3-dimethyl-3-(4-methyl-3-pentenyl)-2-norbornanol	0.052	0.054
49	campherenol	0.065	0.061
50	α-bisabolol	0.081	0.078
51	herniarin	0.229	0.224
52	hexadecanal	0.099	0.096
53	citropten	0.252	0.234
54	bergapten	0.126	0.119
55	isompimpinellin	0.083	0.080

Note
a Coeluted with p-cymene and (Z)-β-ocimene.

INTERACTIVE USE OF MS DATA AND LINEAR RETENTION INDICES

GC-MS technique combine the high separation power of capillary GC with the high identification power of the mass spectrometer. For this reason, GC-MS is today largely employed for peak identification in diverse fields concerning complex mixtures of organic compounds. Unfortunately, the identification of unknown compounds by GC-MS, using commercially available libraries of MS spectra, is often not completely reliable for several reasons. For example, more than one structure for every component under investigation can be suggested by the library, sometimes with very similar degrees of purity. Difficulties increase when a complex mixture is analyzed because of the presence of components with very similar structures corresponding to very similar spectra. It has been demonstrated that the use of a pre-separation technique, either off-line or on-line, before GC-MS analysis of a complex mixture permits to obtain more reliable peak identification.

Mondello et al. (1995b), proposed an other approach that uses Linear Retention Indices, on polar and apolar columns, interactively with a MS-library for a more reliable identification of components of a bergamot essential oil. They built a MS library using about 150 pure standards. This library, called FFC (flavour and fragrance components)

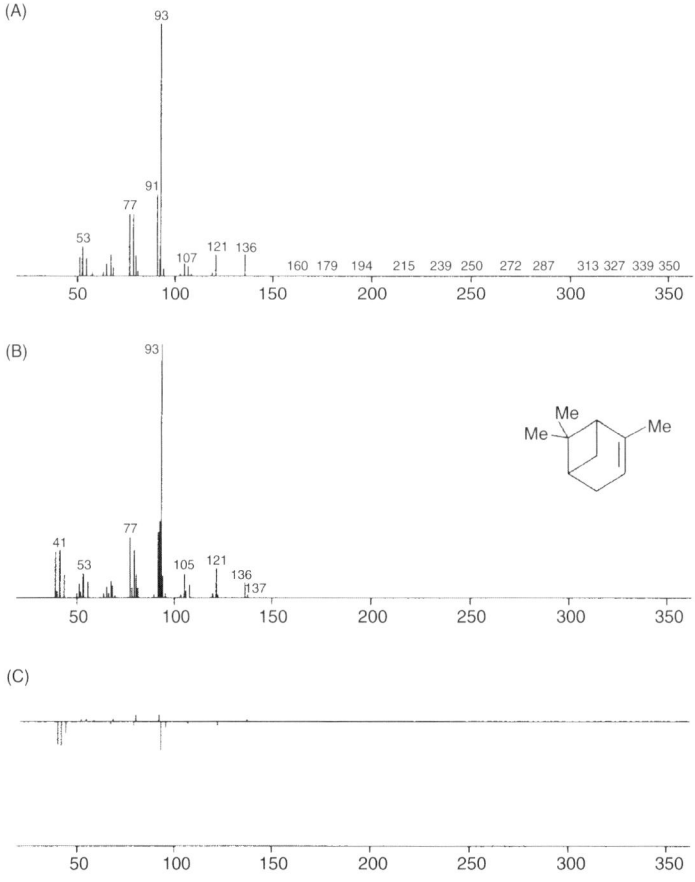

Figure 12.11 Mass spectra of α-pinene obtained from (A) lemon oil and (B) MS library (C) spectra subtraction.

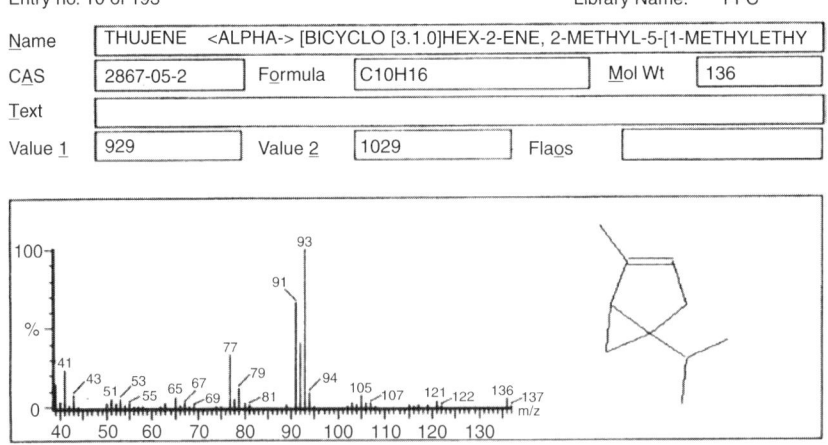

Figure 12.12 Example of the FFC library for α-thujene.

has been augmented with Linear Retention Indices calculated on two columns (SE-52 and Carbowax 20 M). The calculated Linear Retention Indices were used interactively as a filter in conjunction with the spectra library. Figure 12.12 shows an example of the FFC library for α-thujene. This method allowed the identification of 55 components in the analysis of bergamot oil, even when the molecular formula and the molecular weight were identical.

REFERENCES

Anderson, E., Thomason, M.M., Mayfiled, H.T. and Bertsch, W. (1979) Advances in two-dimensional GC with glass capillary columns. *J. High Resolut. Chromatogr.*, 2, 335–342.

Bertsch, W., Anderson, E. and Holzer, G. (1977) Two-dimensional high resolution GLC environmental Analysis, preliminary results. *Chromatographia*, 10, 449–454.

Brotell, H., Rietz, G., Sandqvist, S., Berg, M. and Ehrsson, H. (1982) Two-dimensional capillary gas chromatography without intermediate trapping. Electron capture detector quantitation of an amino alcohol (KABI2128) in serum after trifluoroacetylation. *J. High Resolut. Chromatogr. and Chromatogr. Commun.*, 5, 596–603.

Casabianca, H., Graff, J.-B., Jame, P., Perrucchetti, C. and Chastrette, M. (1995) Application of hyphenated techniques to the chromatographic authentication of flavours in food products and perfumes. *J. High Resolut. Chromatogr.*, 18, 279–285.

David, F., Gere, D.R., Scanlan, F. and Sandra, P. (1999) Instrumentation and applications of fast high-resolution capillary gas chromatography. *J. Chromatogr.*, 842, 309–319.

Deans, D.R. (1968) A new technique for heart cutting in gas chromatography. *Chromatographia*, 1, 19–22.

Deans, D.R. (1981) Use of heart cutting in gas chromatography: a review. *J. Chromatogr.*, 203, 19–28.

Deans, D.R., Huckle, M.T. and Peterson, R.M. (1971) A new system for isothermal gas chromatographic analysis of light gases (H_2, O_2, CO, CH_4, CO_2, C_2H_4, C_2H_6, and C_2H_2) employing a column switch technique. *Chromatographia*, 4, 279–285.

Deans, D.R. and Scott, I. (1973) Gas chromatographic columns with adjustable separation characteristics. *Anal. Chem.*, 45, 1137–1141.

Dugo, G., Cotroneo, A., Del Duce, R., Donato, M.G., Dugo, Giacomo, Dugo, P., Lamonica, G., Licandro, G., Mondello, L., Stagno d'Alcontres, I., Trozzi, A. and Verzera, A. (1994a) The Composition of the volatile fraction of the italian citrus essential oils. *Perfum. Flav.*, 19(6), 29–51.

Dugo, G., Verzera, A., Trozzi, A., Cotroneo, A., Mondello, L. and Bartle, K.D. (1994b) Automated HPLC-HRGC: a powerful method for essential oils analysis. Part I. Investigation on enantiomeric distribution of monoterpene alcohols of lemon and mandarin essential oils. *Essenz. Deriv. Agrum.*, 44, 35–44.

Dugo, G., Verzera, A., Cotroneo, A., Stagno d'Alcontres, I., Mondello L. and Bartle, K.D. (1994c) Automated HPLC-HRGC: a powerful method for essential oils analysis. Part II. Determination of the enantiomeric distribution of linalool in sweet orange, bitter orange and mandarin essential oils. *Flav. Fragr. J.*, 9, 99–104.

Dugo, G., Bartle, K.D., Bonaccorsi, I., Catalfamo, M., Cotroneo, A., Dugo, P., Lamonica, G., McNair, H.M., Mondello, L., Previti, P., Stagno d'Alcontres, I., Trozzi, A. and Verzera, A. (1999a) Advanced analytical technique for the analysis of citrus essential oils. Part. 1. Volatile fraction. HRGC/MS analysis. *Essenz. Deriv. Agrum.*, 69, 79–111.

Dugo, G., Bartle, K.D., Bonaccorsi, I., Catalfamo, M., Cotroneo, A., Dugo, P., Lamonica, G., McNair, H.M., Mondello, L., Previti, P., Stagno d'Alcontres, I., Trozzi, A. and Verzera, A.

(1999b) Advanced analytical technique for the analysis of citrus essential oils. Part. 2. Volatile fraction. LC-HRGC and MDGC. *Essenz. Deriv. Agrum.*, 69, 159–217.

Dugo, G., Mondello, L., Cotroneo, A., Bonaccorsi, I. and Lamonica, G. (2001) Study on the enantiomeric distribution of volatile components of citrus essential oils by multidimensional gas chromatography (MDGC). *Perfum. Flav.*, 26(1), 20–35.

Dugo, P., Mondello, L., Stagno d'Alcontres, I., Cavazza, A. and Dugo, G. (1997) Oxygen heterocyclic compounds of citrus essential oils. *Perfum. Flav.*, 22(1), 25–30.

Faulhaber, S., Hener, U. and Mosandl, A. (1997a) GC/IRMS analysis of mandarin essential oils. 1. $\delta^{13}C_{PDB}$ and $\delta^{15}N_{AIR}$ values of methyl N-methylanthranilate. *J. Agric. Food Chem.*, 45, 2579–2583.

Faulhaber, S., Hener, U. and Mosandl, A. (1997b) GC/IRMS analysis of mandarin essential oils. 2. $\delta^{13}C_{PDB}$ values of characteristic flavor components. *J. Agric. Food Chem.*, 45, 4719–4725.

Gordon, B.M., Rix, C.E. and Borgerding, M.F. (1985) Comparison of state-of-the-art column switching techniques in high resolution gas chromatography. *J. Chromatogr. Sci.*, 23, 1–10.

Grob, K. (1983) Broadening of peaks eluted before the solvent in capillary GC. Part 1: The role of solvent trapping, *Chromatographia*, 17, 357–360.

Grob, K. (1987a) *On-line Coupled LC-GC*. Hüthig, Heidelberg, FRG.

Grob, K. (1987b) On-line coupled HPLC-HRGC *Proceedings 8th Int. Symp. on Capillary Chromatography*. Riva del Garda, Italy. P. Sandra (ed.), Hüthig, Heidelberg, FRG.

Jennings, W.G., Wyllie, S.G. and Alves, S. (1977) WCOT glass capillary columns in flavour chemistry. *Chromatographia*, 10, 426–429.

Jennings, W.G. (1984) State-of-the-art gas chromatography: instrumental design. *J. Chromatogr. Sci.*, 22, 26–33.

Juchelka, D. and Mosandl, A. (1996) Authenticity profiles of bergamot oil. *Pharmazie*, 51, 417–422.

Juchelka, D., Steil, A., Witt, K. and Mosandl, A. (1996) Chiral compounds of essential oils. XX. Chirality evaluation and authenticity profiles of neroli and petitgrain oils. *J. Essent. Oil Res.*, 8, 487–497.

Kaiser, R. (1971) *Preparative gas chromatography*. John Wiley and sons, New York, N.Y.

Martin, R.L. and Winters, J.C. (1963) Adsorption of solutes at the liquid-gas interface as measured by gas chromatography and Gibbs equation. *Anal. Chem.*, 35, 116–117.

Mondello, L., Bartle, K.D., Dugo, G. and Dugo, P. (1994a) Automated HPLC-HRGC: a powerful method for essential oils analysis. Part III. Aliphatic and terpene aldehydes of orange oil. *J. High Resolut. Chromatogr.*, 17, 312–314.

Mondello, L., Dugo, P., Bartle, K.D., Frere, B. and Dugo, G. (1994b) On-line high performance liquid chromatography coupled with high resolution gas chromtography and mass spectrometry (HPLC-HRGC/MS) for the analysis of complex mixtures containing highly volatile compounds. *Chromatographia*, 39, 529–538.

Mondello, L., Bartle, K.D., Dugo, P., Gans, P. and Dugo, G. (1994c) Automated HPLC-HRGC: a powerful method for essential oils analysis. Part IV. Coupled LC-GC/MS (ITD) for bergamot oil analysis. *J. Microcol. Sep.*, 6, 237–244.

Mondello, L., Dugo, P., Bartle, K.D., Dugo, G. and Cotroneo, A. (1995a) Automated HPLC-HRGC: a powerful method for essential oils analysis. Part V. Identification of terpene hydrocarbons of bergamot, lemon, mandarin, sweet orange, bitter orange, grapefruit, clementine and mexican lime oils by coupled HPLC-HRGC/MS (ITD). *Flav. Fragr. J.*, 10, 33–42.

Mondello, L., Dugo, P., Basile, A., Dugo, G. and Bartle, K.D. (1995b) Interactive use of linear retention indices, on polar and apolar columns, with a MS-library for reliable identification of complex mixtures. *J. Microcol. Sep.*, 7, 581–591.

Mondello, L., Dugo, G., Dugo, P. and Bartle, K.D. (1996a) On-line HPLC-HRGC in the analytical chemistry of citrus essential oils. *Perfum. Flav.*, 21(4), 25–49.

Mondello, L., Dugo, G. and Bartle, K.D. (1996b) Coupled HPLC-HRGC-MS: a new method for the on-line analysis of real samples. *American Lab.* (December), 41–49.

Mondello, L., Catalfamo, M., Dugo, P., Proteggente, A.R. and Dugo, G. (1997). Multidimensional GC-GC system for the analysis of real complex samples. Preliminary note. Determination of the enantiomeric distribution of some components of citrus essential oils. *Essenz. Deriv. Agrum.*, 47, 62–85.

Mondello, L., Dugo, P., Dugo, G. and Bartle, K.D. (1998a) On-line HPLC-HRGC-MS for the analysis of natural complex mixtures. *J. Chromatogr. Sci.*, 34, 174–181.

Mondello, L., Catalfamo, M., Dugo, G. and Dugo, P. (1998b) Multidimensional tandem capillary gas chromatography system for the analysis of real complex samples. Part I: development of a fully automated tandem gas chromatography system. *J. Chromatogr. Sci.* 36, 201–209.

Mondello, L., Dugo, P., Cotroneo, A., Proteggente, A.R. and Dugo, G. (1998c) Multidimensional advanced techniques for the analysis of bergamot oil. *EPPOS*, 26, 3–27.

Mondello, L., Catalfamo, M., Dugo, P. and Dugo, G. (1998d). Multidimensional GC-GC system for the analysis of real complex samples. Part II. Enantiomeric distribution of monoterpene hydrocarbons and monoterpene alcohols of cold-pressed and distilled lime oils. *J. Microcol. Sep.*, 10, 203–212.

Mondello, L., Catalfamo, M., Proteggente, A.R., Bonaccorsi, I. and Dugo, G. (1998e). Multidimensional GC-GC system for the analysis of real complex samples. 3. Enantiomeric distribution of monoterpene hydrocarbons and monoterpene alcohols of mandarin oils. *J. Agric. Food Chem.*, 46, 54–61.

Mondello, L., Verzera, A., Previti, P., Crispo, F. and Dugo, G. (1998f). Multidimensional GC-GC system for the analysis of real complex samples. 5. Enantiomeric distribution of monoterpene hydrocarbons, monoterpene alcohols and linalyl acetate of bergamot (*Citrus bergamia* Risso et Poiteau) oils. *J. Agric. Food Chem.*, 46, 4275–4282.

Mondello, L., Dugo, G. and Bartle, K.D. (1998g) A multidimensional HPLC-HRGC system for the analysis of real samples. *LC-GC International*, 11, 26–31.

Mondello, L., Catalfamo, M., Cotroneo, A., Dugo, G., Dugo, Giacomo and McNair, H. (1999). Multidimensional GC-GC system for the analysis of real complex samples. Part IV. Enantiomeric distribution of monoterpene hydrocarbons and monoterpene alcohols of lemon oils. *J. High Resolut. Chromatogr.*, 22, 350–356.

Mondello, L., Zappia, G., Errante, G., Dugo, P. and Dugo, G. (2000) Fast GC and fast GC-MS for the analysis of natural complex matrices. *LC-GC Europe*, 13, 495–502.

Mosandl, A. (1995) Enantioselective capillary gas chromatography and stable isotope ratio mass spectrometry in the authenticity control of flavours and essential oils. *Food Rev. Int.*, 11, 597–664.

Mosandl, A. and Juchelka, D. (1997) Advances in the authenticity assessment of citrus oils. *J. Essent. Oil Res.*, 9, 5–612.

Munari, F., Dugo, G. and Cotroneo, A. (1990) Automated on-line HPLC-HRGC with gradient elution and multiple GC transfer applied to the characterization of citrus essential oils. *J. High Resolut. Chromatogr.*, 13, 56–61.

Rijks, J.A., Van Der Berg, J.H.M. and Diependaal, J.P. (1974) Characterization of hydrocarbons in complex mixtures by two-dimensional precision gas chromatography. *J. Chromatogr.*, 91, 603–12.

Sandra, P., Proot, M., Diricks, G. and David, F. (1987) In P. Sandra and C. Bicchi (eds), *Capillary Gas Chromatography in Essential Oils Analysis*. Huethig, Heidelberg, FRG, p. 38.

Schomburg, G., Kotter, H. and Hack, F. (1973) Improvement of performance and economical utilization of automated preparative scale gas chromatography system by incorporation of cut and backflush devices. *Anal. Chem.*, 45, 1236–1240.

Schomburg, G., Weeke, F., Muller, F. and Oréans, M. (1982) Multidimensional gas chromatography (MDGC) in capillary columns using double oven instruments and a newly designed coupling piece for monitoring detection after pre-separation. *Chromatographia*, 16, 87–91.

Shaw, P.E. (1979) Review of quantitative analyses of citrus essential oils. *J. Agric. Food Chem.*, 27, 246–257.

Simmons, M.C. and Snyder, L.R. (1958) Two stages gas-liquid chromatography. *Anal. Chem.*, 30, 32–35.

13 Composition of the volatile fraction of cold-pressed citrus peel oils

Giovanni Dugo, Antonella Cotroneo, Antonella Verzera and Ivana Bonaccorsi

INTRODUCTION

Citrus essential oils are industrially cold-extracted from the peel of sweet orange, lemon, mandarin, tangerine, grapefruit, Key lime, Persian lime, bitter orange, bergamot and clementine, by mechanical systems. Lime essential oils, however, are most commonly obtained by distillation (Guenther, 1949).

The cold-extraction from the fruit peel, by mechanical systems, consists, independently on the technology used, of three fundamental steps:

1 mechanical action on the peel, in order to cause the utricles breakage, and the consequent release of the oil;
2 the oil is carried by streams of water, that in most cases is recycled;
3 separation, by centrifugation, of the essential oil from the aqueous emulsion.

The volatile fraction of citrus essential oils ranges between 85 per cent, in Key lime oils, and 99 per cent in some sweet orange oils (Di Giacomo and Mincione, 1994). This fraction mostly consists of mono- and sesquiterpene hydrocarbons, and of their oxygenated derivatives, alcohols, aldehydes, esters, ethers, oxides, and also of linear hydrocarbons, alcohols, aldehydes, esters, and acids, and of phenolic compounds and their derivatives. In some cases, as for mandarin, nitrogen esters are present, such as methyl *N*-methylanthranilate. These compounds contribute to the characteristic olfactory note of the oil (Wilson and Shaw, 1981). In citrus essential oil volatile fraction are also present trace amounts of heterocyclic nitrogen containing components (pyrimidines, and pyrazines) (Thomas and Bassols, 1992), and sulphur containing components (Demole *et al.*, 1982). These play a very important role, contributing to the odour character of the oil.

The early studies on the volatile fraction of citrus essential oils date back to the beginning of the nineteenth century, but only since the 1900s have some reliable results been obtained. The reference list is reported by Guenther (1949). The researches of those days allowed with pioneer techniques to identify components that have been mostly confirmed in future.

It was the advent of gas chromatography (GC), first on packed columns, then on capillary columns made of stainless steel at the beginning, then of glass and finally of fused silica, that allowed the study of the composition of complex mixtures of volatile components that are the essential oils. Liberti and Conte (1956) were the first to propose the use gas chromatography to analyse the volatile fraction of citrus essential oils. At the first International Symposium on the Study and Research on Essential Oils they presented the first gas chromatograms of lemon and bergamot essential oils. In these chromatograms 4 and 5 peaks respectively were reported. Immediately after, Bernhard (1957) analysed by GC five monoterpene hydrocarbons standards occurring in cold-pressed lemon oil. Later, the same researcher (Bernhard, 1958) separated five peaks in Californian lemon oil and only two years after the 5 peaks separated by Bernhard become 22, some of which were representative of more than one component; however, 22 different compounds were fully or tentatively identified (Bernhard, 1960).

The preliminary gas chromatographic results obtained, immediately provided information on the complexity of the essential oils and, at the same time, the necessity of some kind of fractionation prior to the gas chromatographic analysis. In those days this was particularly necessary given the limited efficiency of the columns used.

By separating on silica gel columns the hydrocarbons from the oxygenated compounds Clark and Bernhard (1960) found numerous components in lemon oil and Bernhard (1961) underlined the presence, in sweet orange oil, of about 50 components, and most of these were also identified.

At the same time, in Great Britain, Slater (1961) reported the presence of 7 monoterpene hydrocarbons, 9 sesquiterpene hydrocarbons and 24 oxygenated components in lemon and limes oils. Few years later Kovats (1963), and Kugler and Kovats (1963) reported more than 100 components in lime and mandarin oils. Of these, 48 and 44 components were respectively identified.

In the same years Calvarano, M. (1959), and Di Giacomo *et al.* (1962), in Italy, commenced their research on the composition of citrus essential oils, and Di Giacomo and Rispoli (1962) proposed to differentiate lemon oils extracted by the sponge method from those extracted by mechanical systems, using gas chromatographic analysis.

Since the 60s, the capillary columns were introduced for the analysis of essential oils (McFadden *et al.*, 1963; Teranishi *et al.*, 1963; MacLeod *et al.*, 1965; Goretti *et al.*, 1967; Di Giacomo *et al.*, 1971); these columns gradually replaced completely packed columns.

The evolution of the gas chromatographic techniques and the consequent improvement of the information, obtained by their use, on the composition of the volatile fraction of citrus essential oils can be illustrated by comparing Figure 13.1, showing the first gas chromatogram of bergamot oil, obtained by Liberti and Conte (1956), and Figure 13.2, where a bergamot oil gas chromatogram obtained in 1999 is reported (Dugo *et al.*, 1999).

The availability of bench-top mass spectrometer coupled on-line with high resolution gas chromatographs allowed the identification of numerous minor components in citrus essential oils (Mazza, 1986, 1987a,b). It must be noted, however, that the GC/MS identification of components of a complex matrix, as essential oils, by comparison with commercially available mass spectra library is not always possible. The spectra interpretation could be confusing due to peak overlapping and to the structure similarity of many of the components, particularly difficult for those present at trace levels. In order to obtain more reliable information it was convenient to apply an interactive approach of the mass data along with chromatographic retention data, such as linear retention indices (LRI), determined on two columns,

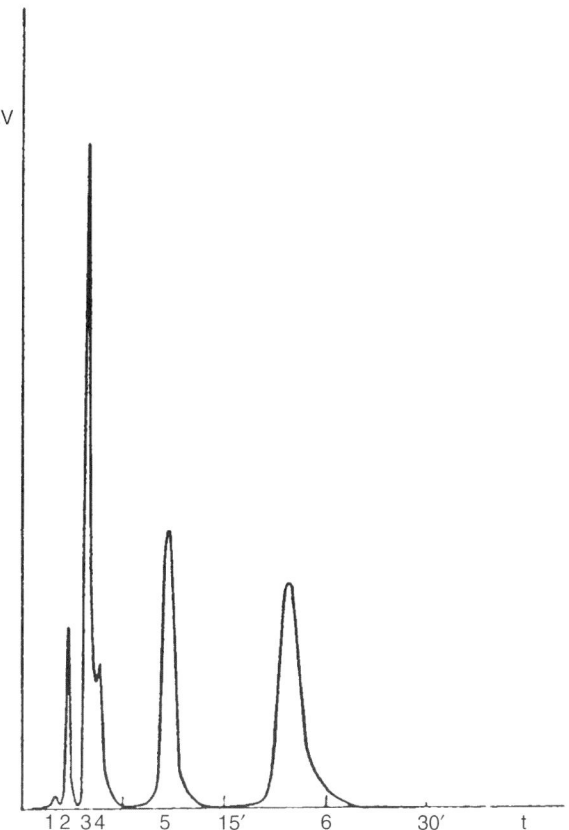

Figure 13.1 Gas chromatogram of bergamot oil. Column: packed column of tricreoylphosphate on celite; column temperature 132 °C (Liberti and Conti, 1956).

one polar and one non-polar (Mondello *et al.*, 1995a). One application reported by these authors on bergamot essential oil proved that the identification of neryl acetate and α-bisabolol using GC/MS equipped with commercial library could result unreliable, while the use of LRI as interactive filter would simplify the identification of these components.

Sebastiani *et al.* (1983), after testing five columns with different stationary phases (OV-1, SE-52, OV-17, UCON and Carbowax), for the separation of lemon essential oil, concluded that, regardless to the efficiency of the chromatographic system used, it was impossible to obtain a complete resolution of all the components of the essential oil. In particular, with polisiloxane stationary phases, as SE-52, some complete or partial peak overlaps were observed between octanal and α-phellandrene, 1,8-cineole and limonene, nerol and citronellol; with polar poliglycol stationary phases co-elution were observed between monoterpene alcohols and esters, and sesquiterpene hydrocarbons. Therefore, in order to obtain satisfactory information on the composition of essential oils, it would be preferable to perform two different GC analyses each on a different stationary phase,

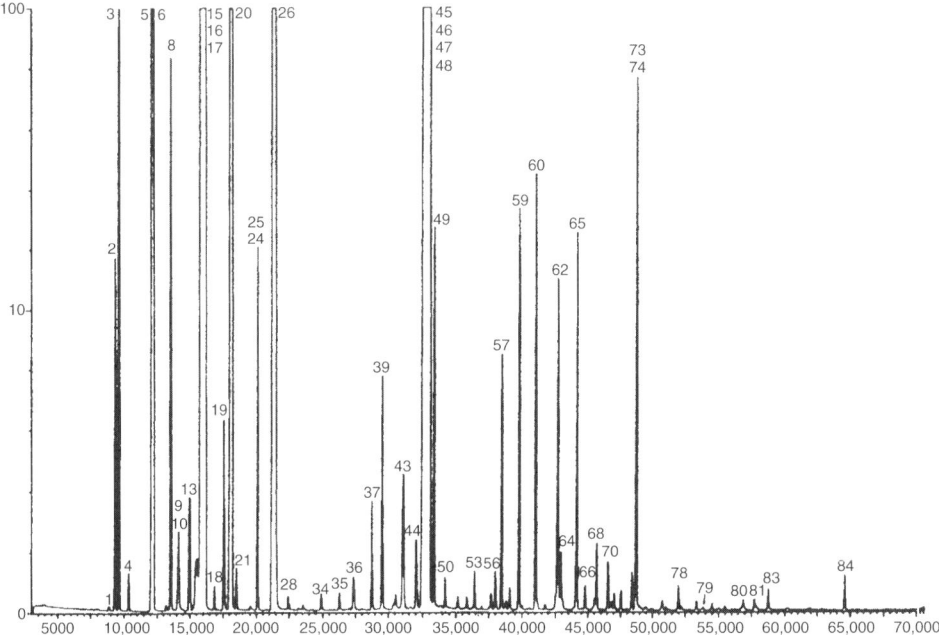

Figure 13.2 Gas chromatogram of bergamot oil. Column: fused silica capillary column 30 m × 0.25 mm i.d. coated with SE-52, 0.25 μm film thickness; column temperature: 45 °C (6 min) to 200 °C at 3 °C/min. (Dugo *et al.*, 1999).

or to pre-separate the oil, obtaining different fractions, simpler in composition, avoiding peak co-elution.

Off-line separation performed by open column liquid chromatography prior to the GC analysis has been largely applied. Chamblee *et al.* (1985), by high performance liquid chromatography (HPLC) separated lime essential oil on three different silica columns in tandem, using as mobile phase a 8 per cent ethyl acetate solution in hexane in a ratio of 1:1 with methylene chloride. The separation allowed to obtain 18 peaks, where the hydrocarbons were concentrated mostly under the first peak, and some under the second, along with oxygenated components that were also the components of all the other peaks. Mazza (1986) used, to separate bergamot oil into 4 fractions, open silica column liquid chromatography, with pure pentane, pentane/diethyl ether mixtures and pure diethyl ether as mobile phases. Cotroneo *et al.* (1985, 1986a) and Dugo *et al.* (1987) used for the separation of kumquat, lemon and bergamot oils, open column liquid chromatography with neutral alumina activity II, and as eluents a system similar to that used by Mazza (1986). These separations allowed to fractionate the oil in classes of compounds, eluting in increasing polarity order: hydrocarbons, esters and ethers, carbonyl compounds and alcohols.

Off-line methods are laborious, time consuming, require sample manipulation, and contamination and loss of components can easily occur.

Recently, for the analysis of essential oils, it has been possible to perform on-line LC pre-fractionation of the oil, so that the fractions can be directly analysed by GC. This procedure does not require manipulations and only sample dilution is necessary, permitting the analysis in a single run (Munari et al., 1990). On-line LC-GC also permits to directly relate the quantitative data obtained by the GC analysis to the amount injected onto the LC system. This technique has been used for the analysis of aldehydes in sweet orange oil (Mondello et al., 1994a). Coupling the LC-GC system to a mass spectrometer detector useful information was obtained on bergamot oil (Mondello et al., 1994b) neroli oil (Mondello et al., 1994c) and petitgrain oil (Mondello et al., 1996) and on the sesquiterpene hydrocarbons of different essential oils (Mondello et al., 1995b). The multidimensional gas chromatography (MDGC) (Mosandl, 1995; Mondello et al., 1998a,b) along with Fast GC (David et al., 1999; Mondello et al., 2000), recently allowed to obtain reliable, and rapid information on the composition of citrus essential oils. The former technique uses the high resolution of GC both during the pre-separation and the final analysis of the transferred fractions; the latter allows the analysis of the volatile fraction and of some non-volatile components of citrus essential oils, with a speed gain of about 5 compared with the conventional GC, without resolution loss.

The quantitative data obtained by the GC analysis of the volatile components of citrus essential oils are usually reported as the relative per cent of the peak areas. In literature are, however, found different papers were the weight per cent of the whole oil are reported for each component. These values are obtained by calibration with internal standard and response factors (Staroscik and Wilson, 1982a,b; Chamblee et al., 1991). The authors of this chapter believe that, even if the results reported as weight per cent give a more correct interpretation of the real composition of the different oils, those reported as relative per cent of the peak area are easily compared and used in quality control laboratories both private and public. The major differences among the two methods can be observed for lime and grapefruit cold pressed oils. These oils in fact contain very high amount of non-volatile residue.

The oldest data reported on the citrus essential oils components have been reviewed, as mentioned previously, by Guenther (1949), those reported later, until 1979, were reviewed by Shaw (1977, 1979). Moreover, since 1976 Lawrence (1976–2000) periodically reviewed the composition of citrus oils.

In this chapter each oil will be singularly described. For each of them, the results reported in literature before 1979, will be briefly reviewed. Then, the most recent data will be summarised in tables, and discussed. In some cases, where the origin of the sample was not certain, the information will be separated from those samples were the origin was reported in detail. The former category will be named 'commercial oils' and the relative data reported in distinct tables. Successively, some information on laboratory extracted oils will be also reported.

The data in the tables, drawn from the original paper, will represent the composition of a single sample, the mean values or, when possible, variability ranges. These are expressed by two decimal figures, or by one if this was the approximation of the original paper. In the tables the single components are grouped by chemical class and, each class, is listed in alphabetical order. In appendix to the tables the following information is provided: geographical origin, the production technology, the number of samples relative to the given data, the analytical technique, the component identification method, how the data are expressed (w per cent or relative peak area per cent).

Those components identified by a single author and their content are also indicated in the appendix.

In both the text and the tables the term tr (trace) indicates percentage equal or less than 0.05 if the results are reported to the first decimal figure; when two decimal figures are used the term indicates percentages equal or less than 0.005; the (+) symbol indicates a component present but not quantitatively determined; cis- and trans-linalool oxide, if not otherwise specified, are in the furanoid form; the asterisk (*) labels those components for which the correct isomer was not characterised in theoriginal paper; the symbol ^{t} signs those components where the identification was tentative.

BITTER ORANGE OIL (*Citrus aurantium* L.)

1959–1979

Only few papers are reported in literature for bitter orange oil in the period between 1959–1979. Shaw (1979), in fact, limited the data reported only to an investigation carried out by Maekawa *et al.* (1967) on laboratory extracted oils, obtained from some varieties of bitter orange (Kabasu, Sudachi, Daidai and Natsudaidai) cultivated in Japan, and on the value of nootkatone (0.01 per cent) reported by MacLeod and Buigues (1964). The results obtained by Maekawa *et al.* are reported below:

camphene	tr–0.89%
limonene	73.82–86.23%
myrcene	1.60–10.90%
β-pinene	0.59–5.50%
nonanal	0.09–0.46%
octanal	1.34–2.17%
citronellal	0.30–0.40%
citronellol	0.23–1.11%
geraniol	0–0.48%
isopulegol	tr–0.41%
linalool	0.42–1.02%
nerol	0–0.29%
α-terpineol	tr–0.23%
citronellyl acetate	0–0.61%
citronellyl formate	0.08–1.08%
geranyl formate	tr–0.57%
linalyl acetate	0.11–0.99%
1,8-cineole	0.77–8.90%

These values cannot be considered representative of the composition of bitter orange oil, industrially cold-pressed, if the origin of the fruits and the extraction technique used are considered.

In these years other studies on bitter orange essential oils were carried out, to determine the qualitative and quantitative composition of the volatile fraction of this oil.

In particular, four papers concerned industrial Italian bitter orange oils. Calvarano M. (1959) reported the presence of limonene, myrcene, α- and β-pinene, γ-terpinene, decanal, dodecanal, nonanal, octanal, geranial, neral, geraniol, linalool, terpineol*, and geranyl acetate. Di Giacomo *et al.* (1963) reported the following composition of the monoterpene hydrocarbons fraction:

p-cymene	tr–0.87%
limonene	94.52–96.20%
myrcene	1.46–1.54%
α-pinene	0.47–0.59%
β-pinene	0.80%
sabinene	tr
γ-terpinene	0.97–1.20%

The year after Di Giacomo *et al.* (1964) determined the average content of monoterpene hydrocarbons ranging between 85 per cent and 87 per cent, indicating the ratios of peak height of decanal, citronellal, linalool, octyl acetate and linalyl acetate as representative in order to evaluate the bitter orange oil purity. Later, Calvarano I. (1966) in addition to components reported in the previously cited papers, identified in this essential oil, camphene, δ-3-carene, p-cymene, phellandrene*, sabinene, terpinolene, citronellol, nerol, terpinen-4-ol, α-terpineol, neryl acetate, terpinyl acetate*. In the same paper Calvarano underlined some quantitative differences between bitter orange oil and sweet orange oil, relatively to the content of β-pinene, sabinene, decanal and linalyl acetate, proposing for genuine bitter orange oil the following limits of some ratios between these components:

β-pinene/sabinene	3.08–3.12	linalyl acetate/linalool	1.67–3.27
linalool/decanal	1.69–4.62	linalyl acetate/decanal	4.40–7.90

Mehlitz and Minas (1965) analysed by gas chromatography oils extracted, from Spanish and Greek bitter oranges, by using different methods (manual, distillation from fresh and stored peels, solvent extraction). They observed that the extraction technique influenced in particular the composition of aliphatic aldehydes and monoterpene and aliphatic alcohols, measured in function of peak height.

Ziegler (1971) reported, for the volatile fraction of bitter orange oil, the presence of the following components: camphene, p-cymene, limonene, myrcene, α-pinene, β-pinene, sabinene, γ-terpinene, δ-cadinene, α- and β-copaene, caryophyllene*, β-elemene, farnesene*, α- and β-humulene, α- and β-ylangene, valencene, decanal, dodecanal, nonanal, octanal, undecanal, geranial, neral, α- and β-sinensal, carvone, nootkatone, citronellol, geraniol, linalool, nerol, α-terpineol, octyl acetate, citronellyl acetate, geranyl acetate, linalyl acetate, neryl acetate. α- and β-copaene and α- and β-ylangene were probably misidentifications, probably as the consequence of the mistaken identification reported by Hunter and Brogden (1965) for some citrus essential oils, and later corrected by Veldhuis and Hunter (1968).

Soulari and Fanghänel (1971) analysed the monoterpene hydrocarbon fraction of bitter orange oil and identified the following components: p-cymene (1.1 per cent), limonene (81.3 per cent), myrcene (1.7 per cent), α-pinene (0.4 per cent), β-pinene (0.6 per cent), α-terpinene (1.9 per cent), γ-terpinene (5.4 per cent), terpinolene (0.8 per cent). The percentages of α-terpinene and γ-terpinene reported in this paper appear, to be, in our opinion, excessively high, as well as the percentage of limonene appears too low, and therefore not compatible with genuine bitter orange oil.

The composition of an Argentine oil, reported by De La Torre and Sardi (1977) appears anomalous due to the low value of limonene, and the high values of almost all the oxygenated compounds:

limonene	71%
α-phellandrene	0.35%
α-pinene	0.24%
β-pinene	0.5%
decanal	4.9%
citronellal	2.7%
geranial + neral	0.6%
geraniol	5%
α-terpineol	3.22%
citronellyl acetate	3.28%
geranyl acetate	7%

Ortiz *et al.* (1978) analysed the laboratory-extracted oils obtained from 17 different cultivars of bitter orange, cultivated in the Citrus Research Centre, Riverside, CA. The quantitative variability range of the components identified in the paper was extremely wide: among the different cultivars studied were chinotto and bergamot, 7 hybrids and some other cultivars that are not usually included for the extraction of bitter orange essential oils. Bergamot oil, in particular, affected the minimum and maximum values obtained by Ortiz *et al.* (1978). The results obtained by Ortiz, with the exclusion of those relative to bergamot, chinotto and hybrids oils, are reported below:

p-cymene	tr–0.01%
limonene	95.26–96.30%
myrcene	1.49–1.66%
α-pinene	0.41–0.47%
β-pinene	0.18–0.61%
sabinene	0.06–0.30%
γ-terpinene	0.22–0.34%
decanal	0.14–0.20%
geranial	0.18–0.29%
neral	0.05–0.12%
citronellol	0.18–0.23%
geraniol	tr
linalool	0.16–0.64%
nerol	0.01–0.06%
α-terpineol	0.05–0.14%
bornyl acetate	tr–0.01%
linalyl acetate	0.02–0.26%

1979–1999

Industrial cold-pressed oils

In Table 13.1 are reported the results published after the review by Shaw (1979), relative to the composition of the volatile fraction of cold-pressed bitter orange essential oils.

As can be observed in Table 13.1 the data found in literature on bitter orange oil, are mostly relative to oils produced in Spain and in Italy. However, in the table are also reported the results obtained for Brazilian oils.

Table 13.1 Relative percentage composition of the volatile fraction of cold-pressed bitter orange oil

	1	2	3(a)	3(b)	4	5	6
HYDROCARBONS							
Monoterpene							
camphene	tr	tr	tr	0.01	tr–0.01	tr	tr–0.01
δ-3-carene	–	tr	–	–	–	0.01	tr
p-cymene	tr	tr	–	–	–	0.01	b
limonene	92.29	92.50	94.34	92.43	91.54–93.70	93.77	93.20–93.86[b]
myrcene	3.10	1.70	1.81	2.07	1.60–1.81	1.73	1.70–1.84
(E)-β-ocimene	–	0.20	0.14	0.37	0.16–0.38	0.27	0.19–0.38
(Z)-β-ocimene	–	tr	–	–	tr	0.05	0.01
α-phellandrene	tr	0.05	0.02	0.01	tr–0.01	0.03	0.04–0.08
β-phellandrene	tr	–	–	–	–	–	b
α-pinene	0.89	0.40	0.45	0.59	0.29–0.45	0.38	0.52–0.58
β-pinene	1.17	0.35	0.29	0.90	0.29–0.39	0.44	0.63–1.28
sabinene	0.23	0.15	0.14	0.27	0.12–0.15	0.16	0.26–0.45
α-terpinene	–	tr	–	–	–	0.02	tr
γ-terpinene	0.07	tr	0.12	0.10	tr–0.01	tr	0.01–0.13
terpinolene	0.07[a]	0.05	0.10	0.09	tr–0.13	0.01	tr–0.01
α-thujene	–	tr	tr	0.02	–	tr	tr–0.01
Sesquiterpene							
β-bisabolene	–	0.02	–	–	tr–0.02	–	tr–0.01
δ-cadinene	–	0.01	–	–	–	tr	–
β-caryophyllene	0.10	0.10	–	–	0.09–0.14	0.10	0.03–0.06
β-elemene	–	0.01	–	–	tr–0.02	0.03	–
δ-elemene	–	0.05	–	–	0.02–0.05	–	0.02
(E)-β-farnesene	–	0.02[c]	–	–	–	0.01	–
(Z)-β-farnesene	–	0.02[c]	–	–	–	0.03	0.01
α-humulene	–	0.04	–	–	0.02–0.04	0.01	tr–0.01
germacrene D	–	0.12	–	–	0.08–0.14	0.13	0.09–0.13
valencene	0.29	0.02	–	–	–	tr	tr
ALDEHYDES							
Aliphatic							
decanal	0.12	0.14	0.14	0.19	0.11–0.20	0.13	0.11–0.19
(E,E)-2,4-decadienal	–	0.01	tr	0.01	–	–	tr
(E,Z)-2,4-decadienal	–	0.01	tr	tr	–	–	–
dodecanal	0.03	0.03	0.02	0.03	0.01.0.04	0.02	0.01–0.03
nonanal	0.02	0.02	0.05	0.11	0.01–0.05	0.02	0.01–0.04
octanal	0.07[a]	0.06	0.13	0.24	0.04–0.15	0.06	0.11–0.19
tetradecanal	–	tr	0.02	0.03	–	–	tr
tridecanal	–	0.01	tr	0.01	–	–	–
undecanal	–	0.01	0.01	0.01	tr–0.01	0.01	tr–0.01

Table 13.1 (Continued)

	1	2	3(a)	3(b)	4	5	6
Monoterpene							
citronellal	0.06[d]	0.01	0.01	0.02	tr–0.01	tr	tr–0.01
geranial	0.06	0.01	0.03	0.09	0.03–0.07	0.04	0.05–0.10
myrtenal	–	tr	0.01	0.01	–	–	–
neral	0.05	0.01	0.02	0.05	tr–0.02	0.04	0.03–0.05
perilla aldehyde	–	0.01	0.01	0.02	tr–0.02	0.02	tr–0.01
Sesquiterpene							
α-sinensal	–	–	tr	tr	–	–	0–tr
β-sinensal	–	–	0.01	tr	–	0.01	–
KETONES							
Monoterpene							
camphor	–	0.01	–	–	tr	–	–
carvone	–	0.02	tr	tr	tr–0.02	–	–
Sesquiterpene							
nootkatone	–	0.13	0.08	0.03	0.05–0.39	0.04	0–tr
7-epi-α-selinen-2-one	–	0.02	–	–	tr–0.02	–	–
ALCOHOLS							
Aliphatic							
decanol	–	0.05	0.01	0.01	tr–0.02	0.01	–
nonanol	0.03	0.01	–	–	–	–	tr[t]
octanol	–	0.03	0.18	0.17	tr–0.04	0.02	tr–0.01
Monoterpene							
citronellol	tr	–	tr	tr	–	–	tr–0.01[e]
geraniol	–	0.02	0.01	0.01	0.01–0.23	0.02	–
linalool	0.21	0.20	0.15	0.37	0.10–0.22	0.22	0.20–0.33
nerol	–	0.01	tr	0.01	tr–0.04	0.01	tr–0.01[e]
terpinen-4-ol	0.03	0.01	tr	tr	tr	tr	tr–0.01
α-terpineol	0.07	0.26	0.57	0.06	0.27–2.94	0.87	0.03–0.06
Sesquiterpene							
(E,E)-farnesol	–	0.02[f]	–	–	tr–0.02[f]	0.01	–
(Z,E)-farnesol	–	0.02[f]	–	–	tr–0.02[f]	tr	tr–0.01
(E)-nerolidol	–	0.23	–	–	0.10–0.23	0.11	0.04–0.08
ESTERS							
Aliphatic							
decyl acetate	–	0.07	0.02	0.02	tr–0.04	tr	0.02–0.04
octyl acetate	0.06[d]	0.05	0.05	0.01	0.01–0.05	0.03	0.03–0.05
Monoterpene							
citronellyl acetate	tr	–	–	–	–	0.05	tr–0.01
geranyl acetate	tr	0.17	0.13	0.09	0.13–0.20	0.14	0.07–0.11
linalyl acetate	0.37	0.86	0.28	0.77	0.19–0.82	0.43	0.64–1.17
1,8(9)-menthadien-10-yl acetate	–	0.02	0.01	0.01	tr–0.02	0.01	–
neryl acetate	0.05	0.05	0.03	0.03	0.02–0.05	0.03	0.02
α-terpinyl acetate	–	0.02	–	–	tr–0.03	0.02	tr–0.01
OXIDES							
Monoterpene							
cis-limonene oxide	–	0.02[g]	0.01[g]	0.01[g]	tr–0.02[g]	0.01	tr–0.01
trans-limonene oxide	–	0.02[g]	0.01[g]	0.01[g]	tr–0.02[g]	0.01	tr–0.01

Notes

t = tentative identification; tr = traces; * correct isomer not characterised; a terpinolene + octanal; b limonene + *p*-cymene + β-phellandrene; c (Z)-β-farnesene + (E)-β-farnesene; d citronellal + octyl acetate; e nerol + citronellol; f (E,E)-farnesol + (E,Z)-farnesol; g *cis*-limonene oxide + *trans*-limonene oxide.

Appendix to Table 13.1

1 Koketsu et al. (1983). Brazil; FMC; one sample; GC on packed columns of Carbowax 20 M and SE-30 and on capillary column coated with Carbowax 20 M; retention times and Kovats indices; relative percentage of peak areas.
2 Boelens and Sindreu (1988). Spain; needle puncturing machine; GC on capillary columns coated with Carbowax 20 M, UCON and SE-54; relative percentage of peak areas. Boelens and Sindreu also found selinenes* (0.01%); (E)-2-decenal (0.02%); (E)-2-dodecenal (0.01%); heptanal (0.01%); hexanal (0.03%); (E)-2-octenal (0.01%); isopiperitone (0.01%); dodecanol (0.01%); dodecyl acetate (0.01%); nonyl acetate (0.01%); geranyl propanate (0.02%); neryl propanate (0.01%); cis- + trans-linalool oxide (0.01%); myrcene oxide (0.01%); decanoic acid (0.05%) and trace amounts of p-cymenene, (E)-2-heptenal, (E)-2-hexenal, (E)-2-nonenal, (E)-2-tetradecenal, (E)-2-tridecenal, (E)-2-undecenal, undecyl acetate, cis- + trans-dehydro linalool oxide, perillene.
3 Boelens and Jimenez (1989). Spain, Andalusia (a), Italy, Sicily (b); needle puncturing machine (a); GC on capillary column coated with SE-54; GC/MS and retention times; relative percentage of peak areas.
4 Boelens (1991). Spain; 7 samples; GC; relative percentage of peak areas.
5 Boelens and Oporto (1991); Spain; needle puncturing machine; Boelens and Oporto also found 1,8(9)-menthadien-10-ol (0.01%).
6 Dugo et al. (1993, 1999). Italy, Sicily; Pelatrice, Torchi, FMC; 10 samples; GC on capillary columns coated with SE-52 and DB-5; GC/MS with interactive use of linear retention indices; relative percentage of peak areas. Dugo et al. also found trans-α-bergamotene (0.01–0.02%), germacrene B (0.01%), β-santalene (tr–0.01%), β-sesquiphellandrene (0.01%), isopulegol (tr–0.01%), cis-sabinene hydrate (tr–0.01%) and trace amounts of bicyclogermacrene, piperitone^t, hexanol, borneol, trans-sabinene hydrate, nonyl acetate and (E)-myroxide; Germacrene B found in the paper of 1993 is probably a misidentification of bicyclogermacrene.

The quantitative data are, generally, in good agreement, especially those relative to the major components. It should be highlighted, however, the value of valencene (0.24 per cent) reported by Koketsu et al. This value does not find correspondence among all the other values reported in literature, where the valencene is reported at trace levels. The high values of α-terpineol and nootkatone, reported in Spanish oils must also be underlined. It is our opinion that such values represent a peculiarity of the Spanish oils; in fact high values of these components were found in several Spanish oils analysed in our laboratories (Dugo, 1997).

In addition to those relative to the results reported in Table 13.1, some other paper are found in literature on some peculiar aspects of the composition of bitter orange oil of secure authenticity.

Micali et al. (1990), using an on-line LC-GC system, identified in bitter orange the linear hydrocarbons C_{22}–C_{29}, C_{31} and the correspondent 'iso' isomers. Later Lanuzza et al. (1991), using the same system identified in the same oil the following sesquiterpene hydrocarbons: β-bisabolene, β-caryophyllene, α-copaene, α-humulene, longifolene. It is our opinion that the identification of longifolene is doubtful: this component is unusual for bitter orange oil.

Mondello et al. (1995b), by using a coupled HPLC-HRGC/MS, identified in the hydrocarbon fraction of Sicilian bitter orange oil the following sesquiterpene hydrocarbons listed in decreasing concentration order: germacrene D, β-caryophyllene, germacrene B, (Z)-β-farnesene, α-humulene, β-elemene, (E,E)-α-farnesene, δ-cadinene^t, and α-muurolene^t.

Commercial oils

Some authors (Haubruge et al., 1989; Inoma et al., 1989; Chouchi et al., 1996) reported the composition of commercial bitter orange oil, of which the origin was unknown.

Even if the first two of these investigations were not detailed, the results obtained for the quantitative composition, shown in Table 13.2, are in good accordance with the data reported in Table 13.1 for genuine oils. The oil analysed by Chouchi *et al.* (1996) resulted different from the others for the low content of limonene and for the relatively high content of β-pinene and some minor components. The content of *p*-cymene, also quite high, could be due to less than optimal storage conditions that may have caused some changes in the composition of the oil.

Table 13.2 Relative percentage composition of the volatile fraction of some commercial bitter orange oils

	1	2	3
HYDROCARBONS			
Monoterpene			
δ-3-carene	tr	–	tr
p-cymene	–	0.1	2.07
limonene	93.50	93.6	82.12
myrcene	2.44	1.8	1.59
(E)-β-ocimene	0.09	–	0.06
α-pinene	0.71	0.5	0.75
β-pinene	0.46	0.5	2.53
sabinene	0.25	0.2	0.05
γ-terpinene	tr	0.3	0.10
Sesquiterpene			
β-caryophyllene	0.06	0.1	–
ALDEHYDES			
Aliphatic			
decanal	0.17	–	0.15
nonanal	tr	tr	0.07
Monoterpene			
perilla aldehyde	0.21	–	0.09
KETONES			
Monoterpene			
carvone	0.06	–	1.21
ALCOHOLS			
Monoterpene			
trans-carveol	0.11	–	0.62
linalool	0.23	0.1	0.52
α-terpineol	0.17	0.2	0.11
ESTERS			
Monoterpene			
geranyl acetate	0.11	0.1	0.13
neryl acetate	0.09	tr	–
OXIDES			
Monoterpene			
trans-linalool oxide	–	tr	0.02

Notes
tr = traces; * correct isomer not characterised.

Appendix to Table 13.2

1 Inoma *et al.* (1989). USA; FMC; one commercial sample; GC on capillary columns coated with OV-101 and Carbowax 20 M; relative percentage of peak areas. Inoma *et al.* also found geranial (0.06%), octanol

(0.10%), limonene oxide* (0.15%), limonene dioxide (0.09%), and trace amounts of valencene, dodecanal, undecanal, neral, geraniol.
2 Haubruge *et al.* (1989). One commercial sample; GC on capillary column coated with CP-TM-WAX 25CB; GC/MS; relative percentage of peak areas. Haubruge *et al.* also found β-phellandrene (1.5%), octanal (0.1%).
3 Chouchi *et al.* (1996). One commercial sample; GC on capillary column coated with DB-5; GC/MS; relative percentage of peak areas corrected using relative response factors. Chouchi *et al.* also found camphene (0.23%), *p*-cymenene (1.36%), (Z)-β-ocimene (0.04%), α-phellandrene (0.27%), *cis-p*-mentha-1,3,8-triene (0.33%), *trans-p*-mentha-1,3,8-triene (0.11%), α-terpinene (0.13%), heptanal (0.08%), *trans*-dihydrocarvone (0.13%), pinocarvone (0.04%), *cis*-carveol (0.34%), iso-dihydrocarveol (0.06%), *cis*-sabinene hydrate (0.66%), *trans*-β-terpineol (0.16%), terpinen-4-ol (0.15%), linalyl acetate (0.68%), eugenol methyl ether (0.76%), *cis*-limonene oxide (0.49%), *trans*-limonene oxide (0.28%), benzaldehyde (0.08%), benzylbenzoate (0.66%), methyl benzoate (0.56%), methyl salicilate (0.14%), and trace amounts of δ-3-carene, terpinolene, α-thujene, α-terpenyl acetate, acetophenone.

Laboratory extracted oils

Numerous researches have been carried out on laboratory extracted bitter orange oils, obtained from different cultivars and different geographic origin (Mediterranean Countries, China and Japan).

El-Samahy *et al.* (1982) and Ashour and El-Kebeer (1983) reported on the composition of oils extracted from Egyptian fruits. The former paper reports the identification of the following components: camphene (14.02 per cent), limonene (95 per cent), α-pinene (22.44 per cent) neral+geranial (0.78 per cent), citronellol (1.03 per cent), geraniol (tr), linalool (18.85 per cent), bornyl acetate (tr), geranyl acetate (tr). The latter reported the following composition for the oil extracted from ripe fruits: α-pinene (0.55 per cent), β-pinene (0.11 per cent), limonene (92.68 per cent), myrcene (2.25 per cent), linalyl acetate (0.12 per cent), terpineol* (0.16 per cent). The oil analysed by El-Samahy appears, based on the content of limonene and linalool, to be a bergamot oil instead of bitter orange oil. The percentage content of camphene is also unusual, not only for bitter orange oil, but, in general, for the most common citrus peels oils. However the chromatograms reported in this papers give the impression of many coeluted peaks, with the consequent difficulty to identify and to quantitatively determine the components. Ashour and El-Kebeer also observed, during the ripening of the fruits, that limonene increased, while β-pinene, terpineol*, linalyl acetate, ocimene* and linalool decreased. In particular the last two components were absent in the oils of ripe fruits.

Tuzcu *et al.* (1985) studied the composition of the bitter orange oil obtained from the Tuzcu cultivar grown in Turkey. The composition of this oil was characterised by the high content of linear alcohols as reported below:

undecane	0.06%
limonene	86.06%
myrcene	0.04%
α-pinene	0.21%
β-pinene	1.61%
γ-terpinene	0.17%
terpinolene	0.72%
β-caryophyllene	0.27%
valencene	0.01%

dodecanol	0.03%
tetradecanal	0.49%
citronellal	0.07%
geranial	0.14%
perilla aldehyde	0.53%
docosanol	0.46%
dodecanol	0.18%
eicosanol	0.02%
eneicosanol	0.03%
heptadecanol	1.70%
hexadecanol	0.07%
nonadecanol	0.01%
pentadecanol	0.17%
tetracosanol	0.24%
tetradecanol	0.01%
tricosanol	0.01%
carvacrol	0.01%
citronellol	0.05%
geraniol	0.03%
cis-myrtanol	0.55%
myrtenol	0.18%
nerol	0.08%
α-terpineol	0.34%
farnesol*	0.55%
nonyl acetate	0.05%
octyl acetate	0.05%
geranyl acetate	0.38%
α-ionone	0.04%
β-ionone	0.35%

In order to determine possible differences in the composition of the volatile fraction of bitter orange oil obtained from raw and ripe fruits, Boelens and Jimenez (1989) directly sampled the oil from the flavedo of living fruits on the tree. These authors observed that the oils sampled from ripe fruits, if compared with the ones extracted from the unripe fruits, produced the following differences: a higher average content of aliphatic aldehydes (0.67 per cent vs 0.48 per cent); a higher content of monoterpene oxygenated compounds (1.02 per cent vs 0.74 per cent), although these varied within a wide range for both oils; a lower content of monoterpene hydrocarbons (96.5 per cent vs 98.0 per cent), a similar content of sesquiterpene hydrocarbons (0.2 per cent) and a higher content of sesquiterpene oxygenated compounds (0.18 per cent vs 0.04 per cent). Some difference in the composition were also due to the presence of acetaldehyde, acetone, nootkatone, 7-epi-α-selinen-5-one, ethanol, methanol, propanol, in the ripe fruit oils, and absent in the unripe fruit oils. Boelens and Jimenez (1989), also noticed that, if the sampling was performed with a syringe rinsed with a methanol solution of triethanolamine, it was possible to avoid possible reactions determined by the low pH of the fruit flesh and pulp, such as: the formation of monoterpene hydrocarbons from linalyl acetate; the formation of α-terpineol by cyclisation; the allylic rearrangement of linalool and linalyl acetate to form nerol and geraniol and their acetates.

Protopapadakis and Papanikolaou (1998) analysed by GC/MS the oils extracted by hydrodistillation from 4 cultivars (Chania, Brazilian, Keen, Bittersweet) cultivated in Crete (Greece). The results of this study are reported below:

limonene	89.65–93.00%
myrcene	1.55–1.62%
(E)-β-ocimene	0.14–0.26%
(Z)-β-ocimene	0.01–0.02%
β-phellandrene	0.25–0.26%
α-pinene	0.37–0.39%
β-pinene	0.23–0.51%
sabinene	0.10–0.13%
α-terpinene	0.01–0.22%
γ-terpinene	0.04%
terpinolene	0.48–0.59%
geraniol	0.09–0.19%
linalool	0.82–1.87%
nerol	0.11–0.13%
terpinen-4-ol	0.08–0.15%
α-terpineol	0.40–0.55%
geranyl acetate	0.22–0.26%
linalyl acetate	0.19–0.47%
neryl acetate	0.05–0.09%
cis-linalool oxide	0.13–0.15%
trans-linalool oxide	0.18–0.24%

Caccioni et al. (1998), while determining the antimicrobial activity of citrus essential oils, analysed, among other oils, a sample of bitter orange oils obtained by hydro-distillation from fruits cultivated in an experimental field in Lentini, Sicily, Italy. The composition of this oil, determined by GC and GC/MS was the following:

δ-3-carene	0.01%
limonene	94.27%
myrcene	1.88%
(E)-β-ocimene	0.23%
α-phellandrene	0.02%
α-pinene	0.40%
β-pinene	0.33%
sabinene	0.08%
γ-terpinene	0.02%
terpinolene	0.02%
β-caryophyllene	0.07%
β-cubebene	0.05%
β-elemene	0.01%
decanal	0.11%
dodecanal	0.01%
octanal	0.08%
geranial	0.11%

neral	0.03%
perilla aldehyde	0.03%
decanol	0.05%
octanol	0.33%
geraniol	0.11%
linalool	0.78%
nerol	0.06%
terpinen-4-ol	0.06%
α-terpineol	0.26%
(E)-nerolidol	0.07%
decyl acetate	0.02%
octyl acetate	0.05%
neryl acetate	0.04%
α-terpinyl acetate	0.03%
trans-linalool oxide	0.05%

Caccioni et al. (1998) also found trace amount of camphene, (Z)-β-ocimene, α-thujene, and cis-linalool oxide.

Lin et al. (1986), Huang et al. (1990), Yang, S. et al. (1992) and Zhu et al. (1995) analysed the essential oils extracted from bitter oranges cultivated in China. The oil studied by Lin et al. (1986) had a composition which was very similar to that commonly observed for bitter orange. Few differences were due to the low content of β-pinene (0.09 per cent) and to the presence of γ-muurolene (0.03 per cent). The oil analysed by Huang et al. (1990) was characterised by a low content of limonene (86.36 per cent), a high content of γ-terpinene (4.25 per cent) and of linalool (1.91 per cent), and by numerous components in the sesquiterpene hydrocarbon fraction (δ-cadinene, γ-cadinene, β-caryophyllene, α-copaene, α-cubebene, β-elemene, δ-elemene, α-humulene), all of them present at levels lower than 0.10 per cent. The samples analysed by Yang, S. et al. (1992) were characterised by a low content of limonene (about 86 per cent) and a high content of α-terpinene (about 4 per cent). The oil analysed by Zhu et al. (1995) presented a very different composition from that commonly observed for bitter orange oil: high amount of linalool (20.69 per cent) and of linalyl acetate (30.72 per cent) that among citrus peel oils could only be compatible with bergamot oil, high amount of geranyl acetate (1.71 per cent), of neryl acetate (4.12 per cent), α-terpinyl acetate (3.10 per cent) and of camphor (4.14 per cent).

Other papers reported on the composition of bitter orange oils extracted from fruits grown in Japan. Kusunose and Sawamura (1980) analysed the diethyl ether extract of bitter orange of the Sumikan cultivar; Njoroge et al. (1994a) analysed the cold-extracted oil obtained from the Daidai cultivar; Sawamura et al. (1999) reported the composition of the cold extracted oil obtained from kabusu bitter orange. The results of these papers are summarised below:

	Kusunose and Sawamura (1980)	Njoroge et al. (1994a)	Sawamura et al. (1999)
limonene	73.8%	94.68%	92.0%
myrcene	24.3%	1.56%	1.7%
α-pinene	0.30%	0.47%	0.6%

β-pinene	0.06%	0.68%	0.9%
γ-terpinene	0.78%	0.05%	–
terpinolene	–	tr	tr
β-caryophyllene	0.04%	0.04%	–
decanal	0.03%	0.09%	0.1%
dodecanal	0.01%	tr	–
nonanal	tr	0.02%	–
octanal	0.01%	0.10%	0.2%
geranial	–	0.03%	0.1%
neral	0.01%	0.02%	0.1%
perilla aldehyde	0.04%	0.01%	–
citronellol	0.02%	tr	–
geraniol	0.02%	tr	–
linalool	0.07%	0.15%	1.1%
nerol	0.01%	tr	–
α-terpineol	0.14%	0.03%	0.1%
octyl acetate	0.01%	0.11%	–
linalyl acetate	–	1.11%	0.1%
geranyl acetate	–	0.08%	0.1%
neryl acetate	tr	0.04%	–

The oil analysed by Kusunose and Sawamura (1980) was characterised by a high content of myrcene and by a low content of limonene; in the oil analysed were also present hexanol (0.01 per cent), cis-hexen-1-ol (0.01 per cent), nonanol (0.06 per cent), β-terpineol (0.01 per cent), 1,8-cineole (0.01 per cent) and trace amounts of hexanal, 6-methyl-5-hepten-2-one, carvacrol, isopulegol, and citronellyl acetate. In addition to the components listed above, Njoroge et al. (1994a) found in the oil analysed the following components: tetradecane (0.02 per cent), (E)-β-ocimene (0.20 per cent), (Z)-β-ocimene (0.01 per cent), sabinene (0.19 per cent), bicyclogermacrene (0.01 per cent), β-farnesene* (0.01 per cent), germacrene D (0.09 per cent), nerolidol* (0.05 per cent), decyl acetate (0.02 per cent), nonyl acetate (0.01 per cent), p-mentha-1,8-dien-10-yl acetate (0.01 per cent), perillyl acetate^t (0.02 per cent), α-terpinyl acetate (0.01 per cent) and trace amounts of camphene, p-cymene, α-copaene, α-humulene, undecanal, citronellal, camphor, carvone, borneol, cis-carveol, thujyl alcohol, thymol, α-bisabolol, cedrol, elemol, dodecyl acetate, heptyl acetate, bornyl acetate, geranyl propanate, p-menth-1-en-9-yl acetate, cis- and trans-limonene oxide.

GRAPEFRUIT OIL (*Citrus paradisi* Macf.)

1960–1979

Shaw (1979) reviewed most of the papers on the composition of grapefruit oil volatile fraction published in the 60s and 70s. Among the papers reviewed by Shaw (1979) only two reported the composition of the whole volatile fraction of grapefruit oil (Attaway et al., 1967; Wilson and Shaw, 1978), while the other authors worked on fractions or single components: aldehydes (Stanley et al., 1961a; Braddok and Kesterson, 1976), monoterpene hydrocarbons (Ikeda et al., 1962a; Ashoor and Bernhard, 1967), nootkatone (MacLeod and Buigues, 1964; Kesterson et al., 1971); Yokoyama (1961) reported the citral content determined by the vanillin-piperidine and barbituric acid condensation method.

It should also be mentioned that the results reported by Ashoor and Bernhard (1967) are not directly comparable with other values found in literature since these are expressed as area per cent of the monoterpene hydrocarbons and not as per cent of volatile fraction, nor of the whole oil.

The limits of the results of the papers reviewed by Shaw, with the exception of Yokoyama's (1961) and those reported by Ashoor and Bernard (1967) are reported below:

limonene	85.60–93.0%
myrcene	1.41–1.9%
myrcene + sabinene	2.12%
ocimene*	0.36%
α-pinene	0.18–0.64%
sabinene	0.68%
δ-cadinene	0.11%
caryophyllene*	0.25%
α-copaene	0.06%
β-copaene	0.01%
β-elemene	0.06%
decanal	0.27–0.81%
dodecanal	0.11–0.19%
heptanal	0.06%
hexanal	tr
nonanal	0.04–0.13%
octanal	0.25–0.71%
tetradecanal	0.10%
undecanal	0.04–0.09%
citronellal	0.14%
geranial	0.11–0.20%
neral	0.03%
perilla aldehyde	0.2%
nootkatone	0.3–0.8%
octanol	0.76%
linalool	0.30–0.36%
terpinen-4-ol	0.08%
α-terpineol	0.16%
elemol	0.04%
decyl acetate	0.15%
octyl acetate	0.09%
neryl acetate	0.22%
cis-linalool oxide	0.03%
trans-linalool oxide	0.03%

The wide range of variation encountered for limonene by Shaw can be explained considering that the value of 85.60 per cent (Wilson and Shaw, 1978) was expressed as w per cent while the value of 93 per cent (Attaway *et al.*, 1967) was expressed as relative peak area per cent. These measures greatly differ since the non-volatile residue of grapefruit is about 7 per cent of the whole oil.

More papers than those reported by Shaw (1979), published in the 60s and 70s, are found in literature. These refer mostly to the qualitative composition of grapefruit oil volatile fraction and most of these papers were reviewed by Shaw (1977) within his first review on the composition of citrus essential oils.

Hunter and Brogden (1964a, 1965) have reported the presence of the following monoterpene and sesquiterpene hydrocarbons: limonene, myrcene, α-pinene, sabinene, γ-terpinene, γ-cadinene, δ-cadinene, α-copaene, β-copaene, caryophyllene*, β-elemene, α- and β-humulene, α-ylangene. Three components, α- and β-copaene and α-ylangene were misidentified, and later corrected as α- and β-cubebene and α-copaene respectively, by Veldhuis and Hunter (1968). Moreover, the presence of β-humulene was not later confirmed with the exception of the paper by Ziegler (1971).

Hunter and Moshonas (1966) reported on the presence of nootkatone, *cis*- and *trans*-linalool oxides, *o*-phenylphenol and on the following alcohols: decanol, dodecanol, nonanol, octanol, *cis*- and *trans*-carveol, citronellol, geraniol, linalool, *cis*- and *trans*-*p*-mentha-2,8-dien-1-ol, *p*-mentha-1,8-dien-9-ol, *p*-menth-8-en-1,2-diol, nerol, α-terpineol, elemol, nerolidol*. The presence of *o*-phenylphenol was probably due to the treatment of the fruits with antibacterial chemicals.

MacLeod (1968) using GC with capillary columns and with programmed flow and temperature, identified in the volatile fraction of grapefruit oil the following components: limonene, myrcene, α-pinene, β-pinene, caryophyllene*, citronellal, geranial, nootkatone, geranyl acetate.

Wenninger *et al.* (1967) using open column liquid chromatography, gas chromatography, fractionated distillation and silver nitrate adduction, separated and identified by IR spectroscopy the following sesquiterpene hydrocarbons in grapefruit oil: β-bisabolene, δ-cadinene, α-cubebene, β-cubebene, α-humulene.

Sulser *et al.* (1971) reported on the isolation and identification by H^1 NMR of a new sesquiterpene alcohol, selin-11-en-4-ol (paradisiol) in grapefruit oil. Huffman and Zalkow (1973) showed that the components previously identified as paradisiol, was its epimer at C-4, intermedeol.

Moshonas (1971) by fractionating on a Florasyl column and purifying by gas chromatography on preparative and analytical columns, isolated and identified the following components: decanal, dodecanal, heptanal, nonanal, octanal, undecanal, citronellal, geranial, neral, perilla aldehyde α-sinensal, β-sinensal, nootkatone, decyl acetate^t, octyl acetate^t, *trans*-carvyl acetate, citronellyl acetate, geranyl acetate^t, *p*-mentha-1,8-dien-2-yl acetate, *p*-mentha-1,8-dien-9-yl acetate, neryl acetate^t, perillyl acetate.

In the same year Ziegler (1971) reported the presence of the following components in the volatile fraction of grapefruit oil: limonene, myrcene, α-pinene, sabinene, γ-terpinene, α- and β-copaene, caryophyllene*, α-and β-humulene, β-elemene, α-ylangene, decanal, dodecanal, nonanal, octanal, undecanal, citronellal, geranial, neral, 6-methyl-5-hepten-2-one, nootkatone, geraniol, α-terpineol, octyl acetate^t, geranyl acetate, neryl acetate. What already stated about α-copaene and α-ylangene referring to the papers by Hunter and Brogden (1964a, 1965), and by Veldhuis and Hunter (1968) is probably valid also in this case.

Di Giacomo *et al.* (1963) determined the average composition of the monoterpene hydrocarbons of five Italian samples. The results obtained by Di Giacomo *et al.* and by Ashoor and Bernhard (1967) are reported below:

	Di Giacomo *et al.*, 1963	Ashoor and Bernhard, 1967
camphene	–	0.17%
p-cymene	–	0.37%
limonene	94.91%	95.0%
myrcene	2.53%	1.41%[a]
α-phellandrene	–	0.84%
β-phellandrene	–	0.66%
α-pinene	0.51%	1.61%
β-pinene	–	1.41%[a]
sabinene	1.03%	–
α-terpinene	–	0.14%
γ-terpinene	1.02%	0.49%

a myrcene + β-pinene.

Kekelidze and Beradze (1974) reported on two oils, obtained in laboratory by distillation from the peel of the fruits of the cvs. Duncan and Marsh, cultivated in Georgia (ex USSR). They provided the following information for the two samples respectively: p-cymene (0.03 per cent, 0.03 per cent); limonene (91.6 per cent, 94.3 per cent); myrcene (1.2 per cent, 1.9 per cent); α-pinene (0.4 per cent, 0.6 per cent); sabinene (0.2 per cent, 1.2 per cent); γ-terpinene (1.2 per cent, 0.1 per cent); terpinolene (0, 0.01 per cent); citronellol (0.1 per cent, 0.2 per cent); linalool (0.1 per cent, 0.1 per cent); α-terpineol (0.1 per cent, 0.3 per cent).

1979–1999

Industrial cold-pressed oils

In Table 13.3 are summarised the data relative to the composition of the volatile fraction of industrial grapefruit oil, found in literature since 1979. These data are mostly in accordance with one another. They provide a good picture of the composition of the volatile fraction of grapefruit oil industrially produced. This is true although the data refer to researches carried out over 20 years, from 1980 to 1999, and none of the data refer to a very large number of samples, on the contrary many refer to a single sample and some to commercial oils (see Appendix to Table 13.3).

Table 13.3 Percentage composition of the volatile fraction of cold-pressed grapefruit oil

	1.2	3	4	5	6	7	8(a)	8(b)	9
HYDROCARBONS									
Monoterpene									
camphene	–	0.01	tr	–	–	–	tr	tr	tr
δ-3-carene	–	0.03	–	–	tr	–	tr	0.01	–
p-cymene	–	0.13	0.10–0.32	0.02	–	–	0.04	0.02	b

limonene	84.84	92.53	92.20–95.40	83.40	92.92	93.7	92.99	93.81	93.44[b]
myrcene	3.67	1.87	1.82–2.63	1.37	1.98	1.9	1.97	1.83	1.92
(E)-β-ocimene	–	–	–	–	0.13	–	0.14	0.16	0.31
α-phellandrene	–	0.03	tr	–	tr	–	0.04	0.04	0.20
β-phellandrene	–	–	tr–0.04	–	–	1.3	–	–	b
α-pinene	0.38	0.54	0.64–0.75	0.38	0.65	0.6	0.59	0.53	0.56
β-pinene	0.05	0.86[a]	0.03–0.04	0.02	tr	0.1	0.26	0.15	0.26
sabinene	1.08	0.86[a]	0.48–0.95	0.42	0.50	0.5	0.38	0.41	1.12
α-terpinene	–	0.01	–	–	–	–	tr	tr	–
γ-terpinene	0.12	0.18	tr–0.04	0.01	–	0.3	0.16	0.10	0.13
terpinolene	–	0.02	–	–	tr	–	0.02	0.01	0.01
α-thujene	–	–	–	–	–	–	0.01	0.01	0.01
Sesquiterpene									
trans-α-bergamotene	–	–	–	–	–	–	0.01	tr	tr
β-bisabolene	–	–	0.21–0.41	–	–	–	0.04	0.03	0.02
γ-cadinene	–	–	–	–	tr	–	–	–	0.01
δ-cadinene	0.12	–	–	–	0.13	–	0.09	0.08	0.09
β-caryophyllene	0.25	0.28	0.18–0.33	0.24	0.46	0.3	0.31	0.27	0.25
α-copaene	0.06	–	–	–	0.16	–	–	–	0.08
β-copaene	0.04	–	–	–	–	–	0.06	0.06	–
β-elemene	0.02	–	–	–	–	–	0.09	0.08	0.05
(E,E)-α-farnesene	–	–	–	–	tr	–	–	–	0.01
β-farnesene	–	–	–	–	–	–	0.01*	0.01*	0.03[f]
α-humulene	0.07	–	–	–	–	–	0.04	0.04	0.03
valencene	–	–	–	tr	tr	–	–	–	–
ALDEHYDES									
Aliphatic									
decanal	0.46	0.33	0.21–0.38	0.40	0.17	–	0.27	0.25	0.27
dodecanal	0.22	–	0.04–0.06	0.02	tr	–	0.02	0.02	0.02
nonanal	0.12	0.05	0.04–0.07	0.07	–	–	0.05	0.01	0.04
octanal	0.79	0.04	–	0.62	–	0.3	0.29	0.32	0.25
undecanal	–	–	0–0.02	–	tr	–	0.01	0.01	0.01
Monoterpene									
citronellal	0.13	0.08	0.06–0.07[c]	0.09	tr	–	0.07	0.05	0.05
geranial	0.11	0.07	0.02–0.07	0.06	tr	–	0.08	0.07	0.10
neral	0.11	0.06	0.03–0.06	0.04	tr	–	0.05	0.04	0.05
perilla aldehyde	0.07	–	–	–	tr	–	0.01	0.01	tr
Sesquiterpene									
β-sinensal	–	–	–	–	–	–	0.01	0.01	0.01
KETONES									
Monoterpene									
carvone	0.02	–	–	0.06	0.21	–	0.01	0.01	0.02
Sesquiterpene									
nootkatone	0.02	–	0.04–0.08	0.10	0.84	–	0.30	0.14	0.02
ALCOHOLS									
Aliphatic									
decanol	–	–	–	–	–	–	tr	tr	tr
octanol	0.04	–	0.05–0.08	0.09	tr	–	0.02	0.04	0.01
Monoterpene									
trans-carveol	–	–	–	–	0.24	–	–	–	tr
citronellol	–	–	0.04–0.06	–	–	–	0.01	tr	–
geraniol	–	–	tr	tr	–	–	tr	tr	tr[d]

Table 13.3 (Continued)

	1.2	3	4	5	6	7	8(a)	8(b)	9
linalool	0.14	0.16	0.06–0.09	0.10	tr	tr	0.09	0.10	0.06
nerol	–	–	–	–	–	–	tr	tr	tr
α-terpineol	0.05	0.05	0.02–0.03	0.04	tr	tr	0.04	0.04	0.03
terpinen-4-ol	–	–	–	–	–	–	tr	tr	0.01
Sesquiterpene									
elemol	0.04	–	–	–	–	–	–	–	0.04[t]
ESTERS									
Aliphatic									
decyl acetate	–	–	–	–	–	–	0.10	0.09	0.01
octyl acetate	0.05	0.01	0.06–0.07[c]	–	–	–	0.03	0.03	0.02
Monoterpene									
citronellyl acetate	0.06	–	tr	–	–	–	0.01	0.01	tr
geranyl acetate	0.04	0.11	0.02–0.06	–	0.13	0.1	0.07	0.06	–
linalyl acetate	–	–	0.03–0.06	–	–	–	–	–	tr[d]
neryl acetate	0.02	–	–	–	tr	tr	0.02	0.02	0.01
α-terpinyl acetate							0.01	0.01	0.01
OXIDES									
Monoterpene									
cis-limonene oxide	0.09	–	–	–	–	–	0.02[e]	0.02[e]	tr
trans-limonene oxide	0.04	–	–	–	–	–	0.02[e]	0.02[e]	tr

Notes

t = tentative; tr = traces; * correct isomer not characterised; a β-pinene + sabinene; b limonene + *p*-cymene + β-phellandrene; c citronellal + octyl acetate; d geraniol + linalyl acetate; e *cis*-limonene oxide + *trans*-limonene oxide; f (Z)-β-farnesene.

Appendix to Table 13.3

1 Wilson and Shaw (1980),
2 Wilson *et al.* (1981). USA; One sample from Duncan and Marsh grapefruit; GC on capillary column coated with Carbovax 20 M; W%.
3 Cappello *et al.* (1981). Argentina; 2 samples; GC on capillary column coated with UCON LB 550 X; relative percentage of peak areas. Cappello *et al.* also found 6-methyl-5-hepten-2-one (0.07%).
4 Koketsu *et al.* (1983). Brazil; FMC; 3 samples; GC on packed columns of Carbowax 20 M and of SE-30 and on capillary column coated with Carbowax 20 M; retention times and Kovats indices; relative percentage of peak areas. Koketsu *et al.* also found nonanol (0.09–0.18%) and methyl N-methyl anthranilate (tr).
5 Myers (1988). USA. Myers also found trace amounts of (E)-2-hexenal, (Z)-3-hexenol, ethyl butyrate.
6 Inoma *et al.* (1989). One commercial sample; GC on capillary columns coated with OV-101 and Carbowax 20 M; GC/MS; relative percentage of peak areas. Inoma *et al.* also found *cis*-carveol (0.12%), limonene oxide* (0.28%), limonene dioxide (0.18%), and trace amounts of *trans*-hexenylbutyrate.
7 Haubruge *et al.* (1989). One commercial sample; GC on capillary column coated with CP-TM-WAX 52 CB; GC/MS; relative percentage of peak areas.
8 Boelens (1991). Israel, white (a) and pink (b); GC; relative percentage of peak areas. Boelens also found mentha-1,8(9)-dien-10-yl acetate (0.01%–tr), and trace amounts of (E,E)-2,4-decadienal, (E,Z)-2,4-decadienal, α-sinensal, perillene.
9 Dugo *et al.* (1999). Italy; one sample; GC on capillary columns coated with SE-52 and DB-5; GC/MS with interactive use of linear retention indices. Dugo *et al.* also found (Z)-β-ocimene (0.01%), bicyclogermacrene (0.03%), germacrene D (0.06%), hexadecanal (0.01%), *cis*-sabinene hydrate (0.01%), *cis*-β-terpineol (0.01%), (E,E)-farnesol (0.01%), (E,Z)-farnesol (0.01%), caryophyllene oxide (0.03%), and trace amounts of tetradecanal, germacrene D-4-ol[t], (E)-nerolidol, *trans*-carvyl acetate[t].

The apparent differences among the quantitative data reported by Wilson and Shaw (1980), by Wilson et al. (1981), by Myers (1988), and those reported by other authors, particularly evident for limonene, are due to the different expression of the quantitative results: the data in rows 1, 2 and 5 are expressed as w per cent of the whole oil, all the others are calculated as relative per cent of peak area obtained by GC analysis of the volatile fraction. The high content of non-volatile residue of grapefruit oil, about 7 per cent, determines these high differences among the data reported in the table.

However, among the results reported in the table, some are not in agreement with all the others: the content of myrcene of 3.67 per cent, reported by Wilson and Shaw (1980) and by Wilson et al. (1981), appears to be high. This value is out of range not only for grapefruit oil, but also for any other cold-pressed citrus peel essential oil. This value could be explained by a possible co-elution of myrcene with another component during the GC analysis performed by these authors. The values reported by Inoma et al. for carvone (0.21 per cent), cis-carveol (0.12 per cent) and trans-carveol (0.24 per cent) are also high.

In addition to the data reported in Table 13.3, in literature other researches on particular aspects of the composition of this oil can be found.

Wilson and Shaw (1984) proposed a GC method on a DB-5 capillary column for the separation of carbonyl compounds (aldehydes and ketones) in various US citrus oils. From the analysis of two samples of Florida grapefruit essential oils they reported the following results:

decanal	0.42%	0.39%
dodecanal	0.05%	0.05%
nonanal	0.08%	0.07%
octanal	0.52%	0.47%
citronellal	0.09%	0.08%
geranial	0.07%	0.09%
neral	0.04%	0.05%
perilla aldehyde	0.02%	0.02%
β-sinensal	0.02%	0.03%
nootkatone	0.24%	0.15%

The data obtained by GC are in agreement with those obtained by the same authors using the method approved by the United States Pharmacopoeia (USP, 1965): 1.55 per cent and 1.46 per cent respectively for the two samples.

In an industrial grapefruit oil produced in Sicily, Mondello et al. (1995b) identified, by on-line HPLC-HRGC/MS, the following sesquiterpene hydrocarbons, listed in decreasing amount order: β-caryophyllene, α-copaene, β-cubebene, δ-cadinene, germacrene D, α-humulene, (Z)-β-farnesene, (E,E)-α-farnesene[t], α-muurolene[t]. The presence of α-copaene, β-cubebene, and germacrene D in grapefruit oil from Florida has been confirmed by Chamblee et al. (1997), using a GC/FT-IR. In this work has also been reported the presence in grapefruit oil, of α-cubebene, bicyclogermacrene and germacrene B.

Schulz *et al.* (1992) proposed a HPLC method for the quantitative determination of nootkatone and the oxygen heterocyclic compounds in grapefruit oil. Three industrial cold-pressed essential oils, produced in Florida (USA), in Italy and in Argentina, were reported to contain the following amounts of nootkatone: 0.7 per cent, 0.2 per cent and 0.6 per cent respectively.

Laboratory extracted oils

In addition to the research carried on industrial grapefruit oils, a literature review has found different papers on the composition of laboratory extracted grapefruit oils, from different geographic origins. Kekelidze *et al.* (1985) reported for 2 different cultivars of grapefruit, Duncan and Marsh, cultivated in Georgia (ex USSR), the following composition:

	Duncan	Marsh
p-cymene	0.40%	0.10%
limonene	88.50%	92.50%
myrcene	0.82%	1.40%
α-phellandrene	0.30%	0.50%
α-pinene	0.15%	0.45%
sabinene	0.81%	0.40%
γ-terpinene	0.05%	0.07%
cadinene*	0.13%	0.05%
β-caryophyllene	0.23%	0.16%
β-copaene	0.02%	0.01%
decanal	0.20%	0.12%
nonanal	0.12%	0.08%
octanal	0.44%	0.23%
citronellal	0.12%	0.07%
neral	0.02%	0.04%
octanol	0.21%	0.12%
linalool	0.13%	0.08%
terpinen-4-ol	0.20%	0.12%
α-terpineol	0.23%	0.12%
decyl acetate	0.23%	–
neryl acetate	0.34%	0.16%
neryl-formate	0.14%	–

Wilson *et al.* (1990) noticed that the content of nootkatone in grapefruit oil was influenced by treatments prior to the harvest of the fruit, with giberellic acid (GA) and with 2-(3,4-dichlorophenoxy)-triethylamine (DCPTA). In particular, the treatment with GA causes a reduction, in function of the dosage, of the rate of increase in nootkatone concentration observed in natural fruit with maturation. The treatment with DCPTA determined an increase of the content of nootkatone. If both chemicals were used simultaneously, the effect of GA was the predominant one.

Dugo et al. (1990), Ruberto et al. (1993, 1997) and Caccioni et al. (1998), analysed some grapefruit oils laboratory extracted from Sicilian fruits. Their results are summarised in Table 13.4.

Sawamura et al. (1991) analysed, by GC/MS after pre-fractionation on a silica gel column, two samples of cold-pressed laboratory extracted grapefruit oil obtained from

Table 13.4 Percentage composition of some italian laboratory extracted grapefruit oils

	1	2	3	4(a)	4(b)
HYDROCARBONS					
Monoterpene					
camphene	tr	tr	tr	tr	tr
limonene	93.54	92.16	93.05	93.59	93.70
myrcene	1.98	2.09	1.76	1.81	1.86
(E)-β-ocimene	0.37	0.22	0.24	0.29	0.21
(Z)-β-ocimene	0.02	–	tr	tr	–
α-phellandrene	0.45[a]	0.03	0.08	0.05	0.07
β-phellandrene	–	tr	–	–	0.18
α-pinene	0.52	0.33	0.42	0.51	0.52
β-pinene	0.04	0.02	0.02	0.07	0.07
sabinene	0.43	0.36	1.20	0.54	0.54
α-terpinene	0.01	–	0.01	0.04	0.03
γ-terpinene	0.01	0.01	0.02	0.07	0.08
terpinolene	0.01	0.31	tr	0.02	0.02
α-thujene	0.01	tr	tr	tr	tr
Sesquiterpene					
β-caryophyllene	0.25	0.34	0.31	0.18	0.17
α-copaene	0.08	0.06	0.06	0.04	0.04
α-cubebene	0.08	0.06	–	–	–
α-humulene	0.03	tr	0.03	0.02	0.02
ALDEHYDES					
Aliphatic					
decanal	0.30	0.25	0.22	0.26	0.26
dodecanal	–	0.02	tr	0.02	0.02
dodec-2-en-1-al*	–	0.05	tr	0.03	0.02
nonanal	0.06	0.07	0.07	0.06	0.05
octanal	0.45[a]	–	0.23	0.58	0.34
undecanal	0.01	tr	–	–	–
Monoterpene					
geranial	0.13	0.11	0.08	0.10	0.07
neral	0.08	–	0.06	0.06	0.05
KETONES					
Sesquiterpene					
nootkatone	0.10	0.74	0.54	0.13	0.07
ALCOHOLS					
Aliphatic					
decanol	–	0.08	0.03	0.01	0.02
nonanol	–	0.02	0.02	tr	tr
octanol	0.01	0.30	0.07	0.14	0.14
Monoterpene					
carveol*	–	0.03	0.10	0.05	0.05
geraniol	0.01	0.05	tr	0.01	0.02
isopulegol	–	0.05	0.04	0.06	0.05

Table 13.4 (Continued)

	1	2	3	4(a)	4(b)
linalool	0.12	0.34	0.45	0.25	0.20
nerol	0.01	0.07	0.03	0.02	0.03
terpinen-4-ol	–	0.07	0.09	0.17	0.18
α-terpineol	0.07	0.18	0.20	0.14	0.15
ESTERS					
Aliphatic					
decyl acetate	–	0.01	tr	tr	tr
Monoterpene					
geranyl acetate	0.06	0.13	0.08	0.05	0.06
neryl acetate	0.01	0.01	tr	tr	0.01
α-terpinyl acetate	–	0.01	0.04	0.01	tr
OXIDES					
Monoterpene					
trans-linalool oxide	–	0.35	–	0.23	0.21

Notes
tr = traces; a correct isomer not characterised; b α-phellandrene + octanal.

Appendix to Table 13.4

1 Dugo Giacomo *et al.* (1990). Sicily, Italy; 2 samples laboratory extracted by applying manual pressure on the rind; GC on capillary column coated with SE-52; relative percentage of peak areas. Dugo *et al.* also found β-cubebene (0.07%), citronellal (0.05%), *cis*-sabinene hydrate (0.01%), citronellyl acetate (0.01%) and trace amounts of *trans*-sabinene hydrate.
2 Ruberto *et al.* (1993). Sicily, Italy; one sample of Marsh seedless grapefruit oil laboratory extracted by simultaneous distillation-solvent extraction; GC on capillary column coated with HP-1; GC/MS, retention times. Ruberto *et al.* also found (E,E)-α-farnesene (0.02%), hex-1-en-3-ol (0.01%), carvone (0.08%) *cis*-linalool oxide (0.76%) and trace amounts of δ-3-carene and carvyl acetate.
3 Ruberto *et al.* (1997). Sicily, Italy; one sample of Duncan grapefruit oil laboratory extracted by simultaneous distillation-solvent extraction; GC on capillary column coated with HP-1; GC/MS, retention times. Ruberto *et al.* also found linalool oxide* (0.66%).
4 Caccioni *et al.* (1998). Sicily, Italy; FMC; one sample of Marsh seedless (a) and Red Blush (b) grapefruit oil laboratory extracted by steam distillation; GC on capillary column coated with HP-1; GC/MS, retention times. Caccioni *et al.* also found in Red Blush grapefruit oil *p*-cymene (0.18%).

fruits provided by the Experimental Station of Kochi (Japan). The results, expressed as w per cent, are reported below:

camphene	tr	–
p-cymene	0.01%	–
limonene	76.14%	83.11%
myrcene	1.32%	1.29%
α-pinene	0.42%	0.47%
β-pinene	0.04%	0.02%
sabinene	0.60%	0.27%
γ-terpinene	0.12%	0.08%
terpinolene	0.01%	0.01%
δ-cadinene	0.15%	0.15%
α-caryophyllene	0.06%	0.04%

α-copaene	0.12%	0.10%
β-cubebene	0.11%	0.16%
β-elemene	0.01%	–
α-muurolene	0.01%	0.03%
valencene	0.04%	0.02%
decanal	0.45%	0.53%
nonanal	0.05%	0.06%
octanal	0.24%	0.45%
undecanal	0.01%	0.01%
2-undecenal*	0.01%	0.01%
citronellal	0.14%	0.12%
geranial	0.13%	0.07%
neral	0.06%	–
nootkatone	0.10%	0.07%
citronellol	0.03%	0.03%
geraniol	0.02%	0.01%
linalool	0.30%	0.17%
terpinen-4-ol	0.40%	0.24%
α-terpineol	0.10%	0.05%
elemol	0.04%	0.02%
nerolidol*	0.02%	0.01%
decyl acetate	0.02%	–
nonyl acetate	0.06%	0.03%
octyl acetate	0.09%	0.04%
geranyl acetate	0.15%	0.10%

Sawamura *et al.* (1991) reported α-caryophyllene and did not report the presence of the β-isomer of this component that is the one reported by all the other authors.

Sun and Petracek (1999) studied the influence on the composition of grapefruit essential oils, of the following processes: waxing, time and temperature storage of the fruit. They obtained the following range of composition of the volatile fraction of the oil:

δ-3-carene	0.04–0.27%
limonene	89.08–92.95%
myrcene	1.89–1.95%
ocimene*	0.19–0.38%
α-phellandrene	0.06–0.19%
α-pinene	0.60–0.63%
β-pinene	0.06–0.09%
sabinene	0.54–0.85%
γ-terpinene	0.04–0.17%
γ-cadinene	0.11–0.12%
β-caryophyllene	0.16–0.18%
α-copaene	0.08–0.09%
α-humulene	0.02–0.03%
decanal	0.47–0.51%

dodecanal	0.07–0.11%
nonanal	0.09–0.15%
octanal	0.51–0.76%
citronellal	0.03–0.06%
neral	0.03–0.05%
nootkatone	0.28–0.64%
octanol	0.01–0.05%
linalool	0.04–0.10%
cis-p-mentha-2,8-dien-1-ol	0.02–0.24%
α-terpineol	0.01–0.05%
cis-linalool oxide	0.01–0.05%
trans-linalool oxide	0.07–0.33%

The results obtained indicated that waxing the fruit caused a decrease of nonanal and nootkatone; limonene decreased with the increasing of storage temperature; the content of β-pinene, α-phellandrene, δ-3-carene, ocimene*, octanol, trans-linalool oxide and cis-p-mentha-2,8-dien-1-ol increased with the storage temperature; the content of α-pinene, limonene, linalool, citronellal, α-terpineol, neral, dodecanal, and α-humulene decreased with the time of storage, while α-phellandrene, δ-3-carene, ocimene*, and trans-linalool oxide presented the opposite behaviour.

LIME OILS

1960–1979

Most of the papers published before 1979 on the volatile fraction of lime cold-pressed oils have been reviewed by Shaw (1979).

Among the papers reviewed by Shaw (1979), one (Yokoyama et al., 1961) reported only the value of the citral content determined by the vanillin-piperidine and barbituric acid condensation method, two (Ikeda et al., 1962a; Ashoor and Bernhard, 1967) only illustrated the monoterpene hydrocarbon fraction of cold-pressed lime oil, one (MacLeod and Buigues, 1964) simply reports on the content of nootkatone, and three (Scora et al., 1968; Shaw et al., 1971; Shaw and Wilson, 1976) studied more extensively the volatile fraction of this oil. The papers by Shaw et al. (1971) and by Shaw and Wilson (1976) are relative to Persian lime oil (*Citrus latifolia* Tanaka); the paper by Scora et al. (1968) was on the Rangpur lime oil laboratory extracted; the papers by Ikeda et al. (1962a), by Ashoor and Bernhard (1967), by MacLeod and Buigues (1964) are relative to oils of not specified botanical origin.

The high values of limonene (64 per cent) and of γ-terpinene (21.7 per cent) reported by Shaw (1979), relative to the paper by Ashoor and Bernhard (1967), do not refer to the percentages of these two compounds in the whole oil, nor to the whole volatile fraction, but are limited to the monoterpene hydrocarbons fraction. The value of 10.3 per cent of myrcene reported by Shaw from the paper by Scora et al. (1968), is too high for a genuine lime oil, but could be due to the peak overlap of myrcene, β-pinene and sabinene. In fact, using a packed column with LAC 446 stationary phase, as the one used by Scora et al. (1968), these components do not separate, as previously reported by Ashoor and

Bernard (1967). The contents of β-pinene and sabinene were, in fact, not reported in the paper by Scora *et al.* (1968).

A summary of the results reported in the papers reviewed by Shaw, with the exception of the paper by Ashoor and Bernhard (1967) which shows values not directly comparable with those obtained by the other authors, the results obtained by Yokoyama *et al.* (1961), and the value of myrcene, reported by Scora *et al.* (1968), are reported in the table below:

p-cymene	0.49–1.3%
limonene	46.78–63.7%
myrcene	0.69%
phellandrene*	0.2%
α-pinene	1.52–2.44%
β-pinene	10.08–11.90%
sabinene	1.58%
γ-terpinene	7.34–16.26%
terpinolene	0.63–1.22%
α-thujene	0.41%
α-bergamotene* + β-caryophyllene	1.51–2.46%
β-bisabolene	2.47–4.02%
decanal	0.1%
nonanal	0.2%
octanal	0.8%
citronellal	1.4%
geranial	5.08–6.81%
neral	0.4–4.63%
nootkatone	0.01%
nonanol	0.1%
linalool	0.09–0.2%
α-terpineol	0.3–1.05%
decyl acetate	0.1%
geranyl acetate	0.5%
geranyl acetate + neryl acetae	3.05%
neryl acetate	2.49%
neryl formate	0.2%

In the period 1960–1979 it was possible to find in literature more papers than those reviewed by Shaw (1979); some of these were previously reviewed by the same author (Shaw, 1977). These are relative to the qualitative composition (Slater, 1961; Hunter and Brogden, 1965; Hunter and Moshonas, 1966; MacLeod, 1968; Ziegler, 1971) and quantitative composition (Haro Guzman and Huet, 1970; Azzouz and Reineccius, 1976; Huet *et al.*, 1978) of lime oil. The results of these papers were as follows:

Slater (1961): camphene, limonene, β-pinene, α-pinene, γ-terpinene, terpinolene, geranial, neral, linalool, 1,4-cineole, 1,8-cineole.

Hunter and Brogden (1965): pentadecane, tetradecane, *p*-cymene, limonene, myrcene, α-pinene, β-pinene, γ-terpinene, terpinolene, α-bergamotene*, β-bisabolene, caryophyllene*, α-elemene, δ-elemene, α-humulene, β-humulene.

Hunter and Moshonas (1966): decanol, nonanol, octanol, geraniol, linalool, nerol, terpinen-4-ol, α-terpineol, and geranial.

MacLeod (1968): camphene, *p*-cymene, limonene, myrcene, α-pinene, β-pinene, sabinene, γ-terpinene, terpinolene, α-thujene, α-bergamotene*, β-bisabolene, caryophyllene*, geranial, neral, terpinen-4-ol, α-terpineol, geranyl acetate, neryl acetate.

Ziegler (1971): camphene, *p*-cymene, limonene, myrcene, α-pinene, β-pinene, sabinene, γ-terpinene, terpinolene, β-bisabolene, caryophyllene*, α-humulene, β-humulene, decanal, dodecanal, octanal, undecanal, neral, geranial, nootkatone, linalool, α-terpineol, geranyl acetate, neryl acetate.

	Haro Guzman and Huet (1970)	Azzous and Reineccius (1976)	Huet et al. (1978)	
			Key	Persian
camphene	0.13%	0.17–0.18%	0.09%	–
p-cymene	1.04%	–	–	0.66–1.39%
limonene	49.20%	34.30–44.17%	48.43%	46.08–48.80%
myrcene	1.74%	1.40–2.10%	1.33%	1.39–1.54%
α-pinene	3.42%	2.57–3.72%	2.45%	2.07–2.91%
β-pinene	21.80%	23.81–24.23%	20.22%	11.76–12.50%
sabinene	–	–	3.07%	2.22–2.23%
γ-terpinene	8.64%	11.88–14.00%	10.10%	19.71–20.17%
terpinolene	0.57%[a]	0.76–0.83%	0.63%	0.99–1.13%
α-bergamotene*	1.67%[b]	1.34–2.16%	0.95%[b]	1.44–1.54%[b]
β-bisabolene	3.98%[c]	3.18–5.70%	–	–
β-caryophyllene	1.17%[d]	0.70–1.36%	–	–
decanal	0.27%	0.48–0.71%	0.15%	–
citronellal	–	–	0.25%	0.08–0.10%
geranial	3.98%[c]	1.70–2.97%	5.06%	4.69–5.75%
neral	0.50%	1.13–1.50%	2.33%	0.04–1.39%
geraniol	–	tr	0.03%	0.10–0.15%
linalool	0.21%	0.25–0.31%	0.31%	0.41–0.45%
nerol	–	tr	–	0.12–0.16%
terpinen-4-ol	1.67%[b]	0.32–0.53%	0.95%[b]	1.44–2.54%[b]
α-terpineol	0.60%	–	0.39%	0.47–0.49%
1,8-cineole	1.17%	0.26–0.35%	–	–

a terpinolene + octanal; b α-bergamotene* + terpinen-4-ol; c β-bisabolene + geranial + decanol; d β-caryophyllene + nonanol + β-terpineol*.

Haro-Guzmán and Huet (1970) also determined in lime essential oil 1,4-cineole (0.23 per cent) and Azzouz and Reineccius (1976) also found the following compounds: nonane (0.02–0.04 per cent) α-thujene (0.54–0.55 per cent), α-elemene (0.54–0.77 per cent), β-elemene (0.33–0.53 per cent), guaiene* (0.42–0.78 per cent), α- + β-humulene (0.18 per cent), geranyl acetate (0.16–0.36 per cent), neryl acetate (0.07–0.16 per cent) and trace amounts of tert-amyl alcohol, decanol, 2-methyl-3-buten-2-ol, octanol, thymol.

1979–1999

Industrial cold-pressed key lime oil (Citrus aurantifolia (Chistm.) Swing.*)*

In Table 13.5 are reported the results relative to the composition of the volatile fraction of Key lime oil industrially processed.

Table 13.5 Percentage composition of the volatile fraction of cold-pressed key lime oil

	1		2		3		4
	Type A	Type B	Type A	Type B	Type A	Type B	Type A
HYDROCARBONS							
Aliphatic							
decane	–	–	–	–	0–tr	tr	tr
dodecane	–	–	–	–	tr–0.01	0.03	0.01
nonane	–	–	–	–	0.01–0.02	0.02	0.01
tridecane	–	–	–	–	0.01–0.03	0.01	0.02
Monoterpene							
camphene	0.10	0.09	0.10	0.10	0.11–0.12	0.11	0.12
δ-3-carene	0.02[a]	0.02[a]	–	–	tr	tr	tr
p-cymene	0.62	0.27	0.38	0.17	0.23–0.41	1.95	0.32
limonene	50.01[b]	47.87[b]	43.89[b]	38.40[b]	49.28–49.42	49.38	49.35[g]
myrcene	1.17[c]	1.22[c]	1.09	1.04	1.24–1.30	1.18	1.26
(E)-β-ocimene	–	–	0.31	0.37	0.38–0.40	0.34	0.39
(Z)-β-ocimene	–	–	–	–	0.13–0.14	0.13	0.14
α-phellandrene	0.04	0.05	0.04	0.04	0.02–0.04	0.03	0.03
α-pinene	2.23	2.16	2.28	2.10	2.29–2.70	2.44	2.45
β-pinene	19.95	19.53	19.50	17.42	21.78–25.45[f]	24.33[f]	23.36[f]
sabinene	3.04	3.28	3.10	3.19	f	f	f
α-terpinene	0.17	0.16	0.14	0.15	0.17–0.35	0.09	0.25
γ-terpinene	7.10[d]	8.23[d]	7.39	7.59	7.70–8.04	6.19	7.79
terpinolene	0.41[a]	0.45[e]	0.35	0.38	0.37–0.49	0.31	0.42
thuja-2,4(10)-diene	–	–	–	–	tr–0.01	tr	tr
α-thujene	0.35	0.37	0.38	0.36	0.36–0.43	0.39	0.39
tryciclene	–	–	–	–	0.01	0.01	0.01
Sesquiterpene							
cis-α-bergamotene	–	–	0.12	0.11	0.08–0.10	0.09	0.09
trans-α-bergamotene	1.27	1.20	1.15	1.01	1.13–1.58	1.35	1.36
(Z)-α-bisabolene	–	–	–	–	0.06–0.20	0.15	0.15
α-bisabolene	–	–	0.05	0.07	–	–	–
β-bisabolene	2.11	1.97	1.54	1.27	2.60–3.79[h]	3.08[h]	3.18[h]
β-caryophyllene	0.93	1.16	0.82	0.89	0.96–1.19	0.98	1.07
β-elemene	–	–	–	–	0.12–0.19	0.16	0.18
δ-elemene	–	–	0.06[t]	0.08[t]	0.25–0.37	0.07	0.31
(E,E)-α-farnesene	–	–	1.24[t]	1.33[t]	2.60–3.79[h]	3.08[h]	3.18[h]
(E)-β-farnesene	–	–	0.11[t]	0.10[t]	0.11–0.16	0.12	0.13
(Z)-β-farnesene	–	–	–	–	0.01–0.02	0.01	0.01
germacrene B	–	–	0.14	0.39	0.43–0.57	0.33	0.50
germacrene D	–	–	–	–	0.26–0.49	0.16	0.35
α-humulene	–	–	0.14[t]	0.13[t]	0.10–0.14	0.11	0.12
β-santalene	–	–	–	–	0.04–0.06	0.05	0.05
α-selinene	–	–	–	–	0.08–0.13	0.08	0.11

Table 13.5 (Continued)

	1		2		3		4
	Type A	Type B	Type A	Type B	Type A	Type B	Type A
ALDEHYDES							
Aliphatic							
decanal	0.20	0.24	0.25	0.29	0.20–0.22	0.14	0.21
dodecanal	0.11	0.12	0.17	0.16	0.10–013	0.08	0.18
hexadecanal	0.06	0.06	–	–	0.04–0.07	0.04	0.06
nonanal	0.41	0.45[e]	0.02	0.02	0.02–0.03	0.02	0.02
octanal	1.17[c]	1.22[c]	0.04	–	0.05–0.06	0.04	0.06
pentadecanal	0.01	0.01	–	–	–	–	tr
tetradecanal	0.04	0.04	0.04	0.04	0.04–0.06	0.05	0.05
undecanal	0.03	0.03	0.04	0.05	0.02–0.04	0.02	0.03
Monoterpene							
citronellal	0.03	0.04	0.03	0.05	0.01	0.03	0.01
geranial	2.36	2.99	2.66	3.65	1.71–2.00[i]	1.97[i]	1.86[i]
neral	1.43	1.82	1.73	2.31	1.06–1.25	1.15	1.15
perilla aldehyde	–	–	–	–	1.71–2.00[i]	1.97[i]	1.86[i]
KETONES							
Aliphatic							
6-methyl-5-hepten-2-one	0.01	0.02	0.02	0.02	0.01–0.03	0.01	0.02
Monoterpene							
cis-pinocamphone	–	–	–	–	0.02	0.02	0.02
piperitone	–	–	–	–	tr	tr	tr[t]
ALCOHOLS							
Monoterpene							
borneol	–	–	0.02	0.03	0.01–0.04	0.02	0.03
geraniol	0.03	0.07	0.04	0.07	0.03–0.07	0.04	0.05
linalool	0.15	0.16	0.21	0.25	0.16–0.18	0.17	0.17
cis-p-menth-2-en-1-ol	–	–	–	–	0.01–0.02	0.01	0.01
trans-p-menth-2-en-1-ol	–	–	–	–	tr–0.01	0.05	0.01
nerol	0.02	0.04	0.03	0.05	0.02–0.05	0.09	0.04
trans-pinocarveol	–	–	–	–	0.01	tr	0.01
cis-sabinene hydrate	–	–	–	–	0.02–0.05	0.01	0.04
trans-sabinene hydrate	–	–	0.05	0.08	0.02–0.04	0.01	0.04
α-terpineol	0.26	0.30	0.23	0.29	0.22–0.35	0.21	0.28
terpinen-4-ol	0.22	0.04	0.23	0.09	0.37–0.70	0.14	0.52
Sesquiterpene							
α-bisabolol	–	–	0.11	0.08	0.07–0.10	0.10	0.09
campherenol	–	–	0.05	0.04	0.05–0.07	0.07	0.06
norbornanol[j]	–	–	0.08[t]	0.06[t]	0.05–0.08	0.07	0.07
ESTERS							
Aliphatic							
bornyl acetate	–	–	0.04	0.04	0.01	0.02	0.01
citronellyl acetate	–	–	0.01	–	–	–	tr
geranyl acetate	0.27	0.25	0.29	0.31	0.21–0.29	0.25	0.24
neryl acetate	0.47[k]	0.51[k]	0.10	0.10	0.07–0.09	0.07	0.08
OXIDES							
Monoterpene							
dehydro-1,8 cineole	–	–	–	–	0–tr[t]	tr[t]	tr[t]

Notes

t = tentative identification; tr = traces; a δ-3-carene + 1,4-cineole; b limonene + 1,8-cineole; c myrcene + octanal; d γ-terpinene + octanol; e terpinolene + nonanal; f β-pinene + sabinene; g limonene + β-phellandrene + 1,8-cineole; h β-bisabolene + (E,E)-α-farnesene; i geranial + perilla aldehyde; j 2,3-dimethyl-3-(4-methyl-3-pentenyl)-2 norbornanol; k neryl acetate + unknown hydrocarbon.

Appendix to Table 13.5

1   Haro and Faas (1985). Mexico; one sample type A and one type B; column chromatography on silica-gel impregnated with 0.7% of Carbowax 1540; GC on capillary column coated with SE-30; relative percentage of peak areas. Haro and Faas also found β-sesquiphellandrene (1.55% in type A and 1.89% in type B), tridecanal (0.01% in type A).
2   Clark and Chamblee (1992). Absolute w%, Clark and Chamblee also found β-copaene[t] (0.33% in type A and 0.30% in type B), heptanal (0.02% in type A), octanol + 1,4-cineole (0.02% in type A).
3   Dugo P. et al. (1997). Four samples type A and one sample type B; GC on capillary columns coated with SE-52 and DB-5; GC/MS; relative percentage of peak areas. Dugo et al. also found exo-fenchol (tr–0.02% in type A and tr in type B), probably a misidentification of endo-fenchol.
4   Dugo et al. (1999). One sample; GC on capillary columns coated with SE-52 and DB-5; GC/MS with interactive use of linear retention indices. Dugo et al. also found α-elemene (0.08%), endo-fenchol (0.01%), β-bisabolol (0.01%), selin-11-en-4-α-ol[t] (0.01%), cis-sesquisabinene hydrate (0.03%), spathulenol (0.02%), trans-pinocarvyl acetate (0.01%), cis-limonene oxide (0.01%), and trace amounts of allo-ocimene[t], (E)-γ-bisabolene, (Z)-γ-bisabolene, α-santalene, cis-piperitol, (E,Z)-farnesol, decyl acetate, trans-limonene oxide, caryophyllene oxide.

The results reported in the table, even if based to a limited number of papers, provide a good description of the real composition of cold-pressed Key lime oil. The results obtained by the different authors are in good agreement. The apparent different values reported by Clark and Chamblee (1992) compared to those obtained by the other authors, particularly for what concerns limonene, are due to the quantification method used. Clark and Chamblee, in fact, reported their values as absolute w per cent, while other authors expressed their results as relative per cent of the peak areas, not adjusted. It should also be noticed that β-sesquiphellandrene identified by Haro and Faas (1985) could have been a misidentification of (E,E)-α-farnesene.

The comparison between lime oil type A and type B shows that these two cold pressed oils do not present substantial composition differences.

In addition to the systemic studies on the composition of industrially processed Key lime oil whole volatile fraction, in literature are found papers on some peculiar aspects on the composition of this oil.

McHale (1980) reported the following information on the quantitative composition of the volatile fraction of ecuelled lime oil:

monoterpene hydrocarbons	74.9%
myrcene	1.2%
limonene	41.8%
β-phellandrene	0.6%
α-pinene	2.9%
β-pinene	17.8%
sabinene	3.1%
γ-terpinene	7.2%
terpinolene	0.4%
sesquiterpene hydrocarbons	3.7%
trans-α-bergamotene	0.6%

β-bisabolene	1.0%
caryophyllene*	0.5%
β-sesquiphellandrene	1.0%
aliphatic aldehydes (C_8–C_{16})	1.0%
geranial + neral	6.2%
α-terpineol	0.3%
other alcohols (fenchol*, linalool, terpininen-4-ol)	0.3%
geranyl acetate + neryl acetate	0.5%

McHale (1980) also reported the presence of camphene, *p*-cymene, isoterpinolene, α-terpinene, δ-elemene, α- and β-humulene. In the same paper McHale also assessed the similarity of the β-sesquiphellandrene structure with that of (E,E)-α-farnesene (isolated and identified by Moshonas and Shaw (1980) in distilled lime oil and in Valencia sweet orange oil), showing how these molecules could not be unequivocally distinguished by IR spectroscopy.

Chamblee *et al.* (1985) after pre-fractionating the oil by normal phase HPLC, analysed the fractions obtained by HRGC-MS. The components identified for the first time in Key lime oil, were the following: farnesal*, 3-hexanone, 6-methyl-5-hepten-2-one, piperitone, *cis*- and *trans*-carveol, citronellol, *p*-cymen-8-ol, γ-isogeraniol, isopiperitenol, *p*-menth-3-en-1-ol, myrtenol, *trans*-pinocarveol, sabinol, β-terpineol*, α-bisabolol, α-cadinol, 1,3- and 2,3-dimethyl-3(4-methyl-3-pentenyl)-2-norbornanol, farnesol*, verbenol, dodecyl acetate.

Clark *et al.* (1987) using HPLC isolated from Key lime oil the sesquiterpene hydrocarbon germacrene B, and confirmed the structure by GC/MS, IR and NMR. The germacrene B is very important to the fresh lime character of the oil.

Mondello *et al.* (1995b) using an on-line coupled HPLC-HRGC/MS system, identified in Key lime oil the following components listed in decreasing amount order: β-bisabolene, *trans*-α-bergamotene, β-caryophyllene, germacrene B, δ-elemene, germacrene D, *cis*-α-bergamotene, β-elemene, (Z)-β-farnesene, α-humulene, α-selinene, β-santalene, (Z)- and (E)-γ-bisabolene, (E)-epi-α-selinene.

Chamblee *et al.* (1997) by using cryofocusing GC/FT-IR confirmed the presence of germacrene B and germacrene D in Key lime oil, and identified for the first time the presence of (Z)-α-bisabolene.

Industrial cold-pressed Persian lime oil (*Citrus latifolia* Tan.)

In Table 13.6 are summarised the literature data relative to the composition of the volatile fraction of cold-pressed Persian lime oil.

Table 13.6 Percentage composition of the volatile fraction of cold-pressed Persian lime oil

	1	2	3	4	5
HYDROCARBONS					
Aliphatic					
decane	–	–	–	tr	tr
dodecane	–	–	–	0.01–0.02	0.01
nonane	–	–	–	tr	tr
tridecane	–	–	–	tr	tr

Monoterpene					
camphene	0.06–0.11	0.06	–	0.06–0.07	0.06
δ-3-carene	–	tr[b]	–	tr–0.01	0.01
p-cymene	1.20–10.43	0.37	2.01	0.11–0.70	0.33
limonene	51.47–59.65	52.95[c]	58.91	51.61–59.81	56.65[i]
myrcene	0.89–1.76	1.35[d]	1.31	1.34–1.50	1.44
(E)-β-ocimene	–	–	–	0.08–0.17	0.11
(Z)-β-ocimene	–	–	–	0.04–0.09	0.06
α-phellandrene	tr	0.05	–	0.01–0.03	0.02
β-phellandrene	tr	–	–	–	i
α-pinene	4.31–5.03	2.01	2.25	1.96–2.21	2.10
β-pinene	14.58–16.04	12.20	11.02	12.04–14.55[g]	12.93[g]
sabinene	0.91–1.40	2.07	1.80	12.04–14.55[g]	12.93[g]
α-terpinene	tr–0.14	0.24	–	0.15–0.33	0.27
γ-terpinene	1.30–8.46	14.45[e]	13.02	12.55–15.65	13.54
terpinolene	0–0.33[a]	0.63[f]	–	0.46–0.70	0.59
thuja-2,4(10)-diene	–	–	–	tr	tr
α-thujene	–	0.52	0.46	0.54–0.60	0.56
tryciclene	–	–	–	tr–0.01	0.01
Sesquiterpene					
cis-α-bergamotene	–	–	–	0.07–0.08	0.07
trans-α-bergamotene	1.03–1.34	1.14	0.86	1.03–1.28	1.11
(Z)-α-bisabolene	–	–	–	0.04–0.16	0.12
β-bisabolene	0.24–0.53	1.75	1.44	1.71–2.25[h]	1.88[h]
β-caryophyllene	0.07–0.40	0.57	0.41	0.48–0.72	0.57
β-elemene	–	–	–	0.05–0.09	0.07
δ-elemene	–	–	–	0.03–0.12	0.07
(E,E)-α-farnesene	–	–	–	1.71–2.25[h]	1.88[h]
(E)-β-farnesene	–	–	–	0.10–0.12	0.10
(Z)-β-farnesene	–	–	–	tr–0.01	0.01
germacrene B	–	–	–	0.10–0.19	0.14
germacrene D	–	–	–	0.06–0.12	0.09
α-humulene	–	–	–	0.04–0.06	0.05
β-santalene	–	–	–	0.04–0.05	0.04
α-selinene	–	–	–	0.03–0.07	0.04
ALDEHYDES					
Aliphatic					
decanal	0.04–0.08	0.09	–	0.06–0.09	0.07
dodecanal	tr	0.05	0.25	0.03–0.05	0.04
hexadecanal	–	0.06	–	0.06–0.12	0.08
nonanal	0.02–0.05	0.63[f]	–	0.01	0.01
octanal	0–0.33[a]	1.35[d]	–	0.04–0.05	0.05
pentadecanal	–	0.01	–	–	tr
tetradecanal	–	0.02	–	0.01–0.05	0.03
undecanal	0.02–0.07	0.02	–	0.01–0.02	0.01
Monoterpene					
citronellal	0.03–0.05[j]	0.05	–	0.03–0.06	0.04
geranial	2.23–3.93	2.77	1.88	1.81–2.77[k]	2.27[k]
neral	0.48–1.17	0.73	1.40	1.06–1.72	1.34
perilla aldehyde	–	–	–	1.81–2.77[k]	2.27[k]
KETONES					
Monoterpene					
cis-pinocamphone	–	–	–	tr	tr
piperitone	–	–	–	tr[t]	tr[t]
6-methyl-5-hepten-2-one	tr–0.07	0.02	–	tr–0.05	tr

Table 13.6 (Continued)

	1	2	3	4	5
ALCOHOLS					
Aliphatic					
octanol	0.05–0.09	e			
Monoterpene					
borneol	–	–	–	0.01–0.03	0.02
geraniol	tr–0.05	0.05	–	0.02–0.10	0.05
linalool	0.11–0.24	0.17	0.19	0.16–0.23	0.18
cis-p-menth-2-en-1-ol	–	–	–	tr–0.01	0.01
trans-p-menth-2-en-1-ol	–	–	–	tr	tr
nerol	tr	0.14	–	0.06–0.20	0.11
trans-pinocarveol	–	–	–	tr–0.02	0.01
cis-sabinene hydrate	–	–	–	0.03–0.05	0.04
trans-sabinene hydrate	–	–	–	0.04–0.07	0.05
α-terpineol	0.16–0.35	0.23	0.23	0.20–0.37	0.28
terpinen-4-ol	0.07–0.17	0.04	–	0.04–0.12	0.08
Sesquiterpene					
α-bisabolol	–	–	–	0.07–0.10	0.08
campherenol	–	–	–	0.05–0.08	0.06
'norbornanol'[l]	–	–	–	0.05–0.07	0.05
ESTERS					
Monoterpene					
bornyl acetate	–	–	–	tr–0.01	0.01
citronellyl acetate	0.03–0.10	–	–	–	0.02
geranyl acetate	0.07–0.27	0.27	0.63	0.17–0.31	0.24
neryl acetate	0.36–0.86	1.53[m]	1.00	0.80–1.26	1.00
OXIDES					
Monoterpene					
dehydro-1,8 cineole	–	–	–	tr[t]	tr[t]

Notes

t = tentative identification; tr = traces; a terpinolene + octanal; b δ-3-carene + 1,4-cineole; c limonene + 1,8-cineole; d myrcene + octanal; e γ-terpinene + octanal; f terpinolene + nonanal; g β-pinene + sabinene; h β-bisabolene + (E,E)-α-farnesene; i limonene + β-phellandrene + 1,8-cineole; j citronellal + octyl acetate; k geranial + perilla aldeyde; l 2,3-dimethyl-3-(4-methyl-3-pentenyl)-2 norbornanol; m neryl acetate + unknown hydrocarbons.

Appendix to Table 13.6

1 Koketsu et al. (1983). Brazil; FMC; 3 samples; GC on packed columns of Carbowax 20 M and SE-30 and on capillary column coated with Carbowax 20 M; retention times and Kovats indices; relative percentage of peak areas: Koketsu et al. also found nonanol (0.12–0.30%), citronellol (tr–0.05%), linalyl acetate (tr–0.12%).
2 Haro and Faas (1985). USA, Florida; one commercial sample; column chromatography on silica-gel impregnated with 0.7% of Carbowax 1540; GC on capillary columns coated with SE-30 and SP-1000; Kovats indices; relative percentage of peak areas. Haro and Faas also found β-sesquiphellandrene (0.44%), tridecanal (0.01%).
3 Lancas et al. (1988). Brazil; FMC; one sample; GC on capillary column coated with SE-54; GC/MS; relative percentage of peak areas.
4 Dugo P. et al. (1997). Six samples; GC on capillary columns coated with SE-54 and DB-5; GC/MS; relative percentage of peak areas. Dugo et al. also found exo-fenchol (tr–0.02%), probably a misidentification of endo-fenchol.
5 Dugo et al. (1999). One sample; GC on capillary columns coated with SE-54 and DB-5; GC/MS with interactive use of linear retention indices. Dugo et al. also found α-elemene (0.09%), β-bisabolol (0.01%), (E,Z)-farnesol (0.01%), cis-sesquisabinene hydrate (0.03%), spathulenol (0.03%),

cis-limonene oxide (0.01%), and trace amounts of allo-ocimene^t, (E)-γ-bisabolene, (Z)-γ-bisabolene, α-santalene, carvacrol^t, endo-fenchol, selin-11-en-4-α-ol^t, decyl acetate, α-terpinyl acetate, *trans*-limonene oxide, caryophyllene oxide.

Although the studies carried out in the 80s (Koketsu *et al.*, 1983; Haro and Faas, 1985; Lancas *et al.*, 1988) were not detailed, the quantitative data reported in Table 13.6 are almost all in good agreement. Although, if compared with the values reported by other authors, the amounts of γ-terpinene and that of β-pinene, reported by Koketsu *et al.* (1983), appear to be low. The high values of *p*-cymene, found by the same authors and by Lancas *et al.* (1988), could be explained by the inappropriate storage conditions of the samples.

The composition of Key lime and Persian lime is very similar. Some differences appear, however, in the quantitative composition, relative to β-pinene, higher in key lime, and γ-terpinene higher in Persian lime. Limonene represents at the most the 50 per cent of the whole volatile fraction in Key lime but its content is lower in Persian lime.

Commercial lime oils

Alessandro *et al.* (1985) carried out a research, by GC-MS, to determine the characteristic composition of lime oil, in order to identify possible adulterations. These authors analysed several samples of cold-pressed lime oils, imported in USA, declared genuine, but of which the botanical origin was not specified. They determined the following composition:

camphene	0–0.5%
p-cymene	0.1–1.3%
myrcene	0.8–1.6%
limonene	43.7–56.2%
α-pinene	2.7–3.1%
(Z)-β-ocimene	0.1–0.3%
β-pinene	4.3–23.0%
sabinene	0.4–4.2%
α-terpinene	0.1–0.4%
γ-terpinene	2.5–16.6%
terpinolene	0.5–0.7%
bergamotene* + terpinen-4-ol	0.7–1.3%

These oils, analysed by Alessandro *et al.* (1985), present wide ranges of variation of the quantitative composition, giving the impression that they were obtained from both Key lime and Persian lime. Moreover, the minima reported for some components, such as β-pinene and γ-terpinene, appear to be incompatible with genuine lime oils.

Laboratory extracted Key lime oils

El-Samahy *et al.* (1982) reported, in cold-pressed laboratory extracted lime oils obtained from Egyptian fruits, the presence of camphene (3.60 per cent), limonene (33.59 per cent), α-pinene (3.10 per cent), citral (as the authors indicated neral) (7.98 per cent), geranial (6.30 per cent), citronellol (tr), linalool (6.60 per cent), α-terpineol (0.90 per cent), bornyl acetate (1.32 per cent), geranyl acetate (0.64 per cent). These results are quite unique for a lime oil. The unusual values reported could be due to the nature of the matrix, but also to poor chromatographic separation, as already mentioned for the bitter orange oil analysed by the same authors.

Khurdiya and Maheshwari (1988) analysed two samples of Key lime oil extracted in laboratory from a stock of fruits harvested in the Delhi area (India). The two oils differed in respect of the treatment used: one obtained by hydrodistillation of the ground and pressed peel; the other from the cut and pressed peel. Both oils presented a high content of monoterpene alcohols (about 19 per cent) and of p-cymene, 7 per cent and 3 per cent respectively, along with some unexplainable differences. In fact, the first oil did not contain γ-terpinene and octanal, while these components were 6 per cent and 2 per cent respectively in the second oil; citronellal was 5 per cent in the first sample and less then 1 per cent in the second one.

Yang, R.H. et al. (1992) in a paper on the odour quality of peel oils of different citrus, determined the composition of one laboratory extracted lime oil. This oil was obtained by solvent extraction, from fruits harvested in Japan. The authors reported its composition as follows:

- the monoterpene hydrocarbons, usually present in lime oil, where the main components were, in decreasing amount order, limonene, γ-terpinene, and β-pinene;
- the sesquiterpene hydrocarbons α-bergamotene*, β-bisabolene, α- and β-caryophyllene, α-cedrene, β- and δ-elemene, (E,E)-α-farnesene, β-farnesene*, γ-muurolene. Unusual appears the presence of α-caryophyllene and α-cedrene.
- numerous oxygenated compounds were present, among which the most abundant were geranial (0.03 per cent), neral (0.02 per cent), neryl acetate (0.02 per cent), geranyl acetate (0.01 per cent) and α-terpineol (0.01 per cent), while were present, at trace levels, the aliphatic aldehydes C_6, C_8–C_{17}, (E,E)-2,4-decadienal, 2,6-dimethyl-5-heptenal, citronellal, perilla aldehyde, 6-methyl-5-hepten-2-one, camphor, fenchone, piperitone, nootkatone, decanol, (Z)-3-hexenol, octanol, borneol, carvacrol, citronellol, geraniol, linalool, nerol, perilla alcohol, trans-sabinene hydrate, terpinen-4-ol, α-terpineol, thymol, nerolidol*, decyl acetate, hexyl acetate, octyl acetate, undecyl acetate, bornyl acetate, citronellyl acetate, geranyl acetate, geranyl formate, methyl geranate, neryl acetate, α-terpinyl acetate, 1,8-cineole, trans-linalool oxide.

Jantan et al. (1996) analysed the peel and leaf oils obtained from different citrus species from Malaysia. The oils extracted by hydrodistillation were analysed by GC using capillary SE-30 and PEG-20M columns, by GC/MS with a SE-30 column equipped with a commercial mass library for the identification; the chromatographic retention index were determined on the two columns, and when possible were co-injected standard compounds. The results obtained by Jantan et al. (1996) for Key lime peel oil are reported below:

camphene	0.16%
δ-3-carene	0.48%
limonene	39.28%
myrcene	1.00%
(E)-β-ocimene	0.47%
(Z)-β-ocimene	0.14%
α-phellandrene	0.08%
α-pinene	1.45%

β-pinene	28.44%
α-terpinene	0.03%
γ-terpinene	0.79%
terpinolene	0.43%
α-thujene	0.07%
α-bergamotene*	0.41%
β-caryophyllene	0.84%
β-elemene	0.45%
δ-elemene	0.44%
(E)-β-farnesene	1.48%
(Z)-β-farnesene	0.41%
α-humulene	0.12%
decanal	0.41%
dodecanal	0.10%
citronellal	0.10%
geranial	2.05%
neral	5.30%
octanol	0.11%
geraniol	7.50%
linalool	1.40%
terpinen-4-ol	2.01%
α-terpineol	2.39%
elemol	0.07%
α-eudesmol	0.06%
β-eudesmol	0.06%
(Z)-nerolidol	0.61%
geranyl acetate	0.59%
neryl acetate	0.21%
sabinyl acetate*	0.10%
cis-limonene oxide	0.11%
hexadecanoic acid	0.06%

Jantan et al. (1996) also found trace amount of β-bisabolene, α-elemene, octanal, isopulegol, and *trans*-limonene oxide. Among the results reported above unusual appear the low value of γ-terpinene as well as the high content of the total amount of monoterpene alcohols.

Sawamura et al. (1999) reported the main components determined in a sample of oil extracted from Key limes cultivated in Japan. The results obtained were as follows:

p-cymene	0.1%
limonene	50.5%
myrcene	1.4%
α-pinene	3.6%
β-pinene	13.4%
α-terpinene	0.3%
γ-terpinene	17.7%
terpinolene	0.7%

citronellal	0.1%
geranial	2.1%
neral	1.0%
geraniol	0.1%
nerol	0.1%
α-terpineol	0.3%
geranyl acetate	0.7%

Sawamura et al. (1999) also found trace amounts of decanal, octanal, nootkatone, citronellol, linalool, and terpinen-4-ol.

Laboratory extracted Persian lime oils

Edwards and Marr (1990), analysing lime oils obtained in laboratory by simultaneous steam distillation–solvent extraction, and oils directly extracted from the utricles, reported the presence of 0.03 per cent and of 0.04 per cent of β-santalene and 0.06 and 0.03 per cent of santalol respectively in the two oils.

Njoroge et al. (1996) and Sawamura et al. (1999) analysed by GC and GC/MS cold-pressed oil obtained from fruits harvested in the Agriculture Research Center in Ehime, Japan.

The composition of the oil, relatively to the main components, is reported below:

limonene	52.2%
myrcene	1.3%
α-pinene	3.2%
β-pinene	13.0%
sabinene	2.0%
α-terpinene	0.3%
γ-terpinene	17.0%
terpinolene	0.7%
α-bergamotene*	0.8%
β-bisabolene	0.8%
β-caryophyllene	0.3%
(E,E)-α-farnesene	0.2%
geranial	2.3%
neral	1.2%
linalool	0.2%
α-terpineol	0.4%
geranyl acetate	1.0%
neryl acetate	1.4%

Njoroge et al. (1996) also found the 0.01 per cent of the following components: camphene, p-cymene, β-elemene, δ-elemene, (E)-β-farnesene, germacrene B, germacrene D, decanal, hexadecanal, citronellal, cis- and trans-sabinene hydrate, α-cadinol, nonyl acetate, and trace amount of pentadecane, δ-3-carene, cis-β-ocimene, α-phellandrene, α-thujene, (E)-α-bisabolene[t], α-cubebene, β-cubebene, γ-elemene, α-humulene, β-santalene[t], sesquiphellandrene*, heptadecanal, nonanal, octadecanal, octanal, tetradecanal, tridecanal, undecanal, perilla aldehyde, (E)-dodec-2-en-1-ol, octanol, borneol, geraniol, *trans*-*p*-mentha-2,8-dien-1-ol^t, thymol, α-bisabolol, γ-eudesmol, (E)-nerolidol, spathulenol, viridiflorol, cinnamyl alcohol, bornyl acetate, citronellyl acetate, geranyl formate, linalyl acetate, *p*-mentha-1, 8-dien-10-yl acetate, perillyl acetate, (E,E)-farnesyl acetate, undecanoic acid.

MANDARIN (*Citrus deliciosa* Ten.) AND TANGERINE OILS (*C. tangerine* Hort. ex Tan.)

Describing mandarin oil is a difficult compared to other citrus oils. From a botanical point of view the mandarin group is a very complex one. It consists of Mediterranean mandarin (*Citrus deliciosa* Ten.), the tangerines (*C. tangerine* Hort. ex Tan.), the clementines (*C. clementina* Hort. ex Tan.) and, the *C. temple* Hort. ex Tan., the *C. nobilis* Lour., the *C. unshiu* (Mark.) as well as the *C. reticulata* Blanco. Often, when the composition of mandarin oil is described, the authors refer to oils obtained from different mandarin species, seldom the botanical origin is not specified, and when it is, sometimes it is not univocal. Mandarins are also commonly described by botanical classifications that differ from the one mentioned above, generating confusion.

The oils obtained from *C. unshiu* (Mark.) and from *C. clementina* Hort. ex Tan. will be described in a different chapter of this book. In this section only the composition of mandarin (*C. deliciosa* Ten.) and tangerine (*C. tangerine* Hort. ex Tan.) oils of ensured origin will be singularly described, as well as the composition of commercial oils that appear to have the same botanical origin.

Most of the literature published before 1979 relating to the composition of mandarin oil has been reviewed by Shaw (1979). Among the papers cited by Shaw (1979) two were on the composition of *C. unshiu* (Yamanishi *et al.*, 1968; Kita *et al.*, 1969), three on the composition of clementine *C. clementina* Hort ex Tan. oils (Scora *et al.*, 1968; Calvarano *et al.*, 1974; Huet and DuPuis, 1969); three were on the tangerine *C. tangerine* Hort. ex Tan. oils (Attaway *et al.*, 1967; Moshonas and Shaw, 1974; Braddock and Kesterson, 1976); two were on the composition of mandarin oil, *C. deliciosa* Ten. (Kugler and Kovats, 1963; D'Amore and Calabrò, 1966) and one was on the hydrocarbon fraction of tangerine and mandarin oils (Ashoor and Bernhard, 1967). In this last paper the quantitative data reported are relative to the hydrocarbon fraction composition, not to the whole essential oils. These values are therefore not comparable with those reported by other authors.

MANDARIN OIL (*Citrus deliciosa* Ten.)

1960–1979

Below are reported the results found in the review by Shaw (1979).

camphene	tr–0.37%
p-cymene	0.58–8.16%
limonene	64.79–67.62%
myrcene	1.19–1.35%

α-phellandrene	0.32%
α-pinene	2.47–2.80%
β-pinene	1.12–1.28%
sabinene	0.45%
γ-terpinene	9.14–17.33%
terpinolene	0.57–0.76%
α-thujene	0.46–0.58%
α-copaene	0.01%
caryophyllene*	0.02%
longifolene	0.01%
α-selinene	0.02%
γ-selinene	0.01%
decanal	0.04%
dodecanal	0.01%
octanal	0.03%
perilla aldehyde	0.05%
carvone	0.03%
decanol	0.04%
heptanol	0.02%
octanol	0.09%
undecanol	tr
cis-carveol[t]	0.02%
trans-carveol	0.04%
citronellol	0.02%
geraniol	0.01%
linalool	0.24%
nerol	0.05%
cis-sabinene hydrate	0.06%
trans-sabinene hydrate	0.11%
terpinen-4-ol	0.11%
α-terpineol	1.11%
benzyl alcohol	0.01%
thymol	0.08%
decyl acetate	tr
geranil acetate	tr
1,8-cineole	tr
methyl N-methylanthranilate	0.85%
decanoic acid	0.03%
dodecanoic acid	0.01%
heptanoic acid	tr
nonanoic acid	0.01%
octanoic acid	0.04%
undecanoic acid	tr

These data only refer to the papers by Kugler and Kovats (1963) and of D'Amore and Calabrò (1966). In particular, the values of the monoterpene hydrocarbons are relative to both papers, while the other values refer uniquely to the paper by Kugler and Kovats

(1963). D'Amore and Calabrò (1966) reported only the qualitative composition relative to the components different from monoterpene hydrocarbons. These components were: bergamotene*, β-bisabolene, γ-cadinene, δ-cadinene, caryophyllene*, β-elemene, longifolene, α-ylangene, the aliphatic linear chained saturated aldehydes C_7–C_{12}, 2-hexen-1-al*, citronellal, cumin aldehyde, geranial, neral, benzyl aldehyde, furfural, 6-methyl-5-hepten-2-one, camphor, carvone, menthone, the aliphatic linear chained saturated alcohols C_6–C_{12}, 3-octanol, citronellol, geraniol, linalool, nerol, α-terpineol, thymol, decyl acetate, nonyl acetate, octyl acetate, undecyl acetate, citronellyl acetate, geranyl acetate, linalyl acetate, neryl acetate, benzyl acetate, methyl N-methylanthranilate, the aliphatic linear chained saturated acids C_7–C_{12} and citronellylic acid.

The remaining values reported by Shaw were eliminated from the table, because as mentioned above, these values refer to different species or are not comparable with the other values as these reported represent the relative percentage of the hydrocarbon fraction (Ashoor and Bernhard, 1967).

Among the data reported above the high value of p-cymene (8.16 per cent) and the low value of γ-terpinene (9.14 per cent) reported by Kugler and Kovats (1963) should be noted. These results could be relative to an aged sample, were p-cymene was formed from γ-terpinene. A different explanation could be given by the formation of this artefact due to excessive analytical stress. The presence of furfural and of menthone reported by D'Amore and Calabrò (1966) is unusual in mandarin oil. α-Ylangene could be a misidentification of α-copaene, as already mentioned for grapefruit oil.

There exist more papers than those reviewed by Shaw (1979) on the composition of mandarin oil: one on the monoterpene hydrocarbon fraction composition by Di Giacomo et al. (1963); one on the qualitative determination (Ziegler, 1971) and one on the quantitative determination (Hussein and Pidel, 1976) on the whole volatile fraction of mandarin oil.

The ranges of variation for the monoterpene hydrocarbons fraction determined by Di Giacomo et al. (1963) from the analysis of six samples along with the results obtained by Ashoor and Bernhard (1967) are reported below:

	Di Giacomo et al. (1963)	Ashoor and Bernhard (1967)
camphene	–	0.14%
δ-3-carene	–	0.06%
p-cymene	0.76–3.60%	1.91%
limonene	62.68–80.52%	78.50%
myrcene	1.47–1.91%	–
myrcene + β-pinene	–	3.01%
α-phellandrene	–	0.59%
β-phellandrene	–	0.51%
α-pinene	2.05–3.26%	1.53%
β-pinene	1.23–1.97%	–
sabinene	0.50–0.99%	–
α-terpinene	–	0.35%
γ-terpinene	13.90–26.76%	13.40%

Ziegler (1971) identified in mandarin oil the following components: camphene, p-cymene, limonene, myrcene, α-pinene, β-pinene, γ-terpinene, terpinolene, δ-cadinene, caryophyllene*, β-copaene, β-elemene, γ-elemene, α- and β-humulene, β-ylangene, decanal, dodecanal, octanal, undecanal, carvone, nootkatone, geraniol, linalool, nerol, α-terpineol, thymol, geranyl acetate, methyl N-methylanthranilate. β-copaene and β-ylangene were probably misidentifications. This mistake was the consequence of the identification reported by Hunter and Brogden (1965) later corrected by Veldhuis and Hunter (1968).

Hussein and Pidel (1976) compared the composition of mandarin oils from Italy and Argentina. The results obtained by these authors are reported below:

	Italy	Argentine
camphene + unknown	0.02%	0.02%
p-cymene + octanal	1.34%	0.90%
limonene	67.10%	71.08%
myrcene	1.80%	1.92%
α-phellandrene	0.03%	0.05%
α-pinene	3.93%	2.81%
β-pinene + unknown	2.16%	1.84%
α-terpinene	0.42%	0.37%
γ-terpinene	20.14%	17.83%
terpinolene	0.89%	0.89%
β-caryophyllene + nonanol	0.07%	0.10%
decanal	0.03%	0.13%
dodecanal + neryl acetate	0.01%	0.06%
nonanal + unknown	0.01%	0.02%
citronellal	0.01%	0.06%
geranial + decanol + citronellol	0.01%	0.01%
neral + α-terpineol	0.15%	0.08%
perilla aldehyde	0.04%	0.04%
sinensal*	0.11%	0.32%
carvone + nerol	0.01%	0.01%
dodecanol	tr	0.01%
heptanol	0.01%	tr
trans-carveol	tr	0.01%
linalool	0.13%	0.09%
sabinene hydrate* + unknown	0.02%	0.02%
thymol	0.03%	0.08%
benzyl alcohol	0.01%	0.02%
decyl acetate	0.01%	0.01%
geranyl acetate + unknown	0.10%	0.35%
methyl N-methylanthranilate	0.33%	0.31%
1,8-cineole	0.54%	0.28%

Hussein and Pidel (1976) also found trace amounts of undecanal, cis-carveol and geraniol in, at least, one type of sample.

1979–2000

Industrial cold-pressed oils

In Table 13.7 are summarised the literature data relative to the composition of mandarin oil published in this period.

Table 13.7 Percentage composition of the volatile fraction of mandarin oil

	1	2	3(a)	3(b)	4(a)	4(b)
HYDROCARBONS						
Monoterpene						
camphene	tr	0.01–0.02	tr	tr	0.02	0.02
δ-3-carene	–	tr	–	–	tr	tr
p-cymene	0.75–1.28	0.12–0.78	0.37	0.45	0.18^d	0.34^d
limonene	70.08–73.89	65.30–74.53^b	66.00	72.27	77.14	71.61
myrcene	1.86–2.27	1.57–1.96	1.75	1.91	1.86	1.86
(E)-β-ocimene	–	0.01–0.03	–	–	tr	tr
α-phellandrene	0.40–0.49	0.03–0.11	–	–	0.05	0.05
β-phellandrene	tr	b	tr	tr	–	–
α-pinene	4.07–5.24	2.00–2.74	3.47	3.31	1.75	2.39
β-pinene	2.17–2.44	1.39–2.10	1.84	1.57	1.15	1.67
sabinene	0.33–0.59	0.23–0.34	0.29	0.24	0.21	0.22
α-terpinene	tr	0.26–0.52	0.51	0.41	0.18^d	0.34^d
γ-terpinene	12.97–15.22	16.23–22.75	22.70	17.39	13.73	18.54
terpinolene	0.54–0.87^a	0.72–1.01	1.04	0.89	0.62	0.86
α-thujene	–	0.72–1.06	tr	tr	0.57	0.86
Sesquiterpene						
β-bisabolene	–	tr	tr	tr	–	–
β-caryophyllene	0.03–0.04	0.07–0.14	0.06	0.05	–	–
α-copaene	–	tr	tr	–	–	–
(E,E)-α-farnesene	–	0.07–0.26	0.12	0.06	–	–
α-humulene	–	tr–0.02	tr	tr	–	–
α-selinene	–	0.02–0.06^c	tr	tr	–	–
ALDEHYDES						
Aliphatic						
decanal	0.05–0.09	0.05–0.12	0.06	0.07	0.06	0.07
(E,E)-2,4-decadienal	–	tr	tr	–	–	–
(E)-dec-2-en-1-al	–	tr–0.03	tr	–	–	–
dodecanal	tr–0.05	0.02–0.04	tr	tr	0.01	0.01
(E)-dodec-2-en-1-al	–	tr–0.03	tr	tr	–	–
hexadecanal	–	tr	tr	tr	–	–
nonanal	0.03–0.04	0.01–0.04	tr	tr	0.06	0.07
octanal	0.54–0.87^a	0.03–0.20	0.05	0.05	0.22	0.17
tetradecanal	–	tr–0.01	tr	tr	–	–
undecanal	0.03–0.05	tr–0.02	tr	tr	tr	tr
Monoterpene						
citronellal	0.03–0.07^e	0.02–0.05	tr	tr	tr	0.02
geranial	0–0.02	tr–0.12^f	tr	–	0.04	0.01
neral	tr	tr–0.03	–	–	0.02	tr
perilla aldehyde	–	tr–0.12^f	tr	0.05	0.01	tr
Sesquiterpene						
α-sinensal	–	0.12–0.53	0.15	0.12	0.19	0.20

Table 13.7 (Continued)

	1	2	3(a)	3(b)	4(a)	4(b)
ALCOHOLS						
Aliphatic						
nonanol	tr	–	tr	–	–	–
octanol	0.06–0.14	tr–0.01	tr	tr	–	–
Monoterpene						
citronellol	0.03–0.04	0.01–0.03^g	tr	tr	–	–
p-cymen-8-ol	–	tr	–	tr	–	–
geraniol	–	tr–0.01	tr	–	0.01	0.01
linalool	0.13–0.31	0.04–0.19	0.18	0.16	0.15	0.10
nerol	–	0.01–0.03^g	tr	tr	tr	tr
perilla alcohol	–	tr	tr	–	–	–
cis-sabinene hydrate	–	0.01–0.06	0.07	0.06	–	–
trans-sabinene hydrate	–	0.01–0.11	tr	tr	–	–
terpinen-4-ol	0.03–0.05	0.01–0.08	0.06	0.05	0.05	0.05
α-terpineol	0.17–0.46	0.04–0.27	0.25	0.20	0.20	0.16
thymol	0.08–0.14	0.01–0.10	0.07	0.08	0.11	0.02
ESTERS						
Aliphatic						
octyl acetate	0.03–0.07^e	tr	–	–	–	–
Monoterpene						
citronellyl acetate	–	tr–0.01	tr	tr	tr	–
geranyl acetate	tr	tr–0.01	tr	tr	tr	tr
linalyl acetate	tr–0.04	tr	tr	tr	–	–
neryl acetate	–	tr–0.01	tr	–	tr	tr
terpinyl acetate*	–	–	–	–	tr	tr
OXIDES						
Monoterpene						
1,8-cineole	–	b	tr	tr	–	–
cis-limonene oxide	–	tr	tr	tr	–	–
trans-limonene oxide	–	tr	tr	–	–	–
OTHERS						
methyl N-methylanthranilate	0.51–0.65	0.26–0.66	0.43	0.38	0.28	0.15

Notes

t = tentative identification; tr = traces; * correct isomer not characterised; a terpinolene + octanal; b limonene + β-phellandrene + 1,8-cineole; c α-selinene + valencene; d p-cymene + α-terpinene; e citronellal + octyl acetate; f geranial + perilla adheyde; g citronellol + nerol.

Appendix to Table 13.7

1 Koketsu *et al*. (1983). Brazil; FMC; 3 samples; GC on packed columns of Carbowax 20 M and SE-30 and on capillary column coated with Carbowax 20 M; retention times and Kovats indices; relative percentage of peak areas. Koketsu *et al*. also found nootkatone (tr).
2 Dugo *et al*. (1984, 1990, 1999); Dugo (1994); Cotroneo *et al*. (1994); Verzera *et al*. (1997). Italy, Sicily and Calabria; Pelatrice, Torchi, FMC; 400 samples; GC on capillary columns coated with SE-52 and DB-5; GC/MS, linear retention indices. These authors also found (Z)-β-ocimene (tr–0.01%), nonyl acetate (tr–0.01%) and trace amounts of p-mentha-1,3,8-triene^t, *trans*-α-bergamotene, bicyclogermacrene, δ-cadinene, β-elemene, germacrene D, camphor, carvone, piperitone^t, *cis*-pinene hydrate^t, (E)-nerolidol.
3 Mazza (1987). Italy; Pelatrice, Torchi; 2 samples, green mandarin (a), red mandarin (b); column chromatography on silica gel; GC on capillary column coated with Carbowax 20 M; GC/MS; relative percentage of peak areas. Mazza also found in one or both the samples analysed trace amounts of

p-cimenene, (E,Z)-2,4-decadien-1-al, heptadecanal, (E)-hexadec-2-en-1-al, (E)-non-2-en-1-al, pentadecanal, (E)-tetradec-2-en-1-al, tridecanal, (E)-tridec-2-en-1-al, (E)-undec-2-en-1-al, cumin aldehyde, 6-methyl-5-hepten-2-one, methylacetophenone, heptanol, carvacrol, *cis*- and *trans*-carveol, γ-isogeraniol^t, *cis*-isopiperitenol, *trans*-isopiperitenol, limonen-4-ol, *p*-mentha-1,8(10)-dien-9-ol, *p*-menth-1-en-9-ol, spathulenol, *p*-mentha-1,8(10)-dien-9-yl acetate, *p*-menth-1-en-4,5-oxide, *p*-menth-4-en-1,2-oxide, methyl anthranilate, acedic acid, octanoic acid.
4 Boelens and Jimenez (1989). Spain, Andalusia (a), Italy, Sicily (b); needle puncturing machine (a); GC on capillary column coated with SE-54; GC/MS and retention times; relative percentage of peak areas.

The data reported are enough homogeneous and well represent, in our opinion, the typical mandarin essential oil composition. It should also be highlighted that, although some authors, cited in the table, refer their data to a small amount of samples, the ranges reported in column 2 are relative to about 400 samples of mandarin industrial essential oils. These oils were produced by the commonly used techniques and within numerous productive seasons between 1982 and 1997.

Among the results published by the different authors, some, and in particular Mazza (1987), reported information on some components (listed in the appendix to the table), present at very low levels. These components were, in a few cases, identified by GC/MS analysing the whole oil or after pre-fractionation by liquid chromatography, and it is our opinion that further confirmation is necessary.

The study carried out by Mazza (1987) was on two samples of fresh mandarin oils and two samples stored for a long period of time. The latter set of sample have been surely subject to some oxidative process that modified the composition of the oils. In Table 13.7 are reported only the data relative to the first set of samples. From the analysis of the oxidised samples Mazza (1987) reported also the presence of the following components: δ-cadinene, *p*-menth-1-en-9-al, β-pinone*, carvone, *cis*- and *trans*-dihydrocarvone^t, isopiperitone, piperitone, α-thujione, cresol, cuminyl alcohol, *cis*- and *trans*-*p*-mentha-1(7),8-dien-2-ol, *cis*- and *trans*-*p*-mentha-2,8-dien-1-ol, myrtenol, *trans*-pinocarveol, terpinen-1-ol, 3,7-dimethyl-octa-1,5-dien-3,7-diol, 3,7-dimethyl-octa-1,7-dien-3,6-diol, caryophyllene diol*, octyl acetate, citronellyl formate, perillene, *cis*- and *trans*-linalool oxide, two *p*-menthene epoxides*, caryophyllene oxide, humulene oxide II, thymol methyl ether.

Dugo *et al.* (1984, 1990), Dugo (1994) noticed that the composition of Italian mandarin oil varied during the productive season. These variation were observed during different productive years. In particular, limonene content increased during the season of production, ranging from the 68 per cent average content in October, to 74 per cent in February; the monoterpene hydrocarbons, such as γ-terpinene, β-pinene, α-pinene, the sesquiterpene hydrocarbons and all the oxygenated compounds decreased during the season of production. Due to such variations it was noticed that, red mandarin, produced at the end of the season, contains only 60 per cent of methyl N-methylanthranilate of the amount present in green mandarin, produced at the beginning of the season, and a total content of alcohols equal to the 25 per cent of the oil produced at the start of the season. If the variation of the components present in mandarin oil are correlated, a good method of evaluation can be set-up, to determine the oil purity. For example (Figure 13.3), if a mandarin oil contains 73 per cent of limonene, 20 per cent of γ-terpinene, and 0.15 per cent of α-terpineol, all these values taken singularly can be considered acceptable for a genuine oil, but this oil cannot be considered genuine: in fact, the values of γ-terpinene and of α-terpineol are typical for

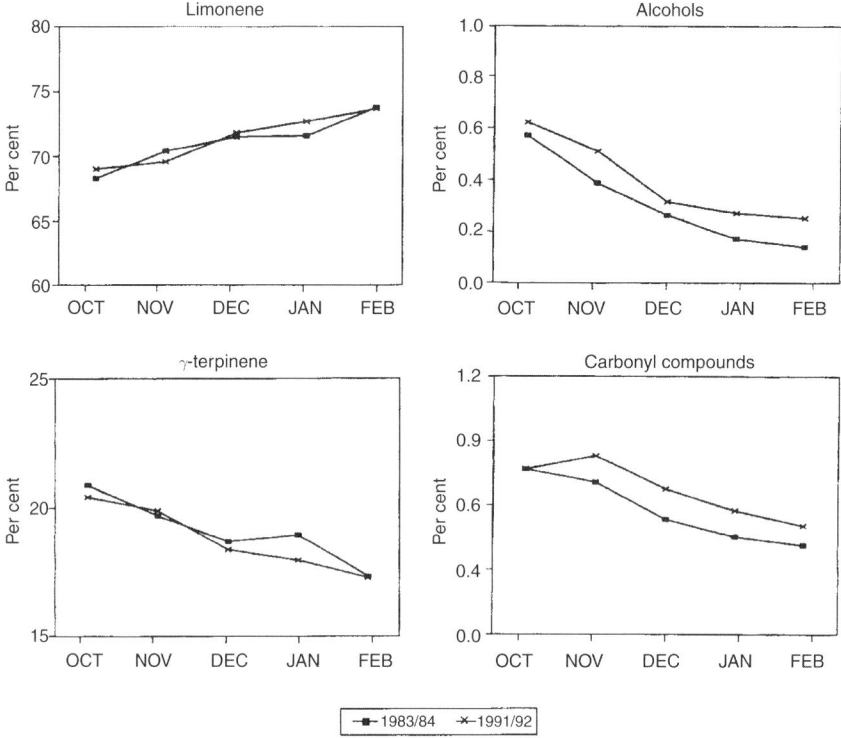

Figure 13.3 Variation in average content of limonene, γ-terpinene, alcohols and carbonyl compounds for mandarin oils produced in Italy during two different seasons (Dugo, 1994).

a green mandarin oil, while the limonene content is characteristic of a red mandarin oil. It is possible that this oil was a green mandarin oil diluted with other product containing high percentages of limonene, such as sweet orange terpene.

Other researches were carried out, other than those reported in Table 13.7. These were on some peculiar aspects of the composition of mandarin oil.

Wilson and Shaw (1981), studying the contribute of some mandarin oil components to its flavour, reported the presence of 1.8 per cent of β-pinene, 14.0 per cent of γ-terpinene, 0.18 per cent of thymol, and 0.65 per cent of methyl *N*-methylanthranilate.

Berger *et al.* (1985) determined in mandarin oil trace amount of 1,3(E),5(Z)-undecatriene, a compound of very likeable odour character.

Calvarano, I. *et al.* (1989), reported the values of some parameters determined by GC for Italian mandarin oil. It was therefore established by these authors that a genuine Italian mandarin oil should present the following characteristics:

β-pinene	1.56–1.91%
sabinene	0.11–0.17%
β-pinene/sabinene	9.58–16.89%
limonene	69.00–73.35%
γ-terpinene	17.14–20.24%
limonene/γ-terpinene	3.41–4.20%

terpinen-4-ol   0.02–0.05%
methyl N-methylanthranilate   0.39–0.56%

They declared that, values of terpinen-4-ol, higher that 0.05 per cent was indicating the presence of distilled oil and that values of methyl N-methylanthranilate lower than 0.38 per cent can reveal fraudulent dilution of the oil. The values determined by Calvarano, I. *et al.* (1989) were determined from analyses carried out during a single productive season, and therefore would be considered by us not surely representative, although these data are in acceptable accordance with those reported in Table 13.7. The value of sabinene reported by these authors, however, is always lower than the minimum reported in Table 13.7. This could be due to the poor chromatographic separation, achieved by Calvarano, I. *et al.* (1989), for the pair sabinene/β-pinene.

Micali *et al.* (1990) and Lanuzza *et al.* (1991), using a coupled HPLC-GC system, identified in mandarin oil the n-alkanes C_{21}–C_{29}, C_{31} and the correspondent 'iso' isomers, and the following sesquiterpene hydrocarbons: β-caryophyllene, α-humulene, (E,E)-α-farnesene.

Using an HPLC-GC/MS system, within a systemic investigation on sesquiterpene hydrocarbons in citrus essential oils, Mondello *et al.* (1995b) identified the following components listed in decresing amount order: (E,E)-α-farnesene, α-copaene, β-caryophyllene, α-selinene, δ-cadinene, α-humulene, germacrene D, β-cubebene, and *trans*-α-bergamotene.

Commercial oils

In Table 13.8 is reported the composition of some mandarin oils that were defined, in the original papers, as commercial oils, or of which the origin was not clearly specified.

Table 13.8 Percentage composition of the volatile fraction of some commercial mandarin oils

	1	2	3
HYDROCARBONS			
Monoterpene			
camphene	0.04	tr–0.04	0.05
δ-3-carene	0.04	0–0.06	–
p-cymene	2.32	1.20–3.65	4.25
limonene	77.02	65.51–68.64	66.41
myrcene	1.62	1.68–2.27	1.73
(E)-β-ocimene	–	tr–0.06	0.03
α-phellandrene	0.04	0.05–0.08	0.48
α-pinene	2.22	2.29–2.87	2.29
β-pinene	1.13	1.73–2.48	1.88
sabinene	0.23	0.29–0.45	–
α-terpinene	0.23	tr–0.30	–
γ-terpinene	11.94	18.46–20.89	19.17
terpinolene	0.49	0.60–0.96	1.38
α-thujene	–	0.34–0.91	0.91
Sesquiterpene			
γ-cadinene	–	tr–0.05	tr
β-caryophyllene	–	0.06–0.11	0.08
(E,E)-α-farnesene	–	0.13–0.18	0.30

Table 13.8 (Continued)

	1	2	3
ALDEHYDES			
Aliphatic			
nonanal	–	tr–0.06	tr
Monoterpene			
perilla aldehyde	–	0.05–0.08	tr
ALCOHOLS			
Monoterpene			
citronellol	–	tr–0.05	tr
linalool	0.25	0.13–025	0.13
terpinen-4-ol	–	tr–0.10	0.04
α-terpineol	–	0.05–0.24	0.12
thymol	–	0.04–0.08	tr
ESTERS			
Monoterpene			
linalyl acetate	0.49	–	tr
OTHERS			
methyl N-methyl anthranilate	–	0.51–0.72	0.50

Appendix to Table 13.8

1 Formàček and Kubeczka (1982). One commercial samples; GC on capillary column coated with WG-11; relative percentage of peak areas. Formàček and Kubeczka also found β-phellandrene (0.21%).
2 Inoma *et al.* (1989). Six commercial samples from France and Italy; GC on capillary columns coated with OV-1 and Carbowax 20 M; GC/MS; relative percentage of peak areas. Inoma *et al.* also found δ-cadinene (0–0.02%), decanal (0.05–0.13%), citronellal (tr–0.02%), α-sinensal (0.15–0.31%), *cis*-carveol (tr–0.05%), (E)-2-hexenyl butyrate (0–0.14%), geranyl acetate (tr–0.09%), neryl acetate (tr–0.03%), limonene oxide* (tr–0.11%), limonene dioxide (tr–0.14%) and in one or more samples analysed trace amounts of α-copaene, dodecanal, undecanal, carvone, *trans*-carveol.
3 Della Porta *et al.* (1997). One commercial sample from France; GC on capillary column coated with DB-5; GC/MS; w%. Della Porta *et al.* also found δ-2-carene (0.03%), (Z)-β-ocimene (0.01%), hexanol (0.02%), decanol (0.03%), 3-octanol (0.12%), γ-terpineol (0.01%), terpinen-4-yl acetate (0.05%), benzaldehyde (0.02%) and trace amounts of *p*-mentha-1,3,8,-triene, α-humulene, β-selinene, geranial, isopulegol, *cis*-*p*-menth-2-en-1-ol, *cis*-sabinene hydrate, *cis*-β-terpineol, *trans*-β-terpineol, acetophenone, 3-methyl benzaldehyde.

The results reported in Table 13.8 are in good agreement with those reported in Table 13.7. The high value of *p*-cymene in all oils reported in Table 13.8, should be highlighted, since this could indicate bad storage conditions of the oils analysed.

Laboratory extracted oils

Recently Caccioni *et al.* (1998), in a paper on the antimicrobial activity of citrus essential oils, analysed a mandarin oil laboratory extracted by hydrodistillation from mandarin of the Avana cultivar, obtained from the experimental field in Lentini, Sicily, Italy. The analysis of the volatile fraction, performed by GC and GC/MS, produced the following results:

camphene	0.02%
p-cymene	0.23%

limonene	72.71%
myrcene	1.71%
(E)-β-ocimene	0.02%
α-phellandrene	0.07%
β-phellandrene	0.52%
α-pinene	1.78%
β-pinene	1.41%
sabinene	0.17%
α-terpinene	0.24%
γ-terpinene	17.17%
terpinolene	0.82%
α-thujene	0.64%
β-caryophyllene	0.06%
(E,E)-α-farnesene	0.05%
α-humulene	0.02%
decanal	0.11%
nonanal	0.02%
octanal	0.06%
citronellal	0.02%
geranial	0.08%
α-sinensal	0.16%
octanol	0.03%
geraniol	0.03%
linalool	0.16%
nerol	0.07%
terpinen-4-ol	0.27%
α-terpineol	0.37%
thymol	0.02%
methyl N-methylanthranilate	0.46%

Caccioni et al. (1998) also found trace amounts of δ-3-carene, (Z)-β-ocimene, and neral.

TANGERINE OIL (*Citrus tangerine* Hort. ex Tan.)

1960–1979

Below are summarised the results reported by Shaw's review (1979) relative to tangerine oil. The papers originating the values reported below, were by Attaway et al. (1967), by Moshonas and Shaw (1974) and by Braddock and Kesterson (1976). The first two were on the analysis of the whole volatile fraction, while that by Braddock and Kesterson was on the total content of aldehydes, esters, alcohols, and only the quantitative composition of the aldehydes was reported in detail. The paper by Attaway et al. (1967) studied the variation of the oil composition during ripening; Shaw (1979) in his review considered only the composition of oils extracted in January from ripe fruits. The same approach has been chosen here.

limonene	87–88.34%
myrcene	1.2–2.26%
ocimene*	0.08%
α-pinene	0.81%
α-pinene + α-thujene	0.54%
β-pinene + sabinene	0.39%
α-terpinene	0.08%
γ-terpinene	2.23%
γ-terpinene + octanal	3.9%
terpinolene	0.31%
β-elemene	0.07%
γ-elemene	0.15%
δ-elemene	0.10%
decanal	0.16–0.26%
dodecanal	0.15%
hexanal	tr
octanal	0.25%
tetradecanal	0.05%
citronellal	0.13%
geranial	0.07–0.31%
neral	0.05%
perilla aldehyde	0.10%
α-sinensal	0.18%
geraniol	0.04%
linalool	0.87–4.2%
terpinen-4-ol + thymol methyl ether	0.27%
α-terpineol	0.06–0.89%
thymol	0.04–0.23%
elemol	tr
thymol methyl ether	0.9%

Among the results listed above those obtained by Ashoor and Bernhard (1967) were not included. These were relative to the composition of the monoterpene hydrocarbon fraction, and therefore not directly comparable with the data calculated relatively to the whole volatile fraction. The results obtained by these Authors (Ashoor and Bernhard, 1967) from the analysis of three tangerine oils, were the following:

camphene	0.01–0.02%
δ-3-carene	0–tr
p-cymene	0.35–0.82%
limonene	92.7–93.6%
myrcene + β-pinene	1.69–2.06%
α-phellandrene	0.03–0.06%
β-phellandrene	0.39–0.42%
α-pinene	0.62–0.99%
α-terpinene	0.08–0.09%
γ-terpinene	2.72–3.75%

During the years before the review by Shaw (1979) some papers, on the qualitative composition of tangerine oil, were reported by the same author in a former review (Shaw, 1977)

Hunter and Brogden (1965) identified the following aliphatic, monoterpene and sesquiterpene hydrocarbons: pentadecane, *p*-cymene, limonene, myrcene, α-pinene, β-pinene, γ-terpinene, terpinolene, δ-cadinene, caryophyllene*, α-, β-, and δ-elemene, α- and β-humulene, α-ylangene (as already mentioned a misidentification of α-copaene).

Hunter and Moshonas (1966) identified the following alcohols: dodecanol, nonanol, octanol, citronellol, *trans*-carveol, geraniol, linalool, *cis*- and *trans-p*-mentha-2,8-dien-1-ol, *p*-mentha-1,8-dien-9-ol, *p*-menth-1-en-9-ol, *p*-menth-8-en-1,2-diol, nerol, α-terpineol, thymol, elemol, *o*-phenyl phenol.

Moreover, Hussein and Pidel (1976) carried out a study for the quantitative determination of the composition of tangerine oil produced in Florida, California, Mexico and Brasil. The results obtained are reported below:

camphene + unknown	0–0.01%
p-cymene + octanal	0.14–0.39%
limonene	89.95–94.17%
myrcene	1.95–2.10%
α-phellandrene	0.03–0.04%
α-pinene	0.70–1.21%
β-pinene + unknown	0.44–1.07%
α-terpinene	0.02–0.06%
γ-terpinene	0.48–3.40%
terpinolene	0.04–0.15%
β-caryophyllene + nonanol	0.02–0.04%
decanal	0.04–0.11%
dodecanal + neryl acetate	0.02–0.04%
nonanal + unknown	0.01–0.03%
citronellal	0.02–0.05%
geranial + decanol + citronellol	0.02–0.03%
neral + α-terpineol	0.04–0.07%
perilla aldehyde	0.03%
sinensal*	tr–0.08%
carvone + nerol	0.03%
trans-carveol	tr–0.01%
linalool	0.30–0.98%
sabinene hydrate* + unknown	0.01%
thymol	tr–0.04%
decyl acetate	tr–0.05%
geranyl acetate + unknown	0.03–0.06%
1,8-cineole	0.63–0.76%

Hussein and Pidel also found, in one or more of the oils analysed, trace amounts of 6-methyl-5-hepten-2-one, dodecanol, heptanol, (Z)-3-hexenol, undecanol, *cis*-carveol, geraniol, thymol, benzyl alcohol, methyl N-methylanthranilate.

1979–2000

In this period there is a scant literature on the volatile fraction of tangerine essential oil. Two papers, Koketsu *et al.* (1983) and Lancas *et al.* (1988) are on the Brazilian tangerine oil, one, Inoma *et al.* (1989) is on commercial tangerine oils. The results relative to these papers are summarised in Table 13.9.

The data reported in Table 13.9, in particular for the major components, are enough homogeneous. However the results relative to the minor components cannot be considered reliable. The latest paper cited in the table is from 1989, and to our knowledge, no later investigations have been done on tangerine oils in the last decade, while the analytical techniques for separation and identification of complex matrices have been greatly improved. In fact, more detailed information could be obtained today on the composition of this oil, if such techniques were applied.

Table 13.9 Percentage composition of the volatile fraction of tangerine oil

	1	2	3
HYDROCARBONS			
Monoterpene			
p-cymene	0.04–1.28	0.22	0.26–0.92
limonene	88.57–91.65	91.44	87.40–91.12
myrcene	2.28–3–16	1.87	2.17–3.11
α-phellandrene	tr–0.02	–	0.03–0.05
α-pinene	1.16–2.01	0.85	0.84–1.19
β-pinene	0.06–0.58	0.31	0.25–0.43
sabinene	0.23–1.44	0.51	0.18–0.24
α-terpinene	tr	–	tr–0.02
γ-terpinene	0.02–2.78	3.41	2.61–4.48
terpinolene	0.07–0.24[a]	0.13	0.14–0.21
α-thujene	–	0.16	0.12–0.22
ALDEHYDES			
Aliphatic			
decanal	0.10–0.23	0.19	0.13–0.18
nonanal	0.05–0.06	–	tr–0.05
undecanal	0.03–0.05	–	0.02–0.03
Monoterpene			
citronellal	0.02–0.04[b]	–	0.05–0.06
geranial	tr–0.05	–	0–0.02
ALCOHOLS			
Monoterpene			
geraniol	0-tr	–	tr–0.02
linalool	0.34–1.17	0.53	0.69–1.22
terpinen-4-ol	tr–0.05	–	tr
α-terpineol	0.04–0.09	–	0.05–0.06
thymol	0–0.07	–	0.04–0.06
ESTERS			
Monoterpene			
citronellyl acetate	0–tr	–	0–0.03
geranyl acetate	tr	–	tr–0.02
neryl acetate	tr–0.02	–	0–0.02

Notes
tr = traces; a terpinolene + octanal; b citronellal + octyl acetate.

Appendix to Table 13.9

1 Koketsu et al. (1983). Brazil; FMC; 4 samples, three Cravo tangerine and one Dancy tangerine oils; GC on packed columns of Carbowax 20 M and SE-30 and on capillary column coated with Carbowax 20 M; retention times and Kovats indices; relative percentage of peak areas. Koketsu et al. also found β-caryophyllene (0.02–0.13%), dodecanal (0.03–0.05%), neral (tr–0.07%), nonanol (tr–0.20%), octanol (tr–0.02%), citronellol (0–0.23%), linalyl acetate (tr–0.07%), methyl N-methyl anthranilate (0–0.07%) and trace amounts of camphene, β-phellandrene, nerol.

2 Lancas et al. (1988). Brazil; FMC; one sample of Cravo tangerine oil; GC on capillary column (narrow-bore) coated with SE-54; GC/MS; relative percentage of paek areas; only those peaks contributing more than 0.1% were taken in account. Lancas et al. also found octanal (0.19%).

3 Inoma et al. (1989). Three commercial samples from Florida; GC on capillary columns coated with OV-101 and Carbowax 20 M; GC/MS; relative percentage of peak areas. Inoma et al. also found (E)-β-ocimene (0.12–0.23%), δ-cadinene (tr–0.02%), α-copaene (tr–0.03%), (E,E)-α-farnesene (tr–0.03%), valencene (0.02–0.05%), dodecanal (0.03–0.04%), perilla aldehyde (0.04%), α-sinensal (0.09–0.10%), trans-carveol (0–0.02%), limonene dioxide (tr–0.07%), limonene oxide* (tr–0.07%), thymol-methyl ether (0.10–0.13%).

The results published by Koketsu et al. (1983) present a maximum value for p-cymene too high, and a minimum value for γ-terpinene too low. It could be possible that these values refer to the same sample which was probably modified at the time of the analysis due to some oxidative processes.

Other researches than those reported in Table 13.9 were on some particular studies on the volatile fraction of tangerine oil.

Moshonas and Shaw (1979) identified the (E,E)-2,4-decadienal; Wilson and Shaw (1981) reported the following quantitative results determined in tangerine oil relative to some components which give the particular flavour note to the tangerine and mandarin oils: β-pinene (0.17 per cent), γ-terpinene (1.74 per cent), thymol (0.02 per cent), dimethyl anthranilate (0.07 per cent); the same authors (1984) also reported the following content of aldehydes determined in two tangerine oils from California:

decanal	0.18–0.21%
dodecanal	0.04–0.05%
nonanal	0.06–0.08%
octanal	0.28–0.46%
citronellal	0.08–0.10%
geranial	0.01–0.02%
neral	0.01–0.02%
perilla aldehyde	0.05–0.06%
α-sinensal	0.14–0.26%

Cartoni et al. (1987) during a research for the optimisation of the separation of volatile complex matrices, using GC columns of different polarities (Carbowax 20 M and SE-54) in series and a multi-step temperature programme, identified in tangerine oil the following components: camphene, p-cymene, limonene, myrcene, α-phellandrene, β-phellandrene, α-pinene, β-pinene, sabinene, γ-terpinene, terpinolene, α-thujene, bergamotene*, β-caryophyllene, humulene*, longifolene (probably misidentification of (E,E)-α-farnesene), decanal, octanal, citronellol, linalool, terpinen-4-ol, α-terpineol, thymol. The polarity of the system could be modified, in function of the composition of the sample to be analysed, changing the order of the columns and the temperature programme. This allowed the separation of critical pairs of components that, otherwise, could not be separated using one or the other column singularly.

COMPARISON OF MANDARIN AND TANGERINE OILS

The data found in literature, particularly for tangerine, are scant, and not recent, therefore it is difficult to identify possible differences among these oils, in particular relatively to the minor components. It is, however, possible to determine some quantitative differences among some major components common to both oils. Mandarin oil can be characterised by a lower content of limonene than tangerine oil (max 77 per cent in mandarin vs min 87 per cent in tangerine); a higher value of γ-terpinene (min 12 per cent in mandarin vs max 4.5 per cent in tangerine); a higher content of α-sinensal and of methyl N-methylanthranilate in mandarin oil. Spanish mandarin oil appear to contain a higher amount of limonene than the Italian mandarin oil, and a lower amount of γ-terpinene. The average content of γ-terpinene in Spanish oil is also similar to the Brazilian oils.

OTHER 'MANDARIN' OILS

More papers than those cited in this chapter, supposed to be on mandarin oils, are found in literature. These papers have not been included because the botanical origin of the fruits was not clear, or because the composition appeared to us incompatible with that normally obtained for *Citrus deliciosa* Ten, or for *C. Tangerine* Hort. ex Tan. oils. Among these were the papers by Blanco Tirado *et al.* (1995) on Columbian oils, by Cappello *et al.* (1981) on oils from Argentina, by Dellacassa *et al.* (1989, 1992), and by Calvarano M. *et al.* (1989) on oils from Uruguay.

SWEET ORANGE OIL (*Citrus sinensis* (L.) Osbek)

Numerous papers are found in literature on the composition of the volatile fraction of sweet orange oil. Most of these are on Italian and American industrial oils, but also on oils produced in Spain, and in South American countries. Moreover, studies on laboratory-extracted oils from fruits of special cultivars from different regions such as Algeria, Libya, Israel, Russia, China and Japan, were published.

1960–1979

Shaw (1979) reviewed many of the papers on the quantitative composition of sweet orange oil, published in this period. Five of these papers contain information on components belonging to different classes of compounds of the volatile fraction (Attaway *et al.*, 1967, 1968; Scora *et al.*, 1968; Ziegler, 1971; Shaw and Coleman, 1974), two are on aldehydes (Stanley *et al.*, 1961a; Braddock and Kesterson, 1976), one is relative to monoterpene aldehydes and hydrocarbons (Lifshitz *et al.*, 1970) and one reports on the nootkatone content (MacLeod and Buigues, 1964). The results considered by Shaw (1979) are reported below:

p-cymene	0.2%
limonene	82.5–97.0%
myrcene	0.9–2.18%
α-pinene	0.09–0.61%
β-pinene + sabinene	0.9%
sabinene	0.08–0.64%
γ-terpinene	0.1%
terpinolene	0.12%
α-bergamotene*	0.06%
δ-cadinene	0.11%
caryophyllene* + β-copaene	0.04–0.09%
β-cubebene	0.10%
β-elemene	0.04–0.06%
farnesene*	0.01–0.07%
valencene	0.04–0.22%
decanal	0.1–0.73%
dodecanal	0.03–0.20%
heptanal	0.02–0.05%
hexanal	tr–0.06%
nonanal	tr–0.17%
octanal	0.16–0.80 (2.8)[a]%
tetradecanal	0.05–0.09%
undecanal	0.01–0.02%
citronellal	tr–0.20%
geranial	0.06–0.23%
neral	0.01–0.21%
perilla aldehyde	0.02–0.03%
α-sinensal	0.02–0.06%
β-sinensal	0.06–0.13%
carvone	0.03–0.13%
nootkatone	0.01–0.02%
nonanol	0.1%
linalool	0.25–1.1 (5.3)[a]%
terpinen-4-ol	0.06–0.2%
α-terpineol	0.1–0.5%
octyl acetate	0.2%
neryl acetate	0.10%
neryl formate	0.1%

[a] unusually high values (both reported by Attaway, 1968) due, as Shaw (1979) indicated, to the calculation method or to a poor chromatographic separation.

In these years more papers than those reviewed by Shaw (1979), report quantitative information on the composition of sweet orange oil, and many are on the qualitative composition of this oil. Shaw (1977) reported some of these in his earlier review.

The qualitative investigations on sweet orange oil, reported in literature in this period (Calvarano, M. 1959; Bernhard, 1961; Teranishi *et al.*, 1963; McFadden *et al.*, 1963; Hunter and Brogden, 1964b,c; 1965; Hunter and Parks, 1964; D'Amore *et al.*, 1968;

Di Giacomo *et al.*, 1971) provided the identification of the following components: n-alkanes C_{21}–C_{31}, camphene, δ-3-carene, *p*-cymene, *p*-cymenene, isoterpinolene, limonene, myrcene, α-phellandrene, α-pinene, β-pinene, sabinene, α-terpinene, γ-terpinene, terpinolene, α-thujene, δ-cadinene, caryophyllene*, α-copaene, β-copaene, α-cubebene, β-cubebene, β-elemene, β-farnesene*, α-humulene, β-humulene, valencene, decanal, dodecanal, nonanal, octanal, undecanal, citronellal, geranial, neral, 6-methyl-5-hepten-2-one, carvone, decanol, hepten-3-ol, nonanol, hexanol, hexen-3-ol, nonanol, octanol, borneol, *trans*-carveol, citronellol, geraniol, isopulegol, linalool, nerol, terpinene-4-ol, α-terpineol, decyl acetate, isoamyl caproate, isobutyl acetate, octyl acetate, bornyl acetate, geranyl acetate, geranyl butyrate, geranyl formate, linalyl acetate, linalyl propanate, terpinyl acetate*, furfural.

Ikeda *et al.* (1962a) in oils from California and Florida found the following monoterpene hydrocarbons contents: limonene 87.5–89.1 per cent; myrcene 1.2–1.3 per cent; α-pinene 0.1–0.3 per cent; sabinene 0.1–0.2 per cent; γ-terpinene 0–0.1 per cent.

Di Giacomo *et al.* (1963) determined in the monoterpene hydrocarbon fraction of Italian oils, the following ranges of variation of some components: limonene 94.31–97.40 per cent; myrcene 1.59–1.96 per cent; α-pinene 0.26–1.28 per cent; sabinene 0.20–1.47 per cent.

Huet and Murail (1972) determined in sweet orange oils from numerous African countries and from Israel, Brazil, Florida and California, the total content of carbonyl compounds and the relative amounts of decanal, nonanal, octanal. The highest content of carbonyl compounds was determined in oils from Cameroon, Guinea, and South Africa.

Kekelidze and Beradze (1974) determined the composition of oils extracted in laboratory by distillation, from orange fruits named by the authors Turkish, Firstling, and Blood. They reported the following values:

limonene	88.4–93.2%
myrcene	1.1–2.8%
α-pinene	0.5–1.0%
sabinene	0.9–1.8%
γ-terpinene	0.5–4.0%
citronellol	tr–0.1%
linalool	0.3–1.4%
α-terpineol	tr–0.4%

Kekelidze and Beradze (1974) also found trace amounts of *p*-cymene and terpinolene.

Shaw and Coleman (1975) studying the most volatile components of sweet orange oil, identified for the first time isoprene and 3-methyl-1-butene, and also determined acetone, benzene, limonene, myrcene, α-pinene, sabinene, octanal, and linalool.

1979–2000

Industrial cold-pressed oils

In Table 13.10 are summarised the results, reported in literature, relative to the composition of the volatile fraction of sweet orange oils industrially produced in different countries, by extraction technologies commonly used by the processing industries. In the table are included oils from the Mediterranean countries (Italy, Spain, Israel, and Algeria), from North America (USA), from South America (Argentina, Brazil, and Uruguay) and from Central America (Mexico and Cuba). These oils were obtained from different fruits varieties, dependently to the different cultivars diffusion in each country.

Although the composition of sweet orange oil depends on the botanical origin of the fruits, and on the period of harvest, as already pointed out by different authors (Lifshitz *et al.*, 1970; Shaw e Coleman, 1974; Braddock e Kesterson, 1976; Dugo *et al.*, 1994; Verzera *et al.*, 1996a), the data in Table 13.10 vary between ranges quite narrow and therefore can represent the quantitative composition of sweet orange oil industrially processed.

The values out of range of myrcene, reported by Koketsu *et al.* (1983) and by Inoma *et al.* (1989) could be due to the not optimised chromatographic separation, and that of neryl acetate (0.24 per cent), reported by Inoma *et al.* (1989) for a Spanish oil, as well as the presence of linalyl acetate in some of the oils analysed by Koketsu *et al.* (1983), also appear unusual.

Further works, not cited in Table 13.10, are found in literature on the composition of sweet orange oil.

Moshonas and Shaw (1979, 1980) isolated and characterised, in cold-pressed oils from Hamlin, Pineapple and Valencia oranges, the following components: (E,E)-2,4-decadienal, the sesquiterpene alcohol intermedeol, and in Valencia oranges the sesquiterpene hydrocarbon (E,E)-α-farnesene. Micali *et al.* (1990) and Lanuzza *et al.* (1991), using a coupled HPLC-HRGC system, identified in sweet orange oil the linear hydrocarbons C_{21}–C_{33}, and the correspondent 'iso' isomers, and the sesquiterpene hydrocarbons α-copaene, β-caryophyllene, and α-humulene. Using a fully automated HPLC-HRGC system, Mondello *et al.* (1995b) were able to pre-fractionate the aldehydes of sweet orange oil, in three fractions (aliphatic, sesquiterpene and monoterpene aldehydes). The fractions were directly transferred into the GC system for the analysis. Avoiding peak overlaps, the authors were able to clearly identify different aldehydes, even some present at trace levels. Using a similar HPLC-HRGC system, coupled to a mass spectrometer, Mondello *et al.* (1995b) identified the following sesquiterpene hydrocarbons, listed in decreasing amount order: valencene, δ-cadinene, (Z)-β-farnesene, α-copaene, β-cubebene, β-caryophyllene, germacrene D, (E,E)-α-farnesene, α-humulene, γ-cadinene, α-muurolene, γ-muurolene, β-elemene, ar-curcumene[t]. Chamblee *et al.* (1997) confirmed the presence, in sweet orange oil, using a GC/FT-IR system, of bicyclogermacrene, α-copaene, α-cubebene, β-cubebene, germacrene B, germacrene D.

Other papers than those reported in Table 13.10 on the composition of sweet orange oils produced by extraction technologies commonly used in the processing industry and from the most common fruits varieties, that contribute for the production of the oil found on the international market, are found in literature on commercial oils, of which the origin is not well established, on oils obtained by semi-industrial processes, on oils obtained from fruits of uncommon varieties and that give small contribute to the international production. In Table 13.11 are the data relative to these papers. The experimental information, along with some short comments on the results obtained by each author, is reported in the appendix to the table.

Table 13.10 Percentage composition of the volatile fraction of sweet orange oil

	1	2	3(a)	3(b)	4	5	6	7	8(a)	8(b)	9(a)	9(b)	10	11	12(a)	12(b)
HYDROCARBONS																
Monoterpene																
camphene	0.01	–	–	–	–	–	tr	–	tr	tr	tr	tr	tr	–	tr	tr
δ-3-carene	0.10	–	–	–	–	–	0.08	0.07–0.21	0.09	0.09	0.09	0.04	0.10	–	0.10	0.14
p-cymene	0.06	tr–0.05	–	–	–	–	e	–	–	–	tr	tr	e	91.4	–	–
limonene	92.25	91.15–93.43	95.17	95.37	96.08	95.66	95.66[e]	91.58–94.88	95.59	95.23	94.34	93.69	95.73[e]	4.3[c]	94.85[h]	95.51[h]
myrcene	1.71	2.95–3.31	1.86[c]	1.81[c]	2.04[c]	1.86	1.72	2.49–3.17	1.85	1.80	1.87	1.91	1.73	–	1.86	1.88
(E)-β-ocimene	–	–	–	–	–	–	0.03	tr–0.07	0.09	0.10	0.03	0.03	0.01	–	0.04	0.03
(Z)-β-ocimene	–	–	–	–	–	–	tr	–	–	–	–	–	0.03	–	0.01	0.01
α-phellandrene	0.03	tr	–	–	–	–	0.14[d]	–	0.07	0.07	0.04	0.04	0.34[d]	–	0.05	0.05
β-phellandrene	–	0–tr	0.05	0.05	0.04	0.05	–	–	–	–	–	–	–	0.2	h	h
α-pinene	0.36	0.89–0.97	0.42	0.43	0.46	0.47	0.47	0.59–0.80	0.54	0.51	0.50	0.53	0.47	1.4	0.51	0.52
β-pinene	0.38[a]	tr–0.06	–	–	–	–	0.04	0.02–0.11	tr	tr	0.04	0.08	0.29[a]	–	0.05	0.03
sabinene	0.38[a]	0.45–0.71	0.24	0.34	0.27	0.28	0.61	0.25–0.58	0.26	0.40	0.29	0.80	0.29[a]	0.4	0.62	0.39
α-terpinene	–	0–tr	–	–	–	–	–	–	–	–	–	–	0.02	–	tr	tr
γ-terpinene	–	tr–0.04	–	–	–	–	0.01	0–0.33	tr	tr	0.03	0.02	0.02	tr	0.03	0.01
terpinolene	0.02	0.14–0.23[b]	–	–	–	–	0.02	–	0.06	0.07	tr	0.01	0.02	tr	0.02	0.03
α-thujene	–	–	–	–	–	–	0.01	–	tr	tr	tr	–	0.01	–	0.01	tr
Sesquiterpene																
γ-cadinene	–	–	–	–	–	–	0.02[t]	0–0.11	–	–	–	–	0.02[t]	–	tr	tr
δ-cadinene	–	–	–	–	–	–	–	tr–0.05	–	–	0.02	0.02	–	–	0.02	0.02
β-caryophyllene	–	0.02–0.04	0.02	0.02	0.03	0.02	0.02	tr–0.04	–	–	0.02	0.02	0.02	tr	0.02	0.02
α-copaene	–	–	–	–	–	–	0.02	tr–0.04	–	–	tr	tr	0.01[t]	–	0.02	0.02
β-copaene	–	–	0.02	0.01	0.02	0.02	–	–	–	–	0.03	0.03	–	tr	–	–
α-cubebene	–	–	–	–	–	–	0.02[tf]	–	–	–	tr	tr	0.02[tf]	–	–	–
β-cubebene	–	–	–	–	–	–	0.02[t]	–	–	–	–	–	–	–	0.02[tg]	0.02[tg]
β-elemene	–	–	–	–	–	–	0.02[f]	–	–	–	0.02	0.02	0.02[tf]	–	0.02[tg]	0.02[tg]
(E,E)-α-farnesene	–	–	–	–	–	–	–	0–tr	–	–	0.10	0.25	–	–	0.01	0.01
β-farnesene*	–	–	0.02	0.01	0.02	0.02	–	–	–	–	0.01	0.02	–	tr	–	–
germacrene D	–	–	–	–	–	–	0.01	tr–0.04	–	–	0.01	0.01	–	–	0.01	0.01
α-humulene	–	–	–	–	–	–	–	tr–0.04	–	–	tr	tr	–	–	tr	tr
valencene	–	0.13–0.38	0.05	0.05	0.05	0.06	0.08	–	–	–	0.01	tr	0.05	0.1	0.08	0.16

ALDEHYDES																
Aliphatic																
decanal	0.36	0.21–0.33	0.28	0.18	0.24	0.26	0.17	0.25–0.43	0.12	0.13	0.35	0.29	0.25	0.4	0.23	0.13
dodecanal	–	tr–0.04	0.07	0.06	0.05	0.07[k]	–	0.04–0.11	0.02	0.02	0.05	0.05	–	0.1	0.04	0.02
(E,E)-2,4-decadienal	–	–	–	–	–	–	–	–	tr	tr	0.01	0.01	–	–	tr[n]	tr[n]
(E,Z)-2,4-decadienal	–	–	–	–	–	–	–	–	tr	tr	0.02	0.01	–	–	–	–
hexadecanal	–	–	–	–	–	–	–	–	–	–	–	–	–	–	tr	tr
nonanal	0.05	0.05–0.07	0.02	0.02	0.06	0.34[p]	0.04	0.05–0.11	0.07	0.07	0.05	0.04	0.04	tr	0.04	0.03
octanal	0.04	0.14–0.23[b]	1.86[c]	1.81[c]	2.04[c]	0.34	0.14[d]	–	0.15	0.13	0.30	0.25	0.34[d]	4.3[c]	0.27	0.13
tetradecanal	–	–	–	–	–	–	–	–	0.01	0.01	–	–	–	–	tr	tr
tridecanal	–	–	–	–	–	–	–	–	tr	0.01	–	–	–	–	tr	tr
undecanal	–	0.02–0.06	0.02	0.01	–	0.02[l]	0.01	tr–0.02	tr	tr	0.01	0.01	0.01	–	0.01	0.01
Monoterpene																
citronellal	0.07	0.03–0.06[i]	0.05	0.04	0.05	0.05	0.03	tr–0.11	0.03	0.03	0.06	0.06	0.06	0.1	0.04	0.04
geranial	0.11	0.04–0.16	010	0.08	0.12	0.11	0.06	0.09–0.22	0.04	0.05	0.08	0.10	0.09	0.2	0.09	0.05
neral	0.06	0.05–0.07	0.07	0.05	0.08	0.09[j]	0.03	0.05–0.15	0.02	0.03	0.05	0.05	0.06	0.1[m]	0.06	0.04
perilla aldehyde	–	–	0.01	0.01	–	0.01	–	tr–0.04	0.02	0.01	0.01	0.02	–	–	0.01	0.01
Sesquiterpene																
α-sinensal	–	–	0.02	0.01	0.02	0.02	–	0–0.04	0.01	0.01	0.02	0.03	0.02	tr	0.02	0.01
β-sinensal	–	–	0.03	0.02	0.04	0.04	0.01	–	tr	0.02	0.04	0.06	–	0.1	0.03	0.02
KETONES																
Monoterpene																
carvone	–	–	–	–	–	–	–	tr–0.09	0.01	0.01	–	–	–	0.1[m]	tr	tr
Sesquiterpene																
nootkatone	–	–	0.01	0.01	0.02	0.01	–	–	tr	0.01	0.02	0.03	–	–	0.01	0.01
ALCOHOLS																
Aliphatic																
decanol	–	–	–	–	–	–	–	–	tr	tr	tr	tr	–	–	–	–
octanol	–	0.05–0.11	0.03	0.03	0.03	–	0.04	0.02–0.12	0.16	0.16	0.07	0.02	–	tr	0.02	0.01
Monoterpene																
trans-carveol	–	–	–	–	–	–	–	0–0.12	–	–	–	–	–	–	–	–
citronellol	–	tr–0.38	–	–	–	–	0.01[o]	–	0.01	0.02	tr	tr	0.02[o]	–	tr	tr
geraniol	–	0–tr	–	–	–	–	–	0–0.02	tr	0.02	0.02	0.02	0.01	–	0.01	tr
linalool	0.47	0.38–0.60	0.25	0.17	0.23	0.34[p]	0.35	0.43–0.73	0.39	0.35	0.41	0.38	0.41	0.8	0.48	0.36

Table 13.10 (Continued)

	1	2	3(a)	3(b)	4	5	6	7	8(a)	8(b)	9(a)	9(b)	10	11	12(a)	12(b)
nerol	–	0–0.05	–	–	–	–	0.01º	–	tr	0.01	0.01	0.01	0.02º	–	0.01	0.01
cis-sabinene-hydrate	–	–	–	–	–	–	0.01	–	–	–	–	–	tr	–	0.01	0.01
terpinen-4-ol	–	–	–	–	–	–	tr	–	tr	tr	tr	tr	tr	tr	0.01	tr
α-terpineol	0.05	0.06–0.09	0.03	0.03	0.02	0.05	0.03	0.05–0.12	0.03	0.04	0.05	0.03	0.03	0.1	0.05	0.03
ESTERS																
Aliphatic																
decyl acetate	–	–	–	–	–	–	–	–	0.01	0.01	0.03	0.03	–	–	0.01	0.01
octyl acetate	0.01	0.03–0.06ⁱ	0.01	0.01	–	0.03	–	–	tr	0.01	0.01	tr	–	tr	0.01	0.01
Monoterpene																
citronellyl acetate	–	tr–0.07	–	–	–	–	tr	–	–	–	tr	tr	–	–	tr	tr
geranyl acetate	–	0–tr	–	–	–	–	0.01	tr–0.03	–	–	tr	0.01	–	–	0.01	0.01
neryl acetate	–	tr–0.06	–	–	–	–	0.01	0–0.24	–	–	0.01	0.01	0.01	–	0.01	0.01
α-terpinyl acetate	–	–	–	–	–	–	–	–	–	–	tr	tr	–	–	tr	tr
OXIDES																
Monoterpene																
cis-limonene oxide	–	–	–	–	–	–	–	–	–	–	0.02ᵠ	0.02ᵠ	–	–	0.01	0.01
trans-limonene oxide	–	–	–	–	–	–	–	–	–	–	0.02ᵠ	0.02ᵠ	–	–	0.02	0.02

Notes

* correct isomer not characterised; tr = traces; t = tentative identification; a β-pinene + sabinene; b terpinolene + octanal; c myrcene + octanal; d α-phellandrene + octanal; e limonene + p-cymene; f α-cubebene + β-elemene; g β-cubebene + β-elemene; h limonene + β-phellandrene; i citronellal + octyl acetate; j neral + unknown; k dodecanal + unknown; l undecanal + unknown; m neral + carvone; n (E,E)-2,4-decadienal + nonyl acetate; o citronellol + nerol; p nonanal + linalool; q cis-limonene oxide + trans-limonene oxide.

Appendix to Table 13.10

1 Cappello et al. (1981). Argentina; one sample; GC on capillary column coated with UCON LB 550 X; relative percentage of peak areas. Capello et al. also found 6-methyl-5-hepten-2-one (0.04%).
2 Koketsu et al. (1983). Brazil; FMC; tree samples of Pera orange and tree samples of Hamlin orange oil; GC on packed columns of Carbowax 20 M and of SE-30 and on capillary column coated with Carbowax 20 M; retention times and Kovats indices; relative percentage of peak areas. Koketsu et al. also found nonanol (0.04–0.34%), linalyl acetate (tr–0.06%).
3 Vora et al. (1983). USA, Florida; FMC; one sample of Valencia (a) and one sample of midseason (b) orange oil; GC on capillary column coated with OV-101; GC/MS; relative percentage of peak areas.
4 Owusu-Yaw et al. (1986). USA; FMC; one sample of Valencia orange oil; GC on capillary column coated with OV-101; GC/MS; relative percentage of peak areas.

5 Ferrer and Matthews (1987). USA; FMC; one sample of Valencia orange oil; GC on capillary column coated with OV-101; GC/MS; w%.

6 Dugo et al. (1988). Italy.; 45 samples; GC on capillary column coated with SE-52; GC/MS; relative percentage of peak areas.

7 Inoma et al. (1989). 9 samples from USA (3 Valencia + 1 midseason), Spain (1 Navel + 1 Blood), Italy (1 Valencia + 1 unknown), Mexico (1); the results relative to one sample obtained from Valencia oranges (USA) have not been included, because it was our opinion that these values were relative to a partially deterpenated oil; GC on capillary columns coated with OV-101 and Carbowax 20 M; GC/MS; relative percentage of peak areas. Inoma et al. also found limonene dioxide (tr–0.20%), limonene oxide* (tr–0.38%).

8 Boelens and Jimenez (1989). Spain, Andalusia (a), Italy, Sicily (b); needle puncturing machine; GC on capillary column coated with SE-54; GC/MS and retention times; relative percentage of peak areas. Boelens and Jimenez also found in both oils myrtenal (0.03%), perillene (0.01%) and trace amounts of menth-1,8-dien-1-yl acetate and farnesols*.

9 Boelens (1991). Israel, (a) Valencia late, (b) Shamouti; GC; relative percentage of peak areas. Boelens also found in both oils β-bisabolene (0.01%), nerolidol* (0.01%) and trace amounts of δ-elemene.

10 Dellacassa et al. (1992). Uruguay; 109 industrial samples; GC; relative percentage of peak areas.

11 Pino et al. (1992). Cuba; 1 sample of Valencia orange oil; GC on capillary columns coated with SPB-1 and PEG 20 M; retention times; relative percentage of peak areas correct for non-volatiles content.

12 Dugo (1994), Dugo et al. (1994, 1999). Italy; Pelatrice and FMC; 86 samples of blond oils (a) and 104 of blood oils (b); GC on capillary column coated with SE-52; GC/MS, linear retention indices; relative percentage of peak areas. Dugo et al. also found in blond and blood oils α-cadinene^t (0.01%), (Z)-β-farnesene (0.01%), γ-muurolene (0.02%), perilla alcohol (0.01%), bornyl acetate^t (0.01%) and trace amounts of bicyclogermacrene, 7-epi-α-selinene, (E)-dec-2-en-1-al^t, hexadecanal, cis-carveol, perilla alcohol, elemol^t, germacrene D-4-ol^t, (E)-nerolidol, nonyl acetate.

Table 13.11 Percentage composition of the volatile fraction of some unusual sweet orange oils

	1	2	3	4	5	6
HYDROCARBONS						
Monoterpene						
δ-3-carene	–	–	–	–	0.03	0–0.17
p-cymene	tr–0.07	–	c	0.2	–	–
limonene	85.16–91.88	89.61	89.53[c]	90.7	94.51	90.57–94.19
myrcene	1.37–2.05	2.00	1.79	2.3	2.09	1.97–2.09
(E)-β-ocimene	–	–	0.21	–	0.21	tr
α-pinene	0.83–0.85	0.85	0.71	0.9	0.62	0.51–0.67
β-pinene	tr	–	0.34	0.3	tr	0–1.25
sabinene	0.30–0.45	2.45	0.13	0.6	0.13	0.78–1.46
γ-terpinene	tr–0.02	–	4.66	0.4	–	0.02–0.07
terpinolene	0.10–0.11[a]	–	0.24	tr	–	–
α-thujene	–	1.72	0.15	–	–	–
Sesquiterpene						
β-bisabolene	1.07–1.50	–	–	–	–	0–tr
γ-cadinene	–	–	0.10[t]	–	–	tr–0.06
δ-cadinene	–	0.07	–	–	tr	–
β-caryophyllene	0.04–0.05	0.11	0.04	tr	tr	tr–0.06
α-copaene	–	0.06	0.08	–	–	0–tr
β-cubebene	–	0.03	0.27	–	–	0–0.10
β-elemene	–	–	0.14[d]	–	–	0–tr
(E,E)-α-farnesene	–	0.25	0.69	–	0.05	–
α-humulene	–	0.04	0.07	–	0.05	–
valencene	tr	0.05	0.03	–	tr	–
ALDEHYDES						
Aliphatic						
decanal	0.19–0.21	–	0.08	–	–	0.19–0.28
dodecanal	0–tr	–	–	–	0.04	0–0.11
nonanal	0.07–0.08	–	0.02	0.1	–	0.05–0.06
octanal	0.10–0.11[a]	0.82	–	0.2	–	0.18–0.35
undecanal	0.04–0.05	0.07	0.01	–	–	–
Monoterpene						
citronellal	0.03–0.07[b]	0.17	0.03	–	–	0–0.04
geranial	0.13–0.25	0.16	0.03	–	–	0.21–0.33
neral	0.17–0.34	0.03	tr	0.1	0.06	0.12–0.30
perilla aldehyde	–	0.04	–	–	0.05	0–tr
Sesquiterpene						
α-sinensal	–	–	–	–	0.06	0–0.15
KETONES						
Monoterpene						
carvone	–	0.07	–	–	0.04	–
ALCOHOLS						
Aliphatic						
nonanol	0.53–0.81	–	–	–	–	0–tr
octanol	0.18–0.25	–	tr	–	–	–
Monoterpene						
trans-carveol	–	–	–	–	0.06	0–tr
citronellol	tr–0.18	–	0.01[e]	–	–	–
geraniol	–	0.08	–	–	–	0–tr

linalool	0.20–0.51	–	0.06	0.9	1.23	0.21–0.70	
nerol	–	0.10	0.01^e	–	–	0–tr	
α-terpineol	0.10–0.18	0.34	0.02	0.1	0.11	0.10–0.22	
ESTERS							
Monoterpene							
citronellyl acetate	tr–0.09	–	0.02	–	–	0–tr	
geranyl acetate	tr–0.17	–	0.07	0.1	tr	tr–0.24	
linalyl acetate	0.12–0.16	–	–	–	–	0–tr	
neryl acetate	0.08–0.09	0.01	0.06	–	0.05	–	

Notes

tr = traces; t = tentative identification; * correct isomer non characterised; a terpinolene + octanal; b citronellal + octyl acetate; c limonene + *p*-cymene; d α-cubebene + β-elemene; e citronellol + nerol.

Appendix to Table 13.11

1 Koketsu *et al.* (1983). Brazil; Piralima orange; three samples extracted by FMC; GC on packed columns of Carbowax 20 M and SE-30 and on capillary columns coated with Carbowax 20 M; retention times and Kovats indices; relative percentage of peak areas. Koketsu *et al.* also found 6-methyl-5-hepten-2-one (0–0.05%).

   The peculiariry of this oil consist of the high content of β-bisabolone (1.07–1.50%) and is also due to the presence of linalyl acetate.

2 Barukadze (1985). Georgia, ex URSS; Washington navel, one sample extracted with a new technology that, by the author, improve the yield of oil; GC on capillary column coated with OV-101. Barukadze also found β-farnesene* (0.15%), eucarvone^t (0.10%), neryl propanate (0.12%), linalool oxide* (0.08%) and trace amounts of (Z)-β-ocimene.

   This oil shows a rather law content of limonene and a high content of α-thujene compared to values generally presented by sweet orange oils for these components.

3 Dugo *et al.* (1988). China, Zhejiang Province; Wenzhou honey orange; Pelatrice; one sample; column chromatography on neutral alumina; GC on capillary column coated with SE-52; GC/MS; relative percentage of peak areas. Dugo *et al.* also found α-phellandrene (0.05%), α-terpinene (0.09%), δ-elemene (0.06%), terpinen-4-ol (0.01%), methyl N-methylantranilate (0.01%) and trace amouns of camphene, *cis*- and *trans*-sabinene hydrate.

   The oil of Wenzhou honey presents some peculiar data relative to its composition; the low content of limonene; the high content of terpinolene and in particular of γ-terpinene; the aliphatic aldehydes and alcohols are present at lower levels than the one commonly observed for sweet orange oils. It should be highlighted the presence of methyl N-methylantranilate, typical component of mandarin oil, that is usually not present in sweet orange, although seldom reported (Thomas and Bassols, 1992).

4 Haubruge *et al.* (1989). One commercial sample; GC on capillary column coated with CP-TM-VAX 52 CB; relative percentage of peak areas. Haubruge *et al.* also found β-phellandrene (1.5%) and trace amounts of *trans*-linalool oxide.

5 Inoma *et al.* (1989). Taiwan; Tankan orange; one commercial sample; GC on capillary columns coated with OV-101 and Carbowax 20 M; relative percentage of peak areas. Inoma *et al.* also found limonene dioxide (0.05%) and limonene oxide* (0.09%).

   The oil extracted from Tankan orange does not present relevant anomalies. It must be noted, however, the absence of decanal and a relatively high content of linalool.

6 Baaliouamer *et al.* (1992). Algery; Semi industrial extractor; 5 samples (Washington navel, Portuguese, Valencia, Hamlin, Sanguigne); preparative GC on packed column of FFAP; GC on capillary column coated with FFAP; GC/MS and linear retention indices; relative percentage of peak areas. Baaliouamer *et al.* also found (E)-β-farnesene (tr–0.02%), hexanal (0–0.02%), β-sinensal (0.04–0.18%), perilla alcohol (0–0.02%), methyl nopinone* (0–0.06%) and trace amounts of *cis*- and *trans*-α-bergamotene, γ-elemene, germacrene D, β-guaiene, β-gurijunene, α-himachalene, (E,E)-2,4-decadienal, heptadecanal, hexadecanal, octadecanal, octadec-9-en-1-al, isopiperitone, *cis*-carveol, *trans*-carveol, δ-cadinol, elemol, farnesol*, (E)-nerolidol, spathulenol, decyl acetate, citronellyl formate, α-terpinyl acetate, *cis*-limonene oxide, caryophyllene oxide and linear chained acids C_8–C_{12} in one or more of the analysed oils.

266 Giovanni Dugo et al.

Laboratory extracted oils

Numerous are the researches carried out on the composition of orange oils obtained from different botanical and geographic origins, extracted in laboratory by different methods. The results of some of these works are summarised in Table 13.12.

Other papers, not reported in Table 13.12, on laboratory-extracted oils, have been published.

El Samay *et al.* (1982) determined, in a laboratory extracted oil, obtained from Egyptian sweet oranges of not specified botanical origin, the following composition:

camphene	3.86%
limonene	28.89%
myrcene	22.40%
α-pinene	14.83%
geranial	1.16%
neral	0.77%
citronellol	0.19%
linalool	4.63%
geranyl acetate	1.11%

El-Samay *et al.* (1982) also found trace amounts of α-terpineol and bornyl acetate.

These values, totally unusual, not only for a sweet orange oil, but for any other citrus cold-pressed peel oil, could be due to the peculiarity of the matrix used, or to the method, probably not adequate, for the analysis of essential oils.

Some Japanese authors (Uchida *et al.*, 1984; Sugisawa *et al.*, 1987, 1989) studied in detail the composition of the oxygenated fraction of oils laboratory extracted from Japanese sweet oranges. They analysed the fractions of these oils separated by silica gel liquid chromatography by GC and GC/MS, and identified about 100 components. The results, expressed as per cent of the main classes of the oxygenated compounds relative to the whole oxygenated fraction, are reported below:

	1	2	3(a)	3(b)	3(c)	3(d)
aldehydes						
aliphatic	27.8	23.8	23.5	28.2	37.0	18.5
monoterpene	2.6	19.5	13.7	14.5	11.4	12.2
sesquiterpene (α-sinensal)	–	2.0	2.1	1.4	0.9	1.9
alcohols						
aliphatic	16.1	1.3	7.0	1.5	5.2	1.7
monoterpene	32.8	34.1	35.0	32.7	29.7	41.3
sesquiterpene	0.5	4.4	6.3	4.0	2.3	5.8
esters						
aliphatic	3.2	2.1	0.8	1.4	1.6	1.0
monoterpene	3.3	2.1	0.8	1.3	2.4	1.1
ketones						
sesquiterpene (nootkatone)	–	0.3	0.6	0.7	0.8	0.9

1. Uchida et al. (1984). Fukuara orange.
2. Sugisawa et al. (1987). Shiroyanagi navel orange.
3. Sugisawa et al. (1989). (a) Shiroyanagi navel, (b) Washington navel, (c) Hukumotobeni, (d) Ohmishima oranges.

The data reported above can provide useful information on the distribution of the different classes of components in the oxygenated fraction of oils extracted from the different sweet orange varieties examined. They indicate, however, that the data obtained from laboratory-extracted oils, referring to a limited number of samples, must be considered with attention. In fact, the same authors, analysing the oils extracted from the Shiroynagi navel orange, in different researches carried out two years apart (columns 2 and 3a), obtained significantly different results for monoterpene aldehydes and aliphatic alcohols.

MacLeod (1988) analysed a sample of a laboratory-extracted oil, obtained by hydrodistillation and solvent extraction with methylene chloride, from oranges cultivated in Libya of unknown botanical origin. This oil presented different anomalies in its composition both qualitative and quantitative. Some examples are the content of limonene (52.0 per cent), linalool (15.8 per cent), β-copaene (4.5 per cent), geranial (3.5 per cent) and carvone (2.5 per cent).

Thomas and Bassols (1992) identified, in oils extracted from Valencia and Pera oranges, some alkyl- and phenyl-pyridines, present at trace levels. They believed that these components were of endogenous nature and not artefacts produced during the extraction process.

Sawamura et al. (1999), during a study intended for the evaluation of the inhibition effect of the formation of N-nitrosodimethylamine *in vitro*, of citrus essential oils, reported the percentage of the major components determined in oils extracted from Valencia and Tarocco oranges cultivated in China. The values reported were the following:

	Valencia	Tarocco
limonene	95.8%	96.6%
myrcene	1.8%	1.8%
α-pinene	0.5%	0.5%
decanal	0.1%	0.1%
octanal	0.3%	0.1%
geranial	0.1%	0.1%
neral	0.1%	tr
linalool	0.4%	0.3%
α-terpineol	0.1%	0.1%
geranyl acetate	tr	0.1%

Sawamura et al. (1999) also found trace amounts of: β-pinene, γ-terpinene, terpinolene, citronellal, octanol, citronellol, geraniol, nerol, terpinen-4-ol.

Table 13.12 Percentage composition of the volatile fraction of laboratory extracted sweet orange oils

	1	2	3	4	5(a)	5(b)	5(c)	5(d)	5(e)	5(f)	5(g)	5(b)	6	7
HYDROCARBONS														
Monoterpene														
camphene	tr	–	tr	0.27–0.52	tr	tr	tr	tr	tr	tr	tr	tr	tr	tr
δ-3-carene	0.01–0.06	–	–	0.04–0.08	0.04	0.41	0.15	0.06	0.20	0.15	0.10	0.06	0–0.22	0.03
p-cymene	–	–	–	–	–	–	–	–	–	–	–	–	tr	tr
limonene	94.99–95.58	96.80	86.18	91.03–94.55	94.01	93.71	94.38	94.28	94.98	94.77	94.45	95.50	91.14–95.29	92.57
myrcene	1.62–1.69	0.93	1.53[b]	0–0.01	1.83	1.86	1.88	2.05	1.82	1.86	1.97	1.82	1.74–1.98	1.95
(E)-β-ocimene	–	–	–	0.02–0.03	0.04	0.03	0.02	0.02	0.03	0.03	0.02	0.02	0.01–0.05	0.06
(Z)-β-ocimene	–	–	–	–	0.01	0.01	tr	tr	0.01	0.01	0.01	tr	tr	tr
α-phellandrene	0.24–0.49[a]	–	0.27	0.02–0.05	0.06	0.09	0.05	0.07	0.04	0.06	0.07	0.07	0.03–0.06	0.08
α-pinene	0.42–0.45	0.44	0.30	0.28–0.53	0.47	0.48	0.47	0.53	0.47	0.48	0.50	0.48	0.41–0.53	0.55
β-pinene	0.02–0.04	–	1.53[b]	0.63–1.05	0.05	0.04	0.04	0.03	0.03	0.03	0.03	0.02	tr–0.06	1.93[e]
sabinene	0.37–0.71	0.24	1.12	0.02–0.03	0.93	0.78	0.52	0.34	0.43	0.60	0.49	0.38	0.13–0.83	1.93[e]
α-terpinene	–	–	–	–	tr	0.01	tr	tr	0.01	0.01	tr	tr	0–0.30	0.01
γ-terpinene	tr–0.01	–	0.03	0.02–1.09	tr	0.01	0.02	0.01	tr	0.01	0.01	tr	0.02–0.18	0.01
terpinolene	tr–0.01	–	–	0.58–2.61	0.01	0.08	0.03	0.02	0.02	0.03	0.02	0.02	0.01–0.07	0.02
α-thujene	tr	–	–	–	0.01	0.01	tr	tr	tr	tr	tr	tr	tr–0.01	0.01
Sesquiterpene														
γ-cadinene	–	–	–	–	0.01	0.01	tr	tr	0.01	0.01	tr	tr	–	–
δ-cadinene	–	0.02	–	0.03–0.04	0.03	0.02	0.03	0.02	0.02	0.03	0.03	0.03	–	0.03
β-caryophyllene	0.01–0.02	tr	0.03	0–0.02	0.02	0.02	0.01	0.01	0.02	0.02	0.03	0.01	tr–0.04	0.01
α-copaene	–	0.02	0.06	–	0.02	0.02	0.02	0.02	0.02	0.02	0.02	0.02	–	0.02
β-cubebene	–	0.01	–	–	0.03[c]	0.02[c]	0.02[c]	0.02[c]	0.03[c]	0.03[c]	0.03[c]	0.03[c]	–	0.02[c]
β-elemene	–	–	–	–	0.03[c]	0.02[c]	0.02[c]	0.02[c]	0.03[c]	0.03[c]	0.03[c]	0.03[c]	–	0.02[c]
(E,E)-α-farnesene	–	–	–	–	0.02	0.01	0.01	0.01	0.01	0.02	0.02	0.01	–	0.03
(E)-β-farnesene	–	–	–	–	0.03	0.01	0.01	0.01	tr	0.02	0.01	0.01	–	0.03
germacrene D	–	–	–	0.02–0.15	0.02	0.02	0.02	0.02	0.02	0.02	0.02	0.02	–	0.01
α-humulene	tr–0.01	–	0.03	–	tr	tr	tr	tr	tr	0.01	0.01	tr	–	0.01
α-muurolene	–	–	–	–	tr	tr	tr	tr	tr	tr	tr	tr	–	–
γ-muurolene	–	0.01	–	–	tr	0.01	tr	0.01	tr	0.01	0.02	0.01	–	–
valencene	0.01–0.09	–	–	–	0.07	0.10	0.04	0.05	0.17	0.18	0.31	0.09	tr–0.03	0.02

ALDEHYDES													
Aliphatic													
decanal	0.20–0.34	0.14	0.89	0.11–0.39	0.32	0.43	0.26	0.40	0.26	0.16	0.27	0.16	0.11–0.26	0.37
dodecanal	–	0.02	0.02	–	0.05	0.06	0.04	0.06	0.04	0.03	0.04	0.03	–	0.05
nonanal	0.03–0.06	0.01	2.45	–	0.07	0.06	0.04	0.06	0.04	0.03	0.04	0.04	tr–0.07	0.10
octanal	0.24–0.49[a]	–	0.77	1.50–1.89	0.33	0.52	0.28	0.42	0.26	0.14	0.17	0.18	0.09–0.53	0.68
tetradecanal	–	–	–	–	tr	tr	tr	tr	tr	tr	tr	tr	–	tr
tridecanal	–	–	–	–	tr[t]	tr[t]	tr[t]	tr[t]	tr[t]	tr[t]	tr[t]	tr[t]	–	tr[t]
undecanal	–	tr	–	–	0.02	0.01	0.01	0.01	0.01	0.01	0.01	0.01	–	0.02
Monoterpene														
citronellal	0.04–0.06	0.02	–	–	0.03	0.06	0.05	0.06	0.04	0.04	0.04	0.03	0.02–0.06	0.03
geranial	0.07–0.10	tr	0.27	0.02–0.23	0.13	0.13	0.12	0.13	0.08	0.08	0.08	0.07	0.07–0.23	0.19
neral	0.06–0.08	tr	0.15	0.08–0.15	0.09	0.09	0.08	0.08	0.05	0.05	0.05	0.05	0.04–0.18	0.13
perilla aldehyde	–	–	–	0–0.03	0.01	0.01	0.01	0.01	0.01	0.01	0.01	0.01	–	–
Sesquiterpene														
α-sinensal	–	0.01	–	–	0.04	0.02	0.02	0.03	0.01	0.03	0.02	0.02	0–0.05	0.06
β-sinensal	–	0.01	–	–	0.04	0.03	0.03	0.05	0.02	0.04	0.02	0.02	0–0.05	0.05
KETONES														
Sesquiterpene														
nootkatone	–	0.01	–	–	0.01	0.02	0.01	0.01	0.03	0.02	0.04	0.01	–	0.01
ALCOHOLS														
Aliphatic														
octanol	tr–0.01	–	0.17	0.02–0.04	0.02	0.03	0.01	0.02	0.01	0.03	0.01	0.01	0.03–0.24	0.01
Monoterpene														
cis-carveol	–	–	–	–	0.01	0.01	0.01	0.01	tr	0.01	0.01	tr	–	0.01
citronellol	tr–0.02[d]	tr	0.05	–	–	–	–	–	–	–	–	–	–	–
geraniol	–	tr	0.01	0.10–0.17	0.01	tr	0.01	tr	0.01	0.01	0.01	tr	0.01–0.09	0.01
linalool	0.51–0.74	0.31	–	–	0.65	0.37	0.86	0.52	0.32	0.54	0.56	0.43	0.41–2.56	0.46
nerol	tr–0.02[d]	–	0.06	0–0.13	0.02	0.02	0.01	0.02	0.01	0.02	0.02	0.01	0.01–0.12	0.02
cis-sabinene-hydrate	–	–	–	–	0.02	0.03	0.02	0.01	0.01	0.02	0.01	0.01	–	0.05
terpinen-4-ol	–	–	–	0.06–0.23	0.01	tr	0.01	0.01	tr	0.01	tr	tr	0.06–0.31	0.01
α-terpineol	0.04–0.06	tr	1.41	–	0.07	0.07	0.07	0.06	0.05	0.05	0.05	0.05	0.12–0.25	0.09

Table 13.12 (Continued)

	1	2	3	4	5(a)	5(b)	5(c)	5(d)	5(e)	5(f)	5(g)	5(b)	6	7
Sesquiterpene														
(E)-nerolidol	–	–	–	–	tr	tr	tr	tr	tr	0.01	tr	tr	–	tr
ESTERS														
Aliphatic														
decyl acetate	–	–	–	–	tr	0.01	0.01	0.02	0.01	0.01	0.03	0.01	–	–
nonyl acetate	–	–	–	–	tr	tr	0.01	0.01	tr	tr	0.01	tr	–	0.01
octyl acetate	–	–	–	–	0.01	tr	tr	0.01	0.01	tr	0.01	tr	–	tr
Monoterpene														
bornyl acetate	–	–	–	–	0.02	0.01	0.04	0.03	0.01	0.02	0.01	0.01	–	0.01
citronellyl acetate	tr	tr	–	–	tr	tr	tr	tr	tr	0.01	tr	tr	–	tr
geranyl acetate	tr–0.01	–	0.03	–	0.01	0.01	0.01	0.01	0.01	0.02	0.01	0.01	–	0.01
neryl acetate	tr–0.01	tr	–	–	0.01	0.01	0.01	0.01	0.01	0.01	0.01	0.01	–	0.01
α-terpinyl acetate	–	–	–	–	tr	tr	tr	tr	0.01	tr	0.01	tr	–	tr
OXIDES														
Monoterpene														
cis-limonene oxide	–	–	–	0–0.37	0.02	0.01	0.02	0.02	0.01	0.01	0.01	0.01	–	0.01
trans-limonene oxide	–	–	–	–	0.02	0.01	0.01	0.01	0.02	0.01	0.02	0.01	–	0.01

Notes

tr = traces; a α-phellandrene + octanal; b myrcene + β-pinene; c β-cubebene + β-elemene; d citronellol + nerol; e β-pinene + sabinene.

Appendix to Table 13.12

1 Arras et al. (1985). Sardinia, Italy; Valencia late orange grafted on different rootstocks (Citrus aurantium, Poncirus trifoliata, P. trifoliata (Roubidoux), P. trifoliata × C. sinensis (Carrizzo), C. macrophylla); 15 samples cold-pressed applying manual pressure on the rind; GC on capillary column coated with SE-52; GC/MS. Arras et al. also found β-bisabolene (0.01–0.02%)

The results reported by Arras et al. prove that the rootstock does not affect the composition of the oil.

2 Lin and Hua (1988). China; one sample; GC on capillary column coated with SE-54; GC/MS; Kovats indices. Liu and Hua also found allo-aromadendrene (0.01%), cis-linalool oxide (0.02%) and trace amounts of cis-α-bergamotene, trans-linalool oxide.

In this paper are compared the composition of sweet orange oil extracted from fresh fruits (results reported in the table) with that obtained from stored fruits. No differences were observed between these oils.

† = tentative identification;

3 Usai *et al.* (1992). Sardinia, Iraly; Thompson navel orange; Three samples extracted with petroleum ether; GC on capillary column coated with NS-54; GC/MS; W%. Usai *et al.* also found formaldeyde (0.44%), hexanal (0.08%), pulegone (0.02%), 3,7-dimethyl-2,6-octadien-1-ol (1.53%), *p*-menth-4(8)-en-9-ol (0.01%), nonanol (0.01%), pentanol (0.01%), acetic acid (0.26%), formic acid (0.13%), octanoic acid (0.01%).

In this paper the composition of the oil obtained from fruits (results reported in the table) is compared with that obtained from fruits stored at different temperature, for a period from 4 to 12 weeks. The most evident effects of the storage are: the increased content of limonene, of aldehydes, of volatile acids (formic and acetic) and of α-terpineol. The oil extracted from fresh fruits presented, however, a content of limonene lower then the one usually determined for genuine sweet orange oil, a very high content of aliphatic aldehydes (4.66%), mostly due to neral (2.45%), and the presence of some minor components reported above.

4 Blanco Tirado *et al.* (1995). Colombia; oil extracted by cold-pressing and by steam distillation from fruits belonging to three different harvesting periods; GC on capillary columns coated with DB-1 and DB WAX; GC/MS; linear retention indices; quantitative calculation performed using internal standard (es. tetradecanal). Blanco Tirado *et al.* also found δ-elemene (0.03–0.09%), isopulegol (0–0.05%), 1,4-cineole (0–0.02%). Qualitative and quantitative differences appear regarding cold-pressed and distilled oils. Myrcene, 1,4-cineole, isopulegol, nerol and perilla aldehyde were absent in the cold-pressed oils; weaver these oils had a lower amount of terpinolene, neral, geranial and a slight higher amount of decanal than distilled oils.

5 Verzera *et al.* (1996a). Sicily, Italy; (a) Biondo comune 31 samples, (b) Naveline 14 samples, (c) Ovale 6 samples, (d) Valencia late 7 samples, (e) Washington navel 6 samples, (f) Moro 51 samples, (g) Sanguinello 11 samples, (h) Tarocco 38 samples cold-pressed by applying manual pressure on the rind; GC on capillary column coated with SE-52; GC/MS; relative percentage of peak areas.

In this work the composition of the sweet orange oils obtained from Sicilian and Calabrian cultivars were compared. The differences of the composition during the productive season were also studied. The blond orange oils, with the exception of those obtained from Washington Navel oranges, present a lower content of monoterpene hydrocarbons and a higher content of oxygenated compounds if compared with blood orange oils. The Washington Navel orange oil is the blond oil that presents more similar composition to the blood orange oils, in particular for the content of aldehydes. The Sanguinello cultivar produces the blood orange oil that is the most similar to the blond oils.

The blood cultivars Moro and Tarocco, during the productive season (December–June) present a variation in composition in particular with a decrease of aliphatic aldehydes and an increase of sesquiterpene hydrocarbons, mainly valencene.

6 Caccioni *et al.* (1998). Sicily, Italy; oils extracted by steam distillation from Sanguinello, Tarocco, Moro, Valencia late and Ovale oranges (one sample from each cultivar); GC on capillary column coated with HP-1; GC/MS; retention times, co-injection with authentic samples; relative percentage of peak areas. Caccioni *et al.* also found β-phellandrene (0–0.20%).

7 Trozzi *et al.* (1999). Calabria, Italy; Maltese orange; 9 samples cold-pressed by applying manual named pressure on the rind; GC on capillary column coated with SE-52; GC/MS; relative percentage of peak areas. Trozzi *et al.* also found bicyclogermacrene (0.01%), β-gurjiunene (0.01%), heptyl acetate (0.01%) and trace amounts of (E)-dec-2-en-1-al¹, carvone, *trans*-sabinene hydrate.

The composition of this oil results to be quite similar to the ones obtained from different varieties cultivated in Italy.

BERGAMOT OIL (*Citrus Bergamia*)

1960–1979

Shaw (1979) in his review on the composition of citrus essential oils, cited only one reference on bergamot oil (Huet and DuPuis, 1969). This paper was on the composition of laboratory extracted oils, by different techniques, from fruits harvested in Corsica. In the review the results on oils produced in Calabria (Italy), were not reported, although the oils produced in this region are the most important, particularly among the industrially processed bergamot oils.

In the 60s and 70s different papers were published (Calvarano, M. 1963, 1965; Calvarano and Calvarano, 1964; Liberti and Goretti, 1974) on the composition of Italian bergamot industrial oils. The limits of the results relative to these papers are reported below:

camphene	0.03–0.10%
δ-3-carene	0.08–0.23%
p-cymene	tr–1.68%
limonene	18.63–48.50%
myrcene	0.40–1.50%
α-phellandrene	0–0.09%
α-pinene	0.61–2.27%
β-pinene	7.75%
β-pinene + sabinene	4.45–11.86%
sabinene	1.26%
γ-terpinene	3.48–11.76%
terpinolene	0.22–0.83%
α-thujene	0.20–0.43%
decanal	0.42–0.69%
nonanal	0.07–0.18%
citronellal	0.04–0.16%
geranial	0.26–0.71%
neral	0.16–0.64%
citronellol	0.25–0.61%
linalool	7.07–29.12%
terpinen-4-ol	0.05–0.09%
α-terpineol	0.10–0.43%
octyl acetate	0–0.12%
linalyl acetate	23.48–35.62%
geranyl acetate	0.27–0.82%
neryl acetate	0.44–1.19%
terpinyl acetate*	0.17–0.54%

Obviously some of the values above are compromised by the limited chromatographic techniques available at that time. In particular, the values of δ-3-carene (0.10–0.23 per cent) are absolutely incompatible with a genuine bergamot oil, where this component is normally present at trace levels; these, therefore, can be explained by some peak overlap.

Di Giacomo et al. (1963) reported the composition of the monoterpene hydrocarbon fraction: p-cymene (tr–2.39 per cent), limonene (61.60–66.33 per cent), myrcene (1.25–2.25 per cent), α-pinene (2.08–2.36 per cent), β-pinene (10.30–11.61 per cent), sabinene (4.56–6.19 per cent), γ-terpinene (13.51–14.00 per cent).

Numerous papers, other than those on the quantitative composition of Italian bergamot oil, have been published in these years on the quantitative or qualitative composition of industrial bergamot oil.

Ikeda et al. (1962b), in a commercial bergamot oil found the following monoterpene hydrocarbons: p-cymene (0.7 per cent); limonene (22.3 per cent); myrcene (0.4 per cent); ocimene* (0.1 per cent); α-pinene (0.6 per cent); β-pinene (3.5 per cent); sabinene (0.5 per cent); γ-terpinene (4.1 per cent); α-thujene (0.1 per cent). MacLeod and Buigues (1964) identified in bergamot oil small amount (0.01 per cent) of nootkatone and Vernin and Vernin (1966) determined trace amount of methyl N-methylanthranilate.

Huet and DuPuis (1968), while analysing bergamot oils produced in Corsica and in Africa (Guinea, Ivory Coast, Algeria, Cameroon, Morocco and Mali), found out that the content of esters varied from 16.0 to 60.1 per cent; that of alcohols from 2.4 to 35.6 per cent. The highest content of esters was determined in the oils from Corsica, that of alcohols in the oils from Guinea and Ivory Coast.

Yoshida et al. (1971) reported for a Japanese oil the following composition: limonene (26.0 per cent); α-pinene (0.6 per cent); β-pinene (5.5 per cent); linalool (29.1 per cent); linalyl acetate (29.3 per cent).

The qualitative analyses on the volatile components in bergamot oil by Wenninger et al. (1967), Dixon et al. (1968), Ziegler (1971), Calabrò and Currò (1973), Hérisset et al. (1973), Di Corcia et al. (1975), provided the identification of other components than those already mentioned: α-terpinene, trans-α-bergamotene, β-bisabolene, caryophyllene*, β-farnesene*, α- and β-humulene, dodecanal, octanal, borneol, geraniol, nerol, bornyl acetate, linalool oxide*, and also the n-alkanes from C_{20} to C_{33} and the relative 'iso' and 'anteiso' isomers from C_{21} to C_{26}.

One study (Mookherjee, 1969), performed by fractionation of bergamot oil by liquid chromatography and by preparative gas chromatography, allowed the isolation and the identification, by spectroscopic techniques, of numerous bifunctional components: 3-acetoxy-3-methylocta-1,5-dien-7-one, 2,6-dimethylocta-1,7-dienyl-3,6-diol, cis- and trans-2,6-dimethylocta-2,7-dienyl acetate, 2,6-dimethyloct-7-enyl acetate, 4-acetoxy-2,6-dimethylocta-2,7-dienyl-6-ol, 1-hydroxydihydrocarveol, 3-hydroxycitrenollyl acetate, 6,7-epoxylinalyl acetate, 7-hydroxylinalyl acetate, dehydro cineole, 8-hydroxy-p-menth-2-en-1-yl methyl ether.

Calvarano, M. (1968) studied the variation of bergamot oil composition during the production season, monitoring the fruits from July (small unripe fruits) up to the end of January (fully mature fruits). The most evident variations concerned the monoterpene hydrocarbons β-pinene and γ-terpinene, and linalool and linalyl acetate. During the season β-pinene content varied from 4 to 10 per cent, γ-terpinene from 6 to 10 per cent; linalyl acetate from 25 to 35 per cent while linalool showed an inverse behaviour varying from 33 to 13 per cent. Other components such as limonene were present at almost constant levels.

Huet and DuPuis (1969) used a similar approach to that used by Calvarano, M. (1968), on bergamot oils, extracted in laboratory by different methods (distillation, solvent extraction and mechanical cold extraction), from fruits harvested in Corsica.

The results, for linalool and linalyl acetate from August to January, determined for the cold pressed oils, were similar to those obtained by Calvarano M. (1968). Linalool varied respectively from 43 to 16 per cent and linalyl acetate from 21 to 38 per cent. Different were the variations observed for β-pinene, γ-terpinene and limonene. The first two hydrocarbons varied randomly during the season, the fruits harvested in August and in January presented almost the same amount (β-pinene about 4 per cent, γ-terpinene about 5 per cent). Limonene varied from 24 to 32 per cent.

In order to provide more complete information, below are summarised the percentage composition determined for the mature fruits by (a) Calvarano, M. (1968), (b) Huet and DuPuis (1969) and also those (c) reported by Ortiz *et al.* (1978) for bergamot oils harvested in California in the period June–July:

	(a)	(b)	(c)
camphene	0.06%	–	–
δ-3-carene	0.05%	0.02%	–
p-cymene	0.43%	–	0.80%
limonene	23.98%	32.35%	48.42%
myrcene	0.95%	0.78%	2.10%
α-phellandrene	0.04%	–	–
α-pinene	1.50%	1.97%	1.25%
β-pinene	9.98%[a]	4.12%[a]	6.15%
sabinene	9.98%[a]	4.12%[a]	1.25%
γ-terpinene	9.89%	4.92%	8.26%
terpinolene	0.30%	–	–
α-thujene	0.02%	–	–
bisabolene*[t]	0.26%	–	–
decanal	0.26%	–	0.04%
nonanal	0.07%	–	–
octanal	0.10%	–	–
citronellal	0.02%	–	–
geranial	0.48%	0.14%	2.58%
neral	0.44%	–	1.75%
citronellol	0.33%[b]	–	0.18%
geraniol	0.02%[t]	–	0.52%
linalool	13.28%	16.50%	12.85%
nerol	0.33%[b]	–	0.24%
terpinen-4-ol	0.02%	0.33%	–
α-terpineol	0.12%	0.30%[c]	3.07%
octyl acetate	0.01%	–	–
bornyl acetate	–	–	0.06%
linalyl acetate	34.84%	37.90%	9.72%
geranyl acetate	0.30%[t]	0.60%	–
neryl acetate[t]	0.98%	–	–
terpinyl acetate*	0.28%	–	–

a β-pinene + sabinene; b nerol + citronellol; c α-terpineol + unknown.

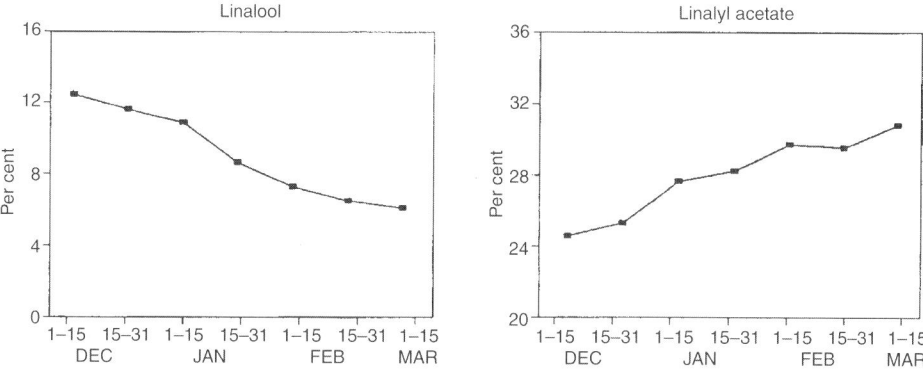

Figure 13.4 Variation in average content of linalool and linalyl acetate for bergamot oils produced in Italy during the productive season (Dugo, 1994).

It should be mentioned that the paper by Ortiz was not meant to be for the determination of bergamot oil composition, but for the comparison between oils obtained by the same extraction method, produced in the same period from numerous cultivars of bitter orange cultivated in the same area.

1979–2000

Industrial cold-pressed oils

In Table 13.13 are reported the results relative to the composition of bergamot oils industrially produced, found in literature in this period.

In this table are not included the results obtained by Zani *et al.* (1991) for a sample of bergamot oil produced in Calabria (Italy). In fact in this sample was determined an unusual amount of camphene (0.93 per cent). This value could be probably due to some contamination of the sample analysed.

The values in Table 13.13 show that bergamot oil is subject to wide ranges of variation in its composition. These changes, as proved by many authors, could be due to the period of harvest of the fruits, (Calvarano, M., 1968; Huet and Puduis, 1969; Dugo *et al.*, 1987, 1991; Dugo, 1994; Verzera *et al.*, 1996b, 1998), to the cultivar of the fruits (Verzera *et al.*, 1996b) and to the area of cultivation of the fruits (Huet and Dupuis, 1969), also for very close fields, as has been noticed in Calabria (Dugo *et al.*, 1987, 1991; Verzera *et al.*, 1998). The most evident changes are, however, noticed during the productive season, in particular for linalool and linalyl acetate. The former, as shown in Figure 13.4 (Dugo, 1994), varied, for a bergamot oil produced in Calabria, during the productive season, from 13 per cent to about 6 per cent; the latter, during the same period varied from an average content of 25 per cent at the beginning of the season, to that of 31 per cent at the end. The ratio linalool/linalyl acetate varied from 0.5 to 0.2. This behaviour has been confirmed for different years of production, although some differences of the average content of these components were determined (Dugo *et al.*, 1991) among different years. As mentioned above, the composition of bergamot oil also depends on the cultivar of the fruits used. For oils produced in Calabria (Italy) it has

Table 13.13 Percentage composition of the volatile fraction of cold-pressed bergamot oil

	1	2(a)	2(b)	2(c)	3	4	5	6	7	8	9
HYDROCARBONS											
Monoterpene											
camphene	–	0.03	0.02	0.04	0.01	0.04–0.08	tr	tr	0.03	0.02–0.05	0.11
p-cymene	2.7	0.23	0.10	0.34	0.16	0.59–0.86	0.54	3.61	0.25–0.32	0.01–0.89	1.29
limonene	33.0	28.38	28.70	35.64	43.29	32.85–34.66	38.35	26.71	38.13–42.54[d]	24.07–54.85[d]	32.14
myrcene	0.9	0.77	0.69	0.90	0.92	0.92–1.32	2.04	0.36	1.02–1.03	0.63–1.81	2.33
allo-ocimene	–	–	–	–	–	–	tr	–	–	tr[t]	–
(E)-β-ocimene	–	–	–	–	–	–	0.23	0.13	0.18–0.19	0.02–0.42	1.06
(Z)-β-ocimene	–	–	–	–	–	–	0.78	–	0.02–0.03	0.01–0.07	0.43
α-phellandrene	–	0.07	0.09	0.04	–	0.13–0.15	–	–	0.04–0.05	0.01–0.06	0.18
β-phellandrene	–	–	–	–	–	0.02–0.04	–	–	d	d	–
α-pinene	1.0	1.56	0.87	1.51	0.88	1.62–1.74	1.38	0.70	1.08–1.14	0.72–1.84	0.99
β-pinene	5.7	7.51	4.11	7.07	5.49	7.63–10.60	5.77	5.11	6.55–6.88[c]	4.81–12.80[c]	6.49
sabinene	1.0	1.26	0.79	1.23	–	1.00–1.69	1.26	0.72	6.55–6.88[c]	4.81–12.80[c]	–
α-terpinene	–	–	–	–	–	–	0.21	–	0.12–0.13	0.08–0.28	–
γ-terpinene	3.7	6.90	4.55	7.63	4.03	3.95–6.73	7.67	1.15	6.19–6.48	5.27–11.38	7.54
terpinolene	–	0.21	0.21	0.17	0.25	0.42–0.50[a]	0.36	tr	0.26–0.27	0.21–0.47	0.72
α-thujene	–	–	–	–	–	–	–	0.15	0.27–0.28	0.19–0.49	0.25
tricyclene	–	–	–	–	–	–	–	–	0–tr	tr–0.01	–
Sesquiterpene											
cis-α-bergamotene	tr	–	–	–	–	–	tr	–	tr	0.02–0.05	–
trans-α-bergamotene	0.23	0.29[t]	0.30[t]	0.32[t]	–	0.09–0.15	0.36	0.20	0.31–0.39	0.16–0.44	–
bicyclogermacrene	–	–	–	–	–	–	–	–	0.04–0.05	0.01–0.08	–
α-bisabolene	tr	–	–	–	–	–	tr	–	–	–	–
β-bisabolene	0.57	0.56[t]	0.73[t]	0.60[t]	–	0.16–0.33	0.46	0.31	0.43–0.56	0.21–0.65	0.52
β-caryophyllene	0.27	–	–	–	–	–	0.27	0.18	0.34–0.50	0.15–0.55	0.45
δ-elemene	–	–	–	–	–	–	tr	–	tr	0–0.06	–
(E,E)-α-farnesene	–	–	–	–	–	–	–	–	tr	0–tr	–
(Z)-β-farnesene	–	–	–	–	–	–	–	–	0.04–0.06	0.03–0.09	0.08
germacrene D	–	–	–	–	–	–	tr	–	0.06–0.10	0.03–0.11	–
α-humulene	tr	–	–	–	–	–	tr	–	0.05–0.07	0.01–0.04	0.03
β-santalene	tr	–	–	–	–	–	tr	–	0.01–0.02	tr–0.02	–

Compound	C1	C2	C3	C4	C5	C6	C7	C8	C9	C10
ALDEHYDES										
Aliphatic										
decanal	—	—	—	—	—	—	—	0.06–0.07	0.04–0.10	0.16
dodecanal	—	0.07	0.06	0.09	—	tr	tr	0.03–0.04	tr–0.05	—
nonanal	—	—	—	—	—	tr–0.06	0.16	0.04–0.05	0.01–0.08	—
octanal	—	0.07	0.03	0.03	—	0.42–0.50[a]	—	0.02–0.03	0.02–0.08	—
tetradecanal	—	—	—	—	—	—	—	tr	0–0.01	—
undecanal	—	—	—	—	—	—	tr	0.01–0.02	0–0.02	—
Monoterpene										
citronellal	—	0.06	0.04	0.02	—	tr[b]	—	0.01	tr–0.03	—
geranial	—	0.58	0.25	0.35	0.23	0.79–1.25	0.16	0.16–0.20[e]	0.19–0.54[e]	0.43
neral	—	0.30	0.12	0.23	0.05	0.20–0.21	0.13	0.11–0.13	0.10–0.72	0.28
perilla aldehyde	tr	—	—	—	—	tr	0.11	0.16–0.20[e]	0.19–0.54[e]	—
KETONES										
Aliphatic										
6-methyl-5-hepten-2-one	—	—	—	—	—	tr	—	—	tr–0.01	—
Monoterpene										
carvone	—	—	—	—	—	—	0.13	—	0–tr	—
Sesquiterpene										
nootkatone	—	—	—	—	—	tr	—	0.04–0.06	0.01–0.10	—
ALCOHOLS										
Aliphatic										
nonanol	—	—	—	—	—	tr	—	—	—	—
octanol	—	—	—	—	—	tr	—	0–tr	0–0.03	—
Monoterpene										
cis-carveol	—	—	—	—	—	tr	tr	—	—	—
trans-carveol	—	—	—	—	—	tr	tr	—	—	—
citronellol	—	—	—	—	0.29	0.17–0.28	tr	—	0.01–0.11[f]	—
hotrienol	tr	—	—	—	—	—	tr	—	—	—
geraniol	0.05	0.02	0.03	0.02	—	0.03–0.07	tr	tr	0–0.01	—
linalool	13.45	18.94	24.22	12.67	15.21	16.76–18.82	9.60	8.91–9.52	1.58–22.68	11.63

Table 13.13 (Continued)

	1	2(a)	2(b)	2(c)	3	4	5	6	7	8	9
nerol	0.10	0.09ᶠ	0.26ᶠ	0.05ᶠ	—	0.04–0.08	tr	—	0.02–0.03	0.01–0.11ᶠ	0.04
perilla alcohol	tr	—	—	—	—	—	tr	—	—	—	—
cis-sabinene-hydrate	tr	—	—	—	—	—	—	—	—	0.01–0.06	0.07
trans-sabinene hydrate	tr	—	—	—	—	—	tr	—	0.02	—	—
terpinen-4-ol	tr	0.21	0.26	0.29	—	tr	tr	tr	0.02	0.01–0.04	tr
α-terpineol	0.13	0.11	0.19	0.08	0.17	0.20–0.27	0.18	0.12	0.03–0.04	0.03–0.13	0.09
Sesquiterpene											
α-bisabolol	tr	—	—	—	—	—	—	—	0.02–0.03	0.01–0.03	—
β-bisabolol	tr	—	—	—	—	—	—	—	—	tr	—
campherenol	—	—	—	—	—	—	—	—	0.02–0.03	0.01–0.02	—
(E)-nerolidol	tr	—	—	—	—	—	tr	—	0.01–0.03	0.01–0.04	—
'norbornanol'ᵍ	—	—	—	—	—	—	—	—	0.01–0.02	0.01–0.02	—
ESTERS											
Aliphatic											
decyl acetate	—	—	—	—	—	—	tr	—	0.02–0.03	tr–0.05	—
heptyl acetate	—	—	—	—	—	—	tr	—	0.01	tr–0.02	—
hexyl acetate	—	—	—	—	—	—	tr	—	—	0–tr	—
nonyl acetate	—	—	—	—	—	—	tr	—	0.02–0.03	0.01–0.05	—
octyl acetate	—	0.03	0.16	0.12	—	trᵇ	tr	—	0.12–0.13	0.06–0.22	—
Monoterpene											
bornyl acetate	—	—	—	—	—	—	tr	—	0.01–0.03	0.01–0.04	—
citronellyl acetate	—	—	—	—	—	—	tr	—	0.01–0.03	tr–0.06	—
geranyl acetate	0.46	0.19	0.16	0.26	0.88	0.15–0.24	0.73	0.64	0.26–0.32	0.11–0.84	0.50
linalyl acetate	31.30	29.48	31.38	32.71	27.40	22.68–27.85	26.88	38.13	28.85–31.83	15.09–41.36	30.06
linalyl propanate	—	—	—	—	—	—	tr	—	0.05–0.06	0.01–0.07	—
methyl geranate	—	—	—	—	—	—	—	—	tr–0.01	tr–0.02	—
neryl acetate	0.42	—	—	—	0.23	tr	0.62	0.47	0.26–0.33	0.13–0.67	0.47
trans-sabinene hydrate acetate	—	—	—	—	—	—	—	—	0.07	0.05–0.13ᶜ	0.07
α-terpinyl acetate	0.21	—	—	—	—	—	0.10	—	0.12–0.15	0.07–0.27	0.18

OXIDES
Monoterpene

1,8 cineole	—	—	—	—	—	tr	—	tr–0.02	—	—
cis-limonene oxide	tr	—	—	—	—	tr	—	tr–0.02	—	—
trans-limonene oxide	tr	—	—	—	—	tr	—	tr–0.01	—	—
cis-linalool oxide	—	—	—	—	—	tr	—	0–tr	—	—
trans-linalool oxide	—	—	—	—	—	tr	—	0–tr	—	—

Sesquiterpene

humulene oxide II	tr	—	—	—	—	tr	—	—	—	—

Notes

tr = traces; t = tentative identification; a terpinolene + octanal; b citronellal + octyl acetate; c β-pinene + sabinene; d limonene + β-phellandrene; e geranial + perilla aldehyde; f citronellol + nerol; g 2,3-dimethyl-3-(4-methyl-3-pentenyl)-2-norbornanol.

Appendix to Table 13.13

1 Schenk and Lamparsky (1981). GC on capillary column coated with UCON. Schenk and Lamparsky also found in bergamot oil traces amount of ar-curcumene, α-santalene, lilial, 2,6-dimethyl-6-acetoxy-oct-7-en-3-one, 2,6-dimethyl-6-acetoxy-octa-1,7-dien-3-one, 2,6-dimethyl-6-acetoxy-octa-1-en-7-one, cumin alcohol, *cis*- and *trans*-2,6-dimethyl-octa-1,5,7-trien-3-ol, β-photosantalol A, *p*-mentha-1,8-dien-9-ol, *p*-menth-1-en-9-ol, *cis*- and *trans*-*p*-menth-2-en-1-ol, α-cadinol, caryophyllenol I and II, cariophyllene alcohol, hotrienyl acetate, *p*-mentha-1,3-dien-7-yl acetate, *p*-mentha-1,7-dien-4-yl acetate, *p*-mentha-1,7(10)-dien-2-yl acetate, *p*-mentha-1,8-dien-9-yl acetate, *p*-menth-1-en-9-yl acetate, terpinen-4-yl-acetate, *cis*- and *trans*-2,3-ocimene oxide, humulene oxide I, isocaryophyllene oxide, 2,3-geranyl acetate oxide, 2,3-neryl acetate oxide.
2 Huet (1981). (a) Corsica, (b) Ivory Coast (Coci and Divo bergamot), (c) Italy. GC on capillary column coated with Carbowax 20 M.
3 Ricciardi *et al.* (1982). Argentina; FMC.
4 Koketsu *et al.* (1983). Brazil; FMC; three sample; GC on packed columns of Carbowax 20 M and SE-30 and on capillary column coated with Carbowax 20 M; retention times and Kovats indices; relative percentage of peak areas.
5 Mazza (1986). Calabria, Italy; three samples; column chromatography on silica gel; GC on capillary column coated with Carbowax 20 M; GC/MS; relative percentage of peak area. Mazza also found traces amount of *p*-cymenene, γ-acoradiene, δ-cadinene, α-muurolene, epi-β-santalene, α-selinene, β-sesquiphellandrene, piperitone, decanol, hexanol, (E)-hex-2-en-1-ol, (Z)-hex-3-en-1-ol, 1 pentanol, 2 pentanol, carvacrol, *p*-cymene-8-ol, limonene-4-ol, 3,7-dimethyl-3-acetoxy-octa-1,5-dien-7-ol, 3,7-dimethyl-3-acetoxy-octa-1,7-dien-6-ol, *cis*- and *trans*-*p*-menth-2,8-dien-1-ol, 3,7-dimethyl-octa-1,5-dien-3,7-diol, 3,7-dimethyl-octa-1,7-dien-3,6-diol, *trans*-pinocarveol, spathulenol, *cis*-hex-3-en-1-yl acetate, 3-(3,4,5-trimethoxyphenyl)-propenyl acetate, geranyl propanate, neryl propanate, perillyl acetate, *trans*-pinocarvyl acetate, sabinyl acetate, *cis*- and *trans*-linalool oxide (pyranoid form), *p*-menth-1-en-4,5-oxide, *p*-menth-4-en-1,2-oxide, linalyl acetate oxide (two isomers), caryophyllene oxide, methyl N-methylanthranilate, acetic acid, octanoic acid.

Table 13.13 (Appendix Continued)

6 Inoma et al. (1989). One commercial sample; GC on capillary columns coated with OV-101 and Carbowax 20 M; GC/MS; relative percentage of peak areas. Inoma et al. also found trace amounts of limonene oxide*.
7 Dellacassa et al. (1997). Uruguay; Sfumatrice; three samples; GC on capillary column coated with SE-52; GC/MS; relative percentage of peak areas.
8 Dugo et al. (1987, 1991, 1999), Lamonica et al. (1990), Verzera et al. (1996b, 1998). Calabria, Italy; Pelatrice; Fantastico, Femminello, Castagnaro bergamot; about 1000 samples from 1984 to 1998; column chromatography on neutral alumina; GC on capillary columns coated with SE-52 and DB-5; GC/MS; linear retention indices; relative percentage of peak areas. These authors also foud dodecane (0–0.01%), δ-3-carene (tr–0.01%), (Z)-α-bisabolene (0–tr), (Z)-γ-bisabolene (0–0.01%), germacrene B (0–0.04%), (E)-dec-2-en-1-al' (0–0.01%), camphor (tr–0.01%), dodecanol (0–0.01%), isopulegol (tr–0.01%), undecyl acetate (tr), indole (0–0.01%).
9 Chouchi et al. (1995). Calabria, Italy; one sample of Pelatrice oil; GC on capillary column coated with DB-5; GC/MS; w%. Chouci et al. also found tridecane (0.05%), δ-2-carene (0.11%), 1,3,8-p-menthatriene (0.17%), aromadendrene (0.36%), γ-muurolene (0.07%), β-selinene (0.04%), isomenthone (0.01%), menthone (0.06%), octen-3-ol (0.08%), dihydrocitronellol (0.05%), menthol (0.16%), neomenthol (0.02%), (E,E)-farnesol (0.01%), (Z)-β-santalol (0.01%), isomenthyl acetate (0.10%), and trace amounts of (E)-β-farnesene, caryl acetate, neomenthyl acetate.

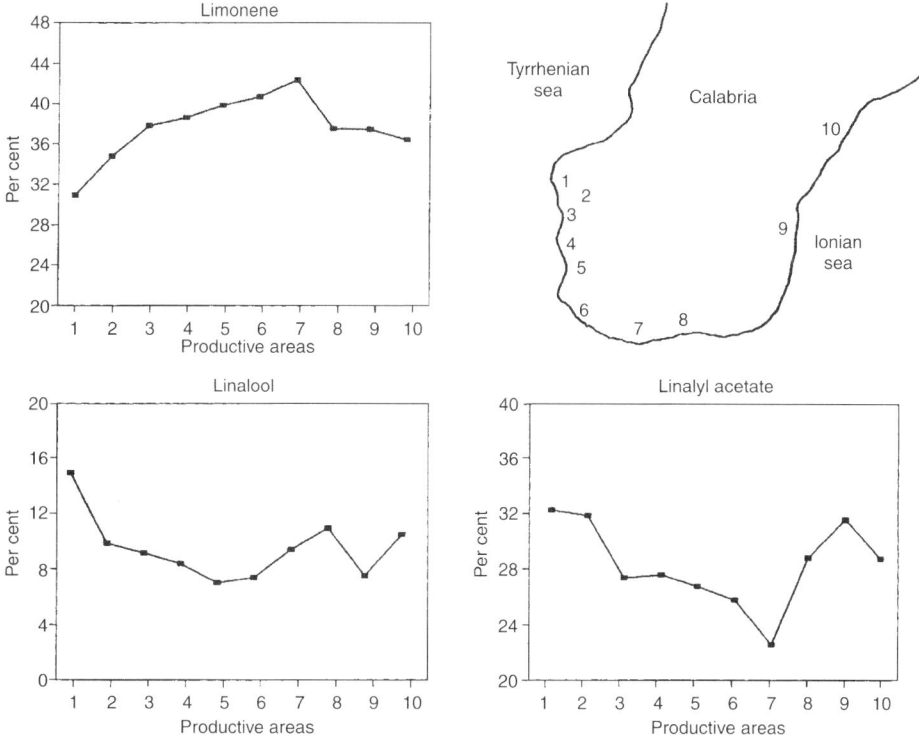

Figure 13.5 Variation in average content of limonene, linalool and linalyl acetate for bergamot oils produced in Calabria, Italy in relations to the production areas (Dugo, 1994).

been assessed (Verzera *et al.*, 1996b) that the *Femminello* cv. produced oils with the highest amounts of oxygenated compounds, while the *Fantastico* cv. provides oils with the lowest amount of these components. The oils produced from the *Castagnaro* cv. present an average content of oxygenated compounds compared to the former two cvs.

It has been noticed that the bergamot oil produced in Calabria provides a good example of how the composition depends on the area of cultivation of the fruits. This oil is subject to evident modifications of its composition even for small changes of the pedo-climatic and weather conditions. In Italy the production of bergamot is limited to a very narrow area on the south part of the Tirrenian and the Jonian Calabrian coasts. Although the limited area, the composition of the oil varies in function of the geographic origin of the fruits, as can be seen in Figure 13.5, where the average content of limonene, linalyl acetate and linalool are correlated with the different productive areas on the Calabrian coast.

More papers than those reported in Table 13.13 have been published on bergamot oil. The information they provide is mostly on the presence of minor components, or they represent applications on bergamot oil analysis of advanced analytical techniques.

Ehret and Maupetit (1982) determined the presence of the sesquiterpene alcohols T-cadinol, β-eudesmol, and farnesol*. Ohloff *et al.* (1986) isolated the isomers (–)-(4S,8R)-8-epi-α-bisabolol, and (–)-(4R,8S)-4-epi-β-bisabolol. Cartoni *et al.* (1987)

described the advantage of using two columns in series, one apolar (SE-54) and one polar (Carbowax 20 M) and a multistep temperature programme for the separation of critical pairs of components in different citrus oils. In particular, using this system, it was possible to separate simultaneously in bergamot oil the critical pairs sabinene/ β-pinene and linalyl acetate/geraniol (difficult to be separated on a SE-54 column) and α-pinene/α-thujene, linalool/geranyl acetate and geraniol/β-bisabolene (difficult to be separated on a Carbowax 20 M column). Lakszner and Szepesy (1988) demonstrated the advantages of a GC detector selective for oxygenated compounds (O-FID) for the quality control of essential oils, bergamot oil included. The technique was quite interesting, but scantly used. Lanuzza et al. (1990) and Micali et al. (1991) used a coupled HPLC-HRGC system to identify in bergamot oil the n-alkanes C_{21}–C_{33} and the correspondent 'iso' isomers and some sesquiterpene hydrocarbons: bergamotene*, bisabolene*, β-caryophyllene, and α-humulene. Mondello et al. (1994b) used a LC-GC system coupled with an ion trap mass detector (ITD), to separate into four fractions bergamot oil and directly transfer each fraction onto the GC system for the GC/MS analysis. These authors highlighted the benefit of the pre-fractionation of the oil prior to the GC/MS analysis, since such technique gave more reliable results than direct GC/MS analysis of the whole oil. The same research group (Mondello et al., 1995b), used the same system for a systemic study of the sesquiterpene hydrocarbons present in citrus oils. In bergamot oil they identified the following components listed in decreasing amount order: β-bisabolene, β-caryophyllene, *trans*-α-bergamotene, germacrene D, (Z)-β-farnesene, *cis*-α-bergamotene, α-humulene, β-santalene, germacrene B, (Z)-γ-bisabolene, (E)-β-farnesene^t and β-sesquiphellandrene^t. In the same year Mondello et al. (1995a) used bergamot oil to provide an example of analysis of natural complex matrices by GC/MS with linear retention index (LRI), calculated on both apolar and polar columns, used as interactive filters for the identification of single components.

Formàcek and Kubeczka (1982) analysed two commercial oils, both Italians, one from Messina and one from Reggio Calabria. Both samples resulted, as revealed by the authors, roughly adulterated. Among the anomalous values the high amounts of α-terpinyl acetate (6 per cent and 4 per cent respectively) was highlighted.

Laboratory extracted oils

In Table 13.14 are summarised the data found in literature relative to bergamot oils, of different origins, extracted in laboratory. Among these results, those relative to δ-3-carene, determined by Huang et al. (1986) and by Kirbaslar et al. (2000) are unusually high for a bergamot oil. This component, in bergamot oil, never exceeds 0.01 per cent of the whole volatile fraction. Too high also appear the values of geraniol and α-terpineol reported in the former paper.

Table 13.14 Percentage composition of the volatile fraction of laboratory extracted bergamot oils

	1	2	3	4	5	6(a)	6(b)	7
HYDROCARBONS								
Monoterpene								
camphene	tr	0.05	0.03	0.01–0.21	0.02	–	–	tr
δ-3-carene	–	0.14^t	2.04	–	–	–	–	0.9
p-cymene	0.2	0.57	1.56	0.10–0.37	0.35	0.3	tr	0.1

limonene	40.2	32.37	42.51	35.42–45.11	32.28	38.8	24.3	23.7[c]
myrcene	0.8	1.17	1.43	0.95–1.97	0.78	0.9	0.7	2.0
(Z)-β-ocimene	–	–	–	0–0.74	0.02	–	–	tr
α-phellandrene	–	–	tr	0.01–0.04	–	–	–	–
β-phellandrene	–	–	–	–	0.15	–	–	[c]
α-pinene	0.7	1.88	0.84	0.52–1.30	0.81	1.6	1.3	0.5
β-pinene	4.5	13.34[b]	4.82	2.90–5.79	3.02	8.9	6.8	3.0
sabinene	–	13.34[b]	0.78	0.58–1.00	0.55	–	–	0.5
α-terpinene	–	–	–	–	0.07	0.2	0.1	0.2
γ-terpinene	4.5	12.60	1.35	4.80–6.02	4.12	8.3	5.6	4.7
terpinolene	0.3	0.20	0.14	0.16–0.44	0.18	0.3	0.2	0.4
α-thujene	–	–	0.18	0–0.03	–	–	–	0.2
Sesquiterpene								
α-bergamotene	–	0.86*	0.02*	0.03–0.15*	0.28*	–	–	0.9[h]
β-bisabolene	–	1.28	0.02	0.35–0.95	1.43[d]	–	–	1.2
β-caryophyllene	–	0.25	0.11	0.21–0.53	0.25	–	–	0.7
(E)-β-farnesene	–	–	0.15	0.20–0.57	–	–	–	–
(Z)-β-farnesene	–	–	–	–	0.07	–	–	0.1
α-humulene	0.8[a]	–	tr	–	–	–	–	0.1
ALDEHYDES								
Aliphatic								
decanal	–	0.29	0.08	0.07–0.12	0.04	0.1	tr	0.1
nonanal	–	0.09	–	0.03–0.04	0.02	–	–	tr
octanal	–	–	–	0.02–0.11	0.03	–	tr	–
Monoterpene								
citronellal	–	0.02	0.02	0.03–0.07	–	–	–	tr
geranial	0.1	0.21	0.48	0.06–0.14	–	0.3	–	tr[g]
neral	0.6	0.16	0.36	0.04–0.10	0.36	0.2	0.2	0.4
KETONES								
Aliphatic								
6-methyl-5-hepten-2-one	–	–	0.02	0.01	tr	–	–	–
Sesquiterpene								
nootkatone	–	–	–	–	–	0.1	tr	0.5
ALCOHOLS								
Monoterpene								
citronellol	0.1	–	tr	0.03–0.06	–	–	tr	0.3[f]
geraniol	–	–	5.67	0.15–1.48	0.04	–	–	–
linalool	7.4	4.39	17.89	7.56–17.38	16.27	4.2	18.2	14.7
nerol	–	–	0.52	0.09–0.56	0.03	tr	0.1	0.3[f]
terpinen-4-ol	–	0.05	0.16	0.03–0.05	–	0.1	0.1	tr
α-terpineol	0.8[a]	–	2.54	0.07–3.11	0.18[e]	–	–	0.1
Sesquiterpene								
(E)-nerolidol	–	–	tr	–	–	–	–	0.1
ESTERS								
Aliphatic								
decyl acetate	–	–	0.02	0.05–0.12	–	–	–	0.1
hexyl acetate	–	–	–	0.01–0.16	–	–	–	0.2
nonyl acetate	–	–	0.03	0.03–0.05	–	–	–	tr
octyl acetate	0.2	0.02	0.11	0.15–0.19	0.09	–	–	0.2
Monoterpene								
citronellyl acetate	–	–	0.15	0.09–0.13	0.05	–	–	0.1
geranyl acetate	0.1	0.58	1.31	0.58–1.93	0.24	0.3	0.2	1.6
linalyl acetate	28.1	23.0	11.37	14.67–31.66	37.39	32.1	39.0	38.7

Table 13.14 (Continued)

	1	2	3	4	5	6(a)	6(b)	7
neryl acetate	0.5	0.40	0.88	0.48–1.28	0.02	–	–	1.6
α-terpinyl acetate	–	–	0.03	0.19–0.31	0.18[e]	–	–	–
ETHERS AND OXIDES								
Monoterpene								
cis-linalool oxide	–	–	0.07	0.02	–	–	–	–

Notes

tr = traces; t = tentative identification; * correct isomer not characterized; a α-humulene + α-terpineol; b β-pinene + sabinene; c limonene + β-phellandrene; d β-bisabolene + 2,7-dimethyl-2,6-octadien-1-ol[t]; e α-terpineol + α-terpinyl acetate; f citronellol + nerol; g geranial + perilla aldehyde; h *trans*-α-bergamotene.

Appendix to Table 13.14

1 Dresher *et al.* (1984). Entre Rios, Argentina; hand cold-pressed samples from the cvs. Monaco, Castagnaro, Femminello; GC on packed columns of Carbowax 20 M and OV-17; relative percentage of peak areas.

2 Calvarano *et al.* (1984). Toscana, Italy; one solvent extracted sample from the cv. Femminello; GC on capillary column coated with UCON LB-550X; relative percentage of peak areas.

3 Huang *et al.* (1986). China; one steam-distilled sample; GC on capillary column coated with SE-54; GC/MS, linear retention indices; relative percentage of peak areas. Huang *et al.* also found pentadecane (0.21%), β-ocimene* (0.27%), hexanol (0.01%), *cis*-hex-3-en-1-ol (0.01%), nonanol (0.02%), dihydrolinalool (0.07%), endofenchol (0.01%), isopulegol (0.01%), lavandulol (0.02%), sabinene hydrate* (0.06), citronellyl formate (0.02%), geranyl formate (0.02%), methyl geranate (0.05%), γ-heptalactone (0.04%) and trace amounts of pulegone, (Z)-nerolidol, butyl acetate, 1,4-cineole.

4 Huang *et al.* (1987). China; four sample extracted by steam distillation, simultaneous steam distillation-solvent extraction, cold solvent extraction, cold-pressing; GC on capillary columns coated with SE-54, PEG-20 M, OV-101; GC/MS, linear retention indices; relative percentage of peak areas. Huang *et al.* also found undecanal (0.07–0.23%), 1,8-cineole (0.16–0.28%).

5 Baser *et al.* (1995). Turkey; one sample cold-pressed extracted using an expeller-type press; GC on capillary column coated with Thermon 600T. Baser *et al.* also found (E)-β-ocimene (0.14%), δ-cadinene (0.33%), γ-muurolene (0.05%), hexanal (0.02%), and trace amounts of (E)-2-hexanal, 6-methyl-3-heptanol.

6 Sawamura *et al.* (1999). One sample of Fantastico bergamot oil from Italy (a) and one sample of Balotin bergamot oil from Japan (b) extracted by hand-pressing of flavedo; GC on capillary column coated with Thermon 600T; GC/MS, retention times, Kovats indices; relative percentage of peak areas. Sawamura *et al.* also found trace amounts of decanol (only in Fantastico oil) and geraniol (in both oils).

7 Kirbaslar *et al.* (2000). Turkey; one sample cold-pressed applying manual pressure on the rind of the fruits; GC on capillary column coated with HP-WAX; GC/MS; relative percentage of peak areas. Kirbaslar *et al.* also found m-cymene (0.3%), *trans*-α-bergamotene (0.9%), germacrene D (0.1%), phenyl ethyl alcohol (0.5%), 1-hydroxy linalool (0.1%), α-bisabolol (0.1%), octadienyl formate (0.1%), and trace amounts of *cis*-α-bergamotene, dodecanal, bornyl acetate, *p*-menth-1-en-8-yl acetate.

LEMON OIL (*Citrus limon* (L.) Burm.)

1956–1979

In his review on the composition of citrus essential oils Shaw (1979) cited nine papers; four of these reported the content, in lemon oil, of components of different classes of compounds (Bernhard, 1960; Günther, 1968; Ziegler, 1971; Fincke and Maurer, 1974; Lund and Bryan, 1976), one was on the aldehyde composition (Stanley *et al.*, 1961a), one on the nootkatone content (MacLeod and Buigues, 1964) and two reported the composition of the monoterpene hydrocarbons (Ikeda *et al.*, 1962a; Ashoor and Bernhard, 1967). The latter paper reported the percentage composition of the components analysed, relative to the monoterpene hydrocarbon fraction and not to the whole oil, therefore these values cannot be compared with those reported by all the other authors. In the paper by Gunther (1968) were reported the results obtained for some commercial oils, some of these were certainly not genuine. In the paper by Ikeda (1962a) were reported the results relative to three samples, one of these was extracted from Meyer lemon.

The range of the results relative to the papers reviewed by Shaw (1979), with the exception of those obtained by Ashoor and Bernhard (1967), those reported by Ikeda *et al.* (1962a), for the Meyer lemon, those obtained by Günther (1968) and those reported in the different papers for co-eluted compounds, are summarised below:

camphene	0.20%
p-cymene	tr–0.61%
limonene	58.4–80.38%
myrcene	0.9–2.13%
α-pinene	1.5–5.0%
β-pinene	5.5–12.4%
sabinene	0.8–1.5%
γ-terpinene	2.92–9.83%
terpinolene	0.62–0.85%
α-bergamotene*	0.38%
β-bisabolene	0.14–0.76%
caryophyllene*	0.33%
α- + β-humulene	0.15%
decanal	0.06–0.22%
heptanal	0.03%
nonanal	0.09–0.27%
octanal	0.10%
undecanal	0.05–0.07%
citronellal	0.1–0.19%
geranial	0.61–2.26%
neral	0.78–1.32%
6-methyl-5-hepten-2-one	0.06%
carvone	0.04%
nootkatone	0.01%
linalool	0.08–0.31%
terpinen-4-ol	0.01–0.35%
α-terpineol	0.18–0.48%
tetrahydrogeraniol	0.11%
decyl acetate	0.05%
octyl acetate	0.04%
citronellyl acetate	0.04–0.17%
geranyl acetate	0.10–0.40%
neryl acetate	0.27–0.44%

In these years other papers than those relative to the data above, have been published on the composition of lemon essential oil. Shaw reviewed most of these in his former paper (Shaw, 1977) on the composition of citrus essential oils. Some were on qualitative analyses performed by GC, including the pioneer papers by Liberti and Conte (1956), Liberti and Cartoni (1958), Bernhard (1958) and those by Di Giacomo and Rispoli (1962), Di Giacomo et al. (1969), Hunter and Bogden (1965), Hunter and Moshonas (1966), MacLeod et al. (1965), Majlath and Takas (1969), Straus and Wolstromer (1974). On the whole these papers reported in lemon oil the presence of other components in addition to those listed above: pentadecane, tetradecane, δ-3-carene, p-cimenene, α- and β-phellandrene, α-terpinene, α-thujene, dodecanal, decanol, nonanol, octanol, citronellol, geraniol, p-mentha-1,8-dien-9-ol, nerol.

Different papers report on the composition of monoterpene hydrocarbons (Clark e Bernhard, 1960; Stanley et al., 1961; Di Giacomo et al., 1962, 1963, 1969). Below are summarised the values reported by these authors together with the composition reported by Ashoor e Bernhard (1967):

	Clark e Bernhard (1960)	Stanley et al. (1961)	Di Giacomo et al. (1962, 1963, 1969)	Ashoor e Bernhard (1967)
camphene	0.08%	–	–	0.49%
p-cymene	1.68%	tr	tr–1.00%	1.08%
limonene	66.59%	69.70%	65.58–74.70%	75.6%
myrcene	1.28%	1.10%	1.37–2.73%	8.02%[a]
α-pinene	2.52%	2.46%	1.85–2.80%	1.30%
β-pinene	20.69%	14.80%	7.93–13.80%	8.02%[a]
sabinene	–	1.67%	1.50–6.60%	–
γ-terpinene	6.88%	10.30%	8.72–12.54%	11.8%
terpinolene	0.14%	–	–	–

a myrcene + β-pinene.

Ashoor and Bernhard also found, α-phellandrene (0.22 per cent), β-phellandrene (0.78 per cent) and α-terpinene (0.67 per cent).

Rispoli and Di Giacomo (1965) and Di Giacomo et al. (1965) reported some information on the sesquiterpene hydrocarbons (in decreasing amount order: β-bisabolene, α-bergamotene*, caryophyllene) and on the oxygenated compounds (in decreasing amount order: geranial, neral, octanal, linalool, nonanal, 6-methyl-5-hepten-2-one, citronellal, decanal, octyl acetate).

Some other papers reported more detailed information on the volatile fraction of lemon oil. Pennisi and Di Giacomo (1965) analysed oils produced by 'sfumatrice' machines from different varieties of lemon cultivated in Sicily; Kekelidze and Beradze (1974) studied the composition of a lemon oil extracted in laboratory by distillation from the peels of lemon of the Novogruzinsky cv. from Russia; Dugo et al. (1977) examined some sample of lemon oils produced by FMC machines in Argentina in the regions of Entre Rios, Corrientes and Misiones; on a Supelco bulletin (1978) was reported the composition of a Californian oil; Huet et al. (1978) analysed the oil extracted by screw press, from lemons of the Eureka cv. cultivated in Corsica. The results of these papers are here reported:

	Pennisi and DiGiacomo (1965)	Kekelidze and Beradze (1974)	Dugo et al. (1977)	Supelco bulletin (1978)	Huet et al. (1978)
camphene	–	0.7%	0.04–0.05%	0.05%	–
p-cymene	–	2.6%	0–0.36%	0.31%	–
limonene	58.86–66.22%	89.0%	70.69–74.25%	69.89%	66.52%
myrcene	1.29–1.92%	1.4%	1.54–1.83%	1.62%	1.52%
α-pinene	2.03–2.34%	0.07%	1.85–2.08%	1.79%	2.02%
β-pinene	11.06–16.66%[a]	0.4%	8.80–10.87%	10.65%	12.88%
sabinene	11.06–16.66%[a]	0.05%	1.54–1.83%	1.75%	2.35%
γ-terpinene	9.15–10.92%	2.6%	7.45–8.29%	8.47%	9.95%
terpinolene	–	–	0.30–0.33%	0.36%	0.70%
α-thujene	–	–	tr–0.01%	0.40%	–
bergamotene	–	–	0.27–0.31%*	0.40%*	0.49%*[b]
bisabolene	–	–	0.40–0.47%*	0.58%[c]	–
caryophyllene	–	–	0.16–0.19%*	0.28%*	–
decanal	0.02–0.06%	–	0.03–0.05%	–	–
nonanal	0.12–0.23%	–	0.08–0.10%	0.07%	0.07%
citronellal	0–0.14%	–	0.05–0.09%	–	0.07%
geranial	1.67–3.37%	–	0.87–1.11%	1.06%	2.43%
neral	0.82–1.52%	–	0.50–0.63%	0.67%	0.89%
linalool	0.15–0.25%	0.1	0.05–0.07%	0.19%	0.24%
terpinen-4-ol	–	–	0.01%	0.11%	0.49%[b]
α-terpineol	0.12–0.32%	0.8	0.06–0.08%	0.16%	0.23%
geranyl acetate	0.29–0.74%	–	0.15–0.27%	0.31%	–
neryl acetate	0.21–1.05%	–	0.28–0.37%	0.41%	–

a β-pinene + sabinene; b bergamotene; * + terpinene-4-ol; c β-bisabolene.

Pennisi and Di Giacomo (1965) also found octanal (0.12–0.30 per cent), 6-methyl-5-hepten-2-one (0–0.24 per cent), and octyl acetate (0.03–0.06 per cent); Dugo et al. (1977) also determined, α-phellandrene (0.03 per cent), α-terpinene (0.13–0.16 per cent), dodecanal (0–0.01 per cent), nerol (0.03–0.05 per cent), citronellyl acetate (0.01–0.02 per cent).

1979–2000

Industrial cold-pressed oils

In Table 13.15 are reported the literature data on the composition of the volatile fraction of lemon oil published since 1979 to February 2000. The values in the table refer to oils produced in different citrus productive areas and by different technologies, as clearly indicated in the appendix to the table.

The results in Table 13.15, although vary between wide ranges, are, at least for the main components, enough homogeneous and can provide a good overview of the composition of lemon oils produced in the different citrus cultivated areas on the globe. The variability of the quantitative data is mostly due to the different regions of origin of the fruits, and the harvest period, with the exception of p-cymene, since the amount

Table 13.15 Percentage composition of the volatile fraction of cold-pressed lemon oil

	1	2	3	4	5(a)	5(b)	6	7	8	9(a)	9(b)	10(a)	10(b)	10(c)	11	12(a)	12(b)	13
HYDROCARBONS																		
Monoterpene																		
camphene	0.07	0.05	0.06	0.05	0.07	0.03	tr–0.09	0.1	—	0.05	0.06	0.04–0.07	0.13	0.05	0.06	0.06	0.06	0.05–0.08
δ-3-carene	0.01	0.01	0.02	tr	tr	tr			—	—	—	0.01	0.01	0.01		tr	tr	tr–0.01
p-cymene	0.12–0.16	0.15	0.98	0.04	0.03	0.02	0.25–1.71	0.3	0.65	0.07	0.14	0.01–0.58	0.05	0.20	g			0.02–0.67
limonene	59.73–62.47	70.29	68.36	68.72	59.58	76.17	64.57–70.09	66.7	70.53	65.16	65.65	65.06–70.36	66.57	71.82	69.70[g]	66.05	67.75	59.57–71.06[k]
myrcene	1.45–1.51	1.62	1.39	1.50	1.28	1.75	1.34–2.23	1.5	1.54	1.61	1.58	1.34–1.93	1.41	1.63	1.60	1.48	1.53	1.05–1.86
(E)-β-ocimene	—	—	0.13	—	—	—	—	—	0.14[h]	0.11	0.09	0.10–0.60	0.14	0.14	0.13	0.13	0.11	0.07–0.20
(Z)-β-ocimene	—	—	0.07	—	—	—	—	—	—	0.05	0.05	—	—	—	0.04	0.07	0.06	0.03–0.15
α-phellandrene	0.04–0.06	0.04	0.03	0.04	0.04	0.04	tr–0.04	—	—	0.04	0.04	0.01–0.10	0.08	0.04	0.18[i]	0.11[i]	0.12[i]	0.01–0.13
β-phellandrene	—	—	0.48	—	—	—	0–tr	0.3	—	—	—	—	—	—	—	—	—	k
α-pinene	0.88–2.00	1.61	2.06	1.72	2.13	1.47	2.70–4.40	2.0	1.83	1.84	1.75	1.23–1.94	1.98	1.65	1.77	1.88	1.87	1.50–2.40
β-pinene	14.69–15.70[a]	11.42[a]	12.27	10.13	16.50	5.96	12.52–16.97	12.1	11.23	10.52	11.13	8.91–12.41	12.68	8.57	11.87[a]	14.35[a]	13.51[a]	9.45–17.79
sabinene	14.69–15.70[a]	11.42[a]	1.92	1.65	2.59	1.12	1.61–2.06	1.9	2.01	1.79	1.85	1.21–1.93	1.60	1.52	11.87[a]	14.35[a]	13.51[a]	1.13–2.79
α-terpinene	0.17–0.23	0.13	0.10	0.32	0.31	0.26	0–0.11	0.2	—	0.20	0.16	0.17–0.21	0.17	0.16	0.25	0.20	0.20	0.05–0.25
γ-terpinene	9.64–11.38	7.50	7.39	8.55	9.67	7.91	2.88–6.21	8.7	7.53	8.86	8.32	8.72–10.55	8.17	7.83	8.84	9.33	8.86	6.59–11.27
terpinolene	0.37–0.49	0.32	0.27	0.39	0.38	0.36	tr–0.27[b]	0.4	0.19	0.39	0.34	0.41–0.50	0.38	0.34	0.37	0.39	0.38	0.20–0.44
α-thujene	0.01	0.01	—	0.38	0.44	0.35	—	0.4	0.37	0.42	0.38	0.27–0.43	0.33	0.37	0.44	0.42	0.41	0.37–0.54
Sesquiterpene																		
cis-α-bergamotene	—	—	—	—	—	—	—	—	—	0.03	0.03	0.03–0.04	0.02	0.03	—	—	—	tr
trans-α-bergamotene	0.54–0.78[h]	0.40[h]	—	0.37	0.38	0.39	0.29–0.45[h]	0.4[h]	0.36[h]	0.36	0.38	0.39–0.53	0.30	0.39	0.38[h]	0.33	0.34	0.21–0.58
β-bisabolene	0.79–0.86[h]	0.56[h]	—	0.56	0.59	0.58	0.08–0.37[h]	0.5	0.50	0.52	0.57	0.65–1.00	0.47	0.66	0.57[h]	0.51	0.52	0.29–0.92
β-caryophyllene	0.28–0.78	0.22[h]	0.24	0.24[h]	0.24	0.23[h]	0.19–0.41[h]	0.2	0.23[h]	0.19[i]	0.22[h]	0.18–0.28[h]	0.16[h]	0.24[h]	0.26[h]	0.25	0.25	0.11–0.33
δ-elemene	—	—	—	—	—	—	—	—	—	—	—	0.01–0.03	0.01	0.02	—	—	—	—
(E)-β-farnesene	—	—	—	—	—	—	—	—	—	0.03[t]	0.03[t]	0.03–0.05	0.03	0.02	—	—	—	0.01
(Z)-β-farnesene	—	—	—	—	—	—	—	—	—	—	—	0.02–0.06	0.02	0.03	—	0.05[i]	0.05[i]	0.01–0.07[i]
α-humulene	—	—	—	0.02	0.02	0.02	—	—	—	0.02[t]	0.02[t]	0.02–0.04	0.02	0.02	0.05[h]	0.02	0.02	0.01–0.03
β-santalene	—	—	—	—	—	—	—	—	—	0.01[t]	0.01[t]	—	—	0.02	—	0.05[i]	0.05[i]	0.01–0.07[i]
valencene	—	—	—	—	—	—	—	—	—	—	—	0.03–0.16	0.06	0.02	tr	0.03	0.02	tr–0.09
ALDEHYDES																		
Aliphatic																		
decanal	0.06–0.10	0.07	—	0.04	0.06	0.04	0–0.06	—	—	0.05	0.05	0.03–0.07	0.05	0.03	0.05	0.04	0.04	0.01–0.08
dodecanal	—	tr	—	0.02	0.03	0.01	tr–0.06	—	—	0.01	0.01[i]	0.01–0.03	0.01	tr	—	—	—	—
heptanal	tr–0.01	—	—	—	tr	tr	—	—	—	—	—	—	—	—	—	—	—	tr

nonanal	0.07–0.09	0.11	–	0.12	0.23	0.06	0.06–0.11	0.1	–	0.12	0.12	0.08–0.16	0.09	0.11	0.09	0.12	0.11	0.04–0.19
octanal	0.35–0.42	0.08	–	0.07	0.13	0.07	tr–0.27[b]	–	–	0.10	0.08	0.03–0.09	0.12	0.05	0.18[f]	0.11[f]	0.12[f]	0.02–0.14
undecanal	–	–	–	–	–	–	0–0.08	–	–	0.03	0.03	0.01–0.02	0.02	0.02	0.02	0.02	0.02	tr–0.05
Monoterpene																		
citronellal	0.12–0.16	0.07	–	0.07	0.08	0.08	0.03–0.14[c]	0.1	–	0.13	0.08	0.08–0.14	0.08	0.08	0.10	0.08	0.07	0.04–0.17
geranial	0.69–0.94	1.03	–	1.22	1.45	0.95	1.66–2.66	1.5	1.30	2.05	1.18	0.86–1.77	1.44	1.18	1.18	1.55	1.45	0.60–2.25
neral	0.34–0.48	0.57	–	0.76	0.90	0.60	0.54–1.20	0.9	0.90	1.26	0.74	0.54–1.06	0.97	0.74	0.71	0.93	0.87	0.45–1.33
perilla aldehyde	–	–	–	–	–	–	–	–	–	0.03	0.03	0.02–0.04	0.05	0.02	–	–	–	tr
KETONES																		
Aliphatic																		
6-methyl-5-hepten-2-one	0.07–0.09	0.36	–	tr	tr	tr	0–0.05	–	–	–	–	–	–	–	–	–	–	tr–0.02
Monoterpene																		
camphor	–	–	–	–	–	–	–	–	–	0.01	0.01	–	–	–	–	–	–	–
carvone	–	–	–	0.01	0.01	7 0.01	–	–	–	–	–	–	–	–	–	–	–	tr–0.01
piperitone	–	–	–	–	–	–	–	–	–	tr	tr	–	–	–	–	–	–	tr–0.01
Sesquiterpene																		
nootkatone	–	–	–	–	–	–	–	–	–	–	–	0.01–0.03	0.02	tr	–	–	–	tr–0.01
ALCOHOLS																		
Aliphatic																		
nonanol	–	–	–	–	–	–	tr–0.21	–	–	–	–	tr–0.01	0.01	0.01	–	–	–	–
octanol	–	–	–	0.01	0.02	0.02	0.04–0.09	–	–	0.05[d]	0.01[d]	0.01–0.03	0.02	0.02	–	tr	tr	tr–0.01
Monoterpene																		
borneol	–	–	–	–	–	–	–	–	–	0.01	0.01	–	–	–	–	–	–	tr–0.02
citronellol	–	–	–	–	–	–	tr–0.07	–	–	0.03[f]	0.06[f]	0.01	0.02	0.02	0.03[f]	0.03[f]	0.03[f]	0.01–0.18[f]
geraniol	–	–	0.05	0.04	0.04	0.02	0.03–0.07	0.2	–	0.02	0.03	0.01–0.12	0.19	0.04	0.04	0.06	0.02	tr–0.06
linalool	0.05–0.09	0.10	0.21	0.14	0.16	0.14	0.08–0.20	0.2	0.11	0.18[e]	0.13[e]	0.08–0.12	0.46	0.11	0.12	0.12	0.10	0.05–0.18
nerol	–	–	–	0.04	0.03	0.03	0.04–0.07	0.1	–	0.03[f]	0.06[f]	0.01–0.16	0.06	0.04	0.03[f]	0.03[f]	0.03[f]	0.01–0.18[f]
cis-sabinene-hydrate	–	–	–	–	–	–	–	–	–	0.18[e]	0.13[e]	–	–	–	0.06	0.04	0.03	0.01–0.07
trans-sabinene hydrate	–	–	–	–	–	–	–	–	–	0.05[d]	0.01[d]	–	–	–	tr	0.04	0.03	0.01–0.07
terpinen-4-ol	0.01–0.05	0.02	–	0.14	0.11	0.06	0.02–0.08	–	–	0.05	0.11	0.02–0.10	0.03	0.03	0.04	0.04	0.04	0.01–0.08
α-terpineol	0.05–0.10	0.12	0.21	0.22	0.23	0.15	0.27–0.54	0.2	0.16	0.19	0.16	0.10–0.84	0.16	0.17	0.13	0.20	0.16	0.06–0.28
Sesquiterpene																		
α-bisabolol	–	–	–	–	–	–	–	–	–	0.04	0.04	–	–	–	–	0.02	0.03	0.01–0.03
campherenol	–	–	–	–	–	–	–	–	–	0.03	0.03	–	–	–	–	0.02	0.02	0.01–0.03

Table 13.15 (Continued)

	1	2	3	4	5(a)	5(b)	6	7	8	9(a)	9(b)	10(a)	10(b)	10(c)	11	12(a)	12(b)	13
'norbornanol'^m	–	–	–	–	–	–	–	–	–	0.02^f	0.03^f	–	–	–	–	0.02	0.02	0.01–0.04
ESTERS																		
Aliphatic																		
decyl acetate	–	–	–	–	–	–	–	–	–	–	–	–	0.01	0.01	–	–	–	tr
nonyl acetate	–	–	–	0.01	0.02	tr	–	–	–	0.01	0.01^1	tr–0.01	0.01	0.01	–	–	–	tr–0.02
octyl acetate	0.01–0.03	0.02	–	tr	0.01	tr	0.03–0.14^c	–	–	–	–	0.01–0.02	0.01	0.01	–	–	–	tr–0.01
												0.01–0.05						
Monoterpene																		
citronellyl acetate	–	0.22	–	0.02	0.02	0.03	0.03–0.10	–	–	0.03	0.03	0.01–0.03	0.02	0.01	0.03	0.02	0.02	tr–0.08
geranyl acetate	0.50–0.86	0.22	0.44	0.43	0.65	0.20	0.06–0.16	0.6	0.18	0.38	0.52	0.25–0.70	0.39	0.20	0.32	0.35	0.27	0.16–0.81
neryl acetate	0.51–0.58	0.26	–	0.50	0.56	0.46	0.07–0.39	0.6	0.24	0.52	0.60	0.26–0.63	0.38	0.32	0.34	0.35	0.35	0.23–0.88
OXIDES																		
Monoterpene																		
cis-limonene oxide	–	–	–	–	–	–	–	–	–	tr	0.01	–	–	–	–	–	–	tr–0.02
trans-limonene oxide	–	–	–	–	–	–	–	–	–	tr	0.01	–	–	–	–	–	–	tr–0.02

Notes

t = traces; t = tentative identification; * correct isomer not characterised; a β-pinene + sabinene; b terpinolene + octanal; c citronellal + octyl acetate; d octanol + *trans*-sabinene hydrate; e linalool + *cis*-sabinene hydrate; f citronellol + nerol; g *p*-cymene + limonene; h the correct isomer was not indicated in the original paper; i (Z)-β-farnesene + β-santalene; j α-phellandrene + octanal; k limonene + β-phellandrene + 1,8-cineole; l dodecanal + decyl acetate; m 2,3-dimethyl-3-(4-methyl-3-pentenyl)-2-norbornanol.

Appendix to Table 13.15

1 Sanchez *et al.* (1980). Spain; 7 sample of Verna lemon; FMC; GC on capillary column coated with UCON LB-550 X; relative percentage of peak areas.
2 Cappello *et al.* (1981). Argentina; one sample; GC on capillary column coated with UCON LB-550 X; relative percentage of peak areas.
3 Formáček and Kubeczka (1982). Italy; one sample; GC on capillary column coated with WG-11; relative percentage of peak areas.
4 Staroscik and Wilson (1982a). USA; one commercial blend of Arizona and California oils; GC on capillary column coated with SE-54; w%.
5 Staroscik and Wilson (1982b). USA; one sample each of early, mid and late season oils from (a) California (coastal area) and (b) Arizona (desert area); GC on capillary column coated with SE-54; w%.
6 Koketsu *et al.* (1983). Brazil; FMC; three sample each of lemon oils cv.s Sicilian and Eureka; GC on packed columns of Carbowax 20 M and of SE-30 and on capillary column coated with Carbowax 20 M; retention times and Kovats indices; relative percentage of peak areas. Koketsu *et al.* also found linanyl acetate (0–0.05%).
7 Analytical Methods Committee (1984). Italy; one sample of 'Sfumatrice' oil; GC on packed columns of Carbowax 20 M and SE-30; relative retention indices; relative percentage of peak areas.
8 Lancas *et al.* (1988). Brazil; one sample of FMC oil from cv. Eureka; GC on microbore capillary column coated with SE-54; GC/MS; relative percentage of peak areas.

9   Chamblee *et al.* (1991). (a) one sample each of Sicilian and (b) California lemon oil; column chromatography on silica gel and GC on capillary column coated with DB-5; GC/MS; w%. These authors also found exo *cis*-4,7-dimethylbicyclo(3.2.1)oct-3-en-6-one (0.01%), undecyl acetate (0.01%), geranyl propanate (0.01%) methyl geranate (tr, 0.01%), neryl propanate (0.01%), 1,8-cineole (0.05%, 0.04%) and trace amounts of nerolidol*, citronellyl propanate.

10 Boelens and Jimenez (1989); Boelens (1991). (a) Spain, cv. Lisbon, needle puncturing machine, 7 samples; (b) Italy, 2 samples; (c) Israel, one sample; GC on capillary column coated with SE-54; GC/MS and retention times; relative percentage of peak areas.

11 Dellacassa *et al.* (1991). Uruguay; FMC; 93 samples from the North productive areas; column chromatography on neutral alumina; GC on capillary column coated with SE-52; relative percentage of peak areas.

12 Dellacassa *et al.* (1995). Uruguay; FMC, Pelatrice; 183 samples from the South productive areas (a) and 52 from the North productive areas (b); column chromatography on neutral alumina, GC on capillary column coated with SE-52; GC/MS; relative percentage of peak areas. These authors also found γ-elemene (0.01–0.02%), tridecanal (tr).

13 Dugo *et al.* (1983, 1999); Dugo (1986, 1994); Cotroneo *et al.* (1986a, 1986b, 1988); Trozzi *et al.* (1993); Verzera *et al.* (1996c). Italy; Pelatrice, Sfumatrice, Torchi, FMC; 1750 samples; column chromatography on neutral alumina, GC on capillary columns coated with SE-52 and IDB-5; GC/MS, linear retention indices; relative percentage of peak areas. These authors also found tricyclene (0.03–0.08%), bicyclogermacrene (0.05–0.07%), (E)-α-bisabolene (0.02%), (Z)-α-bisabolene (0.04%), γ-elemene (0.01–0.03%), germacrene B (0.02–0.12%), germacrene D (tr–0.02%), γ-muurolene (tr–0.02%), tetradecanal (tr–0.02%), β-bisabolol (0.01%), germacrene D-4-ol^t (0.02%), selin-11-en-4-α-ol (0.01%), bornyl acetate (tr–0.01%), methyl geranate (tr–0.01%) and trace amounts of allo-ocimene^t, (Z)-γ-bisabolene.

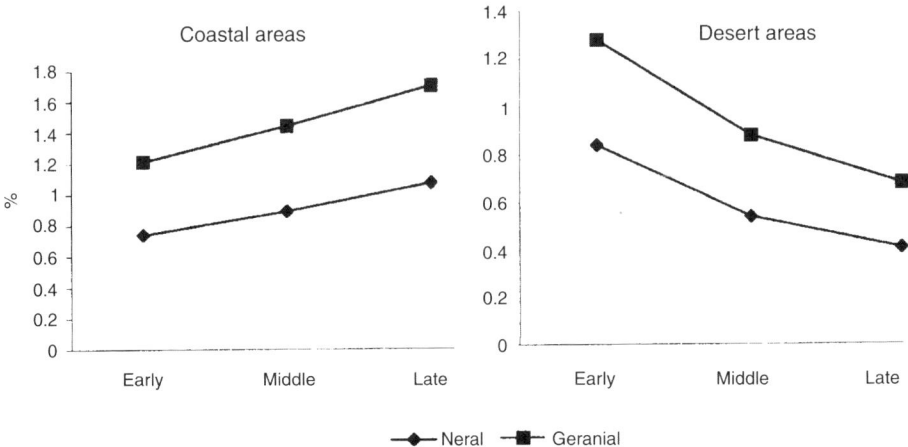

Figure 13.6 Variation of the content of neral and geranial for lemon oils produced in coastal areas (California, USA) and in desertic areas (Arizona, USA) during the productive season (from the results reported by Sharoscik and Wilson, 1982b).

of this component depends on the storage conditions of the oil. The first variability is related to cultivation, environment, climate and weather factors, the second to the maturity stage of the fruits and to the different numbers of blooming that occur during one year, that origin the fruits processed.

Staroscik and Wilson (1982b) underlined the composition differences between oils produced from fruits harvested in the Californian coasts and those from the Arizona desert (average values are reported in columns 5(a) and 5(b) in Table 13.15). The main differences were due to the β-pinene content (15.62–17.29 per cent and 5.38–6.44 per cent respectively); limonene (58.47–60.36 per cent and 75.74–76.48, respectively); to the amount and seasonal variation of neral and geranial that in oils from the coast increased during the season from 1.95 to 2.77 per cent, and decreased in the desert area from 2.12 to 1.09 per cent. These variations are summarised in Figure 13.6.

Licandro *et al.* (1987) observed variation of the composition of oils produced in the same period of the year but from different areas of Sicily. The lemons from the Southeast part gave higher amount of neral and geranial than all the other Sicilian lemons.

Differences of composition in oils produced in the USA, in South America, in Spain, Italy and Israel can be seen from the data reported in Table 13.15. It should be mentioned, however, that the values in Table 13.15 must be carefully considered, since some refer to small numbers of samples, and in some cases the period of production is not specified. Moreover, Sawada and Yamada (1997, 1998) noticed that Sicilian oils, produced with FMC extractor, presented, if compared with Californian oils produced by the same technology, a larger amount of nonanal, citronellal, geranial, neral, α-terpineol, geranyl acetate, neryl acetate, and 1,8-cineole.

The lemon oil from New Zealand shows a characteristic composition, different from lemon oils produced from any other country. Lodge *et al.* (1984) determined only the 52 per cent of limonene in lemon oil from New Zealand produced by the FMC technology.

Dugo et al. (1983, 1984, 1985), Dugo (1986, 1994), Licandro et al. (1984), Cotroneo et al. (1986a,b, 1988), Trozzi et al. (1993), Verzera et al. (1996c) studied in detail the composition of lemon essential oils produced by the common industrial processing technology, in periods that covered entire production seasons, that in the case of lemon last the whole year. They observed that the variation of the composition was reproducible for all the years of production, although there were small differences in the average composition relative to the same period were observed for two productive years. These differences were probably due to the climate and weather factors. In particular it was noticed that the single components and the classes of components slightly changed within the first part of the season of production (October–March), then some of them reached their maximum, others their minimum value in the summer period (June–August), to go back to their initial values at the end of the season. This phenomenon allowed to easily differentiating summer lemon oils, mostly extracted from green lemons (*verdelli*), from the winter lemon oil. Limonene and monoterpene aldehydes presented their minimum values in the summer oils while, in the same period, β-pinene, aliphatic aldehydes and esters were at their maximum values. In Figures 13.7 and 13.8 are plotted the variations of some classes of components and, of some single components, versus the period of production for two different years.

The correlation of the seasonal variation with the different composition parameters can provide, for the evaluation of the genuineness and the quality of lemon oils, a more reliable parameter than the simple ranges of variation of the oils composition (Dugo et al., 1992). As it can be seen in Figure 13.8A, one sample of lemon oil that contains 61 per cent of limonene, 16 per cent of β-pinene and 4 per cent of citral, although these values singularly are compatible with genuine lemon oil, cannot be considered genuine. In fact, if limonene is 61 per cent and β-pinene is 16 per cent the correct amount of citral must be lower than 4 per cent. This sample is probably a summer lemon oil with low citral content which was added with extraneous citral to improve its commercial value, rendering the oil more similar to the winter lemon oil, commercially more valuable. The lemon oil containing 67 per cent of limonene, 0.23 per cent of α-terpineol, and 0.10 per cent of nonanal, Figure 13.8B can provide another example. Again all the values taken singularly are compatible with genuine lemon oil, but this is not the case. This oil was probably a winter lemon oil, that presented a low amount of nonanal and a high content of limonene and that was added with distilled lemon oil. In fact, the oils obtained by distillation present higher content of alcohols, due to hydration of the monoterpene hydrocarbons.

The influence of the extraction technology on the quantitative composition, is limited (Dugo, 1994; Sawada and Yamada, 1997), but it affects the alcohol content of the oil (Dugo, 1994). These components are at higher levels in oils produced by *Pelatrice* and FMC, than those produced by *Sfumatrice* or *Torchi*. The first two extraction techniques permit an easier recycle of the water used to carry the extracted oil. The *sfumatrice* and *torchi* usually require a higher ratio carrier water/oil, since the amount of pectin extracted with such technology is higher and faster, rendering the emulsion water/oil more stable and difficult to break. Therefore the water must be replaced more often with the consequent loss of the oxygenated components, mostly alcohol, that are water-soluble. A similar behaviour was observed comparing the oil manually extracted with the sponge, without water use, and those produced by mechanical processes (Cotroneo et al., 1987). In the oil extracted by the sponge method, the content of

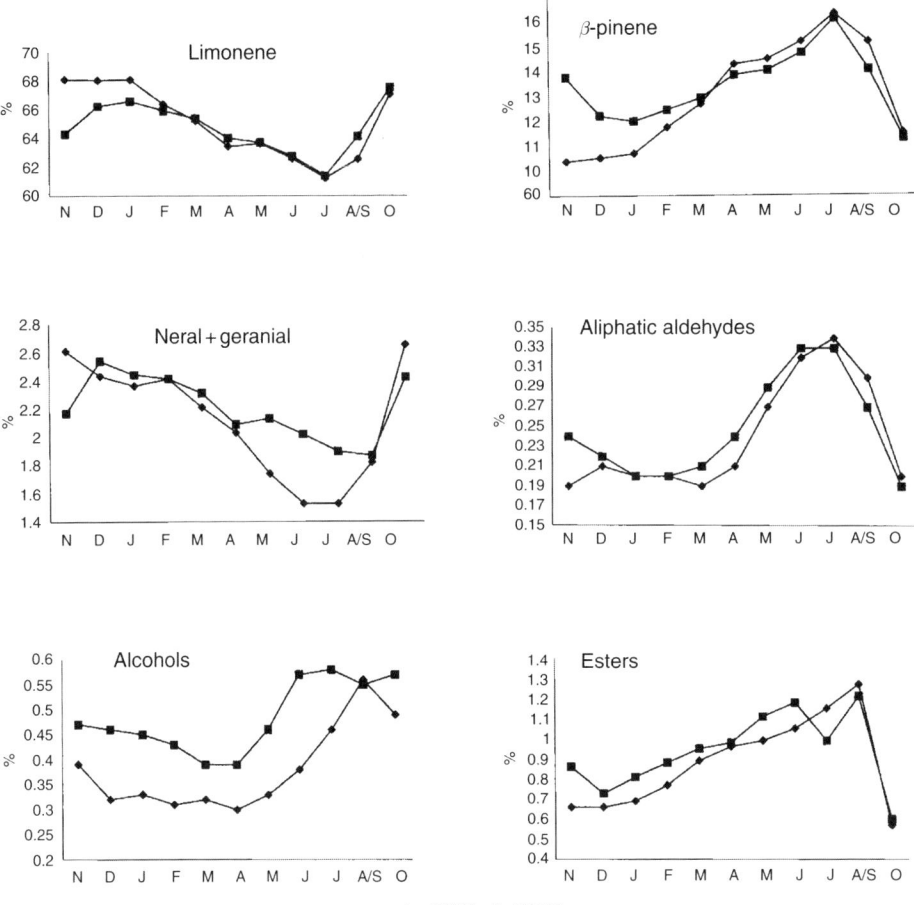

Figure 13.7 Variation in average content of limonene, β-pinene, neral+geranial, aliphatic aldehydes, alcohols and esters for lemon oils produced in Italy during two productive seasons (Dugo et al., 1983, 1984, 1988).

aldehydes and alcohol were higher than in those oils mechanically extracted; in fact, for the latter the loss of water soluble components was inevitable, even if optimised conditions were used. The only exception was terpinen-4-ol present in both oils at almost identical values. This could be explained by hydration phenomena during the mechanical extraction, using carrier water that caused the formation of terpinen-4-ol compensating the amount lost during the process.

Other papers, than those reported in Table 13.15, can be found in literature on the analysis of lemon oil, for the determination of single components or classes of components, or for the application of innovative gas chromatographic set-up for the analysis volatile components present in natural complex matrices.

Nishida and Acree (1984) isolated and characterised from lemon oil two epimeric forms of methyl jasmonate. The methyl epijasmonate was the most abundant.

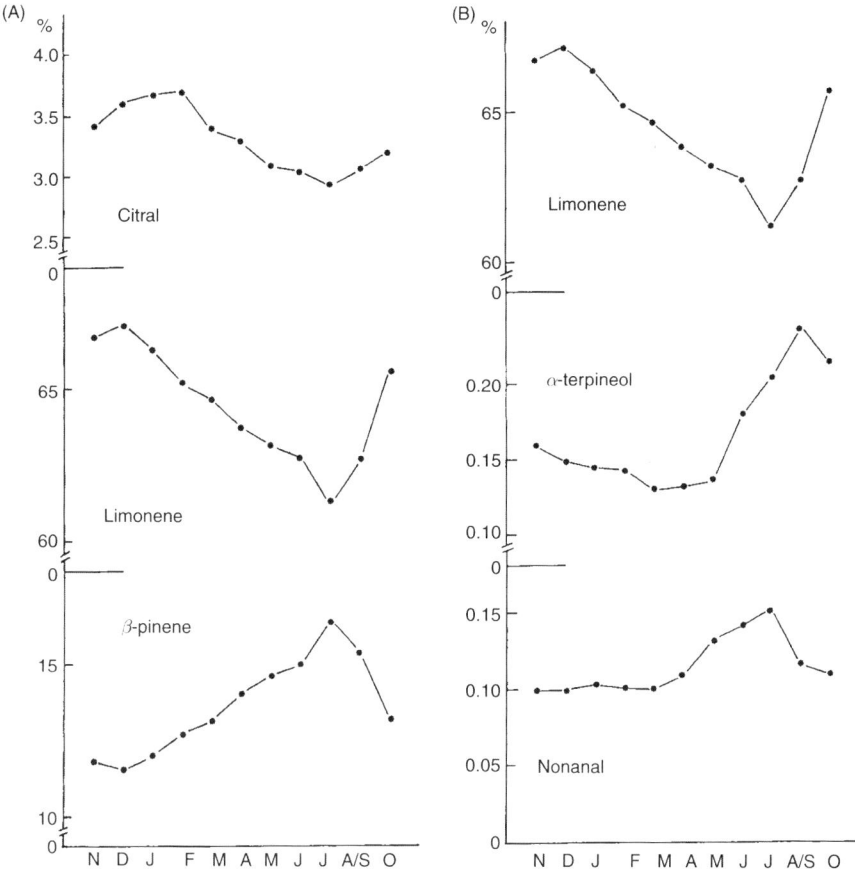

Figure 13.8 Average variation in the citral, limonene and β-pinene content (A) and limonene, α-terpineol and nonanal content (B) for lemon oil for each month of the productive season (Dugo *et al.*, 1992).

Cartoni *et al.* (1986), using two columns, one of PEG-20 M and one of SE-54, in series, obtained a good separation of the critical pairs *p*-cymene/limonene, and sabinene/β-pinene, otherwise not separated by using SE-54 columns, and of the pairs α-thujene/α-pinene and geranial/β-bisabolene, otherwise not separated on a PEG-20 M column.

Mazza (1987b) during a study for the detailed analysis of lemon oil composition, identified by GC/MS, in two samples of lemon oil obtained by FMC, the following components not listed in Table 13.15: *p*-cymenene, β-acoradiene, α-selinene, (E)-2-decenal, hexadecanal, hexanal, (E)-2-nonenal, pentadecanal, (E)-2-tetradecenal, myrtenal, isopiperitone, heptanol, carvacrol, *cis*- and *trans*-carveol, *p*-cymen-8-ol, γ-isogeraniol, *cis*- and *trans*-isopiperitenol, *cis*- and *trans*-*p*-mentha-2,8-dien-1-ol, δ-terpineol, *cis*- and *trans*-*p*-mentha-1(7), 8-dien-2-ol, *p*-mentha-1,8(10)-dien-9-ol, *cis*- and *trans*-1-acetoxy-3,7-dimethyl-2,7-octadien-6-ol, perilla alcohol, thymol; hepthyl acetate, *p*-mentha-1,8(10)-dien-9-yl acetate; ethyl benzoate, 1,4-cineole, acetic acid, octanoic acid.

Lakszner and Szepesy (1988), as mentioned for bergamot oil, proved the possibility to use a selective GC detector for oxygen containing components, for the determination of lemon oil genuineness.

Sawamura et al. (1990) reported the presence of valencene (0.05 per cent).

Munari et al. (1990) were the first who used an automated HPLC-GC system to fractionate lemon oil by HPLC and transfer the fractions onto the GC system. The advantage was that, with such method, the GC analysis of simple fractions avoided peak overlapping.

Micali et al. (1990) and Lanuzza et al. (1991), using a coupled HPLC-GC system identified in a lemon oil the n-alkanes from C_{21} to C_{33} and the correspondent 'iso' isomers, along with β-caryophyllene, bergamotene* (probably the *trans*-α-bergamotene), α-humulene, bisabolene*. Using a similar system Mondello et al. (1995b) identified the following sesquiterpene hydrocarbons listed in decreasing amount order: β-bisabolene, *trans*-α-bergamotene, β-caryophyllene, *cis*-α-bergamotene, valencene[t], (Z)-β-farnesene, β-santalene, α-humulene, γ-curcumene, (Z)-γ-bisabolene, (E)-γ-bisabolene[t], (E)-β-farnesene[t]. Chamblee et al. (1997) using a GC/FT-IR system identified for the first time in lemon oil bicyclogermacrene and (Z)-α-bisabolene.

Commercial oils

Some authors refer to commercial lemon oils of not specified origin. The results of these analyses are reported in Table 13.16. The values obtained by Inoma et al. (1989) are not reported in this table, since, although they referred to lemon oil, the composition reported resulted extremely different from that of such citrus oil. In fact some of the values obtained by these authors were: limonene (90 per cent), other monoterpene hydrocarbons almost absent, linalool (1.95 per cent), decanal (1.56 per cent), valencene (0.53 per cent), γ-cadinene (0.47 per cent).

Table 13.16 Percentage composition of the volatile fraction of some commercial lemon oils

	1	2	3	4	5	6
HYDROCARBONS						
Monoterpene						
camphene	0–0.1	0.03	0.01	–	0.06	–
p-cymene	0.3–1.8	0.29	0.04	0.8	–	0.96
limonene	62.1–74.5	71.80	70.62	67.3	66.80	71.52
myrcene	1.0–2.1	1.80	1.35	1.4	1.44	1.67
β-phellandrene	–	–	6.55	0.8	–	–
α-pinene	1.8–3.6	1.77	1.37	2.4	1.92	1.38
β-pinene	6.1–15.0	9.68	0.59	12.6	12.37	9.98
sabinene	1.5–4.6	–	0.36	2.0	1.98	–
α-terpinene	0–0.5	0.09	–	–	2.53	–
γ-terpinene	6.0–11.6	6.35	8.78	7.6	8.02	9.66
terpinolene	0.3–1.9	0.51	0.34	0.3	–	1.01
α-thujene	–	–	0.30	–	0.44	–
Sesquiterpene						
trans-α-bergamottene	–	–	0.14	0.3	0.40	0.12
β-bisabolene	–	–	–	0.8[a]	0.59	0.60
β-caryophyllene	0–0.5	–	0.05	0.2	0.20	0.42

ALDEHYDES						
Aliphatic						
nonanal	–	0.08	0.04	–	–	–
octanal	0.1–0.5[t]	–	0.31	–	–	–
Monoterpene						
citronellal	–	0.45	0.06	–	0.17	–
geranial	1.0–3.1	1.02	1.94	0.8[a]	1.26	0.82
neral	0.2–0.9	1.89	1.23	0.9	0.80	0.54
ALCOHOLS						
Monoterpene						
geraniol	–	0.01	0.13	–	–	–
linalool	0–0.9	0.07	2.23	0.1	–	0.41
nerol	–	0.01	0.34	–	–	–
terpinen-4-ol	0–0.4	0.01	0.03	–	–	0.14
α-terpineol	0–0.4	–	0.29	0.2	0.12	0.35
ESTERS						
Aliphatic						
octyl acetate	–	0.08	tr	–	–	–
Monoterpene						
geranyl acetate	0–0.3	0.51	0.07	0.3	0.49	0.40
neryl acetate	0–0.5	2.34	0.23	1.6	0.41	0.23

Notes

t = tentative identification; tr = traces; * correct isomer not characterised; a bisabolene* + geranial.

Appendix to Table 13.16

1 Prager and Miskiewicz (1982). Some commercial oils; GC on capillary column coated with Carbowax 20 M; GC/MS; relative percentage of peak areas.
2 Sugiyama and Saito (1988). One commercial sample; GC on capillary column coated with CBP-20; GC/MS. Sugiyama and Saito also found *trans*-sabinene hydrate (0.33%), linalyl acetate (0.09%), limonene oxide* (0.02%).
3 Wen *et al.* (1989). One commercial sample; GC on capillary columns coated with OV-101 and PEG-20 M; GC/MS, retention indices; relative percentage of peak areas. Wen *et al.* also found β-ocimene* (0.04%), α-phellandrene (1.64%), 1,4-cineole (0.29%), *cis*-linalool oxide (0.02%), and trace amounts of α-humulene, decanal, dodecanal, undecanal, 6-methyl-5-hepten-2-one, carvone, citronellol, citronellyl acetate, *trans*-linalool oxide.
4 Haubruge *et al.* (1989). One commercial sample; GC on capillary columns coated with CP-TM-VAX 52 CB and Carbowax 20 M; relative percentage of peak areas.
5 Yamauchi and Saito (1990). One commercial sample; GC on capillary column coated with Ultra-1; GC/MS.
6 Barth *et al.* (1994). One commercial sample from Ivory Coast; GC on capillary column coated with BD-5; GC/MS, retention times; relative percentage of peak areas. Barth *et al.* also found piperitone (0.19%) and trace amounts of γ-bisabolene* and α-bisabolol.

Laboratory extracted oils

The analyses carried out on laboratory extracted oils were meant for the study of the composition of oils extracted from different cultivars of lemon, grown in different geographic regions, and to evaluate the influence of the grafting, the period of harvest and the blooming from which the fruits originated. The extraction systems were: cold-extraction by manual pressure on the fruit peels to break the utricles; extraction with solvents of different polarities; distillation of the ground peels or of the whole fruits, after fine grounding. Of course the results relative to oils extracted by techniques that greatly differ among each other are not directly comparable, in fact different changes can occur depending on the strength of the extraction process. Often these results do not reflect the real composition of the oil contained in the utricles of the flavedo of the fruits. The differences determined by the same authors simultaneously for different varieties of fruits, cannot be considered, unless the reproducibility of extraction process was determined, as variations due to the different botanical origins of the fruits. These variations could be determined by the extraction conditions used (small changes in temperature and time). This is particularly valid for oils obtained by distillation from the whole ground fruits. It is known, in fact, that the conjugated effect of temperature and acidic pH can cause hydration reactions of the unsaturated monoterpene hydrocarbons, with the consequent formation of alcohols, and the degradation of aldehydes, such as neral and geranial, that are the most abundant carbonyl compounds present in lemon oil. These transformations are strictly related to the extraction conditions used.

A group of Spanish researchers studied the composition of lemon oils extracted from different lemon varieties (Verna, Fino, Eureka, Villafranca, Lisbon, Rodrejo) cultivated in Spain (Melendreras *et al.*, 1985; Laencina *et al.*, 1986), and the influence of the grafting on the composition of Verna lemon oil (Melendreras *et al.*, 1986) and on the intermediate grafting (Melendreras *et al.*, 1988). In all these papers the oils were extracted by distillation of the ground and homogenised fruits. From the results obtained the authors could assert that the oils did not differ sensibly in function of the varieties, of the grafting and of the intermediate grafting, although they considered the Fino oil of better quality, for the high aldehydes content; the oil obtained from the Verna variety grafted on *C. macrophyllae*, contained scant amount of aldehydes, therefore seamed to be the least appreciable. These authors also determined a higher yield of oil when the intermediate grafting was used. The four papers reported, for the oils analysed, an average content of α-terpineol, ranging between 2.9 and 8.0 per cent, terpinen-4-ol between 0.8 and 2.0 per cent, neral between 0.2 and 0.6 per cent, geranial between 0.2 and 1.0 per cent. These values appear to be, in our opinion, too high for the alcohols and too low for the aldehydes. These results can be explained by the formation of artefacts during the extraction process, compromising the data obtained. It should also be noted that the authors reported, in all the oils analysed, the presence of methyl anthranilate and methyl N-methylanthranilate.

Germanà *et al.* (1987) analysed the oils extracted with solvent from the triturated flavedo of 44 clones of Feminello, Monachello and Frost Eureka lemons cultivated in an experimental field in the Palermo area in Sicily (Italy). The fruits were obtained, for the different clones, all from the blooming that originates winter lemons, spring lemons (*bianchetti*) and summer lemons (*verdelli*). The authors noticed that winter lemons presented a higher content of limonene compared to bianchetti and verdelli, while β-pinene had an opposite behaviour. The highest amount of oxygenated compounds was determined in the bianchetti lemons, while the lowest was in verdelli lemons. Later the same authors (Crescimano *et al.*, 1988) compared the composition of two clones of Femminello and one clone each of the Monachello and Frost Eureka cultivars. In the Monachello oil limonene was at its highest level (75.6 per cent) and neral + geranial at the lowest (2.3 per cent); the lowest amount of limonene (68.2 per cent) and the highest amount of neral+geranial (4.2 per cent) were determined in one of the two Femminello oils.

Usai *et al.* (1996) compared the composition of the oils, manually extracted by hand pressure from the peel or by solvent extraction with light petroleum, from 12 lemon cultivars: (Verna, Frost Eureka, Interdonato, Frost Lisbon, Santa Teresa (nucellar 165, ICAS 132, ICAS 99, ICAS 146), Massese, Rifiorente (nucellar 19 e 12), Vaniglia) cultivated in Sardinia (Italy) in a field nearby Oristano. The authors found the highest level of monoterpene hydrocarbons (93.6 per cent) in the ICAS S. Teresa 132, of aldehydes (6.6 per cent) and of alcohols (1.7 per cent) in the ICAS S. Teresa 146, of sesquiterpene hydrocarbons (5.3 per cent) and of esters (6.1 per cent) in the Interdonato cv., while the lowest content of these classes of components was, for the monoterpene hydrocarbons (84.5 per cent) in the Interdonato cv., for the alcohols (1.7 per cent) in the Frost Eureka cv., for aldehydes (1.9 per cent), esters (0.8 per cent) and sesquiterpene hydrocarbons (0.8 per cent) in the Rifiorente 19 cv. The statistical analysis of these results elaborated by the principal components analysis (PCA) allowed the authors to include 9 of the 12 cultivars analysed into two groups: (a) Massese, Rifiorente (nucellar 19 e 12), S. Teresa (ICAS 99 e 132), Vaniglia; (b) Verna, Frost Eureka, Frost Lisbon.

Gazea *et al.* (1996) investigated on the composition of the essential oil obtained from the diploid cultivar of Femminello comune nucellar 46-451 and of two selections of the autotetraploids lemons named 'Cavone' and 'Doppio Lentini' cultivated in Italy at the Experimental Institute of Citriculture in Acireale (Sicily). The analysis, performed by GC on the hand-pressed oils or on oils extracted by diethyl ether, allowed to confirm that the Femminello comune cv. presented the highest content of total oxygenated components (6.1 per cent) and of carbonylic compounds (4.0 per cent).

Caccioni *et al.* (1998) analysed three samples of laboratory extracted oils, by hydrodistillation, from lemons of the Femminello cv., harvested in November, February and June respectively, in the experimental field of the Citrus Experimental Institute in Lentini, Sicily. The composition of these oils is reported below:

camphene	0.05–0.10%
δ-3-carene	tr
p-cymene + β-phellandrene	0.17–0.42%
limonene	60.20–71.06%
myrcene	1.39–1.52%
(E)-β-ocimene	0.07–0.15%
(Z)-β-ocimene	0–0.04%
α-phellandrene	0.04–0.06%
α-pinene	1.27–2.27%
β-pinene + sabinene	9.27–19.52%
α-terpinene	0.12–0.30%
γ-terpinene	8.06–9.45%
terpinolene	0.39–0.52%
α-thujene	0.27–0.43%
trans-α-bergamotene	0.17–0.28%
β-bisabolene	0.23–0.47%
β-caryophyllene	0.13–0.19%
α-humulene	0.01–0.03%

decanal	0.03–0.05%
nonanal	0.10–0.13%
octanal	0.06–0.07%
citronellal	0.04–0.08%
geranial	0.56–1.23%
neral	0.38–0.90%
octanol	0.01–0.06%
geraniol	0.18–1.05%
linalool	0.18–0.48%
nerol	0.17–0.86%
α-terpineol	0.41–0.86%
terpinen-4-ol	0.34–0.67%
citronellyl acetate	0.02–0.05%
geranyl acetate	0.23–0.45%
neryl acetate	0.21–0.44%

The sample harvested in February presented a total content of alcohols (3.96 per cent) greatly higher than the samples obtained in November (1.29 per cent) and in June (2.37 per cent). In agreement with what stated in numerous papers (Cotroneo et al., 1986a,b; Dugo, 1986, 1994) on the seasonal variation of the lemon oil composition, the alcohols content in February should be lower or similar to that determined in November and lower than in June. It could be possible that the sample obtained in February, analysed by Caccioni et al. (1998), during the hydrodistillation process was subject to greater variations than those determined on the other two samples.

Kekelidze et al. (1982), in oils obtained from the Novogruzinski variety, cultivated in Russia and extracted by hydrodistillation, determined the following composition:

p-cymene	3.20%
limonene	61.89%
myrcene	1.24%
ocimene*	1.48%
α-pinene	1.47%
β-pinene	13.75%
sabinene	2.32%
γ-terpinene	5.39%
terpinolene	0.50%
bergamotene*	0.20%
β-bisabolene	0.60%
caryophyllene*	0.10%
decanal	0.22%
citronellal	0.18%
geranial	1.21%
neral	0.64%
citronellol	0.15%
geraniol	0.10%
linalool	0.67%
nerol	0.14%
terpinen-4-ol	1.34%

α-terpineol   2.00%
geranyl acetate 0.18%
neryl acetate   0.20%

The oil was characterised by a high content of alcohols (4.40 per cent) and *p*-cymene. It is not easy to determine the reason of such peculiar values, which could be due to the extraction method or to the botanical origin. It is surprising that the composition reported in this paper greatly differ from the one previously reported, for a lemon oil obtained from the same cultivar, by Kekelidze and Beradze (1974).

Sattar *et al.* (1986) determined in a sample of lemon oil extracted from Pakistan fruits the following content of monoterpene hydrocarbons:

camphene   0.87%
p-cymene   1.13%
myrcene   1.36%
limonene   72.36%
α-pinene   0.12%
β-pinene   4.34%
γ-terpinene   8.26%
terpinolene   0.22%

Kumar *et al.* (1992) examined the oil obtained by hydrodistillation from a selection of Kageikalan lemons, named Pant lemon-1 cultivated in India. The composition of this oil, reported below, presented some peculiar characteristics, such as the low value of limonene and the high content of alcohols:

camphene   0.20%
δ-3-carene   0.01%
p-cymene   0.20%
p-cimenene   0.10%
α-fenchene   0.05%
limonene   52.00%
myrcene   1.20%
(E)-β-ocimene   0.30%
(Z)-β-ocimene   0.20%
α-phellandrene + α-terpinene   0.60%
α-pinene   1.10%
β-pinene   12.00%
γ-terpinene   1.10%
terpinolene   2.10%
germacrene D   0.10%
geranial   1.00%
neral   0.30%
fenchone   0.30%
citronellol   0.40%
p-cymen-8-ol   0.02%
geraniol   1.50%
linalool   2.90%

nerol	0.75%
terpinen-4-ol	2.50%
α-terpineol	12.00%
β-terpineol*	0.30%
γ-terpineol	0.10%
3-hexenyl acetate[t]	0.01%
linalool oxide*	0.20%
rose oxide[t]	0.01%

Dellacassa et al. (1994) analysed oils obtained by manual pressure on the peels of fruits of the Verna, Eureka, Mesero and Lisbon cultivars, cultivated in Northern Uruguay, and of the Ferres 1, Ferres 2 and Vila cultivars grown in the South. The results were compared also with those obtained previously for industrially processed lemon oils produced in Uruguay (Dellacassa et al., 1991). Among the varieties studied the highest content of total oxygenated compounds (5.7–5.8 per cent) and of aldehydes (4.2–4.4 per cent) were determined in the Mesero, Lisbon and Ferres 2. It was also determined that the laboratory extracted oils, if compared with the industrial oils, presented a higher content of oxygenated compounds, in particular alcohols and aldehydes, while the esters content did not vary considerably. These authors attributed such behaviour to one or both of the following reasons: the varieties mostly used for the industrial production gave oils of lower quality than those studied; the industrial technology used was not optimised, causing the loss of alcohols and aldehydes.

Wen et al. (1989) examined the composition of the oil extracted by hydrodistillation and simultaneous distillation–solvent extraction, from lemons cultivated in China. The same components identified in the industrial sample used as control (Table 13.16 column 3) were determined. The quantitative composition also resulted very similar to that of the industrial sample, but a higher content of p-cymene, citronellol, geraniol, nerol, terpinene-4-ol, and a lower amount of sesquiterpene hydrocarbons were determined. The amount of 1,4-cineole was higher in the sample obtained by simultaneous distillation–solvent extraction than in the industrial sample and that obtained by hydrodistillation.

Two papers report the results of detailed analysis on the composition of lemon oil extracted in laboratory from fruits cultivated in Japan. Yang et al. (1992) in one oil obtained by solvent extraction from fruits of not specified cultivar, determined 76 components by fractionation of the oil on silica columns and subsequent GC and GC/MS analyses. Njoroge et al. (1994b) in a sample of lemon oil manually pressed from lemons of the Lisbon cv., identified and quantitatively determined 104 components by GC and GC/MS with Kovats indices. Recently, Sawamura et al. (1999), reported the following content of the main components present in two lemon oils, laboratory extracted from the Eureka and Lisbon cvs. cultivated in Japan:

	Eureka	Lisbon
limonene	69.7%	64.6%
myrcene	1.6%	1.4%
α-pinene	2.3%	2.6%
β-pinene	10.5%	14.0%

γ-terpinene	8.2%	10.3%
terpinolene	0.3%	0.4%
octanal	tr	0.1%
citronellal	0.1%	tr
geranial	1.0%	1.1%
neral	1.1%	0.6%
geraniol	tr	0.1%
linalool	0.2%	0.2%
α-terpineol	0.2%	0.3%
geranyl acetate	0.2%	0.2%

The oils analysed by Sawamura *et al.* (1999) also contained trace amounts of *p*-cymene, α-terpinene, decanal, decanol, citronellol, nerol, and terpinen-4-ol.

Baaliouamer and Meklati (1980) determined the quantitative composition of a hand pressed oil obtained from lemons of the Eureka cv. grown in Algeria. The detailed study carried out by GC and GC/MS, using capillary apolar and polar columns, both on the whole oil and on aliquots of this, after derivatisation reactions for the different functional groups (olephines, aldehydes, esters, and acids), separately, permitted the identification of 129 components and 38 of these were confirmed.

Ayedoun *et al.* (1996) determined in a sample of lemon oil extracted by hydrodistillation from lemons cultivated in Benin the following composition:

camphene	0.1%
p-cymene	0.4%
limonene	70.4%
myrcene	1.8%
(E)-β-ocimene	0.3%
(Z)-β-ocimene	0.2%
α-phellandrene	0.1%
α-pinene	1.8%
β-pinene	4.2%
α-terpinene	0.3%
γ-terpinene	11.8%
terpinolene	0.8%
α-thujene	0.3%
trans-α-bergamotene	0.3%
β-bisabolene	0.5%
δ-cadinene	0.1%
β-caryophyllene	0.3%
(E,E)-α-farnesene	0.2%
germacrene D	0.2%
octanal	0.1%
linalool	0.1%
terpinen-4-ol	0.7%
α-terpineol	1.4%
cis-β-terpineol	0.1%
1,8-cineole	3.3%

and trace amounts of sabinene, α-humulene, decanal, geranial neral, borneol, citronellol and endo-phenchol.

Haggag *et al.* (1998) reported, for an Egyptian lemon oil, laboratory extracted both by cold pressing and by steam distillation, an extremely unusual composition. In this oil, in fact, was determined only the 10 per cent of limonene, and very high amounts of α-pinene (27 per cent), β-pinene (18 per cent), monoterpene alcohols (16 per cent), thymol (3 per cent) and eugenol (9 per cent).

REFERENCES

Alessandro, R.T., Adams, J.M. and Miskiewicz, M.A. (1985) Gas chromatographic-mass spectrometric detection of adulteration of natural lime oils. *J. Assoc. Off. Anal. Chem.*, 68, 1154–1159.

Analytical Methods Committee (1984) Application of gas-liquid chromatography to the analysis of essential oils. Part XI. Monographs for seven essential oils. *Analyst*, 109, 1343–1360.

Arras, G., Cotroneo, A., Di Giacomo, A., Picci, V. and Verzera, A. (1985) Sulla genuinità delle essenze agrumarie. Nota XIII. Sulla composizione di essenze di arancia dolce estratte dalla cultivar Valencia late su diversi portainnesti. *Essenz. Deriv. Agrum.*, 55, 374–387.

Ashoor, S.H.M. and Bernhard, R.A. (1967) Isolation and characterization of terpenes from *Citrus reticulata* Blanco and their comparative distribution among other Citrus species. *J. Agric. Food Chem.*, 15, 1044–1047.

Ashour, F.M. and El-Kebeer, M.E. (1983) Studies on physicochemical properties and chemical composition of bitter orange peel oil. *9th International Essential Oil Congress*, Singapore.

Attaway, J.A., Pieringer, A.P. and Barabas, L.J. (1967) The origin of citrus flavor components-III. A study of the percentage variations in peel and leaf oil terpenes during one season. *Phytochemistry*, 6, 25–32.

Attaway, J.A., Pieringer, A.P. and Buslig, B.S. (1968) The origin of citrus flavor components IV. The terpenes of Valencia orange leaf, peel and blossom oils. *Phytochemistry*, 7, 1695–1698.

Ayedoun, A.M., Sossou, P.V., Mardarowicz, M. and Leclercq, P.A. (1996) Volatile constituents of the peel and leaf oils of *Citrus limon* L. Burm. f. from Benin. *J. Essent. Oil Res.*, 8, 441–444.

Azzouz, M.A. and Reineccius, G.A. (1976) Comparison between cold-pressed and distilled lime oils through the application of gas chromatography and mass spectrometry. *J. Food Sci.*, 41, 324–328.

Baaliouamer, A. and Meklati, B.Y. (1980) Analyse qualitative par chromatographie gaz-liquide sur colonnes capillaires de verre de type WCOT des huiles essentielles d'ecorce de fruit et de petitgrain du citronnier 'Eureka' cultivé en Algerie. *Fruits*, 35(9), 561–572.

Baaliouamer, A., Meklati, B.Y., Fraisse, D. and Scharff, C. (1992) The chemical composition of some cold-pressed citrus oils produced in Algeria. *J. Essent. Oil Res.*, 4, 251–258.

Barth, D., Chouci, D., Della Porta, G., Reverchon, E. and Perrut, M. (1994) Desorption of lemon peel oil by supercritical carbon dioxide: deterpenation and psoralen elimination. *J. Supercrit. Fluids*, 7, 177–183.

Barukadze, N.S. (1985) Chemical composition of orange essential oil obtained by a new technique. *Maslo-Zhir. Prom.-st.*, (1), 20–22.

Baser, K.H.C., Özek, T. and Tutas, M. (1995) Composition of cold-pressed bergamot oil from Turkey. *J. Essent. Oil Res.*, 7, 341–342.

Berger, R.G., Drawert, F., Kollmannsberger, H. and Nitz, S. (1985) Natural occurrence of undecaenes in some fruits and vegetables. *J. Food Sci.*, 50, 1655–1656.

Bernhard, R.A. (1957) Separation and identification of some terpenes by gas partition chromatographic analysis. *J. Ass. Off. Agric. Chem.*, 40, 915–921.

Bernhard, R.A. (1958) Examination of lemon oil by gas-partition chromatography. *Food Research*, 23, 213–216.

Bernhard, R.A. (1960) Analysis and composition of oil of lemon by gas-liquid chromatography. *J. Chromatogr.*, 3, 471–476.

Bernhard, R.A. (1961) Citrus flavor. Volatile constituents of the essential oil of the orange. *J. Food Sci.*, 26, 401–411.

Blanco Tirado, C., Stashenko, E.E., Combariza, M.Y. and Martinez, J.R. (1995) Comparative study of Colombian citrus oils by high-resolution gas chromatography and gas chromatography-mass spectrometry. *J. Chromatogr. A*, 697, 501–513.

Boelens, M.H. and Sindreu, R.J. (1988) Essential oils from Seville bitter orange (*Citrus aurantium* L. ssp. *Amara* L.). In B.M. Lawrence, B.D. Mookherjee and B.J. Willis (eds), In *Flavors and Fragrances: a World Perspective*. Elsevier Science BV, Amsterdam, pp. 551–565.

Boelens, M.H. and Jimenez, R. (1989) The chemical composition of some Mediterranean citrus oils. *J. Essent. Oil Res.*, 1, 151–159.

Boelens, M.H. and Jimenez, R. (1989) The chemical composition of the peel oils from unripe and ripe fruits of bitter orange, *Citrus aurantium* L. ssp. *Amara* Engl. *Flav. Fragr. J.*, 4, 139–142.

Boelens, M.H. (1991) A critical review on the chemical composition of citrus oils. *Perfum. Flav.*, 16(2), 17–34.

Boelens, M.H. and Oporto, A. (1991) Natural isolates from Seville bitter orange tree. *Perfum. Flav.*, 16(6), 2–7.

Braddock, R.J. and Kesterson, J.W. (1976) Quantitative analysis of aldehydes, esters, alcohols and acids from citrus oils. *J. Food Sci.*, 41, 1007–1010.

Caccioni, D.R.L., Guizzardi, M., Biondi, D.M., Renda, A. and Ruberto, G. (1998) Relationship between volatile components of citrus fruit essential oils and antimicrobical action on *Penicillum digitatum* and *Penicillum italicum*. *Inter. J. Food Microbiol.*, 43, 73–79.

Calabrò, G. and Currò, P. (1973) Costituenti degli olii essenziali. Nota III. Costituenti sesquiterpenici e paraffinici dell'essenza di bergamotto. *Ann. Fac. Econ. Commer. Univ. Studi Messina*, 10, 68–79.

Calvarano, I. (1966) Contributo all'indagine analitica strumentale dell'essenza di arancio amaro. *Essenz. Deriv. Agrum.*, 36, 5–25.

Calvarano, I., Bovalo, F. and Di Giacomo, A. (1974) L'olio essenziale di clementine. *EPPOS*, 56, 229–231.

Calvarano, I., Calvarano, M. and Di Giacomo, A. (1989) Su alcuni parametri gas-cromatografici dell'essenza di mandarino italiana. *Essenz. Deriv. Agrum.*, 69, 407–416.

Calvarano, M. (1959) Contributo alla conoscenza delle essenze di arancio italiane. *Essenz. Deriv. Agrum.*, 29, 147–159.

Calvarano, M. (1963) La composizione delle essenze di bergamotto. Nota I. Gli idrocarburi monoterpenici. *Essenz. Deriv. Agrum.*, 33, 67–101.

Calvarano, M. and Calvarano, I. (1964) La composizione delle essenze di bergamotto. Nota II. Contributo all'indagine analitica mediante spettrofotometria nell'UV e gascromatografia. *Essenz. Deriv. Agrum.*, 34, 71–92.

Calvarano, M. (1965) La composizione delle essenze di bergamotto. Nota III. *Essenz. Deriv. Agrum.*, 35, 197–211.

Calvarano, M. (1968) Variazioni nella composizione dell'essenza di bergamotto durante la maturazione del frutto. *Essenz. Deriv. Agrum.*, 38, 3–20.

Calvarano, M., Rafanelli, G. and Di Giacomo, A. (1984) Sulle caratteristiche di un'essenza di bergamotto ottenuta da frutti prodotti in Toscana. *Essenz. Deriv. Agrum.*, 54, 220–227.

Calvarano, M., Calvarano, I., Dellacassa, E., Menendez, P. and Di Giacomo, A. (1989) Su alcune essenze di mandarino dell'Uruguay. *Essenz. Deriv. Agrum.*, 69, 397–406.

Cappello, C., Micali, B., Calvarano, M., Retamar, J.A., Rozas De Vottero, L. and Taher, H.A. (1981) Ricerche chimiche sulla composizione dei derivati agrumari argentini. Nota I. Gli olii essenziali. *Essenz. Deriv. Agrum.*, 51, 229–233.

Cartoni, G.P., Goretti, G., Monticelli, B. and Russo, M.V. (1986) Evaluation of capillary gas chromatographic columns in series. Analytical application to lemon oil. *J. Chromatogr.*, 370, 93–101.

Cartoni, G.P., Goretti, G. and Russo, M.V. (1987) Capillary columns in series for the gas-chromatographic analysis of essential oils. *Chromatographia*, 23, 790–795.

Chamblee, T.S., Clark, B.C., Radford, T. and Iacobucci, G.A. (1985) General method for the high performance liquid chromatographic prefractionation of essential oils and flavor mixtures for gas chromatographic-mass spectrometric analysis. Identification of new constituents in cold-pressed lime oil. *J. Chromatogr.*, 330, 141–151.

Chamblee, T.S., Clark, B.C., Brewster, G.B., Redford, T. and Iacobucci, G.A. (1991) Quantitative analysis of the volatile constituents of lemon peel oil. Effects of silica gel chromatography on the composition of its hydrocarbon and oxygenated fractions. *J. Agric. Food Chem.*, 39, 162–169.

Chamblee, T.S., Karelitz, R.L., Radford, T. and Clark, B.C. (1997) Identification of sesquiterpenes in citrus essential oils by cryofocusing GC/FT-IR. *J. Essent. Oil Res.*, 9, 127–132.

Chouchi, D., Barth, D., Reverchon, E. and Della Porta, G. (1995) Supercritical CO_2 desorption of bergamot peel oil. *Ind. Eng. Chem. Res.*, 34, 4508–4513.

Chouchi, D., Barth, D., Reverchon, E. and Della Porta, G. (1996) Bigarade peel oil fractionation by supercritical carbon dioxide desorption. *J. Agric Food Chem.*, 44, 1100–1104.

Clark, B.C., Chamblee, T.S. and Jacobucci, G.A. (1987) HPLC isolation of the sesquiterpene hydrocarbons germacrene B from lime peel oil and its characterization as an important flavor impact constituent. *J. Agric Food Chem.*, 35, 514–518.

Clark, B.C. and Chamblee, T.S. (1992) Acid-catalyzed reactions of citrus oils and other terpene-containing flavors. In G. Charalambous (ed.), *Off-flavors in Food and Beverages*, Elsevier Science Publishers B.V., Amsterdam, Holland 1992, 229–275.

Clark, J.R. and Bernhard, R.A. (1960) Examination of lemon oil by gas-liquid chromatography. II. The hydrocarbon fraction. *Food Research*, 25, 389–394.

Cotroneo, A., Trozzi, A. and Di Giacomo, A. (1985) Sull'olio essenziale di Kumquat. *Essenz. Deriv. Agrum.*, 55, 20–26.

Cotroneo, A., Verzera, A., Lamonica, G., Dugo, G. and Licandro, G. (1986a) On the genuineness of citrus essential oils. Part X. Research on the composition of essential oils produced from Sicilian lemons using 'Pelatrice' and 'Sfumatrice' extractors during the entire 1983/84 production season. *Flav. Fragr. J.*, 1, 69–86.

Cotroneo, A., Dugo, G., Licandro, G., Ragonese, C. and Di Giacomo, G. (1986b) On the genuineness of citrus essential oils. Part XII. Characteristics of Sicilian lemon essential oil produced with the FMC extractor. *Flav. Fragr. J.*, 1, 125–134.

Cotroneo, A., Dugo, G. and Verzera, A. (1987) Sulla genuinità delle essenze agrumarie. Nota XVI. Differenze quantitative nella composizione di essenze di limone estratte a macchina e di essenze estratte manualmente senza uso di acqua. *Essenz. Deriv. Agrum.*, 57, 220–235.

Cotroneo, A., Verzera, A., Dugo, G., Dugo, Giacomo and Licandro, G. (1988) Sulla genuinità delle essenze agrumarie. Nota XXII. La composizione dell'essenza di limone siciliana prodotta industrialmente nell'annata 1984/85. *Industria Bevande*, 17, 209–224.

Cotroneo, A., Mondello, L. and Stagno d'Alcontres, I. (1994) Sulla genuinità delle essenze agrumarie. Nota XLVII. Aggiornamento sulla composizione della frazione volatile dell'olio essenziale di mandarino. *Essenz. Deriv. Agrum.*, 64, 275–285.

Crescimanno, F.G., De Pasquale, F., Germanà, M.A., Bazan, E. and Palazzolo, E. (1988) Influence of the harvesting period on the yield of essential oil from the peel of four lemon cultivars (*C. limon* (L.) Burm f.). *Proceedings of the Sixth International Citrus Congress*, Tel Aviv, Israel, 6–11 March, 589–595.

D'Amore, G. and Calabrò, G. (1966) Costituenti degli olii essenziali. Nota 1. Gascromatografia dell'essenza di mandarino (*Citrus reticulata* Blanco). *Ann. Fac. Econ. Commer. Univ. Studi Messina*, 4(2), 635–661.

D'Amore, G., Calabrò, G. and Currò, P. (1968) Costituenti degli olii essenziali. Nota 2. Idrocarburi sesquiterpenici e paraffinici dell'essenza di arancia dolce. *Ann. Fac. Econ. Commer. Univ. Studi Messina*, 6(3), 3–14.

David, F., Gere, D.R., Scanlan, F. and Sandra, P. (1999) Instrumentation and application of fast high-resolution capillary gas chromatography. *J. Chromatogr. A*, 842, 309–319.

De La Torre, P.C. and Sardi, T.C. (1977) Study on bitter orange (*Citrus aurantium*) essential oil. *Arch. Bioquim. Quim. Farm.*, 20, 69–72. In B.M. Lawrence (1982) Progress in essential oils. *Perfum. Flav.*, 7(5), 43–48.

Dellacassa, E., Menendez, P., Calvarano, M., Calvarano, I. and Di Giacomo, A. (1989) Estudio sobre la esencia de la mandarina malaquina del Uruguay. *Essenz. Deriv. Agrum.*, 69, 266–272.

Dellacassa, E., Rossini, C., Moyna, P., Stagno d'Alcontres, I., Verzera, A., Trozzi, A. and Dugo, G. (1991) Gli olii essenziali dell'Uruguay. Nota II. Primi risultati sulla composizione della frazione volatile dell'olio essenziale di limone. *Essenz. Deriv. Agrum.*, 61, 291–304.

Dellacassa, E., Rossini, C., Menendez, P., Moyna, P., Verzera, A., Trozzi, A. and Dugo, G. (1992) Citrus essential oils of Uruguay. Part I. Composition of oils of some varieties of mandarin. *J. Essent. Oil Res.*, 4, 265–272.

Dellacassa, E., Rossini, C., Lorenzo, D., Moyna, P., Verzera, A., Trozzi, A. and Dugo, G. (1995) Uruguayan essential oils. Part III. Composition of the volatile fraction of lemon essential oils. *J. Essent. Oil Res.*, 7, 25–37.

Dellacassa, E., Lorenzo, D., Moyna, P., Verzera, A. and Cavazza, A. (1997) Uruguayan essential oils. Part V. Composition of bergamot oil. *J. Essent. Oil Res.*, 9, 419–426.

Dellacassa, E., Lorenzo, D., Rossini, C., Moyna, P., Cotroneo, A., Stagno d'Alcontres, I. and Dugo, G. (1994) Gli olii essenziali dell'Uruguay. Nota IV. La composizione della frazione volatile dell'olio di alcune varietà di limone. *Essenz. Deriv. Agrum.*, 64, 45–54.

Della Porta, G., Reverchon, E., Chouci, D. and Barth, D. (1997) Mandarin and lime peel oil processing by supercritical CO_2 desorption: deterpenation and high molecular weight compounds elimination. *J. Essent. Oil Res.*, 9, 512–522.

Demole, E., Eggist, P. and Ohloff, G. (1982) 1-*p*-Menthene-8-thiol: a power flavor impact constituent of grapefruit juice (*Citrus paradisi* Macfayden). *Helvetica Chimica Acta*, 65, 1785–1794.

Di Corcia, A., Liberti, A. and Samperi, R. (1975) Cromatografia gas-liquido-solido: sua applicazione allo studio degli olii essenziali. *Essenz. Deriv. Agrum.*, 45, 89–99.

Di Giacomo, A. and Rispoli, G. (1962) La caratterizzazione delle essenze di limone 'a spugna' rispetto quelle 'a macchina' mediante cromatografia in fase di vapore. *Essenz. Deriv. Agrum.*, 32, 260–273.

Di Giacomo, A., Rispoli, G. and Crupi, F. (1962) Studio della frazione terpenica dell'essenza di limone siciliana mediante cromatografia in fase gassosa. *Essenz. Deriv. Agrum.*, 32, 126–134.

Di Giacomo, A., Rispoli, G. and Tracuzzi, M.L. (1963) Contributo dell'analisi strumentale alla conoscenza dei costituenti terpenici delle essenze agrumarie. *EPPOS*, 45, 269–281.

Di Giacomo, A., Rispoli, G. and Tracuzzi, M.L. (1964) Indagine sulla composizione dell'essenza di arancio amaro e sui metodi atti a determinarne la genuinità. *Essenz. Deriv. Agrum.*, 34, 3–16.

Di Giacomo, A., Pennisi, L. and Rispoli, G. (1965) Studio dei componenti ossigenati dell'essenza di limone siciliana mediante l'impiego di nuove tecniche strumentali. *Essenz. Deriv. Agrum.*, 35, 9–22.

Di Giacomo, A., Pennisi, L. and Raciti, G. (1969) Resa e qualità degli oli essenziali del limone 'femminello comune' in relazione all'ambiente di coltivazione e allo stato nutritivo delle piante. *Industria Conserve*, 44, 110–116.

Di Giacomo, A., Bovalo, F. and Postorino, E. (1971) Sull'essenza di arancia prodotta industrialmente dai frutti della piana di Rosarno. *Essenz. Deriv. Agrum.*, 41, 239–250.

Di Giacomo, A. and Mincione, B. (1994) *Gli Olii Essenziali Agrumari in Italia*. Laruffa Editore, Reggio Calabria, Italy.

Dixon, C.W., Malone, C.T. and Umbreit, G.R. (1968) A scheme for the preparative chromatographic isolation and concentration of trace components in natural products using 4 inch diameter columns. *J. Chromatogr.*, 35, 475–488.

Drescher, R.W., Beñatena, H.N., De Rossin, G.G.T., Ubiergo, G.O. and Retamar, J.A. (1984) Evaluaciòn de le calidad perfumistica del aceite esencial de *Citrus bergamia* Risso (Bergamota) cultivada en la E.E.A. Inta Concordia (Entre Rios). *Essenz. Deriv. Agrum.*, 54, 192–199.

Dugo, G., Ragonese, C. and Licandro, G. (1977) L'essenza di limone della Mesopotamia argentina. *Essenz. Deriv. Agrum.*, 47, 503–514.

Dugo, G., Licandro, G., Cotroneo, A. and Dugo G. (1983) Sulla genuinità delle essenze agrumarie. Nota I. Caratterizzazione di essenze di limone Siciliane. *Essenz. Deriv. Agrum.*, 53, 173–217.

Dugo, G., Cotroneo, A., Licandro, G. and Verzera, A. (1984) Sulla genuinità delle essenze agrumarie. Nota VIII. Caratterizzazione di essenze di mandarino. *Essenz. Deriv. Agrum.*, 54, 62–83.

Dugo, G., Licandro, G., Cotroneo, A. and Verzera, A. (1984) Indagine sulla composizione media dell'essenza di limone siciliana ottenuta industrialmente durante l'intera stagione produttiva 1982/83. *1° Conferenza Nazionale sugli Aromatizzanti*, Salsomaggiore Terme, Parma, Italy, April 2–3, 1984.

Dugo, G. (1986) L'huile essentielle de citron sicilien. *Parfums Cosmétiques Arômes*, 68, 95–105.

Dugo, G., Lamonica, G., Cotroneo, A., Trozzi, A., Crispo, F., Licandro, G. and Gioffrè, D. (1987) Sulla genuinità delle essenze agrumarie. Nota XVII. La composizione della frazione volatile dell'essenza di bergamotto Calabrese. *Essenz. Deriv. Agrum.*, 57, 456–534.

Dugo, G., Cotroneo, A., Trozzi, A., Barbeni, M. and Di Giacomo, A. (1988) On the genuineness of citrus essential oils. Part XXV. The essential oil from 'Wenzhou honey' oranges. *Flav. Fragr. J.*, 3, 161–166.

Dugo, G., Rouzet, M., Verzera, A., Cotroneo, A. and Merenda I.F. (1990) La pureté des essences d'agrumes. Note XXIV. Composition de l'huile essentielle italienne de mandarine. *Parfums Cosmétiques Aromes*, (93), 77–84.

Dugo, G., Cotroneo, A., Verzera, A., Donato, M.G., Del Duce, R., Licandro, G. and Crispo, F. (1991) Genuineness characters of Calabrian bergamot essential oil. *Flav. Fragr. J.*, 6, 39–56.

Dugo, G., Lamonica, G., Cotroneo, A., Stagno d'Alcontres, I., Verzera, A., Donato M.G. and Dugo, P. (1992) High resolution gas chromatography for detection of adulterations of citrus cold-pressed essential oils. *Perfum. Flav.*, 17(5), 57–74.

Dugo, G., Verzera, A., Stagno d'Alcontres, I., Cotroneo, A. and Ficarra, R. (1993) On the genuineness of citrus essential oil. Part XLI. Italian bitter orange essential oil: composition and detection of contamination and additions of oils and terpenes of sweet orange and of lemon. *Flav. Fragr. J.*, 8, 25–33.

Dugo, G. (1994) The composition of the volatile fraction of the Italian citrus essential oils. *Perfum. Flav.*, 19(6), 29–51.

Dugo, G., Verzera, A., Stagno d'Alcontres, I., Cotroneo, A., Trozzi, A. and Mondello, L. (1994) On the genuineness of citrus essential oils. Part XLIII. The composition of the volatile fraction of Italian sweet orange oils (*Citrus sinensis* (L) Osbeek). *J. Essent. Oil Res.*, 6, 101–137.

Dugo, G. (1997) Unpublished results.

Dugo, G., Bartle, K.D., Bonaccorsi, I., Catalfamo, M., Cotroneo, A., Dugo, P., Lamonica, G., McNair, H., Mondello, L., Previti, P., Stagno d'Alcontres, I., Trozzi, A. and Verzera, A. (1999) Advanced analytical techniques for the analysis of citrus essential oils. Part I. Volatile fraction: HRGC/MS analysis. *Essenz. Deriv. Agrum.*, 69, 79–111.

Dugo, Giacomo, Cotroneo, A., Verzera, A., Dugo, G. and Licandro, G. (1990) Mapo tangelo essential oil. *Flav. Fragr. J.*, 5, 205–210.

Dugo, P., Mondello, L., Lamonica, G. and Dugo, G. (1997) Characterization of cold-pressed Key and Persian lime oils by gas chromatography, gas chromatography/mass spectroscopy, high-performance liquid chromatography and physicochemical indices. *J. Agric. Food Chem.*, 45, 3608–3616.

Edwards, D.J. and Marr, I.M. (1990) Previously unreported sesquiterpenes of lime oil (*Citrus latifolia* Tanaka). *J. Essent. Oil Res.*, 2, 137–138.

Ehret, C. and Maupetit, P. (1982) Two sinapyl alcohol derivatives from bergamot essential oil. *Phytochemistry*, 21, 2984–2985.

El-Samahy, S.K., Askar, A. and Abd El-Fadeel, M.G. (1982) Quantitative analysis of some citrus peel oils. *Riechstoffe, Aromen, Kosmetica*, 3, 68–70.

Ferrer, O.J. and Matthews, R.F. (1987) Terpene reduction in cold-pressed orange oil by frontal analysis-displacement adsorption chromatography. *J. Food Sci.*, 52, 801–805.

Fincke, A. and Maurer, R. (1974) Lemon oil content after production and storage of lemon oil-containing sweets. *Deut. Lebensm-Rundsch.*, 70, 100–104

Formàček, V. and Kubeczka, K.H. (1982) *Essential Oils Analysis by Capillary Gas Chromatography and Carbon-13 NMR Spectroscopy*. Whiley, Chichester, UK.

Gazea, F., Calvarano, I., Calvarano, M., Russo, G. and Starrantino, A. (1996) Gli olii essenziali del flavedo del limone 'Femminello comune' NUC 46-451 e di due selezioni autotetraploidi. *Essenz. Deriv. Agrum.*, 66, 321–328.

Germanà, M.A., De Pasquale, F., Bazan, E. and Palazzolo, E. (1987) Ricerche sugli olii essenziali del flavedo di 44 cloni di limone. *Essenz. Deriv. Agrum.*, 57, 421–455.

Goretti, G., Nota, G. and Zoccolillo, L. (1967) Impiego di colonne gascromatografiche ad alta risoluzione per la valutazione di olii essenziali di agrumi. *Essenz. Deriv. Agrum.*, 37, 209–220.

Guenther, E. (1949) *The Essential Oils*. D. Van Nostrand Co., New York, N.Y.

Günther, H. (1968) Gas-chromatographic and infrared-spectroscopic studies of lemon oils. *Deutsche Lebensmittel-Rundsch.*, 64, 104–111.

Haggag, E.G., Wahab, S.M.A., El-Zalabany, S.M., Moustafa, E.A.A., El-Kherasy, E.M. and Mabry, T.J. (1988) Volatile oils and pectins from *Citrus aurantifolia* (Lime) and *Citrus limonia* (Lemon). *Asian J. Chem.*, 10, 828–833.

Haro, L. and Faas, W.E. (1985) Comparative study of the essential oils of Key and Persian limes. *Perfum. Flav.*, 10(5), 67–72.

Haro-Guzmán, L. and Huet, R. (1970) L'huile essentielle de lime au Mexique (*Citrus aurantifolia* Swingle). *Fruits*, 25, 887–899.

Haubruge, E., Lognay, G., Marlier, M., Danhier, P., Gilson, J.C. and Gaspar, Ch. (1989) Etude de la toxicite de cinq huiles essentielles extraites de Citrus sp. a l'égard de *Sitophilus zeamais* Motsch (Col., Curculionidae), *Prostephanus truncatus* (Horn) (Col., Bostrychidae) et *Tribolium castaneum* Herbst (Col., Tenebrionidae). *Med. Fac. Landbouww. Rijksuniv. Gent*, 54, 1083–1093.

Hérisset, A., Jolivet, J., Rey, P. and Lavault, M. (1973) Differénciation de quelques huiles essentielles présentant une constitution voisine. Essences de divers *Citrus. Plant. Med. Phytoter.*, 7, 306–318.

Huang, M.-B., Zhang, D.-X., Yan, C.-T. and Yang, S.-B. (1990) Identification and determination of the constituents of essential oil extracted from Fructus Citri aurantii by GC and GC/MS. *Sepu*, 8, 321–325. In B.M. Lawrence, (2000) Progress in essential oils. *Perfum. Flav.*, 15(6), 45–66.

Huang, Y., Wen, M., Xiao, S., Zhao, H., Ren, W., Chen, Q., Liu, X. and Guo, T. (1986) Studies on the chemical constituents of the steam-distilled leaf and peel essential oil from *Citrus bergamia*. *Acta Botanica Yunnanica*, 8, 471–476.

Huang, Y., Wen, M., Xiao, S., Zhao, H., Ren, W., Xiao, S., Chen, Q., Liu, X. and Guo, T. (1987). Chemical constituents of the essential oil from the peel of *Citrus Bergamia* Risso. *Acta Botanica Sinica*, 29, 77–83.

Huet, R. and DuPuis, C. (1968) L'huile essentielle de bergamote en Afrique et en Corse. *Fruits*, 23, 301–311.

Huet, R. and DuPuis, C. (1969) Evolution de la composition chimique de l'huile essentielle de clementine (hybride de *Citrus reticulata* Blanco) et de l'huile essentielle de bergamote (*Citrus aurantium* LIN. Subsp. *Bergamia* Risso et Poiteau Engler) au cours de la croissance du fruit. *La France et ses Parfums*, 12(63), 123–130.

Huet, R. and Murail, M.C. (1972) L'huile essentielle d'órange de type Guineé, *Fruits*, 27, 297–301.

Huet, R., Cassin, J., Tisseau, R., Dalnic, R. and Terrien, A. (1978) Citrons et limes acides. Les limes à gros fruits, production d'avenir pour les DOM-TOM. *EPPOS*, 60, 511–524.

Huet, R. (1981) Etude comparative de l'huile essentielle de bergamote provenant d'Italie, de Corse et de Côte d'Ivoire. *EPPOS*, 63, 310–313.

Huffman, J.W. and Zalkow, L.H. (1973) The structure of paradisiol and its identity to intermedeol. *Tetrahedron Letters*, 10, 751–754.

Hunter, G.L.K. and Brogden, W.B. (1964a) A rapid method for isolation and identification of sesquiterpene hydrocarbons in cold-pressed grapefruit oil. *Anal. Chem.*, 36, 1122–1123.

Hunter, G.L.K. and Brogden, W.B. (1964b) 2,4-p-Menthadiene, a new monoterpene from Valencia orange oil. *J. Org. Chem.*, 29, 498.

Hunter, G.L.K. and Brogden, W.B. (1964c) β-Ylangene, a new sesquiterpene hydrocarbon from orange oil. *J. Org. Chem.*, 29, 2100.

Hunter, G.L.K. and Parks, G.L. (1964) Isolation of β-elemene from orange oil. *J. Food Sci.*, 29, 25.

Hunter, G.L.K. and Brogden, W.B. (1965) Analysis of the terpene and sesquiterpene hydrocarbons in some citrus oils. *J. Food Sci.*, 30, 383–387.

Hunter, G.L.K. and Moshonas, M.G. (1966) Analysis of alcohols in essential oils of grapefruit, lemon, lime and tangerine. *J. Food Sci.*, 31, 167–171.

Hussein, M.M. and Pidel, A.R. (1976) Gas chromatographic analysis of tangerine peel oils and characterization of their geographic origin. 36th *National I.F.T. Meeting*, Paper No. 261, Anaheim, California, June 1976.

Ikeda, R.M., Stanley, W.L., Rolle, L.A. and Vannier, S.H. (1962a) Monoterpene hydrocarbons composition of citrus oils. *J. Food Sci.*, 27, 593–596.

Ikeda, R.M., Stanley, W.L., Vannier, S.H. and Spitler, S.H. (1962b) The monoterpene hydrocarbons of some essential oils. *J. Food Sci.*, 27, 455–458.

Inoma, S., Miyagi, Y. and Akieda, T. (1989) Characterization of citrus oils (mainly orange oil, mandarin oil, tangerine oil). *Kanzei Chuo Bunsekishoho*, 29, 87–97.

Jantan, I., Ahmad, A.S., Amhad, A.R., Mohd Ali, N.A. and Ayop, N. (1996) Chemical composition of some citrus oils from Malaysia. *J. Essent. Oil Res.*, 8, 627–632.

Kekelidze, N.A. and Beradze, L.V. (1974) Essential oils of fruit of some citrus plants growing in the Georgian SSR. *International Congress of Essential Oils*, Paper No. 104, San Francisco, September 1974.

Kekelidze, N.A., Dzhanikashvili, M.I. and Kutateladze, V.V. (1982) Essential oil of the Navogruzinskii variety of lemon. *Maslo-Zhir. Prom-st.*, (2), 25–26.

Kekelidze, N.A., Dzhanikashvili, M.I. and Fishman, G.M. (1985) Essential oil of grapefruit *Citrus paradis* cultivated in Georgian SSR. *Khim. Prir. Soedin*, (1), 119–120.

Kesterson, J.W., Hendrickson, R. and Braddock, R.J. (1971) Florida citrus oils. *Univ. Fl. Inst. Food Agric. Bull.*, 749.

Khurdiya, D.S. and Maheshwari, M.L. (1988) Lime peel oil and its composition. *Pafai J.* (April/June), 25–28.

Kirbaslar, S.I., Kirbaslar, F.G. and Dramur, U. (2000) Volatile constituents of Turkish bergamot oil. *J. Essent. Oil Res.*, 12, 216–220.

Kita, Y., Nakatani, Y., Kobayashi, A. and Yamanishi, T. (1969) Composition of peel oil from *Citrus (nobilis) unshiu*. *Agric. Biol. Chem.*, 33, 1559–1565. In P.E. Shaw, (1979) Review on quantitative analysis of citrus essential oils. *J. Agric. Food Chem.*, 27, 246–257.

Koketsu, M., Magalhaes, M.T., Wilberg, V.C. and Donaliso, M.G.R. (1983) Oleos essenciais de frutos citricos cultivados no Brazil. *Bol. Pesqui EMBRAPA Cent. Technol. Agric. Aliment.*, 7, 3–21.

Kovats, E. (1963) Essential oils. IV. The so called 'distilled' oil of lime (*Citrus medica* L., var. *acida*, Brandis; *Citrus auratifolia* Swingle). *Helvetica Chimica Acta.*, 46, 2705–2731.

Kugler, E. and Kovats, E. (1963) Essential oils. I. The oil of mandarin peels (*Citrus reticulata* Blanco, bzw. *Citrus nobilis var. deliciosa* Swingle, Mandarin). *Helvetica Chimica Acta.*, 46, 1480–1513.

Kumar, U., Ram, B., Pant, A.K., Gupta, K.C. and Brophy, J.J. (1992) Volatile constituents of the distilled leaf and peel oils of *Citrus limon* Burm cv. 'Pant lemon-1'. *J. Essent. Oil Res.*, 4, 643–644.

Kusunose, H. and Sawamura, M. (1980) Aroma constituents of some sour citrus oil. *Nippon Shokuhin Gakkaishi*, 27, 517–521.

Laencina, J., Goretti, G., Melendreras, F.A. and Flores, J. (1986) Kaolin plot columns in *Citrus limon* essential oils analysis. *Anal. Bromatol.*, 37, 315–326.

Lakszner, K. and Szepesy, L. (1988) Application of the O-FID oxygenes analyzer in the cosmetic industry. *Chromatographia*, 26, 91–96.

Lamonica, G., Stagno d'Alcontres, I., Donato, M.G. and Merenda, I. (1990) On the Calabrian bergamot essential oil. *Chimica Oggi*, May, 59–63.

Lancas, F., David, F. and Sandra, P. (1988) CGC analysis of the essential oils of citrus fruits on 100 µm i.d. columns. *J. High Resolut. Chromatogr. and Chromatogr. Comm.*, 11, 73–75.

Lanuzza, F., Micali, G., Currò, P. and Calabrò, G. (1991) On-line HPLC-HRGC coupling in the study of citrus oils: sesquiterpene and paraffin hydrocarbons. *Flav. Fragr. J.*, 6, 29–37.

Lawrence, B.M. (1976–2000) Progress in essential oils. Bimontlhy reviews published on *Perfum. Flav.*

Liberti, A. and Conte, G. (1956) Possibilità di applicazione della cromatografia in fase gassosa allo studio delle essenze. *Atti I° Congresso Internazionale di Studi e Ricerche sulle Essenze*, Reggio Calabria, Italy, Marzo 1956.

Liberti, A. and Cartoni, G.P. (1958) Cromatografia in fase gassosa degli idrocarburi terpenici. *La Ricerca Scientifica*, 28, 1192–1198.

Liberti, A. and Goretti, G. (1974) Moderni criteri sulla valutazione degli olii essenziali: l'olio di bergamotto. *Essenz. Deriv. Agrum.*, 44, 197–208.

Licandro, G., Dugo, G., Lamonica, G. and Cotroneo, A. (1984) Sulla genuinità delle essenze agrumarie. Nota VI. Caratterizzazione di essenze di limone Siciliane (Parte II). *Essenz. Deriv. Agrum.*, 54, 22–48.

Licandro, G., Cotroneo, A. and Trozzi, A. (1987) Sulla genuinità delle essenze agrumarie. Nota XX. Particolarità riscontrate nella composizione di essenze di limone prodotte durante il periodo estivo da frutti provenienti da una zona della provincia di Siracusa. *Essenz. Deriv. Agrum.*, 57, 620–627.

Lifshitz, A., Stanley, W.L. and Stepak, Y. (1970) Comparison of Valencia essential oil from California, Florida and Israel. *J. Food Sci.*, 35, 547–548.

Lin, Z.-K., Hua, Y.-F. and Gu, Y.-H. (1986) The chemical constituents of the essential oil from the flowers, leaves and peels of *Citrus aurantium*. *Acta Botanica Sinica*, 28, 635–640.

Lin, Z.-K. and Hua, Y.-F. (1988) Changes in the volatile flavor constituents of *Citrus sinensis* (L.) Osbeck during storage. *Acta Botanica Sinica*, 30, 623–628.

Lodge, N., Paterson, V.J. and Young, H. (1984) The physico-chemical composition of New Zealand lemon oil. *J. Sci. Food Agric.*, 35, 447–451.

Lund, R.D. and Bryan, W.L. (1976) Composition of lemon oil distilled from commercial mill waste. *J. Food Sci.*, 41, 1194–1197.

MacLeod, A.J., MacLeod, G. and Subramanian, G. (1988) Volatile aroma constituents of orange. *Phytochemistry*, 27, 2185–2188.

MacLeod, W.D. and Buigues, N.M. (1964) Sesquiterpenes. I. Nootkatone, a new grapefruit flavor costituent. *J. Food Sci.*, 29, 565–568.

MacLeod, W.D., Mc Fadden, W.H. and Buigues, N.M. (1965) Lemon oil analysis. II. Gas-liquid chromatography on a temperature-programmed, long, open tubular column. *J. Food. Sci.*, 30, 591–594.

MacLeod, W.D. (1968) Rapid analysis of natural essences by combined flow and temperature-programmed capillary gas chromatography. *J. Agric. Food Chem.*, 16, 884–886.

Maekawa, K., Kodama, M., Kushii, M. and Mitamura, M. (1967) Essential oils of some orange peels. *Agr. Biol. Chem.*, 31, 373–377.

Majláth, P. and Takács, J. (1969) Gas-chromatographic analysis of the lemon oil. *Herba Hungarica*, 8, 181–188.

Mazza, G. (1986) Etude sur la composition aromatique de l'huile essentielle de bergamote (*Citrus aurantium* subsp. *Bergamia* Risso et Poiteau Engler) par chromatographie gazeuse et spectrometrie de masse. *J. Chromatogr.*, 362, 87–99.

Mazza, G. (1987a) Gas-chromatography and mass spectrometry study on the aromatic composition of mandarin essential oil. *Sci. Alim.*, 7, 459–479.

Mazza, G. (1987b) Identificazione di nuovi composti negli olii agrumari mediante GC/MS. *Essenz. Deriv. Agrum.*, 57, 19–33.

McFadden, W.H., Teranishi, R., Black, D.R. and Day, J.C. (1963) Use of capillary gas-chromatography with time-of-flight mass spectrometer. *J. Food Sci.*, 28, 316–319.

McHale, D. (1980) Effect of processing conditions on composition of distilled oil of lime. *VIII International Congress of Essential Oils*, Cannes, France, October 1980.

Mehlitz, A. and Minas, Th. (1965) Orange peel oil. III. Oxigen-containing components of orange peel oil by gas chromatography. *Riechstoffe, Aromen, Koerperpflegemittel*, 15, 365–374.

Melendreras, F.A., Laencina, J., Flores, J. and Guzmán, G. (1985) Aceites esenciales de frutos de variedades de limonero (*Citrus limon* L. Burm. f.). *Rev. Agroquím Tecnol. Aliment.*, 25, 133–143.

Melendreras, F.A., Laencina, J., Flores, J. and Guzmán, G. (1986) Influencia del portainjerto sobre el aceite esencial de frutos de limonero Verna. *Agrochimica*, 30, 311–324.

Melendreras, F.A., Laencina, J., Flores, J. and Guzmán, G. (1988) Efecto de la madera intermedia sobre el aceite esencial de limon Verna. *Agrochimica*, 32, 1–15.

Micali, G., Lanuzza, F., Currò, P. and Calabrò, G. (1990) Separation of alkanes in citrus essential oils by on-line coupled high performance liquid chromatography–high-resolution gas chromatography. *J. Chromatogr.*, 514, 317–324.

Mondello, L., Bartle, K.D., Dugo, G. and Dugo, P. (1994a) Automated HPLC-HRGC: a powerful method for essential oil analysis. Part III. Aliphatic and terpene aldehydes of orange oil. *J. High Resolut. Chromatogr.*, 17, 312–314.

Mondello, L., Bartle, K.D., Dugo, P., Gans, P. and Dugo, G. (1994b) Automated LC-GC: a powerful method for essential oils analysis. Part IV. Coupled LC-GC-MS (ITD) for bergamot oil analysis. *J. Microcol. Sep.*, 6, 237–244.

Mondello, L., Dugo, P., Bartle, K.D., Frere, B. and Dugo, G. (1994c) On-line high performance liquid chromatography coupled with high resolution gas chromatography and mass spectrometry (HPLC-HRGC-MS) for the analysis of complex mixtures containing highly volatile compounds. *Chromatographia*, 39, 529–538.

Mondello, L., Dugo, P., Basile, A., Dugo, G. and Bartle, K.D. (1995a) Interactive use of linear retention indices, on polar and apolar columns, with a MS-library for reliable identification of complex mixtures. *J. Microcol. Sep.*, 7, 581–591.

Mondello, L., Dugo, P., Bartle, K.D., Dugo, G. and Cotroneo, A. (1995b) Automated HPLC-HRGC: a powerful method for essential oils analysis. Part V. Identification of terpene hydrocarbons of bergamot, lemon, mandarin, sweet orange, bitter orange, grapefruit, clementine and Mexican lime oils by coupled HPLC-HRGC-MS (ITD). *Flav. Fragr. J.*, 10, 33–42.

Mondello, L., Dugo, P., Dugo, G. and Bartle, K.D. (1996) On-line HPLC-HRGC-MS for the analysis of natural complex mixtures. *J. Chromatogr. Sci.*, 34, 174–181.

Mondello, L., Catalfamo, M., Dugo, P. and Dugo, G. (1998a) Multidimensional capillary GC-GC for the analysis of real complex samples. Part II. Enantiomeric distribution of monoterpene hydrocarbons and monoterpene alcohols of cold-pressed and distilled lime oils. *J. Microcol. Sep.*, 10, 203–212.

Mondello, L., Catalfamo, M., Proteggente, A.R., Bonaccorsi, I. and Dugo, G. (1998b) Multidimensional capillary GC-GC for the analysis of real complex samples. 3. Enantiomeric distribution of monoterpene hydrocarbons and monoterpene alcohols of mandarin oils. *J. Agric. Food Chem.*, 46, 54–61.

Mondello, L., Zappia, G., Errante, G. and Dugo, G. (2000) Fast GC and fast GC/MS for the analysis of natural complex matrices. *LC-GC Europe*, 13, 495–502.

Mookherjee, B.D. (1969) Occurrence of bifunctional monoterpene compounds in bergamot oil. *158th Amer. Chem. Soc. Meeting*, Paper No. 37. In B.M. Lawrence (1979) Progress in essential oils. *Perfum. Flav.*, 4(3), 50–52.

Mosandl, A. (1995) Enantioselective capillary gas chromatography and stable isotope ratio mass spectrometry in the authenticity control of flavors and essential oils. *Food Rev. Int.*, 11, 597–664.

Moshonas, M.G. (1971) Analysis of carbonyl flavor constituents from grapefruit oil. *J. Agric. Food Chem.*, 19, 769–770.

Moshonas, M.G. and Shaw, P.E. (1974) Quantitative and qualitative analysis of tangerine peel oil. *J. Agric. Food Chem.*, 22, 282–284.

Moshonas, M.G. and Shaw, P.E. (1979) Isolation of *trans, trans*-2,4-decadienal and intermedeol from cold-pressed citrus oils. *J. Agric. Food Chem.*, 27, 210–211.

Moshonas, M.G. and Shaw, P.E. (1980) Characterization of new Citrus components, trans, trans-α-farnesene. Isolation of α-farnesene isomers from dehydration of farnesol. *J. Agric. Food Chem.*, 28, 680–681.

Munari, F., Dugo, G. and Cotroneo, A. (1990) Automated on-line HPLC-HRGC with gradient elution and multiple GC transfer applied to the characterization of citrus essential oils. *J. High Resolut. Chromatogr.*, 13, 56–61.

Myers, F.S. (1988) Unpublished work. In R.L. Swaine and R.L Swaine (1988) Citrus oils: processing, technology and application. *Perfum. Flav.*, 13(6), 2–20.

Nishida, R. and Acree, T.E. (1984) Isolation and characterization of methyl epijasmonate from lemon (*Citrus limon* Burm). *J. Agric. Food Chem.*, 32, 1001–1003.

Njoroge, S.M., Ukeda, H., Kusunose, H. and Sawamura, M. (1994a) Volatile components of the essential oils from Kabosu, Daidai and Yuko sour *Citrus* fruits. *Flav. Fragr. J.*, 9, 289–297.

Njoroge, S.M., Ukeda, H., Kusunose, H. and Sawamura, M. (1994b) Volatile components of japanese Yuzu and lemon oils. *Flav. Fragr. J.*, 9, 159–166.

Njoroge, S.M., Ukeda, H., Kusunose, H. and Sawamura, M. (1996) Japanese sour *Citrus* fruits. Part IV. Volatile compounds of Naoschichi and Tahiti lime essential oils. *Flav. Fragr. J.*, 11, 25–29.

Ohloff, G., Giersch, W., Näf, R. and Delay, F. (1986) The absolute configuration of β-bisabolol. *Helvetica Chimica Acta*, 69, 698–703.

Ortiz, J.M., Kumamoto, J. and Scora, R.W. (1978) Possible relationship among sour oranges by analysis of their essential oils. *Internat. Flav. Food Addit.*, 9, 224–226.

Owusu-Yaw, J., Matthews, R.F. and West, P.F. (1986) Alcohol deterpenation of orange oil. *J. Food Sci.*, 51, 1180–1182.

Pennisi, L. and Di Giacomo, A. (1965) Contributo alla conoscenza delle essenze di limone di alcune cultivar italiane. *EPPOS*, 47, 370–377.

Pino, J., Sanchez, M., Sanchez, R. and Roncal, E. (1992) Chemical composition of orange oil concentrates. *Die Nahrung*, 36, 539–542.

Prager, M.J. and Miskiewicz, M.A. (1982) Gas chromatographic-mass spectrometric analysis, identification and detection of adulteration of natural and concentrated lemon oils. *J. Assoc. Off. Anal. Chem.*, 65, 166–171.

Protopapadakis, E. and Papanikolaou, X. (1998) Characterization of *Citrus aurantium* and *C. Taiwanica* rootstoks by isoenzyme and essential oil analysis. *J. Hort. Sci. Biotechnol.*, 73, 81–85.

Ricciardi, A.I.A., Agrelo de Nassif, A.E., Olivetti de Bravi, M.G., Peruchena de Godoy, N.M. and Moll, E. (1982) *SAIPA*, 8–13. In R.W. Dresher, H.N. Beñatena, G.G.T. De Rossin, G.O. Ubiergo and J.A. Retamar (1984) Evaluaciòn de le calidad perfumistica del aceite esencial de *Citrus bergamia* Risso (Bergamota) cultivada en la E.E.A. Inta Concordia (Entre Rios). *Essenz. Deriv. Agrum.*, 54, 192–199.

Rispoli, G. and Di Giacomo, A. (1965) Studio dei composti sesquiterpenici dell'essenza di limone siciliana. *EPPOS*, 47, 650–655.

Ruberto, G., Biondi, D., Piattelli, M., Rapisarda, P. and Starrantino, A. (1993) Profiles of essential oils of new citrus hybrids. *Flav. Fragr. J.*, 8, 179–184.

Ruberto, G., Biondi, D., Rapisarda, P., Renda, A. and Starrantino, A. (1997) Essential oil of Cami, a new citrus hybrid. *J. Agric. Food Chem.*, 45, 3206–3210.

Sanchez, J.L., Calvarano, M. and Micali, B. (1980) El aceite esencial de limon Verna. *Essenz. Deriv. Agrum.*, 50, 311–323.

Sattar, A., Mahmud, S. and Khan, S.A. (1986) Citrus oils. Part I. Composition of the monoterpenes of the peel oils of oranges, Kinnows and lemons. *Pakistan J. Sci. Ind. Res.*, 29(3), 196–198.

Sawada, M. and Yamada, T. (1997) The analysis and sensory evaluation of lemon oils prepared by three extraction methods. *Nippon Shokuhin Kagaku Kogaku Kaishi.*, 44, 243–247.

Sawada, M. and Yamada, T. (1998) The analysis of oxygenated compounds fraction of lemon oils prepared by different extraction methods and application to a beverage. *Nippon Shokuhin Kagaku Kogaku Kaishi.*, 45, 134–143.

Sawamura, M., Kuwahara, S., Shichiri, K. and Aoki, T. (1990) Volatile constituents of several varieties of pummelos and comparison of the nootkatone levels in pummelos and other citrus fruits. *Agric. Biol. Chem.*, 55, 803–805.

Sawamura, M., Shichiri, K., Ootani, Y. and Zheng, X.H. (1991) Volatile constituents of several varieties of pummelos and characteristics among citrus species. *Agric. Biol. Chem.*, 55, 2571–2578.

Sawamura, M., Sun, S.H., Oraki, K., Ishikawa, J. and Ukeda, H. (1999) Inhibitory effects of citrus essential oils and their components on the formation of N-nitrosodimethylamina. *J. Agric. Food Chem.*, 47, 4868–4872.

Schenk, H.P. and Lamparsky, D. (1981) Analysis of citrus oils, especially bergamot oil. *Seifen, Öle, Fette, Wachse*, 107, 363–369.

Schulz, H., Albroscheit, G. and Nowak, D. (1992) Characterization of grapefruit oil and juice by HPLC. *Z. Lebensm. Unters Forsch.*, 195, 254–258.

Scora, R.W., Cameron, J.W. and England, A.B. (1968) Rind oil components of intergeneric *Citrus-Poncirus* hybrids and their parents. *Proc. Am. Soc. Hortc. Sci.*, 92, 221–226.

Sebastiani, E., Dugo, G. and Cotroneo, A. (1983) Sulla genuinità delle essenze agrumarie. Nota V. Valutazione di alcuni tipi di fasi stazionarie per l'analisi della frazione volatile degli olii essenziali di limone mediante gascromatografia ad alta risoluzione. *Essenz. Deriv. Agrum.*, 53, 501–514.

Shaw, P.E., Coleman, R.L. and Moshonas, M.G. (1971) Quantitative analysis of Persian lime oil. *Proc. Fla. State Hort. Soc.*, 84, 187–189.

Shaw, P.E. and Coleman, R.L. (1974) Quantitative composition of cold-pressed orange oils. *J. Agric. Food Chem.*, 22, 785–787.

Shaw, P.E. and Coleman, R.L. (1975) Composition and flavour evaluation of a volatile fraction from cold-pressed Valencia orange oil. *Flavour*, (May/June), 190.

Shaw, P.E. and Wilson, C.W. (1976) Comparison of extracted peel oil composition and juice flavor for Rough lemon, Persian lime and a lemon-lime cross. *J. Agric. Food Chem.*, 24, 664–666.

Shaw, P.E. (1977) Essential oils. In S. Nagy, P.E. Shaw and M.K. Veldhuis (eds), *Citrus Science and Technology*. The Avi publishing Company, Westport, Connecticut, Vol. I, pp. 427–462.

Shaw, P.E. (1979) Review of quantitative analyses of citrus essential oils. *J. Agric. Food Chem.*, 27, 246–257.

Slater, C.A. (1961) Composition of natural lime oil. *Chem. Ind.*, 17, 833–835.

Soulari, M. and Fanghänel, E. (1971) Analisis parcial del aceite esencial de naranja agria (*Citrus aurantium* L.) producido en Cuba. *Revista CENIC*, 3, 125–128.

Stanley, W.L., Ikeda, R.M., Vannier, S.H. and Rolle, L.A. (1961a) Determination of the relative concentrations of the major aldehydes in lemon, orange and grapefruit oils by gas chromatography. *J. Food Sci.*, 26, 43–48.

Stanley, W.L., Ikeda, R.M. and Cook, S. (1961b) Hydrocarbon composition of lemon oils and its relationship to optical rotation. *J. Food Technol.*, 15, 381–385.

Staroscik, J.A. and Wilson, A.A. (1982a) Quantitative analysis of cold-pressed lemon oil by glass capillary gas chromatography. *J. Agric. Food Chem.*, 30, 507–509.

Staroscik, J.A. and Wilson, A.A. (1982b) Seasonal and regional variation in the quantitative composition of cold-pressed lemon oil from California and Arizona. *J. Agric. Food Chem.*, 30, 835–837.

Straus, D.A. and Wolstromer, R.J. (1874) The examination of various essential oils. *Sixth Internatiol Congress of Essential Oils*, Paper No. 94, San Francisco, September 1974. In B.M. Lawrence (1978) Progress in essential oils, *Perfum. Flav.*, 3(6), 54–58.

Sugisawa, H., Yang, R.-H, Matsuo, T. and Tamura, H. (1987) Volatile compounds in peel oil of Shiroyanagi navel. *Nippon Nogeikagaku Kaishi*, 61, 1101–1106.

Sugisawa, H., Yamamoto, M., Tamura, H. and Takagi, N. (1989) The comparison of volatile components in peel oil from four species of Navel oranges. *Nippon Shokuin Kogyo Gakkaishi*, 36, 455–462.

Sugiyama, K. and Saito, M. (1988) Simple microscale supercritical fluid extraction system and its application to gas chromatography-mass spectrometry of lemon peel oil. *J. Chromatogr.*, 442, 121–131.

Sulser, H., Scherer, J.R. and Stevens, K.L. (1971) The structure of paradisiol, a new sesquiterpene alcohol from grapefruit oil. *J. Org. Chem.*, 36, 2422–2426.

Sun, D. and Petracek, P.D. (1999) Grapefruit gland oil composition is affected by wax application, storage temperature and storage time. *J. Agric. Food Chem.*, 47, 2067–2069.

Supelco Bull. N.776 (1978) GC Analysis of citrus oils. Bellefonte, PA, USA. In B.M. Lawrence (1978) Progress in essential oils, *Perfum. Flav.*, 3(6), 54–58.

Teranishi, R., Schultz, T.H., McFadden, W.H., Lundin, R.E. and Black, D.R. (1963) Volatiles from oranges. I. Hydrocarbons. Identified by infrared, nuclear magnetic resonance and mass spectra. *J. Food Sci.*, 28, 541–545.

Thomas, A.F. and Bassols, F. (1992) Occurrence of pyridines and other bases in orange oils. *J. Agric. Food Chem.*, 40, 2236–2243.

Trozzi, A., Verzera, A., Del Duce, R. and Cotroneo, A. (1993) Sulla genuinità delle essenze agrumarie. Nota XLIV. Aggiornamento sulla composizione della composizione volatile dell'olio essenziale di limone. *Essenz. Deriv. Agrum.*, 63, 375–394.

Trozzi, A., Verzera, A. and Lamonica, G. (1999) Essential oil composition of *Citrus sinensis* (L.) Osbeck cv. Maltese. *J. Essent. Oil Res.*, 11, 482–488.

Tuzcu, O., Neubeller, J. and Buchloh, G. (1985) Essential oil contents in the rinds eastern Mediterranean sour oranges (*Citrus aurantium* L.). *Doga Bilim Dergisi*, 9(1), 34–39.

Uchida, K., Kobayashi, A. and Yamanishi, T. (1984) Composition of oxigenated compounds in the peel oil of Fukuhara oranges. *Nippon Nogeikagaku Kaishi*, 58, 691–694.

Usai, M., Arras, G. and Fronteddu, F. (1992) Effects of cold storage on essential oils of peel of Thompson navel oranges. *J. Agric. Food Chem.*, 40, 271–275.

Usai, M., Picci, V. and Arras, G. (1996) Influence of cultivar on the composition of lemon peel oil. *J. Essent. Oil Res.*, 8, 149–158.

Veldhuis, M.K. and Hunter, G.L.K. (1968) Nomenclature of ylangene, copaene and cubebene. *J. Food Sci.*, 32, 697.

Vernin, G. and Vernin, G. (1966) Détection et évaluation de l'anthranilate de méthyle et de ses dérivés méthyles dans différents échantillons naturels et de syntése par C.C.M. et G.L.C. *France et ses Parfumes*, 9, 429–448.

Verzera, A., Trozzi, A., Stagno d'Alcontres, I. and Cotroneo, A. (1996a) On the genuineness of citrus essential oils. Part XLVIII. The composition of volatile fraction of some varieties of sweet orange oils. *J. Essent. Oil Res.*, 8, 159–170.

Verzera, A., Lamonica, G., Mondello, L., Trozzi, A. and Dugo, G. (1996b) The composition of bergamot oil. *Perfum. Flav.*, 21(6), 19–34.

Verzera, A., Trozzi, A., Licandro, G. and Scuderi, L. (1996c) Sulla genuinità delle essenze agrumarie. Nota LVI. La composizione dell'olio essenziale di limone FMC prodotto nella stagione 1995. *Essenz. Deriv. Agrum.*, 56, 353–367.

Verzera, A., Proteggente, A.R. and Lamonica, G. (1997) Sulla genuinità delle essenze agrumarie. Nota LV. La composizione dell'olio essenziale di mandarino prodotto nella stagione 1996/97. *Essenz. Deriv. Agrum.*, 67, 154–170.

Verzera, A., Trozzi, A., Stagno d'Alcontres, I., Mondello, L., Dugo, G. and Sebastiani, E. (1998) The composition of the volatile fraction of calabrian bergamot essential oil. *EPPOS*, (25), 17–38.

Vora, J.D., Matthews, R.F., Crandall, P.G. and Cook, R. (1983) Preparation and chemical composition of orange oil concentrates. *J. Food Sci.*, 48, 1197–1199.

Wen, M., Xiao, S., Zhao, H., Ren, W. and Huang, Y. (1989) A study on the chemical components of the essential oil from the peel of *Citrus limon* (L.) Burm f. *Tianran Chanwu Yanjiu Yu Kaifa*, 1(2), 11–17.

Wenninger, J.A., Yates, R.L. and Dolinsky, M. (1967) High resolution infrared spectra of some naturally occurring sesquiterpene hydrocarbons. *J. Ass. Off. Agric. Chem.*, 50, 1313–1335.

Wilson, C.W. and Shaw, P.E. (1978) Quantitative composition of cold-pressed grapefruit oil. *J. Agric. Food Chem.*, 26, 1432–1434.

Wilson, C.W. and Shaw, P.E. (1980) Glass capillary gas chromatography for quantitative determination of volatile constituents in cold-pressed grapefruit oil. *J. Agric. Food Chem.*, 28, 919–922.

Wilson, C.W. and Shaw, P.E. (1981) Importance of thymol, methyl N-methylanthranilate and monoterpene hydrocarbons to the aroma and flavor of mandarin cold-pressed oils. *J. Agric. Food Chem.*, 29, 494–496.

Wilson, C.W., Shaw, P.E. and Berry, R.E. (1981) Analysis of cold-pressed grapefruit oil by glass capillary gas chromatography. In B.D. Mookherjee and C.J. Mussian (eds), *Essential Oils*, Allured Pub. Co., Wheaton, IL. 1981.

Wilson, C.W. and Shaw, P.E. (1984) Quantitation of individual and total aldehydes in citrus cold-pressed oils by fused silica capillary gas chromatography. *J. Agric. Food Chem.*, 32, 399–401.

Wilson, C.W., Shaw, P.E., McDonald, R.E., Greany, P.D. and Yokoyama, H. (1990) Effect of gibberellic acid and 2-(3,4-dichlorophenoxy) triethylamina on nootkatone in grapefruit peel oil and total peel oil content. *J. Agric. Food Chem.*, 38, 656–659.

Yamanishi, T., Kobayashi, A., Mikumo, Y., Nakasone, Y., Kita, M. and Hattori, T. (1968) Composition of peel oil from *Citrus unshiu*. *Agric. Biol. Chem.*, 32, 593–598.

Yamauchi, Y. and Saito, M. (1990) Fractionation of lemon-peel oil by semi-preparative supercritical fluid chromatography. *J. Chromatogr.*, 505, 237–246.

Yang, R.-H, Sugisawa, H., Nakatani, H., Tamura, H. and Takagi, N. (1992) Comparison of odor quality in peel oils of acid citrus. *Nippon Shokuhin Kogyo Gakkaishi*, 39, 16–24.

Yang, S., Wang, Q., Zhang, D., Huang, M. and Zhang, D (1992) Constituents of essential oil extracted from Seville orange (*Citrus aurantium*) and its processed products. *Zhongcaoyao*, 23, 14–15.

Yokoyama, F., Levi, L., Laughton, P.M. and Stanley, W.L. (1961) Determination of citral in citrus extracts and citrus oils by conventional and modern chemical methods of analysis. *J. Assoc. Off. Agric. Chem.*, 44, 647–653.

Yoshida, J., Ikawa, S. and Yasunaga, K. (1971) On the production of bergamot oil in the district of the Seto Inland Sea. *Jap. J. Trop. Agric.*, 14, 199–203.

Zani, F., Massimo, G., Benvenuti, S., Bianchi, A., Albasini, A., Melegari, M., Vampa, G., Bellotti, A. and Mazza, P. (1991) Studies on the genotoxic properties of essential oils with *Bacillus subtilis rec.* Assay and *Salmonella* microsome reversion assay. *Planta Medica*, 57, 237–241.

Zhu, L.F., Li, H.-Y., Li, B.-L., Lu, B.-Y. and Zhang, W.-L. (1995) *Aromatic plants and essential constituents* (supplement 1). Peace Book Co. Ltd., Hong Kong.

Ziegler, E. (1971) The examination of citrus oils. *Flavour Ind.* (November), 647–653.

14 The oil composition of less common *Citrus* species

Brian M. Lawrence

INTRODUCTION

Although numerous *Citrus* species, varieties, cultivars and hybrids are grown for a variety of reasons, a few of the less common species are somewhat popular as a source of rare oils or as fruit. The species that will be discussed are as follows: *Citrus aurantium* L. var. *myrtifolia* Ker-Gawl (Chinotto), *C. clementina* Hort. ex Tanaka (clementine), *C. grandis* (L.) Osbeck (pummelo or shaddock), *C. hystrix* DC (combava), *C. jambhiri* Lush. (rough lemon), C. *junos* Sieb. ex Tanaka (yuzu), *C. madurensis* Lour. (calamandin), *C. medica* L. (citron etrog and fingered citron) and *C. unshiu* Marcovitch (mikan).

Each of the species above was reviewed. A summary of the pertinent literature was conducted by species and can be found in the following section of this chapter. To simplify the quantitative data, all results were rounded up or down to one decimal place only, as most of the literature data were found as area per cent, without the use of an internal standard. All data designated as 0.04 per cent or less were grouped together with all other low level identifications as trace constituents (tr). Rather than report on each of the analyses by listing all analytical data from previously published studies separately, it was decided to either combine the data into one list or present it in tabular form. If a component was only found in an amount of 0.1 per cent in only one analysis while it was not found in other analyses, it was listed as 0–0.1 per cent (as shown for the β-phellandrene content of *C. clementina*). Throughout the review, if a compound appeared to this reviewer to be a misidentification, an asterisk (*) was placed against it. This means that the compound identification in that particular oil is in question and needs confirmation before acceptance by the scientific community.

Each of the nine species will be treated separately. Also, the use of *cis*- and *trans*-linalool oxide in the analysis described will refer to the furanoid form unless otherwise stated.

OILS COMPOSITION

Citrus aurantium L. var. *myrtifolia* Ker-Gawl

C. aurantium var. *myrtifolia*, which is known as myrtle-leafed orange or chinotto, is an important sour orange with small fruit. It has been classified by Tanaka (1961) to be in the Section Aurantium, subsection Aurantioides, and Racemosa group. Both the leaf and peel oils have been the subject of limited study.

Peel oil

The composition of the peel oil of *C. aurantium* var. *myrtifolia* has been studied by Ortiz *et al.* (1978), Chialva and Doglia (1990) and Verzera *et al.* (1991). A summary of the compounds found in the oil can be seen in Table 14.1. The same authors listed above also found trace amounts of dehydro-1,8-cineole, 9-decenol, (E)-2-hexenal, hexyl acetate, α-fenchyl alcohol, piperitone, hexyl hexanoate, *cis*-dihydrocarvone, (E)-2-decenal, undecanal, nonyl acetate, α-terpinyl acetate, bornyl acetate, 2,4-decadienal, and methyl eugenol in one or more of the chinotto peel oil studied.

Leaf oil

The leaf oil composition, which has been studied by Ortiz *et al.* (1978) and Huang *et al.* (2000), is reported in Table 14.1.

Huang *et al.* also found trace amounts of linalool oxide and β-farnesene in the leaf oil but the correct isomers of these two components were not reported.

Citrus clementina Hort. ex Tanaka

C. clementina, which is commonly known as clementine, is a member of the section Acrumen, subsection Microacrumen, group Citriodora, and subgroup Megacarpa (Tanaka, 1961). Numerous cultivars of clementine are grown mainly in Spain and Italy, as well as several North African Mediterranean countries. Although the leaf oil has not been the subject of intense study, a number of relevant studies on the peel oil have been reported.

Peel oil

The composition of the peel oil of *C. clementina* has been determined by Calvarano *et al.* (1974), Dugo *et al.* (1988), Baaliouamer *et al.* (1992), Ruberto *et al.* (1993), Ruberto *et al.* (1994), Mondello *et al.* (1995), Ruberto *et al.* (1997), Verzera *et al.* (1997), Verzera *et al.* (1998), Gazea *et al.* (1998) and Ruberto *et al.* (1999). A summary of the components found in the oil can be seen in Table 14.1.

The authors listed above also found trace amounts of heptyl acetate, decanol, perillyl alcohol, nonyl acetate, undecanal, (E)-2-decenal, *trans*-caren-3-ol*, β-gurjunene, (E)-2-dodecenal, isopiperitenone, bicyclogermacrene, (E,E)-2,4-decadienal, γ-muurolene, α-muurolene, tridecanal, tetradecanal and tetradecanol in one or more of the clementine peel oil studies.

Leaf oil

In 1983, Ortiz Marcide *et al.* analysed the leaf oil obtained from the following cultivars: 'Arrufatina,' 'Clemenules,' 'Clemenvilla,' 'Esbal,' 'Fina,' 'Guillermina,' 'Hernandina,' 'Monreal,' 'Oroval' and 'Tomatera.' More recently, Huang *et al.* (2000) also analysed the leaf oil of *C. clementina*. A summary of their results can be seen in Table 14.2.

Table 14.1 Percentage composition of the peel oils of seven less common *Citrus* species

Compound	Chin	Clem	Gran	Hyst	Juno	Medi	Unsh
ethanol	–	–	–	–	–	–	0–0.1
ethyl acetate	–	–	–	–	–	–	0–0.1
(Z)-3-hexenal	–	tr	–	–	0–0.7	–	–
hexanol	tr	–	–	–	0–0.4	–	tr
α-thujene	tr	tr	tr	0.1–0.2	0–0.3	tr–1.0	0.1–0.3
α-pinene	0.2–0.5	0.1–2.0	0.2–1.7	2.0–3.2	1.0–1.8	0.1–2.3	0.1–1.3
camphene	tr	tr–0.3	0–0.1	0.1–0.2	tr	tr–0.3	tr
6-methyl-5-hepten-2-one	tr	–	–	–	tr	0–0.2	tr
sabinene	0.2–0.4	0.1–1.1	tr–6.2	13.0–25.9	tr–0.2	0–2.4	tr–0.2
β-pinene	0.4–1.6	tr–0.3	tr–11.3	20.4–42.2	0.3–0.7	0.2–14.8	0.2–2.3
myrcene	1.3–2.0	1.5–2.1	1.2–4.4	1.3–1.6	1.4–2.2	0.4–1.6	0.9–2.0
δ-2-carene	–	–	0–0.2	–	–	–	–
octanal	0.1–0.4	tr–0.6	0–0.4	tr	tr	0–0.1	0–0.2
α-phellandrene	0–0.1	tr–0.1	0–0.1	0–0.1	tr–0.5	tr–0.2	tr
δ-3-carene	–	tr–0.1	tr	0–1.4	–	0.01	tr
α-terpinene	tr	tr	0–0.3	0–0.4	tr–0.2	tr–4.2	tr–0.4
p-cymene	tr	tr–0.2	tr–0.4	0–0.3	tr–0.4	tr–3.3	tr–26.9
limonene	80.1–95.8	83.0–95.0	48.9–95.6	2.8–14.2	60.4–82.4	51.2–93.6	41.2–90.7
1,8-cineole	–	–	0–2.4	tr	0–8.3	tr	0–1.6
(Z)-β-ocimene	tr	tr	tr	tr	tr	tr–1.5	–
β-phellandrene	0–0.3	0–0.1	–	0–0.8	tr–2.5	–	0–0.3
(E)-β-ocimene	0.2–0.3	tr–0.1	0–0.1	0–0.1	–	0–2.1	0–0.9
isoamyl butyrate	–	–	–	–	–	–	0–0.1
γ-terpinene	tr–0.5	tr–2.1	tr–6.9	0.9–2.4	7.6–10.7	tr–26.2	0.3–5.7
octanol	tr	tr–0.1	–	–	tr	0–0.1	0–0.1
cis-sabinene hydrate	tr	0–0.1	tr	0–0.3	tr	0–0.1	–
cis-linalool oxide	0–0.4	–	0–0.3	0.3–1.9	–	–	tr
terpinolene	tr	tr–0.2	0–0.7	1.6–3.0	0–0.5	0.3–1.1	tr–1.2
trans-linalool oxide	0–0.2	–	tr	0–0.9	tr	–	tr
trans-sabinene hydrate	tr	tr	tr	0–1.0	tr	0–0.1	–

linalool	0.1–5.5	0.2–1.5	tr–0.7	1.9–2.8	0.9–5.6	0.1–4.8	0.1–3.3
nonanal	tr	–	0–0.1	–	tr–0.1	tr–0.2	0–0.1
cis-limonene oxide	tr	tr	0–0.1	0–0.4	tr	tr	tr
trans-limonene oxide	tr	tr	0–0.2	tr	tr	–	%
β-terpineol	–	–	0–0.1	0–0.3	0–0.2	–	–
isopulegol	–	0–0.1	–	0.3–2.5	0–0.1	–	tr
citronellal	tr–0.1	tr–0.2	0–0.2	3.4–16.8	tr	tr–0.2	tr–0.1
nonanol	–	0–0.1	–	–	0–0.2	–	0–0.1
menthol	–	–	–	0–0.4	–	–	–
terpinen-4-ol	tr–0.3	tr–0.1	tr–0.1	3.8–8.9	tr	tr–3.9	tr–0.2
naphthalene	–	–	–	–	–	–	0–0.1
α-terpineol	tr–1.3	tr–0.2	tr–1.0	1.7–7.4	0.1–0.9	tr–3.9	tr–0.6
cis-piperitol	–	–	–	0–0.2	–	–	–
decanal	0.1–0.2	0.2–0.5	0–0.6	tr	tr	0–0.1	tr–0.4
octyl acetate	0–0.1	tr	0–0.4	–	tr	0–0.1	tr–0.1
limonen-4-ol	–	–	–	–	tr	0–0.6	–
citronellol	tr–0.1	tr	tr	1.7–3.0	tr	0–0.1	tr–0.3
nerol	tr	tr–0.1	0–0.3	–	tr	0–0.6	–
cis-carveol	tr	tr	tr	–	tr–0.1	–	tr
limonen-10-ol	–	–	–	–	tr	–	0–0.1
trans-carveol	0–0.1	–	–	–	–	–	–
methyl thymol	–	–	–	–	tr	–	tr
neral	tr–0.1	tr–0.1	0–0.3	tr	tr	0.4–7.4	tr–0.1
carvone	–	0–0.1	tr	–	tr	0–1.1	tr
geraniol	tr–0.6	tr	0–0.2	0–0.7	tr	0–0.7	0–0.1
linalyl acetate	0.2–0.3	–	–	–	tr	0–5.4	–
geranial	0.1–0.3	tr–0.2	tr–0.8	tr	tr	0.7–13.3	tr
perillaldehyde	0–0.1	tr	tr	–	tr	0–0.1	0–0.1
guaiacol	–	–	–	–	0–1.5	0–0.4	–
bicycloelemene	–	–	–	–	tr–0.3	–	–
δ-elemene	–	tr	tr	tr	–	–	tr–0.2
α-terpinyl acetate	–	tr	–	–	–	0–0.1	tr
α-cubebene	–	tr–0.1	–	0–0.1	tr	–	0–0.3

Table 14.1 (Continued)

Compound	Chin	Clem	Gran	Hyst	Juno	Medi	Unsh
citronellyl acetate	tr	tr	tr	0.4–1.5	tr	0–2.2	tr–0.1
citronellic acid	–	–	–	0–0.3	–	–	–
neryl acetate	tr–0.1	tr	0–0.1	0–0.1	0–0.3	–	0–0.2
α-copaene	–	tr	0–0.4	0–0.1	0–0.3	–	0–0.2
geranyl acetate	0.1–0.2	tr	0–0.3	0.4–0.5	tr	tr–0.9	tr–1.1
β-cubebene	–	tr	0–1.7	0–0.1	–	0–0.3	–
β-elemene	–	tr	0–0.1	0–0.3	0–0.4	tr	0–3.4
methyl N-methyl anthranilate	–	0–0.1	–	–	–	–	–
dodecanal	tr–0.1	0–0.1	tr	–	0–0.2	–	0–0.1
decyl acetate	tr–0.1	–	0–0.1	–	tr	–	tr
β-caryophyllene	tr–0.1	tr	tr–0.2	0–0.4	0.1–0.3	0.1–0.6	tr–1.2
β-copaene	–	–	0–0.2	–	tr	–	0–1.2
γ-elemene	–	–	–	tr	tr–0.5	–	–
β-bisabolene	tr–0.1	–	–	–	–	–	–
trans-α-bergamotene	–	tr	–	–	–	0.1–0.6	–
(Z)-β-farnesene	–	0–0.1	0–0.3	0–0.2	–	tr	–
citronellyl propionate	–	0–0.1	–	–	–	–	–
α-humulene	tr	tr	0–0.1	0–0.1	tr	tr	tr–0.2
(Z)-nerolidol	tr–0.1	–	–	–	–	–	–
(E)-β-farnesene	tr	tr	–	0–0.1	0–2.0	–	–
germacrene D	–	0–0.1	0–1.1	0–0.4	0–0.2	0–0.1	0–0.5
trans-β-bergamotene	–	–	–	–	–	0–1.1	–
(E)-β-ionone	–	tr–0.5	–	–	–	–	0–0.1
valencene	–	tr	–	–	0–1.0	0–0.1	0–0.3
bicyclogermacrene	–	–	–	–	–	0–0.9	–
α-bisabolene	–	tr–0.1	–	–	–	–	–
(E,E)-α-farnesene	–	0–0.5	–	–	tr	0.2–3.2	0–2.5
β-bisabolene	tr	–	–	–	tr	–	–
γ-cadinene	tr	–	–	–	tr	–	0–0.3

Compound	Chin	Clem	Gran	Hyst	Jamo	Medi	Unsh
δ-cadinene	—	0–0.1	—	tr–0.4	0.5–0.7	tr	tr–0.9
hedycaryol	—	—	—	—	0–0.3	tr	—
elemol	—	—	—	tr	0.1–0.3	tr	0–2.7
(E)-nerolidol	0.1–0.2	tr	—	—	tr–0.1	tr	0–0.1
β-eudesmol	—	—	—	—	0–0.2	tr	—
α-eudesmol	—	—	—	—	tr–0.2	—	—
β-sinensal	—	tr–0.2	—	—	—	tr	tr
(Z,E)-farnesol	tr–0.5	—	—	—	—	—	—
α-sinensal	—	0.2–0.6	—	—	—	—	—
nootkatone	tr–0.7	tr	—	tr–1.0	—	—	tr
phytol	—	—	—	0–0.1	tr	—	0–0.6
hexadecanoic acid	—	—	—	0–0.1	0–0.1	tr	—

Notes

Chin = C. aurantium var. myrtifolia, Clem = C. clementina, Gran = C. grandis, Hyst = C. hystrix, Jamo = C. junos, Medi = C. medica, Unsh = C. unshiu;

Compounds are listed in elution order from a non-polar column;

tr = < 0.05%.

Table 14.2 Percentage composition of the leaf oils of seven less common Citrus species

Compound	Chin	Clem	Gran	Hyst	Juno	Medi	Unsb
(Z)-2-pentenol	–	–	0–0.8	–	0.2	–	0–0.5
(E)-2-hexenol	–	–	0–0.4	–	0.3	tr	–
(E)-2-hexenal	–	–	0–1.4	–	2.4	tr	0–3.7
(Z)-3-hexenol	0.8–1.3	–	0–0.2	–	0.9	0–0.3	0–5.3
hexanol	–	–	0–0.1	–	0.2	–	–
α-thujene	tr	0.5	0–0.5	tr	0.2–2.2	0–0.2	1.6–2.0
α-pinene	0–0.1	1.5–2.6	0.1–4.3	0.1–0.4	0.6–6.5	0.2–1.2	3.8–4.8
camphene	tr	0.1	tr–0.5	–	tr–0.1	tr	1.2–3.0
6-methyl-5-hepten-2-one	–	0.2	0–0.8	tr	0–0.1	0.1–5.0	–
sabinene	0.2–0.6	33.9–42.5	0.1–9.4	0–4.8	0.2–2.8	0.2–1.6	6.2–10.3
β-pinene	0.5–2.5	1.9–2.3	4.9–59.1	0.2–1.9	0.8–12.5	0–0.2	0.2–1.3
myrcene	1.1–2.3	2.5–3.6	tr–1.5	0.4–1.4	0.3–1.9	0.5–1.6	–
octanal	–	–	–	0–0.2	–	0–0.2	–
α-phellandrene	tr	0.1	0–0.9	tr	0.1–1.8	0–1.8	0.1–0.6
δ-3-carene	tr	3.4–6.3	tr–3.9	tr–0.1	tr	0–0.6	–
α-terpinene	tr	0.5–1.9	tr–0.1	tr–0.1	0.1–1.1	tr	0.1–0.7
p-cymene	tr	0.3	tr–2.3	tr–0.1	0.7–17.9	0–3.9	14.8–22.7
limonene	0.3–0.6	3.6–12.5	1.4–19.1	tr–0.6	3.3–19.0	15.4–56.6	2.9–5.6
1,8-cineole	tr	–	0–2.2	–	tr–0.2	0.1–2.2	0–0.1
(Z)-β-ocimene	0–1.0	0.6	0–0.8	tr–0.1	0.1–0.3	0.1–0.5	0–0.1
β-phellandrene	tr	0.5	0.2–0.7	tr–0.1	0.2–14.7	0–0.7	0–0.6
(E)-β-ocimene	1.4–2.5	3.5–4.7	4.9–31.2	0.2–0.5	2.3–5.9	0.2–6.5	0–6.4
γ-terpinene	0–0.1	1.6–2.9	tr–2.0	0.1–1.1	7.8–53.2	0–0.3	19.5–36.3
octanol	tr	–	tr	0–0.1	tr	0–0.1	–
cis-sabinene hydrate	–	1.0	0–0.1	0.1–1.5	tr	tr	0–0.1
cis-linalool oxide	–	0.2	–	tr–0.4	tr	–	tr
terpinolene	0.1–0.5	1.2–1.3	tr–1.6	0.1–0.2	0.3–2.5	tr–0.4	0–2.7
trans-linalool oxide	–	–	–	tr–0.2	–	–	–
p-cymenene	–	–	tr	0–0.3	–	–	1.8–3.6
2,6-dimethyl-5-heptenol	–	–	–	0–0.1	0–9.6	tr	–
trans-sabinene hydrate	0–0.1	–	–	0–0.1	–	–	–
linalool	37.7–63.0	14.9–22.6	0.1–1.1	1.7–4.7	0.1–25.4	0.8–1.6	4.1–22.0

Compound						
nonanal	0–0.1	0.2			tr	tr
β-terpineol	—	—			—	—
isopulegol	tr	—		0–0.2	0–0.9	—
citronellal	0–0.1	2.6	tr–6.8	58.9–81.5	0–2.3	0.2–1.9
menthol	—	—	—	0–0.3	—	—
terpinen-4-ol	0.1–0.5	3.0–4.2	0.1–0.6	0.1–0.9	0.1–0.2	0.3–0.4
α-terpineol	4.2–9.8	0.9	0.1–1.3	tr–0.1	tr–2.2	0.2–5.8
decanal	tr–0.9	0.1	0–0.2	0–0.2	0–2.5	tr–0.3
citronellol	tr–0.9	1.9	tr–8.7	6.0–8.2	0–1.2	0–0.1
nerol	2.0–3.5	0.6	0.1–1.5	tr	0.5–6.0	0–2.0
methyl thymol	—	—	0–0.1	—	—	tr
neral	tr–0.4	0.3	tr–14.5	tr	8.2–28.5	0–0.5
piperitone	—	—	—	—	0–0.1	—
geraniol	1.5–5.5	0.7	0.1–1.6	0.1–0.3	0.8–6.6	tr–0.5
linalyl acetate	—	—	0–0.1	—	tr	—
geranial	18.2–26.1	0.4	tr–17.9	0–0.1	13.5–44.0	tr–0.5
safrole	0.1–1.0	—	—	0–0.2	—	—
thymol	tr	tr	0–0.2	—	—	tr
carvacrol	—	tr	0–0.8	—	—	—
undecanal	0–0.1	—	0–0.1	tr	0–0.3	—
methyl geranate	tr	0.2	—	—	tr	—
δ-elemene	—	—	0–0.7	0.1–0.2	0–0.2	—
α-cubebene	—	—	—	—	0–2.1	—
citronellyl acetate	0–0.1	0.1	0.2–1.8	0.9–5.1	0–1.3	0–0.2
α-terpinyl acetate	0–0.1	—	—	—	—	tr
citronellic acid	—	—	—	0–0.1	—	—
neryl acetate	0–2.9	0.1	0.1–0.9	tr–0.1	0.4–1.1	tr
α-ylangene	—	—	—	—	0–13.4	—
α-copaene	—	—	0–0.2	0–0.1	—	—
geranyl acetate	0–5.4	0.6	0–8.1	0.1–0.8	tr–1.9	0.1–3.1
β-cubebene	—	—	0–0.1	0–0.2	—	—
β-elemene	tr	—	0–0.2	tr	tr–2.5	0–0.3
methyl N-methyl anthranilate	—	—	—	—	0–64.1	—
dodecanal	—	—	0–0.1	—	0–0.2	0.9–7.2
cis-α-bergamotene	—	—	0–0.2	tr	—	—

Table 14.2 (Continued)

Compound	Cbtn	Clem	Gran	Hyst	Juno	Medi	Unsb
β-cedrene	–	–	–	–	–	–	0–1.4
β-caryophyllene	0–0.2	0.2	0.9–15.4	0.2–0.9	tr–3.5	0.1–1.1	0.4–2.1
(Z)-β-farnesene	–	tr	tr–2.2	0–0.1	tr–0.3	–	–
α-humulene	0–0.1	tr	0.2–1.8	tr–0.2	tr–0.2	0–0.1	0.2–0.4
(E)-β-farnesene	–	–	–	0–0.2	–	0.5	–
γ-muurolene	–	–	0–0.4	–	tr–1.9	–	–
trans-β-bergamotene	–	–	–	–	–	0–0.2	–
β-selinene	–	–	0–0.4	–	0–0.8	0–0.3	0–4.7
valencene	–	–	0–0.4	–	–	–	–
bicyclogermacrene	–	tr	0–4.1	0–0.1	tr–0.6	–	0–0.1
(E,E)-α-farnesene	tr	tr	0–0.9	–	0–0.1	–	0–0.1
β-bisabolene	–	–	0–0.4	–	–	tr–0.2	–
γ-cadinene	–	–	0–0.9	–	–	–	–
δ-cadinene	tr	0.1	0–0.6	tr–0.4	0–0.2	tr	0–0.1
α-cadinene	–	–	0–7.1	0–0.4	–	–	–
hedycaryol	–	0	0–2.4	0–0.3	–	–	–
elemol	–	0.1	0–0.2	–	0–0.1	–	–
germacrene B	0–0.1	–	–	–	–	–	–
(E)-nerolidol	0–0.1	0.6	0.2–2.8	0.2–0.9	0–0.3	0–0.2	0–0.2
spathulenol	0–0.1	–	0–0.7	tr	0–0.3	–	0–0.1
caryophyllene oxide	0–0.1	tr	0–0.5	0–0.1	–	0–0.1	–
β-eudesmol	–	tr	0.2–1.6	0–0.2	0–0.2	–	–
α-eudesmol	–	–	0–1.8	0–0.2	–	–	–
β-sinensal	–	1.5	–	–	0–1.0	–	–
α-sinensal	–	0.6	–	–	0–1.1	–	–
nootkatone	–	–	0–0.1	–	–	–	–
phytol	–	0	0–23.1	–	–	0–10.2	–
hexadecanoic acid	–	–	0–3.4	–	–	–	–

Notes

Cbtn = C. aurantium var. myrtifolia, Clem = C. clementina, Gran = C. grandis, Hyst = C. bystrix, Juno = C. junos, Medi = C. medica, Unsb = C. unsbiu; Compounds are listed in elution order from a non-polar column; tr = < 0.05%.

Citrus grandis (L.) Osbeck

Pummelo or shaddock is known botanically as *C. grandis* [syn. *C. maxima* (J. Burman) Merrill]; *C. decumana* L.). According to Tanaka (1961), *C. grandis* is a member of the subsection Decumana within the Cephalocitrus section. The peel, leaf and flower oils have been subject of much study.

Peel oil

Numerous studies on the peel oil of *C. grandis* have been done by Yeh (1955), Gao and Zheng (1986), Sawamura and Kuriyama (1988), Sawamura *et al.* (1989), Lin and Hua (1989), Sawamura *et al.* (1990), Sawamura *et al.* (1991), Mondello *et al.* (1996), Jantan *et al.* (1996), Asano (1997), Bordoloi *et al.* (1999) and Sawamura *et al.* (1999). A summary of the components identified is shown in Table 14.1.

In addition, the above listed authors also found trace amounts of nonane, tricyclene, undecane, hexyl acetate, heptyl acetate, butyl acetate, butyl butyrate, 4-methyl-2-pentanone*, *trans*-carveol, perillyl alcohol, nonyl acetate, carvotanacetone, dodecyl acetate, 2-decenal, 2-undecenal, undecanal, β-terpinyl acetate, bornyl acetate, perillyl acetate, α-guaiene, *cis*-α-bergamotene, germacrene B, α-selinene, 2-dodecenal, α-muurolene, γ-muurolene, aromadendrene, allo-aromadendrene, α-cadinene, tetradecanal, hexadecanal, (Z)-nerolidol, elemicin, α-cadinol, farnesol, farnesyl acetate and octadecanol in the peel oil.

In 1995, Zhu *et al.* analysed a peel oil of *C. grandis* of Chinese origin and found that it contained the following constituents:

(Z)-3-hexenol	0.9%
6-methyl-5-hepten-2-one	0.2%
myrcene	0.2%
(Z)-3-hexenyl acetate	0.2%
octanal	0.1%
limonene	15.5%
trans-linalool oxide	1.0%
linalool	12.7%
terpinen-4-ol	1.3%
α-terpineol	9.1%
nerol	5.1%
cis-carveol	3.5%
neral	11.0%
geraniol	3.8%
geranial	9.2%
perillaldehyde	1.5%
indole	0.9%
perillyl alcohol	0.6%
geranyl acetate	0.5%
palmitic acid	0.9%

A trace amount of *cis*-linalool oxide was also found by Zhu *et al.* in this peel oil. As can be seen, this peel oil is totally dissimilar to the analytical results listed above in the summary of the peel oil studies.

Leaf oil

In 1973, Cameron and Scora determined that the leaf oil of an open pollinated *C. grandis* was found to possess the following composition:

α-pinene	0.9%
β-pinene	2.5%
myrcene	2.4%
limonene	15.0%
γ-terpinene	0.8%
ocimene	7.0%
p-cymene	0.5%
terpinolene	0.3%
2-methyl-2-heptanone*	1.4%
citronellal	1.4%
decanal	0.4%
linalool	2.4%
β-caryophyllene	1.9%
neral	0.4%
geraniol	2.7%
citroncllol	19.0%
nerol	31.9%
geraniol	8.6%

Perez Zayas *et al.* (1980) also analysed a leaf oil of *C. grandis*, which they called *C. decumana* or shaddock petitgrain. Although they did not present much quantitative data, they did determine that the oil possessed the following components: α-pinene; camphene; β-pinene; sabinene (40 per cent); δ-3-carene; myrcene; α-terpinene; limonene; (E)-β-ocimene; γ-terpinene; p-cymene; terpinolene; β-elemene; β-caryophyllene; β-farnesene; α-humulene; α-selinene*; α-farnesene; δ-cadinene; 1,8-cineole; 6-methyl-5-hepten-2-one; citronellal; linalool (10 per cent); linalyl acetate (21 per cent); terpinen-4-ol; α-terpineol; neral; geranial; neryl acetate; geranyl acetate; nerol; geraniol; p-cresol*; carvacrol.

Further studies on the leaf oil of *C. grandis* have been completed by Kamiyama and Amaha (1972), Scora *et al.* (1976), Cheng and Lee (1981) and Jantan *et al.* (1996). Also, Huang *et al.* (2000) examined the composition of the leaf oils of the following cultivars: 'Shatianyou', 'Wubuhongxingyou', 'Chumenweidanyou', 'Xianluomiyou', 'Dianjiangbaiyou', 'Cuba', 'Jinxiangyou' and 'Madoumendanyou'.

A summary of the components identified in all of the above mentioned studies on the leaf oil of *C. grandis* can be seen listed in Table 14.2.

Flower oil

In 1979, Wang determined that the flower oil of *C. grandis* was found to contain the following constituents:

α-pinene	0.3–0.6%
camphene	1.0%
limonene	28.8–31.6%

linalool	24.4–25.6%
α-terpineol	tr–0.2%
camphor	tr–0.2%
nerol	30.0–30.4%
neryl acetate	0.3–0.4%
geraniol	0.9–1.0%
eugenol	tr–0.3%
methyl anthranilate	9.2–9.3%
indole	tr–0.5%
2-phenethyl alcohol	0.4–0.5%
phenyl *p*-cresol ether	0.8–1.3%
methyl β-naphthyl ketone*	0.3–0.4%

Trace amounts of linalyl acetate and geranyl acetate were also found by Wang (1979) in this flower oil.

In 1991, Dung *et al.* analysed a flower oil of *C. maxima* (syn. *C. grandis*) of Vietnamese origin. They found that this oil possessed the following composition:

α-pinene	0.4%
sabinene	0.7%
β-pinene	3.4%
myrcene	3.9%
α-phellandrene	1.3%
p-cymene	0.1%
limonene	18.2%
(Z)-β-ocimene	0.1%
(E)-β-ocimene	4.4%
γ-terpinene	0.1%
linalool	16.4%
nerol	1.1%
neral	0.5%
geraniol	0.6%
geranial	0.5%
methyl anthranilate	0.3%
β-caryophyllene	0.4%
geranyl acetone	0.2%
selina-4(14), 11-diene	0.3%
methyl *N*-formylanthranilate	0.1%
(E)-nerolidol	29.3%
(Z)-farnesol	0.3%
(E)-farnesol	15.7%
(E)-farnesal	0.55%
farnesyl acetate	0.2%

Dung *et al.* (1991) also found trace amounts of heptanal, α-thujene, camphene, β-phellandrene, terpinen-4-ol, (Z)-3-hexenyl butyrate, α-terpineol, neryl acetate and geranyl acetate in this flower oil.

From a headspace study of the flowers of *C. grandis*, Toyoda *et al.* (1993) determined that it comprised of the following constituents: methacrolein, 1-buten-3-one, sabinene, myrcene, 3-methyl-2-butenal, limonene, β-phellandrene, 1,8-cineole, β-ocimene, *p*-cymene, 6-methyl-5-hepten-2-one, allo-ocimene, myrcene-6,7-oxide, linalool oxide, β-ocimene-6,7-oxide, benzaldehyde, 2-methyl-6-methylene-1,7-octadien-3-one, linalool, 3,7-dimethyl-1,7-octadien-3-ol, β-caryophyllene, 3-methylene-7-methyl-1,7-octadien-7-ol, 3-methylene-7-methyl-1,7-octadien-6-ol, α-terpineol, nerol, geraniol, citronellol, geranial, neral, 3,7-dimethyl-1,5-octadiene-3,7-diol, 3,7-dimethyl-1,7-octadiene-3,6-diol, linalool, oxide (pyranoid), anthranilaldehyde, methyl anthranilate, indole.

Citrus hystrix DC

According to Tanaka (1961), *C. hystrix* is a member of the section Papeda. It is known in France as combava, and for many years a leaf oil has been produced in limited quantities in Madagascar and other Indian Ocean islands. *C. hystrix* leaf oil became popular because of its richness in citronellal. As a result, the peel and leaf oils have been the subject of some study. In addition, a study of the juice oil of *C. hystrix* by Sato *et al.* (1990) has been conducted.

Peel oil

The peel oil of *C. hystrix* has been analysed by Schwob and Reignier (1964), Lawrence *et al.* (1971), Schwob (1977), Pieribattesti (1982), Sato *et al.* (1990), Jantan *et al.* (1996) and Pudil *et al.* (1998). The oil composition can be seen summarised in Table 14.1.

Trace amounts of acetone, methanol, ethanol, *p*-mentha-1,4(8)-diene, 3-hydroxycyclopentanone*, 4-methyl-(2-methyl-1-propenyl)-tetrahydropyran*, 2,2,6-trimethylcyclohexanone*, octahydro-3a-methyl-*cis*-2H-inden-2-one*, citronellyl formate, α-bergamotene, α-cadinene and cadinol were also found by the above listed authors in this peel oil.

Leaf oil

Analyses of the leaf oil of *C. hystrix* by Lawrence *et al.* (1971), Moreuil and Huet (1973), Schwob (1977), Pieribattesti (1982), Sato *et al.* (1990) and Jantan *et al.* (1996) can be seen summarised in Table 14.2.

Trace amounts of rose oxide, menthone, isomenthone, neryl acetone, β-elemene, isocaryophyllene, (Z, Z)-α-farnesene, α-muurolene, germacrene D, ar-curcumene, calamenene, cubenol, epi-cubenol and *p*-menthan-1,8-diol were also found by the above authors in combava leaf oil.

Juice oil

Sato *et al.* (1990) also examined the composition of a juice extract of *C. hystrix*. They found that it contained the following constituents:

α-pinene	4.6%
camphene	0.4%
β-pinene	39.5%
sabinene	4.5%

myrcene	1.1%
α-terpinene	1.5%
limonene	2.8%
β-phellandrene	1.4%
β-ocimene	0.1%
γ-terpinene	4.0%
p-cymene	0.4%
terpinolene	1.1%
cis-linalool oxide	1.0%
α-copaene	1.2%
linalool	0.8%
terpinen-4-ol	17.6%
β-caryophyllene	0.8%
citronellyl acetate	0.5%
α-terpineol	7.1%
germacrene D	0.9%
geranyl acetate	2.0%

Citrus jambhiri Lush.

C. jambhiri has been classified by Tanaka (1961) as being a member of the section Citrophorum, subsection Limonoides. It is commonly known as rough lemon and it originates from India. It is known to exhibit a wide range of variation both in shape and size. Although it is used to some extent as a substitute for lemon it has become much more valuable as a rootstock (Reuter *et al.*, 1967).

A review of the literature indicates that only one study has been conducted on the peel oil, while a number of studies have been completed on the leaf oil.

Peel oil

The peel oil of *C. jambhiri* was analysed by Shaw and Wilson (1976) and was found to contain:

α-pinene	0.5%
β-pinene + myrcene	4.9%
limonene	92.2%
γ-terpinene	0.1%
linalool	0.2%
citronellal	0.3%
α-terpineol	0.1%
decanal	0.2%
neral	0.2%
geraniol	0.2%
neryl acetate	0.4%
geranyl acetate	0.1%
β-caryophyllene + α-bergamotene	0.3%
β-bisabolene	0.3%

A trace amount of hexane was also found in the peel oil.

Leaf oil

The composition of the leaf oil of *C. jambhiri* has been examined by Scora *et al.* (1969), Lund *et al.* (1981), Agarwal *et al.* (1989), Nemec and Lund (1990) and Huang *et al.* (2000). It is obvious that the leaf oil composition is extremely variable, as can be seen from the compositional summary shown as follows:

2-pentanol	0–0.2%
2-pentenol	0–0.3%
(E)-2-hexenal	0–2.0%
α-thujene	tr–0.6%
α-pinene	tr–0.9%
camphene	tr–1.6%
6-methyl-5-hepten-2-one	0–2.1%
sabinene	1.2–35.5%
β-pinene	0.1–2.4%
myrcene	1.2–31.4%
α-phellandrene	0–0.1%
α-terpinene	tr–0.2%
p-cymene	tr–3.3%
p-mentha-1,3,8-triene	0–0.2%
limonene	13.1–41.0%
1,8-cineole	0–3.4%
(Z)-β-ocimene	0–0.7%
β-phellandrene	0–0.4%
(E)-β-ocimene	tr–3.6%
γ-terpinene	0.4–7.4%
octanal	0–0.6%
octanol	0–0.1%
acetic acid	0–1.2%
sabinene hydrate	0–0.6%
nonanol	0–2.2%
p-cymenene	0–0.4%
terpinolene	tr–0.2%
nonanal	0–0.3%
linalool	1.3–34.1%
iso(iso)pulegol	0–0.6%
isopulegol	0–4.6%
citronellal	0.8–17.0%
carvone	0–1.0%
terpinen-4-ol	0.4–1.6%
α-terpineol	0–2.7%
decanal	0.1–0.7%
citronellol	0.1–7.9%
sabinol	0–0.9%
nerol	0.1–2.7%
neral	1.2–6.5%
geraniol	tr–3.0%
linalyl acetate	0–5.0%

geraniol	1.1–4.2%
octadecanol	0–0.5%
α-terpinyl acetate	0.5–3.5%
neryl acetate	0.5–1.3%
δ-elemene	0–0.2%
geranyl acetate	0.3–1.4%
geranyl formate	0–1.8%
dodecanal	0–0.1%
neryl formate	0–1.8%
pyridine	0–2.7%
2,4-pentadiene nitrile	0–7.0%
N,N'-dimethyl formamide	0–0.6%
α-collidine	0–1.2%
methyl N-methyl anthranilate	7.7–4.0%
β-elemene	0–0.8%
α-copaene	0–0.6%
β-bourbonene	0–0.3%
β-caryophyllene	0.2–13.3%
α-bergamotene	0–0.2%
α-humulene	0.1–0.7%
β-selinene	0–1.9%
γ-muurolene	0–0.1%
α-farnesene	0–0.1%
germacrene B	0–0.4%
β-bisabolene	0–0.7%
δ-cadinene	0–0.1%
(E)-nerolidol	0–0.1%
spathulenol	0–0.1%
caryophyllene oxide	0–0.2%
methyl 14-methyl pentadecanoate	0–0.7%

Trace amounts of (Z)-3-hexenol, δ-3-carene, linalool oxide, methyl geranate and β-farnesene were also found by Nemec and Lund (1990), and Huang et al. (2000).

Citrus junos Sieb. ex Tanaka

In 1961, Tanaka reported that *C. junos* was a member of section Osmocitrus subsection Eusomocitrus. It is a very popular sour orange in Japan where it is known as yuzu.

The fruits are reported to have a pleasant aroma and are used as a seasoning and as a raw material for vinegar manufacture. It was reported (Matsuura et al., 1977) that ca. 3700 tonnes of fruit are harvested annually in Japan. Limited quantities of a CP peel oil have been produced irregularly. The peel, leaf and juice oils have been the subject of previous study.

Peel oil

The presence of germacrene D, germacrene B and a new germacrene, which was named bicyclogermacrene, was characterised in the sesquiterpenoid fraction of *C. junos* by Nishimura et al. (1969). The configurations of the endocyclic double bonds were determined by Nishimura et al. (1973) to have an (E,E) orientation. In addition, the authors determined that the 10 membered ring possessed a fused 1,1'-dimethyl cyclopropane ring.

The peel oil of C. junos has been analyzed by Shinoda et al. (1970), Kunose and Sawamura (1980), Sawamura et al. (1989), Yang et al. (1992), Njoroge et al. (1994), Njoroge et al. (1996) and Sawamura et al. (1999) and a summary of these findings can be seen in Table 14.1.

A number of trace constituents were also identified by the above authors and Watanabe et al. (1983), Escher (1985), Tajima et al. (1990) and Yukawa et al. (1994) in the peel oil. These trace compounds found were camphene, p-cymenene, undecatriene, tetradecane, p-mentha-1,4,8-triene, α-bergamotene, α-ylangene, β-ylangene, aromadendrene, allo-aromadendrene, germacrene B, α-muurolene, δ-guaiene, β-sesquiphellandrene, ar-curcumene, calamenene, butyl acetate, butyl butyrate, ethyl hexanoate, hexyl acetate, (Z)-3-hexenyl acetate, heptyl acetate, hexyl butyrate, (Z)-3-hexenyl butyrate, citronellyl formate, perillyl acetate, geranyl formate, limonen-10-yl acetate, cis-carvyl acetate, methyl palmitate, ethyl palmitate, methyl jasmonate, 4-methylpentan-2-one*, 2-nonanone, fenchone, camphor, dihydrocarvone, cryptone, piperitone, hexanal, heptanal, 2-octenal, 6-methyloctanal, 2-nonenal, 8-methylnonanal, undecanal, (Z)-4-decenal, 8-methyldecanal, dihydrolimonen-10-al, (E)-2-decenal, (E)-2-undecenal, dodecanal, (E)-2-dodecenal, (E,E)-2,4-decadienal, tridecanal, pentadecanal, benzaldehyde, cuminaldehyde, 2-pentanol, amyl alcohol, 2-pentenol, 2,5-dimethyl-2-vinylhexanol, (E)-2-hexenol, heptanol, decanol, undecanol, 2-pinanol*, cis-p-menth-2-en-1-ol, trans-p-menth-2-en-1-ol, trans-p-mentha-2,8-dien-1-ol, 3,7-dimethyl-5-oxo-octen-3-ol, cis-p-mentha-2,8-dien-1-ol, borneol, trans-carveol, trans-piperitol, trans-p-mentha-1(7), 8-dien-2-ol, p-cymen-8-ol, benzyl alcohol, cuminyl alcohol, cis-p-mentha-1(7), 8-dien-2-ol, perillyl alcohol, juniper camphor, globulol, spathulenol, T-cadinol, T-muurolol, α-cadinol, limonene diol, viridiflorol, 2,6-dimethyl-2,7-octadien-1,6-diol, eugenol, carvacrol, (Z,E)-farnesol, (E,E)-farnesyl acetate, mesityl oxide, myrcene-6,7-epoxide, β-ocimene-6,7-epoxide, methyl carvacrol, isovaleric acid, 2-methylbutyric acid, linalool oxide (pyranoid), ascaridole*, caryophyllene oxide, 9-cyclodecanolide, hexanoic acid, decanoic acid, dodecanoic acid, tetradecanoic acid and octadecanoic acid.

In 1980, Kitahara et al. found two unusual monoterpene ketols as minor components of a peel oil of C. junos of Chinese origin. The two compounds, which were structurally elucidated using IR, MS and ^{1}H-NMR and confirmed by synthesis, were determined to be 2,6-dimethyl-4-oxo-oct-7-en-6-ol (also known as tagetonol) and 2,6-dimethyl-4-oxo-octa-2,7-dien-6-ol, respectively.

In 1994, Yukawa et al. determined that the sulphurous note associated with yuzu peel oil was caused by p-menth-1-en-8-thiol.

Leaf oil

Kamiyama (1970), Kekelidze and Dzhanikashvili (1982), Kekelidze et al. (1984) and Huang et al. (1993) examined the composition of the leaf oil of C. junos; the latter study over two seasons. Huang et al. (2000) further analyzed the following cultivars: Xiecheng, Luohancheng, Wangchangxiancheng, Mitaoxiancheng, Tangcheng, Nippon-Xiancheng, Jiangbeixiancheng, Chanju, Zhencheng and Bazhongxiancheng of C. junos. The results of these various studies are summarised in Table 14.2.

Trace amounts of α-terpinyl formate and nonanal were also found by Kekelidze *et al.* (1984) and Huang *et al.* (1993) in the leaf oils. The three cultivars of *C. junos* examined by Huang *et al.* (2000) that contained methyl N-methyl anthranilate were 'Zhencheng' (0.4 per cent), 'Xiecheng' (64.1 per cent) and 'Luohancheng' (63.6 per cent). These latter two leaf oils could have economic potential as a natural source of methyl N-methyl anthranilate.

Juice oil

Both Ohta and Osajima (1983) and Miyake *et al.* (1990) reported that an oil extracted from *C. junos* juice contained:

α-pinene	0.3–1.3%
β-pinene	0.3–0.7%
myrcene	1.5–2.4%
α-terpinene	0.2–0.6%
limonene	77.1–78.9%
γ-terpinene	8.7–10.8%
p-cymene	tr–1.2%
terpinolene	0.2–0.6%
linalool	0.7–4.6%
α-fenchol	0–0.1%
terpinen-4-ol	0.6–0.8%
β-caryophyllene	0–0.3%
β-farnesene	0.6–0.7%
α-terpineol	0.3–0.5%
cis-β-terpineol	0–1.8%
carvone	0–0.2%
citronellol	0–0.3%
geraniol	0–0.1%
geranyl acetate	0–0.2%
T-cadinol	0–0.1%
thymol	0–0.3%
ethyl hexadecanoate	0–0.2%
ethyl oleate	0–0.1%
ethyl linoleate	0–0.2%
ethyl linolenate	0–0.2%

Ohta and Osajima (1983), and Miyake *et al.* (1990) also found trace amounts of 1,4-cineole, 1,8-cineole, nonanal, citronellal, decanal, octyl acetate, *p*-menth-2-en-1-ol, *trans*-β-terpineol, ocimenol, nerol, α-eudesmol, and β-eudesmol in the juice oil.

Citrus madurensis Lour.

Calamondin or calamansi is the common name given to this *Citrus* species that possesses small fruit. According to Tanaka, cited in (Reuther *et al.*, 1967) *C. madurensis* (syn. *C. mitis* Blanco; *C. macrocarpa* Bunge) is classified as being the lone member of section

Pseudofortunella. From a literature survey it can be seen that the peel, leaf and juice oils have been the subject of limited study.

Peel oil

In 1958, Nigam *et al.* performed a preliminary analysis of *C. madurensis* (which they named *C. macrocarpa*). More recently, Moshonas and Shaw (1996) also analysed the peel oil of *C. madurensis* (which they called *C. mitis*). A summary of the analyses can be seen as follows:

α-pinene	0.1%
sabinene	0.1%
β-pinene	1.4%
myrcene	0.3%
α-phellandrene	0.1%
limonene	91.3–91.9%
(Z)-β-ocimene	0.1%
(E)-β-ocimene	0.2%
octanal	0.1%
octanol	0.2%
nonanal	0.3%
linalool	0.4–2.5%
α-terpineol	0.4–1.4%
decanal	0.3%
cis-dihydrocarvone	0.2%
carvone	0.3%
undecanal	0.1%
neral and geranial	tr–1.1%
linalyl acetate	0–1.4%
geraniol	0–0.6%
nonyl acetate	0.1%
decyl acetate	0.2%
geranyl acetate	0.5%
neryl propionate	0.3%
dodecanal	0.1%
methyl anthranilate	0–0.2%
indole	0.3%
β-elemene	0.1%
valencene	0.7%
elemol	0.3%
eudesmol	0.3%

Trace amounts of acetaldehyde, acetone, heptane, octane, 3-hexanone, 2-hexanone, 3-hexenol, α-thujene, 2-octenal, γ-terpinene, heptyl acetate, *cis-p*-mentha-1,8-dien-1-ol, *trans-p*-mentha-1,8-dien-1-ol, citronellal, 2-nonenal, methyl salicylate, 2,4-nonadienal, *trans*-dihydrocarvone, verbenone, nonanoic acid, perillaldehyde,

2,4-decadienal, citronellyl acetate, neryl acetate, dodecanoic acid, tetradecanoic acid and hexadecanoic acid were also found by Nigam *et al.* (1958) and Moshonas and Shaw (1996).

Leaf oil

The leaf oil of *C. madurensis* was analysed by Gupta and Gupta (1957) and Huang *et al.* (2000). A summary of their results is shown as follows:

α-thujene	0.4%
α-pinene	1.3%
sabinene	26.7%
β-pinene	1.6%
myrcene	1.7%
α-phellandrene	0.1%
α-terpinene	0.3%
p-cymene	0.3%
limonene	1.3%
(Z)-β-ocimene	0.2%
β-phellandrene	0.3%
(E)-β-ocimene	3.5%
γ-terpinene	2.5%
sabinene hydrate	0.7%
p-cymenene	0.2%
terpinolene	0.3%
nonanal	0.2%
linalool	50.0%
citronellal	0.1%
terpinen-4-ol	0.9%
α-terpineol	0.6%
decanal	0.1%
citronellol	0.1%
thymol	2.1%
β-elemene	1.2%
β-caryophyllene	0.4%
α-humulene	0.2%
germacrene B	0.1%
(E)-nerolidol	0.1%
β-sinensal	1.5%
α-sinensal	0.6%

Trace amounts of camphene, linalool oxide, methyl anthranilate, nerol, methyl thymol, neral, geranial, carvacrol, citronellyl acetate, neryl acetate, geranyl acetate, β-bisabolene, δ-cadinene, elemol, spathulenol, caryophyllene oxide and β-eudesmol were also found by Huang *et al.* (2000).

Juice oil

In 1992, Nisperos-Carriedo et al. examined the juice volatiles of *C. madurensis*. They found that the ppm level of each of the detected volatiles was:

acetaldehyde	2.4
decanal	1.1
nonanal	0.5
octanal	1.2
perillaldehyde	0.4
ethanol	11.0
linalool	0.9
methanol	9.0
terpinen-4-ol	0.2
α-terpineol	20.6
geranyl acetate	2.5
neryl acetate	0.3
δ-3-carene	0.2
α-pinene	5.6
β-pinene	2.0
γ-terpinene	3.0
terpinolene	0.3
valencene	0.2

Citrus medica L.

C. medica, which is known as citron, has been classified by Tanaka (1961) as being in the subsection Citroides within the section Citrophorum. According to Reuter et al. (1967), this was the first citrus fruit to reach the Mediterranean region from Asia. Citron peel oil has been an occasional item of commerce for many years, although currently little is available annually. Two distinct varieties, *C. medica* var. *sarcodactylis* (Noot.) Swingle and *C. medica* var. *ethrog* Engl. are also known. A review of the literature has revealed that the peel, leaf and flower oils have been previously studied.

Peel oil

The composition of the oil of *C. medica* has been studied by Capello et al. (1982), Cotroneo et al. (1986), Huet et al. (1986), Dugo et al. (1988), Fleisher and Fleisher (1991 and 1998), Poiana et al. (1998) and Lota et al. (1999). A summary of the components identified in the oil by these various groups can be seen in Table 14.1. Also, trace amounts of allo-ocimene, undecanal, β-ionone, germacrene B, spathulenol and α-bisabolol were found by Dugo et al. (1988), Fleisher and Fleisher (1991 and 1998), Poiana et al. (1998) and Lota et al. (1999) in the peel oil.

According to Dugo et al. (1988) the peel oil of *C. medica* can be differentiated from other commercial peel oils by the following ratios:

α-thujene/sabinene: 0.226
α-thujene/β-pinene: 0.034

α-pinene/β-pinene: 0.139
β-pinene/γ-terpinene: 1.245

Finally, an unusual strain of *C. medica* produced in China was found Zhu *et al.* (1995) to possess a peel oil, which contained the following, components:

limonene	1.5%
1,8-cineole	0.5%
cis-linalool oxide	2.8%
trans-linalool oxide	1.4%
linalool	8.4%
terpinen-4-ol	7.7%
α-terpineol	22.1%
nerol	1.4%
neral	18.5%
geraniol	1.3%
geranial	27.6%
geranyl acetate	0.3%

Based on this analysis, it would appear that either this is a different chemotypic form of *C. medica* or the plant was incorrectly taxonomically classified.

Leaf oil

The leaf oil of *C. medica* has been analysed by Kamiyama and Amaha (1972), Sun *et al.* (1984), Fleisher and Fleisher (1998), Lota *et al.* (1999) and Huang *et al.* (2000). The oil components found by each of these authors can be seen summarised in Table 14.2. Trace amounts of *cis*-limonene oxide, *trans*-limonene oxide, limonen-4-ol, *cis*-carveol, *trans*-carveol, perillaldehyde, carvone, and geranic acid were also found by Fleisher and Fleisher (1998), and Lota *et al.* (1999) in the leaf oil.

Flower oil

Altenburger and Matile (1990) examined the headspace of the flowers of *C. medica* over a 24-hour period. Although the authors did not present any useable quantitative data, they did show that an extract, which was rich in linalool, also contained the following components: α-pinene, β-pinene, limonene, γ-terpinene, α-terpineol, indole, methyl anthranilate, nerolidol.

Citrus medica L. var. *ethrog* Engl.

According to Reuther *et al.* (1967), *C. medica* var. *ethrog* (known as etrog) is probably an ancient form of citron from which the modern citron was developed by selection. Fleisher and Fleisher (1991) stated that an important characteristic of the etrog fruit is that it possesses a violet-like aroma. Limited studies on the peel and leaf oil are reported in the literature.

Peel oil

Analyses of *C. medica* var. *ethrog* have been reported by Fleisher and Fleisher (1991 and 1998) and Lota *et al.* (1999). The components identified in this oil are as follows:

α-thujene	0–0.8%
α-pinene	1.0–2.0%
sabinene	0.2–0.4%
β-pinene	0.4–2.1%
myrcene	1.6–2.0%
α-phellandrene	0.1%
α-terpinene	0.1–0.5%
p-cymene	0–0.3%
limonene	46.9–81.3%
(Z)-β-ocimene	0.4–0.6%
β-phellandrene	0–0.2%
(E)-β-ocimene	0–0.8%
γ-terpinene	4.5–30.7%
trans-sabinene hydrate	T–0.1%
cis-sabinene hydrate	0–0.1%
terpinolene	0–1.3%
linalool	0.2–0.3%
citronellal	0.2%
terpinen-4-ol	0–0.1%
α-terpineol	0.2–0.3%
nerol	0.1–0.2%
neral	2.6–2.8%
geraniol	0.1–0.2%
geranial	4.2–5.4%
neryl acetate	0.2–0.3%
geranyl acetate	0.5–0.7%
β-caryophyllene	0.2–0.3%
trans-α-bergamotene	0–0.4%
trans-β-bergamotene*	0–0.2%
α-bisabolene	0–0.1%
β-bisabolene	0–0.6%

In addition, Fleisher and Fleisher (1991 and 1998) and Lota *et al.* (1999) also found trace amounts of toluene, camphene, ethyl benzene, δ-3-carene, allo-ocimene, 1,8-cineole, limonene oxide, hexanal, octanal, nonanal, decanal, undecanal, tetradecanal, pentadecanal, heptadecanal, perillaldehyde, benzaldehyde, 6-methyl-5-hepten-2-one, β-ionone, hexyl acetate, ethyl hexanoate, ethyl myristate, ethyl palmitate, ethyl heptadecanoate, ethyl stearate, ethyl linoleate, ethyl linolenate, citronellyl propionate, hexanol, 4-methylhexanol, nonanol, (Z)-3-hexenol, terpinen-4-ol, γ-isogeraniol*, citronellol, *trans*-carveol, *cis*-carveol, benzyl alcohol, 2-phenethyl alcohol, (E)-nerolidol, α-bisabolol, β-bisabolol, phytol, anethole, α-humulene, *cis*-α-bergamotene, aromadendrene, germacrene D, neric acid and 14-pentadecanoic acid in the peel oil.

Leaf oil

The leaf oil of *C. medica* var. *ethrog* has been the subject of analyses by Fleisher and Fleisher (1991 and 1998) and Lota *et al.* (1999). A summary of the findings of these studies can be seen as follows:

(Z)-3-hexenol	0–0.1%
α-pinene	0.1–0.3%
6-methyl-5-hepten-2-one	0–6.7%
5-methyl-5-hepten-2-one	0–2.8%
sabinene	0.1–0.6%
β-pinene	tr–0.1%
myrcene	1.0%
p-cymene	0–0.1%
limonene	27.8–37.9%
1,8-cineole	1.0–1.8%
β-phellandrene	tr–1.0%
(Z)-β-ocimene	tr–0.4%
(E)-β-ocimene	0.5–0.6 %
γ-terpinene	tr–0.5%
(Z)-3-hexenyl acetate	0–0.4%
ethyl hexanoate	0–0.1%
benzaldehyde	0–0.1%
trans-sabinene hydrate	0–0.1%
nonanal	0.1–0.2%
nonanol	0–0.1%
linalool	1.2–1.6%
citronellal	0.5–1.5%
terpinen-4-ol	0–0.2%
α-terpineol	0.7–1.1%
citronellol	0.3–1.4%
trans-carveol	0–0.1%
nerol	1.3–7.3%
neral	5.1–14.8%
geraniol	6.2–6.8%
linalyl acetate	0–0.2%
geranial	8.2–21.3%
anethole	0–0.5%
neryl acetate	0–0.7%
geranyl acetate	2.4–3.7%
β-caryophyllene	0.1–4.8%
cis-β-bergamotene	0–0.2%
trans-β-bergamotene	0–0.7%
α-humulene	0–0.4%
(E)-nerolidol	tr–0.1 %
caryophyllene oxide	tr–0.4%
β-bisabolol	0–0.1%
ethyl myristate	0–0.2%

ethyl palmitate	0–0.1%
neric acid	0–1.0%
ethyl stearate	0–0.2%
4-pentadecanoic acid	0–0.3%
ethyl linoleate	0–1.0%
ethyl linolenate	0–3.0%
phytol	0–1.9%

Fleisher and Fleisher (1991 and 1998), and Lota *et al.* (1999) also found trace amounts of α-thujene, camphene, δ-3-carene, α-phellandrene, α-terpinene, ethyl benzene, terpinolene, 1,1-diethoxy-2-methylbutane*, linalool oxide, limonene oxide, 2-ethylfuran, pentanal, hexanal, octanal, decanal, undecanal, dodecanal, 2,4-heptadienal, 2-cyclohexenone*, piperitenone, 2-methyl-3-buten-2-ol, 1-penten-3-ol, octanol, undecanol, *cis*-sabinene hydrate, *cis*-carveol, 2-phenethyl alcohol, ethyl octanoate, ethyl pentadecanoate, citronellyl acetate, (E)-β-farnesene, β-bisabolene, bicyclogermacrene, humulene oxide and geranic acid in the leaf oil.

Citrus medica L. var. *sarcodactylis* (Noot.) Swingle

C. medica var. *sarcodactylis*, which is the most unusual *Citrus* species, is known as fingered citron because of the shape of the fruit. The peel and leaf oils have been the subject of limited study.

Peel oil

Analyses of *C. medica* var. *sarcodactylis* by Shiota (1990), Zhu *et al.* (1993), Dung *et al.* (1996) and Lota *et al.* (1999) revealed that the peel oil had the following composition:

α-thujene	tr–1.3%
α-pinene	0.1–2.9%
6-methyl-5-hepten-2-one	tr–0.4%
sabinene	0–0.4%
β-pinene	0.2–2.7%
myrcene	0.3–1.7%
δ-3-carene	0–0.4%
α-terpinene	tr–0.7%
p-cymene	0.1–33.7%
limonene	11.2–55.5%
(Z)-β-ocimene	0.1–0.2%
(E)-β-ocimene	0.1–0.5%
γ-terpinene	1.5–32.1%
cis-linalool oxide	tr–0.2%
trans-linalool oxide	tr–0.2%
terpinolene	tr–1.4%
linalool	0.2–1.4%
camphor	0–0.2%
citronellal	tr–0.2%
cis-limonene oxide	0–0.6%

trans-limonene oxide	0–0.8%
limonen-10-ol	0–0.5%
terpinen-4-ol	0.1–7.6%
α-terpineol	0.2–7.6%
nerol	tr–6.4%
neral	1.2–19.2%
geraniol	tr–9.3%
geranial	1.3–22.0%
limonene diepoxide	0–0.6%
2-undecanone	tr–0.2%
citronellyl acetate	0–0.1%
geranyl acetate	tr–0.3%
neryl acetate	0–0.3%
geranyl isobutyrate	tr–0.3%
germacrene D	0–0.4%
β-cubebene	0–0.2%
β-caryophyllene	tr–0.2%
trans-α-bergamotene	0.1–0.2%
β-bisabolene	0.1–0.4%
lauric acid	0–0.1%

The above listed authors also found trace amounts of (Z)-3-hexenol, camphene, 1,8-cineole, octanal, nonanal, fenchone, camphor, decanal, piperitone, undecanal, 2-tridecanone, β-ionone, epoxy-β-ionone, methyl acetophenone, dihydroactinodiolide, hexanol, sabinene hydrate, borneol, *p*-cymen-8-ol, *trans*-carveol, isopulegol, β-ionol, α-humulene, α-bisabolene, geranyl propionate, (E)-nerolidol, spathulenol, α-bisabolol, caryophyllene oxide, tetradecanoic acid and pentadecanoic acid in the peel oil.

Leaf oil

Lota *et al.* (1999) and Huang *et al.* (2000) analysed the leaf oil of fingered citron and found it to contain the following components:

α-pinene	0.3%
camphene	0–0.1%
6-methyl-5-hepten-2-one	0.5–3.8%
sabinene	0.3–1.2%
β-pinene	0.1–0.2%
myrcene	1.2–1.3%
α-terpinene	0–0.1%
p-cymene	0.1–0.2%
limonene	38.1–43.8%
1,8-cineole	0.2–1.7%
(Z)-β-ocimene	0.1–0.7%
β-phellandrene	0–0.1%
(E)-β-ocimene	0.2–1.1%
γ-terpinene	0.1–0.4%
linalool oxide	0–0.1%

terpinolene	0.1%
nonanal	0.2–0.4%
linalool	0–0.6%
isopulegol	0–0.9%
citronellal	2.1%
terpinen-4-ol	0.2–0.3%
α-terpineol	0.2–1.0%
decanol	0.1%
citronellol	0.2–1.1%
nerol	2.1–4.2%
neral	14.0–17.2%
geraniol	2.2–3.8%
geranial	20.1–23.6%
undecanal	0.1–0.2%
methyl geranate	0–0.1%
citronellyl acetate	tr–0.5%
neryl acetate	0.2–2.4%
geranyl acetate	0.3–3.1%
β-caryophyllene	0.2–0.5%
α-bergamotene	0–0.5%
α-humulene	0–0.1%
β-bisabolene	0–0.1%
δ-cadinene	0–0.1%
(E)-nerolidol	0–0.2%
spathulenol	0–0.2%
caryophyllene oxide	0–0.1%

In addition, Lota *et al.* (1999) and Huang *et al.* (2000) also found trace amounts of α-thujene, α-phellandrene, δ-3-carene, octanol, *cis*-limonene oxide, sabinene hydrate, terpinolene, thymol, dodecanal and β-elemene in the leaf oil.

Citrus unshiu Marcovitch

C. unshiu has been classified as being a member of section Acrumen, subsection Euacrumen according to Tanaka (1961). It is known in Japan as Unshiu Mikan. Both the peel, leaf and flower and juice oils have been the subject of numerous analytical studies.

Peel oil

The composition of the peel oil of *C. unshiu* has been determined by Kato (1974), Yajima *et al.* (1979), Sawamura (1983), Namba *et al.* (1985), Kekelidze and Dzhanikashvili (1985), Gao and Zheng (1986), Inoma *et al.* (1989), Ifuku *et al.* (1992), Sakamoto *et al.* (1994), Sakamoto *et al.* (1997) and Sawamura *et al.* (1999). A summary of the findings of these analyses can be seen in Table 14.1.

Trace amounts of acetic acid, octanoic acid, 2-butanol, isobutanol, 2-pentanol, isoamyl alcohol, amyl alcohol, 1-octen-3-ol, heptanol, *cis*-p-mentha-2,8-dien-1-ol, *trans*-*p*-mentha-2,8-dien-1-ol, borneol, *cis*-isopiperitenol, *trans*-isopiperitenol, decanol, *trans*-carveol, dihydrolimonen-10-ol, perillyl alcohol, α-cadinol, hexanal, heptanal,

(E)-2-hexanal, (Z)-4-heptenal, 2,4-hexadienal, (E)-2-nonenal, (E)-2-decenal, (Z)-4-decenal*, undecanal, (E,E)-2,4-decadienal, cuminaldehyde, anisaldehyde, ethyl formate, ethyl propionate, ethyl butyrate, amyl formate, methyl hexanoate, heptyl acetate, nonyl acetate, undecyl acetate, dodecyl acetate, citronellyl formate, geranyl propionate, bornyl acetate, limonen-10-yl acetate, dihydrolimonen-10-yl acetate, perillyl acetate, 2-butanone, camphor, menthone, dihydrocarvone, geranyl acetone, anethole, isosafrole*, phenol, carvacrol, eugenol, β-sesquiphellandrene, β-selinene, germacrene B, and indole were also found in the peel oil by the above listed authors.

Leaf oil

The composition of *C. unshiu* leaf oil has been studied by Kamiyama (1967), Kamiyama (1968), Kekelidze et al. (1981), Kekelidze et al. (1984), Keklidze et al. (1985) and Huang et al. (2000). A summary of their findings can be seen in Table 14.2.

In 1990, Ogihara et al. compared the leaf oil composition of diploid and tetraploid *C. unshiu*. They found that the oil compositions of the diploid and tetraploid forms were very similar as can be seen as follows:

camphene	$1.2\%^a$, $0.6\%^b$
limonene	$2.6\%^a$, $3.2\%^b$
p-cymene	7.55^a, $9.1\%^b$
γ-terpinene	$21.8\%^a$, $19.6\%^b$
linalool	$2.3\%^a$, $2.1\%^b$
β-elemene	$4.4\%^a$, $3.6\%^b$
α-elemene	$4.0\%^a$, $2.5\%^b$
β-caryophyllene	$8.4\%^a$, $6.4\%^b$
β-bisabolene	$0.4\%^a$, $0.8\%^b$

a = diploid; b = tetraploid.

Flower oil

Analyses of the flower oil and absolute of *C. unshiu* have been performed by Sakurai et al. (1979), Kekelidze et al. (1985), Kharebeva and Tsertsvadze (1986) and Toyoda et al. (1993). The results of these analyses are summarised as follows:

α-pinene	tr–2.8%
camphene	tr–6.9%
β-pinene	tr–3.7%
δ-3-carene	tr–1.5 %
myrcene	0.1–2.4%
α-terpinene	tr–2.9%
limonene	0.1–32.2%
γ-terpinene	tr–0.1%
(E)-β-ocimene	tr–1.9%
p-cymene	0–0.3%
tridecane	0–0.1%

nonanol	0–0.1%
linalool	1.9–2.9%
linalyl acetate	0.1–1.8%
terpinen-4-ol	0.1–4.8%
methyl thymol	0–2.6%
neral	tr–0.3%
α-terpineol	0.7–2.4%
heptadecane	0–0.1%
geranial	0.1–0.8%
citronellol	0–1.1%
geranyl acetate	tr–1.6%
ethyl phenylacetate	0–0.1%
octadecane	0–0.1%
hexadecanoic acid	0–0.1%
geraniol	0.1–0.6%
benzyl alcohol	0.1–6.9%
nonadecane	tr–4.6%
2-phenethyl alcohol	1.1–5.1%
phenylacetonitrile	2.8–4.7%
(Z)-jasmone	0.4–0.8%
eicosane	0–0.1%
methyl myristate	0–0.1%
nerolidol	0–0.1%
m-cresol	0–0.1%
heneicosane	0–0.3%
ethyl cinnamate	0–0.3%
thymol	tr–12.2%
docosane	0–0.1%
methyl anthranilate	1.8–2.0%
ethyl palmitate	0–0.1%
decanoic acid	0–0.1%
tricosane	0–0.6%
(E,Z)-farnesol	0–0.1%
(E,E)-farnesol	3.9–13.0%
benzoic acid	0–0.2%
indole	0.3–12.5%
methyl oleate	0.1–0.2%
phenylacetic acid	0–0.2%
myristic acid	0–0.5%
pentadecanoic acid	0–0.2%
palmitic acid	0–13.7%
linoleic acid	0–6.0%

Trace amounts of sabinene, β-phellandrene, terpinolene, allo-ocimene, tetradecane, pentadecane, hexadecane, heptadecane, heptadecene, nonadecane, tetracosane, β-elemene, β-caryophyllene, methacrolein, valeraldehyde, 2-methyl-2-butenol, nonanal, 3,7,11-trimethyl-dodecatrienal, benzaldehyde, phenylacetaldehyde, jasmone, 6-methyl-5-hepten-2-one, piperitone, 6,10,14-trimethyl-2-pentadecanone, geranyl acetone, 2-methyl-6-methylene-

1,7-octa-dien-3-one, 5-(3-furyl)-2-methyl-4-penten-3-one, 1,8-cineole, *cis*-linalool oxide, *trans*-linalool oxide, β-ocimene-6,7-oxide, myrcene-6,7-oxide, caryophyllene oxide, hexanol, (Z)-3-hexenol, *trans*-sabinene hydrate, *cis*-sabinene hydrate, piperitol, menthol, *p*-cymen-8-ol, nerol, 3-methylene-7-methyl-1,5-octadien-7-ol, 3-methylene-7-methyl-1,7-octadien-6-ol, 2,6-dimethyl-1,5,7-octatrien-3-ol, tridecyl alcohol, geranyl linalool, elemol, 2-phenethyl alcohol, vanillin, anethole, carvacrol, 4-heptenoic acid, levulinic acid, octanoic acid, 2-nonenoic acid, undecanoic acid, pentadecanoic acid, lauric acid, palmitic acid, sebacic acid, stearic acid, oleic acid, linoleic acid, suberic acid, benzoic acid, *p*-toluic acid, cinnamic acid, *p*-anisic acid, 3-phenylpropionic acid, 4-isopropylbenzoic acid, (Z)-3-hexenyl acetate, neryl acetate, farnesyl acetate, methyl (E)-jasmonate, methyl (Z)-jasmonate, methyl pentadecanoate, methyl stearate, methyl palmitate, methyl linoleate, methyl benzoate, methyl salicylate, methyl phenylacetate, methyl cinnamate, ethyl decanoate, ethyl stearate, ethyl myristate, ethyl oleate, ethyl linoleate, ethyl benzoate, 2-phenethyl formate, 2-phenethyl acetate, anthranilaldehyde, methyl anthranilate, methyl *N*-methyl anthranilate, methyl *N*-formylanthranilate, ethyl anthranilate, benzyl cyanide, phenylaldoxime, 2-phenylnitromethane and 3-ethyl-4-methylpyridine were also found in the flower oil by the above listed authors.

Juice oil

According to Hattori (1959), after World War II the juice of *Citrus natsudaidai* Hayata and that of *C. unshiu* were blended together to make some of the popular citrus drinks in Japan. The author also stated that 'Unshiu Mikan' (*C. unshiu*) was the most popular citrus fruit in Japan.

The composition of oils isolated from the juice of *C. unshiu* has been examined by Imagawa *et al.* (1974) and Yajima *et al.* (1979), Kim *et al.* (1980) and Araki and Sakakibara (1991), and found to contain the following components:

2-butanone	0–0.1%
ethanol	0–0.8%
butanol	0–1.8%
isobutanol	0–0.1%
isoamyl alcohol	0.2–10.0%
amyl alcohol	0–1.1%
hexanol	0–1.3%
(E)-2-pentenol	0–1.1%
(Z)-3-hexenol	0–1.9%
(E)-2-hexenol	0–0.5%
hexanal	0.2–5.1%
heptanal	0–0.1%
1-penten-3-ol	0–0.1%
(E)-2-pentenal	0–0.8%
(E)-2-hexenal	0–9.2%
(E)-2-octenal	0–0.4%
(E)-2-nonenal	0–0.1%
octanol	1.0%
butyl ethyl ether	0–1.8%

4-hydroxy-2-butanone	0–1.1%
benzaldehyde	0–0.1%
α-pinene	0.2–1.0%
sabinene	0–0.2%
β-pinene	0–0.4%
myrcene	0.9–1.3%
α-terpinene	0–0.2%
p-cymene	0.2–0.7%
1,8-cineole	0–0.1%
limonene	51.6–70.8%
β-phellandrene	0–0.8%
γ-terpinene	2.9–4.7%
cis-linalool oxide	0.1–0.3%
terpinolene	0.1–0.7%
octanal	0–0.4%
linalool	0.6–11.9%
citronellol	0–0.5%
terpinen-4-ol	0.3–1.5%
α-terpineol	1.1–2.1%
nerol	0–0.1%
dihydrocarvone	0–0.1%
nonanol	0–0.1%
carvone	0–0.1%
benzyl alcohol	0–0.1%
cis-p-mentha-2,8-dien-1-ol	0–0.1%
2-phenethyl alcohol	0–0.1%
perillaldehyde	0–0.4%
citronellyl acetate	0–0.1%
neryl acetate	0–0.4%
geranyl acetate	0.2–0.4%
limonen-10-yl acetate	0–0.1%
α-copaene	0–0.2%
limonen-10-ol	0–0.2%
β-elemene	0–0.2%
β-caryophyllene	0.1–0.2%
β-sesquiphellandrene	0–4.9%
(E,E)-α-farnesene	0–0.1%
germacrene D	0–0.1%
β-selinene	0–0.1%
δ-cadinene	0–0.1%

The above listed authors also found trace amounts of hexane, α-thujene, camphene, α-phellandrene, isopropyl ethyl ether*, acetaldehyde, propionaldehyde, furfuraldehyde, nonanal, acetone, 2-propanol, 2-butanol, 3-methyl-2-butenol, 3-methyl-2-butanol, heptanol, cyclohexanol, α-fenchyl alcohol, trans-sabinene hydrate, β-terpineol, borneol, terpinen-1-ol, geraniol, p-cymen-8-ol, dihydrolimonen-10-ol, perillyl alcohol, nerolidol, α-cadinol, octyl acetate, perillyl acetate, α-terpinyl acetate and longifolene in the juice oil.

REFERENCES

Altenburger, R. and Matile, P. (1990) Further observations on rhythmic emission of fragrance in flowers. *Planta*, 180, 194–197.

Araki, C. and Sakakibara, H. (1991) Changes in the volatile flavor compounds by heating Satsuma mandarin (*Citrus unshiu* Marcov.) juice. *Agric. Biol. Chem.*, 55, 1421–1423.

Agarwal, S.G., Lal, S., Thappa, R.K., Kapahi, B.K. and Sarin, Y.K. (1989) Seasonal studies of India *Citrus jambhiri* Lush. leaf oil an new chemotype. *Flav. Fragr. J.*, 4, 33–36.

Asano, K.-I. (1997) Thailand in view of her agricultural products particularly aromatic plants, spices and a number of *Citrus Koryo*, 195, 53–64.

Baaliouamer, A., Meklati, B.Y., Fraisse, D. and Scharff, C. (1992) The chemical composition of some cold-pressed citrus oils produced in Algeria. *J. Essent. Oil Res.*, 4, 251–258.

Bordoloi, A.K., Pathak, M.G., Sperkova, J. and Leclercq, P.A. (1999) Volatile constituents of the fruit peel oil of *Citrus maxima* (J. Burman) Merrill from Northeast India. *J. Essent. Oil Res.*, 11, 629–632.

Calvarano, I., Bovalo, F. and Di Giacomo, A. (1974) L'olio essenziale di clementine. *Essenz. Deriv. Agrum.*, 44, 117–123.

Cameron, J.W. and Scora, R.W. (1973) Leaf oils and fruit characters in relation to genetic identity among twin and triplet *Citrus* hybrids. *Lloydia*, 36, 410–415.

Cappello, C., Calvarano, M., Di Giacomo, A. and Gioffre, D. (1982) Sull' olio essenziale di cedro (*Citrus medica* L.). *Essenz. Deriv. Agrum*, 52, 59–66.

Cheng, Y.-S. and Lee, C.-S. (1981) Composition of leaf essential oils from ten *Citrus* species. *Proc. Natl. Sci. Counc. B. Roc.*, 5, 278–283.

Chialva, F. and Doglia, G. (1990) Essential oil constituents of chinotto (*Citrus aurantium* L. var. *myrtifolia* Guill.). *J. Essent. Oil Res.*, 2, 33–35.

Cotroneo, A., Verzera, A., Alfa, M. and Dugo, G. (1986) Sulla genuinità delle essenze agrumarie Nota XIV. L'olio essenziale di cedro. *Essenz. Deriv. Agrum.*, 56, 105–120.

De Rocca Serra, D., Lota, M.L., Tomi, F. and Casanova, J. (1998) Essential oils and taxonomy among *Citrus* example bergamot. *EPPOS* (Spec. Num.), 38–43.

Dugo, G., Cotroneo, A., De Filippo, V. and Daghetta, A. (1988) Sulla genuinità delle essenze agrumarie. Nota XXI gli alcolati di Cedro. *Industria Bevande*, 17, 17–31.

Dugo, G., Cotroneo, A., Trozzi, A., Barbeni, M. and Di Giacomo, A. (1988) On the genuineness of citrus essential oils. Part XXV. The essential oil from 'Wenzhou Honey' oranges. *Flav. Fragr. J.*, 3, 161–166.

Dung, N.X., Pha, N.M., Lo, V.N., An, N.T.K. and Leclercq, P.A. (1991) The essential oil from the flowers of *Citrus maxima* (J. Burman) Merrill from Vietnam. *J. Essent. Oil Res.*, 3, 359–360.

Dung, N.X., Pha, N.M., Lo, V.N., Thien, N.H. and Leclercq, P.A. (1996) Chemical investigation of the fruit peel oil of *Citrus medica* L. var. *sarcodactylis* (Noot.) Swingle from Vietnam. *J. Essent. Oil Res.*, 8, 15–18.

Escher, S., cited in Flament, I. and Ohloff, G. (1985) Volatile constituents of algae. Odoriferous constituents of seaweeds and structure of nor-terpenoids identified in asakusa–nori flavour. *Progress in Flavour Research 1984*. J. Adda (ed.), Elsevier Science Publ., Amsterdam.

Fleisher, Z. and Fleisher, A. (1991) Aromatic plants of the Holy Land and the Sinai. Part VI. The essential oils of etrog (*Citrus medica* L. var. *ethrog* Engl.). *J. Essent. Oil Res.*, 3, 377–379.

Fleisher, Z. and Fleisher, A. (1998) Ethrog: the first citrus of the Western world. *Perfum. Flav.* 21(6), 11–16.

Gao, S.Y. and Zheng, H.-J. (1986) Determination of the main constituents of the essential oils from the peels and mature fruits of *Citrus* by gas chromatography. *Yaowu Fenxi Zazhi*, 6(2), 83–85.

Gazea, F., Calvarano, I. and Calvarano, M. (1998) Characteristics of a new citrus hybrid essential oil *Citrus clementina* x *C. limon*. *J. Essent. Oil Res.*, 10, 235–239.

Gupta, G.N. and Gupta, J.C. (1957) Petitgrain oil of *Citrus microcarpa* Bunge or Hazara. *Indian Perfum.*, 1(1), 36–38.

Hattori, S. (1959) Biochemical studies of natsudaidai. *Amer. Perfum. Aromat.*, 73(4), 30–35.

Huang, Y.-Z., Pu, Z.-L. and Chen, Q.-Y. (2000) The chemical composition of leaf oils from 110 *Citrus* species, varieties, cultivars and hybrids of Chinese origin. *Perfum. Flav.* 25(1), 53–66.

Huang, Y.-Z., Cao, Y.-H., Chen, Q.-Y. and Wu, W-L. (1993) Studies on the chemical components of essential oil from the leaves of *Citrus junos* (Sieb.) Tan. cv. luohancheng Hort. *Chemistry Industry Forest Prod.*, 13, 165–168.

Huet, R., Dalnic, R., Cassin, J. and Jacquemond, C. (1986) Le cedrat Mediterranéen. Le cedrat de Corse. *Fruits*, 41(2), 113–119.

Ifuku, Y., Yonei, H., Takahashi, Y. and Yutaka, J. (1992) Separation of volatile components in citrus peel utilization of membranes. *Shokohin Sangyo Senta Gijutsu Kenkyu Hokoku*, 18, 31–42.

Imagawa, K., Yamanishi, T. and Koshika, M. (1974) Changes in volatile flavor constituents during manufacturing the concentrated juice from *Citrus unshiu*. *Nippon Nogeikagaku Kaishi*, 48, 561–567.

Inoma, S., Miyagi, Y. and Akieda, T. (1989) Characterization of citrus oil (mainly orange oil, mandarin oil, tangerine oil). *Kanzei Chuo Busekishoho*, 29, 87–97.

Jantan, I., Ahmad, A.S., Ahmad, A.R., Ali, N.A.M. and Ayop, N. (1996) Chemical composition of some citrus oils from Malaysia. *J. Essent. Oil Res.*, 8, 627–632.

Kamiyama, S. (1967) Studies of the leaf oils of *Citrus* species. Part I. Composition of leaf oils from *Citrus unshiu, Citrus natsudaidai, Citrus kokitsu* and *Citrus limon*. *Agric Biol. Chem.*, 31, 1091–1096.

Kamiyama, S. (1968) Studies on the leaf oils of *Citrus* species. Part II. An examination of the seasonal variation of leaf oil composition. *Bull. Brew. Sci.*, 14(12), 43–48.

Kamiyama, S. (1970) Studies on leaf oils of *Citrus* species. Part IV. Composition of leaf oils from funadoko-mikan, sanbokan, kawabata-mikan, shiikuwasha, yuzu and otaheite-orange. *Agric. Biol. Chem.*, 34, 1561–1568.

Kamiyama, S. and Amaha, M. (1972) Studies on the leaf oils of *Citrus* species VI. Composition of leaf oils from ten *Citrus* taxa and some intrageneric hybrids. *Bull Brew. Sci.*, 18, 17–27.

Kato, Y. (1974) Organoleptic assessment of the flavor in *Citrus unshiu*. *Kaseigaku Zasshi*, 25, 436–442.

Kekelidze, N.A. and Dzhanikashvili, M.I. (1982) Essential oil from *Citrus junos* leaves. *Khim. Prir. Soedin.*, (6), 786.

Kekelidze, N.A. and Dzhanikashvili, M.I. (1985) Essential oils in the fruit of early maturing varieties of *Citrus unshiu*. *Khim. Prir. Soedin.*, (4), 572–573.

Kekelidze, N.A., Dzhanikashvili, M.I. and Kutateladze, V.V. (1981) Petitgrain essential oil of the Mandarin variety Unshiu. *Maslo-Zhir. Prom-St.*, (2), 35–36.

Kekelidze, N.A., Dzhanikashvili, M.I. and Fishman, G.M. (1985) Production of absolute essential oil from flowers of Mandarin (*Citrus unshiu*). *Maslo Zhir. Prom. St.*, (12), 18–19.

Kekelidze, N.A., Dzhanikashvili, M.I. Tartarishvili, A.N. and Bagdosvili, T.P. (1984) Essential oils from the leaves of *Citrus unshiu*. *Khim. Prir. Soedin.*, (5), 607–610.

Kharebeva, L.G. and Tsertsvadze, V.V. (1986) Volatile compounds of flowers of *Citrus unshiu* Marc. *Subtrop. Kul't.*, 1, 119–121.

Kim, H., Jo, D.H., Park, Y.H., Lee, C.Y. and Lee, Y.H. (1980) Quantitative determination of flavor constituents of Korean milgam (*Citrus unshiu*) juice. *J. Korean Agric. Chem. Soc.*, 23, 106–114.

Kitahara, T., Yoshikazu, T. and Matsui, M. (1980) Structure and the synthesis of novel constituents of Yudzu peel oil and their conversion to related monoterpenes. *Agric. Biol. Chem.*, 44, 897–901.

Kunose, Y. and Sawamura, M. (1980) Aroma constituents of some sour *Citrus* oils. *Nippon Shokuhin Kogyo Gakkaishi*, 27, 517–521.

Lawrence, B.M., Hogg, J.W., Terhune, S.J. and Podimuang, V. (1971) Constituents of the leaf and peel oils of *Citrus hystrix* DC. *Phytochemistry*, 10, 1404–1405.

Lin, Z.-K. and Hua, Y.-F. (1989) A study on the chemical composition of the essential oil from peel of *Citrus grandis* Osbeck native to China. *Zhiwu Xuebao*, 31(1), 73–76.

Lota, M.L., de Rocca Serra, D., Tomi, F., Bessiere, J.M. and Casanova, J. (1999) Chemical composition of peel and leaf essential oils of *Citrus medica* and *C. limonimedica* Lush., *Flav. Fragr. J.*, 14, 161–166.

Lund, E.D., Shaw, P.E. and Kirland, C.L. (1981) Composition of rough lemon leaf oil. *J. Agric. Food Chem.*, 29, 490–494.

Matsuura, Y., Hata, G., Abe, S., Sakai, I. and Abe, A. (1977) A novel 13-membered macrolide. A flavour component of yuzu (*Citrus junos*) oil. *Paper No. 147, 7th International Essential Oil Congress*, Kyoto, 1997.

Miyake, M., Inaba, N., Maeda, H. and Ifuku, Y. (1990) Quality characteristics of Jabara (*Citrus jabara* Hort. ex Tanaka) and Yuzu (*Citrus junos* Sieb. ex Tanaka) fruit juice. *Nippon Shokuhin Kogyo Gakkaishi*, 37, 346–354.

Mondello, L., Dugo, P., Bartle, K.D., Dugo, G. and Cotroneo, A. (1995) Automated HPLC-HRGC: a powerful method for essential oil analysis. Part V. Identification of terpene hydrocarbons of bitter orange, grapefruit, clementine and Mexican lime oils by coupled HPLC-HRGC-MS(ITD). *Flav. Fragr. J.*, 10, 33–42.

Mondello, L., Dugo, P., Cavazza, A. and Dugo, G. (1996) Characterization of essential oil of Pummelo (c.v. Chandler) by GCLMS, HPLC and physicochemical indices. *J. Essent. Oil Res.*, 8, 311–314.

Moreuil, C. and Huet, R. (1973) Le combava culture et débouchés à Madagascar. *Fruits*, 28, 703–708.

Moshonas, M.G. and Shaw, P.E. (1996) Volatile components of calamondin peel oil. *J. Agric. Food Chem.*, 44, 1105–1107.

Namba, T., Araki, T., Mikage, M. and Hattori, M. (1985) Fundamental studies on the evaluation of crude drugs. VIII. Monthly variation in anatomical characteristics and chemical components of the dried peels of *Citrus unshiu, C. aurantium* and *C. natsudaidai*. *Shoyakugaku Zasshi*, 39, 52–62.

Nemec, S. and Lund, E. (1990) Leaf volatiles of mycorrhizal and nonmycorrhizal *Citrus jambhiri* Lush. *J. Essent. Oil. Res.*, 2, 287–297.

Nigam, M.C., Dhingra, D.R. and Gupta, G.N. (1958) Essential oil from the peels of *Citrus macrocarpa* Bunge. *Indian Perfum*, 2(2), 36–38.

Nishimura, K., Horibe, I. and Tori, K. (1973) Conformations of 10-membered rings in bicyclogermacrene and isobicyclogermacrene. *Tetrahedron*, 29, 271–274.

Nishimura, K., Shinoda, N. and Hirose, Y. (1969) New sesquiterpene, bicyclogermacrene. *Tetrahedron Lett.*, 36, 3097–3100.

Nisperos-Carriedo, M.O., Baldwin, E.A., Moshonas, M.G. and Shaw, P.E. (1992) Determination of volatile flavor components, sugars and ascorbic, diascorbic and other organic acids in calamondin (*Citrus mitis* Blanco). *J. Agric. Food Chem.*, 40, 2464–2466.

Njoroge, S.M., Ueda, H. Kusunose, H. and Sawamura, M. (1994) Volatile components of Japanese yuzu and lemon oils. *Flav. Fragr. J.*, 9, 159–166.

Njoroge, S.M., Ukeda, H. and Sawamura, M. (1996) Changes in the volatile composition of yuzu (*Citrus junos* Tanaka) cold-pressed oil during storage. *J. Agric. Food Chem.*, 44, 550–556.

Ogihara, K., Munesada, K. and Suga, T. (1990) Variation in leaf terpenoids with ploidy level in *Citrus* cultivars and hybrids. *Phytochemistry*, 29, 1889–1891.

Ohta, H. and Osajima, Y. (1983) Glass capillary gas chromatographic analysis of oil components extracted from yuzu (*Citrus junos*) juice. *J. Chromatogr.*, 268, 336–340.

Ortiz Marcide, J.M., Tadeo Lluch, J.L., Dias Llanos, F.J. and Estelles Adam, A. (1983) Etude des huiles essentielles de feuilles du groupe de la Mandarine clementine. Utilisation Taxonomique. *Fruits*, 38, 125–131.

Oritz, J.M., Kumamoto, J. and Scora, R.W. (1978) Possible relationships among sour oranges by analysis of their essential oils. *Int. Flav. Food Addit.*, 9, 224–226.

Perez Zayas, J., Baluja, R., Tapanes, R., Schmidt, W. and Rosado, A. (1980) On the chemical composition of Cuban petigrain oils III. Analytical study of shaddock petitgrain oil. *Revista CENIC*, 11(1/2), 9–16.

Pieribattesti, J.C. (1982) Contibution a l'etude de quelques huiles essentielles da la Reunion. *Phd. Thesis, Fac. Sci. St. Jerome, Univ. D'Aix-Marseille*, 189–195.

Poiana, M., Sicari, V. and Mincione, B. (1998) A comparison between the chemical composition of the oil, solvent extract and supercritical carbon dioxide extract of *Citrus medica* cv. Diamante. *J. Essent. Oil Res.*, 10, 145–152.

Pudil, F, Wijaya, H., Janda, V., Volfová, J., Valentová, H. and Pokorny, J. (1998) Changes in *Citrus hystrix* oil during oxidation. In E.T. Contis, C.-T. Ho, C.J. Mussinan, T.H. Parliament, F. Shahidi and A.M. Spanier (eds), *Food Flavors: Formation, Analysis and Packaging Influences*, Elsevier Sciencie Publ., B.V. Amsterdam., pp. 707–718.

Reuther, W., Webber, H.J. and Batchelor, L.D. (1967) *The Citrus Industry*. University of California, Div. Agric. Sci., Riverside, CA.

Ruberto, G., Biondi, D., Piatelli, M., Rapisarda, P. and Starrantino, A. (1994) Essential oil of the new citrus hybrid *Citrus clementina* x *C. limon. J. Essent. Oil Res.*, 6, 1–8.

Ruberto, G., Biondi, D. Piatelli, M., Rapisarda, P. and Starrantino, A. (1993) Profiles of essential oils of new *Citrus* hybrids. *Flav. Fragr. J.*, 8, 179–184.

Ruberto, G., Starrantino, A. and Rapisarda, P. (1999) Essential oils from new citrus fruits. *Essenz. Deriv. Agrum.*, 69, 15–26.

Ruberto, G., Bondi, D., Rapisarda, P., Renda, A. and Starrantino, A. (1997) Essential oil of Cami, a new citrus hybrid. *J. Agric. Food Chem.*, 45, 3206–3210.

Sakamoto, K., Inoue, A., Yoshiwa, T., Morimoto, K., Nakatani, M., Kozuka, H., Ohta, H. and Osajima, Y. (1997) Composition of peel oils from citrus hybrids 'Hayaka,' 'Southern red' and 'Shiranui.' *Nippon Nogeikagaku Kaishi*, 71, 403–412.

Sakamoto, K., Inoue, A., Yoshiwa, T., Morimoto, K., Nakatani, S. and Kozuku, H. (1994) Proportions of peel oil components in hybrid citrus seedlings. *Nippon Nogeikagaku Kaishi*, 68, 815–820.

Sakurai, K., Toyoda, T., Muraki, S. and Yoshida, T. (1979) Odorous constituents of the absolute from flower of *Citrus unshiu* Marcovitch. *Agric. Biol. Chem.*, 43, 195–197.

Sato, A., Asano, K. and Sato, T. (1990) The chemical composition of *Citrus hystrix* DC (Swangi). *J. Essent. Oil Res.*, 2, 179–183.

Sawamura, M. and Kuriyama, T. (1988) Quantitative determination of volatile constituents in the pummelo (*Citrus grandis* Osbeck form a Tosa-buntan). *J. Agric. Food Chem.*, 36, 567–569.

Sawamura, M., Kuwahara, S., Shichiri, K.-I. and Aoki, T. (1990) Volatile constituents of several varieties of pummelos and a comparison of the nootkatone levels in pummellos and other citrus fruits. *Agric. Biol. Chem.*, 54, 803–805.

Sawamura, M., Hattori, M., Yanogawa, K., Manabe, T., Akita, T. and Kusunose, H. (1983) Chemical constituents and sensory evaluation of satsuma mandarin oranges grown in vinyl houses and in open fields. Part I. Quality of satsuma mandarin fruits in vinyl houses cultures. *Nippon Nogeikagaku Kaishi*, 57, 757–763.

Sawamura, M., Kuriyama, T., Li, Z.-F. and Kusunose, H. (1989) Composition of peel oil components of yuzu (*Citrus junos* Tanaka) extracted by supercritical carbon dioxide and those of traditional ones. *Nippon Shokuhin Kogyo Gaikkaishi*, 36, 34–38.

Sawamura, M., Shichiri, K.-I., Ootani, Y. and Zheng, X.-H. (1991) Volatile constituents of several varieties of pummelos and characteristics among *Citrus* species. *Agric. Biol. Chem.*, 55, 2571–2578.

Sawamura, M., Tetsuya, T. and Kuwahara, S. (1989) Studies on the essential oils of pummelo. Part II. Changes in the volatile constituents of pummelo (*Citrus grandis* Osbeck form a Tosa-buntan) during storage. *Agric. Biol. Chem.*, 53, 243–246.

Sawamura, M., Yanogama, K., Hattori, M. and Kusunose, H. (1983) Aroma components of satsuma mandarin oranges grown in vinyl houses and in open fields. Part II. Quality of satsuma mandarin fruits from vinyl houses cultures. *Nippon Nogeikagaku Kaishi*, 57, 863–871.

Sawamura, M., Sun, S.H., Ozaki, K., Ishikawa, J. and Ueda, H. (1999) Inhibitory effects of citrus essential oils and their components on the formation of N-nitrosodimethylamine. *J. Agric. Food Chem.*, 47, 4868–4872.

Schwob, R. (1977) L'essence de combava, produit naturel original. *Parfums Cosmetiques Arômes*, (16), 48.

Schwob, R. and Reignier, R. (1964) L'essence de zeste de combava. *Premier Congrès Internat. Industries Agric. Aliment. Zones Trop. Subtrop.* Abidjan, December 1994.

Scora, R.W., Esen, A. and Kumamoto, J. (1976) Distribution of essential oils in leaf tissue of an F_2 population of *Citrus. Euphitica*, 25, 201–209.

Scora, R.W., England, A.B. and Chang, D. (1969) Taxonomic affinities within the rough lemon group (*Citrus jambhiri* Lush.) as aided by gas chromatography of their essential leaf oils. *Proceed. 1st International Citrus Sympos.* Riverside, CA, 1969.

Shaw, P.E. and Wilson, C.W. (1976) Composition of extracted peel oil composition and juice flavor for rough lemon, Persian lime and a lemon-lime cross. *J. Agric. Food Chem.*, 24, 664–666.

Shinoda, N., Shiga, M. and Nishimura, K. (1970) Constituents of yuzu *Citrus junos* oil. *Agric. Biol. Chem.*, 34, 234–242.

Shiota, H. (1990) Volatile components in the peel oil from fingered citron (*Citrus medica* L. var. *sarcodactylis* Swingle). *Flav. Fragr. J.*, 5, 33–37.

Sun, H.-D., Ding, L.-S. and Wu, Y. (1984) The chemical constituents of the essential oil from the leaves of *Citrus medica. Yunnan Zhiwu Yanjiu (Acta Bot. Yunnan.)*, 6, 457–460.

Tajima, K., Tanaka, S., Yamaguchi, T. and Fujita, M. (1990) Analysis of green and yellow yuzu peel oils (*Citrus junos* Tanaka). Novel aldehyde components with remarkably low odor thresholds. *J. Agric. Food Chem.*, 38, 1544–1548.

Tanaka, T. (1961) *Citrologia. Semi-centennial Commemoration Papers on Citrus Studies*. Citrologia Supporting Foundation, Osaka.

Toyoda, T., Nohara I. and Sato, T. (1993) Headspace analysis of volatile compounds emitted from various *Citrus* blossoms. In R. Teranishi, R.G. Buttery and H. Sagisawa (eds), *Bioactive volatile compounds from plants*. Amer. Chem. Soc., Washington, DC, pp. 205–218.

Verzera, A., Mondello, L., Trozzi, A. and Dugo, P. (1997) On the genuineness of Citrus essential oils. Part LII. Chemical characterization of essential oil of three cultivars of *Citrus clementina* Hort. *Flav. Fragr. J.*, 12, 163–172.

Verzera, A. Stagno D'Alcontres, I., Trozzi, A. and Saitta, M. (1991) Sulla genuinità delle essenze agrumarie. Nota XXXVIII. La composizione della frazione volatile dell'olio essenziale di chinotto. *Essenz. Deriv. Agrum.*, 61, 232–330.

Verzera, A., Trozzi, A., Mondello, L., Dellacassa, E. and Lorenzo, D. (1998) Uruguayan essential oils. Part X. Composition of the oil of *Citrus clementina* Hort. *Flav. Fragr. J.*, 13, 189–195.

Wang, D.-J. (1979) Studies on the constituents of the essential oils of four aromatic flowers. *National Sci. Council Monthly (Taiwan)*, 7, 1036–1048.

Watanabe, I., Yanai, T., Furuhata, A., Awano, K., Kogami, K. and Hayashi, K. (1983) Volatile components of yuzu (*Citrus junos*). *Proceedings of IXth International Congress of Essential Oils*, Singapore, March 1983.

Yajima, I., Yanai, T., Nakamura, M. Sakakibara, H. and Hayashi, K. (1979) Composition of the volatiles of peel oil and juice from *Citrus unshiu. Agric. Biol. Chem.*, 43, 259–264.

Yang, R., Sugisawa, H., Nakatani, H., Tamura, H. and Takagi, N. (1992) Comparison of odor quality in peel oils of acid citrus. *Nippon Shokuhin Kogyo Gakkaishi*, 39, 16–24.

Yeh, P.-H. (1955) Oil of Yuh Tsu *Citrus maxima* (Burm. f.) Merrill essential oils. II. *Amer. Perfum. Essent. Oil Rev.*, 65(2), 29–30.

Yukawa, C., Osaki, K. and Iwabuchi, H. (1994) Study on the volatile components of yuzu (*Citrus junos* Sieb. ex Tanaka). *Japan J. Food Chem.*, 1, 46–49.

Zhu, L.-F., Li, Y.-H., Li, B.-L., Lu, B.-Y. and Xia, N.-H. (1993) *Aromatic Plants and Essential Constituents*. Hai Feng Publ. Co., Peace Book Co., Hong Kong.

Zhu, L.-F., Li, Y.-H., Li, B.-L., Lu, B.-Y. and Zhang, W.-L. (1995) *Aromatic Plants and Essential Constituents (Supplement 1)*. Hai Feng Publ. Co., Peace Book Co., Hong Kong.

15 The oxygen heterocyclic compounds of citrus essential oils

Paola Dugo and David McHale

INTRODUCTION

The early studies into citrus essential oils revealed that they gave significant amounts of non-volatile residue on distillation. The isolation of pure crystalline compounds, such as bergapten (5-methoxypsoralen) from bergamot oil (Pomerantz, 1891) and limettin (5,7-dimethoxycoumarin) from lime oil (Tilden and Burrows, 1902), provided the first indication that these residues contained oxygen heterocyclic compounds. Further studies over the years have established that coumarins, psoralens (furanocoumarins) and polymethoxyflavones are mainly present. Some 70 of these compounds have been reported to occur in the cold-pressed peel oils of the *Citrus* species (Gray and Waterman, 1978; Murray *et al.*, 1982). Without exception they are all derivatives of the three basic structures shown in Figure 15.1. The majority of the coumarins and psoralens are substituted in the numbered positions by methoxy and/or isopentenyl or geranyl related alkoxy groups. In certain cases, the terminal double bond of a prenyl substituent has undergone oxidation forming an epoxide which may subsequently react further to yield a side chain containing a diol, enol or carbonyl. Methoxy substitutents predominate in the flavones but hydroxy groups are occasionally present.

Although bergapten and limettin (also known as citropten) were identified as constituents of citrus oils as early as the turn of the nineteenth century, it was not until nearly fifty years later that other psoralens and coumarins were found to be present. Späth and Kainrath (1937) isolated bergaptol and bergamottin from bergamot oil and Caldwell and Jones (1945) showed that a pale yellow solid which had separated from West Indian lime oil on standing, consisted mainly of isopimpinellin and limettin, together with a small amount of 5-geranyloxy-7-methoxycoumarin.

A major advance in knowledge of the oxygen heterocyclic compounds of citrus followed from a chromatographic study of lemon oil on powdered silicic acid. This was undertaken by Stanley and Vannier (1957a) to explain the difference between the position of its UV maximum (315 nm) and that of limettin (326 nm). The results established that the UV maximum of lemon oil was not due to a single component but, to a mixture of closely related compounds with UV maxima dependent on their aromatic substitution patterns. Later, D'Amore and Calapaj (1965), using thin layer chromatography (TLC) on silica gel, were able to demonstrate that the peel oils of lemon, bergamot, mandarin and sweet and bitter orange contained a range of substances that were strongly fluorescent under UV light. Isolation and extraction of the individual bands and measurement of

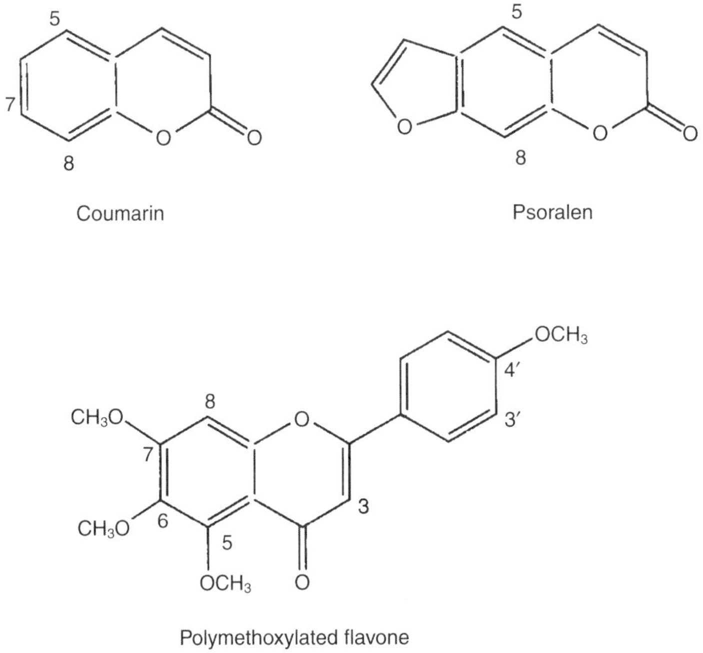

Figure 15.1 Structures of oxygen heterocyclic compounds present in citrus oil.

their UV spectra suggested that the substances responsible were coumarins, psoralens and polymethoxyflavones.

The recognition that citrus peel oils contained complex mixtures of oxygen heterocyclic compounds led Stanley and Jurd (1971) to conclude that knowledge of the identity and distribution of such compounds offered a means of detecting the adulteration of one oil with another. However, while the separation procedures allowed the structures of the individual compounds to be established, the analytical procedures were not sufficiently precise to provide reliable information on the concentrations at which individual compounds occurred. It is only within the last three decades that new instrumental techniques have been developed that incorporate highly sensitive detectors which allow more accurate determinations of the concentrations of these components to be made, even when present in trace amounts.

Due to the non-volatile nature of the oxygen heterocyclic compounds of citrus, liquid chromatography is without doubt the technique of choice for their analysis. Some gas chromatographic (GC) methods have been developed (Berahia *et al.*, 1994; Chouchi and Barth, 1994), but TLC and high-performance liquid chromatography (HPLC) in the normal or the reversed phase mode, are the most used techniques (Dugo *et al.*, 2000). Ultraviolet and fluorescence detectors are also widely used. Recently, the development of MS interface technology has allowed rapid advances in the on-line coupling of LC and MS, and affords a very powerful technique for the identification of natural components (Dugo *et al.*, 2001; Careri *et al.*, 1998).

In contrast to the wealth of analytical data available for the volatile fraction of the different citrus peel oils, there is little published quantitative information on the composition of the corresponding oxygen heterocyclic fraction. Moreover, data often relate to a limited number of samples whose geographical origin, date of production and method of extraction are not reported. The early analytical data for citrus coumarins and psoralens were reviewed by Lawrence (1982) and that for polymethoxy-flavones in (1984). For convenience and because of their wide acceptance, the trivial names for these compounds have been included together with their chemical equivalent in Table 15.10.

THE DISTRIBUTION OF THE DIFFERENT OXYGEN HETEROCYCLIC COMPOUNDS THROUGHOUT THE CITRUS PEEL OILS

Lemon

The first coumarin to be reported in lemon oil was citropten, a crystalline solid isolated from the waxy residue remaining after the distillation of the oil. Schmidt (1901) established that it was identical to the limettin that Tilden and Beck (1890) had obtained from lime oil. From structural studies, Tilden and Burrows (1902) concluded that limettin was a dimethoxycoumarin and this was confirmed when Schmidt (1904) proved that it was 5,7-dimethoxycoumarin by its synthesis from phloroglucinol aldehyde.

No further work on the oxygen heterocyclic compounds of lemon oil appeared in the literature until Stanley and Vannier (1957a) reported the isolation of seven crystalline substances having UV spectra characteristic of coumarins or psoralens. The identities of four of these were established unequivocally as limettin, 5-geranyloxy-7-methoxycoumarin, 5-geranyloxypsoralen and byakangelicin. A fifth compound was tentatively identified as 8-geranyloxypsoralen. Stanley (1963) then reported the occurrence in lemon oil of a further five psoralens and another coumarin. Among these, imperatorin (8-isopentenyl-oxypsoralen) was recognised by comparison with an authentic sample, and the structures of isoimperatorin (5-isopentenyloxypsoralen) and 5-isopentenyloxy-7-methoxycoumarin were deduced by synthesis. The remaining compounds were tentatively identified as oxypeucedanin hydrate, phellopterin and 5-geranyloxy-8-methoxypsoralen.

The first quantitative data, other than chromatographic isolation yields, appeared in 1969 when Cieri reported the analysis of two samples of lemon oil by TLC on silica gel. The bands corresponding to citropten, bergamottin and 5-geranyloxy-7-methoxy-coumarin were eluted and their concentration measured by UV spectophotometry. A year later, Madsen and Latz (1970) analysed five lemon oils of different geographical origin. Quantitative results for citropten and 5-geranyloxy-7-methoxycoumarin were obtained by *in situ* fluorimetry of their thin layer chromatograms. Using a similar technique, Calabrò and Currò (1976) measured the amounts of citropten, bergamottin and 5-geranyloxy-7-methoxycoumarin present in twenty samples of Italian genuine lemon oils.

Fisher and Trama (1979) used normal phase HPLC to investigate the coumarins and psoralens in a sample of lemon oil isolated hypodermically from the oil glands of the fruit. Their system showed eight peaks but only five of these were identified (two as coumarins and three as psoralens), and no quantitative data were reported. Several years later, McHale and Sheridan (1988) were able to resolve lemon oil into fourteen peaks by applying gradient elution. Detection was by UV absorption with a photodiode array and the identities of the eluting peaks were checked by scanning their UV absorption

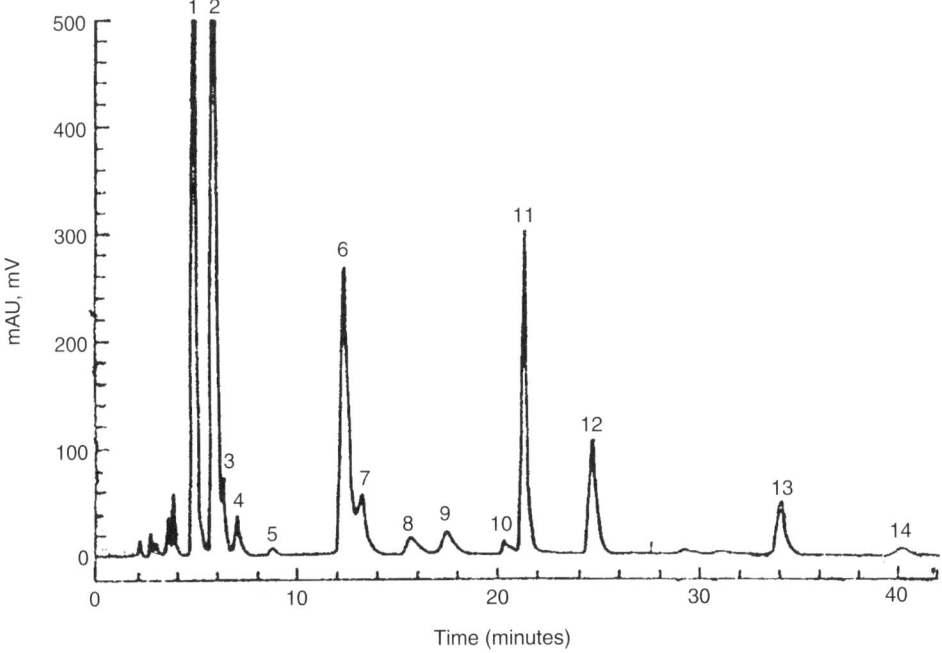

Figure 15.2 Normal phase HPLC chromatogram of oxygen heterocyclic components of genuine cold-pressed lemon oil. (1) bergamottin; (2) 5-geranyloxy-7-methoxycoumarin; (3) isoimperatorin; (4) isopentenyloxy-7-methoxycoumarin; (5) unidentified; (6) citropten; (7) 8-geranyloxypsoralen; (8) phellopterin + imperatorin; (9) 5-isopent-2'-enyloxy-8-(2',3'-epoxyisopentyloxy)psoralen; (10) unidentified; (11) oxypeucedanin; (12) byakangelicol; (13) oxypeucedanin hydrate; (14) byakangelicin. Reproduced from McHale and Sheridan, 1988.

spectra. Where necessary, the compounds responsible were isolated by preparative HPLC and their identities confirmed by MS and ^{1}H-NMR.

The resolution of the coumarins and psoralens of lemon oil is shown in Figure 15.2. The UV spectra of ten of the peaks suggested that they represented single components and could be assigned to previously reported constituents of lemon oil. The compound responsible for peak nine was identified as 5-isopent-2'-enyloxy-8-(2',3'-epoxyisopentyloxy)-psoralen. It was apparent from the spectral data that the fraction contained minor amounts of two other psoralens. These were tentatively identified as phellopterin and imperatorin on the basis of the presence of $(M+1)^+$ ions at m/z 301 and 271 in the desorption chemical ionisation mass spectrum, using ammonia as reagent gas.

Ziegler and Spiteller (1992) undertook a detailed study of the non-volatile components of Sicilian lemon oil. The residue remaining after removal of non-polar compounds was fractionated by preparative TLC on silica gel. Eight fractions were collected and these were further separated by preparative reversed phase HPLC. Individual isolates were identified by MS and, where possible, by ^{1}H-NMR. Figure 15.3 shows the reversed phase HPLC chromatogram for the whole oil. Twenty-five coumarins and psoralens were identified, of which thirteen were already known to occur in lemon oil. With the exception of cnidicin, the level of each of the other twelve new compounds was

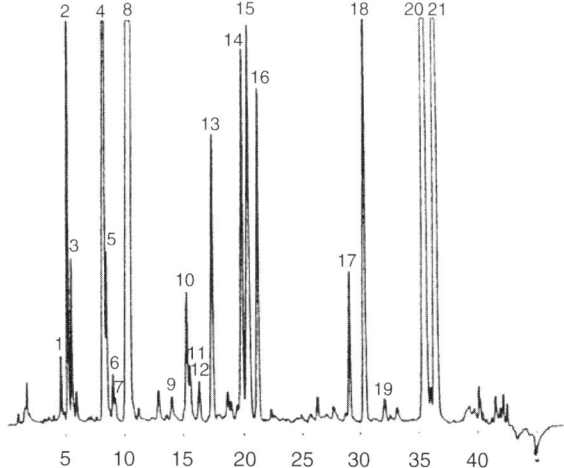

Figure 15.3 Reversed phase HPLC chromatogram of oxygen heterocyclic components of cold-pressed lemon oil. (1) 5-(2',3'-dihydroxyisopentyloxy)-7-methoxycoumarin; (2) oxypeucedanin hydrate; (3) byakangelicin; (4) citropten; (5) heraclenin; (6) pabulenol/gosferol; (7) neobyakangelicol; (8) oxypeucedanin; (9) 5-isopentenyloxy-8-(2',3'-dihydroxyisopentyloxy)psoralen; (10) imperatorin; (11) 7-isopentenyloxycoumarin; (12) 8-(6',7'-epoxygeranyloxy)psoralen; (13) phellopterin; (14) isoimperatorin; (15) 5-isopentenyloxy-7-methoxy coumarin; (16) 5-isopentenyloxy-8-(2',3'-epoxyisopentyloxy)psoralen; (17) cnidicin; (18) 8-geranyloxypsoralen; (19) aurapten; (20) bergamottin; (21) 5-geranyloxy-7-methoxycoumarin. Reproduced from Ziegler and Spiteller, 1992.

estimated not to exceed 10 mg/kg. Surprisingly, no quantitative data were included for the previously known constituents.

In contrast, a study by Dugo *et al*. (1998) of 37 samples of genuine industrial cold-pressed Italian lemon oils produced during the 1994–95 season has provided reliable quantitative data for the seven major oxygen heterocyclic compounds. The analytical resolution of the three coumarins and the ten psoralens known to be present was achieved by normal phase HPLC. In addition, all of these compounds were isolated by a combination of column chromatograhy, TLC and semi-preparative HPLC, and their identities confirmed by ^{1}H-NMR and MS. The presence of trace amounts of herniarin was also detected unequivocally in these samples, and oil extracted from the summer 'verdelli' fruit showed a significantly higher concentration than that from winter lemons.

The qualitative and quantitative data from the work discussed above, together with the other reports on the composition of lemon oil that have appeared in the literature since 1957, are summarised in Table 15.1. Except where the analytical procedure was inappropriate, bergamottin, citropten, and 5-geranyloxy-7-methoxycoumarin have always been found to be present. Irrespective of geographical origin, period of production, or lemon variety, these compounds are undoubtedly the major components of the oxygen heterocyclic fraction. Certain workers have established that oxypeucedanin and byakangelicol are, frequently, also significant components of this fraction. However, since they both contain an epoxy group that is prone to hydrolysis, their occurrence is very dependent on the conditions prevalent during the isolation of the oil. The resulting diols (oxypeucedanin hydrate and byakangelicin) are poorly soluble in the oil and tend to

Table 15.1 Qualitative and quantitative (ppm) results found in literature for oxygen heterocyclic compounds of lemon oil

	1	2	3	4	5	6	7	8	9	10	11	12	13	14*	15	16
Citropten	X	X	X	500–700	297–606	X	1200	X	1800	X	650	X	X	X	X	520–1420
5-Geranyloxy-7-methoxycoumarin	X	X	X	300	1290–1580	X	2000	X	1700	X	1600		X	X		1800–2500
5-Isopentenyloxy-7-methoxycoumarin		X				X		X	60		80		X	X		X
Herniarin															t	X
Umbelliferone													X		X	
Scoparone																
Scopoletin																
Bergamottin	X	X		1000–1200	X	X	2700	X	2000	X	2200	X	X	X		1600–2910
Bergapten			X						100						X	
Bergaptol			X	X									X			
Oxypeucedanin		X	X			X			1000		1100			X		890–1570
Oxypeucedanin hydrate		X	X		X	X			300		260		X	X		X
Isoimperatorin											180			X		X
Byakangelicol									500		450		X	X		660–1230
Byakangelicin	X	X	X		X	X			100		70		X	X		X
Phellopterin		X			X	t				X	90		X	X	X	X
5-Geranyloxy-8-methoxypsoralen		X														
5-Methoxy-8-isopentenyloxypsoralen								X								
5-Isopentenyloxy-8-(2',3'-epoxyisopentyloxy)psoralen											220			X		190–370
Isopimpinellin												X	X			
8-Geranyloxypsoralen	X	X	X		X	X			1000	X	750		X	X		190–360
Imperatorin		X									60	X	X	X		X

Notes

1 Stanley and Vannier, 1957; *2* Stanley, 1963; *3* D'Amore and Calapai, 1965; *4* Cieri, 1969; *5* Madsen and Latz, 1970; *6* Stanley and Jurd, 1971; *7* Calabrò and Currò, 1976; *8* Latz and Ernes, 1978; *9* Glandian *et al.*, 1978; *10* Fisher and Trama, 1979; *11* McHale and Sheridan, 1988; *12* Benincasa *et al.*, 1990; *13* Di Giacomo, 1990; *14* Ziegler and Spiteller, 1992; *15* Chouchi and Barth, 1994; *16* Dugo *et al.*, 1998.

t = tentative; *Ziegler and Spiteller (1992) identified also: aurapten; 7-(isopent-2'-enyloxy)coumarin; 5-(2',3'-epoxyisopentyloxy)-7-methoxycoumarin; 5-(2',3'-dihydroxyisopentenyloxy)-7-methoxycoumarin; 8-(6',7'-epoxygeranyloxy)psoralen; heraclenin; heraclenol; pabulenol/gosferol; 5-methoxy-8-geranyloxypsoralen; cnidicin; 5-Isopentenyloxy-8-(2',3'-dihydroxyisopentyloxy)psoralen; neobyakangelicol.

crystallise out on standing, particularly at sub-ambient temperatures. Consequently, the concentrations of these four substances found in a lemon oil can vary widely.

Other compounds that are generally accepted as being present, although at a lower concentration, include 5-isopentenyl-7-methoxycoumarin, isoimperatorin, phellopterin, 8-geranyloxypsoralen, and imperatorin. Glandian *et al.* (1978) detected bergapten in the Eureka variety of lemon oil from the Ivory Coast and, after isolation, confirmed its identity by IR, MS and ^{1}H-NMR. Apart from an absence of isoimperatorin, phellopterin and imperatorin, the other components comprised the usual constituents of lemon oil. Using GC-MS, Chouchi and Barth (1994) also found bergapten in the Eureka variety of lemon oil from the Ivory Coast, and identified herniarin and scoparone (tentatively) as components of lemon oil for the first time. Citropten, oxypeucedanin and phellopterin, were readily detected but no 5-geranyloxy-7-methoxycoumarin or bergamottin was observed, presumably due to the significant increase in molecular weight imposed by the geranyl group of these molecules, rendering difficult the GC analyses. The MS spectra were checked with three commercial data banks. In addition, retention times were compared with those of authentic compounds, where available.

In general, the quantitative results given in Table 15.1 show good overall agreement with perhaps the exception of the low figure for 8-geranyloxypsoralen (Dugo *et al.*, 1998) as compared to (McHale and Sheridan, 1988) and (Glandian *et al.*, 1978). The difference in the figures for byakangelicol may reflect the method of processing or the age of the oil.

Lime

Prior to 1963 only two coumarins and one psoralen had been reported to occur in lime oil (Caldwell and Jones, 1945). However, Stanley (1963), in a review of recent developments in coumarin chemistry, listed the coumarins and psoralens that occurred in the various citrus oils and recorded that lime oil contained an additional seven psoralens. Following this, Stanley and Vannier (1967) reported the isolation of ten crystalline solids from expressed West Indian lime oil (*Citrus aurantifolia* Swingle) by column chromatography on silicic acid. Compounds identified, but not previously found in lime oil, were bergamottin, isoimperatorin, 8-geranyloxypsoralen, imperatorin and phellopterin. Structural studies suggested that two other compounds were 5-geranyloxy-8-methoxypsoralen and oxypeucedanin hydrate. Quantitative data, based on the yields of individual compounds recovered, were also recorded. Soon after, Cieri (1969) carried out the first direct estimation of the coumarins and psoralens of lime oil. Samples of three commercial oils from Florida, Mexico and Guinea and one from the Florida Citrus Station were analysed by TLC on silica gel. After development, the appropriate bands were detected under UV-light, extracted into solvent and determined spectrophotometrically.

In order to study the total luminescence of the coumarins and psoralens present in expressed lime oil, Latz and Madsen (1969) carried out an isolation of the individual compounds in a similar manner to Stanley and Vannier (1967). Eleven compounds were isolated of which two did not correspond to any previously reported constituent of lime oil. On the basis of the UV spectrum and the luminescence data, one of these compounds was tentatively identified as 5-isopentenyloxy-8-methoxypsoralen. The other compound was not identified, but by TLC it was more polar than all other compounds, except oxypeucedanin hydrate. Isoimperatorin was not detected by these workers. In a later paper in which the resolution by TLC had been improved, Madsen and Latz (1970) succeeded also in identifying begapten and byakangelicin in expressed lime oil.

McHale (1980) investigated the coumarin and psoralen fraction of écuelled West Indian lime oil. The whole oil was separated by TLC on silica gel and, after visualisation under UV-light, the coumarin and psoralen bands were eluted individually. Ten fractions were collected and examined spectrophotometrically. The results showed some significant differences from those of Stanley and Vannier (1967), Latz and Madsen (1969), and Madsen and Latz (1970). Thus, 5-isopentenyloxy-7-methoxycoumarin was identified as a minor constituent (70 mg/100 g). It was also established that oxypeucedanin (300 mg/100 g), and not the hydrate, was present. The latter observation raised the question was any oxypeucedanin hydrate previously reported as a natural constituent of the oil, an artefact of the isolation procedure.

Using more sophisticated analytical techniques, McHale and Sheridan (1989) investigated the oxygen heterocyclic fraction of the West Indian or Key variety of acid lime and that of Persian lime (*Citrus latifolia* Tanaka), both with samples from different geographical origins. Normal phase gradient HPLC of the oils yielded thirteen peaks. All but one of the peaks had a UV spectrum characteristic of a single component and ten were readily assignable to previously reported coumarins and psoralens of lime oil. The UV spectrum of one peak corresponded to that of a coumarin monosubstituted at C-7. This was unexpected as no such coumarins had been reported to be lime oil constituents. The compound responsible for the peak was isolated and shown by ^{1}H-NMR and MS to be 7-methoxycoumarin (herniarin). While herniarin had been reported to occur in many plant families, there had been no corroborated evidence at that time for its occurrence in *Citrus*.

A minor late running peak with a characteristic 5,8-disubstuted psoralen UV spectrum was shown by MS and ^{1}H-NMR to be due to 5-(2',3'-epoxyisopentyloxy)-8-methoxypsoralen, a previously unknown isomer of byakangelicol. It was given the trivial name 'isobyakangelicol'. Byakangelicol was also identified for the first time as a constituent of lime oil. It co-eluted with an 8-substituted psoralen which was tentatively identified as heraclenin on the basis of the physical data. The failure to detect isoimperatorin may well have been due to a too rapid change in the gradient of the solvents in the initial stages of chromatography. The peak would have been expected to elute between bergamottin and 5-geranyloxy-7-methoxycoumarin where the window is quite narrow. Since only a trace amount was likely to be present, it could well have been masked by one or other of these two major peaks.

Although the same pattern of coumarins and psoralens was present in the two varieties of lime oil, the concentration of individual compounds was lower in the Persian oil. Analysis of the crystalline residue, which separated from the oil samples on standing, confirmed that it was primarily citropten and isopimpinellin.

Nigg *et al.* (1993) studied the composition of the phototoxic coumarin fraction in the peel and the pulp of limes. After a long and laborious method of sample preparation and extraction, reversed phase HPLC analyses were carried out. Compounds present in the samples and in the fractions isolated by TLC were identified by GC-MS. The retention times of psoralen, xanthotoxin, bergapten, citropten and isopimpinellin matched those of authentic standards in both the GC and the HPLC analyses. Psoralen and xanthotoxin had not been previously reported to be constituents of lime oil and, together with bergapten, they were considered to be the most phototoxic of the lime furanocoumarins.

Nowadays, two varieties of cold-pressed Key lime oil, known as Type A and Type B, are produced (Haro-Guzman, 1980). Dugo *et al.* (1997b) compared the qualitative and the quantitative composition of these oils with those of Persian lime oil. The normal phase HPLC chromatograms for the Key lime Type A and the Persian oil are shown in

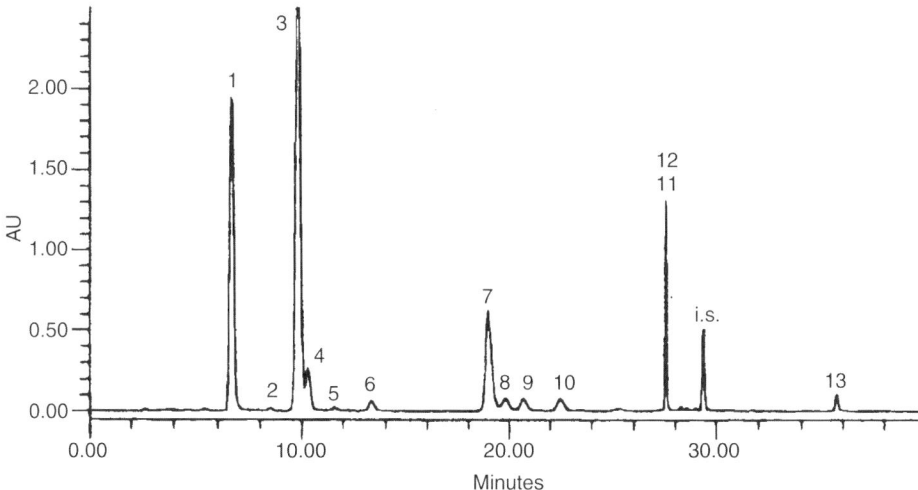

Figure 15.4 Normal phase HPLC chromatogram of a cold-pressed Key lime oil type A. (1) bergamottin; (2) isoimperatorin; (3) 5-geranyloxy-7-methoxycoumarin; (4) 5-geranyloxy-8-methoxypsoralen; (5) 5-isopentenyloxy-7-methoxycoumarin; (6) cnidilin; (7) citropten; (8) 8-geranyloxypsoralen; (9) herniarin; (10) bergapten; (11) isopimpinellin; (12) oxypeucedanin; (13) oxypeucedanin hydrate; (i.s.) tangeritin (internal standard). Reproduced from Dugo *et al.*, 1997.

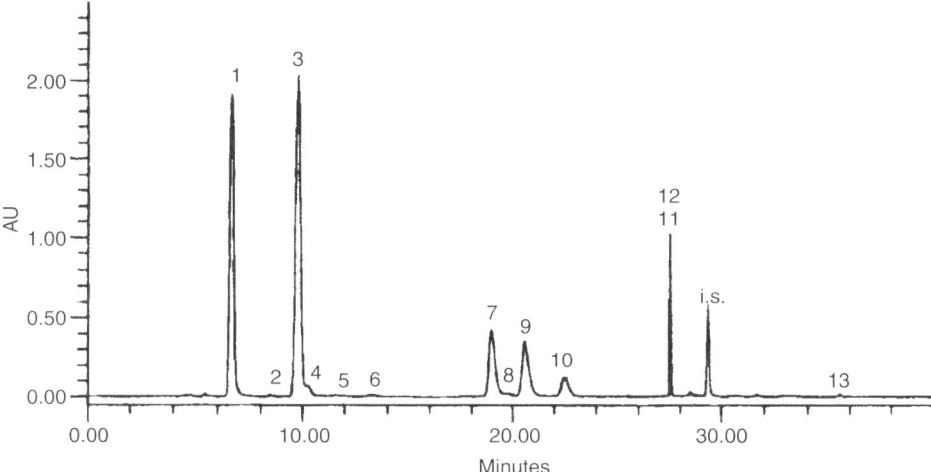

Figure 15.5 Normal phase HPLC chromatogram of a cold-pressed Persian lime oil. Peak identification: see Figure 15.4. Reproduced from Dugo *et al.*, 1997.

Figures 15.4 and 15.5, respectively. Qualitatively the chromatograms are similar, but there are quantitative differences. Of the thirteen compounds identified, oxypeucedanin was not present in the Key lime type A oil, probably due to the intimate mixing of the oil and juice during extraction. In contrast to the results of McHale and Sheridan (1989), no heraclenin, byakangelicol or isobyakangelicol was detected, but isoimperatorin and oxypeucedanin hydrate were present in all three oils.

Table 15.2 Qualitative and quantitative (mg/100g) results found in literature for oxygen heterocyclic compounds of lime oil

	1	2	3	4	5	6	7	8	9		10*		11		
	Key lime								Key lime	Persian lime	Key lime	Persian lime	Key lime type A	Key lime type B	Persian lime
Citropten	X	464	760–1700	X	203–842	X	X	X	440–670	280–340	291.1	310.1	581	484	443
5-Geranyloxy-7-methoxycoumarin	X	1725	2200–5200	X	690–4660	X	X	X	1770–1960	1490–1520			3703	4093	3111
5-Isopentenyloxy-7-methoxycoumarin									13–17	2–7			X	X	X
Herniarin									50–390	170–270			91	74	435
Psoralen												3.9			
Bergamottin		3025	2200–2500	X	X	X	X	X	1500–1600	1360–1420			3408	3154	3067
Bergapten			170–330		X	X	X	X	120–240	140–220	20.9	128.7	113	89	217
Bergaptol	X		60–100												
Oxypeucedanin									210–430	120–210			–	144	272
Oxypeucedanin hydrate		25		X	X	X	X	X					X	X	X
Isoimperatorin		33				X							X	X	X
Heraclenin									**	**					
Byakangelicol									66–100	8–16					
Byakangelicin					X	X	X	X							
Phellopterin		11		t	X	X	X	X							
5-Geranyloxy-8-methoxypsoralen		945		X	X	t	X	X	330–450	55–88			X	X	X
Cnidilin							X	X	8–15				31	24	7
Isobyakangelicol				t	X				27–46	5–9					
Isopimpinellin	X	508	0–100	X	X	X	X	X	300–570	100–210	22.0	53.7	356	331	217
Xanthotoxin												5.9			
8-Geranyloxypsoralen		105		X	X	X	X	X	160–350	58–100			X	X	X
Imperatorin		18		t	X	X	X	X							

Notes

1 Caldwell and Jones, 1945; 2 Stanley and Vannier, 1967; 3 Cieri, 1969; 4 Latz and Madsen, 1969; 5 Madsen and Latz, 1970; 6 Stanley and Jurd, 1971; 7 Calvarano and Gallino, 1975; 8 Latz and Ernes, 1978; 9 McHale and Sheridan, 1989; 10 Nigg et al., 1993; 11 Dugo et al., 1997.

t = tentative; * Results expressed as µg/g of rind fresh weight; ** Co-eluted with byakangelicol.

Table 15.2 summarises the qualitative and quantitative data available in the above papers. In all cases where quantitative data have been provided, the main oxygen heterocyclic compounds present in the oil are bergamottin and 5-geranyloxy-7-methoxycoumarin. There are quite significant differences in the values reported by the various workers for the concentrations of individual compounds. In general, the figures reported by Dugo *et al.* (1997b) for the concentrations of the major compounds exceed those of other workers. McHale and Sheridan (1989) and Dugo *et al.* (1997b) both found higher concentrations of 5-geranyloxy-7-methoxycoumarin, isopimpinellin and citropten in Key lime oil than in Persian oil, but only the latter workers observed that Persian oil had the higher concentrations of herniarin and bergapten.

Care must be taken in drawing firm conclusions from the results of analyses of different oils by different workers using a variety of procedures. Particularly, because the short chain coumarins and psoralens are much less soluble in the oil than those containing geranyl substituents and, have a greater tendency to crystallise out on prolonged standing. Moreover, the concentration of epoxy compounds remaining in the oil will be very dependent on the method of production used. The geographical origin, storage conditions and age may also influence considerably the composition of the oxygen heterocyclic fraction.

Bitter orange

Apart from the confusion over the presence of aurapten in *Citrus aurantium* Natsudaidai which is detailed under grapefruit oil, Van Os and Dykstra (1937) provided the first indication that bitter orange oil contained coumarins, by showing that it had a UV spectrum similar to other citrus peel oils. The isolation of the compound named 'aurapten' from the stearoptenes of bitter orange oil (Böhme and Pietsch, 1938) and its subsequent identification as 7-methoxy-8-(2',3'-epoxyisopentyl)coumarin (Böhme and Pietsch, 1939; Böhme and Schneider, 1939), confirmed that such compounds were present.

Patnayak *et al.* (1942) were the first to report the occurrence of methoxyflavones in citrus peel oils when they identified auranetin (3,4',6,7,8-pentamethoxyflavone) in a sample of Indian bitter orange oil. Several years later Sarin and Seshadri (1960) isolated auranetin and desmethylnobiletin (5-hydroxy-3',4',6,7,8-pentamethoxyflavone) from the peel of Nagpur oranges (*Citrus aurantium*). Stanley (1963) commented that most of the solid product obtained from bitter orange oil comprised the three polymethoxyflavones, tangeritin, nobiletin and heptamethoxyflavone. However, bergapten and the coumarin (osthol) were also present. Another 7,8-disubstituted coumarin was isolated by Stanley *et al.* (1965) when commercial Seville bitter orange oil was chromatographed on silicic acid. It was identified as 7-methoxy-8-(2'-hydroxyisopent-3'-enyl)coumarin and named 'auraptenol'. The structure of this compound was unusual in that, like osthol, the derived prenyl group was directly substituted on the coumarin ring and furthermore, at the time of this work, no other known coumarin or psoralen present in *Citrus* possessed a 2-hydroxyisopent-3-enyl substituent.

D'Amore and Calapaj (1965) detected seventeen fluorescent spots when ten samples of Italian bitter orange oil were analysed by TLC. Nine compounds were identified as oxygen heterocyclic compounds from their UV spectra and a tenth compound, as methyl anthranilate. It was not until Shu *et al.* (1975) developed an HPLC method to determine the bergapten content of citrus oils that any quantitative data became available. Fisher and Trama (1979) used reversed phase HPLC to examine bitter orange oil isolated hypodermically from the oil glands of the fruit. Detection was by UV and identification was by peak enrichment. Surprisingly, apart from osthol and bergapten, the only other compound they identified was bergaptol.

Subsequently, more details of the composition of the oxygen heterocyclic fraction were reported by McHale and Sheridan (1983, 1989), and Dugo *et al.* (1996a). Having established that the major coumarins and psoralens of bitter orange oil were unstable when subjected to column chromatography on neutral alumina, McHale and Sheridan (1983) used normal phase HPLC to examine four samples of authentic bitter orange oil hand-pressed directly from the peel of Seville oranges and one sample of commercial oil. The latter differed from the hand-pressed oils in that it showed two additional peaks. Structural studies established that the peaks of the genuine oils in order of elution were: osthol and 6',7'-epoxybergamottin, bergapten, tangeritin, unidentified, heptamethoxyflavone, meranzin, and nobiletin. The two additional peaks in the commercial oil were due to aurapten and 6',7'-epoxyaurapten. Tetra-O-methylscutallerein was also detected in this oil. The results were strongly indicative that the commercial oil had been extended with grapefruit and sweet orange oil. Meranzin was the main compound present in the genuine oils.

McHale and Sheridan (1989) obtained reliable compositional data on analysing Seville bitter orange oil using normal phase HPLC with photodiode array detection. Close examination of the nobiletin peak revealed that isomeranzin (7-methoxy-8-(2'-oxoisopentyl)coumarin) was the major contributor. A comparison of the composition of Italian and Spanish bitter orange oils was made by Dugo *et al.* (1996a). They observed that meranzin was always the main component and was present at a significantly higher concentration in the Italian oils, whereas the Spanish samples had the higher content of osthol. The results for the Spanish oils were in good agreement with those obtained by McHale and Sheridan (1989).

The normal phase HPLC chromatogram of a genuine Italian bitter orange oil is shown in Figure 15.6 (Dugo *et al.*, 1996a). In order to obtain complete separation of all the components of the fraction, Dugo *et al.* (1996a) used a switching system between two columns of different selectivity. In particular, working under normal phase conditions, the separation of meranzin from isomeranzin proved to be the most critical. Using reversed phase HPLC with a C18 column, this pair were well resolved as can be seen in Figure 15.7 (Dugo *et al.*, 2001).

McHale and Sheridan (1989) and Dugo *et al.* (1996a) isolated the individual components by preparative HPLC and identified them by ^{1}H-NMR and MS spectrometry. Tetra-O-methylscutellarein was not detected by McHale and Sheridan (1989), nor were the diols derived from meranzin and epoxybergamottin. Dugo *et al.* (1996a) concluded that meranzin hydrate forms during the industrial process of extraction of the oil, since they failed to find it in laboratory extracted oils obtained by cold-pressing the peel of the fruits and avoiding contact of the oil with water or juice.

Although meranzin hydrate may not be a natural constituent of bitter orange oil, McHale *et al.* (1987) found clear evidence for its occurrence in the flavedo. Examination of an aqueous extract of the flavedo by HPLC with dual UV/fluorescence detection revealed the presence of a number of components. Four intensely fluorescent substances were isolated from the extract and found to be closely related 7,8-disubstituted coumarins. Two of these were recognised as meranzin hydrate and isomeranzin, while the two more polar components isolated were both glycosides. One was identified as the 13-O-β-D-glucopyranoside of meranzin hydrate on the basis of its facile conversion to

The oxygen heterocyclic compounds of citrus essential oils 367

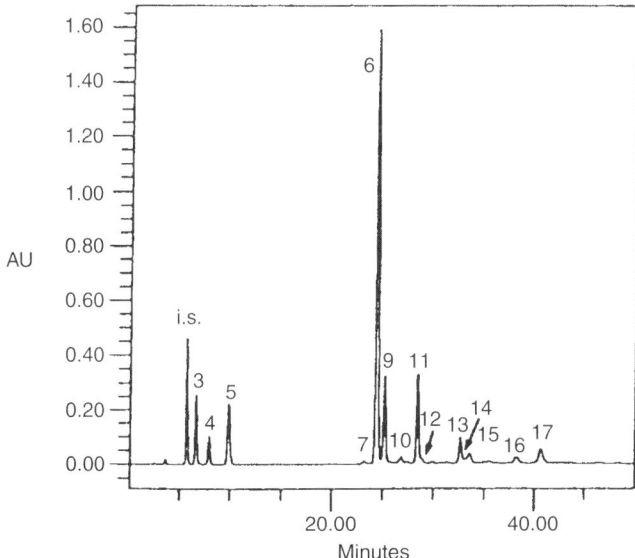

Figure 15.6 Normal phase HPLC chromatogram of oxygen heterocyclic components of cold-pressed bitter orange oil obtained using a column switching system. (3) osthol; (4) bergapten; (5) epoxybergamottin; (7) unidentified coumarin; (8) meranzin; (9) isomeranzin; (10) unidentified coumarin; (11) tangeritin; (12) heptamethoxyflavone; (13) nobiletin; (14) tetra-O-methylscutellarein; (15) unidentified coumarin; (16) epoxybergamottin hydrate; (17) meranzin hydrate. Reproduced from Dugo *et al.*, 1996a.

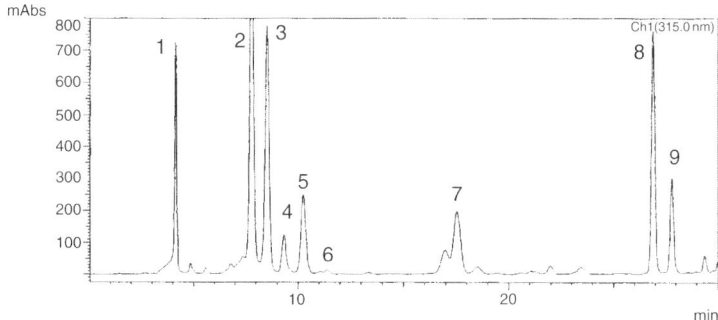

Figure 15.7 Reversed phase HPLC chromatogram of oxygen heterocyclic components of cold-pressed bitter orange oil. (1) meranzin hydrate; (2) meranzin; (3) isomeranzin; (4) nobiletin; (5) heptamethoxyflavone+bergapten; (6) tetra-O-methylscutellarein; (7) tangeritin; (8) osthol; (9) epoxybergamottin. Reproduced from Dugo *et al.*, 2000.

meranzin hydrate on treatment with an enzyme possessing β-glucosidase activity and on spectral evidence.

The results of the above studies are summarised in Table 15.3. Dugo *et al.* (1996a) also analysed certain Italian and Spanish commercial oils which they called 'anomalous'. Their results are given in Table 15.4. As can be seen, meranzin, usually the main component

Table 15.3 Qualitative and quantitative (ppm) results found in literature for oxygen heterocyclic compounds of bitter orange oil

	1	2	3	4	5	6	7	8	9	10	11*	12	13	14 Italian	Spanish
Aurapten	X														
Auraptenol											X		X		
Umbelliferone					X	X	X						X		
Meranzin		X									X	3000	X	7880–11720	3070–3320
Isomeranzin												1800		1540–2110	2080–2130
Osthol					X			X		X		620	X	1540–1840	3660–3710
Meranzin hydrate											X			100–700	180–390
Citropten							X						X		
Isoimperatorin							X						X		
Bergapten					X		X	X	690–730	X	X	1000	X	520–730	710
Bergaptol										X					
Epoxybergamottin											X	820		1880–3220	3040–3280
Epoxybergamottin hydrate											X			130–450	240–420
Auranetin			X	X			t								
Desmethylnobiletin				X											
Tangeritin					X		X				X	890	X	590–1560	950–990
Nobiletin					X		X				X	500	X	340–870	760–850
Sinensetin							t								
Heptamethoxyflavone					X						X	310		50–140	230–250
Tetra-O-methylscutellarein														100–180	50–80

Notes

1 Kamatsu *et al.*, 1930; *2* Böhme and Pietsch, 1938; *3* Patnayak *et al.*, 1942; *4* Sarin and Seshadri, 1960; *5* Stanley, 1963; *6* Stanley *et al.*, 1965; *7* D'Amore and Calapaj, 1965; *8* Stanley and Jurd; 1971; *9* Shu *et al.*, 1975; *10* Fisher and Trama, 1979; *11* McHale and Sheridan, 1983; *12* McHale and Sheridan, 1989; *13* Di Giacomo, 1990; *14* Dugo *et al.*, 1996a. t = tentative; * McHale and Sheridan (1983) also isolated 5-(6'-hydroxy-3',7'-dimethylocta-2',7'-dienyl)oxypsoralen.

Table 15.4 Oxygen heterocyclic compounds in 'anomalous' Italian and Spanish bitter orange oils (ppm) (Dugo et al., 1996a)

Compound	Italian oils	Spanish oils
Osthol	1410–1780	2320–2390
Bergapten	50–400	750–770
Epoxybergamottin	0–180	70–80
Unknown coumarin 1	50–110	20–40
Meranzin	–	–
Isomeranzin	1720–2010	2140–2290
Unknown coumarin 2	0–100	60–70
Tangeritin	560–940	700–730
Heptamethoxyflavone	0–70	120–190
Nobiletin	170–650	630–750
Tetra-O-methylscutellarein	50–80	tr–160
Unknown coumarin 3	–	0–tr
Epoxybergamottin hydrate	230–410	490–610
Meranzin hydrate	700–1320	60–70

of the fraction, was absent, and epoxybergamottin was only present at a low concentration. The samples seemed not to be contaminated or adulterated on the basis of other parameters considered indicative of authenticity, such as CD value, (−)/(+) linalool ratio and δ-3-carene and α-terpineol content. The reduction in concentration of the two epoxy compounds was considered to be a consequence of their hydrolysis during the extraction process. Two factors were thought to be involved: contact of the oil with juice during the release from the peel in the 'Torchi' machines; and contact with the aqueous citric acid used to break the resulting oil in water emulsion. The diols formed under such conditions would be preferentially soluble in the aqueous phase and much would be lost from the oil.

Grapefruit

It would appear that Komatsu et al. (1930) were the first to recognise that grapefruit oil, like lemon and lime oil, contained oxygen heterocyclic compounds. In a paper with the conflicting title 'Biochemical Studies on Grapefruits, Citrus aurantium L', they reported that a crystalline solid was obtained when a sample of cold-pressed peel oil from Citrus aurantium Natsudaidai was chilled. The solid was called 'aurapten(e)' in accord with the use of the name citropten for the first coumarin isolated from lemon oil, and was identified as 4-heptyloxycoumarin. Unfortunately, Böhme and Pietsch (1938) also used the name 'aurapten' for a compound isolated from the stearoptenes of bitter orange oil and which they identified initially as 7-geranyloxycoumarin. After further study (Böhme and Pietsch, 1939; Böhme and Schneider, 1939), the identity was reassessed and shown to be 7-methoxy-8-(2',3'-epoxyisopentyl)coumarin, a compound which had previously been named 'meranzin' by Dodge (1938). Nomura (1950) repeated the work of Komatsu et al. (1930) and established that aurapten from Citrus aurantium Natsudaidai differed from meranzin. Kariyone and Matsuno (1953) finally identified it as 7-geranyloxycoumarin. Since, nowadays, the occurrence of this compound appears to be restricted to grapefruit oil, it must be assumed that Citrus aurantium Natsudaidai is a variety of grapefruit and not bitter orange. Nomura (1950) isolated another coumarin from this fruit and named it 'auraptin'. It was later shown by Matsuno (1956) to be isoimperatorin.

The occurrence of aurapten in grapefruit oil was confirmed by Stanley and Vannier (1957b) and, in a general review of coumarin chemistry, Stanley (1963) named six other compounds as occurring in this oil, but no details were given. Using column chromatography on alumina, Fisher and Nordby (1965) isolated five coumarins and four psoralens from Florida grapefruit oil. Three of the coumarins were reported to be aurapten, osthol and citropten, while the psoralens included bergaptol, bergapten and bergamottin. The remaining psoralen and one of the unidentified coumarins were shown to be unusual in that their side chains contained an aldehyde function (Fisher and Nordby, 1966). They were identified as 5-(3',6'-dimethyl-6'-formylhept-2'-enyloxy)-psoralen (aurantiumal) and 7-methoxy-8-(2'-formyl-2'-methylpropyl)coumarin (citrusal). Further work (Fisher et al., 1967) established that the other unknown coumarin was 7-(6',7'-dihydroxygeranyloxycoumarin) (marmin).

Dreyer (1967) suggested that aurantiumal and citrusal could be formed as the result of a pinacol rearrangement but Stanley and Jurd (1971) pointed out that the appropriate precursor diols were not known. The matter was resolved when Tatum and Berry (1979) investigated the solid that separated from the oil on prolonged standing at $-20\,^{\circ}\mathrm{C}$. After fractionation by TLC on silica gel using a variety of solvent systems and multiple development, four coumarins, four psoralens and two methoxyflavones were isolated and subjected to structural study. Three of the compounds were identified as known constituents of grapefruit oil (aurapten, bergamottin and bergapten). The two methoxyflavones (nobiletin and tangeritin) had been previously found in orange and mandarin oils. The remaining compounds were epoxybergamottin, epoxybergamottin hydrate, epoxyaurapten, meranzin and meranzin hydrate. Of these, epoxybergamottin was a new natural compound. The two hydrates were of particular interest because they were the required precursors for the formation of aurantiumal and citrusal. Neither of the latter two compounds were detected in this study, nor was the previously identified marmin (epoxyaurapten hydrate), possibly because it co-eluted with epoxybergamottin hydrate. Although no quantitative analysis was undertaken, it was estimated that aurapten and epoxybergamottin were the major compounds present and that meranzin hydrate and epoxybergamottin hydrate were minor constituents of the mixture.

Using both normal and reversed phase HPLC, Fisher and Trama (1979) examined oil taken hypodermically from the glands of the fruit. They identified three coumarins and three psoralens by peak enrichment, but there were other peaks present which were not identified. As well as aurapten and osthol, they reported marmin to be present. The psoralens were said to be bergamottin, bergapten and bergaptol.

Shu et al. (1975) provided the first quantitative data on the oxygen heterocyclic compounds of grapefruit oil when they measured the bergapten content by HPLC. It was not until several years later that any detailed compositional data became available. McHale and Sheridan (1989) examined the oil by normal phase HPLC under isocratic conditions with detection by UV absorbance at 315 nm. The UV spectra of eluting peaks were monitored with a photodiode array. Nine peaks were obtained (Figure 15.8) of which eight gave UV spectra of a single component and all but one of these were assignable to known constituents of grapefruit oil. The unknown peak was identified as 3,3',4',5,6,7,8-heptamethoxyflavone, which occurred in both sweet and bitter orange oil but had not been associated with grapefruit oil. The incompletely resolved peak was shown to contain nobiletin and isomeranzin, a coumarin present in the aqueous extract

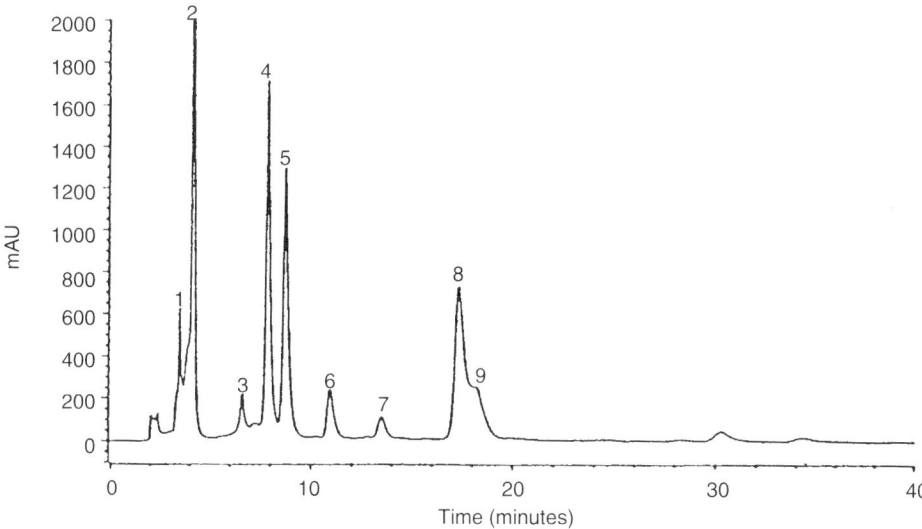

Figure 15.8 Normal phase HPLC chromatogram of oxygen heterocyclic components of cold-pressed grapefruit oil. (1) Bergamottin; (2) Aurapten; (3) Osthol; (4) Epoxybergamottin; (5) Epoxyaurapten; (6) Tangeritin; (7) Heptamethoxyflavone; (8) Meranzin; (9) Isomeranzin + Nobiletin. Reproduced from McHale and Sheridan, 1989.

of grapefruit flavedo but not previously found in the oil (McHale *et al.*, 1987). Although this coumarin was known to be formed from meranzin in the presence of acid, it was concluded that it was a natural component of grapefruit oil, since the mild HPLC conditions used were unlikely to have effected the isomerisation. Dugo *et al.* (1997a) improved the resolution of some of the peaks by using column switching and gradient elution. They detected seventeen peaks of which three were not characterised but were attributed to coumarins. The compounds responsible for the remaining fourteen were identified and quantified. The four substances additional to those observed by McHale and Sheridan (1989) were bergapten, the hydrates of meranzin and epoxybergamottin and the flavone, tetra-O-methylscutallerein. The latter had not been previously found in grapefruit oil.

The detailed qualitative and quantitative results for the studies on grapefruit oil are given in Table 15.5. A comparison of the two sets of quantitative data show that the values for meranzin, osthol, epoxybergamottin and the three major polymethoxyflavones are in reasonable agreement, while those for aurapten, epoxyaurapten, isomeranzin and bergamottin differ significantly. This may reflect differences in the origin of the oils and their method of production. The partial co-elution of certain compounds may well also have influenced the individual values.

Sweet orange and mandarin

Unlike the other citrus oils, the oxygen heterocyclic fractions of sweet orange, tangerine and mandarin consist almost entirely of polymethoxyflavones. However, much of the earlier work contained uncorroborated reports of the presence of coumarins and psoralens

Table 15.5 Qualitative and quantitative (ppm) results found in literature for oxygen heterocyclic compounds of grapefruit oil

	1	2	3	4	5	6	7	8	9	10	11
Citrusal (artefact)	X				X						
Aurapten	X	X	X		X		X	X	X	7200	11240
Meranzin								X		4900	5100
Isomeranzin										1500*	810
Osthol	X		X		X			X	X	700	580
Meranzin hydrate											120
Citropten	X				X		X		X		
Marmin (synthetic)	X		t	X	X						
5-Geranyloxy-7-methoxycoumarin							X				
Epoxyaurapten	X				X			X		4300	9300
Bergamottin	X				X		X	X	X	2000	970
Aurantiumal (artefact)	X	X			X						
Bergapten	X				X	120–130		X	X		110
Bergaptol	X		X		X			X	X		
Epoxybergamottin								X		9500	11260
Epoxybergamottin hydrate											700
Byakangelicin			X		X						
Tangeritin								X		600	680
Nobiletin								X		400*	460
Heptamethoxyflavone										570	370
Tetra-O-methylscutellarein											20

Notes
1 Fisher and Nordby, 1965; 2 Fisher and Nordby, 1966; 3 Stanley and Vannier, 1967; 4 Fisher and Nordby, 1966; 3 Stanley and Vannier, 1967; 4 Fisher et al., 1967; 5 Stanley and Jurd, 1971; 6 Shu et al., 1975; 7 Latz and Ernes, 1978; 8 Tatum and Berry, 1979; 9 Fischer and Trama, 1979; 10 McHale and Sheridan, 1989; 11 Dugo et al., 1997a.
* Nobiletin and isomeranzin are coeluted.

in such oils. Thus, Böhme and Pietsch (1938) reported the occurrence of 'aurapten' in the stearoptenes of orange oil. But, as mentioned previously, it was subsequently shown that the compound isolated was, in fact, meranzin (Böhme and Pietsch, 1939; Böhme and Schneider, 1939). On the basis of the R_f value and a blue fluorescence on a TLC plate, D'Amore and Calapaj (1965) concluded that citropten occurred in both sweet orange and mandarin oil. According to Fisher and Trama (1979), HPLC showed the presence of bergaptol in an oil taken hypodermically from the glands of Valencia orange, however, the only confirmation of identification was by peak enrichment. Buiarelli *et al.* (1991) indicated that aurapten was detected in a mandarin oil analysed by HPLC. In contrast to the earlier unconfirmed reports, more recently, Ziegler and Spiteller (1992) have isolated low levels of osthol from the non-volatile residue of sweet orange oil and established its identity by MS.

The earliest observation that sweet orange oil contained polymethoxyflavones was provided by Böhme and Wölcker (1959), who reported that the residue from Californian orange oil contained tangeritin and 3,3',4',5,6,7,8-heptamethoxyflavone. Shortly after this, Born (1960) found sinensetin (3',4',5,6,7-pentamethoxyflavone) in the orange peel juice that separated during commercial recovery of oil, and nobiletin was isolated from the juice obtained from Valencia orange peel by Swift (1960). Using column chromatography on Celite, Swift (1965a) separated a fifth polymethoxyflavone from orange peel juice and identified it as tetra-O-methylscutellarein. Following a complex purification of a benzene extract of peel juice and separation of individual polymethoxyflavones by column chromatography, Swift (1965b) found the relative proportions of the individual compounds to be tangeritin (6.5 per cent), heptamethoxyflavone (6.2 per cent), tetra-O-methylscutellarein (8.7 per cent), nobiletin (36 per cent), sinensetin (23.2 per cent) and unidentified (19.4 per cent). The complexity of the polymethoxyflavone patterns in *Citrus* was emphasised when Tatum and Berry (1972) reported the isolation of thirteen flavonoids from Valencia orange and Robinson tangerine peel.

Bianchini and Gaydou (1980) developed a normal phase HPLC procedure capable of resolving polymethoxyflavones. The method was used to quantify those compounds present in commercial sweet orange oil (Bianchini and Gaydou, 1981). McHale and Sheridan (1983) used TLC to determine the relative contribution of the individual polymethoxyflavones to the absorption maximum of sweet orange oil. The results with oils from five different countries were in good agreement and showed that nobiletin was the major contributor followed in order by tangeritin, heptamethoxyflavone and tetra-O-methylscutellarein. Subsequently, McHale and Sheridan (1989) used a normal phase isocratic HPLC method to quantify the individual polymethoxyflavones present in both sweet orange and tangerine oil. Using a similar procedure, Dugo *et al.* (1994) studied the variations in the concentration of the individual polymethoxyflavones with respect to cultivar type and method of oil extraction over a whole production season.

Packed column carbon dioxide SFC (supercritical fluid chromatography) has been also used for the rapid analysis of the polymethoxyflavones in sweet orange and mandarin oils (Morin *et al.*, 1991; Dugo *et al.*, 1996b). In both cases the best results were obtained using a bare silica stationary phase and carbon dioxide modified with methanol. Figure 15.9 shows the SFC chromatogram for a sweet orange oil. Qualitative and quantitative results obtained by SFC were in good agreement with those obtained by HPLC.

The qualitative and quantitative data described above for sweet orange oil are summarised in Table 15.6. Overall, three of the four sets of quantitative data are in good agreement, however, the results of Bianchini and Gaydou (1981) show a number

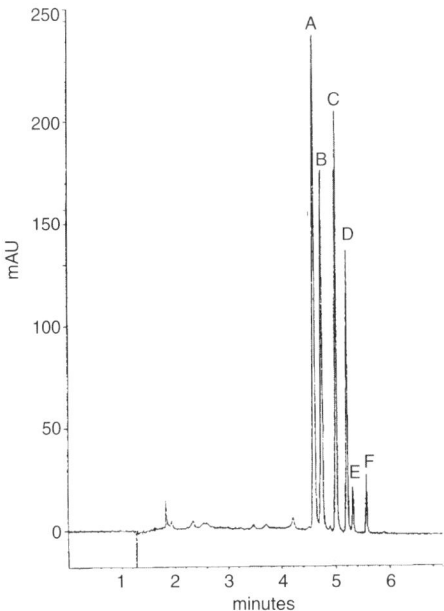

Figure 15.9 Packed SFC chromatogram of polymethoxylated flavones of sweet orange oil. (A) Tangeritin; (B) Heptamethoxyflavone; (C) Nobiletin; (D) Tetra-O-methylscutellarein; (E) 3,3',4',5,6,7-Hexamethoxyflavone; (F) Sinensetin. Reproduced from Dugo *et al.*, 1996b.

of significant differences. In particular, the sample of sweet orange from Guinea has the highest value for each of the individual compounds.

Tangeritin was isolated from the peel of tangerine oranges by Nelson (1934) and Goldsworthy and Robinson (1937) confirmed its structure by synthesis. Later, Riganesis (1956) observed that it was present in the residue that separated from mandarin oil on standing. Nobiletin was first obtained from the Chinese drug 'Chen pi' which was derived from dried mandarin peel (Tseng, 1938) and it was eventually found in the oil by Venturella *et al.* (1961). Iinuma *et al.* (1980) used GLC to study the polymethoxyflavones present in the peel of *Citrus reticulata* and reported the occurrence of thirteen compounds, but no quantitative data were given. A comparative study of sweet orange and mandarin peel oil by Gaydou *et al.* (1987) led to the identification of three compounds not previously reported as constituents of mandarin oil. Again no quantitative data were reported. McHale and Sheridan (1989) analysed mandarin oil by HPLC and confirmed the presence of one of these new compounds (heptamethoxyflavone). They also examined tangerine oil and found it to have a pattern of polymethoxyflavones similar to that of sweet orange oil, but the tangeritin content was higher. In a more detailed study over a whole production season, Dugo *et al.* (1994) investigated the variations in the concentration of the individual polymethoxyflavones in mandarin oil with respect to method of oil extraction.

The results of the above studies are summarised in Table 15.7. Apart from the high concentration of nobiletin reported by McHale and Sheridan (1989), the quantitative

Table 15.6 Qualitative and quantitative (g/l) results found in literature for oxygen heterocyclic compounds of sweet orange oil

	1	2	3	4	5	6	7	8	9	10	11	12	13	14
							Guinea	Israel	Brazil					
Citropten	X													
Aurapten			X											
Bergaptol					X									
Nobiletin		X	X	X		X	0.98	0.25	0.34			X	0.52	0.39
Tangeritin			X	X		X	0.84	0.36	0.68			X	0.48	0.60
5,8-Dihydroxy-3,7,3',4'-tetramethoxyflavone			X											
Tetra-O-methylscutellarein				X		X	0.50	0.16	0.18		X	X	0.31	0.24
Heptamethoxyflavone				X		X	1.24	0.42	0.36		X	X	0.84	0.71
Sinensetin				X		X	0.16	0.05	0.04			X	0.09	0.06
3,5,6,7,3',4'-Hexamethoxyflavone				X			0.20	0.06	0.04			X	0.13	0.07
Auranetin				X							X			
3,5,7,8,3',4'-Hexamethoxyflavone				X										
5,7,8,4'-Tetramethoxyflavone				X										
5,7,8,3',4'-Pentamethoxyflavone				X										
5-Hydroxy-3,7,8,3',4'-pentamethoxyflavone				X										
5-Hydroxy-3,6,7,8,3',4'-hexamethoxyflavone				X										

Notes
1 Böhme and Pietsch, 1938; 2 Swift, 1960; 3 D'Amore and Calapaj, 1965; 4 Tatum and Berry, 1972; 5 Fisher and Trama, 1979; 6 Bianchini and Gaydou, 1980; 7 Bianchini and Gaydou, 1981; 8 Gaydou et al., 1987; 9 McHale and Sheridan, 1989; 10 Di Giacomo et al., 1990; 11 Buiatelli et al., 1991; 12 Morin et al., 1991; 13 Dugo et al., 1994; 14 Dugo et al., 1996b.

Table 15.7 Qualitative and quantitative (g/l) results found in literature for oxygen heterocyclic compounds of mandarin oil

	1	2	3	4	5	6	7	8
Citropten	X				X			
Aurapten						X		
Nobiletin	X	X	X	1.2		X	0.74	0.66
Tangeritin	X	X	X	2.0		X	2.14	2.18
5,7,4'-Trimethoxyflavone		X						
3',4',5,7,8-Pentamethoxyflavone		X						
4',5,7,8-Tetramethoxyflavone		X						
5-Hydroxy-4',7,8-trimethoxyflavone		X						
4',5-Dihydroxy-7,8-dimethoxyflavone		X						
Sinensetin		X	X				0.02	0.01
5-Hydroxy-3',4',6,7-tetramethoxyflavone		X						
5-Hydroxy-3',4', 6,7,8-pentamethoxyflavone		X						
4',5,7-Trihydroxy-3',6,8-trimethoxyflavone		X						
4',5,6,7,8-Pentamethoxyflavone		X						
4'-Hydroxy-5,6,7,8-tetramethoxyflavone		X						
Tetra-O-methylscutellarein			X				0.05	0.04
Heptamethoxyflavone			X	0.5		X	0.37	0.50
3,5,6,7,3',4'-Hexamethoxyflavone			X					

Notes
1 D'Amore and Calapaj, 1965; 2 Iinuma, 1980; 3 Gaydou et al., 1987; 4 McHale and Sheridan, 1989; 5 Di Giacomo et al., 1991; 6 Buiarelli et al., 1991; 7 Dugo et al., 1994; 8 Dugo et al., 1996b.

data are in good agreement. However, Dugo et al. (1994, 1996) also detected low levels of sinensetin and tetra-O-methylscutellarein in mandarin oil, thus confirming some of the observations of Gaydou et al. (1987).

Bergamot

The oxygen heterocyclic fraction of bergamot oil has undoubtedly received more study than that of any other citrus oil. Bergamot oil is a very valuable product, largely used in perfumery as the basis of eau de Cologne, and in cosmetics as a constituent of suntan products. It is known to have strong photosensitising activity, due to the presence of psoralens and, in particular, bergapten. This has prompted the development of anaytical procedures for the determination of bergapten in bergamot oil and in cosmetic products containing the oil. There have also been many investigations into the pharmacological effects of the oil.

The early studies on the composition of the oxygen heterocyclic fraction of bergamot oil have been reviewed by Mossman and Bogert (1941) and, more recently, by Di Giacomo and Calvarano (1978) and by Lawrence (1982). The isolation of colourless needles from bergamot oil was first reported by Mulder (1839) just before Ohme (1839), working independently, gave the name 'begapten' to crystals obtained from a similar source. The composition of the crystals was studied by Crismer (1891) and by Pomerantz (1891) who found their formula to be $C_{12}H_8O_4$. Subsequently, Thoms and Baetecke (1912) proved that the compound was 5-methoxypsoralen. A second crystalline compound was isolated from bergamot oil by von Soden and Rojahn (1901) and named 'bergaptin'. In the years between 1934 and 1937, Späth and co-workers confirmed the presence of the previously isolated components, and identified new compounds: bergaptol (Späth and Socias, 1934) and citropten (Späth and Kainrath, 1937). In addition the latter workers established that 'bergaptin' (von Soden and Rojahn, 1901) was identical with their compound bergamottin (5-geranyloxypsoralen). Rodighero and Caporale (1954) also found bergapten, citropten and bergaptol by distilling the residue of bergamot oil under high vacuum, but they believed that the bergaptol was derived from the decomposition of bergamottin. Guenther (1949) reported the tendency of bergamottin to decompose into bergaptol during some analytical operations.

The qualitative and quantitative data on the composition of the oxygen heterocyclic fraction of bergamot oil available in the literature are summarised in Table 15.8. Apart from four papers and others where bergapten alone had been determined, the two coumarins and two psoralens of bergamot oil have always been detected. The research group from the Experimental Station for the Industries of essential oils and citrus derivatives of Reggio Calabria, using a normal phase HPLC method, report having identified in bergamot oil significant amounts of certain compounds usually found in lemon and lime oils (Calvarano et al., 1995; Gionfriddo et al., 1997). Identifications were based on the comparison of UV spectra and retention times with those of authentic compounds. However, it is worth pointing out that the bergamottin content of the oil analysed by Gionfriddo et al. was considerably lower than any other oil in Table 15.8. The HPLC chromatogram is shown in Figure 15.10. Dugo et al. (1999) were the first to report finding traces of polymethoxyflavones present in bergamot oil. They confirmed this result by the highly sensitive HPLC-MS analysis of a genuine bergamot oil.

Bergamottin is clearly the main component of the oxygen heterocyclic fraction of bergamot oil. Values reported in Table 15.8 range from about 1.0 to 2.7 per cent, and

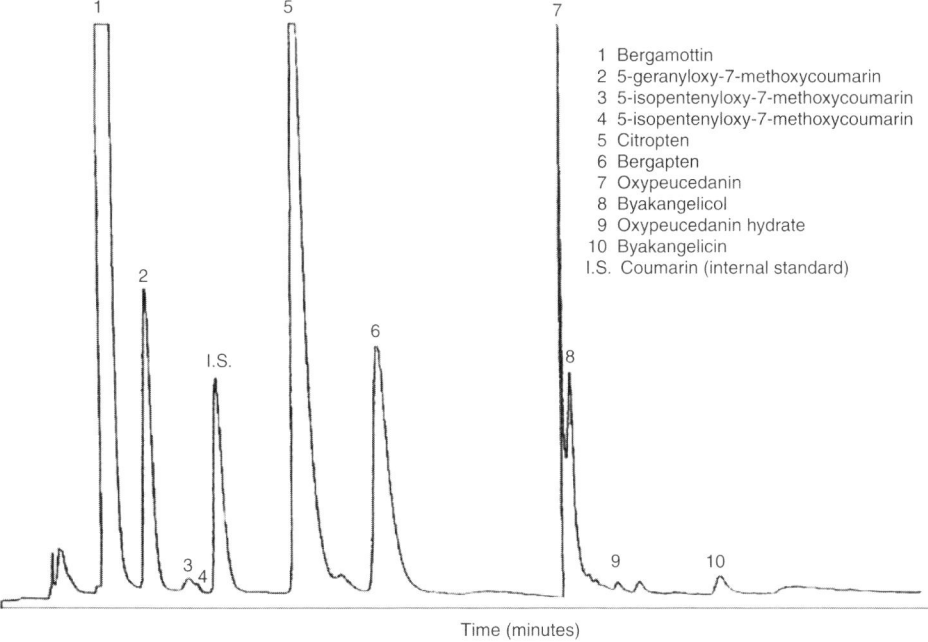

Figure 15.10 Normal phase HPLC chromatogram of oxygen heterocyclic compounds of bergamot oil. Reproduced from Gionfriddo et al., 1997.

Table 15.8 Qualitative and quantitative (ppm) results found in literature for oxygen heterocyclic compounds of bergamot oil

	1	2	3	4	5	6	7	8	9	10
Bergapten	X	2400–3600	X		1620–3570	2800–3300	1500–3300	2110–4560	X	6000–8700
Citropten	X	2100–3200	X	X			1400–2400		X	5000–7500
Bergamottin	X	11000–19000	X				14000–22000		X	
5-Geranyloxy-7-methoxycoumarin	X	400–600	X	X			1000–1500		X	
Bergaptol				X						
5-Isopentenyloxy-7-methoxycoumarin										
5-isopentenyloxy-8-methoxypsoralen*										
Oxypeucedanin										
Byakangelicol										
Oxypeucedanin hydrate										
Byakangelicin										
Tetra-O-methylscutellarein										
Sinensetin										

	11	12	13	14	15	16	17	18	19
Bergapten	1560–4040	2300–3000	2700	X	X	1100–3200		1893–3310	1100–2900
Citropten	1600–3000		2100	X	X	1400–3500		1176–3207	1200–3500
Bergamottin	11400–27300		18300	X		10200–27500		534–1072	10000–19400
5-Geranyloxy-7-methoxycoumarin			1100		X	800–2200		6327–8263	900–2700
Bergaptol					X				
5-Isopentenyloxy-7-methoxycoumarin					X		20–80	40–169	
5-isopentenyloxy-8-methoxypsoralen*							tr–60	0–15	
Oxypeucedanin							440–640	180–475	
Byakangelicol							150–410	165–895	
Oxypeucedanin hydrate							120–190	22–59	
Byakangelicin							60–80	23–90	
Tetra-O-methylscutellarein									X
Sinensetin									X

Notes

1 D'Amore and Calapaj, 1965; *2* Cieri, 1969; *3* Stanley and Jurd, 1971; *4* Di Giacomo and Calvarano, 1974; *5* Porcaro and Shubiak, 1974; *6* Shu *et al.*, 1975; *7* Calabrò and Currò, 1973; *8* Di Giacomo and Calvarano, 1978; *9* Latz and Ernes, 1978; *10* Vernin *et al.*, 1979; *11* Calvarano *et al.*, 1979; *12* Analytical Methods Committee, 1987; *13* McHale and Sheridan, 1989; *14* Benincasa *et al.*, 1990; *15* Di Giacomo, 1990; *16* Mondello *et al.*, 1993; *17* Calvarano *et al.*, 1995; *18* Gionfriddo *et al.*, 1997; *19* Dugo *et al.*, 1999.

* reported as 5-isopentenyloxy-7-methoxypsoralen in Calvarano *et al.*, 1995 and Gionfriddo *et al.*, 1997.

Table 15.9 Bergapten and other oxygen heterocyclic compounds in bergamot oil 'bergapten free' (ppm)

	1	2	3	4	
				NaOH treatment	Distillation
Bergapten	170–180	tr–420	1.7–5.7	0–91	0–41
Citropten				0–52	0–47
Bergamottin				11726–16250	0–3017
5-Geranyloxy-7-methoxycoumarin				1539–1975	0–349

Notes
1 Porcaro and Shubiak, 1974; 2 Di Giacomo and Calvarano, 1978; 3 Analytical Methods Committee, 1987; 4 Dugo et al., 1999; tr = traces.

are in good agreement, independent of the technology used. Data for bergapten, citropten and 5-geranyloxy-7-methoxycoumarin also agree well. However, Vernin et al. (1979) found very high values for citropten and bergapten, but these were obtained by GC analysis and may have been subject to interference from other peaks.

Many of the studies on the composition of the oxygen heterocyclic compounds of bergamot oil are linked to the problem of its phototoxicity and, in particular, to the amount of bergapten present in the oil. There are available on the market bergapten-free oils that have been produced by different methods. The procedures commonly used to remove bergapten from the oil involve either distillation or treatment with NaOH. Obviously, it is essential that the olfactory and physico-chemical characteristics of the oil are not affected by the treatment except, perhaps, any contribution from the oxygen heterocyclic compounds themselves.

Literature data on the composition of bergapten-free bergamot oil, obtained from the market or laboratory prepared, are included in Table 15.9. Porcaro and Shubiak (1974) and Di Giacomo et al. (1978) analysed commercially available samples, but reported values for bergapten only. Di Giacomo et al. (1978) specified that their samples were obtained by NaOH treatment and found that the bergapten content varied from a trace to 420 ppm, while the two samples analysed by Porcaro and Shubiak (1974) contained 170–80 ppm. Dugo et al. (1999) analysed eight samples of NaOH treated oil, and eight samples of distilled oil obtained on the market. It is apparent that the samples treated with NaOH had contents of 5-geranyloxy-7-methoxy-coumarin and bergamottin comparable to those of genuine cold-pressed bergamot oil, whereas the more polar compounds (bergapten and citropten) were almost completely extracted by the NaOH. In contrast, distillation had a major effect on all four compounds.

ISOLATION ARTEFACTS

Much of the confusion in the past over the natural occurrence of certain coumarins and psoralens in citrus oils has arisen because of the lability of particular modifications of the isoprenoid side chains that are present in the oxygen heterocyclic compounds of this species. An epoxy function in the isoprenyl side chain would appear to be especially prone to undergo hydration under the relatively mild acidic conditions imposed by contact with juice or during chromatography on silicic acid. Thus, Hata et al. (1963) established that byakangelicol was converted into byakangelicin under the latter conditions.

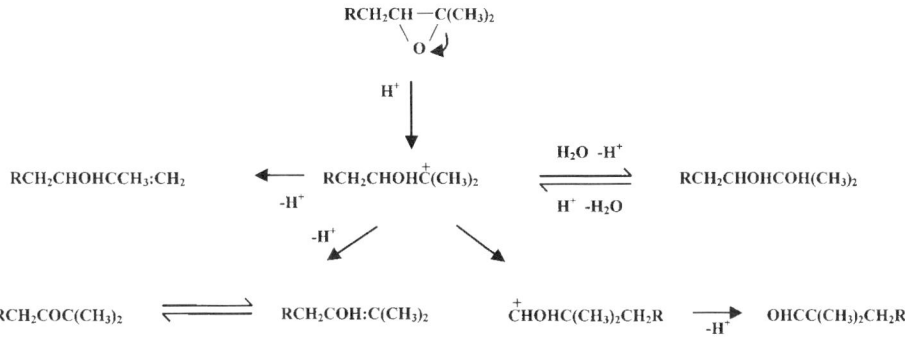

Figure 15.11 Acid catalysed reactions of an epoxy function in the prenyl side chain of coumarins and psoralens.

In addition to forming a diol, the epoxy compound may undergo acid catalysed change to yield a 1,3-enol by initial protonation of the epoxide, ring fission to a tertiary carbonium ion and elimination of a proton. The only instance of such a reaction among the citrus coumarins is the known formation of auraptenol from meranzin. However, in the case of the psoralens, McHale and Sheridan (1983) identified 5-(6'-hydroxy-3',7'-dimethylocta-2',7'-dienyl)oxypsoralen among the compounds obtained when bitter orange oil was fractionated on alumina, and Ziegler and Spiteller (1992) reported the isolation of neobyakangelicol and pabulenol/gosferol from TLC fractions of Sicilian lemon oil. It is likely that these compounds were formed during the chromatography as a result of the isomerisation of the epoxy group present in epoxybergamottin, byakangelicin and oxypeucedanin. Stanley *et al.* (1965) discovered that under acid conditions auraptenol underwent further isomerisation to yield the ketone, isomeranzin.

On prolonged contact with acid, the diols may be subject to bond migration similar to the pinacol rearrangement. It has been proposed that the two aldehydes (citrusal and aurantiumal) that Fisher and Nordby (1965, 1966) isolated from grapefruit oil by column chromatography on silicic acid were artefacts formed from 5-(6',7'-dihydroxygeranyloxy)psoralen and meranzin hydrate.

In addition to the above reactions, which are depicted schematically in Figure 15.11, it should be recognised that many of the oxygen heterocyclic compounds are aromatic 2,3-enyl ethers and may undergo thermal cleavage or rearrangement. They will also undergo facile cleavage to the parent phenol in the presence of acid. Thus under mild acidic conditions bergamottin can be readily converted into bergaptol (Stanley and Vannier, 1957a).

BIOGENESIS

It is generally accepted that phenylpropanoids, such as coumarins, psoralens and flavonoids, are formed in plants via the shikimate-chorismate biosynthetic pathway. In the case of *Citrus*, all the coumarins contain oxygen at C-7 and three are also oxygenated in the 5-position. For coumarins oxygenated at C-7 it is well established that the precursors are *cis*- and *trans*-4'-hydroxycinnamic acid. The biosynthesis of lignin and many other phenolic products of secondary plant metabolism follow this same initial

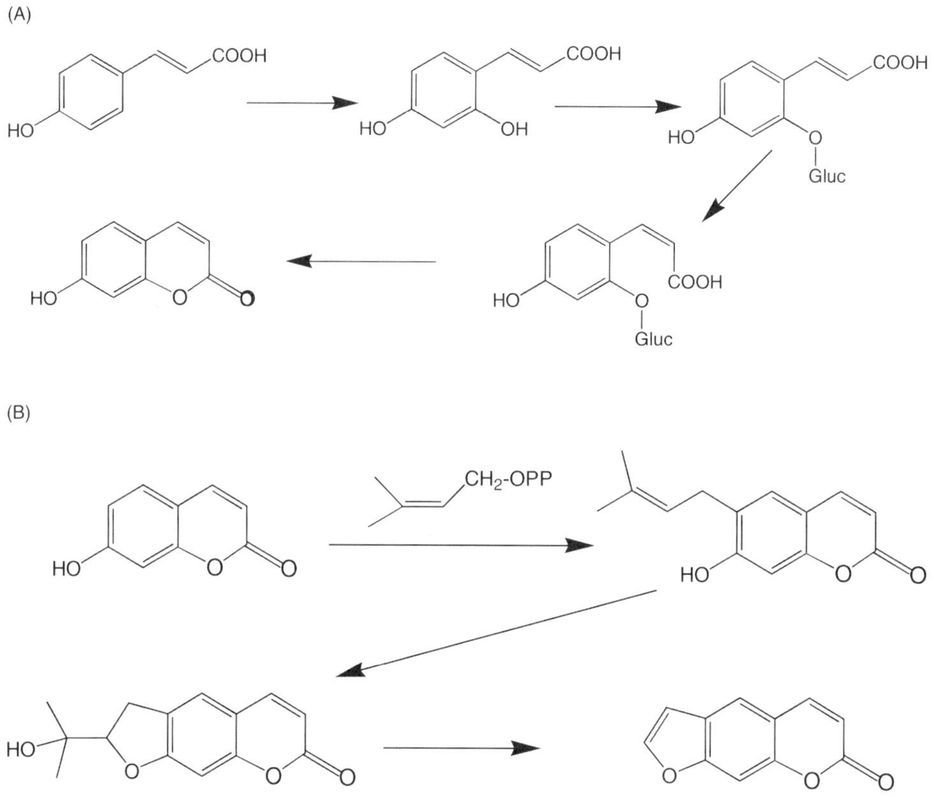

Figure 15.12 The accepted biosynthetic pathways to (a) 7-oxygenated coumarins and (b) linear psoralens.

pathway. There is evidence that the 4'-hydroxy group undergoes glucosylation prior to the introduction of the 2'-hydroxy group (Murray *et al.*, 1982). The resulting 2'-hydroxy group is itself glucosylated before the isomerisation of the *trans*-acid to the *cis*-acid. Such processes may involve both enzyme catalysis and UV-mediated isomerisation. The stage at which methylation of the 7-hydroxy group occurs has not yet been determined. It is assumed that any C- or O-prenylation takes place after ring formation and involves mevalonate. The synthesis up to the hydroxycoumarin stage is shown in Figure 15.12A.

Since two major citrus coumarins are also oxygenated at C-5, the question arises as to when the additional oxygen is incorporated. According to Murray *et al.* (1982) it has not been possible to demonstrate the synthesis of coumarins from cinnamic acids containing oxygen in other than the position *para* to the three carbon chain, although the existence of such routes cannot be excluded. One should perhaps point out in this particular case that the additional oxygen occupies a position originally equivalent to the 2'-hydroxy group and, as a consequence, this position may also be hydroxylated by the 2'-hydroxylase available to the system.

Explaining the synthesis of the 7-oxygenated-8-isoprenyl derived substituted coumarins is less of a problem since prenylation after ring formation is a recognised procedure.

However, it is interesting to note that the only furanocoumarins found in *Citrus* are psoralens, and yet a major coumarin of grapefruit and bitter orange has a structure closely related to the possible precursor for the angular furanocoumarins. Thus, the free phenol corresponding to meranzin or osthol could well cyclise to columbianetin, the likely immediate precursor of angelicin.

Umbelliferone is considered to be the precursor of the linear furanocoumarins and it has been established that the carbons at C-1 and C-2 of an isoprenyl group provide the carbons at C-6 and C-7. There is now evidence to suggest that after prenylation of umbelliferone at C-6, the resulting demethylsuberosin is converted into the dihydrofuranocoumarin (marmesin) by a concerted reaction catalysed by a cytochrome-P_{450} dependent mono-oxygenase and not via an epoxide intermediate (Wendorff and Matern, 1986; Hamerski and Matern, 1988). The mechanism by which the isopropyl residue is eliminated and the furan ring double bond introduced has not been fully established but it may involve oxygenation at C-6 prior to the elimination of water and acetone (Birch *et al.*, 1969). Stanjek *et al.* (1999) have now suggested that a second cytochrome-P_{450} dependent mono-oxygenase catalyses this reaction. The overall scheme is summarised in Figure 15.12B. Since the psoralens of *Citrus* all have an oxygen containing substituent at C-5 and/or C-8, it is possible that this oxygen is introduced early in the synthesis. However, it would seem more probable that hydroxylation and ether formation occur after the psoralen structure has been generated. The recent evidence is that a third cytochrome-P_{450} dependent mono-oxygenase is involved (Stanjek *et al.*, 1999).

The biosynthesis of polymethoxyflavones is well documented and involves the coupling of a malonate derived C_6 unit with 4'-hydroxycinnamoyl CoA. The resulting chalcone-flavanone isomeric pair then undergo a series of oxidation and methylation reactions to give the many polymethoxyflavones.

CHEMOTAXONOMIC CONSIDERATIONS

Tatum *et al.* (1974, 1975) reported that coumarins and polymethoxyflavones were useful as taxonomic markers in the *Citrus* species. For example, they observed that *C. paradisi* contained three unidentifed coumarins, while hybrids of *C. paradisi* × *C. reticulata* contained from 0 to 3 of these compounds. Somewhat unexpectedly, certain of these compounds were also observed in hybrids of *C. sinensis* × *C. sinensis*. Such results do little to encourage the view that coumarins, psoralens and polymethoxyflavones offer a clear cut method of dividing a genus into distinct species.

Scora and Kumamoto (1983) investigated the presence of certain enzymes and substrates in the *Citrus* species and concluded that bitter orange is a hybrid of pummelo (*C. grandis*) and mandarin (*C. reticulata*) whereas grapefruit is a hybrid of pummelo and sweet orange (*C. sinensis*). Other evidence for a close relationship between pummelo, grapefruit and bitter orange is the occurrence of the bitter flavanone, neohesperidose, in these species but, in no other *Citrus* species except the Pondorosa lemon (Maier, 1983). Further support for this view was obtained by McHale *et al.* (1987) when they detected meranzin and a glycoside of meranzin hydrate in aqueous extracts of the flavedo from bitter orange and pummelo. Neither compound was present in an extract of sweet orange flavedo.

Although there appear to be certain oxygen heterocyclic compounds unique to one *Citrus* species (McHale and Sheridan, 1989), the taxonomic significance is not clear. Thus, despite aurapten being widely found in the Rutaceae, it and the closely related 6',7'-epoxyaurapten are unique to grapefruit oil. Furthermore, apart from these two compounds and herniarin, which occurs in both lime and lemon oil, no other coumarins substituted only at C-7 are found in *Citrus*. There are other noticeable trends in the substitution patterns of the coumarins of *Citrus*. All the coumarins of lemon, lime and bergamot, with the exception of herniarin, are 5,7-oxygenated. In contrast, the disubstituted coumarins of grapefruit and bitter orange are only 7-oxygenated but, also contain an isopentenyl derived substituent at C-8. The substitution patterns of the psoralens for lime and lemon include all three possibilities (C-5 and/or C-8) while only C-5 substitution is observed with bergamot, grapefruit and bitter orange. Finally, bergamot is the only species that does not contain a coumarin or psoralen with a prenyl substituent embracing an epoxy function, suggesting that this species lacks the appropriate epoxidase.

Another point of note concerns the fact that most *Citrus* is grown on a foreign rootstock, such as rough lemon, bitter orange, grapefruit etc. This introduces the question as to where the coumarin, psoralen and polymethoxyflavone syntheses occur. Is it possible that different patterns of oxygen heterocyclic formation can arise as a result of the synthesis taking place at more than one site? Could some of the anomalous analytical results be explained in this way? At this time there is no simple answer to such questions.

CONCLUSIONS

The oxygen heterocyclic compounds that have been reported to occur in citrus essential oils are listed in Table 15.10. It is obvious from the earlier detailed discussion that many of these identifications are incorrect or, have not been corroborated. Knowledge of the distribution of such compounds throughout the *Citrus* species has advanced rapidly in the last decade as a result of the advent of modern efficient separation techniques and reliable instrumental means of identification. In fact the analytical procedures are sometimes too sensitive and may raise questions as to whether trace constituents are natural components or contaminants (accidental or deliberate).

The role of the oxygen heterocyclic compounds has been neglected in this article. This is due to an absence of any factual data rather than an oversight. Mention has been made of the photosensitising properties of bergamot oil, but nothing was said about the incidence of dermatitis among the workers who produced écuelled lime oil in the West Indies. Psoralens are known to be active against many fungi and it is possible that they may be employed by the *Citrus* species to combat diseases such as fungal withertip. There has been a suggestion that nobiletin and tangeritin might give a degree of protection to the leaves of citrus plants against the fungus 'mal-secco' (Pinkas *et al.*, 1968). As all the oxygen heterocyclic compounds found in the oil glands of the flavedo are intensely fluorescent, this may well be the characteristic determining the natural role they have. However, it is difficult to explain why so many different structures are required. Until further research indicates their true role, we should perhaps just continue to use them as a criteria of quality of the oils.

Table 15.10 Oxygen heterocyclic compounds identified in citrus essential oils

	BO	SO	Lem	Lim	B	M	G
Aurapten (7-geranyloxycoumarin)	X	X	X			X	X
Auraptenol (7-[2'-hydroxyisopent-3'-enyloxy]coumarin)	X						
Marmin (7-[6',7'-dihydroxygeranyloxy]coumarin)							X
Umbelliferone (7-hydroxycoumarin)	X						
Herniarin (7-methoxycoumarin)			X	X			
Epoxyaurapten (7-[6',7'-epoxygeranyloxy]coumarin)							X
7-Isopentenyloxycoumarin			X				
Meranzin (7-methoxy-8[2',3'-epoxyisopentenyl]coumarin)	X						X
Osthol (7-methoxy-8-isopentenylcoumarin)	X						X
Meranzin hydrate (7-methoxy-8-[2',3'-dihydroxyisopentyl]coumarin)	X						X
Isomeranzin (7-methoxy-8-[2'-oxoisopentyl]coumarin)	X						X
Citrusal (7-[methoxy-8-[2'-formyl-2'-methylpropyl]coumarin)							X
Citropten (5,7-dimethoxycoumarin)	X	X	X	X	X	X	X
5-Isopentenyloxy-7-methoxycoumarin			X	X	X		
5-(2',3'-Epoxyisopentyloxy)-7-methoxycoumarin			X				
5-Geranyloxy-7-methoxycoumarin			X	X	X		X
5-(2',3'-Dihydroxyisopentyloxy)-7-methoxycoumarin			X				
Scoparone (5,6-dimethoxycoumarin)			X				
Scopoletin (5-methoxy-6-hydroxycoumarin)			X				
Psoralen				X			
Bergapten (5-methoxypsoralen)	X		X	X	X		X
Epoxybergamottin (5-[6',7'-epoxygeranyloxy]psoralen)	X						X
Epoxybergamottin hydrate (5-[6',7'-dihydroxygeranyloxy]psoralen)	X						X
5-(6'-Hydroxy-3'-7'-dimethylocta-2'-7'-dienyloxy)psoralen	X						
Bergamottin (5-geranyloxypsoralen)			X	X	X		X
Bergaptol (5-methoxypsoralen)	X	X	X		X		X
Oxypeucedanin (5-[2',3'-epoxyisopentyloxy]psoralen)			X	X	X		
Oxypeucedanin hydrate (5-[2',3'-dihydroxyisopentyloxy]psoralen)			X	X	X		
Pabulenol/Gosferol (5-[2'-hydroxyisopent-3'-enyloxy]psoralen)			X				
Isoimperatorin (5-isopentenyloxypsoralen)	X		X	X			
Aurantiumal (5-[3,6-dimethyl-6-formylhept-2'-enyloxy]psoralen)							X
Xanthotoxin (8-methoxypsoralen)				X			
8-Geranyloxypsoralen			X	X			
Heraclenin (8-[2',3'-epoxyisopentyloxy]psoralen)			X	X			
Heraclenol (8-[2',3'-dihydroxyisopentyloxy]psoralen)			X				
Imperatorin (8-isopentenyloxypsoralen)			X	X			
8-(6',7'-Epoxygeranyloxy)psoralen			X				
5-Methoxy-8-isopentenyloxypsoralen			X				
5-Geranyloxy-8-methoxypsoralen			X	X			
Byakangelicin (5-methoxy-8-[2',3'-dihydroxyisopentyloxy]psoralen)			X	X	X		X
Isobyakangelicol (5-[2',3'-epoxyisopentyloxy]-8-methoxypsoralen)				X			
5-Isopentenyloxy-8-[2',3'-epoxyisopentyloxy]psoralen			X				
Phellopterin (5-methoxy-8-isopentenyloxypsoralen)			X	X			
Byakangelicol (5-methoxy-8-[2',3'-epoxy-isopentyloxy]psoralen)			X	X	X		
Isopimpinellin (5,8-dimethoxypsoralen)			X	X			

Table 15.10 (Continued)

	BO	SO	Lem	Lim	B	M	G
Cnidilin (5-Isopentenyloxy-8-methoxypsoralen)				X	X		
Neobyakangelicol (5-methoxy-8-[2'-hydroxyisopent-3'-enyloxy]psoralen)			X				
5-Isopentenyloxy-8-[2',3'-dihydroxyisopentyloxy]psoralen			X				
Cnidicin (5,8-diisopentenyloxypsoralen)			X				
5-Methoxy-8-geranyloxypsoralen			X				
Auranetin (3,4',6,7,8,-pentamethoxyflavone)	X	X					
Desmethylnobiletin (5-hydroxy-3',4,6,7,8-pentamethoxyflavone)	X						
Tangeritin (4',5,6,7,8-pentamethoxyflavone)	X	X				X	X
Nobiletin (3',4',5,6,7,8-hexamethoxyflavone)	X	X				X	X
3,3',4',5,6,7,8-Heptamethoxyflavone	X	X				X	X
Sinensetin (3',4',5,6,7-pentamethoxyflavone)	X	X			X	X	
3,3',4',5,6,7-Hexamethoxyflavone		X				X	
Tetra-O-methylscutellarein (4',5,6,7-tetramethoxyflavone)	X	X			X	X	X
5,8-Dihydroxy-3,3',4',7-tetramethoxyflavone		X					
3,3',4',5,7,8-Hexamethoxyflavone		X					
4',5,7,8-Tetramethoxyflavone		X				X	
3',4',5,7,8-Pentamethoxyflavone		X				X	
5-Hydroxy-3,3',4',7,8-pentamethoxyflavone		X					
5-Hydroxy-3,3',4',6,7,8-hexamethoxyflavone		X					
4',5,7-Trimethoxyflavone						X	
5-Hydroxy-4',7,8-trimethoxyflavone						X	
4,5-Dihydroxy-7,8-dimethoxyflavone						X	
5-Hydroxy-3',4',6,7-tetramethoxyflavone						X	
4',5,7-Trihydroxy-3',6,8-trimethoxyflavone						X	
4'-Hydroxy-5,6,7,8-tetramethoxyflavone						X	

Notes

BO = Bitter Orange; SO = Sweet Orange; M = Mandarin; Lem = Lemon; Lim = Lime; G = Grapefruit; B = Bergamot; Isopentenyloxy = 3'-methylbut-2'-enyloxy except where otherwise designated; Geranyloxy = 3',7'-dimethylocta-2',6'-dienyloxy.

REFERENCES

Analytical Methods Committee (1987) Applications of high performance liquid chromatography to the analysis of essential oils. Part 1. Determination of bergapten (4-methoxyfuro[3,2-g] chromen-7-one)(5-methoxypsoralen) in oils of bergamot. *Analyst*, 112, 195–198.

Benincasa, M., Buiarelli, F., Cartoni, G.P. and Coccioli, F. (1990) Analysis of lemon and bergamot essential oils by HPLC with microbore columns. *Chromatographia*, 30, 271–276.

Berahia, T., Gaydou, E.M., Cerrati, C. and Wallet, J. (1994) Mass spectrometry of polymethoxylated flavones. *J. Agric. Food Chem.*, 42, 1697–1700.

Bianchini, J.P. and Gaydou, E.M. (1980) Separation of polymethoxylated flavones by straight-phase high-performance liquid chromatography. *J. Chromatogr.*, 190, 233–236.

Bianchini, J.P. and Gaydou, E.M. (1981) Role of water in qualitative and quantitative determination of polymethoxylated flavones by straight-phase high-performance liquid chromatography: application to orange peel oils. *J. Chromatogr.*, 211, 61–78.

Birch, A.J., Maung, M. and Pelter, A. (1969) Biosynthesis. XL. Some aspects of the chemistry of *o*-isopentenylphenols and related compounds. *Aus. J. Chem.*, 22, 1923–1932.

Born, R. (1960) 3',4',5,6,7-Penta-O-methylflavone in orange peel. *Chem. &Ind.*, 264–265.

Böhme, H. and Pietsch, G. (1938) Über das Stearopten des Pomeranzenschalenöls. *Arch. Pharm.*, 276, 482–488.

Böhme, H. and Pietsch, G. (1939) Zu Kenntnis des Auraptens. *Ber.*, 72B, 773–779.

Böhme, H. and Schneider, E. (1939) Oxidativer Abbau und Konstitution des Auraptens. *Ber.*, 72B, 780–784.

Böhme, H. and Völcker, P.E. (1959) Zur Kenntnis der nichtflüchtigen Anteil des Pomeranzenschalenöl. *Arch Pharm.*, 292, 529–536.

Buiarelli, F., Cartoni, G.P., Coccioli, F. and Ravazzi, E. (1991) Analysis of orange and mandarin essential oils by HPLC. *Chromatographia*, 31, 489–492.

Calabrò, G. and Currò, P. (1975) Spectrofluorimetric determination of the coumarins of bergamot essential oil. *Essenz. Deriv. Agrum.*, 45, 246–262.

Calabrò, G. and Currò, P. (1976) Spectrofluorimetric determination of coumarins from lemon essential oil. *Essenz. Deriv. Agrum.*, 46, 215–223.

Caldwell, A.G. and Jones, E.R.H. (1945) The constituents of expressed West Indian lime oil. *J. Chem. Soc.*, 540–543.

Calvarano, I., Ferlazzo, A. and Di Giacomo, A. (1979) The coumarin and furocoumarin composition of bergamot oil. *Essenz. Deriv. Agrum.*, 49, 12–21.

Calvarano, I., Calvarano, M., Gionfriddo, F., Bovalo, F. and Postorino, E. (1995) HPLC profile of citrus essential oils from different geographic regions. *Essenz. Deriv. Agrum.*, 65, 488–502.

Calvarano, M. and Gallino, M. (1975) Contributo allo studio del limao Galego (*Citrus aurantifolia* Sw.) coltivato in Brasile. *Essenz. Deriv. Agrum.*, 45, 279–291.

Careri, M., Mangia, A. and Musci, M. (1998) Overview of the applications of liquid chromatography-mass spectrometry interfacing systems in food analysis: naturally occurring substances in food. *J. Chromatogr. A*, 794, 263–297.

Chouchi, D. and Barth, D. (1994) Rapid identification of some coumarin derivatives in deterpenated citrus peel oil by gas chromatography. *J. Chromatogr. A*, 672, 177–183.

Cieri, U.R. (1969) Characterization of the steam non-volatile residue of bergamot oil and some other essential oils. *J. Assoc. Off. Anal. Chem.*, 52, 719–728.

Crismer, M.L. (1891) Sur produits crystallisés des essences de citron et de bergamote. *Bull. Soc. Chim.*, 6, 30–33.

Croud, V.B., Michaelis, J.R. and Pindar, A.G. (1983) Isolation of bergapten and limettin from bergamot oil. *Analyst*, 108, 1532–1534.

D'Amore, G. and Calapaj, R. (1965) Le sostanze fluorescenti contenute nelle essenze di limone, bergamotto, mandarino, arancio dolce, arancio amaro. *Rassegna Chimica*, 6, 264–269.

Di Giacomo, A. (1990) Valutazione della qualità delle essenze agrumarie 'cold-pressed' in relazione al contenuto in composti cumarinici e psoralenici. *Essenz. Deriv. Agrum.*, 60, 313–334.

Di Giacomo, A. and Calvarano, M. (1978) The bergapten content in cold-extracted bergamot essential oil. *Essenz. Deriv. Agrum.*, 48, 51–83.

Dodge, F.D. (1938) Lactones of the citrus oils. *Amer. Perf.*, 37(6), 34–36.

Dreyer, D.L. (1967) Chemotaxonomy of the Rutaceae. II. Extractives of *Severinia buxifolia* (Poir.) Ten. *Tetrahedron*, 23, 4613–4646.

Dugo, P., Mondello, L., Cogliandro, E., Stagno d'Alcontres, I. and Cotroneo, A. (1994) On the genuineness of citrus essential oils. Part 46. Polymethoxylated flavones of the non-volatile residue of Italian sweet orange and mandarin essential oils. *Flav. Fragr. J.*, 9, 105–111.

Dugo, P., Mondello, L., Cogliandro, E., Verzera, A. and Dugo, G. (1996a) On the genuineness of citrus essential oils. 51. Oxygen heterocyclic compounds of bitter orange oil (*Citrus aurantium* L.). *J. Agric. Food Chem.*, 44, 544–549.

Dugo, P., Mondello, L., Dugo, G., Heaton, D.M., Bartle, K.D., Clifford, A.A. and Myers, P. (1996b) Rapid analysis of polymethoxylated flavones from citrus oils by supercritical fluid chromatography. *J. Agric. Food Chem.*, 44, 3900–3905.

Dugo, P., Mondello, L., Stagno d'Alcontres, I., Cavazza, A. and Dugo, G. (1997a) Oxygen heterocyclic compounds of citrus essential oils. *Perfum. Flav.*, 22(1), 25–30.

Dugo, P., Mondello, L., Lamonica, G. and Dugo, G. (1997b) Characterization of cold-pressed key and Persian lime oils by gas chromatography, gas chromatography/mass spectroscopy, high-performance liquid chromatography and physicochemical indices. *J. Agric. Food Chem.*, 45, 3608–3616.

Dugo, P., Mondello, L., Cogliandro, E., Cavazza, A. and Dugo, G. (1998) On the genuineness of citrus essential oils. Part LIII. Determination of the composition of the oxygen heterocyclic fraction of lemon essential oils (*Citrus limon* L. Burm. f.) by normal phase high performance liquid chromatography. *Flav. Fragr. J.*, 13, 329–334.

Dugo, P., Mondello, L., Proteggente, A.R. and Dugo, G. (1999) Oxygen heterocyclic compounds of bergamot essential oils. *EPPOS*, (27), 31–41.

Dugo, P., Mondello, L. and Dugo, G. (2000) Citrus oils: Liquid chromatography. *Encyclopedia of Separation Science*, 3, 2441–2454, Academic Press, London.

Dugo, P., Mondello, L., Dugo, L., Stancanelli, R. and Dugo, G. (2001) LC-MS for the identification of oxygen heterocyclic compounds in citrus essential oils. *J. Pharm. Biomed. Anal.*, 24, 147–154.

Fischer, J.F. and Nordby, H.E. (1965) Isolation and spectral characterization of coumarins in Florida grapefruit peel oil. *J. Food Sci.*, 30, 869–873.

Fischer, J.F. and Nordby, H.E. (1966) Two new coumarins from grapefruit peel oil. *Tetrahedron*, 22, 1489–1493.

Fischer, J.F., Nordby, H.E., Waiss, A.C., Jr and Stanley, W.L. (1967). A new coumarin from grapefruit peel oil. *Tetrahedron*, 23, 2523–2528.

Fisher, J.F. and Trama, L.A. (1979) High-performance liquid chromatographic determination of some coumarins and psoralens found in citrus peel oils. *J. Agric. Food Chem.*, 27, 1334–1337.

Gaydou, E.M., Bianchini, J.P. and Randriamiharisoa, R.P. (1987) Orange and mandarin peel oils differentiation using polymethoxylated flavone composition. *J. Agric. Food Chem.*, 85, 525–529.

Gionfriddo, F., Postorino, E. and Bovalo, F. (1997) On the authenticity of bergamot oil: HPLC profile of heterocyclic components. *Essenz. Deriv. Agrum.*, 67, 342–352.

Glandian, R., Corneteau, H., Drouet, S. and Rouzet, M. (1978) Composition of the non-volatile fraction of the essential oil of lemon from the Ivory Coast. *Labo-Pharma Probl. Tech.*, 26, 503–507.

Goldsworthy, L.J. and Robinson, R. (1937) A synthesis of tangeritin. *J. Chem. Soc.*, 46–49.

Guenther, E. (1949) *The essential oils*. D. Van Nostrand Company, New York.

Hamerski, D. and Matern, U. (1988) Elicitor-induced biosynthesis of psoralens in *Ammi majus* L. suspension cultures. Microsomal conversion of demethylsuberosin into (+)-marmesin and psoralen. *Eur. J. Biochem.*, 171, 369–375.

Haro-Guzman, L. (1980) Les essences de lime mexicaine. Changements dans la composition dus a la technique d'extraction. *Essenz. Deriv. Agrum.*, 50, 332–344.

Hata, K., Kozawa, M. and Yen, K. (1963) Pharmacognostical studies on umbelliferous plants. XIX. Chinese drug byakushi. 4. Coumarins of the roots of *Angelica dahurica* var. *dahurica* and *A. dahurica* var. *pai-chi*. *J. Pharm. Soc. Japan*, 83, 606–610.

Iinuma, M., Matsuura, S., Kurogochi, K. and Tanaka, T. (1980) Studies on the constituents of useful plants. 5. Multisubstituted flavones in the fruit peel of *Citrus reticulata* and their examination by gas-liquid chromatography. *Chem. Pharm. Bull.*, 28, 717–722.

Kariyone, T. and Matsuno, T. (1953) Studies on the constituents of orange oil. I. On the structure of auraptene. *Pharm Bull. Japan*, 1, 119–122.

Komatsu, S., Tanaka, S., Ozawa, S., Kubo, R., Ono, Y. and Matsuda, Z. (1930) Biochemical studies on grapefruits, *Citrus aurantium* L. *J. Chem. Soc. Japan*, 51, 478–498.

Latz, H.W. and Ernes, D.A. (1978) Selective fluorescence detection of citrus oil components separated by high-pressure liquid chromatography. *J. Chromatogr.*, 166, 189–199.

Latz, H.W. and Madsen, B.C. (1969) Total luminescence of coumarin derivatives isolated from expressed lime oil. *Anal. Chem.*, 41, 1180–1185.

Lawrence, B.M. (1982) Progress in essential oils. Coumarins and psoralens in citrus oils. *Perfum. Flav.*, 7(3), 57–65.

Lawrence, B.M. (1984) Progress in essential oils. *Perfum. Flav.*, 9(6), 61–70.

Madsen, B.C. and Latz, H.L. (1970) Qualitative and quantitative *in situ* fluorimetry of citrus oil thin-layer chromatograms. *J. Chromatogr.*, 50, 288–303.

Maier, V.P. (1983) Chemistry and significance of selected citrus limonoids and flavonoids. In P.G. Waterman and M.F. Grunden (eds), *Chemistry and Chemical Taxonomy of the Rutales*, Academic Press, London, 319–342.

Matsuno, T. (1956) Components of *Citrus* species. IV. Constituents of orange oil. *J. Pharm. Soc. Japan*, 76, 1136–1138.

McHale, D. (1982) Effect of processing conditions on composition of distilled oil of lime. Paper No. 67, *Proceedings of the VIII International Congress of Essential Oils*, Cannes-Grasse, October 12–17, 1980, Fedarom, Grasse, pp. 177–180.

McHale, D., Khopkar, P.P. and Sheridan, J.B. (1987) Coumarin glycosides from *Citrus* flavedo. *Phytochemistry*, 26, 2547–2549.

McHale, D. and Sheridan, J.B. (1983) Detection of adulteration of cold-pressed bitter orange oil. Paper presented at *IXth International Essential Oil Congress*, Singapore. *Technical paper book* 3, 43–48.

McHale, D. and Sheridan, J.B. (1988) Detection of adulteration of cold-pressed lemon oil. *Flav. Fragr. J.*, 3, 127–133.

McHale, D. and Sheridan, J.B. (1989) The oxygen heterocyclic compounds of citrus peel oils. *J. Ess. Oil Res.*, 1, 139–149.

Mondello, L., Stagno d'Alcontres, I., Del Duce, R. and Crispo, F. (1993) On the genuineness of citrus essential oils. Part 40. Composition of the coumarins and psoralens of Calabrian bergamot essential oil (*Citrus bergamia* Risso). *Flav. Fragr. J.*, 8, 17–24.

Morin, Ph., Gallois, A., Richard, H. and Gaydon, E. (1991) Fast separation of polymethoxylated flavones by carbon dioxide supercritical fluid chromatography. *J. Chromatogr.*, 586, 171–176.

Mossman, D.D. and Bogert, M.T. (1941) *Bergamot oil*, The American Pharmaceutical Association, Washington.

Mulder, G.J. (1839) Über die Zusammensetzung einiger Stearopten und ätherischen Öle. *Annalen*, 31, 67–72.

Murray, R.D.H., Mendez, J. and Brown, S.A. (1982) *The Natural Coumarins. Occurrence, Chemistry and Biochemistry*, John Wiley & Sons, Chichester.

Nelson, E.K. (1934) The occurrence of a pentamethyl flavonoid in tangerine peel. *J. Am. Chem. Soc.*, 56, 1392–1393.

Nigg, H.N., Nordby, H.E., Beier, R.C., Dillman, A., Macias, C. and Hansen, R.C. (1993) Phototoxic coumarins in limes. *Fd. Chem. Toxic.*, 31, 331–335.

Nomura, D. (1950) Coumarin derivatives found in the fruit of *Citrus aurantium* Natsudaidai. *Kagaku No Ryoiki*, 4, 561–564.

Ohme, C. (1839) Über die Zusammensetzung des Bergamottöls. *Annalen*, 31, 316–321.

Patnayak, K.C., Rangaswani, S. and Seshadri, T.R. (1942) Chemical investigation of Indian fruits. III. Characteristic crystalline components of certain citrus fruits (oranges of the circars). *Proc. Indian Acad. Sci.*, 16A, 10–15.

Pinkas, J., Lavie, D. and Chorin, M. (1968) Fungistatic constituents in citrus varieties resistant to mal-secco disease. *Phytochemistry*, 7, 169–174.

Pomerantz, C. (1891) Über das Bergapten. *Monatsh. f. Chem.*, 12, 379–392.

Porcaro, P.J. and Shubiak, P. (1974) Liquid chromatographic determination of bergapten content in treated or natural bergamot. *J. Assoc. Off. Anal. Chem.*, 57, 145–147.

Riganesis, M. (1956) Chemical, photometric and chromatographic research of tangerine oil. *Essenz. Deriv. Agrum.*, 26, 107–121.

Rodighero, G. and Caporale, G. (1954) The coumarins obtained from extracts of *Citrus bergamia*. *Atti Ist. Veneto Sci. Nat.*, 112, 97–102.

Sarin, P.S. and Seshadri, T.R. (1960). New components of *Citrus aurantium*. *Tetrahedron*, 8, 64–66.

Schmidt, E. (1901) Über das Citropten. *Apoth. Z.*, 16, 619–620.

Schmidt, E. (1904) Über das Citropten (Citronenölstearopten, Citroenkampfer, Limettin). *Arch. der Pharm.*, 242, 288–295.

Scora, R.W. and Kumamoto, J. (1983) Chemotaxonomy of the genus *Citrus*. In P.G. Waterman and M.F. Grunden (eds), *Chemistry and Chemical Taxonomy of the Rutales*, Academic Press, London, 343–351.

Shu, C.K., Walradt, J.P. and Taylor, W.I. (1975) Improved method for bergapten determination by high-performance liquid chromatography. *J. Chromatogr.*, 106, 271–282.

Späth, E. and Socias, E.L. (1934) Über Bergaptol einen neuen Inhaltstoff des calabrischen Bergamottöles (VIII. Mitteil über naturliche Cumarine). *Ber.*, 67B, 59–61.

Späth, E. and Kainrath, P. (1937) Über Bergamottin und über die Auffindung von limettin in Bergamottöl (XXXIV. Mitteil über naturliche Cumarine). *Ber.*, 70B, 2272–2276.

Stanjek, V., Miksch, M., Lüer, P., Matern, U. and Boland, W. (1999) Biosynthesis of psoralen: mechanism of a cytochrome P450 catalyzed oxidative bond cleavage. *Angew. Chem. Int. Ed.*, 38, 400–402.

Stanley, W.L. (1963) Recent development in coumarin chemistry. *Aspects Plant Phenolic Chem. Proc. Symp. 3rd. Univ. Toronto*, 79–102.

Stanley, W.L. and Jurd, L. (1971) Citrus coumarins. *J. Agric. Food Chem.*, 19, 1106–1110.

Stanley, W.L. and Vannier, S.H. (1957a) Chemical composition of lemon oil. Isolation of a series of substituted coumarins. *J. Am. Chem. Soc.*, 79, 3488–3491.

Stanley, W.L. and Vannier, S.H. (1957b) Analysis of coumarin compounds in citrus oils by liquid solid partition. *J. Assoc. Off. Anal. Chem.*, 40, 582–588.

Stanley, W.L. and Vannier, S.H. (1967) Psoralens and substituted coumarins from expressed oil of lime. *Phytochemistry*, 6, 585–596.

Stanley, W.L., Waiss, A.C. Jr, Lundin, R.E. and Vannier, S.H. (1965) Auraptenol, a coumarin compound in bitter (Seville) orange oil. *Tetrahedron*, 21, 89–92.

Swift, L.J. (1960) Nobiletin from the peel of the Valencia orange (*Citrus sinensis* L.) *J. Org. Chem.*, 25, 2067–2068.

Swift, L.J. (1964) Isolation of 5,6,7,3',4'-pentamethoxyflavone from orange peel juice. *J. Food. Sci.*, 29, 766–767.

Swift, L.J. (1965a) Tetra-O-methylscutallerein in orange peel. *J. Org. Chem.*, 30, 2079–2080.

Swift, L.J. (1965b) Flavones of the neutral fraction of the benzene extractables of an orange peel juice. *J. Agric. Food Chem.*, 13(5), 431–433.

Tatum, J.H. and Berry, R.E. (1972) Six new flavonoids from *Citrus*. *Phytochemistry*, 11, 2283–2288.

Tatum, J.H. and Berry, R.E. (1979) Coumarins and psoralens in grapefruit peel oil. *Phytochemistry*, 18, 500–502.

Tatum, J.H., Berry, R.E. and Hearn, C.J. (1974) Characterisation of citrus cultivars and separation of nucellar and zygotic seedlings by thin layer chromatography. *Proc. Fla. State Hortic. Soc.*, 87, 75–81.

Tatum, J.H., Hearn, C.J. and Berry, R.E. (1975) Characterisation of citrus cultivars by chemical differentiation. *J. Am. Soc. Hortic. Sci.*, 103, 492–496.

Thoms, H. and Baetcke, E. (1912) Die Konstitution des Bergaptens. *Ber.*, 45, 3705–3712.

Tilden, W.A. and Beck, C.R. (1890) XXIII. Some crystalline substances from the fruit of various citrus. Part 1. *J. Chem. Soc.*, 57, 323–328.

Tilden, W.A. and Burrows, H. (1902) The constitution of limettin. *J. Chem. Soc.*, 508–512.

Tseng, K.F. (1938) Nobiletin. Part I. *J. Chem. Soc.*, 1003–1004.

Van Os, D. and Dykstra, K. (1937) Examination of essential oils by measurement of absorption in the ultraviolet. *J. Pharm. Chim.*, 25, 437–454; 485–501.

Venturella, P., Bellino, A. and Cusmano, S. (1961) Flavonoid components of *Citrus*. I. Isolation of 5,6,7,8,3',4'-hexamethoxy-flavone (nobiletin) from *Citrus deliciosa* Tenore. *Ann. Chim. (Rome)*, 51, 101–115.

Vernin, G., Bianchini, J.P. and Siouffi, A. (1979) Methods for determining bergapten and citropten in bergamot essential oils. Comparative study. *Parfums, Cosmetiques, Aromes*, 30, 49–50; 53–55.

Von Soden, H. and Rojahn, W. (1901) Über Bergaptin einen neuen Inhaltstoff des Bergamottöles. *Pharm. Ztg.*, 46, 778–779.

Wendorff, H. and Matern, U. (1986) Differential response of cultured parsley cells to elicitors from two non-pathogenic strains of fungi. *Eur. J. Biochem.*, 161, 391–398.

Ziegler, H. and Spiteller, G. (1992) Coumarins and psoralens from Sicilian lemon oil (*Citrus limon* (L.) Burm. f.). *Flav. Fragr. J.*, 7, 129–139.

16 Terpeneless and sesquiterpeneless oils

David Moyler

INTRODUCTION

Many of the applications of citrus oils involve incorporation into water based food and beverage consumer products.

The natural high total terpene (consisting of both mono and sesquiterpene hydrocarbons) contents of citrus oils as they are produced by cold expression and or distillation, detract from the oils solubility in the water based product and limit the amount of flavour that can be incorporated into consumer beverages. Terpene reduction and or removal is therefore a key process and the commercial methods, each having their own advantages and disadvantages, are reviewed.

A significant proportion of deterpenated citrus oils are now used for top notes in reconstituted fruit juices and juice blends which are from the named fruit. These beverages are made from concentrated juice which, for the sensible commercial practices of stability and minimisation of shipment and transport costs, lose volatile aromatic components during their processing from pressed fruit juice to concentrate.

The terpeneless citrus oil made from the appropriate fruit fraction (and country of origin where labelling of product requires) can be used to replace some or all of those components that are lost during the concentration.

This reflects the realism of today's beverage industry, and when done correctly, justifies the non-labelling of the aroma fraction on the packaging of the beverage.

In this author's opinion this should be a part of the reconstitution of the juice to compensate for components lost during processing, not to additionally flavour the juice to give it more aroma than naturally provided at the dilution normally consumed.

In Table 16.1 the yield of terpeneless, sesquiterpeneless peel oils and essences are reported for some citrus fruits.

PROCESSES

The processes covered in this chapter include those known to be used at present, or that have been used in the past, or have potential for the future.

Table 16.1 Yields of terpeneless oils from peel oils and essences

	Terpeneless peel oil	Sesquiterpeneless peel oil	Terpeneless essence
Sweet orange	2.5%	1.9%	1.8%
Bitter orange	3.0%	2.0%	N/A
Mandarin	4.0%	2.8%	N/A
Lemon	4.8%	3.8%	3.0%
Lime distilled	15.0%	12.0%	3.0%
Grapefruit	5.0%	2.5%	2.0%
Bergamot	60.0%	N/A	N/A

Note

N/A = not applicable or available. The yield of cold pressed bergamot oil to furanocoumarin free oil (FCF), for perfumery with the skin sensitising psorelens removed is 92%.

Washing

The traditional and time honoured method with food grade solvents, usually ethanol (but sometimes other alcohols) and water mixtures, are used to extract the flavouring components, rejecting the terpene hydrocarbons.

This process is a simple operation. The citrus oil and solvent are placed in a covered stainless steel tank with a stirrer inserted, usually set at an angle to the vertical position, and a tap at the bottom. After stirring sufficiently to allow for the optimum of flavour extraction, the stirrer is switched off and the contents of the tank allowed to separate over night, or up to 48 hours for better separation of the emulsions which can form at the interface. Many of the large international flavour companies have a centrifugal system built into the process to break any emulsions and speed up the separation and minimise the dead time of standing for separation between batches.

The extract is a solution of the flavour compounds in ethanol, which can be standardised for flavour by dilution with solvent and then used directly in finished beverages by adding to the bottling syrup containing sugar, sweeteners, preservative, acidulant etc.

The secondary product, washed oil, is composed of the water immiscible terpenes and still has about half of the flavour components partitioned in it. As such, it has real commercial value for use as a base for reconstituted essential oils by adding ingredients which are either natural or nature identical (NI). This practice is relevant in Europe only, as the NI classification is not recognised in the USA, where it is called artificial and the addition of natural components not derived from the named fruit gives the blend a WONF (natural–with other natural flavoring) status in the USA.

Examples of this type of reconstitution of citrus oils are:

- lemon oil, based upon desolventised washed lemon oil with added citral ex litsea cubeba oil and terpene alcohols and esters ex citronella oil, to make up for those components that were removed by the ethanol during the washing process;
- lime oil distilled, based upon desolventised washed lime oil with added α-terpineol fraction ex pine oils to replace those components extracted.

Such reconstituted oils are offered to users at a fraction of the cost of pure oils and are a viable alternative for the flavouring of budget priced consumer products. An alternative to blending washed oil is to distil it into citrus terpenes, removing traces of ethanol and water so giving a colourless, low peroxide content ingredient, which forms an excellent base for both flavour and fragrance products.

A seminal paper (Owusu-Yaw et al., 1986) from the University of Florida, reported the analysis of cold pressed Valencia orange oil washed with varying ratios of aqueous ethanol, at around the strengths considered to be optimum for the extraction of flavour components. Full quantitative GLC and GC/MS analysis results were given, clearly showing the optimum ratios of oil and solvent and relative ethanol strength needed to get the maximum of aldehydes and alcohols with the minimum of hydrocarbons. Optimum conditions for the removal of terpenes were reported to be with 70 per cent ethanol and a ratio of oil to solvent of 1:3. The results obtained by Owusu-Yaw et al. (1986) are reported in Table 16.2.

This analysis clearly shows how the oxygenated fraction of orange oil is enriched by the washing process. In this author's experience, similar enrichment results are obtained with all of the other citrus oils.

For optimisation of the dynamics of the stirrer rotating in the tank of the mixture of citrus oil and solvent, the reader is referred to the Sourcebook of Flavors (Heath, 1980). Here the positioning of the angled stirrer is shown to significantly alter the turbulence of the system. If vigorous agitation is used, the system is best protected from oxidation

Table 16.2 Percentage composition of cold-pressed valencia orange and of its extract with 70% aqueous ethyl alcohol (Owusu-Yaw et al., 1986)

	Cold-pressed	Ratio of oil to alcohol		
		1:15	1:7	1:3
Monoterpene hydrocarbons[a]	98.69	84.88	2.83	1.93
Sesquiterpene hydrocarbons	0.12	0.26	0.32	0.27
octanal[a]	0.20	2.70	11.35	15.65
nonanal	0.06	0.35	2.29	2.10
decanal	0.24	3.53	24.10	18.27
undecanal	nd	0.18	0.89	1.18
dodecanal	0.05	0.77	4.61	2.21
citronallal	0.05	0.40	2.63	2.27
geranial	0.12	0.91	4.97	6.31
neral	0.08	0.66	4.61	5.17
perilla aldehyde[b]	nd	0.12	0.76	0.99
α-sinensal	0.02	0.26	1.02	0.44
β-sinensal	0.04	0.52	2.09	0.91
nootkatone	0.02	0.24	1.08	0.75
octanol	0.03	0.40	3.24	4.72
linalool	0.23	1.81	18.45	25.02
α-terpineol	0.02	0.53	3.70	4.30

Notes

nd = not detected; a In the analysis of Owusu-Yaw myrcene and octanal were coeluted; in the cold-pressed oil, values of 1.84% and of 0.20% have been assigned to myrcene and octanal respectively. These values have been given by comparison with the composition of sweet orange oil reported in literature; in the alcoholic extracts, values for myrcene and octanal have been deduced by comparison with the behaviour of the other monoterpene hydrocarbons and of decanal; b The perilla aldehyde and any appreciable level of carvone in alcoholic extracts of cold-pressed oils, are believed by this author to be oxidation products caused by processing.

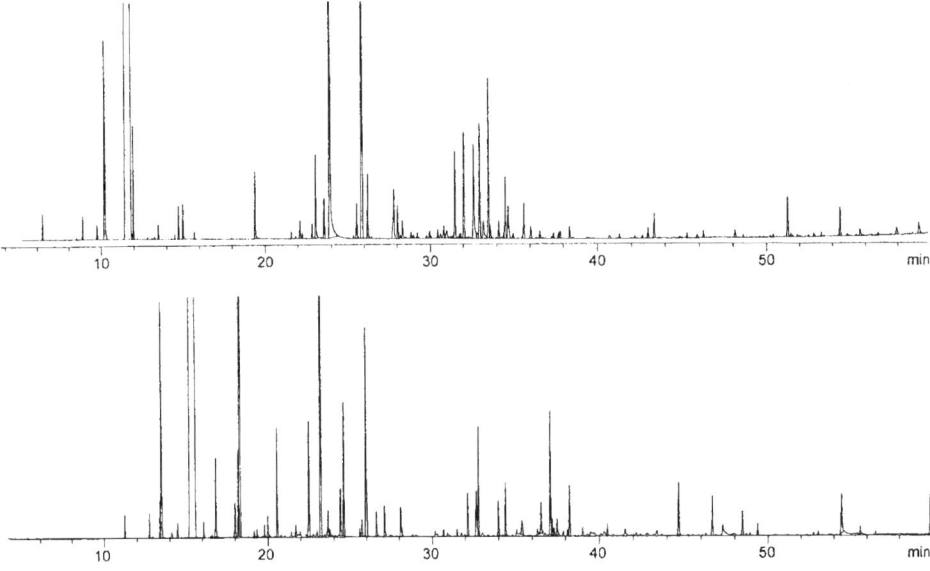

Figure 16.1 GLC analysis of folded Pera orange oil from Brazil (Moyler, 1999). Top trace: Silica fused capillary column, 60 m × 0.25 mm, coated with carbowax, 0.25 μm film thickness. Lower trace: Silica fused capillary column, 60 m × 0.25 mm, coated with polymethylsiloxane 0.25 μm film thickness.

with inert gas. Solid CO_2 can be utilised to blanket the system and help cool the mass to 2 °C, the optimum temperature for the citrus deterpenation process.

Advantages of the washing method are those of simplicity and hence low cost of the operation. Disadvantages of the process are the high volumes of flammable solvent, relative slowness and 50 per cent efficiency of flavour extraction and if the processor does not have a ready use for the washed oil, a large waste steam of terpenes.

Despite these drawbacks, the washing method is without doubt, still the most important commercial deterpenation method for citrus oils. This popularity is in some part due to the simplicity of the system but also the good stability of the extract in soft drinks which is due to its low peroxide value. The peroxides, presumed to be mostly limonene epoxide and oxide, seem to be left in the by product washed oil after the extraction.

Folding

The term folding is used to mean concentration by distilling the monoterpene hydrocarbons out of the remaining concentrated oil by vacuum distillation.

The folded orange oil is a complex oil as illustrated in Figure 16.1, where the chromatograms of a folded Brasilian Pera orange oil obtained on a polar and an apolar column are reported (Moyler, 1999).

This system was studied and reported (Vora *et al.*, 1983) using the 10 fold concentration of Valencia orange oil as the model. The results given are interpreted here and used in a comparison with those of Owusu-Yaw *et al.*, 1986 (Table 16.3). They are a clear

Table 16.3 Percentage content of some flavouring components in cold-pressed and ten fold orange oils

	Vora et al., 1983				Osuwu-Yaw et al., 1986			
	Cold-pressed	Ten fold			Cold-pressed	Ten fold		
		Experimental	Calculated	Loss %		Experimental	Calculated	Loss %
octanal	0.20	0.11	2.0	94.5	0.20	0.30	2.0	95.0
nonanal	0.02	0.03	0.2	85.5	0.06	0.07	0.6	88.3
decanal	0.28	1.48	2.8	47.1	0.24	1.04	2.4	56.7
undecanal	0.02	0.16	0.2	20.0	nd	0.10	–	–
dodecanal	0.07	0.92	0.7	0	0.05	0.41	0.5	18.0
citronallal	0.05	0.14	0.5	72.0	0.05	0.11	0.5	78.0
geranial	0.10	0.68	1.0	32.0	0.12	0.68	1.2	43.3
neral	0.07	0.41	0.7	41.4	0.08	0.41	0.8	48.7
α-sinensal	0.02	0.37	0.2	0	0.02	0.16	0.2	25.0
β-sinensal	0.03	0.55	0.3	0	0.04	1.32	0.4	0
nootkatone	0.01	0.26	0.1	0	0.02	0.16	0.2	25.0
octanol	0.03	0.03	0.3	90.0	0.03	0.02	0.3	93.3
linalool	0.25	0.52	2.5	79.2	0.23	0.62	2.3	73.0
α-terpineol	0.03	0.17	0.3	43.3	0.02	0.18	0.2	10.0
valencene	0.05	0.72	0.5	0	0.05	1.01	0.5	0

Note
nd = not detected.

illustration of the losses of useful flavouring components that are lost by evaporation with the monoterpene hydrocarbons.

The distillation of orange oil using a thin film evaporator (molecular still) to produce a partly deterpenated oil was reported by Tateo (1990). This method has the advantage of a short distillation residence time of orange oil in the heated zone, thus minimising any thermal degradation. However in general, molecular stills are good separators but poor fractionators of oils. In the case of orange oil, the waxes and carotenoids are well separated on a molecular still, but the terpenes are not easy to separate from the octanal and decanal.

In the case of orange oil, there is a ready use for orange terpenes as a natural industrial solvent to replace the ozone depleting organo-chloro and fluoro solvents and as a feedstock to make (−)-carvone (the dextro-rotary hydrocarbon forms the laevo-rotary ketone in the industrial process). The five fold and ten fold oils are therefore plentiful and cheap and, despite the drawbacks of unbalanced flavour profile, are a useful flavouring source.

With the other citrus oils, where the demand for the terpenes is less than that for orange, the alternative methods of flavour extraction are more attractive.

Chromatography

This method has been popular with some processors in the past, using columns packed with activated alumina or silica gel to deterpenate orange and lemon oils.

The method reported by Tzamtzis *et al.* (1990) involves passing the citrus oil through a column, wet slurry packed with non-polar solvent to expel air. They reported this wet packing to be an improvement on the dry pack method of Braverman *et al.* (1957), which reduces the risk of decomposition, isomerisation or polymerisation of the sensitive components of the oil.

The citrus oil may be dissolved in the same solvent if desired and the hydrocarbons are washed through the column with more non-polar solvent. The oxygenated flavouring components are adsorbed onto the active sites of the activated packing and then eluted from the sites by displacement with the more polar solvent ethyl acetate. The solvent is then evaporated under vacuum to give the terpeneless flavouring components.

Modern commercial practice does not use this method because of the large volumes of organic solvents required. Solvents pose problems for economic recovery, static electrical charges, flammability, emissions of voc's (volatile organic compounds) to the environment and their inhalation by operators, potentially causing long term health problems.

PTFE packed column

A series of American articles (Fleisher et al., 1987; Fleisher, 1990, 1994) reported the use of chromatographic columns packed with PTFE=poly tetra fluor ethylene powder which they called Poroplast. This is the familiar non-stick plastic used for coating kitchen pans.

These packed columns were used to deterpenate citrus and other essential oils. The results published by Fleisher et al. (1987), Fleisher (1990, 1994), showed almost complete removal of the terpene hydrocarbons under laboratory and pilot scale conditions. These authors claimed that this is a cold process which gives high percentage recoveries of flavouring components. However, such columns do get warm at the active sites in the middle of the column, and such heat generation increases as the process is scaled up. Perhaps this helps to explain the relatively high levels of carvone which were reported in the concentrate of orange oil made by this method or explained if older oil was used. Carvone is not a natural component of fresh orange oil but can form by the degradation of the limonene, the main component of the terpene hydrocarbon fraction of citrus oils.

This PTFE method has the same limitations of high volume solvent use and recovery as the other column chromatographic techniques. A comparison with the results obtained by this method compared to the counter current method is made in Table 16.4 (orange terpeneless) and Table 16.5 (lemon terpeneless), later in this chapter.

Counter current

This is the name given to the multistage washing extraction of oils, effectively and efficiently removing the monoterpene hydrocarbons.

The method was reputedly used by the Shell Motor oil company for extracting sulphur compounds from gasoline products by extracting with liquid sulphur dioxide under mild pressure conditions.

The Naarden Company in Holland (now part of Quest International) utilised this concept for citrus oils, which they patented in the 1950's. They used a horizontal extractor, diluted the citrus oil in pentane to lower its density and partitioned it against a counter flow of alcohol/water mixture. The alcoholic solution of the citrus concentrate was then standardised with more solvent and incorporated directly in flavours.

The problems associated with handling, recovering and evaporation losses for pentane must have been considerable. The boiling point is 36 °C and pentane has a flash point of minus 49 °C, which makes it a diffusive and highly flammable processing aid.

At the Reigate UK laboratories of Pauls (now part of Universal Flavors) in the early 1980's, an existing citrus extraction process which used diacetin, (glyceryl diacetate)

Table 16.4 Comparison of percentage composition of terpeneless orange oils obtained by counter current and poroplast techniques

	Counter current	Poroplast			Counter current	Poroplast	
	(Moyler and Stephens, 1992)	(Fleisher, 1990)	(Fleisher, 1994)		(Moyler and Stephens, 1992)	(Fleisher, 1990)	(Fleisher, 1994)
hexanal	–	–	0.04	cis-carveol	–	0.09	0.05
heptanal	tr	–	0.09	trans-carveol	–	0.08	0.38
octanal	11.6	6.85	14.17	citronellol	11.6	1.23	0.63
nonanal	2.3	1.87	2.57	geraniol	0.2	0.96	0.43
decanal	17.0	15.91	12.72	linalool	28.0	34.55	29.99
undecanal	0.8	0.31	0.94	nerol	1.2	1.55	0.25
dodecanal	3.3	1.96	1.93	perilla alcohol	–	–	0.10
(E)-2-nonenal	–	–	0.22	trans-sabinene hydrate	–	0.18	–
2-decenal*	0.1	–	–	terpinen-4-ol	0.4	1.02	–
2,4-decadienal*	0.1	–	0.29	α-terpineol	3.0	5.06	2.90
citronellal	3.2	3.59	2.87	elemol	0.4	–	0.48
geranial	6.5	6.47	4.21	nerolidol*	0.2[a]	0.24*	0.12*
neral	4.0	3.92	2.81	hexyl acetate	–	–	0.06
perilla aldehyde	1.1	–	1.89	heptyl acetate	tr	–	0.01
α-sinensal	0.6	0.56	0.24	heptyl formate	–	–	–
β-sinensal	0.9	0.83	0.31	octyl acetate	0.4	0.21	–
6-methyl-5-hepten-2-one	tr	–	–	nonyl propanate	–	–	0.13
carvone[b]	<0.01	0.25	1.07	geranyl acetate	0.1	1.79	–
cis-dihydrocarvone[b]	–	0.07	–	neryl acetate	0.2	1.52	–
trans-dihydrocarvone[b]	–	0.05	0.10*	1,8-cineole	–	–	0.09
piperitone	0.2	–	–	cis-limonene oxide	–	0.25	0.23
nootkatone	0.5	0.53	0.73	trans-limonene oxide	0.2	0.08	0.63
(E)-2-hexenol	tr	–	–	octanoic acid	–	–	0.12
(Z)-3-hexenol	–	–	0.02	decanoic acid	–	–	0.47
heptanol	–	–	0.11	limonene	0.2	–	–
octanol	2.5	2.87	2.25	valencene	0.1	–	–
nonanol	–	–	0.26	total aldehyde acetals	8.0	–	–
decanol	0.3	–	0.94				

Notes
tr = traces; *correct isomer not characterized in the original paper; a (E)-nerolidol; b lower levels of carvone and derivates indicate less decomposition of the orange oil components.

was altered to use some of the principles of the Naarden process but with a vertical column. The principles of the technique were reported (Moyler and Stephens, 1992) and shown to be based on having a difference in density of the two phases being partitioned of 0.04 g/ml. This can be achieved in several ways; reducing the density of the citrus oil with non-polar solvent, thus increasing the density of the aqueous alcohol with soluble salts or weighing agents. A combination of diluting the citrus oil in hexane and high levels of water in the alcohol is one way to achieve the required difference in density.

The choice of which alcohol can also have an effect on the citrus concentrate composition. The three most practical alcohols are listed and their attributes compared:

1. Methanol (Boiling Point 64°C and Flash Point 11°C) is the lowest boiling, so to remove at the lowest temperature hence minimising loss of topnotes. During processing, the prolonged contact of the alcohol with the aliphatic aldehydes form acetal compounds. These acetals do help to stabilise the aldehydes, not only in the concentrate but also in the finished beverage or food that the concentrate is added to. These acetals will eventually decompose back to the corresponding aldehyde in the beverage and this aldehyde will eventually decompose completely. The net effect is to prolong the beverage flavour's shelf life. A minute trace of nature identical methanol will also appear in the beverage.
2. Ethanol (Boiling Point 78°C and Flash Point 14°C) is the most acceptable food solvent internationally. In some countries (UK customs, USA BATF) the problem of recovery of ethanol by distillation is an issue to be considered. Acetals are also formed and improve the stability and shelf life of the flavour. A trace of ethanol will appear in the beverage.
3. Propan-2-ol (Boiling Point 80°C and Flash Point 12°C) is traditionally not accepted in some countries as a food solvent. Therefore, if any extract is to be made for a broad international market, it is better to avoid the use of this solvent. Unlike the other alcohols, it does not form acetals, which may be undesirable for some applications. This lack of acetal formation is believed to be caused by the steric hindrance of the molecular shape of this alcohol. Propan-2-ol is nature identical, but not in citrus oils.

As an illustration, Moyler *et al.* (1992) published the results of the capillary GLC analysis of cold pressed sweet orange oil made using the counter current technique, which in Table 16.4 are compared to the published results from the Poroplast method (Fleisher, 1990, 1994).

A similar behaviour for lemon oil is reported in Table 16.5. The counter current lemon oil tested was made from a Sicilian spring crop yellow cold pressed oil extracted by the screw press method. The lemon oil used for the Poroplast publication was unspecified.

A range of concentrated citrus peel and oil phase essence oils made in the UK using this method are commercially available and include: bitter orange oil cold pressed peel, West Indian; orange essence oil phase, both Florida and Brazil; lemon essence oil phase; lime oil distilled, West Indian; grapefruit oil cold pressed peel.

The citrus concentrates made by this counter current method have a true tasting flavour profile and this is revealed in the confirming GLC of the deterpenated oil compared to the folded citrus oils.

Table 16.5 Comparison of percentage composition of terpeneless lemon oils obtained by counter current and poroplast techniques

	Counter current (Moyler, 1999)	Poroplast (Fleisher, 1990)
octanal	0.4	0.53
nonanal	0.9	0.94
decanal	0.6	0.59
undecanal	0.3	0.15
citronellal	1.0	0.24
geranial	37.2	41.29
neral	23.0	22.16
6-methyl-5-hepten-2-one	0.04	0.09
piperitone	0.1	–
nootkatone	0.05	0.15
octanol	0.03	0.18
cis-carveol	–	0.07
trans-carveol	–	0.05
geraniol	0.7	0.74
linalool	2.6	1.58
nerol	0.7	0.75
cis-sabinene hydrate	–	0.09
trans-sabinene hydrate	–	0.07
terpinen-4-ol	1.0	1.42
α-terpineol	3.3	3.10
α-bisabolol	1.2	0.64
citronellyl propanate	–	0.55
geranyl acetate	6.9	8.37
geranyl butyrate	–	0.21
neryl acetate	9.1	9.11
1,4-cineole	0.03	0.06
1,8-cineole	0.05	0.06
limonene	0.5	0.07
γ-terpinene	0.1	–
β-bisabolene	0.2	–
total aldehyde acetals	2.1	–

Carbon dioxide

Publications have appeared on the extraction of citrus oils, using CO_2 to dissolve and remove the terpenes in a high pressure SFE (super critical fluid) system. The fundamentals of CO_2 extraction and its relation to the flavour industry were published by King and Bott (1993). Chapters on applications in the various aspects of the food and beverages industries were included, chapter six by this author is a review of the extraction of flavour and fragrance materials.

The system as relating to citrus oils is fundamentally different from the conventional counter current method. Alcohol is more soluble in CO_2 than it is in aliphatic hydrocarbon solvents which involves careful control of the pressure of the system, hence density of the CO_2.

Perhaps, to overcome the difficulties with the control of CO_2 solubility, the system used by the University of Leeds UK (Dugo et al., 1995) utilised a silica gel packed column. The citrus oil was dissolved in supercritical CO_2 and pumped through the column. The oxygenated components were absorbed and later eluted by changing the solvent. They experimented with both orange and lemon peel oils and compared the deterpenated oils with those made by vacuum distillation. They reported that the SFE process gave concentrates which showed qualitative and quantitative ratios of the oxygenated compounds near to those of the original oils (see this author's comments earlier about losses during distillation).

Bitter orange oil was the subject of a similar study by Couchi et al. (1996). Using pressures between 77 and 120 bar and 40°C conditions for processing, they reported that the oxygenated compounds in the deterpenated fraction was 6.6 times higher than in the crude oil and they were able to separate a residue fraction containing the waxes, psoralens and colour.

There is clearly a future in the use of carbon dioxide to replace the flammable solvents currently used in citrus deterpenation processes. Industrially, the use of CO_2 to replace the highly flammable hexane has to be welcomed, even if the downside is a substantial capital cost of high pressure equipment.

At present, no processors offering such CO_2 concentrated citrus oils are commercially available, although there may be some made for in-house use by flavour companies.

CONCLUSIONS

A variety of viable commercial methods are suitable for the manufacture of terpeneless and sesquiterpeneless citrus oils.

A comparison of all of the deterpenation methods is summarised in Table 16.6, which shows the temperatures to which the oil and concentrate are exposed during processing and their relative operating efficiency. From Table 16.6 it can be seen that the most efficient deterpenation methods are d, e and f which operate at the lowest temperatures. The significance of the lower temperatures is due to:

- the minimal evaporation of topnote volatiles from the concentrate;
- the minimal oxidation of terpenes;
- the formation of less acetals.

Although less efficient at deterpenation, method c distillation is a method which prevents acetal formation, because no alcohols are present and the use of any flammable

Table 16.6 Citrus deterpenation method comparison

	Max (°C) with terpenes	Max (°C) with concentrate	Extract efficiency(%)	Residual terpenes(%)	Capital cost
a Washing	25	40	50	10	low
b Folding and washing	60	40	50	10	low
c Distillation	60	60	50	10–40	low
d Chromatographic	40	40	95	<1	low
e Counter current	25	40	95	<1	medium
f Carbon dioxyde	40	40	95	<1	high

solvents is completely avoided. Some deficiencies in the full flavour profile are to be expected because of the loss of all components of the citrus oil with low boiling points e.g. octanal, with boiling points below that of limonene. Despite the flavour differences, distillation is without doubt the most cost effective of all of the methods. It gives a commercially acceptable product, which can be modified in most flavour systems to supplement the flavour profile with natural flavouring substances or food grade aromachemicals, depending on the legislative status of the flavour.

The comparison of methods for cost, efficiency and profile of flavouring components shows that the lower capital cost methods give incomplete profiles due to losses during processing. However, for flavour compounds, these incomplete profiles for e.g. 10 fold orange oil, are extremely cost effective in blends.

The more complete flavour profile concentrates made from the named fruit variety and origin, are best suited for the direct incorporation into fruit juice blends. They form part of the reconstitution of the ready to drink juice made from high Brix concentrated juice where they add only a minimal cost of 0.01 USD per litre.

REFERENCES

Braverman, J.B.S. and Solomiansky, L. (1957) Separation of terpeneless essential oils by chromatographic method. *Perfumery Essent. Oil Record*, 6, 284–287.

Chouchi, D., Barth, D., Reverchon, E. and Della Porta, G. Bigarade peel oil fractionation by supercritical carbon dioxyde desorption. *J. Agric. Food Chem.*, 44, 1100–1104.

Dugo, P., Mondello, L., Bartle, K.D., Clifford, A.A., Breen, G.P.A. and Dugo, G. (1994) Deterpenation of sweet orange and lemon essential oils with supercritical carbon dioxyde using silica gel as an adsorbent. *Flav. Fragr. J.*, 10, 51–58.

Fleisher, A. (1990) The Poroplast extraction technique in the flavor and fragrance industry. *Perfum. Flav.*, 15(5), 27–36.

Fleisher, A. (1994) Citrus hydrocarbons from essential oils. *Perfum. Flav.*, 19(1), 40–44.

Fleisher, A., Biza, J., Secord, N. and Dono, J. (1987) Ultra-Tech citrus concentrates. A new series of deterpenified citrus oils. *Perfum. Flav.*, 12(2), 57–61.

Heath, H.B. (1982) *Sourcebook of Flavors*. AVI Publishing Co., Westport, Connecticut, US.

King, M.B. and Bott, T.R. (1993) *Extraction of Natural Products Using Near Critical Solvents*. Blackie/Chapman and Hall, London, UK.

Moyler, D.A. and Stephens, M.A. (1992) Counter current deterpenation of cold pressed sweet orange peel oil. *Perfum. Flav.*, 17(2), 37–38.

Moyler, D.A. (1999) Unpublished results.

Owusu-Yaw, J., Matthews, R.F. and West, P.F. (1986) Alcohol deterpenation of orange oil *J. Food Sci.*, 51, 1180–1182.

Tateo, F. (1990) Production of concentrated orange oils using a thin film evaporator. *J. Essent. Oil Res.*, 2, 7–13.

Tzamtzis, N.E., Liodakis, S.E. and Parissakis, G.K. (1990) The deterpenation of orange and lemon oils using preparative adsorption chromatography. *Flav. Fragr. J.*, 5, 56–67.

Vora, J.D., Matthews, R.F., Crandall, P.G. and Cook, R. (1983) Preparation and chemical composition of orange oil concentrate. *J. Food Sci.*, 48, 1197–1199.

17 Composition of distilled oils

Luis Haro-Guzmán

COMPOSITION OF DISTILLED KEY AND PERSIAN LIME OILS

Distilled lime oils differ very noticeably in aspect and odour from centrifuged oils, also known as cold-pressed oils. Almost colourless or pale yellow with a sharp and terpenic odour, distilled oils have few points in common with centrifuged oils, which are brownish green and have a fruit fragance. Arctander (1960) describes the odour of distilled oil as: 'sharp, fresh, terpene-like, somewhat perfumey-fruity citrus-type'.

During distillation lime oil undergoes important modifications in its chemical composition. Placed during several hours at a temperature equivalent to the boiling point of juice and in a very acidic environment, many constituents are degraded, generating other more stable compounds but no longer keeping the same odour. Finally non-volatile compounds such as wax, colouring matters, coumarins and psoralens of course do not enter in the distillate. Table 17.1 shows the relative amounts of the different groups of compounds that constitute the distilled oil of Mexican lime. The composition of distilled oils of Mexican and Persian limes is shown in Table 17.2. A typical Gas Chromatogram of distilled Key lime oil is shown in Figure 17.1. The oils obtained from Key and Persian limes show remarkable quantitative differences, large enough to explain those found in odour and flavour (Haro and Faas, 1985). The ratio of two compounds: neryl acetate/δ-elemene has been used to differentiate distilled Key lime oil from Persian. The relative amounts of these compounds vary little with the change in distillation conditions. The values of the ratio are less than 1.2 for Key and more than 10.0 for Persian. This ratio is very helpful in the detection of Persian lime oil being added to Key lime oil (Haro-Guzmán, 1999).

COMPOSITION OF DISTILLED (PERATONER) LEMON, MANDARIN AND GRAPEFRUIT OILS

Distilled and cold-pressed essential oils show notable differences though not so marked as in the case of lime oil. The distillation temperature is lower, the contact time shorter and the emulsion less acidic. There are also clear differences in the organoleptic characteristics.

These products show evident differences in composition: the distilled oils practically do not contain the components of the fixed residue. Sesquiterpenes and some aldehydes are in smaller amounts in distilled oils. Some alcohols such as terpinen-4-ol and α-terpineol

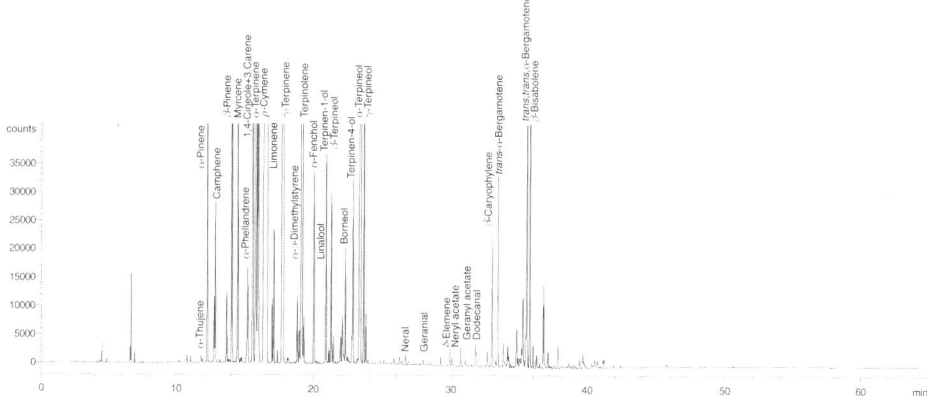

Figure 17.1 GC profile of distilled Key lime oil on a Non-Polar Phase (SE-30) (Haro-Guzmán, 1999).

Table 17.1 Quantitative composition of distilled key lime oil by groups (Clark and Chamblee, 1992)

Group	Weight (%)
Monoterpene hydrocarbons	
Monocyclic	72.13
Bycyclic	3.31
Sesquiterpene hydrocarbons	3.48
Aldehydes	0.48
Alcohols	13.41
Esters	0.15
Ethers, ketones	3.96
Unidentified	2.38

are probably formed under the distillation conditions. Table 17.3 shows the composition of a distilled Peratoner oil of Italian lemon compared with its cold-pressed counterpart. Table 17.4 lists the same information for Italian mandarin oils and Table 17.5 describes the composition of a distilled grapefruit oil.

Table 17.2 Composition of distilled lime oils

	Key lime (%)	Persian lime (%)
Paraffins		
nonane	0.06[a], 0.02–0.04[d], 0.03[f]	
tridecane	0.02[a], tr.–0.02[d]	
undecane	0.03[a], 0.01–0.03[d], 0.19[f] (incl. linalool)	
Monoterpene hydrocarbons		
bornylene	X[c], 0.02[f]	
camphene	0.77[a], 0.42[b], 0.68–0.79[d] (incl. α-fenchene), 0.53[f]	0.25[c]
δ-3-carene	0.21[b], tr.–0.03[d], 0.01[f]	0.39[e] (incl. 1,4-cineol)
p-cymenene	0.08–0.16[d]	
α-fenchene	X[c], 0.68–0.79[d] (incl. camphene), 0.21[f]	

Table 17.2 (Continued)

	Key lime (%)	Persian lime (%)
limonene	59.95[a], 38.99[b], 38.95–42.24[d], 46.86[f] (incl. 1,8-cineole)	59.02[e] (incl. 1,8-cineol)
p-mentha-3,8-diene	0.01–0.02[d]	
p-mentha-1,3,8-triene	tr.–0.01[d]	
Myrcene	0.77[a], 1.20[b], 0.85–1.10[d], 0.34[f] (incl. 2, 3-dehydro-1,8-cineole)	1.56[e]
allo-ocimene	tr[d]	
neo-allo-ocimene	tr[d]	
(E)-β-ocimene	0.20–0.50[d], 0.68[f] (incl. 2,2-dimethyl-5 (1-methylpropenyl)-tetrahydrofuran)	
(Z)-β-ocimene	0.15–0.20[d], 0.20[f]	
α-phellandrene	0.34–0.48[d], 0.46[f]	0.20[e]
α-pinene	0.84[a], 0.85[b], 0.94–1.26[d], 1.02[f]	2.25[e]
β-pinene	0.90[a], 0.42[b], 0.77–1.90[d], 1.34[f]	6.04[e]
sabinene	tr.[d], 0.02[f]	
α-perpinene	2.47[b], 1.47–2.35[d], 2.84[f]	1.05[e]
γ-perpinene	0.63[a], 12.41[b], 8.51–13.35[d], 11.02[f]	16.09[e]
terpinolene	0.82[a], 9.52[b], 6.91–9.70[d], 8.43[f] (incl. trans-linalool oxide and monoterpene hydrocarbon)	2.69[e] (incl. nonanal)
α-phujene	tr.–0.02[d], 0.01[f]	0.21[e]
tricyclene	0.02–0.03[d], 0.02[f]	
Sesquiterpene hydrocarbons		
cis, α-bergamotene	0.04–0.09[d], 0.06[f]	
trans, α-bergamotene	0.52[a], 0.45–0.80[d], 0.70[f]	0.56[e]
(Z)-α-bisabolene	tr.–0.14[d], 0.17[f] (incl. α-selinene)	
β-bisabolene	0.94[a], 1.09–2.47[d] (incl. (E,E)-α-farnesene), 1.04[f]	0.63[e]
(E)-γ-bisabolene	0.01–0.04[d], 0.01[f]	
(Z)-γ-bisabolene	tr.–0.01[d], 0.04[f]	
δ-cadinene	0.03–0.08[d], 0.05[f] (incl. epi-α-selinene)	
γ-cadinene	X[c]	
β-caryophyllene	0.27[a], 0.42–0.57[d], 0.60[f]	0.29[e]
γ-curcumene	0.07–0.11[d]	
β-elemene	0.04–0.08[d], X[f]	
δ-elemene	X[c], 0.04–0.08[d], 0.06[f]	0.54[e] (incl. neryl acetate)
γ-elemene	tr.[d]	
trans, trans, α-farnesene	X[c]	0.18[e]
(E,E)-α-farnesene	1.09–2.47[d] (incl. β-bisabolene), 0.68[f]	
β-farnesene	X[c]	
(Z)-β-farnesene	tr.–0.09[d], 0.10[f]	
germacrene-B	0.01–0.08[d], 0.03[f]	
cis-β-guaiene	0.08–0.16[d]	
α-humulene	0.08–0.15[d], 0.09[f]	
isocaryophyllene	tr.–0.09[d]	
α-muurolene	0.03[f]	
β-santalene	X[c], 0.02–0.08[d], 0.04[f]	
selina-3,7(11)-diene	0.03–0.17[d], 0.05[f]	
selina-4,11-diene	0.13[f]	
α-selinene	0.06–0.15[d], 0.17[f] (incl. (Z)-α-bisabolene)	
epi-α-selinene	0.05[f] (incl. δ-cadinene)	
β-selinene	0.01[f]	

δ-selinene	0.20^f	
trans-trans-sesquicitronellene	0.20^a	
Alcohols		
α-bisabolol	X^c, 0.02–0.06^d, 0.06^f	
epi-α-bisabolol	0.01–0.06^d	
borneol	0.60^a, 0.59–0.82^d, 0.69^f (incl. nonanol)	0.10^e
α-cadinol	0.01–0.03^d, 0.04^f	
epi-α-cadinol	0.02^f (incl. epi-α-muurolol)	
campherenol	X^c, 0.01–0.05^d, 0.04^f	
trans-carveol	tr.^{f}	
caryophyllene alcohol	0.02–0.09^d, 0.07^f	
citronellol	0.02^f	
p-cymen-8-ol	0.11–0.28^d, 0.13^f (incl. p-methyl-aceto phenone)	
decanol	0.06^a, tr.–0.04^d, 0.09^f (incl. geranial)	
dihydrocarveol	1.07–1.31^d (incl. γ-terpineol)	
dodecanol	tr.–0.04^d, 0.04^f	
elemol	tr.–0.04^d	
γ-eudesmol	0.03–0.10^d, 0.05^f	
α-fenchol	1.19^a, 0.66–0.83^d, 0.79^f	0.30^e
β-fenchol	tr.–0.02^d	
geraniol	tr.–0.04^d, 0.02^f	
isoborneol	0.05–0.15^d (incl. (E)-ocimenol), 0.08^f	
isopulegol	tr.–0.08^d	
linalool	0.10^a, 0.11–0.22^d, 0.19^f (incl. undecane)	0.05^e
cis-p-menth-2-en-1-ol	tr.^{f}	
trans-p-menth-2-en-1-ol	0.02^f	
epi-α-muurolol	0.02^f (incl. epi-α-cadinol)	
myrcenol	tr.–0.08^d, 0.08^f	
nerol	tr.–0.02^d	
nerolidol	0.01^f	
nonanol	0.01^a, 0.69^f (incl. borneol)	
(E)-ocimenol	0.05–0.15^d (incl. isoborneol), 0.06^f	
(Z)-ocimenol	0.60–0.91^d (incl. cis-β-terpineol), 0.09^f	
octanol	0.01^a, 0.01–0.02^d, 0.02^f	
trans-pinene hydrate	tr.–0.03^d	
selin-7(11)-en-4-ol	0.01^f	
terpinen-1-ol	0.68^a, 0.74–1.27^d, 1.20^f	0.12^e
terpinen-4-ol	1.61^a, 0.56–0.88^d, 0.97^f	0.46^e
α-terpineol	5.86^a, 7.00–8.04^d, 7.9^f	2.15^e
β-terpineol	0.70^a	0.19^e
cis-β-terpineol	0.60–0.91^d (incl. (Z)-ocimenol), 0.22^f	
trans-β-terpineol	0.10–0.20^d, 0.81^f	
δ-terpineol	X^c	
γ-terpineol	X^c, 1.07–1.31^d (incl. dihydrocarveol), 1.21^f	0.19^e
α, α, p-trimethylbenzyl alcohol	0.56^a	
undecanol	tr.–0.01^d, 0.03^f	
Aldehydes		
decanal	0.09^a, 0.02–0.27^d, 0.20^f	
dodecanal	0.01^a, tr.–0.10^d, 0.08^f	
furfural	0.01^a	
geranial	0.02–0.04^d, 0.09^f (incl. decanol)	0.12^c
neral	tr.–0.03^d, 0.02^f	0.09^c
nonanal	0.01–0.04^d, 0.03^f	2.69^c (incl. terpinolene)
octanal	0.03^a, 0.02–0.05^d, 0.04^f	

Table 17.2 (Continued)

	Key lime (%)	Persian lime (%)
perillaldehyde	X[c], 0.01–0.05[d], 0.02[f]	
α-terpinen-7-al	tr.–0.02[d]	
tetradecanal	0.01–0.04[d], 0.02[f]	
undecanal	0.01–0.06[d], 0.03[f]	
Esters		
bornyl acetate	tr.–0.02[d]	
citronellyl acetate	tr.–0.01[d]	
decyl acetate	tr.–[d]	
geranyl acetate	0.28[a], 0.06–0.13[d], 0.10[f]	0.09[e]
neryl acetate	0.01[a], 0.02–0.07[d], 0.04[f] (incl. decanoic acid)	0.54[e] (incl. δ-elemene)
trans-pinocarvyl acetate	tr.–0.04[d]	
sabinyl acetate	0.01[f]	
Ketones		
carvone	tr.–0.02[d], 0.01[f]	0.01[e]
methylheptenone	0.01[a], tr.–0.01[d]	
p-methylacetophenone	tr.–0.02[d], 0.13[f] (incl. *p*-cymen-8-ol)	
cis-pinocamphone	0.01–0.05[d]	
trans-pinocamphone	0.02–0.06[d]	
piperitone	tr.[d], 0.01[f]	
Oxides		
caryophyllene oxide	tr.–0.05[d]	
1,4-cineole	1.75[a], 2.43–3.72[d], 2.32[f]	0.39[e] (incl. δ-3-carene)
1,8-cineole	0.70[a], 4.35–9.47[d], 46.86[f] (incl. limonene)	59.02[e] (incl. limonene)
2,3-dehydro-1,8-cineole	0.19–0.29[d], 1.34[f] (incl. myrcene)	
dihydrocaryophyllene oxide	tr.–0.01[d]	
4,8-epoxy-*p*-menth-1-ene	tr.–0.03[d]	
cis-limonene oxide	tr.–0.03[d]	
trans-limonene oxide	tr.–0.04[d]	
cis-linalool oxide	0.02[f]	
trans-linalool oxide	8.43[f] (incl. monoterpene hydrocarbon and terpinolene)	
Others		
p-cymene	11.59[a], 3.88[b], 1.81–4.28[d], 1.89[f]	1.51[e]
decanoic acid	0.04[f] (incl. neryl acetate)	
2,2-dimethyl-5-(1-methyl-1-propenyl)-tetrahydrofuran	0.30[a], 0.02–0.06[d], 0.68[f] (incl. (E)-β-ocimene)	
α,*p*-dimethylstyrene	0.49[a], 0.14[b]	
p-menth-1-en-8-yl vinyl ether	tr.–0.01[d]	
1,1,3a,7-tetramethyl-1*H-1* a,2,3,3a, 4,5,6,7b-octahydro-cyclopropa [a]-naphtalene	0.13–0.24[d]	
2,6,6-trimethyl-tetrahydropyran	X[c]	
2,6,6-trimethyl-2-ethenyl-tetrahydropyran	0.16[a], 0.18–0.24[d], 0.22[f]	

Notes
a Kovats, 1963; b Tapanes *et al.*, 1971. Values reported (monoterpenes = 100%) were corrected estimating 70.5% of monoterpenes in the oil; c Chamblee *et al.*, 1985; d Dugo *et al.*, 1998; e Haro and Faas, 1985; f Chamblee and Clark, 1997 (Figures in weight per cent); X = present.

Table 17.3 Composition of lemon oils

	Cold-pressed[a] area (%)	Distilled[b] area (%)
α-thujene	0.43	0.24–0.49
α-pinene	1.95	1.10–2.11
camphene	0.06	0.04–0.06
sabinene	2.02	0.78–1.73
β-pinene	13.01	8.22–13.99
myrcene	1.44	1.36–1.58
limonene	65.23	64.01–70.49
γ-terpinene	9.54	8.37–10.72
terpinolene	0.38	0.37–0.51
linalool	0.11	0.15–0.31
nonanal	0.11	0.06–0.13
citronellal	0.09	0.01–0.04
terpinen-4-ol	0.04	0.22–0.88
α-terpineol	0.17	0.27–0.70
decanal	0.04	0.02–0.06
nerol + citronellol	0.04	0.02–0.37
neral	0.83	0.51–1.45
geraniol	0.02	0.02–0.43
geranial + perillaldehyde	1.39	0.69–2.20
neryl acetate	0.40	0.23–0.52
geranyl acetate	0.42	0.18–0.59
β-caryophyllene	0.23	0.10–0.23
trans-α-bergamotene	0.34	0.12–0.33
α-humulene	0.02	0.01
β-bisabolene	0.51	0.16–0.50

Notes
a Verzera *et al.*, 1996; Average values for oil produced with FMC machines, 1984/93; b Dugo *et al.*, 1983.

Table 17.4 Composition of mandarin oils (Verzera *et al.*, 1992)

	Cold-pressed area (%)	Distilled area (%)
α-thujene	0.73–1.06	0.47–1.86
α-pinene	2.03–2.74	1.41–4.87
camphene	0.01–0.02	0.01–0.04
sabinene	0.23–0.34	0.09–0.33
β-pinene	1.39–2.10	1.17–2.60
myrcene	1.57–1.84	1.56–2.01
octanal	0.06–0.20	0.09–0.29
α-phellandrene	0.02–0.09	0.03–0.10
α-terpinene	0.26–0.52	0.25–0.60
p-cymene	0.10–1.38	0.20–0.94
limonene	65.30–74.26	66.28–73.33
γ-terpinene	16.40–22.75	16.37–21.79
terpinolene	0.72–1.01	0.69–1.02
linalool	0.04–0.19	0.15–0.34
nonanal	0.01–0.04	0.02–0.06
citronellal	0.02–0.05	tr.–0.04
terpinen-4-ol	0.01–0.08	0.18–0.48
α-terpineol	0.03–0.27	0.20–0.49

Table 17.4 (Continued)

	Cold-pressed area (%)	Distilled area (%)
decanal	0.05–0.12	0.05–0.12
nerol + citronellol	tr.–0.05	0.02–0.06
neral	tr.–0.05	0.01–0.05
geraniol	tr.–0.01	tr.–0.01
geranial	0.01–0.10	0.03–0.11
thymol	0.01–0.11	0.01–0.13
undecanal	tr.–0.02	tr.–0.01
methyl N-methyl-anthranilate	0.26–0.66	0.14–0.79
β-caryophyllene	0.07–0.14	0.03–0.09
α-humulene	tr.–0.02	tr.–0.01
α-farnesene	0.06–0.32	0.02–0.16
α-sinensal	0.12–0.53	0.01–0.21

Note
tr = trace value.

Table 17.5 Composition of distilled grapefruit oil (Pino et al., 1999)

	Area %
α-pinene	3.8
sabinene	0.2
myrcene	13.6
limonene	70.9
octanol	0.9
cis-linalool oxide*	1.0
trans-linalool oxide*	0.4
linalool	1.6
cis-limonene oxide	0.1
trans-limonene oxide	tr
cis-β-terpineol	0.1
terpinen-4-ol	0.8
α-terpineol	2.3
dihydrocarveol	0.1
neodihydrocarveol	tr
decanal	0.1
trans-carveol	0.1
cis-carveol	tr
neral	0.3
carvone	tr
carvotanacetone	0.1
geranial	0.3
neryl acetate	0.1
α-copaene	0.2
longifolene	0.1
β-caryophyllene	0.6
α-humulene	0.1
δ-cadinene	0.1

Notes
tr = trace (<0.01%) value; * furanoid form.

CHEMICAL TRANSFORMATIONS OCCURRING DURING THE DISTILLATION

Subject to high temperatures (95–105 °C) for long periods of time (6–12 hours) in a very acidic environment (pH 2.2–2.4) the lime oil suffers deep modifications in its composition: neral, geranial and sabinene almost disappear, α- and β-pinene react to different extents. On the other hand 1,4- and 1,8-cineole, terpinolene, α-terpineol, fenchol, borneol, α-p-dimethylstyrene, α- and γ-terpinene and p-cymene are formed.

α- and β-Pinene

Haro and Huet (1970) obtained important proportions of 1,4-cineole, terpinolene and α-terpineol, as well as lesser quantities of 1,8-cineole and p-cymene when treating β-pinene under conditions similar to those of a typical distillation. Clark and Chamblee (1992) mention as products: limonene, terpinolene, α-terpineol, borneol, isoborneol, 1,4- and 1,8-cineol, camphene, α-fenchol and fenchene. It was calculated that β-pinene reacts approximately 10 times faster than α-pinene which itself reacts more quickly than limonene by a factor of approximately 16.

Sabinene and α-thujene

The major hydration products are: terpinen-4-ol, γ- and α-terpinene, terpinolene, and sabinene hydrate. McHale (1980) mentions 1,4-cineole as product of sabinene. Spontaneous ring opening and proton elimination yield monoterpene hydrocarbons: γ-terpinene, terpinolene and α-terpinene (Clark and Chamblee, 1992).

p-Cymene

This compound is stable under normal distillation conditions (Clark and Chamblee, 1992).

d-Limonene

Slater and Watkins (1964) have shown that d-limonene which makes up around 50 per cent of the oil can, under distillation conditions, give out α-terpineol, terpinen-4-ol, 1,4- and 1,8-cineole, α and γ-terpinene and terpinolene.

γ-Terpinene

It has been found that p-cymene is formed from γ-terpinene during deterioration of lemon oil (Ikeda et al., 1961). A similar situation can be expected to occur in lime oil. Table 17.2 shows abnormal amounts for some components as determined by Kovats (1963): α-terpinene was not found, γ-terpinene content is 0.63 per cent and p-cymene is 11.59 per cent. This situation make us think he worked with a very deteriorated sample of oil. In fact, the content of p-cymene is used as an indicator of the degree of deterioration of the oil. A maximum of 2.6 per cent is considered acceptable.

Citronellal

It has long been known to be unstable under acidic conditions. It cyclizes to form *cis*-and *trans-p*-menthan-3,8-diols as the major products plus a small amount of the corresponding isopulegols (Clark and Chamblee, 1992).

Terpinen-4-ol

1,4-Cineole is also a secondary product probably formed from terpinen-4-ol (Clark and Chamblee, 1992).

Citral

It readily cyclises in acidic conditions yielding some *p*-cymene, α-*p*-dimethylstyrene and a wide range of other products (Loori and Cover, 1964; McHale, 1980; Clark and Chamblee, 1992).

Neryl and geranyl acetates

The major products of terpene ester hydrolysis are diols, along with linalool and geraniol which will partially cyclize to α-terpineol and a small amount of Kovats ether 'tetrahydropyran' (Clark and Chamblee, 1992).

Table 17.6 compares the composition of an average Mexican distilled Key lime oil with that of a cold-pressed type A of the same origin thus permitting to observe the deep changes that happen in the composition during the distillation.

Table 17.6 Composition of key lime oils

	Cold-pressed type A[a] area (%)	Distilled[b] area (%)
α-thujene	0.39	0.01
α-pinene	2.45	1.03
α-fenchene	–	0.71[c]
camphene	0.12	
sabinene + β-pinene	23.36	1.39
myrcene	1.26	0.98
α-phellandrene	0.03	0.42
δ-3-carene	0.01	0.01
1,4-cineole	–	2.84
α-terpinene	0.26	1.89
p-cymene	0.32	2.97
limonene	49.35	41.50
1,8-cineole	–	7.27
γ-terpinene	7.80	10.22
terpinolene	0.42	8.01
linalool	0.17	0.15
nonanal	0.02	0.02
fenchol	0.01	0.73
citronellal	0.01	–
borneol	0.03	0.64
terpinen-4-ol	0.52	0.75

α-terpineol	0.28	7.43
γ-terpineol	–	1.16
decanal	0.21	0.12
nerol	0.04	0.01
neral	1.15	0.01
geraniol	0.05	0.03
geranial + perillaldehyde	1.86	0.04
undecanal	0.03	0.03
δ-elemene	0.31	0.06
neryl acetate	0.08	0.04
geranyl acetate	0.24	0.08
β-elemene	0.18	0.06
dodecanal	0.12	0.03
cis-α-bergamotene	0.09	0.06
β-caryophyllene	1.07	0.55
trans-α-bergamotene	1.36	0.65
(Z)-α-bisabolene	0.18	0.11
(E,E)-α-farnesene + β-bisabolene	3.18	1.81
germacrene B	0.50	0.06

Notes
a Dugo et al. (1997); b Dugo et al. (1998); c α-fencene + camphene.

We can conclude with Slater and Watkins (1964) that: 'the differences in flavour and aroma between distilled and écuelled (cold-pressed) lime oils and between fresh and old lime juices are mainly due to the chemical changes in natural lime oil brought about by contact with citric acid solutions' and, with Clark and Chamblee (1992): 'model experiments with expressed lime oil constituents and intermediates, similar to those conducted in early work by Slater and Watkins, would allow an even greater understanding of the chemistry of this complex process'.

It can be seen that the production of a good quality lime oil depends mostly in carrying the reactions to the necessary extent to obtain the correct balance of constituents.

REFERENCES

Arctander, S. (1960) *Perfume and Flavor Materials of Natural Origin*. Author's Edition, Elizabeth, New Jersey.

Chamblee, T.S., Clark, B.C., Radford, T. and Iacobucci, A. (1985) Citrus essential oil analyses by HPLC and capillary GC. *ACS Meeting*, Chicago, IL., September 1985.

Chamblee, T.S. and Clark, B.C. (1997) Analysis and chemistry of distilled lime oil (*Citrus aurantifolia* Swingle). *J. Essent. Oil Res.*, 9, 267–274.

Clark, B.C. and Chamblee, T.S. (1992) Acid-catalyzed reactions of citrus oils and other terpene-containing flavors. In G. Charalambous (ed.), *Off-Flavors in Foods and Beverages*, Elsevier, Holland, pp. 229–285.

Dugo, G., Licandro, G., Cotroneo, A. and Dugo, G. (1983) Sulla genuinità delle essenze agrumarie. Nota II–Individuazione di aggiunta di essenze ottenute per distillazione alle essenze di limone estratte a freddo. *Essenz. Deriv. Agrum.*, 53, 218–257.

Dugo, P., Cotroneo, A., Bonaccorsi, I. and Mondello, L. (1998) On the genuineness of citrus essential oils. Part LVII. The composition of distilled lime oil. *Flav. Fragr. J.*, 13, 93–97.

Dugo, P., Mondello, L., Lamonica, G. and Dugo, G. (1997) Characterization of cold-pressed Key and Persian lime oils by gas chromatography, gas chromatography/mass spectroscopy, high-performance liquid chromatography and physicochemical indices. *J. Agric. Food Chem.*, 45, 3608–3616.

Haro-Guzmán, L. and Huet, R. (1970) L'huile essentielle de lime au Mexique (*Citrus aurantifolia* Swingle). *Fruits*, 25, 887–899.

Haro, L. and Faas, W.E. (1985) Comparative study of the essential oils of Key and Persian limes. *Perfum. Flav.*, 10(5), 67–72.

Haro-Guzmán, L. (1999) Unpublished work.

Ikeda, R.M., Stanley, W.L., Vannier, S.H. and Rolle, L.A. (1961) Deterioration of lemon oil. Formation of p-cymene from γ-terpinene. *Food Tech.*, 15, 379–380.

Kovats, E. sz. (1963) Zur Kenntnis des sog. 'destillierten' Limetten-Öls. *Helvetica Chimica Acta.*, 46, 2705–2731.

Loori, J.J. and Cover, A.R. (1964) The mechanism of formation of p, α-dimethylstyrene in the essential oil of distilled lime (*Citrus aurantifolia*). *J. Food Sci.*, 29, 576–582.

Pino, J.A., Acevedo, A., Rabelo, J., Gonzales, C. and Escadon, I. (1999). Chemical composition of distilled grapefruit oils. *J. Essent. Oil Res.*, 11, 75–76.

Slater, C.A. and Watkins, W.T. (1964) Chemical transformations of lime oil. *J. Sci. Food Agric.*, 15, 657–664.

Tápanes, R., Pérez Zayas, J. and Fanghänel, E. (1971) Análisis de la fracción terpénica del aceite esencial de limeta destilado (*Citrus aurantifolia* Swingle) producido en Cuba. *Revista CENIC*, 3, 99–110.

Verzera, A., Trozzi, A., Licandro, G. and Scuderi, L. (1996) Sulla genuinità delle essenze agrumarie. Nota LVI. La composizione dell'olio essenziale di limone FMC prodotto nella stagione 1995. *Essenz. Deriv. Agrum.*, 66, 353–367.

Verzera, A., Cotroneo, A., Stagno d'Alcontres, I. and Donato, M.G. (1992) On the genuineness of essential oils. Part XXX. Detection of distilled essential oils added to cold-pressed mandarin essential oils. *J. Essent. Oil Res.*, 4, 273–280.

18 Extracts from the bitter orange flowers (*Citrus aurantium* L.): composition and adulteration

Louis Peyron and Ivana Bonaccorsi

INTRODUCTION

This chapter describes the composition of some of the products obtained from bitter orange flowers and their possible adulterations.

The products described will be:

- neroli oil obtained by steam distillation from the flowers;
- bitter orange flower water absolute obtained from bitter orange flowers by solvent extraction of the water-phase produced by steam distillation;
- bitter orange flower absolute obtained by extraction in alcohol from the solvent extract of the flowers.

COMPOSITION OF NEROLI OIL

In the text and tables will be used the same symbols and criteria applied in the chapters on the peel cold pressed oils and on petitgrain oils.

Guenther (1949) reported the physicochemical indices of neroli oils of different origins, and, for some of these, the esters content (expressed as linalyl acetate), the free alcohols content (expressed as linalool) and content of methyl anthranilate. These values were the following:

	Esters	Free alcohols	Methyl anthranilate
France	6.7–21.0%	–	0.4–1.2%
Italy	7.3–14.7%	49.9–67.5%	0.6–1.4%
Spain	12.3–32.3%[a]	–	–
Haiti	8.7–18.7%	–	–

a this value was considered by Guenther too high, and probably due to the presence of leaves in the raw material used for distillation.

For the neroli oil produced in France, Guenther (1949) also reported the following approximated composition:

Hydrocarbons (α-pinene, camphene, limonene, heptacosane)	35%
(−)-linalool	30%
geraniol + nerol	4%
(+)-α-terpineol	2%
linalyl acetate	7%
geranyl acetate + neryl acetate	4%
(+)-nerolidol	6%
indole	< 0.1%
methyl anthranilate	0.6%
acetic acid + palmitic acid	0.1%
Others (decanal^t, esters of phenyl acetic acid and of benzoic acid, jasmone*, farnesol*)	11.2%

Guenther also found β-ocimene, phenyl ethyl alcohol, and trace amounts of phenols. The same quantitative data reported by Guenther were later included by Gildmeister and Hoffmann (1959) and by Bigi (1962) in their reviews respectively.

Calvarano, M. (1963) determined, by classical methods, the following composition, relative to six samples of neroli oils:

esters (as linalyl acetate)	11.98–25.99%
free alcohols (as linalool)	49.14–58.75%
aldehydes (as decanal)	0.62–1.06%
methyl anthranilate	0.45–0.73%
indole	0.09–0.10%

The gas chromatographic analysis, carried out on a stainless steel capillary column, coated with UCON LB-550X, on a neroli sample obtained by mixing some of the six samples, put in evidence the presence of p-cymene (2.53 per cent), limonene (5.25 per cent), myrcene (0.8 per cent), α-pinene (0.15 per cent), β-pinene (6.09 per cent), γ-terpinene (0.15 per cent) and also, not quantified, decanal, geraniol, linalool, nerol, α-terpineol, farnesol*, nerolidol*, phenyl ethyl alcohol, geranyl acetate, linalyl acetate, neryl acetate, terpenyl acetate*.

Peyron (1965) determined in neroli oils, using gas chromatography on packed columns, the presence of the following components, listed in decreasing amount order: linalool, β-pinene, linalyl acetate, limonene, β-ocimene*, nerolidol*, α-terpineol, neryl acetate, geranyl acetate, α-pinene.

Few years later McHale (1971) determined that the β-ocimene, identified in neroli oil, was the trans isomer, and described the alkaline isomerisation to the conjugated trien of this component. Corbier and Tiesseire (1974) isolated and identified in neroli oil the *cis*-heptadec-8-ene, and the 2,5-dimethyl-2-vinyl-4-hex-4-enal.

Since 1981 numerous papers have been published on the systemic analysis on the composition of neroli oil. The results of these papers are reported in Table 18.1.

In the period 1980–1990 papers other than those relative to systemic studies on the composition of neroli oil were published. These are on particular aspects of the composition of neroli oils, or limited to qualitative information.

Toyoda *et al.* (1993) identified in neroli oil p-cymene, limonene, myrcene, β-ocimene*, β-caryophyllene, benzyl aldehyde, geraniol, linalool, nerol, α-terpineol, nerolidol*, 2-phenyl ethyl alcohol, linalool oxides*, indole, methyl anthranilate, phenylacetonitrile.

Table 18.1 Percentage composition of neroli oil

	1	2	3(a)	3(b)	3(c)	3(d)	3(e)	4	5	6	7	8	9
HYDROCARBONS													
Monoterpene													
camphene	–	–	–	–	–	–	–	0.02	0.05	–	–	0.05	0.04
δ-3-carene	–	–	–	–	–	–	–	–	0.1	–	–	0.05	0.52
p-cymene	–	–	–	0.15	0.17	0.56	0.05	–	tr	–	–	0.22	1.04
limonene	16.1[a]	9–17	13.35	18.12	15.05	15.60	12.58	9.13	16.6	1.06	7.15	12.88	24.57
myrcene	1.6	1.1–3.2	2.07	3.76	9.34	7.87	1.97	6.26	1.7	0.07	1.34	2.49	2.33
(E)-β-ocimene	6.0	3.2–8.2	–	–	–	–	–	–	6.0	0.40	–	5.60	3.60
(Z)-β-ocimene	16.1[a]	0.5–1.0	1.02[c]	0.98[c]	0.86[c]	0.66[c]	0.78[c]	–	0.5	–	–	0.82	0.34
α-phellandrene	–	–	–	–	–	–	–	–	0.05	–	–	0.04	0.09
β-phellandrene	–	–	1.02[c]	0.98[c]	0.86[c]	0.66[c]	0.78[c]	–	0.4	–	–	–	–
α-pinene	0.8	0.7–1.9	0.95	0.86	1.26	1.18	1.35	0.28	1.1	–	0.25	0.75	1.31
β-pinene	15.0	11–23	9.31	7.42	8.22	9.97	14.08	0.53	11.8	0.13	4.85	0.52	20.22[e]
sabinene	–	1.0–3.3	7.24	5.27	9.12	7.93	9.53	–	2.8	–	0.66	2.26	20.22[e]
α-terpinene	–	–	–	0.59	0.74	–	–	–	0.5	–	5.64	0.18	0.51
γ-terpinene	–	–	7.92	8.34	7.34	2.75	6.98	–	1.0	–	–	0.33	3.71
terpinolene	–	–	0.45	0.71	0.66	0.1	0.43	–	0.4	–	–	0.42	0.53
α-thujene	–	–	–	–	–	–	–	tr	0.1	–	–	0.05	0.25
Sesquiterpene													
δ-cadinene	–	–	–	–	–	–	–	–	tr	–	–	–	0.03
caryophyllene	–	0.3–1.6[b]	–	–	–	–	–	–	0.5	–	1.23	–	–
β-caryophyllene	–	–	–	–	–	–	–	0.13	–	–	–	0.54	0.72
β-elemene	–	–	–	–	–	–	–	–	0.1	–	–	0.05	1.63
cis-β-farnesene	–	–	–	–	–	–	–	–	–	–	–	0.08	0.14
germacrene D	–	–	–	–	–	–	–	–	0.1	–	–	0.05	0.05
α-humulene	–	–	–	–	–	–	–	tr	0.1	0.06	–	0.18	0.10
valencene	–	–	–	–	–	–	–	–	tr	–	–	0.05	–
ALDEHYDES													
Aliphatic													
nonanal	–	–	–	–	–	–	–	tr	–	–	–	0.01	–

Table 18.1 (Continued)

	1	2	3(a)	3(b)	3(c)	3(d)	3(e)	4	5	6	7	8	9
Monoterpene													
citronellal	–	–	–	–	–	–	–	–	–	–	–	0.01	0.06
geranial	–	–	–	–	–	–	–	0.04	tr	–	1.05	0.10	0.65
neral	–	–	–	–	–	–	–	0.58	tr	–	3.7	0.03	0.41
ALCOHOLS													
monoterpene													
citronellol	0.2	–	4.08	2.21	2.81	3.59	3.08	–	–	–	–	2.18	–
geraniol	2.0	1.2–2.9	1.99	2.00	1.71	1.50	4.09	tr	4.25	–	0.45	8.93	15.59
linalool	30.6	31–41	25.48	29.99	31.23	38.81	29.00	50.86	37.5	73.72	53.19	8.93	15.59
nerol	0.5	0.5–1.2	0.79	0.64	0.81	0.48	0.77	tr	0.5	0.19	0.98	0.82	0.69
terpinen-4-ol	–	0.3–1.6[b]	0.17	1.59	1.90	0.43	0.53	0.11	0.75	–	–	0.42	1.20
α-terpineol	3.0	2.8–4.9	2.26	2.59	2.96	3.09	3.35	4.75	2.0	–	9.76	3.30	1.79
β-terpineol*	–	–	0.60	0.32	0.50	0.85	0.55	–	–	–	–	–	–
Sesquiterpene													
farnesol*	4.0	0.9–2.8	–	0.72	–	1.01	2.17	tr	1.0[d]	2.46	7.55	2.58	–
(E,E)-farnesol	–	–	–	–	–	–	–	–	–	–	–	1.48	0.01
(Z,E)-farnesol	–	–	–	–	–	–	–	–	–	–	–	0.04	0.98
nerolidol*	4.0	1.4–5.6	6.88	2.76	3.03	4.61	4.33	–	2.6	–	–	–	–
(E)-nerolidol	–	–	–	–	–	–	–	0.64	–	–	–	–	1.76
(Z)-nerolidol	–	–	–	–	–	–	–	1.09	–	1.18	–	–	tr
ESTERS													
Monoterpene													
citronellyl acetate	–	–	–	–	–	–	–	2.15	1.7	0.21	1.08	0.09	0.03
geranyl acetate	2.9	2.5–4.1	–	–	–	–	–	18.39	2.8	16.53	–	2.65	–
linalyl acetate	9.1	3.4–7.6	5.39	9.45	4.90	2.48	3.84	1.09	0.9	0.45	0.48	6.37	9.76
neryl acetate	1.7	1.3–2.0	–	–	–	–	–	tr*	0.2	–	–	1.36	0.92
α-terpinyl acetate	–	–	–	–	–	–	–	–	–	–	–	0.06	0.05
ETHERS and OXIDES													
caryophyllene oxide	–	–	–	–	–	–	–	–	–	–	–	0.01	0.04
cis-linalool oxide	–	–	0.05	0.05	0.08	0.08	0.11	–	–	–	–	0.02	0.02

trans-linalool oxide	–	–	0.05	0.02	0.05	0.24	0.02	0.15	–	–	–	–	0.07	–
perillene	–	–	–	–	–	–	–	–	tr	–	–	–	0.02	–
OTHERS														
geranyl acetone	–	–	–	–	–	–	–	–	0.05	–	–	–	0.05	–
indole	0.1	–	–	–	–	–	–	–	0.1	0.83	0.16	–	–	0.06
cis-jasmone	–	–	–	–	–	–	–	–	0.05	–	–	–	0.05	–
methylanthranilate	0.3	0–0.5	0.31	0.85	0.36	0.13	0.26	tr	–	0.25	0.10	–	–	0.11
methyl N-methylanthranilate	–	–	–	–	–	–	–	–	0.1	–	–	–	0.10	3.18
methyl jasmonate	–	–	–	–	–	–	–	–	tr	–	–	–	0.01	–
2-phenyl ethanol	–	–	–	–	–	–	–	–	0.2	–	–	–	0.20	–

Notes

t = tentative identification; tr = traces; * correct isomer not characterised; a limonene + (Z)-β-ocimene; b β-caryophyllene + terpinen-4-ol; c (Z)-β-ocimene + β-phellandrene; d farnesol isomers sum; e β-pinene + sabinene.

Appendix to Table 18.1

1   Buccellato (1981). One sample.
2   Prager and Miskiewicz (1982). 16 samples from Mediterranean Countries; GC on capillary column coated with Carbowax 20 M; GC/MS; relative percentage of peak areas. These authors also analysed 3 commercial samples considered adulterated.
3   Srinivas (1986). Oils produced in Algeria, Egypt, Italy, Spain and Tunisia; GC/MS. Srinivas also found farnesil acetate* (0.03%).
4   Lin et al. (1986). China; one sample; column chromatography on silica gel, GC on capillary column coated with SP 2305 and SE 30; GC/MS, retention indices, IR; relative percentage of peak areas. Lin et al. also found β-ocimene* (3.91%); cis-linalool oxide, pyranoid form (0.06%); trans-linalool oxide, pyranoid form (0.09%). Lin et al. also observed that during the storage the linalool decreased and that the linalool oxides increased up 13.3%.
5   Boelens and Sindreu (1988). Spain. GC on capillary columns coated with Carbowax 20 M, UCON and SE 54; GC/MS; relative percentage of peak areas. Boelens and Sindreu also found bisabolene (0.1%), cis + trans-farnesene (0.1%), cis + trans-anhydrolinalool oxide (0.05%), cis + trans-linalool oxide (0.1%) and trace amounts of p-cymenene, selinene*, methyl N-methylanthranilate.
6   Ma et al. (1988). China; one sample; GC/MS and retention indices on two columns of differing polarity; relative percentage of peak areas. Ma et al. also found aromadendrene (0.30%), isocaryophyllene (0.09%), benzyl alcohol (0.15%), 3-cycloexenyl carbinol* (0.43%), (E)-2-hexenol (0.09%), hexadecanoic acid (0.36%), phtalic acid* (0.09%).
7   Germanà et al. (1990). Sicily, Italy; one sample; GC on packed column of EAS; relative percentage of peak areas.
8   Boelens and Oporto (1991). Spain. Boelens and Oporto also found δ-cadinene (0.03%), trans-β-farnesene (0.13%), decanol (0.03%), octanol (0.02%), δ-cadinol (0.02%), 1,8(9)-menthadienyl acetate (0.02%), cis-limonene oxide (0.03%), trans-limonene oxide (0.03%), 2,2,6-trimethyl-6-vinyl tetrahydropyrane (0.04%).
9   Mondello et al. (1994). Sicily, Italy; one commercial sample; GC on capillary column coated with SE 52; LC-GC/MS; relative percentage of peak areas. Mondello et al. also found tricyclene (0.01%), trans-α-bergamotene (0.02%), (E,E)-α-farnesene (0.07%), benzaldehyde (0.01%), 6-methyl-5-hepten-2-one (0.11%), cis-p-menth-2-en-1-ol (0.09%), α-cadinol (0.02%), globulol (0.01%).

Boelens (1997) compared the composition of Spanish and Tunisian neroli oils, obtained by hydrodistillation, and one extracted by supercritical CO_2, from bitter orange flowers from Morocco. The main differences between the two types of oils are reported below:

	Hydrodistilled oil	Supercritical CO_2 extracted oil
monoterpene hydrocarbons	38%	28%
linalyl acetate	3–5%	24%
linalool	38%	35%
nitrogen derivatives	<0.5%	2%
sesquiterpene alcohols	4%	<2%

Boelens (1997) explained the different contents between the two oils, in monoterpene hydrocarbons and in linalyl acetate, to some chemical change of linalyl acetate during the distillation process, leading to the formation of monoterpene hydrocarbons and other monoterpenoids. The higher content of nitrogen containing components in the oil extracted by supercritical CO_2 could be due to the water solubility of these components and to their consequent loss during the hydrodistillation process. Boelens also asserted that a group of perfumers preferred the oils extracted by supercritical CO_2 for their odour character and intensity (double if compared with the hydrodistilled ones).

COMPOSITION OF THE BITTER ORANGE FLOWERS WATER ABSOLUTE

The quantitative composition of the flower water absolute was reported by Buccellato (1981), Prager and Mishiewicz (1982) and by Remy *et al.* (1993). The results reported in these papers are summarised in Table 18.2.

Remy *et al.* (1993) monitored the chemical changes of the bitter orange water absolute in function of the time and the conditions used for the storage of the distillation water obtained after the separation of the neroli oil, during the first five days of storage. They determined a decrease of the free monoterpene alcohols content and of those esterified (geraniol, linalool, nerol) and of the sesquiterpene alcohols (farnesol* and nerolidol*) and the increase of α-terpineol, 6-methyl-5-hepten-2-one and of another ketone not identified. These phenomena were more evident after longer periods of storage if the water was kept in contact with air. In these conditions the racemisation of linalool and of α-terpineol, would also occur. Moreover, the odour would turn to a green note. Such behaviour would therefore suggest that, in order to obtain a good quality flower water absolute, it would be preferable to perform the extraction in short time after the distillation.

COMPOSITION OF BITTER ORANGE FLOWERS ABSOLUTE

The quantitative composition of the bitter orange flower absolute has been studied by Buccellato (1981) and by Prager and Mishiewicz (1982). Their results are summarised in Table 18.3.

Table 18.2 Percentage composition of flower water absolute

	1	2	3
limonene	0.5[a]	0–2.2	–
(E)-β-ocimene	0.2	–	–
(Z)-β-ocimene	0.5[a]	–	–
β-pinene	1.1	0–11	–
β-caryophyllene	–	0.2–2.0[b]	–
citronellol	0.2	–	–
eugenol	0.5	–	–
geraniol	6.4	3.7–7.6	8.25
linalool	44.2	40–53	47.80
nerol	2.6	1.7–2.7	2.86
terpinen-4-ol	–	0.6–2.0[b]	–
α-terpineol	18.5	13–22	19.80
farnesol*	0.5	0–1.9	1.00
nerolidol*	1.7	0–1.5	0.70
phenylethyl alcohol	1.9	0.9–3.9	2.54
geranyl acetate	0.5	0–1.1	0.70
linalyl acetate	–	0–2.4	0.18
neryl acetate	0.5	0–0.6	0.33
cis-linalool oxide	–	0.6–2.3	5.10[c]
trans-linalool oxide	–	1.3–3.7	5.10[c]
phenylacetonitrile	–	0.8–5.0	–
indole	0.1	0–3.2	1.50
methylanthranilate	4.1	2.0–11.4	2.97

Notes
* correct isomer not characterised; a limonene + (Z)-β-ocimene; b β-caryophyllene + terpinen-4-ol; c cis- + trans-linalool oxide.

Appendix to Table 18.2

1 Buccellato, 1981;
2 Prager and Miskiewicz, 1982. Eleven samples imported from France and four samples obtained from perfume companies; GC on capillary column coated with Carbowax 20 M; relative retention times, MS of separated components; relative percentage of peaks area.
3 Remy et al., 1993. Tunisian samples from the productive seasons 1990–1992.

Kaiser and Lamparsky (1982) studied the nitrogen containing components present in bitter orange flower absolute and in its headspace. The components determined were: methyl anthranilate (6.5 per cent), indole (5.0 per cent), phenylacetaldoxime (2.0 per cent), phenylacetonitrile (1.5 per cent), 1-nitro-2-phenylethane (0.4 per cent), methyl N-acetylanthranilate (0.03 per cent), ethylanthranilate (tr), and methyl N-formylanthranilate (tr); were also identified the following components: methyl N-methylanthranilate, decanalaldoxime, nonanalaldoxime, undecanalaldoxime, geranialaldoxime, neralaldoxime, (6E)-farnesaloxime, methylheptenone oxime, geranylcetone oxime, methylheptenone isooxazoline.

Yang and Lee (1988), during a study on the toxicity of bitter orange flower water absolute, isolated by a method that applied distillation, solvent extraction and adsorption chromatography, from 3 samples, Egyptian, French and from Morocco, 2.650, 545 and 680 ppm respectively of bergapten, and a not determined amount of epoxibergamotin. The authors attributed to bergapten the phototoxic activity of bitter orange flower water absolute.

Table 18.3 Percentage composition of bitter orange flower absolute

	1	2
limonene	5.1[a]	0–3.6
myrcene	0.1	0–0.5
(E)-β-ocimene	0.6	0–2.2
(Z)-β-ocimene	5.1[a]	–
α-pinene	tr	–
β-pinene	0.4	0–2.7
β-caryophyllene	–	0–0.8[b]
citronellol	0.5	–
eugenol	0.3	–
geraniol	1.6	0–2.0
linalool	32.0	34–48
nerol	0.9	0–1.1
terpinen-4-ol	–	0–0.8[b]
α-terpineol	2.4	1.5–3.7
farnesol*	7.7	3.6–15.4
nerolidol*	7.6	4.8–8.9
phenylethyl alcohol	4.5	0–2.1
citronellyl acetate	0.1	–
geranyl acetate	0.6	0–0.3
linalyl acetate	16.8	14–21
neryl acetate	0.8	0–0.7
indole	1.0	2.9–9.9
methylanthranilate	30	0–1.1
phenylacetonitrile	–	1–4.3

Notes
* correct isomer not characterised; a limonene+(Z)-β-ocimene; b β-caryophyllene+terpinen-4-ol.

Appendix to Table 18.3

1 Buccellato, 1981;
2 Prager and Miskiewicz, 1982. Eleven samples imported from France and four samples obtained from perfume companies; GC on capillary column coated with Carbowax 20 M; relative retention times, MS of separated components; relative percentage of peaks area.

Toyoda *et al.* (1993) identified in bitter orange flower water absolute. the same components identified in neroli oil and previously reported in this chapter.

During the same year, Surburg *et al.* (1993) identified the heterocyclic nitrogen containing components listed following: 0.01 per cent of each: 3-phenylquinoline, and 2-phenyl-2-propylpyridine; 0.001 per cent of each: 4-methyl-3-phenylpyridine, 2-methyl-5-phenylpyridine, and 2-ethyl-5-phenylpyridine; and 0.0002 per cent of 4-ethyl-3-phenylpyridine.

ADULTERATION

The bitter orange flower oil (or neroli) is essential for perfumery. In these days its production is limited and is expensive. The price of neroli produced in Tunisia in 1998 was about 1850 US$/Kg.

Due to this high cost, during the years, the producers and the users of natural flowers have tried to obtain less expensive products, with odour characters close to that of the neroli oil to be used as substitute and sometimes as adulterants of the genuine oil.

In the last decade the adulteration became more subtle along with the development of the analytical technique and the increased availability of the raw material. These are mainly of the following types:

- the first more or less rough, consist of the adjustment of the main components composition and of the correction of the physico-chemical indices, to imitate the genuine oils;
- the second, more refined, also consists of the reproduction of the odour character of the natural oil.

The products commonly used as adulterants are numerous:

- citrus essential oils, different from neroli, that are present on the market at a lower price;
- single components natural or synthetic;
- reconstituted essential oils that present chemical composition and sensorial properties close to those of genuine neroli oils.

A detailed list of the most common ingredients used to prepare a synthetic neroli oil has been reported by Anonis (1983). The adulteration can also be performed by distilling along with bitter orange flowers, sweet orange flowers, bitter orange foliage and any other vegetable material different from bitter orange flowers, but containing components of the same nature as those normally present in neroli oil. Obviously the presence of vegetable material different from the bitter orange flowers can occur by accident and be simply due to the poor selection of the raw material. Some anomalies of the composition and the odour character can be also due to the technology and the inadequate condition used, to the uncleanness of the stills and containers, and to the modification of the natural composition during the storage of the oil.

The techniques and the analytical methods used for the detection of adulterations in neroli oil, as it is also common for other essential oils, consist of the determination of classical indices (refractive index, optical rotation, density, solubility, acidity number, esters number etc.), mentioned by the International Regulations and by the Pharmacopoeias, indicating for each oil the minimum and maximum values or the ranges variability. The modern instrumental methods, in particular the chromatographic and spectroscopic ones, are also applied, in order to determine the content of each single component and specific characters of some of these, as the enantiomeric distribution and the isotopic ratios.

Guenther (1949) asserted that the adulterations, that are most commonly used for neroli oils, are the additions of bitter orange petitgrain, either as whole or deterpenated. These additions could be detected measuring the total amount of esters. In fact, this adulterant is characterised by amounts of linalyl acetate much higher than neroli oil. Guenther suggested, however, to proceed by an accurate sensorial test, since the chemical analyses at that time could be inadequate for the comprehension of the adulteration. Guenther also reported the comments published by Naves (1948), who asserted that the enantiomeric distribution of linalool, and the amount and composition of the free alcohols fraction are influenced by the duration of the distillation.

Moreover, analysing the esterified alcoholic fraction after saponification, the addition of synthetic linalyl acetate (presence of partially racemic linalool) and of terpinyl acetate could be revealed.

The ISO (1975), AFNOR (1976), and the Swiss regulations (1984), provide the following limits for neroli oil:

		AFNOR (1976) ISO (1975)	Manuel Suisse des denrées alimentaires (1984)
Relative density (at 20/20 °C)			0.866–0.879
	France	0.866–0.871	
	Italy	0.866–0.879	
	North Africa	0.866–0.876	
Refractive index (at 20 °C)			1.469–1.474
	France	1.4690–1.4740	
	Italy	1.4690–1.4740	
	North Africa	1.4700–1.4740	
Optical rotation (at 20 °C)			from $+1°$ to $+11°$
	France	from $+1.5°$ to $+7°$	
	Italy	from $+2.5°$ to $+11.5°$	
	North Africa	from $+6°$ to $+11°$	
Alcohol solubility[a] (at 20 °C)			1/2 (alcohol 80%)
	France	1/2 (alcohol 80%)	
	Italy	1/2 (alcohol 80%)	
	North Africa	1/3.5 (alcohol 85%)	
Acidity index			max 2
	France	max 2	
	Italy	max 2	
	North Africa	max 2	
Esters index			16–50
	France	25–44	
	Italy	20–44	
	North Africa	28–50	
Total alcohols (as linalool)			50%

[a] oil value/max value of aqueous alcohol.

Prager and Mishiewicz (1981) esteemed adulterate same samples of neroli oil due to their low content of monoterpene hydrocarbons (23 per cent, 15 per cent and 22 per cent respectively) compared with the average content of 39 per cent determined by these author in genuine oils, and for the high content of farnesol* (8.3 per cent vs 3 per cent average in genuine samples); almost doubled values of linalool and of linalyl acetate if compared with the genuine ones; presence of 2.5–3 per cent of *cis*- and *trans*-linalool oxides. The same authors considered adulterated some samples of bitter orange flower water absolute for the high content of linalyl acetate (5.9–9.4 per cent vs the average value of 1 per cent determined in genuine samples); and/or of terpinen-4-ol (2.6–3.2 per cent); of nerolidol* (2.9–3.5 per cent), of farnesol* (2.8–4.6 per cent), and of indole (3.2–5.4 per cent) that were determined in genuine sample at 2.0 per cent, 1.5 per cent, 1.9 per cent, 2.7 per cent respectively. In the same paper were also considered adulterated those samples of bitter orange flower water that presented a high level of phenyl ethyl alcohol (4.7–12.2 per cent vs the range of variation of 0.3–2.1 per cent normally determined in genuine samples); of limonene (6.2–9.6 per cent vs the range of 0.3–3.6 per cent determined in genuine oils), and a high level of one linear chained alcohol of 20 or 22 carbon atoms. Some of the oils considered adulterated also did not contain nerolidol, which in genuine samples was present at an average value of 7 per cent. The high amount of indole (9.9 per cent) in one sample was not considered sign of adulteration, but a consequence of the processing conditions. In fact, it was Guenther's opinion (1949) that, if the flowers were packed too tightly or processed too slowly, an increase of the indole content would result.

Juchelka et al. (1996), and Mosandl and Juchelka (1997) from the results obtained analysing numerous samples of neroli oil extracted in laboratory by distillation and solvent extraction, and samples of industrially processed oils of ensured origin, and of commercial oils, come to the conclusion that the presence of the R-(−)-linalyl acetate (>95 per cent), of S-(−)-β-pinene (>98 per cent) can be considered tools of extreme importance for the quality control of neroli oils. Even if the enantiomeric distribution of limonene ranged widely, it could be used as a characteristic marker of genuineness. For this compound was indicated a value of the enantiomer R-(+) > 93 per cent for a genuine neroli oil. The enantiomeric distributions of linalool, α-terpineol, terpinen-4-ol, and α-pinene were not considered by these authors reliable for the neroli oil characterisation. In fact, the enantiomeric distribution of linalool presented a ratio of the R-(−) and S-(+) isomers of about 90:10 in the solvent cold extracted flowers, but this ratio shifted toward racemization up to 70:30 in the oils obtained by distillation, depending on the length of the process. The α-terpineol present in neroli oil ranged between 2 and 7 per cent, representing an artefact generated during the distillation process. This is in fact almost absent in the extracts obtained by solvent cold extraction. The enantiomeric distribution of the S-(−) and R-(+) isomers was about 70:30. Terpinen-4-ol is usually present at trace levels as racemic mixture. Least, the enantiomeric distribution of α-pinene varied between wide ranges.

REFERENCES

Association Française de Normalisation (AFNOR) (1976) Huile essentielle de neroli, T 75–202.

Anonis, D.P. (1985) Neroli in perfumery. *Perfum. Flav.*, 9(1), 7–10.

Bigi, B. (1962) Neroli and orange flower absolute. *Amer. Perfum.*, 77(9), 21–24.

Boelens, M.H. and Sindreu, R.J. (1988) Essential oils from Seville bitter orange (*Citrus aurantium* L. ssp. *Amara* L.). In B.M. Lawrence, B.D. Mookherjee and B.J. Willis (eds), *Flavors and Fragrances: a World Perspective*. Elsevier Science Publishing B.V., Amsterdam, pp. 551–565.

Boelens, M.H. and Oporto, A. (1991) Natural isolates from Seville bitter orange tree. *Perfum. Flav.*, 16(6), 2–7.

Boelens, M.H. (1997) Differences in chemical and sensory properties of orange flower and rose oils obtained from hydrodistillation and from supercritical CO_2 extraction. *Perfum. Flav.*, 22(3), 31–35.

Buccellato, F. (1981) Orange blossom. *Perfum. Flav.*, 6(2), 31–34.

Calvarano, M. (1963) Sulla composizione dell'essenza di neroli di Calabria. *Essenz. Deriv. Agrum.* 33, 5–21.

Corbier, B. and Teisseire, P. (1974) Contribution to the knowledge of neroli oil from Grasse. *Recherches*, 19, 289–290.

Germanà, M.A., De Pasquale, F., Bazan, E. and Palazzolo, E. (1987) Ricerche sugli oli essenziali del flavedo di 44 cloni di limone. *Essenz. Deriv. Agrum.*, 57, 421–455.

Gildemeister, E. and Hoffmann, F. (1959) *Die Ätherischen Öle*, Band V, 608–624. Akademie Verlag, Berlin.

Guenther, E. (1949) The essential oils. D. Van Nostrand Co., New York, N.Y.

Juchelka, D., Steil, A., Witt, K. and Mosandl, A. (1996) Chiral compounds of essential oils. XX. Chirality evaluation and authenticity profiles of neroli and petitgrain oils. *J. Essent. Oil Res.*, 8, 487–497.

Kaiser, R. and Lamparsky, D. (1982) Constituents en trace azotes de quelques absolues de fleurs et leurs headspaces correspondents. *Proceeding VII International Essential Oil Congress*, paper. No. 90, 287–294, Fedarom, Grasse.

Lin, Z.-K., Hua, Y.-F. and Gu, Y.-H. (1986) The chemical constituents of the essential oil from the flowers, leaves and peels of *Citrus aurantium*. *Acta Botanica Sinica*, 28, 635–640.

Ma, L., Zheng, Y.Q., Sun, Y.L., Lin, M.X. and Wu, Z.P. (1988) Aroma volatile constituents of *Citrus aurantium* L. ssp. *amara* Engl. *Beijing Daxue Zuebao Ziran Kexueban*, 24, 687–694.

Manuel Swisse des denrées alimentaires (1978, revisè partiellement en 1984) *Substances aromatisantes*, Vol. 2, Chapter 43.

McHale, D. (1971) The base catalysed rearrangement of *cis*- and *trans*-β-ocimene to *trans*-4-trans-6-alloocimene. *Tetrahedron*, 27, 4843–4844.

Mondello, L., Dugo, P., Bartle, K.D., Frere, B. and Dugo, G. (1994) On line high performance liquid chromatography coupled with high resolution gas chromatography and mass spectrometry (HPLC–HRGC–MS) for the analysis of complex mixtures containing highly volatile compounds. *Chromatographia*, 39, 529–538.

Mondello, L., Dugo, G., Dugo, P. and Bartle, K.D. (1996) On-line HPLC–HRGC in the analytical chemistry of citrus essential oils. *Perfum. Flav.*, 21(6), 25–49.

Mosandl, A. and Juchelka, D. (1997) Advances in the authenticity assessment of citrus oils. *J. Essent. Oil Res.*, 9, 5–12.

Normes Internationales (1975) ISO, 3517.

Peyron, L. (1965) Petitgrain oils in perfumery. *Soap Perfum. Cosmet.*, 38, 769–780.

Prager, J.M. and Miskievicz, M. (1981) Gas chromatographic mass spectrometric analysis, identification and detection of adulteration of perfumery products from bitter orange trees. *J. Assoc. Off. Anal. Chem.*, 64, 131–138.

Remy, M., Bayle, J.C., Derbesy, M. and Uzio, R. (1993) The Tunisian orange flower water absolute LMR, *EPPOS* (numero speciale), 593–602.

Sirinivas, S.R. (1986) Atlas of essential oils. *Published by author*, Bronx, NY.

Surburg, H., Guentert, M. and Harder, H. (1993) Volatile compounds from flowers. Analytical and olfactory aspects. In R. Teranishi, R.G. Buttery and H. Sugisawa (eds), *Bioactive Volatile Compounds from Plants. ACS Symposium Series 525*, Am. Chem. Soc., Washington, DC., pp. 168–186.

Toyoda, T., Nohara, I. and Sato, T. (1993). Headspace analysis of volatile compounds emitted from various Citrus blossoms. In R. Teranishi, R.G. Buttery and H. Sugisawa (eds), *Bioactive Volatile Compounds from Plants. ACS Symposium Series n. 525*, Am. Chem. Soc. Washington, DC, pp. 205–219.

Yang, H.J. and Lee N.A. (1988) A study of phototoxicity of orange flower absolute. In B.M. Lawrence, B.D. Mookherjee and B.M. Willis (eds), *Flavors and Fragrance: a World Perspective*. Elsevier Science Publishing B.V., Amsterdam, pp. 1039–1044.

19 Composition of petitgrain oils

Giovanni Dugo, Luigi Mondello and Ivana Bonaccorsi

INTRODUCTION

Petitgrain oils are obtained by steam distillation of the leaves, buds and small branches of the different citrus species.

Due to its odour properties, bitter orange petitgrain is the most important and the most appreciated among these oils. It is produced mainly in the Mediterranean countries and in Paraguay. Bitter orange petitgrain oils produced in the Mediterranean areas present better odour properties if compared with those produced in Paraguay, usually obtained from an hybrid named 'Apepu-Jhai' (Di Giacomo, 1974; Huet, 1991).

Lemon, mandarin, bergamot petitgrain oils are produced at small quantities, almost exclusively in Italy. Grapefruit petitgrain is rare, due to its flat odour properties (Peyron, 1965), however small amounts of this oil are seldom found on the market (Lawrence, 1993). Sweet orange petitgrain, only occasionally produced industrially, is considered the least valuable among these oils, and some time it is used to adulterate the more valuable ones. Because of its low cost, the production of this oil lacks in attention, and scant selection of the raw material along with the uncleanness of the production lines compromises the purity of the product. It is common, therefore, to find sweet orange petitgrain oils industrially produced, contaminated by other citrus species, rendering difficult to find pure industrial samples (Dugo *et al.*, 1996).

It has been noticed that the quality of the raw material available for distillation, in different countries is a problem, compromising the production of the different petitgrain oils. The increased cost of agriculture specialised workers, the consequent use of occasional workers, the contraction of some cultivated areas, are the reasons why the raw material that is used for the industrial production is not homogeneous.

Most of the components in citrus petitgrain oils are the same as those present in the volatile fraction of the correspondent citrus peel oil. Substantial qualitative and quantitative differences among the two types of oil, exist. The qualitative differences are determined by the different matrix used for the extraction and by the different process used to obtain these oils. The extraction by distillation of petitgrain oils can determine, some transformation of the components naturally present in the leaves, and causes the absence of the non volatile components present in the peel oils obtained by cold-extraction. Petitgrain oils present substantial differences in the quantitative composition relatively to the cold-pressed peel oils. These are mainly due to the higher content of oxygenated compounds, highly represented in bitter orange and bergamot petitgrain oils, if compared to the hydrocarbons content. Limonene, the main component of almost all the peel oils, is present at levels even lower than 1 per cent in bitter orange petitgrain oil. This oil is characterised by linalyl acetate and linalool, that together can exceed the 80 per cent of the whole oil. In mandarin petitgrain oil, methyl N-methylanthranilate is equal to about the 50 per cent of the whole oil. Among the monoterpene hydrocarbons of this petitgrain oil the most abundant is γ-terpinene. In sweet orange petitgrain oil the main component is sabinene (about 40 per cent), and among the oxygenated components linalool is the predominant. In lemon petitgrain oil the main component is limonene, but with pecentages greatly lower than the correspondent peel oil (about 30 per cent vs 65 per cent).

Most of the papers on the composition of citrus petitgrain have been reviewed by Lawrence (1993) and by Mondello *et al.* (1996, 1997a,b,c). The composition of the industrially processed petitgrain oils have been the subject of very few papers and the grapefruit and bergamot oils have been scantly studied. There is more abundant research on laboratory extracted oils, obtained by distillation or solvent extraction. These last studies were focused on the biogenesis and the ontogenesis of the volatile flavouring components in citrus and on the characterisation and classification of the different citrus species.

Due to the complexity of petitgrain oils, but mostly due to the nature of the sample hitherto analysed that differ for the geographic origin, the cultivar, the extraction technology used, and the age and freshness of the leaves, the composition of citrus petitgrain is not as well defined as the correspondent peel oils.

In this chapter will be described the composition of bitter orange (*C. aurantium* L.), mandarin (*C. deliciosa* Ten.), tangerine (*C. tangerine* Hort. ex Tan.), lemon (*C. limon* L. Burm), sweet orange (*C. sinensis* L. Osbek), bergamot (*C. bergamia*), Key lime (*C. aurantifolia* (Christm.) Swing.), Persian lime (*C. latifolia* Tan.) and grapefruit (*C. paradisi* Macf.) petitgrain oils.

The composition of those petitgrain that are not of industrial interest are described in a separate chapter dedicated to the minor citrus species, along with the corresponding peel and flower oils.

The literature results relative to the industrial products and to the laboratory extracted oils, will be summarised in separate tables, and discussed in the text. The same criteria and symbols, used in the chapter for cold-pressed citrus oil, will be here applied.

BITTER ORANGE PETITGRAIN OIL (*Citrus aurantium* L.)

Bitter orange petitgrain, often called bigarade petitgrain, produces the most appreciated odour character among all the citrus petitgrain.

Although there is extensive literature on its composition, most papers are on the oils extracted in laboratory by distillation or by solvent extraction. Only few papers have been published on the industrial oils.

Industrial oils

Since 1965, studies by gas chromatographic techniques, on the composition of bitter orange petitgrain industrially processed oils, or at least so denominated, were carried out. The determinations were qualitative or semiquantitative (Peyron, 1965, 1973; De Vottero *et al.*, 1978) and quite numerous quantitative studies (Calvarano, I., 1968; Lawrence, 1980; Prager and Miskiewicz, 1981; Formáček and Kubeczka, 1982; Boelens and Sindreu, 1988; Haubruge *et al.*, 1989; Boelens and Oporto, 1991; Mondello *et al.*, 1996).

The qualitative studies named above, provided all together the identification of the following components: *p*-cymene, limonene, myrcene, (E)-β-ocimene, (Z)-β-ocimene, α-pinene, β-pinene, γ-terpinene, octanal, geraniol, linalool, nerol α-terpineol, citronellyl acetate, geranyl acetate, linalyl acetate, neryl acetate, terpinyl acetate*.

The results of the quantitative studies are summarised in Table 19.1.

Table 19.1 Percentage composition of industrially extracted bitter orange petitgrain oil

	1	2	3	4	5	6	7	8
HYDROCARBONS								
Monoterpene								
camphene	tr	–	–	0.11	–	–	0.01	tr–0.01
δ-3-carene	0.31	–	–	0.39	tr	–	0.03	0.21–0.67
p-cymene	0.76	1.0–2.7	–	0.07	tr	tr	0.05	0.03–0.08
p-cymenene	–	–	–	–	tr	–	–	0.01–0.08
limonene	1.07	0.7–1.1	0.3–2.3	1.05	4.0	4.4	5.43	0.44–2.17
myrcene	1.59	1.3–5,5	0–2.4	1.96	2.5	1.5	2.60	0.56–1.24
(E)-β-ocimene	–	tr–3.3	0–2.7	1.29	2.5	–	2.44	0.57–1.76
(Z)-β-ocimene	–	tr–1.1	0–0.8	0.52	0.1	–	0.84	0.20–0.44
α-phellandrene	0.03	tr–0.2	–	–	–	–	0.01	tr–0.03
β-phellandrene	–	–	–	0.10	tr	–	–	0.03–0.04
α-pinene	0.11	tr	–	0.22	0.2	0.2	0.19	0.03–0.30
β-pinene	0.83	0.7–1.7	0.3–2.7	1.57	2.5	2.4	2.53	0.65–1.15
sabinene	0.10	tr–0.4	0–1.4	0.30	0.4	0.4	0.40	0.13–0.23
α-terpinene	0.02	tr	–	–	0.3	–	0.06	tr–0.02
γ-terpinene	1.56	0.5–1.1	–	–	0.1	–	0.06	0.01–0.09
terpinolene	0.46	tr–0.1	–	0.19	0.4	0.2	0.29	0.08–0.22
α-thujene	tr	–	–	–	tr	–	0.02	tr–0.01
Sesquiterpene								
δ-cadinene	–	–	–	–	–	–	0.07	0.02–0.03
β-caryophyllene	–	–	0.3–0.9[a]	0.67	1.6	–	1.77	0.48–0.61
β-elemene	–	–	–	–	–	–	0.03	tr–0.02
(E)-β-farnesene	–	–	–	–	–	tr	0.46	–
(Z)-β-farnesene	–	–	–	–	–	–	0.08	0.04–0.07
germacrene D	–	–	–	–	0.1	–	0.04	–
α-humulene	–	–	–	–	0.2	0.1	0.18	0.04–0.06
ALDEHYDES								
Aliphatic								
decanal	0.12	tr	–	–	–	–	0.02	–
nonanal	0.03	–	–	–	–	–	–	0.02–0.05
octanal	0.02	–	–	–	–	–	0.01	–
Monoterpene								
citronellal	0.03	–	–	–	–	–	0.05	0.01–0.04
geranial	0.11	tr	–	–	tr	–	0.07	0.38–0.64
neral	0.03	tr	–	–	tr	–	0.03	0.21–0.43
ALCOHOLS								
Monoterpene								
citronellol	0.47	tr–0.02	–	–	–	–	–	–
geraniol	0.34	2.0–3.5	0.5–3.0	2.24	1.8	2.3	3.00	0.71–0.95
linalool	17.66	19.9–26.9	12.3–33.7	26.62	24.1	17.0	20.20	21.70–32.55
nerol	1.22	1.0–1.5	0.4–1.2	0.95	0.8	0.8	1.00	0.75–0.99

Table 19.1 (Continued)

	1	2	3	4	5	6	7	8
terpinen-4-ol	0.74	0.5–0.8	0.3–0.9[a]	–	0.1	–	0.15	0.05–0.08
α-terpineol	4.16	4.6–7.6	2.1–6.8	5.10	5.2	3.7	4.00	3.09–5.63
Sesquiterpene								
nerolidol	–	–	–	–	0.15*	0.1*	0.12*	0.05–0.08[b]
ESTERS								
Monoterpene								
citronellyl acetate	–	–	–	–	–	0.1	0.07	tr
geranyl acetate	–	–	1.9–4.5	2.89	4.2	3.2	3.92	1.90–3.16
linalyl acetate	–	46.3–55.0	38.4–71.0	50.81	45.5	56.8	48.85	50.68–62.57
neryl acetate	–	2.1–2.6	1.1–3.0	1.69	2.2	–	2.15	1.04–1.73
α-terpinyl acetate	–	0.2–2.2	–	–	0.1	–	0.10	0.08–0.16
OXIDES								
Monoterpene								
cis-linalool oxide	–	–	–	–	0.2[c]	tr	0.06	0.03–0.09
trans-linalool oxide	–	–	–	–	0.2[c]	tr	0.04	0.01–0.03
perillene	–	–	–	–	tr	–	0.01	–
Sesquiterpene								
Caryophyllene oxide	–	–	–	–	–	–	0.04	0.02–0.07
OTHERS								
Methyl N-methyl-anthranilate	–	–	–	–	–	–	0.05	tr-0.14

Notes

tr = traces; * correct isomer not characterised; a β-caryophyllene + terpinene-4-ol; b (E)-nerolidol; c *cis*-linalool oxide + *trans*-linalool oxide.

Appendix to Table 19.1

1 Calvarano, I. (1968). Calabria, Italy; 5 samples; Chemical and TLC fractionation; GC on capillary column coated with UCON LB 550X; relative percentage of peak areas. Calvarano also found a β-ocimene (0.21%), heptanal (0.02%), methyl N-methylanthranilate (2.51%) and a total content of esters of 62.92%. The last two results were obtained by conventional laboratory procedures.
2 Lawrence (1980). Several samples of commercial oils. GC, IR. Lawrence also found trace amounts of (Z)-3-hexenol and thymol.
3 Prager and Miskiewicz (1981). 10 samples (5 from France, 4 from Paraguay and one from Egypt); GC on capillary column coated with Carbowax 20 M; GC/MS; relative percentage of peak areas. From the original author's results, one France sample with a high limonene level (8%), that can be considered adulterated, has not been included in the table.
4 Formáček and Kubeczka (1982). One sample; GC on capillary column coated with WG-11; relative percentage of peak areas.
5 Boelens and Sindreu (1988). Spain; GC on capillary columns coated with Carbowax 20 M, UCON and SE-54; GC/MS; relative percentage of peak areas. Boelens and Sindreu also found farnesols* (0.05%), *cis*-anhydrolinalool oxide + *trans*-anhydrolinalool oxide (0.1%).
6 Haubruge *et al.* (1989). One commercial sample; GC on capillary column coated with CP-TM-VAX 52 CB; relative percentage of peak areas. Haubruge *et al.* also found cyclophenchene (0.1%), m-cymene (0.5%) and trace amounts of β- or γ-elemene, p-menth-6-en-2-one.
7 Boelens and Oporto (1991). Spain. Boelens and Oporto also found valencene (0.03%), perilla aldehyde (0.01%), nootkatone (0.03%), δ-cadinol (0.01%), mentha-1,8(9)-dienyl acetate (0.01%), methyl-anthranilate (0.10%), 2 phenylethanol (0.20%), 2,2,6-trimethyl-6-vinyl tetrahydropyrane (0.01%).
8 Mondello *et al.* (1996). Sicily, Italy; 5 samples; GC on capillary columns coated with Carbowax 20 M and SE-52; GC/MS, LC-GC/MS; relative percentage of peak areas. Mondello *et al.* also found o-cymene (0–0.01%), isoterpinolene (0.02–0.06%), *trans*-α-bergamotene (tr–0.01%), bicyclogerma-crene (0.04–0.30%), β-bisabolene (0–0.01%), α-copaene (tr–0.01%), α-cubebene (tr–0.01%),

δ-elemene 0.01%), (E,E)-α-farnesene (0.01–0.06%), 6-methyl-5-hepten-2-one (0.01–0.10%), cis-p-menth-2-en-1-ol (tr–0.01%), trans-p-menth-2-en-1-ol (tr–0.02%), cis-sabinene hydrate (tr–0.01%), spathulenol (0.03–0.13%), linalyl propanate (0.02–0.04%), methyl geranate (tr–0.03%), 1,8-cineole (0.02–0.05%) and trace amounts of α-fenchene, tricyclene, hexanal.

The data in Table 19.1 provide good information on the basic composition of bitter orange petitgrain oil. The oxygenated components are present at higher amounts than hydrocarbons and in some cases reach the 90 per cent of the whole oil. The most represented fraction is that of esters, followed by alcohols. Aldehydes are present at very low levels and none of these components ever exceeds 1 per cent. The main components are linalyl acetate, almost never less than 40 per cent, and that reaches values of 71 per cent, and linalool that varies between 12 and 34 per cent. The sesquiterpene hydrocarbons, as could been foreseen for a distilled oil, are under represented, both by number of components and by amount of the single components; the most represented in this class is β-caryophyllene, identified in this oil by almost all the authors of the papers reported in Table 19.1. In Italian oils (Mondello et al., 1996), among monoterpene hydrocarbons, only limonene, myrcene, (E)-β-ocimene and β-pinene reach the value of 1 per cent but never exceed 2 per cent. In Spanish oils (Boelens and Sindreu, 1988; Boelens and Oporto, 1991) the main monoterpene hydrocarbons were the same as the Italian oil, but present at higher levels: limonene about 4.5 per cent, and myrcene, (E)-β-ocimene, and β-pinene are about 2.5 per cent each. The value, reported by Calvarano, I. (1968), of γ-terpinene (1.56 per cent) for Italian petitgrain oil, could be due, in our opinion, to a contamiantion with mandarin petitgrain oil, where this component is present at more than 20 per cent. This explanation is also prooved by the fact that Calvarano reported for the same oil the presence of 2.5 per cent of methyl N-methylanthranilate, present in mandarin at 40–50 per cent. The values of γ-terpinene and methyl N-methylanthranilate reported for this sample are compatible with a bitter orange petigrain contaminated with about 5 per cent of mandarin petitgrain oil.

Not reported in Table 19.1 are the results listed below, obtained by Vernin and Vernin (1966) on bitter orange petitgrain of different geographic origins:

	Spain	Tunisia	Morocco	Paraguay
limonene	10%	3.5%	7%	3.5%
linalool	28%	26%	30%	29%
terpineol*	6.5%	3%	5%	7%
linalyl acetate	32%	46%	30%	36%
methyl N-methylanthranilate	tr	0.15%	1%	0.5%

Maurer and Hauser (1992), identified in the alkaline extract of a commercial sample of bitter orange petitgrain oil, trace amounts of (Z)-3-(but-1-enyl)pyridine.

Laboratory extracted oils

In Table 19.2 are reported the results relative to papers on the composition of oils extracted from bitter orange leaves of different geographic origins by distillation or solvent extraction.

Table 19.2 Percentage composition of laboratory extracted bitter orange petitgrain oils

	1	2	3	4	5	6	7	8	9
HYDROCARBONS									
Monoterpene									
camphene	1.15	0.1	tr	–	–	tr	–	–	–
δ-3-carene	–	–	tr	–	0.20	–	–	–	–
p-cymene	1.50	4.1	0.2	0.20–0.35	0.16	–	–	–	–
p-cymenene	–	0.3	tr	–	–	–	–	–	–
limonene	26.79	0.9	0.4	0.14–0.58	1.25	0.78	0.97	–	6.82–11.13
myrcene	0.92	1.6	1.0	0.51–1.24	1.82	1.02	0.04	tr	2.03–2.29
β-ocimene	1.84	–	–	1.01–2.56	1.92	0.30	–	tr[g]	2.07–2.61[i]
(E)-β-ocimene	–	–	1.5	–	–	–	–	tr	–
α-phellandrene	–	–	–	0.76–2.66	–	–	–	0.12	–
β-phellandrene	–	0.1	tr	–	0.01	–	–	–	–
α-pinene	1.66	4.4[b]	0.5	–	0.19	0.25	–	–	–
β-pinene	2.65	9.4[c]	3.0	tr–2.51	1.76	tr	0.40	–	1.71–2.25
sabinene	0.95	9.4[c]	0.5	0.13–0.63	1.58	–	0.03	–	–
α-terpinene	–	–	tr	tr	–	–	0.02	–	–
γ-terpinene	–	2.7	1.0	tr	0.08	–	–	–	–
terpinolene	–	–	tr	0.08–0.37	0.30	–	–	0.18	–
α-thujene	–	4.4[b]	0.1	–	–	tr	–	–	–
Sesquiterpene									
β-caryophyllene	–	0.8	0.6	–	0.21	0.11	0.10	–	–
β-elemene	–	0.1[d]	tr	–	+	–	–	–	–
β-farnesene*	–	–	tr	–	+	–	–	–	–
humulene	–	0.9*[f]	tr*	–	+[h]	1.35[h]	–	–	–
ALDEHYDES									
Aliphatic									
decanal	–	–	tr	tr–0.02	–	–	–	–	–
(E)-2-hexen-1-al	–	0.2	0.3	–	–	–	–	–	–
nonanal	–	–	tr	–	–	0.45	–	–	–
Monoterpene									
citronellal	0.87	0.2	tr	–	+	–	–	–	–
geranial	2.43	–	0.2	1.00–2.10[e]	+	tr	0.93	0.99	–
neral	1.84	–	–	tr–2.21	0.12	tr	0.78	–	–
ALCOHOLS									
Aliphatic									
cis-hex-3-en-1-ol	–	0.3	tr	0.80–2.34	–	–	–	tr	–
cis-pent-2-en-1-ol	–	0.1	tr	–	–	–	–	–	–
Monoterpene									
citronellol	–	–	–	1.00–2.10[e]	–	–	–	2.33	–
geraniol	1.38	–	0.2	1.07–2.13	3.97	0.98	0.08	6.60	4.85–8.43
linalool	4.95	36.0	38.4	59.23–71.04	36.10	11.72	94.1	66.10	24.24–26.32
nerol	6.72[a]	–	0.1	1.13–4.33	1.51	tr	0.36	–	1.70–1.93
terpinen-4-ol	–	0.1[d]	tr	0.17–0.61	0.42	tr	–	20.86	–
α-terpineol	1.96	0.9[f]	1.0	3.02–5.87	6.80	1.26	0.38	0.30	7.11–7.90
Sesquiterpene									
farnesol*	6.72[a]	–	–	–	–	0.23	0.16	–	–
ESTERS									
Monoterpene									
geranyl acetate	3.45	0.8	0.5	–	3.54	1.73	0.49	–	5.19–5.50
linalyl acetate	3.45	35.4	49.8	13.59–20.51	28.84	44.08	–	–	37.39–38.97

neryl acetate	–	0.6	0.3	–	+	0.83	0.82	0.53	1.70–2.78
α-terpinyl acetate	–	–	–	–	+	1.11	–	–	–
OXIDES									
Monoterpene									
cis-linalool oxide	–	–	–	–		0.10	5.40	–	–
trans-linalool oxide	–	–	–	–		0.07	7.13	–	–
OTHERS									
methyl anthranilate	2.42	–	–	–	–	3.49	0.11	–	–

Notes

tr = traces; + = present, not quantified; t = tentative identification; * correct isomer not characterized; a nerol + farnesol* + unknown; b α-pinene + α-thujene; c β-pinene + sabinene; d β-elemene + terpinen-4-ol; e geranial + citronellol; f humulene* + α-terpineol; g (Z)-β-ocimene; h α-humulene.

Appendix to Table 19.2

1 Karawya *et al.* (1970). Egypt; several steam-distilled samples; column chromatography, TLC, GC on stainless steel capillary column coated with nujol. Karawya *et al.* also found terpinene* (2.53%).

2 Kamiyama (1970). Japan; one sample of Daidai bitter orange leaf oil; extraction with methylene chloride, concentration of extract, steam-distillation of the residue and extraction with methylene chloride; column chromatography on silica gel, analytical and semipreparative GC on packed columns of PEG 20 M, Apiezon M, β, β'-oxydipropionitrile, IR; relative percentage of peak areas.

3 Kamiyama and Amaha (1972). Japan; one sample of Daidai bitter orange leaf oil extracted as reported by Kamiyama (1970); GC on packed columns of PEG 20 M, Apicron M, β, β'-oxydipropionitrile, UCON 50HB 280X, LAC-3R-728; relative percentage of peak areas. Kamiyama and Amaha also found trace amounts of β-selinene, (E)-hex-2-en-1-ol and isopulegol.

4 Ortiz *et al.* (1978). USA; one sample each of Bittersweet, Bouquet de fleurs, Daidai, Granitos, Myrtifolia, Paraguay, Salicifolia, Seville, Four, Standard, Stow n. 15 bitter orange leaf oil; steam-distillation; GC on packed columns of LAC 446; relative percentage of peak areas. Ortiz *et al.* also found trace amounts of thymol.

5 Baaliouamer and Meklati (1986). Algeria; one sample steam-distilled; GC on capillary columns coated with FFAP and CP-Sil-5; GC/MS; relative percentage of peak areas. Baaliouamer and Meklati also found cyclophenchene (0.57%), bicycloelemene (0.20%), geranyl formate (2.00%), farnesyl acetate* (0.09%), phytol (0.23%) and the following components not determined quantitatively: aromadendrene, δ-cadinene, γ-patchoulene, terpinen-2-ol, nerolidol*, neryl formate.

6 Lin *et al.* (1986). China; one sample; column chromatography on silica gel; GC on capillary columns coated with SP-2305 and SE-30; GC/MS, retention indices, IR; relative percentage of peak areas. Liu *et al.* also found γ-muurolene (0.45%), 6-mentha-1,4-dien-7-ol (0.07%), *cis*-linalool oxide (pyranoid) (0.17%), *trans*-linalool oxide (pyranoid) (0.60%) and trace amounts of camphor, *trans*-pinocarveol, (E)-nerolidol.

7 Germanà *et al.* (1990). Sicily, Italy; one sample steam-distilled; GC on packed column of EAS; relative percentage of peak areas. Germanà *et al.* also found benzaldeyde (0.07%).

8 Gurib-Fakin and Demarne (1995). Mauritius; one sample steam-distilled; GC on capillary column coated with HP-101; GC/MS, Kovats indices; relative percentage of peak areas. Gurib-Fakin and Demarne also found β-copaene (0.23%), octanol (0.34%), *cis*-carveol (0.98%), geranyl formate (0.20%).

9 Protopapadakis and Papanikolau (1998). Creta, Greece; four samples obtained by hydrodistillation from leaves of the cv.s Chania, Brazilian, Keen and Bittersweed; GC; GC/MS.

The composition relative to all the oils reported in Table 19.2 present some peculiar character, more or less accentuated, if compared with the industrial sample described in Table 19.1. These differences, are surely related to the botanical origin of the vegetable matrix used, and also to the extraction process, to the storage conditions, to the period of harvest of the leaves and to their age, along with the analytical capabilities of the methods used.

The Egyptian oil analysed by Karawya et al. (1970) was characterised by the high content of limonene (about 27 per cent) and monoterpene aldehydes (about 5 per cent), values that are similar to those of lemon petigrain oil (Mondello et al., 1997c). This oil also presented a content of linalool (about 5 per cent) and linalyl acetate (about 3.5 per cent) extremely lower that those normally determined for bitter orange petitgrain oils.

In the oil extracted with methylene chloride from leaves of C. aurantium L. of the Kabusu form (Daidai), cultivated in Japan, Kamiyama (1970) determined amounts of monoterpene hydrocarbons greatly higher than those usually determined in bitter orange leaf oil. The total amount of these components was about the 23 per cent of the whole oil analysed. Strangely two years later, Kamiyama and Amaha (1972) analysing one oil obtained from the same matrix, determined a total amount of monoterpene hydrocarbons of 8.1 per cent.

The oils analysed by Ortiz et al. (1978), obtained from numerous cultivars of bitter orange (see appendix to Table 19.2), were characterised by a high amount of α-phellandrene (0.76–2.66 per cent) and a slightly high amount of geranial (1.00–2.10 per cent), while the Algerian oil, analysed by Baaliouamer and Meklati (1986) contained trace amount of sesquiterpene hydrocarbons (bicyclogermacrene and γ-patchoulene) and of esters (geranyl and neryl formate) unusual in bitter orange petitgrain oil.

The Chinese oil analysed by Lin et al. (1986) showed a high content of of α-humulene (1.35 per cent), usually present in such oil at levels lower than 0.2 per cent, of γ-muurolene (0.45 per cent) and mostly strange a total content of cis- and trans-linalool oxides, both in the furanoid and pyranoid forms, of about 13 per cent. These data gave the idea that the stored conditions used for the oil were inappropriate.

In the Italian oil analysed by Germanà et al. (1990) the linalyl acetate was absent, while linalool was present at a very high level (94.1 per cent).

Linalyl acetate was also absent in the oil from Mauritius, analysed by Gurib-Fakim and Demarne (1995). This oil was also characterised not only by the presence of high amounts of linalool, but also by the high levels of other monoterpene alcohols such as terpinen-4-ol (20.86 per cent) and of geraniol (6.60 per cent), and by the absence of limonene, that is always present in this petitgrain oil, although at levels less than 1 per cent.

LEMON PETITGRAIN OIL (*Citrus limon* (L.) Burm.)

Industrial oils

The number of papers found in literature on the composition of lemon petitgrain oils industrially produced, is small: one paper by Peyron (1965) on a semi quantitative determination; one that reports on the dosage of very few components (Vernin and Vernin, 1966); one on the qualitative analysis on Argentinean oils (De Vottero et al., 1978); four papers on the quantitative determinations on Italian industrial production (Calvarano, I., 1967; Di Giacomo et al., 1985; Goretti et al., 1986; Mondello et al., 1997c); and one paper by Cartoni et al. (1987) where lemon petitgrain oil was used,

Composition of petitgrain oils 433

with other essential oils, to prove the use of two columns in series, one apolar and one polar, for the GC analysis of essential oils.

Peyron (1965) identified in lemon petitgrain oils the following components listed in decreasing amount order: limonene, β-pinene, geranyl acetate+neryl acetate, geranial, neral+nerol, linalool, γ-terpinene, α-pinene. De Vottero *et al.* (1978), during a study on GC analysis on packed columns with Carbowax 20 M, identified in lemon petitgrain oil, the following components in addition to those reported by Peyron: *p*-cymene, myrcene, citronellal, citronellol, geraniol, α-terpineol. Cartoni *et al.* (1987) also identified camphene, α- and β-phellandrene, sabinene, terpinolene, α-thujene, bergamotene*, β-bisabolene, β-caryophyllene, humulene*, decanal, octanal, 6-methyl-5-hepten-2-one, terpinen-4-ol, citronellyl acetate, linalyl acetate, 1,8-cineole. The GC system used by Cartoni *et al.* as already described for bergamot and tangerine peel oils, in the relative chapter, permitted to separate some critical pairs otherwise co-eluted using one or the other column separately.

Vernin and Vernin (1966) determined in lemon petitgrain oil, by GC and TLC the following composition: limonene (30 per cent), geranial+neral (7–10 per cent), linalool (24 per cent), geraniol+nerol+geranyl acetate+neryl acetate (15–20 per cent), terpineol* (0.1 per cent), linalyl acetate (tr), methyl anthranilate (tr), methyl N-methyl-anthranilate (0.05 per cent).

The results relative to the quantitative studies on industrial petitgrain oils are reported in Table 19.3.

Table 19.3 Percentage composition of industrially extracted lemon petitgrain oil

	1	2	3	4
HYDROCARBONS				
Monoterpene				
camphene	0.09	0.08–0.11	0.12	0.05–0.07
δ-3-carene	0.63	0.11–0.15	–	0.63–1.08
o-cymene	–	0.04–0.06	–	tr–0.01
p-cymene	1.28	0.67–0.82	0.91	0.04–0.51
limonene	30.67	30.18–35.57	30.71	28.41–34.82
myrcene	1.01	1.02–1.80	1.49	0.79–1.60
β-ocimene	1.37*	–	–	1.50–2.43^d
α-phellandrene	0.08	–	0.16	0.03–0.09
α-pinene	2.29	1.12–1.48	1.21^a	0.84–1.13
β-pinene	11.53	15.63–19.86	13.59	11.86–16.03
sabinene	2.85	1.05–1.88	3.56	2.99–3.81
α-terpinene	0.12	–	–	0.04–0.11
γ-terpinene	1.80	2.16–3.19	2.32	0.34–0.70
terpinoplene	0.30	0.18–0.22	0.31	0.19–0.31
Sesquiterpene				
bisabolene	–	0.09–0.10*	–	0.07–0.43^e
caryophyllene	–	3.19–6.20*	0.95^f	0.60–1.54^f
ALDEHYDES				
Aliphatic				
decanal	0.09	0.01–0.05	0.06	0.03–0.09
nonanal	0.16	0.29–0.35	–	0.08–0.22
octanal	0.41	–	–	0.01–0.04
Monoterpene				
citronellal	0.34	1.11–1.89	1.48	0.61–1.41

Table 19.3 (Continued)

	1	2	3	4
geranial	10.30	10.91–17.27	10.93	9.87–14.07
neral	5.80	8.30–12.63	6.48	6.64–10.78
KETONES				
Monoterpene				
6-methyl-5-hepten-2-one	–	0.33–0.51	0.26	0.67–1.61
ALCOHOLS				
Monoterpene				
citronellol	0.84	0.02–0.19[b]	0.22	tr
geraniol	1.95	0.43–1.05	0.51	0.87–6.25
linalool	6.30	0.87–1.24	1.77	0.88–3.87
nerol	2.20	2.66–3.10	1.30	1.90–3.14
terpinen-4-ol	0.91	0.15–0.20	0.24	0.25–0.59
α-terpineol	2.65	0.38–0.45	0.37	0.53–1.00
ESTERS				
Aliphatic				
octyl acetate	–	tr	0.04	–
Monoterpene				
citronellyl acetate	–	–	0.25	0.13–0.23
geranyl acetate	10.12[c]	0.19–0.88	2.91	2.17–2.92
linalyl acetate	10.12[c]	0.02–0.19[b]	6.50	0.31–0.42
neryl acetate	10.12[c]	–	7.44	3.75–6.74
terpinyl acetate*	10.12[c]	1.77–2.23	–	–
OTHERS				
methyl N-methylanthranilate	–	0.03–0.07	0.78	tr–0.39

Notes

tr = traces; * correct isomer not characterised; a α-pinene + β-thujene; b citronellol + linalyl acetate; c α-terpinyl acetate + geranyl acetate + linalyl acetate + neryl acetate (listed in decreasing concentration order); d (E)-β-ocimene; e β-bisabolene; f β-caryophyllene.

Appendix to Table 19.3

1 Calvarano, I. (1967). Calabria, Italy; 9 samples; conventional laboratory procedures; GC on capillary column coated with UCON LB 550X; relative percentage of peak areas. Calvarano, I. also found heptanal (0.09%).
2 Di Giacomo *et al.* (1985). Calabria, Italy; 3 samples; GC on capillary column coated with UCON LB 550X; relative percentage of peak areas.
3 Goretti *et al.* (1986). Calabria, Italy; one sample; GC on capillary column coated with FFAP; relative percentage of peak areas.
4 Mondello *et al.* (1997c). Sicily, Italy; 6 samples; GC on capillary columns coated with SE 52 and Carbowax 20 M; GC/MS LC-GC/MS; relative percentage of peak areas. Mondello *et al.* also found isoterpinolene (0.01–0.03%), (Z)-β-ocimene (0.30–0.44%), β-phellandrene (2.22–2.60%), α-thujene (0.06–0.09%), tricyclene (tr–0.01%), *trans*-α-bergamotene (0.03–0.20%), bicyclogermacrene (0.06–0.23%), δ-cadinene (0.02–0.05%), β-elemene (tr–0.03%), (E,E)-α-farnesene (0.03–0.14%), (Z)-β-farnesene (tr–0.03%), α-humulene (0.06–0.13%), undecanal (0.02–0.08%), octanol (tr–0.01%), isopulegol (tr–0.03%), *cis-p*-menth-2-en-1-ol (0.01–0.03%), *cis*-sabinene hydrate (0.02–0.06%), α-bisabolol (0.01–0.03%), campherenol (0.01–0.02%), 2,3-dimethyl-3-(4-methyl-3-penthenyl)-2-norbornanol (0.02–0.06%), (E)-nerolidol (0.01–0.03%), spathulenol (0.01–0.09%), geranyl propanate (0.03–0.04%), 1,8-cineole (1.12–2.13%), *cis*-limonene oxide (0.01–0.05%), *trans*-limonene oxide (0.01–0.06%), *cis*-linalool oxide (tr–0.01%), caryophyllene oxide (0.04–0.14%) and trace amounts of α-fenchene, *cis*-α-bergamotene and iso-(iso)pulegol.

Among these results, those reported by Calvarano, I. (1967) have been calculated from the relative composition of each class of components, and from the total amount of each of these reported in the paper.

The data relative to Di Giacomo *et al.* (1985), reported in Table 19.3, were extrapolated from the results found on the original paper relative to the composition of fractions obtained by distillation at different times (from 30 minutes to 3 hours), and considering the yield of each fraction. It is noteworthy that the fractions collected during the first hour of distillation were much richer of aldehydes, in particular neral and geranial, and of esters, while the monoterpene hydrocarbons were not present at high levels, especially limonene and β-pinene. In the later fractions, collected after 90–180 minutes, it was observed a decrease of the aldehydes and esters content, and the increase of monoterpene hydrocarbons. The authors explained this with transformations occurred on the naturally occurring components, during the hot distillation, more or less evident in function of the time. In the same paper was extracted at 40 °C by solvent an aliquot of the same leaves used for distillation. The composition determined for these oils, considered by the authors representative of the natural oil composition, resulted very similar to that observed for the fractions distilled in the first hour. The authors, therefore, concluded that the distillation prolonged for a long time, produced higher yields of oil, but also caused visible changes of the composition, while shorter time would allow to avoid formation of such evident alterations.

The composition of the petitgrain oils reported in Table 19.3, although relative to samples analysed in a range of time of about 30 years, are sufficiently homogeneous. The ranges of variation of some components are quite wide. These variations are probably due to the different durations of the distillation process, that of course, determine the presence of lower or higher amounts of the least volatile components. Should however be highlighted the percentage values, sensibly high, reported by Calvarano, I. (1967) and by Di Giacomo *et al.* (1985) of terpinyl acetate*. It could be possible that this component was misidentified, in fact only Melendreras *et al.* (1984), and Scora *et al.* (1968) reported the presence terpinyl acetate*. The former determined in *C. limon* leaves laboratory distilled oils only trace amounts of this component; the latter found in the samples analysed about 2–3 per cent of terpinyl acetate*.

Laboratory extracted oils

Numerous studies have been carried out on the laboratory extracted oils, and the relative results are summarised in Table 19.4.

In the appendix to Table 19.4 are reported detailed information on the different botanical and geographic origins, of the oils analysed by the different authors, either obtained by solvent extraction or by distillation. The each vegetable material used for extraction, by the different authors, probably differed for the age of the tree and the period of harvest. These variables could be the explanations to the diversity of the quantitative composition among the oils that are described in Table 19.4. All the oils are however characterised by a high content of limonene, with the exception of that analysed by Ayedoun *et al.* (1996), and by a high amount of neral and geranial. These values are in agreement with those determined above for the oils industrially processed.

Table 19.4 Percentage composition of laboratory extracted lemon leaf oil

	1	2	3	4	5	6	7(a)	7(b)	8	9	10	11	12	13
HYDROCARBONS														
Monoterpene														
camphene	0.1	0.01	–	0.08	tr–0.16	0.08	0.10	0.06	–	–	–	–	0.1	–
δ-3-carene	–	0.3–0.6	–	1.08	0.57–0.98	0.39	–	–	–	–	–	–	–	–
p-cymene	0.5	–	–	0.04	tr	0.12	–	–	–	0.25	–	–	0.1	0.29
limonene	23.1	22.2–28.9	13.00–14.70	38.63	25.84–34.55	25.90	26.07	22.71	12.05–17.47	9.84	28.39	38.20	40.8	20.05
myrcene	1.7	1.6–2.2	4.77–7.36	1.61	1.15–1.64	0.94^a	1.03	0.98	–	–	1.04	1.20	1.5	–
β-ocimene*	–	–	–	2.90	2.05–2.57	1.99	1.96	1.67	–	–	–	–	–	–
(E)-β-ocimene	–	–	–	–	–	–	–	–	–	0.75	–	0.15	3.1	–
(Z)-β-ocimene	–	–	–	–	–	–	–	–	–	–	–	0.10	0.8	–
α-phellandrene	–	–	–	–	–	0.94^a	–	–	–	–	–	0.19^b	–	–
β-phellandrene	–	–	–	0.55	–	–	–	–	–	9.33	–	0.35	–	–
α-pinene	3.4^c	2.1–3.6^c	0.19–0.27	1.58	0.83–2.15^c	1.38^c	1.35	0.84	1.63–2.23	–	0.29	0.16	1.5	3.95
β-pinene	12.4^d	12.1–14.4^d	–	18.75	9.79–26.86	17.11	18.43	13.32	3.44–6.85	–	6.72	0.10	18.5	9.72
sabinene	12.4^d	12.1–14.4^d	–	2.06	2.24–3.96	4.53	–	–	–	–	0.58	0.32	3.8	–
α-terpinene	tr	–	–	0.08	tr–0.28	0.05	–	–	–	–	3.41	0.19^b	tr	–
γ-terpinene	3.3	2.1–3.5	2.39–2.49	0.22	0.85–1.24	0.26	0.28	0.21	0.30–2.51	0.06	–	0.50	0.8	–
terpinolene	–	–	–	0.29	0.25–0.31	0.16	–	–	–	0.04	–	0.16	0.1	–
α-thujene	3.4^c	2.1–3.6^c	–	–	0.83–2.15^c	1.38^c	–	–	–	–	–	–	0.1	–
Sesquiterpene														
α-bergamotene	–	–	–	–	–	0.79*	0.30*	0.53*	–	0.06*	–	–	0.1^e	–
β-bisabolene	–	–	–	–	–	2.78	–	–	–	–	–	–	0.3	–
δ-cadinene	–	–	–	–	–	0.03	–	–	–	–	–	–	tr	–
caryophyllene	0.2*	0.2–2.3*	–	0.66^m	0.57–1.44^m	0.79^m	–	–	0.24–1.44*	0.25*	0.22*	–	0.8^m	–
β-elemene	0.4^f	0.4–0.7^f	–	–	–	–	–	–	–	–	–	–	–	–
α-farnesene	–	–	–	–	–	0.03*	0.07	0.15	–	–	–	–	tr^g	–
α-humulene	–	–	–	0.21	–	9.38^h	0.07	0.10	–	0.05	–	–	0.1	–
β-selinene	1.9	1.4–2.0	–	1.03	–	–	–	–	–	–	–	–	–	–
ALDEHYDES														
Aliphatic														
(E)-2-butenal	–	–	–	–	tr–0.11	0.23	–	–	–	–	–	–	–	–
decanal	–	–	–	–	–	0.02	–	–	–	0.01	–	–	0.4	–

Compound													
dodecanal	–	–	–	–	–	–	–	–	–	–	–	–	–
(e)-2-hexenal	2.0	–	–	–	–	–	–	–	–	–	–	–	–
nonanal	–	–	–	–	0.21–0.35	–	0.14	–	0.39	–	–	0.02	–	0.2	–
octanal	–	–	–	–	tr	–	–	–	–	–	–	–	–	0.1	–
undecanal	–	–	–	–	tr–0.12	–	–	–	–	–	–	0.05	–	0.1	–
Monoterpene													
citronellal	24.3	24.2–29.6	2.03–3–12	0.91	1.06–1.90	1.07	–	–	–	2.63	–	0.06	16.5	2.16
geranial	16.4	16.0–18.2	27.06–31.77	9.73	9.40–15.19	11.11	12.71	19.17	–	21.31	19.27	15.20	0.3	+
neral			18.14–22.54	5.97	7.60–12.10	1.56	12.03	14.79	–	13.96	12.22	10.20	0.3	+
KETONES													
Aliphatic													
6-methyl-5-hepten-2-one	–	–	–	2.30	–	1.06	–	–	–	–	–	3.20	–	–
ALCOHOLS													
Aliphatic													
(Z)-3-hexenol	0.9	–	1.17–2.39	–	–	0.03	0.20	0.12	–	0.92	–	–	–	–
nonanol	–	–	–	–	–	–	–	–	–	–	–	–	–	–
Monoterpene													
borneol	–	–	–	0.02	0.10–0.30	–	–	–	–	–	–	–	–	–
citronellol	–	1.3–2.8	–	0.02	0.27–0.39	0.02	–	–	–	–	–	1.40	2.3	+
geraniol	2.8	–	–	1.63	0.99–1.63	1.49	–	–	11.40–15.33	0.17	–	3.00	tr	3.20
isopulegol	–	–	–	–	–	0.02	–	–	–	0.52	–	–	0.3	–
linalool	3.1	1.7–3.2	1.75–2.11	1.20	1.24–1.46	1.60	1.03	0.89	–	17.26	1.64	8.00	0.9	1.50
nerol	2.0	1.7–2.2	6.58–7.39	2.18	1.66–2.13	1.54	–	–	–	4.47	0.07	3.50	–	1.24
terpinen-4-ol	0.4	0.4–0.7	–	0.21	0.43–0.91	0.15	0.31	0.27	–	0.28	–	0.80	0.6	–
α-terpineol	–	–	–	0.30	0.63–1.00	9.38	0.27	0.31	–	0.07	4.59	3.00	0.6	–
trans-β-terpineol	–	–	–	–	–	0.10	–	–	–	–	–	0.10*	–	–
Sesquiterpene													
farnesol	–	–	–	–	–	0.01*	0.04*	0.11*	–	–	–	–	–	+*
nerolidol	–	–	–	–	–	0.05*	0.02*	0.03*	–	–	–	–	tr	–
ESTERS													
Aliphatic													
Octyl acetate	–	–	–	–	–	0.01	–	–	–	0.01	–	–	–	–

Table 19.4 (Continued)

	1	2	3	4	5	6	7(a)	7(b)	8	9	10	11	12	13
Monoterpene														
citronellyl acetate	–	–	–	0.11	tr	9.38[h]	0.02	0.02	–	0.37	–	–	0.7	–
geranyl acetate	–	–	5.38–7.98	2.58	2.30–3.31	1.75	0.26	0.59	2.71–4.03	1.66	16.37	–	0.3	+
geranyl formate	–	–	2.57–3.32	–	–	0.21	–	–	–	–	–	–	–	–
linalyl acetate	–	–	–	0.21	–	0.23	–	–	–	2.22	–	–	–	+
neryl acetate	–	–	1.32–1.75	–	4.12–8.18	1.56[i]	0.18	0.39	3.48–5.67	2.82	4.46	–	0.2	–
α-terpinyl acetate	–	–	2.48–2.97	–	tr	–	–	–	–	–	–	–	–	–
OXIDES														
Monoterpene														
1,8-cineole	–	–	–	0.70	–	4.36	–	–	–	–	–	–	2.9	–
OTHERS														
toluene	–	–	–	–	–	0.01	0.02	0.02	–	–	–	–	–	–

Notes

tr = traces; * correct isomer not characterised; + present, not quantified; a myrcene + α-phellandrene; b α-phellandrene + α-terpinene; c α-pinene + α-thujene; d β-pinene + sabinene; e *trans*-α-bergamotene; f β-elemene + terpinen-4-ol; g (E,E)-α-farnesene; h α-humulene + α-terpineol + citronellyl acetate; i neral + neryl acetate; j *trans*-β-terpineol; k (E)-nerolidol; l octyl acetate + *cis*-limonene oxide; m β-caryophyllene.

Appendix to Table 19.4

1 Kamiyama (1967). Japan; one sample; extraction with methylene chloride, steam-distillation of the extract and further extraction with methylene chloride; TLC, GC on packed columns of Carbowax 20M, Reoplex 400, Apiezon M, Succinate polyester and β,β'-oxydipropionitrile; relative percentage of peak areas. Kamiyama also found trace amounts of (E)-2-hexenol.

2 Kamiyama (1968). Japan; four samples from leaves collected at four different season (from June to January); extraction as in above reference; GC on packed column of Carbowax 20 M; relative percentage of peak areas.

3 Scora *et al*. (1969). USA; samples extracted from the cultivars Eureka, Lisbon and San Fernando. Scora *et al*. also found neryl formate (0.55–2.39%).

4 Cheng and Lee (1981). China. Cheng and Lee also found thymol methyl ether (0.03%).

5 Melendreras *et al*. (1984). Spain; samples extracted from the cultivars Verna, Fino, Eureka, Villafranca and Lishon. Melendreras *et al*. also found methyl N-methylanthranilate (0.14–0.19%).

6 Baaliouamer *et al*. (1985). Algeria; one sample extracted by hydrodistillation from Eureka lemon leaves; preparative and analytical GC on packed and capillary columns of FFAP; GC/MS; relative percentage of peak areas. Baaliouamer *et al*. also found methylciclopentane (0.01%), δ-4-carene (0.01%), cyclocamphene (0.33%), α-fenchene (0.01%), tricyclene (0.01%), bicycloelemene (0.62%), γ-guriunene (0.03%), perilla aldehyde (0.01%), camphor (0.02%), epicamphor (0.03%),

p-menth-3-en-1-one (0.01%), pulegone (0.44%), 2-butyloctanol (0.04%), (E)-3-hexenol (0.03%), *trans*-carveol (0.02%), iso-geraniol (0.03%), cedrenol (0.04%), 2,3-dihydro farnesol (0.01%), epiglobulol (0.03%), spathulenol (0.04%), (Z)-3-hexenyl acetate (0.01%), octyl acetate + *cis*-limonene oxide (0.01%), geranyl propanate (0.03%), farnesyl acetate* (0.03%), methyl salicilate (0.01%), phytol (0.22%), *p*-tolyl aldehyde (0.05%) and trace amounts of cedrol, citronellyl propanate, 2,2,3-thrimethyl butane.

7   Adeishvili and Kharebava (1987). Ex URSS; two samples extracted by distillation from the cultivars Georgian (a) and Monachello(b); GC on capillary column coated with OV-101; relative percentage of peak areas. Adeishvili and Kharebava also found hexanol (0.09–0.02%), octanol (0.02–0.01%), *p*-cymenol* (0.12–0.11%), dibutyl phtalate (0.08–0.09%, probably a contaminant), eneicosane (0.20–0.19%) and trace amounts of furfural and methylanthranilate.

8   Crescimanno *et al*. (1988). Sicily, Italy; several samples extracted by petroleum ether from cultivars Femminello, San Giuseppe Larena, Femminello Favazzina, Monachello non refloring, Frost Eureka; four samples from November–September for each cultivar; GC on packed column of WEAS; relative percentage of peak areas. The authors observed the highest citral content in the winter samples and in the Monachello cultivar.

9   Wen *et al*. (1989). China; one sample of steam-distilled lemon leaf oil; GC on capillary columns coated with OV-101 and PEG-20M; GC/MS relative retention indices. Wen *et al*. also found carvone (2.02%), *cis*-linalool oxide (0.01%), *trans*-linalool oxide (0.17%).

10 Kumar *et al*. (1992). India; one sample of hydrodistilled Pant lemon-1 leaf oil; GC on capillary column coated with FFAP; GC/MS; relative percentage of peak areas. Kumar *et al*. also found *p*-cymenene (0.50%), linalool oxide* (1.80%).

11 Germanà *et al*. (1990). Sicily, Italy; one sample extracted by hydrodistillation; GC on packed column of WEAS; relative percentage of peak areas. Germanà *et al*. also found benzaldehyde (0.50%).

12 Ayedoun *et al*. (1996) Benin; one sample extracted by steam-distillation; GC on capillary column coated with HP-1 and BB-5; GC/MS; relative percentage of peak areas. Ayedoun *et al*. also found (E,E)-α-farnesene (0.2%), caryophyllene oxide (0.1%) and trace amounts of germacrene D.

13 Haggag *et al*. (1998) Egypt; cold-pressing, steam distillation; TLC, GC; relative percentage of peak areas. Haggag *et al*. also found carvacrol (0.12%), a terpineol* (18.93%), thymol (0.23%), eugenol (0.21%).

MANDARIN PETITGRAIN OIL (*Citrus deliciosa* Ten.)

In this section will be considered the petitgrain oils that present high content of methyl N-methylanthranilate obtained from leaves of Mediterranean plants of mandarin, which botanical name is *Citrus deliciosa* Ten., although in the original papers are some times indicated by a different botanical name. These oils are also characterised by levels of γ-terpinene higher than 10 per cent.

Industrial oils

The information on industrially processed mandarin petitgrain oils are quite scant. The data found in literature are relative to two papers by Peyron (1965, 1966), one by Vernin and Vernin (1966) on the main components of this oil, one qualitative study by De Vottero *et al.* (1978) on oils produced in Argentina, and two quantitative studies on Italian oils: the first by Calvarano, I. (1967) the second, more recent and extensive, by Mondello *et al.* (1997a). In the first of his papers Peyron (1965) identified in mandarin petitgrain oil the following components, listed in decreasing amount order: limonene, linalool, methyl N-methylanthranilate, α-pinene, linalyl acetate, β-pinene, γ-terpinene; in the second paper Peyron (1966) also identified small amounts of p-cymene, and considered the high content of γ-terpinene one peculiarity of mandarin petitgrain oil. In this paper Peyron also asserted the existence of mandarin petitgrain oils that presented a content of methyl N-methylanthranilate not higher than 15 per cent, differing from what previously determined for mandarin petitgrain oils obtained from the market, that contained 45–60 per cent of methyl N-methylanthranilate. Peyron indicated as possible reason of such great differences, the long lasting periods of distillation, the lack of homogeneity of the vegetable material used for distillation (presence of clementine and tangerine foliage, that did not contain methyl N-methylanthranilate), or fraudulent preparation of reconstituted oils from different citrus products or from synthetic compounds.

Vernin and Vernin (1966) found in mandarin petitgrain oil limonene (35 per cent), linalool (2 per cent), terpineol* (1 per cent), linalyl acetate (5 per cent), methyl anthranilate (tr), and methyl N-methylanthranilate (10–15 per cent).

De Vottero *et al.* (1978), in addition to all the compounds already mentioned, identified myrcene, citronellol, α-terpineol, and geranyl acetate.

Calvarano, I. (1967), during a study on eleven samples of mandarin petitgrain oils, by fractionation using classical methods, and gas chromatographic analysis of the whole oils and of the fractions, determined the following average composition:

camphene	0.02%
δ-3-carene	0.11%
p-cymene	3.43%
limonene	8.22%
myrcene	0.88%
β-ocimene*	0.68%
α-phellandrene	0.12%
α-pinene	2.16%
β-pinene	1.86%

α-terpinene	0.17%
γ-terpinene	13.74%
terpinolene	0.59%
α-thujene	0.01%
decanal	0.01%
octanal	0.01%
citronellal	0.13%
geranial	0.01%
citronellol	1.14%
geraniol	0.10%
linalool	7.92%
terpinen-4-ol	0.09%
α-terpineol	0.01%
methyl anthranilate	1.68%
methyl N-methylanthranilate	55.19%

Calvarano, I. (1968), also found trace amounts of sabinene, heptanal, nonanal, neral, nerol, and indicated the presence of geranyl, linalyl and terpinyl acetate, but did not report the percentage amount.

The most recent data on the composition of the industrially produced mandarin petitgrain is found in one paper by Mondello et al. (1997a). The research was carried out on five samples, produced in Sicily (Italy), by GC and GC/MS on capillary SE-52 and Carbowax 20 M columns, and by an on-line HPLC-HRGC/MS fully automated system that allowed the fractionation of the oils into the hydrocarbons and oxygenated compounds, and the direct GC/MS analysis of the fractions. The results obtained from this study are reported in Table 19.5.

The results obtained by Mondello et al. (1997a) on Italian mandarin petitgrain industrially produced, differed from the former obtained by Calvarano, I. (1967), not only for the extension of the investigation, but also for the quantitative values of some of the components reported in both papers, such as those relative to methyl anthranilate. The value reported by Calvarano, I. (1967) of methyl anthranilate (1.68 per cent) is not compatible with the composition of a genuine mandarin petitgrain oil, and should be explained by the inadequate analytical techniques used at that time.

Table 19.5 Percentage composition of mandarin petitgrain oil (Mondello et al., 1997a)

HYDROCARBONS			
Monoterpene		α-phellandrene	0.03–0.06
camphene	0.01–0.02	β-phellandrene	0.03–0.05
δ-3-carene	0.01–0.10 (0.29)[a]	α-pinene	1.75–2.30
p-cymene	2.96–5.19	β-pinene	1.90–2.45
p-cymenene	0.10–0.18	sabinene	0.22–0.90 (2.33)[a]
α-fenchene	tr–0.01	α-terpinene	0.19–0.33
limonene	7.18–12.59	γ-terpinene	23.94–28.48
p-mentha-1,3,8-triene	tr–0.02	terpinolene	0.71–0.88
myrcene	0.62–0.82	α-thujene	0.78–1.04
(E)-β-ocimene	0.42–0.92	*Sesquiterpene*	
(Z)-β-ocimene	0.15–0.20	bicyclogermacrene	0.03–0.13

Table 19.5 (Continued)

β-bisabolene	tr–0.02	cis-sabinene hydrate	0.01–0.05
β-caryophyllene	0.92–1.40	terpinen-4-ol	0.20–0.42
β-elemene	tr–0.01	α-terpineol	0.16–0.26
(E,E)-α-farnesne	tr–0.03	thymol	0.11–0.17
α-humulene	0.07–0.13		
α-selinene	tr–0.02	*Sesquiterpene*	
		spathulenol	tr–0.02
ALDEHYDES		**ESTERS**	
Aliphatic		*Monoterpene*	
decanal	0.01–0.02	linalyl acetate	0.02–0.10 (0.96)[b]
nonanal	tr–0.01	geranyl acetate	tr–0.02
		neryl acetate	tr–0.05
Monoterpene			
citronellal	0.02–0.08	**OTHERS**	
geranial	tr–0.10	caryophyllene oxide	0.01–0.08
neral	tr–0.06	1,8-cineole	0.01–0.02
		6-methyl-5-hepten-2-one	tr–0.01
ALCOHOLS		(Z)-3-hexenyl benzoate	tr–0.02
Monoterpene		methyl anthranilate	tr–0.03
carvacrol	0.01	methyl	41.61–51.93
p-cymen-8-ol	0.01–0.02	N-methylanthranilate	
linalool	0.27–1.10	methyl	tr–0.04
trans-p-menth-2-en-1-ol	0.02–0.03	N-dimethylanthranilate	
nerol	0.01–0.03	thymol methyl ether	0.10–0.16

Notes
a The value in parentheses indacates the percentage of δ-3-carene and of sabinene determined in the same sample obtained from foliage that probably contained 3% sweet orange leaves; b The value in parentheses indacates the percentage of linalyl acetate determined in a sample obtained from foliage that probably contained 2% of bitter orange leaves.

Appendix to Table 19.5

Mondello *et al.* (1997a) also found trace amounts of tricyclene, *o*-cymene, δ-cadinene, octanal, octanol, citronellol, geraniol, *cis-p*-menth-2-en-1-ol, α-terpinyl acetate, *cis*-linalool oxide.

Laboratory extracted oils

The information relative to the laboratory extracted oils from mandarin foliage, are limited to one paper by Germanà *et al.* (1990) on the Tardivo di Ciaculli cultivar grown in Sicily (Italy) and to the papers by Karawya and Hyfnawy (1979) and by Fleisher and Fleisher (1990, 1991) on the Balady cultivar grown in Egypt and Israel. In the first two papers were identified and quantitatively determined few components. The results are listed below:

	Karawya and Hyfnawy (1979)	Germanà *et al.* (1990)
p-cymene	4%	–
limonene	6%	24.24%
myrcene	1%	8.03%

α-phellandrene	12%	–
α-pinene	3%	0.96%
β-pinene	2%	0.96%
sabinene	–	0.54%
α-terpinene	–	4.07%
caryophyllene*	–	0.06%
β-farnesene*	tr	–
geranial	–	1.99%
neral	–	0.03%
geraniol	–	1.11%
linalool	–	4.37%
nerol	–	1.27%
terpineol*	–	0.71%
farnesol*	–	0.40%
geranyl acetate	–	0.78%
neryl acetate	–	0.17%
methyl anthranilate	3%	–
methyl N-methylanthranilate	60%	50.10%
benzaldehyde	–	0.11%

Both samples, relative to the composition reported above, although the amount of methyl N-methylanthranilate is in accordance with that of a genuine mandarin petitgrain oil, resulted anomalous: for both oils the presence of γ-terpinene, usually the most represented monoterpene hydrocarbon in such oil, was not reported. The oil analysed by Karawya and Hyfnawy (1979) presented the 3 per cent of methyl anthranilate, probably misidentified. The presence of this components at levels higher than 0.1 per cent in mandarin petitgrain oil is considered a sign of adulteration. Methyl anthranilate, in fact, was not identified in the oil extracted from the foliage of same cultivar, analysed by Fleisher and Fleisher (1990, 1991) by more updated analytical techniques.

The results obtained by Fleisher and Fleisher (1990, 1991) are reported in Table 19.6.

TANGERINE PETITGRAIN OIL (*Citrus tangerine* Hort. ex Tan.)

In this section will be shortly described some tangerine petitgrain oils extracted from foliage obtained from plants classifiable, in our opinion, as *C. tangerine* Hort. ex Tan., although if in the original paper were indicated by a different botanical name, and characterised by a high content of linalool.

Attaway *et al.* (1966) identified in Murcott and Dancy tangerine leaf oils, cultivated in Florida, camphene[t], *p*-cymene, limonene, myrcene, β-ocimene*, α- and β-pinene, sabinene, α-terpinene, terpinolene, α-thujene[t] and linalool; in the Murcott tangerine oil were identified also, β-caryophyllene, geraniol, nerol, and terpinen-4-ol while in the Dancy tangerine also found *p*-cymenene, isoterpinolene[t], α-terpineol, thymol, and thymol methyl ether.

The quantitative data on the composition of tangerine leaf oils, of different geographic origins, reported in papers published between 1967 and 1995, are summarised in Table 19.7.

Table 19.6 Percentage composition of mandarin leaf oil laboratory extracted (Fleisher and Fleisher, 1990, 1991)

HYDROCARBONS		linalool	0.11
Monoterpene		*trans*-sabinene hydrate	0.01
p-cymene	3.10	terpinen-1-ol	0.01
limonene	4.91	terpinen-4-ol	1.63
myrcene	0.31	α-terpineol	0.24
(Z)-β-ocimene	0.04	thymol	0.12
α-phellandrene	0.05		
β-phellandrene	0.03	**ESTERS**	
α-pinene	1.34	*Aliphatic*	
β-pinene	0.72	ethyl decanoate	0.03
α-terpinene	0.15	ethyl heptadecanoate	0.04
γ-terpinene	12.60	ethyl linoleate	0.74
terpinolene	0.53	ethyl linolenate	2.64
		ethyl myristate	0.18
Sesquiterpene		ethyl nonanoate	0.01
β-caryophyllene	0.14	ethyl octanoate	0.01
α-humulene	0.08	ethyl palmitate	1.83
valencene	0.05	ethyl pentadecanoate	0.03
		ethyl stearate	0.06
ALCOHOLS		(Z)-3-hexenyl formate	0.11
Aliphatic			
(Z)-3-hexenol	0.06	**OTHERS**	
		ethyl phenyl acetate	0.06
Monoterpene		methyl salicitate	1.00
p-cymen-8-ol	0.10	methyl N-methylanthranilate	65.71
cis-carveol	0.04	phytol	0.14
trans-carveol	0.04		

Notes
Fleisher and Fleisher (1990, 1991) also found trace amounts of viridiflorene, (Z)-2-pentenol, caryophyllene oxide.

Table 19.7 Percentage composition of tangerine leaf oil

	1	2	3(a)	3(b)	3(c)	4
HYDROCARBONS						
Monoterpene						
p-cymene	–	–	6.54	4.09	1.43	2.35
limonene	0.8–1.9	–	1.00	0.39	0.23	8.32
myrcene	0.3–0.6	–	0.34	0.11	0.25	0.05
β-ocimene	2.8–8.2*	5.3*	–	0.10[e]	2.31[e]	7.87[f]
β-phellandrene	0.2–0.5	–	–	–	0.06	–
α-pinene	1.1–2.7[a]	–	1.83	1.31	–	2.66
β-pinene	1.4–2.6	–	1.51	1.18	0.11	0.86
sabinene	–	4.1	–	0.41	1.41	2.81
α-terpinene	0.1–0.3	–	0.11	0.21	0.07	1.95
γ-terpinene	4.3–1.0	6.6	4.57	3.11	2.12	1.20
terpinolene	0.7–1.3	–	0.34	–	–	1.49
α-thuyene	1.1–2.7[a]	–	0.78	0.65	–	1.26
Sesquiterpene						
γ-cadinene	–	–	0.06	0.02	0.06	–
β-caryophyllene	–	–	5.44	0.24	2.00	0.24
β-elemene	–	–	–	0.10	0.04	–

(E,E)-α-farnesene	–	–	0.71	0.50	1.00	–
(E)-β-farnesene	–	–	–	tr	tr	–
α-humulene	–	–	0.51	0.25	0.11	0.08
α-muurolene	–	–	tr	tr	0.02	0.08
viridiflorene	–	–	0.31	0.18	–	–
ALDEHYDES						
Monoterpene						
neral	0.2–0.6[b]	–	–	–	–	6.05
Sesquiterpene						
α-sinensal	–	0.9[d]	0.93	0.24	0.72	0.50
β-sinensal	–	0.9[d]	–	0.41	2.98	–
ALCOHOLS						
Aliphatic						
(Z)-3-hexen-1-ol	–	–	0.13	0.05	0.33	–
Monoterpene						
cis-carveol	–	–	tr	0.01	–	–
p-cymen-8-ol	–	–	0.05	tr	tr	–
geraniol	0.1–0.2	–	0.02	tr	0.03	0.16
linalool	52–78	35.9	45.12	50.73	55.10	52.66
myrtenol	–	–	tr	tr	tr	–
terpinen-1-ol	–	–	tr	tr	–	–
α-terpineol	0.2–0.6[b]	–	0.33	0.58	1.30	0.06
thymol	2.6–6.9	4.9	14.43	11.70	8.21	–
Sesquiterpene						
nerolidol*	–	–	0.04	tr	0.04	–
ESTERS						
Aliphatic						
ethyl eptadecanoate	–	–	0.09	0.15	0.07	–
ethyl laurate	–	–	0.02	tr	0.03	–
ethyl linoleate	–	–	0.78	0.39	1.06	–
ethyl linolenate	–	–	2.86	0.68	6.88	–
ethyl myristate	–	–	0.11	0.14	0.05	–
ethyl palmitate	–	–	2.31	2.79	6.08	–
ethyl octanoate	–	–	tr	–	0.02	–
ethyl stearate	–	–	0.09	0.25	0.87	–
ETHERS AND OXIDES						
Monoterpene						
cis-linalool oxide	–	12.2[c]	0.06	0.44	–	–
trans-linalool oxide	–	12.2[c]	–	0.21	–	–
Thymol methyl ether	1.1–16	–	–	4.63	–	–
Sesquiterpene						
caryophyllene oxide	–	–	0.31	0.33	–	–
humulene oxide*	–	–	0.03	tr	–	–
OTHERS						
anethole*	–	–	–	0.02	0.01	–
β-phenyl ethanol	–	–	tr	0.02	0.03	–
phytol	–	–	0.19	0.12	1.29	0.13

Notes

tr = traces; * correct isomer not characterised; a α-thujene + α-pinene; b neral + α-terpineol; c cis + trans-linalool oxide; d α + β-sinensal; e (Z)-β-ocimene; f (E)-β-ocimene.

Table 19.7 (Continued)

Appendix to Table 19.7

1. Attaway *et al.* (1967). Florida, USA; Dancy tangerine; changes in composition from the first flesh of growth (Marsh) and the time of fruit harvest (January); steam-distillation; GC on packed columns of Carbowax 20 M; relative percentage of peak height. Attaway *et al.* also found *p*-cymenene (0.6–1.5%).
2. Lin and Hua (1992). China; Dahongpao tangerine; GC on capillary columns coated with SE-54; GC/MS, Kovats indices.
3. Fleisher and Fleisher (1990, 1991). Israel; Yussuf Effendy(a), Dancy(b) and Maya(c) tangerine; water distillation; GC on capillary column; GC/MS, retention indices. Fleisher and Fleisher also found trace amounts of α-phellandrene and ethyldecanoate in Yussuf Effendy oil; trace amounts of sabinol in Dancy oil; *trans*-α-bergamotene (0.02%), bicyclogermacrene (0.06%), α-copaene (0.02%), decanal (0.03%), *trans*-sabinene hydrate (2.79%), ethyl-2-butenoate (0.04%), benzaldehyde (0.02%), ethyl phenyl acetate (0.02%) and trace amounts of citronellol in Maya oil.
4. Blanco Tirado *et al.* (1995). Colombia; one sample steam-distilled; GC on capillary columns coated with DB-1 and Carbowax 20 M; GC/MS, Kovats indices; relative percentage of peak areas. Blanco Tirado *et al.* also found camphene (0.37%), δ-3-carene (0.04%), α-bergamotene* (0.11%), δ-cadinene (0.08%), δ-elemene (0.22%), germacrene B (0.10%), germacrene D (0.17%), longifolene (0.23%), β-selinene (1.01%), geranial (7.45%), nonanol (0.19%), 1,2-dihydrolinalool (0.03%), *cis*-sabinene hydrate (0.08%), α-bisabolol (0.29%), δ-cadinol (0.06%), citronellyl acetate (0.43%), 1,4-cineole (0.27%), 1,8-cineole (0.70%), *cis*-limonene oxide (0.59%).

SWEET ORANGE PETITGRAIN OIL (*Citrus sinensis* L. Osbeck)

Industrial oils

Sweet orange petitgrain oil is of poor commercial value and therefore scantly produced. For this reason not enough care is given to the selection of the foliage to be used for the distillation, and the cleanliness of the stills, often used for the distillation of more valuable oils, is not controlled. The possible contamination by bitter orange, lemon and mandarin leaves, are not, in fact, considered undesirable, and therefore it is quite difficult to find pure sweet orange petitgran oils (Mondello *et al.*, 1997b).

The analyses on sweet orange petitgrain have been almost completely carried out on laboratory extracted oils, therefore the information on the composition of oils produced industrially are limited to the semiquantitative, only of the major components, reported by Peyron (1965) and to those, relative to a more recent paper on Italian industrial sweet orange oils (Mondello *et al.*, 1997b).

Peyron (1965) identified in sweet orange petitgrain the following components listed in decreasing amount order: limonene, linalool, α-terpineol, β-pinene and β-ocimene*.

The results obtained by Mondello *et al.* (1997b) using modern analytical techniques, GC/MS on SE-52 and Carbowax 20 M columns, and by HPLC-GC/MS, on four samples industrially produced of Sicilian sweet orange petitgrain oils are reported in Table 19.8.

As it can be seen from the values reported in Table 19.8, the variability ranges of some components are quite wide. The cause can be assigned to different factors, such as the distillation conditions used; the cultivar of *C. sinensis* from which originated the foliage: in Italy are cultivated numerous varieties of both blond and blood oranges; the poor selection of the foliage, leading to the contamination by the presence of foliage of citrus species different from the sinensis. The amount of methyl

N-methylanthranilate (1.26–10.29 per cent), characteristic of mandarin petitgrain oils, could be due to the poor selection of the foliage. This component was, in fact, determined only at trace levels in laboratory extracted oils from carefully selected leaves of some sweet orange varieties grown in Italy. The composition of these oils was reported in the same paper (Mondello *et al.*, 1997b) that generated the results summarised in Table 19.8.

Table 19.8 Percentage composition of sweet orange petitgrain oil (Mondello *et al.*, 1997b)

HYDROCARBONS		*Sesquiterpene*	
Aliphatic		(α)-sinensal	0.04–0.32
camphene	0.03–0.04	(β)-sinensal	0.23–1.29
δ-3-carene	3.46–5.91	**KETONES**	
o-cymene	0.03–0.05	*Monoterpene*	
p-cymene	0.59–2.89	6-methyl-5-hepten-2-one	0.04–0.06
α-fenchene	0.01		
limonene	5.12–8.37	**ALCOHOLS**	
p-menta-2,4(8)diene	0.09–0.22	*Monoterpene*	
myrcene	2.66–3.85	citronellol	0.10–0.22
(E)-β-ocimene	4.53–9.21	*p*-cymen-8-ol	tr–0.09
(Z)-β-ocimene	0.21–0.33	geraniol	0.06–0.29
α-phellandrene	0.23–0.50	linalool	4.34–10.71
β-phellandrene	0.78–1.06	*cis-p*-menth-2-en-1-ol	0.04–0.11
α-pinene	0.99–1.51	*trans-p*-menth-2-en-1-ol	0.03–0.07
β-pinene	1.89–2.62	nerol	0.13–0.26
sabinene	37.64–41.93	*cis*-piperitol	0.01–0.07
α-terpinene	0.27–0.96	*trans*-piperitol	tr–0.05
γ-terpinene	2.19–2.98	*cis*-sabinene hydrate	0.13–0.42
terpinolene	0.98–1.52	terpinen-4-ol	0.55–2.59
α-thujene	0.21–0.39	α-terpineol	0.21–0.30
		thymol	0.01–0.05
Sesquiterpene			
bicyclogermacrene	0.01–0.24	*Sesquiterpene*	
β-bisabolene	0–0.09	(E)-nerolidol	0.01–0.05
β-caryophyllene	0.13–2.47	**ESTERS**	
α-copaene	tr–0.01	*Monoterpene*	
β-cubebene	tr–0.10	citronellyl acetate	0.21–0.65
β-elemene	0.04–3.80	linalyl acetate	0.02–0.12
(E,E)-α-farnesene	0.02–0.13	geranyl acetate	0.15–0.28
(Z)-β-farnesene	0.01–0.58	methyl geranate	0.05–0.07
α-humulene	0.05–0.60	neryl acetate	0.29–0.38
ALDEHYDES		**OXIDES**	
Aliphatic		*Monoterpene*	
decanal	0.02–0.07	(1,8)-cineole	0–0.05
nonanal	tr–0.03	*cis*-linalool oxide	tr–0.01
octanal	0.03–0.05		
		Sesquiterpene	
Monoterpene		caryophyllene oxide	0.02–0.24
citronellal	0.43–4.44	**OTHERS**	
geranial	0.59–3.11	methyl *N*-methylanthranilate	1.26–10.29
neral	0.28–2.18		

Notes
Mondello *et al.* also found trace amounts of tricyclene, *cis*-α-bergamotene, valencene, α-selinene, hexanal, hexanol, thymol methyl ether, (Z)-3-hexenyl benzoate.

Laboratory extracted oils

The analyses carried out on laboratory extracted oils from sweet orange leaves, are quite numerous. These are relative to many cultivars of different geographic origin. Due to the vast number of papers, on different cultivars, that presented greatly different qualitative and quantitative composition, it resulted very difficult to summarise all the data available in one table. The papers will, therefore, be described singularly, and for each will be reported the major information on the composition of the oils extracted from each cultivar.

Most of these papers (Attaway et al., 1967; Kamiyama and Amaha, 1972; Cheng and Lee, 1981; Baaliouamer et al., 1988; Ekundayo et al., 1990; Fleisher and Fleisher, 1990; Lin and Hua, 1992; Germanà et al., 1990; Blanco Tirado, 1995; Mondello et al., 1997b) report as the major component, sabinene ranging between 16 and 58 per cent. Some papers (Ogihara et al., 1990; Gurib-Fakim and Demarne, 1995), however, do not report the presence of sabinene. Only linalool (1.5–2.4 per cent) was determined in all oils of which the composition was reported in literature.

Attaway et al. (1967) examined the variation of the composition of the oil extracted by hydrodistillation from the leaves of Hamlin oranges during the period March (first flush of growth) to January (fruits ripening). During this period the composition of the oil randomly changed, and in general, the variations of linalool content were correspondent to opposite variations of the total monoterpene hydrocarbons content. Sabinene (52–58 per cent) was quite stable during the whole period, while evident variations were determined for linalool, ranging from 16 per cent in March, to its minimum value, 1.5 per cent, in September; δ-3-carene, varied from 2 per cent in March, to about the 8 per cent in December and January.

Kamyama and Amaha (1972) analysed the oil extracted from leaves of the Fukuara and Washington navel cultivars, grown in China. In both oils the main component was sabinene, 51.0 per cent and 42.4 per cent respectively, while linalool was the oxygenated component mostly represented, 8.7 per cent and 14.5 per cent respectively.

Cheng and Lee (1981) studied the oil of the Snow, Pineapple and of the ordinary round orange leaves. For all the varieties the major components were: sabinene (41.80–54.75 per cent), β-ocimene* (4.73–8.88 per cent), δ-3-carene (4.66–5.23 per cent), linalool (3.90–8.55 per cent) and β-caryophyllene (4.18–5.95 per cent).

Baaliouamer et al. (1988) extensively studied, by GC/MS, the oils extracted from the Washington navel, Valencia, Sanguine, and Portuguese cultivars, grown in Algeria. The authors determined, among these oils, numerous qualitative and quantitative differences. The most significant are reported below:

	Washington navel	Valencia	Sanguine	Portuguese
sabinene	10.20%	22.32%	36.47%	9.02%
δ-3-carene	0.76%	1.66%	6.47%	0.63%
myrcene	0.36%	0.40%	4.18%[a]	0.54%
limonene	14.39%	9.76%	5.00%	16.63%
γ-terpinene	0.12%	0.11%	2.15%	0.04%
(E)-β-ocimene	–	–	5.39%	–
terpinolene	–	0.02%	1.52%	–
α-sinensal	–	–	0.67%	–
β-sinensal	–	–	1.30%	0.46%
linalool	11.26%	12.09%	3.79%	15.78%

a myrcene + α-phellandrene.

The oil extracted from the Sanguine cultivar contained the highest amount of monoterpene hydrocarbons, and the lowest amount of oxygenated compounds.

Ekundayo et al. (1990) analysed the oils extracted from the leaves of six cultivars (Agege I, Bendel, Meran, Umudike, Etinam, Hamlin) of sweet oranges grown in Nigeria. The percentage composition of these oils is reported below:

	Agege I	Bendel	Meran	Umudike	Etinam	Hamlin
δ-3-carene	7.53%	11.68%	9.97%	6.09%	11.58%	9.32%
p-cymene	3.38%	3.89%	4.94%	12.31%	2,73%	–
limonene	3.71%	5.53%	5.68%	6.30%	7.38%	6.94%
myrcene	2.09%	3.09%	2.50%	1.37%	3.84%	2.38%
(E)-β-ocimene	1.99%	3.76%	1.76%	–	6.36%	–
β-phellandrene	0.41%	0.64%	–	–	0.85%	–
α-pinene	1.48%	2.25%	1.72%	3.03%	2.41%	3.09%
β-pinene	1.28%	1.83%	1.58%	2.49%	1.88%	2.48%
sabinene	16.03%	23.88%	18.84%	21.47%	24.19%	29.48%
γ-terpinene	0.72%	1.16%	0.64%	–	2.13%	–
terpinolene	0.77%	0.96%	0.40%	–	1.73%	–
β-caryophyllene	0.94%	0.92%	1.30%	1.67%	1.81%	–
α-humulene	5.10%	–	–	0.71%	0.29%	0.50%
geranial	1.95%	2.16%	2.30%	1.22%	0.69%	7.90%
neral	1.92%	1.65%	–	1.28%	0.38%	–
citronellol	1.58%	1.85%	3.44%	2.77%	2.93%	1.63%
geraniol	2.63%	2.86%	4.84%	2.50%	2.06%	1.98%
linalool	12.98%	13.99%	17.59%	14.48%	9.67%	15.68%
nerol	2.72%	3.03%	4.92%	1.79%	2.78%	2.35%
terpinen-4-ol	7.20%	8.99%	9.77%	–	8.81%	11.43%
α-terpineol	1.78%	1.92%	2.24%	2.04%	0.62%	1.87%
octenyl acetate*	–	–	0.57%	0.56%	–	–

The oil extracted from the Agege I cv. also contained camphor (1.44 per cent) and bornyl acetate (9.00 per cent); that of the Umudike cv. trans-sabinene hydrate (0.63 per cent); that of the Etinam cv. α-phellandrene (0.84 per cent), geranyl acetate (0.64 per cent), and linalyl acetate (0.26 per cent).

Fleisher and Fleisher (1990, 1991) reported the composition of oils obtained from the leaves of the three most common cvs. in Israel: Valencia, Jaffa (Shamouti) and Ruby. The three oils presented the same qualitative composition, but different quantitative composition; the quantitative difference most significant are here reported:

	Valencia	Jaffa	Ruby
δ-3-carene	2.52%	4.34%	3.06%
limonene	1.70%	1.25%	2.96%
myrcene	2.79%	5.63%	3.64%
sabinene	18.66%	32.58%	15.81%
γ-terpinene	7.52%	7.41%	10.50%
terpinolene	1.31%	0.75%	1.51%
β-caryophyllene + terpinen-4-ol	7.87%	3.62%	6.95%

β-elemene	4.13%	2.38%	4.00%
α-humulene	0.11%	0.98%	1.19%
β-sinensal	1.22%	0.43%	2.35%
linalool	8.21%	5.13%	20.92%

These authors noticed that the composition of the oils obtained from the Valencia cv., grown in Israel, was very similar to that obtained from the same cv. grown in Florida (Attaway et al., 1966; Moshonas and Shaw, 1986), while greatly differed from that of the Valencia cv. grown in Algeria (Baaliouamer, 1988).

Lin and Hua (1992) determined the concentration of the major components of the oils extracted from the leaves of three cultivars of sweet orange (Glorious, Washington Navel, Hamlin) cultivated in China. All the three oils were characterised by a high content of sabinene and linalool. The oil extracted from the Hamlin orange leaves also contained a high amount of β-pinene (27.2 per cent). This value was not determined before in oils extracted from the leaves of the same cv. analysed by Attaway (1967) and by Ekundayo et al. (1990). All the results obtained by Lin and Hua are reported below:

	Glorious	Washington navel	Hamlin
sabinene	46.6%	45.5%	27.2%
limonene	4.1%	3.8%	4.6%
myrcene	4.1%	4.3%	4.6%
β-pinene	–	–	27.2%
geranial + neral	6.4%	8.8%	3.2%
α-sinensal + β-sinensal	0.9%	1.3%	1.5%
linalool	12.7%	9.5%	7.7%

Germanà et al. (1990) analysed the oil obtained from the Valencia orange leaves, cultivated in Sicily (Italy). This oil presented peculiar composition. Although the oil contained high amount of sabinene (27.86 per cent), and of linalool (18.28 per cent), as it is common for sweet orange leaf oils, it showed a high content of geranyl acetate and of neryl acetate (about 10 per cent totally). The latter value results unusual for other sweet orange leaf oils. Also the presence of 4.18 per cent of methyl anthranilate and the 2.37 per cent of one farnesol* is peculiar, since these components have never been determined in sweet orange petitgrain oils.

Mondello et al. (1997b), using coupled analytical techniques (GC/MS) and multi-dimensional LC-GC/MS, studied some samples of sweet orange petitgrain industrially processed (the results are reported in Table 19.8), and some laboratory extracted oils, obtained from the Valencia, Biondo comune, and Moro cultivars, grown in Sicily (Italy). From the comparison of the composition determined for the three oils, it was determined:

- the almost identical qualitative composition; the few differences were related to the sesquiterpene hydrocarbons fraction, where cis-α-bergamotene was present at trace levels in Valencia and Moro leaf oils, and absent in the Biondo comune oils, β-santalene was present (0.04 per cent) only in Moro oil and absent in the others, while bicyclogermacrene (0.05 per cent and 0.09 per cent respectively) and (E,E)-α-farnesene (0.01 per cent and 0.05 per cent respectively) were present in Valencia and Biondo comune leaf oils, and absent in Moro;

- in all oils the main components were the three monoterpene hydrocarbons sabinene, δ-3-carene and (E)-β-ocimene, and the two monoterpene alcohols linalool and terpinen-4-ol;
- the three oils presented numerous quantitative differences, of which the most important are below reported:

	Valencia	Biondo comune	Moro
δ-3-carene	4.45%	10.28%	5.51%
(E)-β-ocimene	6.14%	9.73%	4.99%
sabinene	40.66%	38.46%	48.52%
β-caryophyllene	0.28%	0.83%	0.27%
β-elemene	0.49%	1.07%	0.36%
α-sinensal	0.27%	0.46%	0.08%
linalool	15.12%	6.29%	6.52%
terpinen-4-ol	7.33%	3.75%	6.14%

The composition of a sweet orange leaf oil, of not specified cultivar, cultivated in Columbia, was reported by Blanco Tirado *et al.* (1995). The main components, among the monoterpene hydrocarbons were, listed in amount order: sabinene (47.68 per cent), (E)-β-ocimene (8.31 per cent), δ-3-carene (7.38 per cent), limonene (5.95 per cent), and myrcene (3.73 per cent); among the oxygenated compound, were linalool (4.38 per cent) and the monoterpene aldehydes: citronellal (2.83 per cent), geranial (2.61 per cent) and neral (1.87 per cent). In the oil were also present δ-elemene (0.63 per cent), germacrene B (0.12 per cent) and longifolene (0.04 per cent) never reported by other authors in sweet orange leaf oils.

Two papers on the composition of sweet orange leaf oils do not report the presence in such oils of sabinene (Ogihara *et al.*, 1990; Gurib-Fakim and Demarne, 1995). The first by Ogihara *et al.* (1990) was carried out on oils obtained from the foliage of diploid and tetraploid plant of three varieties of sweet orange, Morita navel, Fukuara, and Hamlin, grown in Japan. These oils were obtained by extraction in pentane, from the methanol extract of the foliage. The components identified in these oils were: camphene, p-cymene, limonene, α-phellandrene, γ-terpinene, β-bisabolene, caryophyllene*, α-elemene, β-elemene and linalool. The main components were: camphene (19.4–42.0 per cent), α-phellandrene (5.5–41.6 per cent) and linalool (5.4–8.6 per cent). In these oils was also determined the presence of 12.8–21.4 per cent of phytol and β-amirin totally. Gurib-Fakim and Demarne (1995) determined for a sweet orange leaf oil, of not specified cultivar, from the Mauritius, a composition extremely different from that commonly reported in other papers. In the oil, where 53 components were identified, were determined, only the 2.80 per cent of monoterpene hydrocarbons ((E)- and (Z)-β-ocimene, β-phellandrene, α- and β-pinene); a high amount of alcohols, among which the most abundant were *cis*-piperitol (26.42 per cent), and 1-octen-3-ol (16.97 per cent); and some esters, in particular, cinnamyl formate (0.56 per cent), citronellyl tiglate (0.37 per cent) and geranyl tiglate (1.42 per cent) unusual in citrus petitgrain oils.

In addition to the papers already mentioned Moshonas and Shaw (1986) compared the oil extracted from Valencia foliage, cultivated in Florida, and obtained by plant treated with an abscission agent (5-Cl-3-methyl-4-nitro-1-H-pirazole) and that obtained from not treated plants. The author did not determined relevant qualitative and quantitative differences among the oils. The paper reported only qualitative information. The components identified were: δ-3-carene, m-cymene, p-cymene, limonene, myrcene, ocimene*, α-pinene, sabinene, γ-terpinene, terpinolene, α-thujene, β-caryophyllene, β-elemene, humulene*, citronellal, geranial, neral, α-sinensal, β-sinensal, 6-methyl-5-hepten-2-one, hexanol, (Z)-3-hexen-1-ol, (E)-2-hexen-1-ol, citronellol, p-cymen-8-ol, geraniol, isopulegol, linalool, cis-p-menth-2-en-1-ol, nerol, terpinen-4-ol, α-terpineol, 4-vinyl guaiacol, citronellyl acetate, geranyl acetate, neryl acetate.

BERGAMOT PETITGRAIN OIL (*Citrus bergamia*)

The literature data on the industrial bergamot petitgrain composition are limited to the semiquantitative results obtained by Peyron (1965) and to the results relative to oils produced in Calabria (Italy) reported by Calvarano, I. (1968).

Peyron reported for petitgrain bergamot oil the presence of the following components listed in decreasing amount order: linalyl acetate, linalool, limonene, α-terpineol, β-pinene, geranyl acetate+neryl acetate, and γ-terpinene.

Calvarano, I. (1968), after pre-fractionation of the oil, and by GC analysis of the fractions, on stainless steel capillary column, coated with a film of UCON LB 550X, identified in Calabrian bergamot petitgrain oil 33 components determined at the following relative percentages:

camphene	0.01%
δ-3-carene[t]+myrcene	2.04%
p-cymene	1.02%
limonene	2.80%
α-phellandrene	0.04%
α-pinene	0.16%
β-pinene	1.11%
β-ocimene*	0.18%
sabinene	0.12%
α-terpinene	0.09%
γ-terpinene	1.42%
terpinolene	0.42%
α-thujene	tr
decanal	0.07%
heptanal	0.01%
nonanal	0.02%
octanal	0.01%
citronellal	0.02%
geranial	0.14%
neral	0.03%
citronellol	0.50%
geraniol	0.09%
linalool	18.82%
nerol	1.74%
terpinen-4-ol	0.21%
α-terpineol	5.69%

ESTERS (octyl acetate, citronellyl acetate, geranyl acetate, linalyl acetate, neryl acetate, terpenyl acetate*)	51.02%
methyl N-methylanthranilate	7.53%

The information on the composition of bergamot leaf oil laboratory extracted are also scant. These are limited to researches carried out on foliage grown in Egypt (Karawya et al., 1970), in California (Ortiz et al., 1978), in Taiwan (Cheng and Lee, 1981; Huang et al., 1986).

In the oil analysed by Karawya et al. (1970), the main components, among those identified, were limonene (10.91 per cent), β-pinene (8.24 per cent), and geranial (5.33 per cent), while linalool and linalyl acetate were only the 1.33 per cent and 0.83 per cent respectively.

In the oils analysed by Ortiz et al. (1978), by Cheng and Lee (1981) and by Huang et al. (1986) were present high amounts of linalool (55.16 per cent; 22.39 per cent; 41.24 per cent respectively), of linalyl acetate (22.30 per cent; 51.64 per cent; 11.31 per cent respectively) and of α-terpineol (5.17 per cent; 4.21 per cent; 7.89 per cent respectively). These results were in accordance with what determined by Peyron (1965) and by Calvarano, I. (1968). The oil analysed by Huang et al. (1986) contained also high amount of geraniol (22.51 per cent).

GRAPEFRUIT PETITGRAIN OIL (*Citrus paradisi* Macf.)

From the literature data at that time available, Peyron (1965) observed that grapefruit leaf oil contained from 65 to 73 per cent of monoterpene hydrocarbons.

Attaway et al. (1966) during a qualitative study on different citrus leaf oils found that in the laboratory extracted oil obtained from Duncan and Marsh grapefruit foliage, cultivated in Florida, were present the following components: camphene^t, δ-3-carene, p-cymene, limonene, myrcene, β-ocimene*, α- and β-pinene, sabinene, α-terpinene, β-caryophyllene, citronellal, geranial, neral, 6-methyl-5-hepten-2-one, linalool and terpinen-4-ol. In the oil obtained from the Duncan variety were also present terpinolene, cadinene*^{t}, geraniol and nerol while in the Marsh variety were also present isoterpinolene^t, humulene*^{t}, α-terpineol and neryl acetate.

The results on the quantitative analysis on grapefruit petitgrain, found in literature, all relative to laboratory extracted oils, are not enough homogeneous, and therefore do not permit to define the typical composition of this oil. Some paper report, as the main component, sabinene, present at levels up to 62 per cent of the whole oil; in other papers sabinene was present at small amounts or totally absent, and in these cases the main components were p-cymene or linalool. The results relative to these papers are summarised in Table 19.9.

KEY LIME PETITGRAIN OIL (*Citrus aurantifolia* (Chistm.) Swing.)

Kumar and Banerijee (1957), determined, in a oil extracted by steam distillation from the leaves obtained from Indian trees, the presence of: limonene (20.5 per cent), geranial + neral (36.0 per cent), geraniol + linalool (13.2 per cent), esters calculated as linalyl acetate (23.8 per cent), acids calculated as acetic acid (2.0 per cent), citropten (2.0 per cent). The analyses were carried out by using the common laboratory procedures on the fractions obtained by distillation of the oil.

Table 19.9 Percentage composition of grapefruit leaf oil

	1	2	3	4	5	6
HYDROCARBONS						
Monoterpene						
camphene	–	0.2	0.06	–	–	–
δ-3-carene	0.1–0.2	6.1	0.12	–	–	–
p-cymene	–	1.4	0.05	21.32	–	–
limonene	1.6–3.3	6.4	4.38	11.37	2.12	0.14
myrcene	2.7–4.3	4.9	4.02	–	1.25	–
β-ocimene	8.1–13*	4.6[a]	7.18*	–	–	0.18[b]
β-phellandrene	0.7–2.0	2.3	0.88	–	–	0.21
α-pinene	1.6–2.7	2.4	2.98	3.09	–	–
β-pinene	–	3.1	4.50	3.17	–	–
sabinene	42–59	18.5	61.91	2.49	10.22	–
α-terpinene	0.7–1.8	10.7	0.55	–	0.49	tr
γ-terpinene	1.2–3.8	12.0	0.82	–	–	–
terpinolene	0.4–0.8	2.2	0.23	1.56	–	0.33
Sesquiterpene						
caryophyllene	–	0.9*	3.14[c]	1.83[c]	0.99*	–
humulene	–	0.5*	0.28[d]	1.60[d]	–	–
ALDEHYDES						
Aliphatic						
nonanal	–	tr	–	–	–	0.36
Monoterpene						
citronellal	0.1–12.0	3.1	0.60	–	–	–
geranial	0.1–3.3	0.6	0.20	1.70	5.33	0.10
neral	0.4–1.2[e]	1.2	0.06	–	3.07	0.81
ALCOHOLS						
Aliphatic						
(Z)-3-hexanol	–	0.1	–	–	–	tr
Monoterpene						
citronellol	–	1.4	–	3.83	–	8.60
geraniol	–	0.3	0.04	1.69	3.74	2.73
linalool	5.9–24.0	4.4	3.33	12.78	26.66	22.93
nerol	–	0.3	0.05	1.11	3.49	–
terpinen-4-ol	1.4–14.0	0.8	0.99	17.02	–	20.00
α-terpineol	0.4–1.2[e]	0.3	0.12	1.61	13.56	0.30
ESTERS						
Monoterpene						
citronellyl acetate	–	–	0.12	–	–	4.84
geranyl acetate	–	–	0.09	0.53	7.65	0.29
linalyl acetate	–	–	0.08	0.57	–	–

Notes
tr = traces;   t = tentative identification;   * correct isomer not characterised;   a (E)-β-ocimene;   b (Z)-β-ocimene;
c β-caryophyllene;   d α-humulene;   e neral + α-terpineol.

Appendix to Table 19.9

1. Attaway *et al.* (1967). Florida, USA; Marsh grapefruit; change in composition from the first flush of growth (March) and time of fruit harvest (January); steam-distillation; GC on packed columns of Carbowax 20 M; relative percentage of peak height.
2. Kamiyama and Amaha (1972). Japan; CH_2Cl_2 extraction; GC on packed columns of PEG 20 M, Apiezon M, β,β'-oxidypropionitrile, UCON 50 HB, LAC-3R-728; relative percentage of peak areas. Kamiyama and Amaha also found α-thuyene (4.0%), β-farnesene (0.7%), β-selinene (1.2%), decanal (0.1%), (E)-2-hexanal (0.3%), isopulegol (0.6%) and trace amounts of *p*-cymenene, β-elemene, dodecanal, hexanol, (E)-2-hexanol, (Z)-2-pentenol.
3. Cheng and Lee (1981). Taiwan; prefractionation techniques; GC. Cheng and Lee also found 6-methyl-5-hepten-2-one (0.15%), 1,8-cineole (0.36%), thymol methyl ether (0.16%).
4. Ekundayo *et al.* (1991a). Nigeria; Duncan and Marsh grapefruit; water distillation; GC on capillary columns coated with DB-wax; relative percentage of peak areas. Ekundayo *et al.* Also found *trans*-sabinene hydrate (0.85%), caryophyllene oxide (2.28%).
5. Germanà *et al.* (1990). Sicily, Italy; Marsh grapefruit; water distillation; GC on packed column of WEAS; relative percentage of peak areas. Germanà *et al.* also found farnesol (5.63%, correct isomer not characterised), neryl acetate (3.92%), benzaldehyde (2.94%), methyl anthranilate (7.97%).
6. Gurib-Fakim and Demarne (1995). Mauritius; water distillation; GC on capillary column coated with HP 101; GC/MS, Kovats indices; relative percentage of peak areas. Gurib-Fakim and Demarne also found β-copaene (0.62%), α-cubebene (0.30%), menthone (0.96%), verbenone (0.36%), octanol (0.64%), carvacrol (0.22%), *cis*-carveol (0.18%), isomenthol (0.10%), *trans*-piperitol (7.40%), *cis*-β-terpineol (6.98%), γ-terpineol (0.64%), bornyl acetate (1.17%), geranyl formate (0.23%), *cis*-linalool oxide (0.13%), *trans*-rose oxide (0.82%), cinnamyl aldehyde (11.11%), cinnamyl formate (0.18%), citronellic acid (0.41%) and trace amounts of α-phellandrene, *trans*-β-bergamotene, nerolidol*, citronellyl tiglate.

From the literature available at that time, Peyron (1966) observed that key lime petitgrain oil contained from 20 to 50 per cent of monoterpene hydrocarbons, and from 30 to 60 per cent of citral. Peyron (1965, 1966) also studied the composition of the petitgrain oil obtained from a citrus (*limonette*) belonging to the group of lime (*C. aurantifolia*), largely diffused in Morocco. This author noticed that this oil resembled the bitter orange petitgrain, for the high level of linalyl acetate, the main component of the ester fraction (59 per cent of the whole oil). In the oil were identified the following components: camphene, cymene*, limonene, myrcene, β-ocimene*, α- and β-pinene, γ-terpinene, saturated linear aldehydes C_2, C_5, C_6, C_8–C_{12}, isopentenal, citronellal, geranial, neral, benzaldehyde, carvone, citronellol, geraniol, linalool, nerol, α-terpineol, farnesols*, nerolidol*, citronellyl acetate, geranyl acetate, linalyl acetate, linalyl formate, cineol*, methyl N-methylanthranilate, the carboxylic acids C_1–C_3, C_5, C_6, iso-C_6, C_8, C_{10}, C_{12}, C_{14}, C_{16}.

The quantitative results on key lime petitgrain oil composition, reported in papers published from 1972 to 1998 are summarised in Table 19.10.

Table 19.10 Percentage composition of key lime leaf oil

	1	2	3	4	5	6
HYDROCARBONS						
Monoterpene						
camphene	tr	–	tr	–	0.16	–
δ-3-carene	0.4	–	–	0.67	0.48	–
p-cymene	tr	1.4	0.68	–	–	0.28
limonene	24.7	21.6	28.93	33.76	39.28	32.07
myrcene	1.0	0.6	0.99	0.56	1.00	–

Table 19.10 (Continued)

	1	2	3	4	5	6
β-ocimene	–	0.14*	1.58*	–	–	–
(E)-β-ocimene	1.4	–	–	–	0.47	–
β-phellandrene	0.1	–	–	1.81	–	–
α-pinene	0.3	–	0.20	0.61	1.45	4.11
β-pinene	0.3	–	0.31	5.83	28.44	0.14
sabinene	2.3	–	0.38	–	–	–
γ-terpinene	0.3	–	2.20	–	0.79	–
terpinolene	0.1	–	0.05	–	0.43	–
α-thujene	0.2	–	0.01	–	0.07	–
Sesquiterpene						
β-caryophyllene	0.7	–	–	–	0.84	–
ALDEHYDES						
Aliphatic						
decanal	0.2	–	0.81	–	0.41	–
nonanal	tr	–	tr	–	–	–
octanal	–	–	0.10	–	tr	–
Monoterpene						
citronellal	0.5	0.03	2.50	–	0.10	2.64
geranial	40.0	17.4	25.34	21.54	2.05	+
neral	23.8	12.4	19.72	–	5.30	+
ALCOHOLS						
Monoterpene						
citronellol	–	0.04[a]	–	0.42	–	21.39
geraniol	0.6	–	0.25	3.95	7.50	10.38
isopulegol	0.1	0.09	–	–	tr	–
linalool	0.8	0.4	1.24	1.93	1.40	3.97
nerol	0.2	–	1.80	6.15	–	1.48
terpinen-4-ol	0.1	–	0.15	0.81	2.01	–
α-terpineol	0.2	–	0.34	1.25	2.39	–
ESTERS						
Monoterpene						
geranyl acetate	0.5	–	5.49	0.98	0.59	+
linalyl acetate	tr	–	0.55	–	–	+
neryl acetate	–	6.9	0.62	–	0.21	–

Notes
tr = traces; * correct isomer not characterised; + present, not quantified; a citronellol + unidentified.

Appendix to Table 19.10

1 Kamiyama and Amaha (1972). Japan; CH_2Cl_2 extraction; GC on packed columns of PEG 20 M, Apiezon M, β,β'-oxidypropionitrile, UCON 50 HB 280X, LAC-3R-728; relative percentage of peak areas. Kamiyama and Amaha also found β-selinene (0.1%) and trace amounts of p-cymenene, (E)-2-hexen-1-al, hexanol, (E)-2-hexen-1-ol, (Z)-3-hexen-1-ol.
2 Lund *et al.* (1982). Florida, USA; steam distillation; prefraction by preparative GC on packed column of DBGS; GC on packed columns of Carbowax; IR, GC/MS; relative percentage of peak areas. Lund *et al.* also found 6-methyl-5-hepten-2-one (0.02%), iso isopulegol (0.01%).
3 Calvarano, M. *et al.* (1982). Brazil; Galego lime; steam distillation; GC on packed column of UCON LB 550X; retention times, co-chromatography with standard; relative percentage of peak areas. Calvarano, M. *et al.* also found terpinyl acetate* (0.74%).

4 Ekundayo *et al.* (1991b). Nigeria; water distillation; GC on capillary columns coated with DB-wax; relative percentage of peak areas. Ekundayo *et al.* also found caryophyllene oxide (1.03%).
5 Jantan *et al.* (1996). Malaysia; water distillation; GC on capillary columns coated with SE-30 and PEG 20 M; GC/MS, retention indices and co-chromatography with standard compounds; relative percentage of peak areas. Jantan *et al.* also found (Z)-β-ocimene (0.14%), α-phellandrene (0.08%), α-terpinene (0.03%), α-bergamotene* (0.41%), β-elemene (0.45%), δ-elemene (0.44%), (E)-β-farnesene (1.48%), (Z)-β-farnesene (0.41%), α-humulene (0.12%), dodecanal (0.10%), octanal (0.11%), sabinyl acetate* (0.10%), *cis*-limonene oxide (0.11%) and trace amounts of *trans*-limonene oxide.
6 Haggag *et al.* (1998). Egypt; steam distillation; TLC, GC; relative percentage of peak areas. Haggag *et al.* also found carvacrol (0.38%), thymol (0.50%), eugenol (0.70%).

PERSIAN LIME PETITGRAIN OIL (*Citrus latifolia* Tanaka)

Peyron (1965), determined, as was reported in literature, that Persian lime leaf oil contained from 20 per cent to 50 per cent of monoterpene hydrocarbons and from 30 per cent to 60 per cent of aldehydes.

Calvarano *et al.* (1982) reported the composition of a leaf oil sample, obtained in laboratory by hydrodistillation, from Persian lime foliage grown in Brazil. The qualitative composition of this oils resulted identical and the quantitative composition was very similar to the Key lime oil reported in the same paper (see Table 19.10). The most evident quantitative differences were due to a higher content, in Persian lime petitgrain oil than in Key lime petitgrain oil, of limonene (34.47 per cent vs 28.93 per cent), and of nerol^t (6.48 per cent vs 1.80 per cent), and a lower amount of neral (14.95 per cent vs 19.72 per cent) and of geranial (19.54 per cent vs 25.34 per cent).

REFERENCES

Adeishvili, N. and Khrarebava, L.G. (1987) Volatile components of lemon leaves. *Subtropicheskie Kul'tury*, 6, 63–66.

Attaway, J.A., Pieringer, A.P. and Barabas, L.J. (1966) The origin of citrus flavor components. I. The analysis of citrus leaf oils using gas-liquid chromatography, thin layer chromatography and mass spectrometry. *Phytochemistry*, 5, 141–151.

Attaway, J.A., Pieringer, A.P. and Barabas, L.J. (1967) The origin of citrus flavor components. III. A study of the percentage variations in peel and leaf oil terpenes during one season. *Phytochemistry*, 6, 25–32.

Ayedoun, A.M., Sossou, P.V., Mardarowicz, M. and Leclerq, P.A. (1996) Volatile constituents of the peel and leaf oils of *Citrus limon* L. Burm. from Benin, *J. Essent. Oil Res.*, 8, 441–444.

Baaliouamer, A., Meklati, B.Y., Fraisse, D. and Scharff, C. (1985) Qualitative and quantitative analysis of petitgrain Eureka lemon essential oil by fused silica capillary column gas chromatography mass spectrometry. *J. Sci. Food Agric.*, 36, 1145–1154.

Baaliouamer, A. and Meklati, B.Y. (1986) Analysis of bitter orange petitgrain essential oil by combined gas chromatography-mass spectrometry. *Agric. Biol. Chem.*, 50, 2111–2114.

Baaliouamer, A., Meklati, B.Y., Fraisse, D. and Scharff, C. (1988) Analysis of leaf oils from four varieties of sweet orange by combined gas chromatography-mass spectrometry. *Flav. Fragr. J.*, 3, 47–52.

Blanco Tirado, C., Stashenko, E.E., Combariza, M.Y. and Martinez, J.R. (1995) Comparative study of Colombian citrus oils by high-resolution gas chromatography and gas chromatography-mass spectrometry. *J. Chromatogr. A*, 697, 501–513.

Boelens, M.H. and Sindreu, R.J. (1988) Essential oils from Seville bitter orange (*Citrus aurantium* L SSP *Amara* L). In B.M. Lawrence, B.D. Mookherjee and B.J. Willis (eds), *Flavors and Fragrances: A World Perspective*, Amsterdam: Elsevier Sci. Publ. BV, pp. 551–565.

Boelens, M.H. and Oporto, A. (1991) Natural isolates from Seville bitter orange tree. *Perfum. Flav.*, 16(6), 2–7.

Calvarano, I. (1967) Le essenze italiane di petitgrain. Nota I. I petitgrain limone e mandarino. *Essenz. Deriv. Agrum.*, 37, 27–54.

Calvarano, I. (1968) Le essenze italiane di petitgrain. Nota II. I petitgrain bigarade e bergamotto. *Essenz. Deriv. Agrum.*, 38, 31–48.

Calvarano, M., Salnitro, F. and Sacco, T. (1982) L'olio essenziale delle foglie di limáo Tahiti (*Citrus latifolia* Tanaka) e di limáo Galego (*Citrus aurantifolia* Sw.) del Brasile. *Essenz. Deriv. Agrum.*, 52, 52–58.

Cartoni, G.P., Goretti, G. and Russo, M.V. (1987) Capillary columns in series for the gas chromatographic analysis of essential oils. *Chromatographia*, 23, 790–795.

Cheng, Y.S. and Lee, C.S. (1981) Composition of leaf essential oils from ten *Citrus* species. *Proc. Natl. Sci. Counc. B. ROC*, 5, 278–283. In B.M. Lawrence, *Perfum. Flav.*, 18(5), 43–48 (1993).

Crescimanno, F.G., De Pasquale, F., Germanà, M.A., Bazan, E., and Palazzolo, E. (1988) Annual variation of essential oils in the leaves of four lemon (*Citrus limon* (L.) Burm f.) cultivars. *Proceeding Sixth Int. Citrus Congress, Tel Aviv, Israel, March 6–11*, 583–588, Margraf Scientific Books, Weikersheim, Germany.

De Vottero, L.R., Tonarelli, G., De Krasnogor, E.R. and Retamar, J.A. (1978) Petitgrains de hojas de naranjo amargo, limonero y mandarinero. *Essenz. Deriv. Agrum.* 48, 150–156.

Di Giacomo, A. (1974) *Gli Olii Essenziali degli Agrumi*. EPPOS, Milan, Italy.

Di Giacomo, A., Calvarano, I., Calvarano, M. and Belmusto, G. (1985) Petitgrain di limone: composizione e tecnologia. *Essenz. Deriv. Agrum.*, 55, 251–267.

Dugo, G., Mondello, L., Cotroneo, A., Stagno D'Alcontres, I., Basile, A., Previti, P., Dugo, P. and Bartle, K.D. (1996) Characterization of Italian citrus petitgrain oils. *Perfum. Flav.*, 21(3), 17–28.

Ekundayo, O., Bakare, O., Adesomoju, A. and Stahl-Biskup, E. (1990) Nigerian sweet orange leaf oil composition. *J. Essent. Oil Res.*, 2, 199–201.

Ekundayo, O., Bakare, O., Adesomoju, A. and Stahl-Biskup, E. (1991a) Composition of the leaf oil of grapefruit (*Citrus paradisi* Macf.). *J. Essent. Oil Res.*, 3, 55–56.

Ekundayo, O., Bakare, O., Adesomoju, A. and Stahl-Biskup, E. (1991b) Volatile constituents of the leaf oil of Nigerian lime (*Citrus aurantifolia*). *J. Essent. Oil Res.*, 3, 119–120.

Fleisher, Z. and Fleisher, A. (1990) Mandarin leaf oil (*Citrus reticulata* Blanco). Aromatic plants of the Holy Land and the Sinai. Part III. *J. Essent. Oil Res.*, 2, 331–334.

Fleisher, Z. and Fleisher, A. (1991) Citrus petitgrain oils of Israel. *Perfum. Flav.*, 16(1), 43–46.

Formàček, K. and Kubeczka, K.H. (1982) *Essential Oils Analysis by Capillary Chromatography and C-13 NMR Spectroscopy*. Wiley & Sons, New York.

Germanà, M.A., De Pasquale, F., Bazan, E. and Palazzolo, E. (1990) Indagine sugli oli essenziali contenuti nei fiori, nelle foglie e nei germogli di cinque specie di citrus. *Essenz. Deriv. Agrum.*, 60, 297–312.

Goretti, G., Russo, M.V., Liberti, A. and Belmusto, G. (1986) Valutazione gas cromatografica degli olii essenziali mediante campionamento dello spazio di testa. *Essenz. Deriv. Agrum.*, 56, 345–358

Gurib-Fakim, A. and Demarne, F. (1995) Aromatic plants of Mauritius: volatile constituents of the leaf oils of *Citrus aurantium* L., *Citrus paradisi* Macfad. and *Citrus sinensis* (L.) Osbeck. *J. Essent. Oil Res.*, 7, 105–109.

Haggag, E.G., Wahab, S.M.A., El-Zalabany, S.M., Moustafa, E.A.A., El-Kherasy, E.M. and Mabry, T.J. (1998) Volatile oils and pectins from *Citrus aurantifolia* (lime) and *Citrus limonia* (lemon). *Asian J. Chem.*, 10, 828–833.

Haubruge, E., Lognay, G., Marlier, M., Danhier, P., Gilson, J.-C. and Gaspar, C. (1989) Etude de la toxicite de cinq huiles essentielles extraites de *Citrus* sp. a l'egard de *Sitophilus zeamais* Motsch (Col. Curculionidae), *Prostephanus truncatus* (Horn) (Col. Bostrychidae) et *Tribolium castaneum* Herbst (Col. Tenebrionidae). *Med. Fac. Landbouww Rijksuniv Gent.*, 54, 1083–1093.

Huang, Y., Wen, M., Xiao, S., Zhao, H., Ren, W., Chen, Q., Liu, X. and Gui, T. (1986) Studies on the chemical constituents of the steam-distilled leaf and peel essential oil from *Citrus bergamia*. *Acta Botanica Yunnanica*, 8, 471–476.

Huet, R. (1991) Les huiles essentielles d'agrumes. *Fruits*, 46, 551–576.

Jantan, I., Ahmad, A.S., Ahmad, A.R., Mohd Ali, N.R. and Ayop, N. (1996) Chemical composition of some citrus oils from Malaysia. *J. Essent. Oil Res.*, 8, 627–632.

Kamiyama, S. (1967) Studies on the leaf oils of *Citrus* species. Part I. Composition of leaf oils from *Citrus unshiu*, *Citrus natsudaidai*, *Citrus kokitsu* and *Citrus limon*. *Agric. Biol. Chem.*, 31, 1091–1096.

Kamiyama, S. (1968) Studies on the leaf oils of *Citrus* species. Part II. An examination of the seasonal variation of leaf oil composition. *Bull. Brew. Sci.*, 14, 43–47.

Kamiyama, S. (1970) Studies on the leaf oils of *Citrus* species. Part III. Composition of leaf oils from Hassaku, Daidai, Tachibana and Kishu-mikan. *Agric. Biol. Chem.*, 34, 540–546.

Kamiyama, S. and Amaha, M. (1972) Studies on the leaf oils of *Citrus* species. Part VI. Composition of leaf oils from ten citrus taxa and some intrageneric hybrids. *Bull. Brew. Sci.*, 18, 17–27.

Karawya, M.S., Balbaa, S.I. and Hifnawy, M.S. (1970) Study of the leaf essential oils of bitter orange and bergamot growing in Egypt. *American Perfumer and Cosmetics*, 85(11), 29–32.

Karawya, M.S. and Hifnawy, M.S. (1979) Leaf essential oils of three different varieties of *Citrus reticulata* Blanco growing in Egypt. *Perfum. Flav.*, 4(2), 27–30.

Kumar, S. and Banerjee, B.C. (1957) Chemical examination of petitgrain oil from lime (*Citrus aurantifolia*). *Indian Perfumer*, 1, 25–29.

Kumar, U., Ram, B., Pant, A.K., Gupta, K.G. and Brophy, J.J. (1992) Volatile constituents of the distilled leaf and peel oils of *Citrus limon* Burm. cv. 'Pant lemon-1', *J. Essent. Oil Res.*, 4, 643–644.

Lawrence, B.M. (1980) Progress in essential oils. *Perfum. Flav.*, 5(6), 27–32.

Lawrence, B.M. (1993) Progress in essential oils. *Perfum. Flav.*, 18(5), 43–58.

Lin, Z., Hua, Y. and Gu, Y. (1986) The chemical constituents of the essential oil from the flowers, leaves and peels of *Citrus aurantium*. *Acta Botanica Sinica*, 28, 635–640.

Lin, Z. and Hua, Y. (1992) Systematic evolutional relation of chemical components of the essential oils from 11 taxa of citrus leaves, *Acta Botanica Sinica*, 34, 133–139.

Lund, E.D., Shaw, P.E. and Kirkand, C.L. (1982) Components of Key lime leaf oil. *J. Agric. Food Chem.*, 30, 94–95.

Maurer, B. and Hauser, A. (1992) New pyridine derivates from essential oils. *Chimia*, 46, 93–95.

Melendreras, F.A., Laencina, J., Flores, J. and Guzman, G. (1984) Aceites esenciales en hojas de variedades de limonero (*Citrus limon* L. Burm.). *An. Edafol. Agrobiol.*, 43, 1161–1180. In B.M. Lawrence, *Parfum. Flav.*, 18(5), 51–54 (1993).

Mondello, L., Dugo, P., Dugo, G. and Bartle, K.D. (1996) Italian citrus petitgrain oils. Part I. Composition of bitter orange petitgrain oil. *J. Essent. Oil Res.*, 8, 597–609.

Mondello, L., Basile, A., Previti, P. and Dugo, G. (1997a) Italian citrus petitgrain oils. Part II. Composition of mandarin petitgrain oil. *J. Essent. Oil Res.*, 9, 255–266.

Mondello, L., Cotroneo, A., Stagno d'Alcontres, I. and Dugo, G. (1997b) Italian citrus petitgrain oils. Part III. Composition of sweet orange petitgrain oil. *J. Essent. Oil Res.*, 9, 379–392.

Mondello, L., Cotroneo, A., Dugo, P. and Dugo, G. (1997c) Italian citrus petitgrain oils. Part IV. Composition of lemon petitgrain oil. *J. Essent. Oil Res.*, 9, 495–508.

Moshonas, M.G. and Shaw, P.E. (1986) Valencia orange leaf oil composition. *J. Agric. Food Chem.*, 34, 818–820.

Ogihara, K., Munesada, K. and Suga, T. (1990) Variations in leaf terpenoids with ploidy level in citrus cultivars and hybrids. *Phytochemistry*, 29, 1889–1891.

Ortiz, J.M., Kumamoto, J. and Scora, R.W. (1978) Possible relationships among sour oranges by analysis of their essential oils. *Int. Flav. Food Addit.*, **5**, 224–226.

Peyron, L. (1965) Petitgrain oils in perfumery. *Soap Perfum. Cosmet.*, **38**, 769–780

Peyron, L. (1966) Some little-known essential oils of potential interest in perfumery. *Soap Perfum. Cosmet.*, **39**, 633–643.

Peyron, L. (1973) Sur quelques essences en provenance du Mato Grosso. *Parf. Cosm. Sav. France*, **3**, 371–378.

Prager, M.J. and Miskiewicz, M.A. (1981) Gas chromatographic-mass spectrometric analysis, identification and detection of adulteration of perfumery products from bitter orange trees. *J. Assoc. Off. Anal. Chem.*, **64**, 131–138

Protopapadakis, E. and Papanikolau, X. (1998) Characterization of *Citrus aurantium* and *taiwanica* rootstocks by isoenzyme and essential oil analysis. *J. Hort. Sci. Biotechnol.*, **73**, 81–85.

Scora, R.W., England, A.B. and Chang, D. (1968) Taxonomic affinities within the rough lemon group (*Citrus jambhiri* Lush.) as aided by gas chromatography of their essential leaf oils. *Proceeding First Int. Citrus Symp.*, **1**, 441–450, Riverside, Calif. 1969. In B.M. Lawrence (1993) *Perfum. Flav.*, **18**(5), 51–54.

Vernin, G. and Vernin, G. (1966) Détection et évaluation de l'anthranilate de méthyle et de ses dérives méthyles dans différents échantillons naturales et de synthèse par C. C. M. et G. L. C.. *La France et ses Parfums*, **9**, 429–448.

Wen, M., Xiao, S., Zhao, H., Ren., W. and Huang, Y. (1989) A. study on the chemical components of the essential oil from leaves of *Citrus limon* (L.) Burm f., *Tianran Chanwu Yanjiu Yu Kaifa*, **1**(2), 18–22.

20 The chiral compounds of citrus essential oils

Luigi Mondello, Paola Dugo and Giovanni Dugo

INTRODUCTION

The enantiomeric distribution of the components of essential oils can provide useful information on the determination of authenticity, quality, extraction technique used, geographic origin and biogenesis of the oils. Moreover, it has been proved that enantiomers can possess different biological activity. The most suitable method for determining the enantiomeric ratio of volatile compounds is high resolution gas chromatography (HRGC). Modified cyclodextrins (CD), either pure or diluted in polysiloxanes, are today the most widely used chiral stationary phases in HRGC.

Essential oils are usually complex mixtures. Direct GC separation is sometimes possible, mainly for those components present in high amount, but is usually difficult. Experimental conditions have to be chosen in order to avoid overlaps between the peaks of the enantiomers and those of other components. Often, a sample clean-up prior to chiral GC analysis is necessary to obtain chiral volatiles at high chemical purity, ready for the direct resolution into their enantiomers by GC analysis. Pre-separation using high performance thin layer chromatography (HPTLC) or high performance liquid chromatography (HPLC), off-line or on-line coupled to GC, have been performed. In particular, multidimensional GC with the combination of a non-chiral precolumn and a chiral main column has been demonstrated to be a powerful technique for the determination of the enantiomeric distribution of volatile components in complex mixtures, without problems of peak overlapping.

In this chapter, data present in literature on the enantiomeric distribution of some components of the volatile fraction of citrus oils are reported. Many papers are related to more than one oil, or sometimes the same authors published a series of papers where the same (or a similar) analytical method was used to determine the enantiomeric distribution of some components in the different oils. Since in this chapter, for an easier and clearer approach, the single citrus oils will be treated separately, a brief summary of the most used analytical techniques and experimental conditions used is cited in the introduction section. In this way, the section dedicated to the single oils will focus mainly on the results reported in literature and their discussion.

Mosandl's research group (Mosandl *et al.*, 1990; Hener *et al.*, 1990a,b; Kreis *et al.*, 1991; Mosandl, 1995) used a fully automated multidimensional gas chromatographic system (MDGC) (Siemens Sichromat 2) with a 'live switching' coupling piece ('live-T-piece') to determine the enantiomeric distribution of α-pinene, β-pinene and

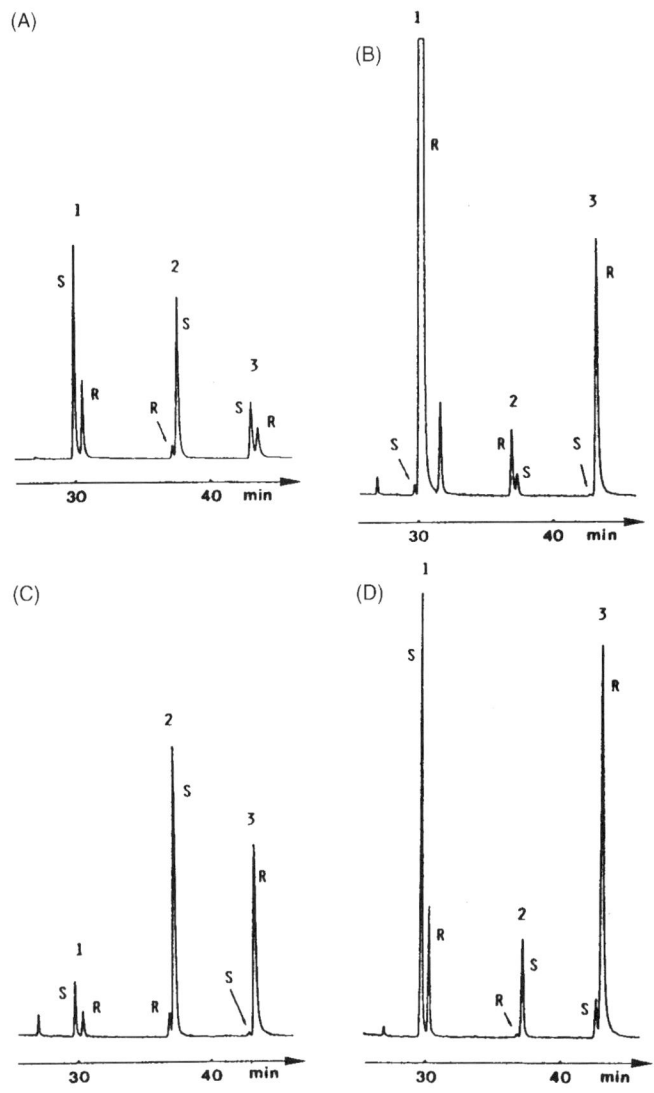

Figure 20.1 (A) MDGC analysis of α-pinene (1), β-pinene (2), limonene (3) using heptakis (2,3,6-tri-O-methyl)-β-cyclodextrin as the chiral stationary phase; (B) MDGC analysis of self-prepared orange oil. (C) MDGC analysis of self-prepared lemon oil; (D) MDGC analysis of lime oil. Reproduced with permission from Mosandl *et al.* (1990).

limonene in different citrus oils, as shown in Figure 20.1. The double oven system was equipped with a Carbowax 20 M pre-column and a heptakis (2,3,6-tri-O-methyl)-β-cyclodextrin coated main column (10 per cent of CD and 90 per cent OV-1701 vinyl), and the heart-cutting technique was used to transfer the fractions from the non chiral precolumn to the chiral main column.

Casabianca and his co-workers (Casabianca *et al.*, 1995; Casabianca and Graff, 1994, 1996) also used a Siemens Sichromat 2 MDGC system for their study on the enantiomeric distribution of some components in citrus oils.

The group from the University of Messina has extensively studied the enantiomeric distribution of some components of citrus essential oils employing different analytical techniques. The first studies were carried out using a GC system and a combination of columns to obtain reliable results on the enantiomeric distribution of components present in high percentage in the oils (Dugo *et al.*, 1992a,b, 1993; Cotroneo *et al.*, 1992). Later on, they used a fully automated HPLC-HRGC system (Dualchrom 3000 Series, Fisons) with an on-column type interface and partially concurrent eluent evaporation for the stereodifferentiation of monoterpene alcohols present in small amounts in citrus oils (Dugo *et al.*, 1994a,b; Mondello *et al.*, 1996). The HPLC-HRGC system permits the transfer and subsequent GC analysis of LC fractions that contain compounds of the same polarity, but each LC fraction subsequent to the first can be transferred and analysed by GC only after the end of the previous GC run. Recently, the enantiomeric distribution of sabinene, β-pinene, limonene, linalool, terpinen-4-ol and α-terpineol in different citrus oils, have been determined using a fully automated multidimensional GC system based on the use of mechanical valves, developed in their laboratory. With this system, linalyl acetate has been also analysed in bergamot oil (Mondello *et al.*, 1997, 1998a,b,c,d, 1999; Dugo *et al.*, 2000). Figure 20.2 shows the GC chromatograms obtained for a bergamot oils on the precolumn and the multidimensional system in the standby position, the same chromatogram with the six heart-cuts, and the chiral chromatogram of the transferred components.

LEMON OIL

Table 20.1 summarises data of the literature on the enantiomeric distribution of the components of lemon oil.

The research group of Mosandl (Mosandl *et al.*, 1990; Mosandl, 1995; Hener *et al.*, 1990a) used a multidimensional GC system to obtained direct stereodifferentiation of α-pinene, β-pinene and limonene for lemon oils hand-squeezed in laboratory. Moreover, the same stereodifferentiation has been carried out on distilled and solvent extracted lemon oils (Kreis *et al.*, 1991). The enantiomeric distribution of the same components have been determined by Rocca *et al.* (1992) using MDGC and a modified β-cyclodextrin main column. Results obtained are in perfect agreement with those of Mosandl's group. Dugo *et al.* (1992a, 1993) used GC techniques to determine the enantiomeric ratio of limonene in cold-pressed lemon oils of sure origin, employing a non-chiral (SE-52) and a chiral (permethylated-β-ciclodextrin) columns coupled on-line. This analysis allows the detection of addition as low as 5 per cent of reconstituted lemon oils to genuine ones. Figure 20.3 shows the GC chromatograms of a mixture of cold-pressed lemon oil with 5 per cent and 10 per cent of reconstituted lemon oil. In fact, such mixtures presented values of the ratio $(-)/(+)$-limonene outside the 99 per cent confidence limit allowed for genuine oils. The authors underlined that the method is easy and quick, but, since no pre-separation of the oil occurred prior to GC analysis, the possible occurrence of peak overlapping between the limonene enantiomers with those of other components of lemon oil has to be checked. In particular, the overlapping with $(-)$-limonene.

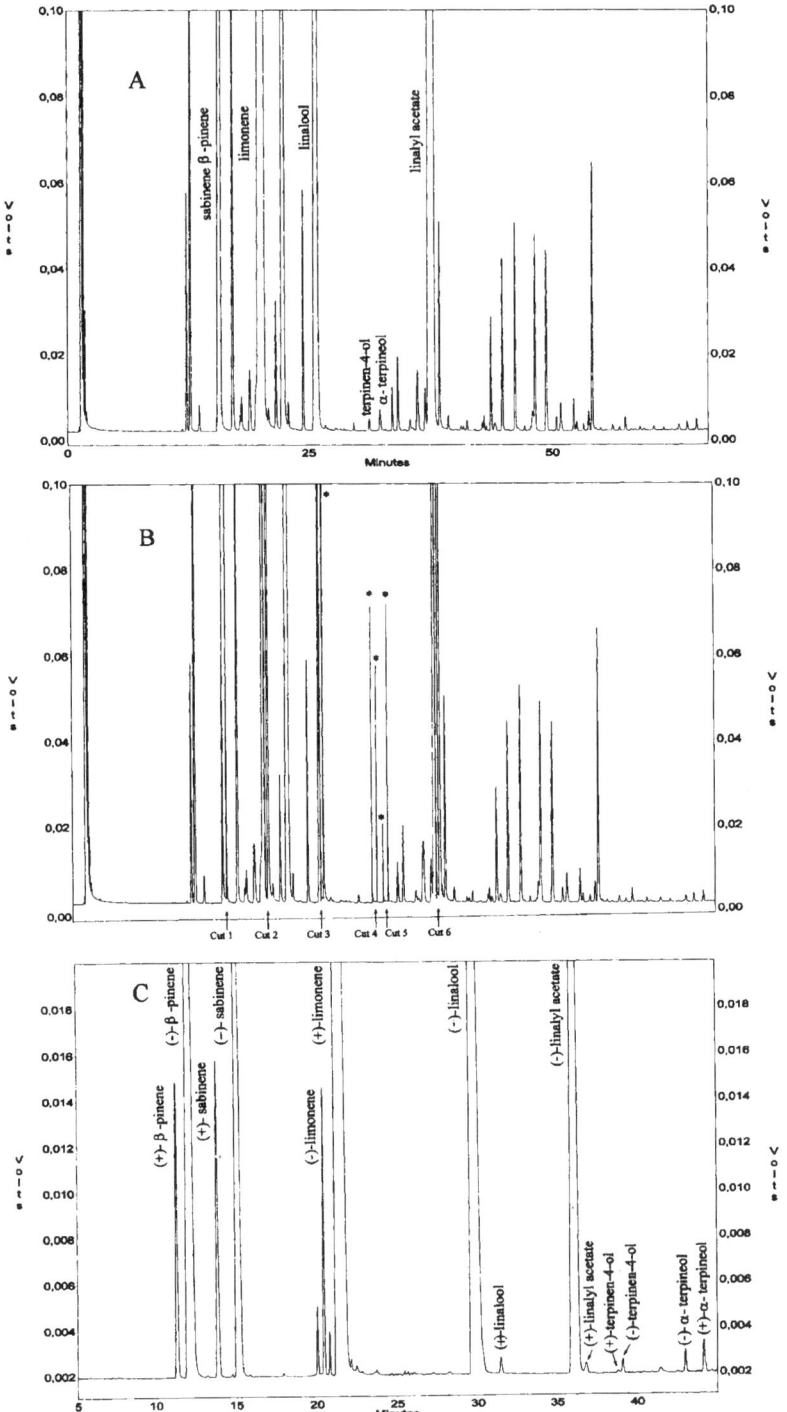

Figure 20.2 GC chromatogram of a cold-pressed bergamot oil obtained with (A) the SE-52 column, and (B) the SE-52 column with the six heart-cuts, and (C) GC-GC chiral chromatogram of the transferred components. Reproduced with permission from Mondello *et al.* (1998d).

Table 20.1 Enantiomeric distribution of some components of lemon oils reported in the literature

		1,2,3	4		5	6,7	8		9	10	11	12	13	
		c.p.	a	b	c.p.	c.p.	c.p.	a	c	d	e	e	c.p.	a
α-pinene	1R,5R(+)	33–38	28–34	23–33	30									
	1S,5S(−)	67–62	72–66	77–67	70						33–35			
β-pinene	1R,5R(+)	5–6	5–6	4–6	5						67–65			
	1S,5S(−)	95–94	95–94	96–94	95						tr–7			
sabinene	1R,5R(+)										100–93			
	1S,5S(−)													
limonene	4S(−)	1	1–2	1–2	2	1.5–2.1					0.5–5			
	4R(+)	99	99–98	99–98	98	98.5–97.9					99.5–95			
linalool	3R(−)						51–58	51–56	56.8–57.3			32–46	54	53
	3S(+)						49–42	49–44	43.2–42.7			68–54	46	47
terpinen-4-ol	4S(+)						18–22	27–31	38.4				20	28
	4R(−)						82–78	73–79	61.6				80	72
α-terpineol	4S(−)								75.8–76.3	67				
	4R(+)								24.2–23.7	33				

		14			15	16	17	18,19	18,19	18,19	18,19
		f	f	f	c.p.	c.p.	c.p.	c.p.	g	a	
α-pinene	1R,5R(+)										
	1S,5S(−)										
β-pinene	1R,5R(+)	95.1	0.7	11.2	5.8–6.4	5.0–7.0	6.3	4.2–7.0	3.5–7.8	6.4–6.6	
	1S,5S(−)	4.9	99.3	88.8	94.2–93.6	95.0–93.0	93.7	95.8–93.0	96.5–92.2	93.6–93.4	
sabinene	1R,5R(+)	73.2	21.2	23.4	14.2–15.5	13.3–15.3	14.9	12.5–15.3	16.9–14.0	12.7–14.6	
	1S,5S(−)	26.8	78.8	76.6	85.8–84.5	86.7–84.7	85.1	87.5–84.7	83.1–6.0	87.3–85.4	
limonene	4S(−)	1.5	0.7	1.1	1.5–1.6	1.7–1.9	1.6	1.5–2.0	5.0–14.5	1.7	
	4R(+)	98.5	99.3	98.9	98.5–98.4	98.3–98.1	98.4	98.5–98.0	95.0–85.5	98.3	
linalool	3R(−)	65.7	50.4	52.4	49.5–56.8	57.4–68.7	58.0	50.8–71.5	5.7–23.3	53.1–60.0	

Table 20.1 (Continued)

		14			15	16	17	18, 19	18, 19	18, 19
		f	f	f	c.p.	c.p.	c.p.	c.p.	g	a
terpinen-4-ol	3S(+)	34.3	49.6	47.6	50.5–43.2	42.6–31.3	42.0	49.2–28.5	94.3–76.7	46.9–40.0
	4S(+)	11.2	70.5	30.1	28.4–32.5	15.9–26.7	24.7	13.7–26.9	28.9–75.4	28.4–28.5
	4R(−)	88.8	29.5	69.9	71.6–67.5	84.1–73.3	75.3	86.3–73.1	71.1–24.6	71.6–71.5
α-terpineol	4S(−)	75.0	44.7	60.3	74.5–79.0	73.5–80.9	75.2	64.2–82.0	8.8–67.1	76.4–77.0
	4R(+)	25.0	55.3	39.7	25.5–21.0	26.5–19.1	24.8	35.8–18.0	91.2–32.9	23.6–23.0

Notes

1 Mosandl et al. (1990); 2 Hener et al. (1990a); 3 Mosandl (1995); 4 Kreis et al. (1991); 5 Rocca et al. (1992); 6 Dugo et al. (1992a); 7 Dugo et al. (1993); 8 Dugo et al. (1994a); 9 Bicchi et al. (1994); 10 Ravid et al. (1995); 11 Casabianca et al. (1995); 12 Casabianca and Graff (1996); 13 Mondello et al. (1996); 14 Starrantino et al. (1997); 15 Dellacassa et al. (1997a); 16 Mondello et al. (1997); 17 Mondello et al. (1998a); 18 Mondello et al. (1999); 19 Dugo et al. (2001); c.p. cold-pressed; a distilled; b solvent extracted; c results obtained for the same sample using three different stationary phases; d concentrated; e supercritical fluid extracted; f new hybrids; commercial oils.

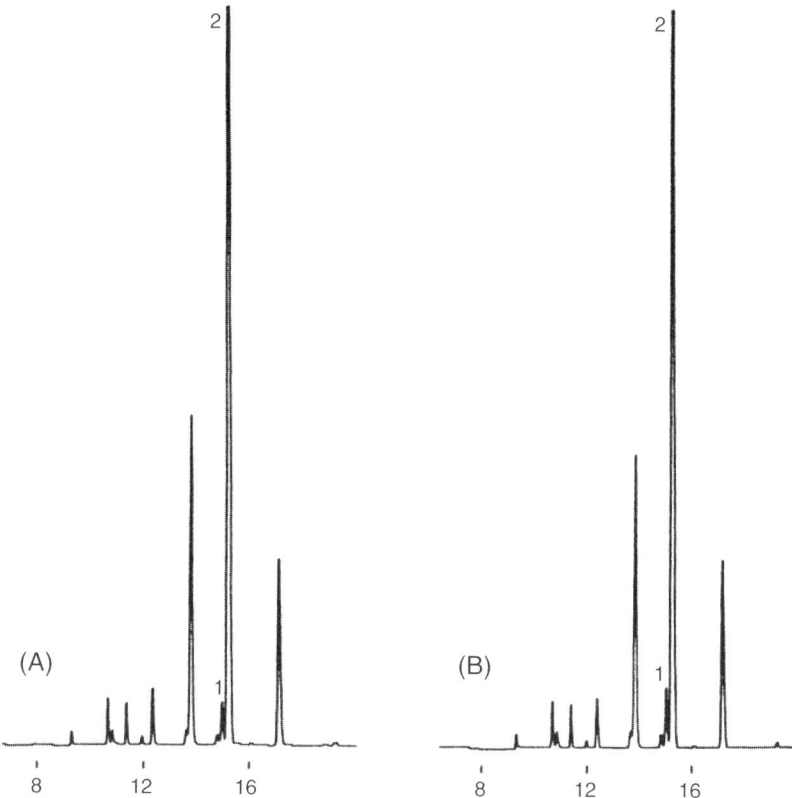

Figure 20.3 GC chromatogram of a mixture of cold-pressed lemon oil with (A) 5 per cent of reconstituted lemon oil and (B) 10 per cent of reconstituted lemon oil. (1) (−)-limonene; (2) (+)-limonene. Reproduced with permission from Dugo *et al.* (1993).

Obviously, the direct GC analysis becomes very difficult for components present in small quantities. Dugo *et al.* (1994a) and Mondello *et al.* (1996) used a fully automated HPLC-HRGC system with an on-column type interface and partially concurrent solvent evaporation for the stereodifferentiation of monoterpene alcohols in cold-pressed and distilled lemon oils. Linalool represents 0.1 per cent of the volatile fraction of cold-pressed oil, and 0.19 per cent of distilled oil; terpinen-4-ol, 0.02 per cent of cold-pressed lemon oils, and 0.35 per cent of distilled oil. The fraction containing linalool and terpinen-4-ol was pre-separated by HPLC, as shown in Figure 20.4, and analysed by GC for the determination of the enantiomeric distribution of these alcohols without interferences. Figure 20.5 shows the enantiomeric distribution of linalool and terpinen-4-ol in cold-pressed and distilled lemon oils obtained by HPLC-HRGC. As can be seen from results reported in Table 20.1 the enantiomeric ratio of linalool is approximately the same in both type of oils, while the enantiomeric distribution of terpinen-4-ol is different. This is due to the technology used to obtain the samples. In this way, it is possible to differentiate the more valuable cold-pressed lemon oils from the distilled ones.

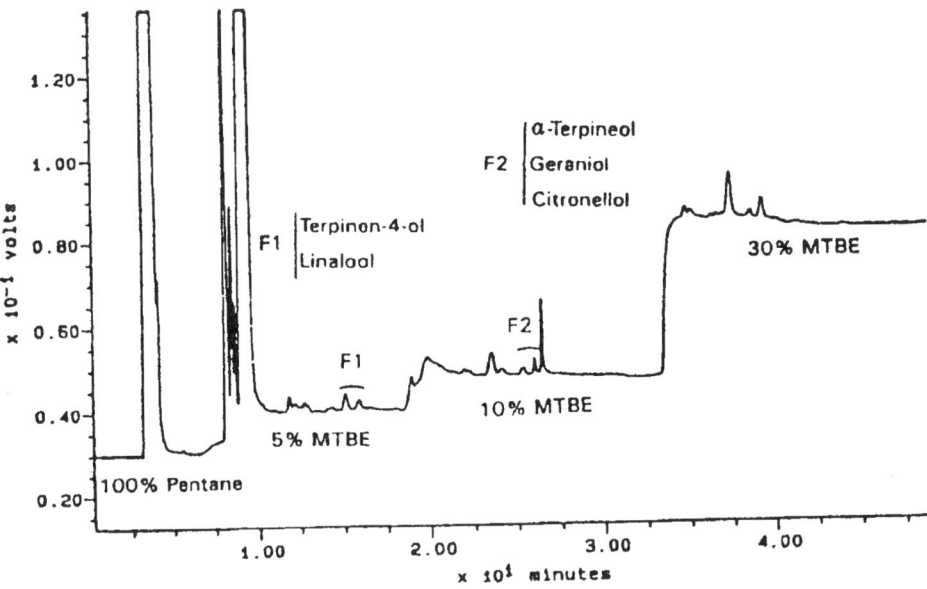

Figure 20.4 HPLC chromatogram of a lemon oil. Reproduced with permission from Dugo *et al.* (1994a).

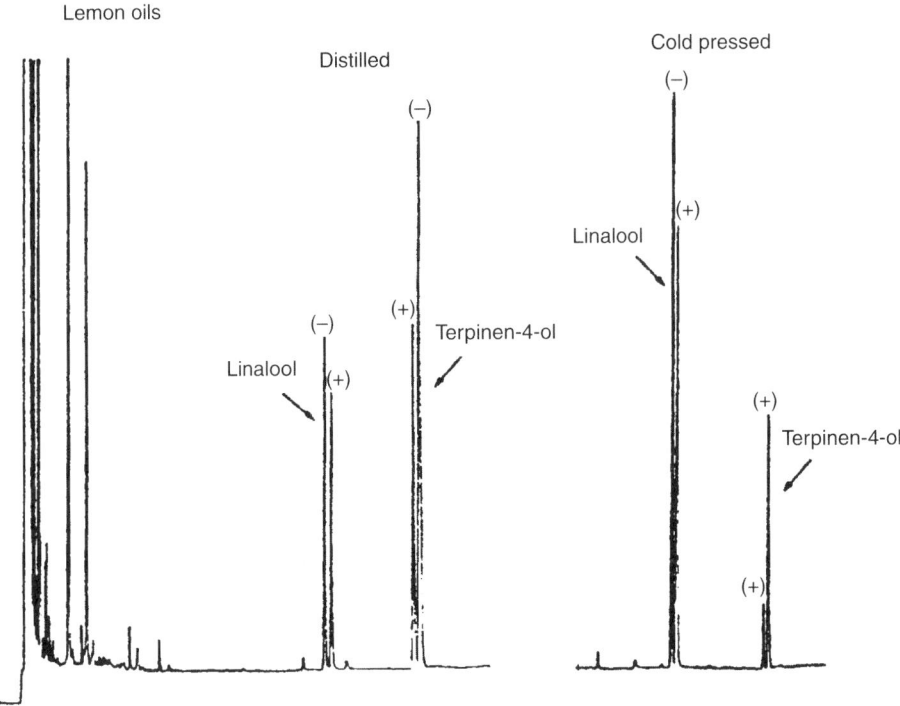

Figure 20.5 Enantiomeric distribution of linalool and terpinen-4-ol for lemon oil obtained by HPLC-HRGC. Reproduced with permission from Dugo *et al.* (1994a).

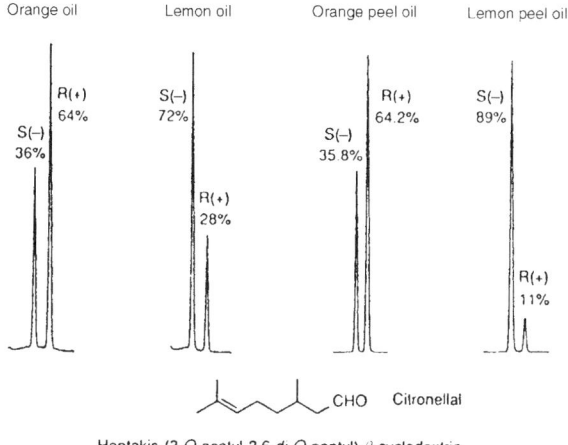

Figure 20.6 Stereodifferentiation of natural citronellal in orange and lemon samples. Reproduced with permission from Werkhoff *et al.* (1993).

Bicchi *et al.* (1994) determined the enantiomeric distribution of linalool, terpinen-4-ol and α-terpineol by GC analysis, using a chiral column of modified cyclodextrin (2,6-di-O-methyl-3-O-pentyl-β-cyclodextrin) diluted in different polysiloxane phases of different polarity (PS-347.5, PS-086 and OV-1701). As can be seen, results obtained for linalool and terpinen-4-ol are sensibly different from those of Dugo *et al.* (1994a) and Mondello *et al.* (1996), mainly those for terpinen-4-ol. These differences may be due to the analytical technique used. In fact, direct GC analysis of components present in low amount, like terpinen-4-ol, may give problems of peak overlappings. However, the aim of the work of Bicchi *et al.* was to demonstrate that it is possible to obtain different interactions and as a consequence different retention times of components, simply changing the polarity of the diluting phase. Werkhoff *et al.* (1993) presented for the first time data on the enantiomeric distribution of citronellal in lemon oil. Figure 20.6 reports the chromatograms obtained for lemon and orange oil samples. The samples analysed demonstrated an enantiomeric excess of (−)-(S)-citronellal (ratio between 3S(−) and 3R(+) was 72–89:28–11). In the same paper the authors reported also data on the enantiomeric distribution of methyl jasmonate and methyl epijasmonate in lemon peel concentrate. They found the following values:

(−)-(1R,2R)/(+)-(1S,2S) methyl jasmonate: >99.0/<1.0
(+)-(1R,2S)/(−)-(1S,2R) methyl epijasmonate: >99.0/<1.0.

Ravid *et al.* (1995) determined the enantiomeric distribution of α-terpineol in a commercial sample of lemon oil concentrate, from Israel, by GC analysis using a modified β-cyclodextrin. Casabianca *et al.* (1995) and Casabianca and Graff (1996) analysed lemon oils obtained by SFE (supercritical fluid extraction). In the first paper, the authors reported the enantiomeric distribution of α-pinene, β-pinene and limonene, while in 1996 the enantiomeric distribution of linalool was studied.

Mondello *et al.* (1997; 1998a) developed a MDGC system, double oven, fully automated, that allowed multitransfers during the same analysis. Using a SE-52 pre-column and a modified β-cyclodextrin main column (diethyl, ter-butyl-β-cyclodextrin), they separated sabinene, β-pinene, limonene, terpinen-4-ol, linalool, and α-terpineol in lemon essential oil. Limonene was only partially transferred, due to its high concentration, while the other components were totally transferred from the pre-column to the main column. With the same MDGC system, enantiomeric distribution of these six components has been determined for Uruguayan lemon oils (Dellacassa et al., 1997a) and for three laboratory extracted lemon oils from new lemon hybrids (Starrantino et al., 1997). In the following years, Mondello et al. (1999) and Dugo et al. (2000) presented a slightly modified version of the MDGC system, and analysed with this system samples of lemon oil from a whole productive season. Figure 20.7 shows the variation of the enantiomeric excess of limonene, β-pinene, sabinene, terpinen-4-ol, α-terpineol and linalool in lemon oils during the productive season. Limonene, sabinene and β-pinene showed values of the enantiomeric ratio practically constant all around the year. Values of the enantiomeric distribution of terpinen-4-ol and α-terpineol during the production season showed similar values and trends. The ratio (−)/(+)-linalool varied over a wide range. At the beginning of the season, the enantiomeric excess of (−)-linalool is about 33; this value decreases from October to May, and then increases again reaching in September values comparable to those of the beginning of the season. Mondello et al. (1999) and Dugo et al. (2000) also analysed distilled lemon oils and commercial oils of unknown origin. As can be seen, distilled oils show very similar values to cold-pressed oil, with the exception of terpinen-4-ol that tend to racemise. This behaviour has been already observed in the past (Dugo et al., 1994a; Mondello et al., 1996). For the commercial samples, only β-pinene shows values similar to those of genuine oils. The other components show very different enantiomeric distributions, and the ranges were so large that the authors were not able to detect the kind of products used for the reconstitution of the oils.

Comparing results reported in Table 20.1, it is possible to see that all the data related to monoterpene hydrocarbon α-pinene, are in agreement. Also values reported for the other monoterpene hydrocarbons are generally in a very limited range, but values reported for β-pinene by Casabianca et al. (1995) extend beyond the range. As previously noted by Mondello et al. (1999), the value of 100 per cent reported for (−)-β-pinene by Casabianca et al. (1995) seemed inexplicable. This difference may be due to the technology used for the extraction of the oil. Again, values reported by Casabianca et al. (1995) for limonene differ from the others. In particular, the lowest value of the ratio (−)/(+)-limonene could be due to the presence of orange oil in the lemon oil analysed (Mondello et al., 1997) while the highest value could be due to the presence of reconstituted oils (Dugo et al., 1993). Regarding to monoterpene alcohols, values reported for the enantiomeric distribution of linalool in cold-pressed and distilled lemon oils are in the same range, as well as values for α-terpineol. Terpinen-4-ol, in the distilled samples, shows a slight tendency to racemise. For samples of lemon oil obtained from hybrids, the enantiomeric ratio of the six components analysed show values which are almost always very different for the three oils, and different again from those of industrial lemon oils.

MANDARIN OIL

Table 20.2 presents data reported in the literature on the enantiomeric distribution of some components of mandarin oil. As can be seen, the enantiomeric distribution of some monoterpene hydrocarbons have been studied by Mosandl's research group

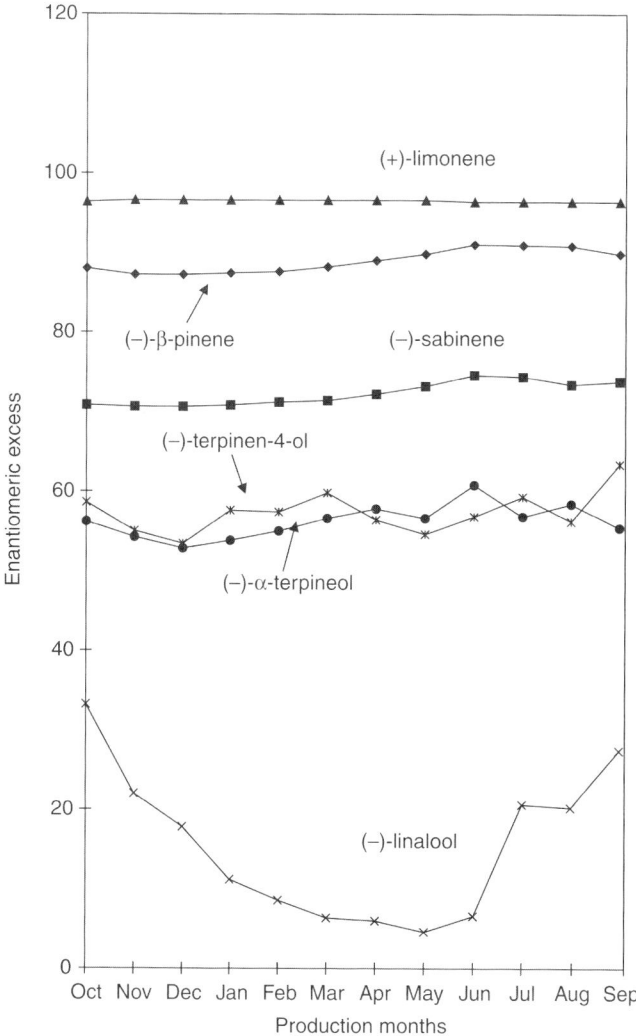

Figure 20.7 Variation of the enantiomeric excess of 4R-(+)-limonene, 1S,5S-(−)-β-pinene, 1S,5S-(−)-sabinene, 4R-(−)-terpinen-4-ol, 4S-(−)-α-terpineol and 3R-(−)-linalool in lemon oils during the productive season. Reproduced with permission from Mondello *et al.* (1999).

(Mosandl *et al.*, 1990; Hener *et al.*, 1990a; Mosandl, 1995; Kreis *et al.*, 1991), Dugo *et al.* (1992b, 1993), Rocca *et al.* (1992) and Casabianca *et al.* (1995). As can be seen observing the results, data are not in good agreement for all the components with the exception of limonene. These differences may be due to several factors: many authors report results of a research carried out on only one sample or a very limited number of samples. These samples can be of unknown origin, and data are not related to the extraction technology or to the harvest time. Moreover, different analytical procedures have been applied. In the case of the chiral analysis of monoterpene hydrocarbons in mandarin oil, Dugo *et al.* (1992b, 1993) used a GC system to stereodifferentiate limonene. With a simple and

Table 20.2 Enantiomeric distribution of some components of mandarin oils reported in the literature

		1, 2, 3	4			5	6, 7	8	9		10	11		
		c.p.	c.p.	a	b	c.p.	c.p.	c.p.	c.p.	b	c	c.p.	b	c
α-pinene	1R,5R(+)	57–68	45–100	66–68	67–60	52								
	1S,5S(−)	43–32	55–tr	34–32	33–30	48								
β-pinene	1R,5R(+)	84–97	37–1	84	88–76	96								
	1S,5S(−)	16–3	63–99	16	12–24	4								
sabinene	1R,5R(+)													
	1S,5S(−)													
limonene	4S(−)	tr–1	tr–1	<1	<1–tr	2	1.8–2.6							
	4R(+)	100–99	100–99	>99	>99–100	98	98.2–97.4							
linalool	3R(−)							16.1–16.8	16–18	16–20	12–30	17	17	12–30
	3S(+)							83.9–93.2	84–82	84–80	88–70	83	83	88–70
terpinen-4-ol	4S(+)							11.3–11.9	12–14	25–30		13	27	
	4R(−)							88.7–88.1	88–86	75–70		87	73	
α-terpineol	4S(−)							74.8–76.7						
	4R(+)							25.2–23.3						

		12	13	14	15, 16		
		c.p.	c.p.	c.p.	c.p.	d	e
α-pinene	1R,5R(+)	68–96					
	1S,5S(−)	32–4					
β-pinene	1R,5R(+)	76–78		97.0–98.8	97.0–98.8	29.3–94.6	98.3–98.8
	1S,5S(−)	24–22		3.0–1.2	3.0–1.2	70.7–5.4	1.7–1.2
sabinene	1R,5R(+)			76.2–80.5	76.2–83.4	37.6–96.6	79.9–83.4
	1S,5S(−)			23.8–19.5	23.8–16.6	62.4–3.4	20.1–16.6
limonene	4S(−)	tr–0.5		2.0–2.3	1.5–2.3	0.7–5.0	1.5–1.9
	4R(+)	100–99.5		98.0–97.7	98.5–97.7	99.3–95.0	98.5–98.1

linalool	3R(−)	4.0–6.5	13.1–19.8	12.7–19.8	16.3–20.4	3.5–26.5	12.7–19.3
	3S(+)	96.0–93.5	86.9–80.2	87.3–80.2	83.7–79.6	96.5–73.5	87.3–80.7
terpinen-4-ol	4S(+)		10.0–19.2	10.0–19.2	25.3–28.7	12.4–38.9	10.5–17.5
	4R(−)		90.0–81.8	90.0–81.8	74.7–71.3	87.6–61.1	89.5–82.5
α-terpineol	4S(−)		69.6–76.8	67.8–76.8	61.2–73.4	24.2–57.4	67.8–75.1
	4R(+)		30.4–23.2	32.2–23.2	38.8–26.6	75.8–42.6	32.2–24.9

Notes

1 Mosandl et al. (1990); 2 Hener et al. (1990a); 3 Mosandl (1995); 4 Kreis et al. (1991); 5 Rocca et al. (1992); 6 Dugo et al. (1992b); 7 Dugo et al. (1993); 8 Bicchi et al. (1994); 9 Dugo et al. (1994a); 10 Dugo et al. (1994b); 11 Mondello et al. (1996); 12 Casabianca et al. (1995); 13 Casabianca and Graff (1996); 14 Mondello et al. (1997); 15 Mondello et al. (1998b); 16 Dugo et al. (2001).

c.p. cold-pressed oils; a solvent extracted oils; b distilled oils; c Uruguayan oils; d commercial oils; e laboratory-extracted oils.

rapid procedure analogous to those already reported for lemon oil (Dugo et al., 1992a), the authors were able to detect presence of reconstituted mandarin oil (as low as 5 per cent) in cold-pressed genuine mandarin oil.

The distribution of monoterpene alcohols was studied by Bicchi et al. (1994), Dugo et al. (1994a,b), Mondello et al. (1996), Casabianca and Graff (1996). Bicchi et al. (1994) used a GC system to determine the enantiomeric distribution of linalool, terpinen-4-ol and α-terpineol. They carried out the analysis using the same cyclodextrin derivative and different diluting phases. In this way it is possible to produce sufficient variations in the retention indices to identify successfully each enantiomer of the components under investigation in the total chromatogram. The authors pointed out also that retention times differences are more significant than those obtained using different CD derivatives and the same diluting phase. Moreover, using the same CD, no inversion of the elution order of enantiomers is obtained. Dugo et al. (1994a,b) and Mondello et al. (1996) used HPLC-HRGC system to determine linalool and terpinen-4-ol in cold-pressed Italian and Uraguayan mandarin oils and in distilled oils. Results obtained by Bicchi et al. (1994), Dugo et al. (1994a,b) and Mondello et al. (1996) for linalool are in good agreement. Values for terpinen-4-ol are still in good agreement for cold-pressed oils, while distilled oils show very different values. Distilled oils are obtained from the acid aqueous residue of cold extraction, under conditions of temperature and pH that cause the hydration of monoterpene hydrocarbons and the consequent increase of alcohols, especially terpinen-4-ol and α-terpineol. Since this reaction is not stereospecific, the ratio of terpinen-4-ol goes towards the racemate, as can be seen comparing results obtained for these oils with those of cold-pressed oils. Casabianca and Graff (1996) used an MDGC system to stereodifferentiate linalool in French and Spanish mandarin oils obtained by SFE. Results obtained are significantly different from the others found in literature. As observed by Mondello et al. (1998b), these data seem to be closer to data reported for clementine, tangerine or sweet orange oil than to those of mandarin oil.

Mondello et al. (1997, 1998b) and Dugo et al. (2000) determined during the same analysis both monoterpene hydrocarbons and monoterpene alcohols, using a home-made MDGC system which allowed multitransfers during the same run. Figure 20.8 reports the chromatogram of a cold-pressed mandarin oil obtained with the SE-52 precolumn and the system in stand-by position, the chromatogram of the same oil obtained with the SE-52 column and the system in the cut position and the chromatogram obtained with the chiral column from the fraction transferred from the SE-52 precolumn. They carried out an extensive research on a large number of genuine, cold-pressed oils produced during a whole productive season with the most commonly used extraction techniques. Moreover, Figure 20.9 shows the variation of the enantiomeric excesses of the components under investigation plotted as a function of the month of production of the oils. Laboratory extracted from two varieties (Avana Comune and Tardivo di Ciaculli), distilled and commercial oils of unknown origin have been also analysed. The authors found that laboratory extracted oils of the two varieties showed similar values of enantiomeric distribution and therefore industrial samples' values are not affected by the proportion of each variety. In fact, data reported for cold-pressed industrial samples and laboratory extracted oils are in perfect agreement. Again, distilled mandarin oils present values similar to those of cold-pressed oil, with the exception of terpinen-4-ol. This behaviour is in accordance with that observed by Dugo et al. (1994b). Commercial oils (11 samples) presented values out of ranges for genuine mandarin oils. Mondello et al. (1998b) also analysed mixture of lemon, clementine and sweet orange oils with

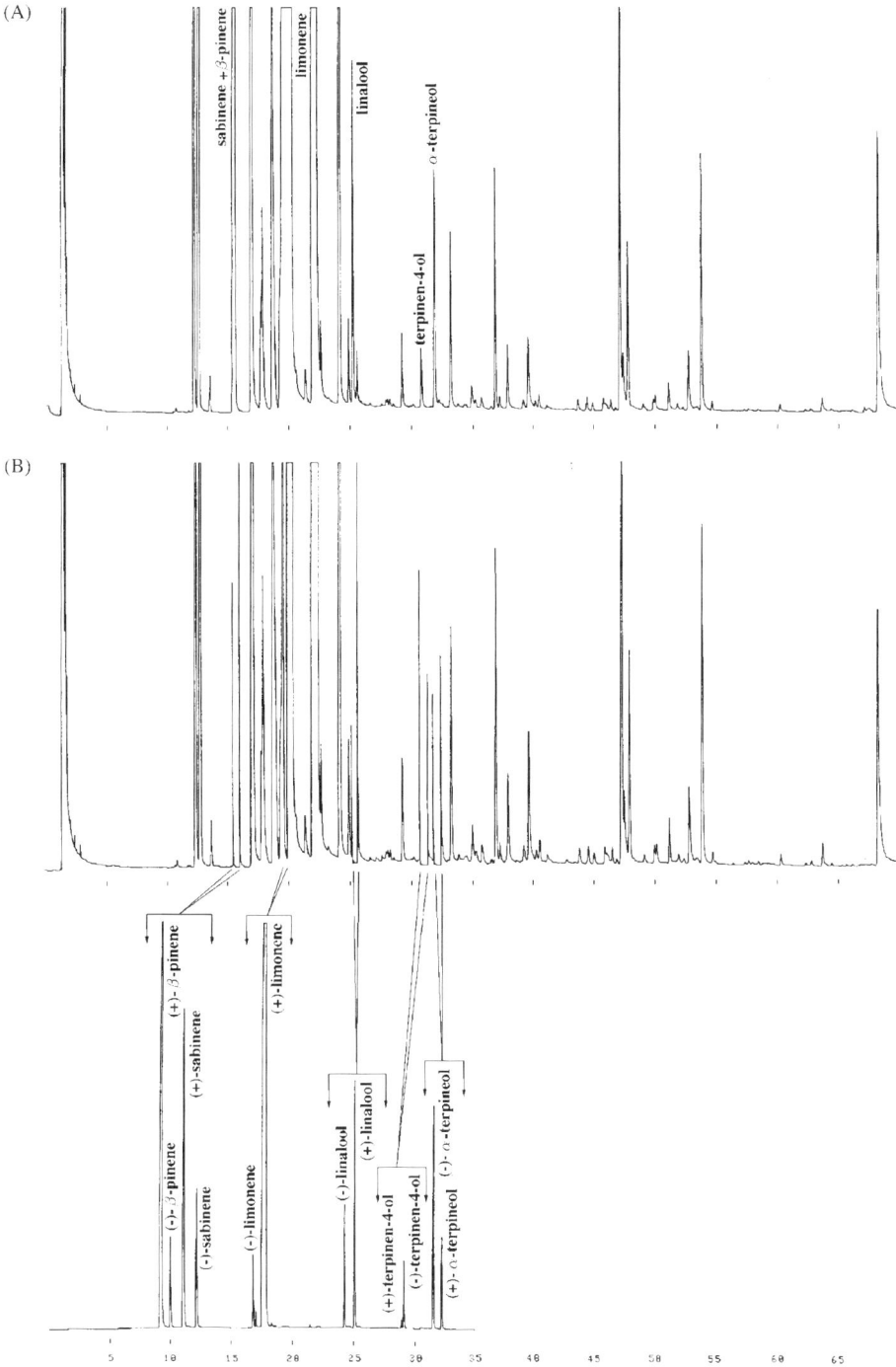

Figure 20.8 GC chromatogram of a cold-pressed mandarin oil obtained with (A) the SE-52 column, and (B) the SE-52 column with the five heart-cuts, and (C) GC-GC chiral chromatogram of the transferred components. Reproduced with permission from Mondello *et al.* (1998b).

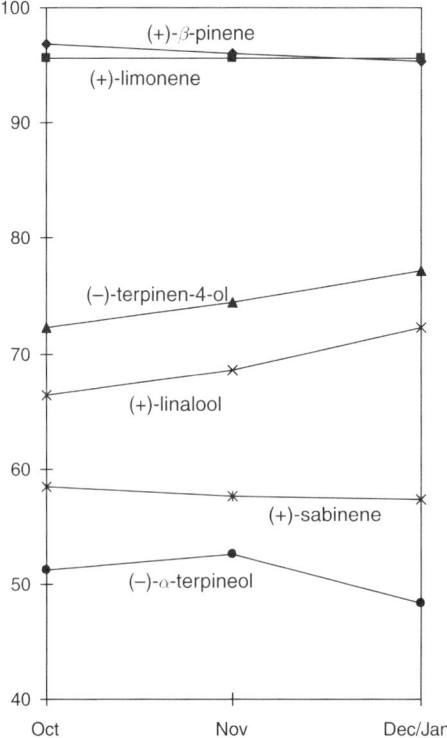

Figure 20.9 Variation of the enantiomeric excess of 1R,5R-(+)-β-pinene, 4R-(+)-limonene, 4R-(−)-terpinen-4-ol, 3S-(+)-linalool, 1R,5R-(+)-sabinene, and 4S-(−)-α-terpineol in mandarin oils during the productive season. Reproduced with permission from Mondello *et al.* (1998b).

mandarin oils. The addition of sweet orange terpenes is the most common adulteration of mandarin oils, while the presence of lemon or clementine oils is the most common cause of contamination. The results are reported in Table 20.3. The presence of lemon and clementine may be detected by observing the enantiomeric ratio of β-pinene, while the presence of sweet orange terpenes by observing the enantiomeric ratio of sabinene.

SWEET ORANGE OIL

Table 20.4 summarises data of the literature on the enantiomeric distribution of some components of sweet orange oil. Some papers report the stereodifferentiation of α-pinene, β-pinene and limonene (Mosandl *et al.*, 1990; Hener *et al.*, 1990a; Kreis *et al.*, 1991; Rocca *et al.*, 1992; Casabianca *et al.*, 1995; Mosandl, 1995). Data obtained for the enantiomeric distribution of α-pinene are practically coincident. Only Casabianca *et al.* (1995) found a slightly wide range for this compound. Data for β-pinene and limonene are in perfect agreement. It should be noted that the ratio of 54:46 reported by Mosandl's group for (+)/(−)-β-pinene was found for commercial oils, while a value of

Table 20.3 Enantiomeric distribution of some components of cold-pressed lemon, sweet orange and clementine oils and of their mixtures with cold-pressed mandarin oil (Mondello et al., 1998b)

		Lemon	Mandarin (90%) Lemon (10%)	Mandarin (99%) Lemon (1%)	Sweet orange	Mandarin (80%) Sweet orange (20%)	Clementine	Mandarin (90%) Clementine (10%)	Industrial and laboratory-extracted cold-pressed mandarin oils Range
β-pinene	1R,5R(+)	5.1	50.2	90.3	42.2	97.7	22.9	95.8	97.0–98.8
	1S,5S(−)	94.9	49.8	9.7	57.8	2.3	77.1	4.2	3.0–1.2
sabinene	1R,5R(+)	15.1	45.6	75.0	96.8	85.8	90.1	82.5	76.2–83.4
	1S,5S(−)	84.9	54.4	25.0	3.2	14.2	9.9	17.5	23.8–16.6
limonene	4S(+)	1.9	2.2	2.0	0.6	1.7	0.7	1.9	1.5–2.3
	4R(+)	98.1	97.8	98.0	99.4	98.3	99.3	98.1	98.5–97.7
linalool	3R(−)	71.5	26.9	19.3	4.9	12.4	7.5	17.1	12.7–19.8
	3S(+)	28.5	73.1	80.7	95.1	87.6	92.5	82.9	87.3–80.2
terpinen-4-ol	4S(+)	19.7	15.7	14.4	64.9	15.0	46.4	15.1	10.0–19.2
	4R(−)	80.3	84.3	85.6	35.1	85.0	53.6	84.9	90.0–81.8
α-terpineol	4S(−)	77.4	74.4	73.5	8.2	69.2	13.1	71.8	67.8–76.8
	4R(+)	22.6	25.6	26.5	91.8	30.8	86.9	28.2	32.2–23.2

73:27 was determined for peel laboratory extracted oils. Mondello et al. (1997) and Dugo et al. (2000) also reported values for the enantiomeric distribution of limonene, that are still in accordance with results previously cited. Moreover, they reported values for the enantiomeric distribution of sabinene; no other data on the enantiomeric distribution of sabinene in sweet orange oils are available up to now. However, results reported by Mondello et al. (1997) and Dugo et al. (2000) refer to twelve samples of cold-pressed genuine sweet orange oil. The method proposed by the authors allowed the determination, together with limonene and sabinene, of the enantiomeric distribution of β-pinene, linalool, terpinen-4-ol and α-terpineol. Results obtained for these components are not reported since they varied within a very wide range, therefore a larger number of samples should be analysed to obtain enough data for a good interpretation. These large variations may be due to numerous varieties cultivated and used for the industrial processing of blond and blood sweet orange fruits. Since the volatile composition of sweet orange oil is influenced by the different varieties of the fruit processes, it is possible to foresee a similar behaviour for the enantiomeric distribution of its components. Takeoka et al. (1990) and König et al. (1994) reported values of the enantiomeric distribution of α-copaene. This sesquiterpene hydrocarbon is reported to be an effective attractant for the male Mediterranean fruit fly. (+)-enantiomer seems to be more effective than the (−)-enantiomer. Sweet orange oil shows an abundance of levorotatory form. König et al. (1994) also reported results on the stereodifferentiation of an other sesquiterpene hydrocarbon, δ-cadinene. No other papers report results on this compound. Also the enantiomeric distributions of carvone (Konig et al., 1990) and citronellal (Werkhoff et al., 1993) have been reported in only one work.

Linalool is an important fragrance component found in many essential oils, spice and herbs. It has been found that the two enantiomers strongly differ in odour: (−)-S-linalool has the fragrance of petitgrain oil, while (+)-R-linalool is characterised by a woody fragrance and a typical lavander note (Werkhoff et al., 1993). Numerous studies reported results on the enantiomeric distribution of linalool (Bernreuther and Schreier, 1991; Werkhoff et al., 1993; Dugo et al., 1994b; Wang et al., 1995; Mosandl, 1995; Mondello et al., 1996; Casabianca and Graff, 1996). Berneuther and Schreier (1991), Werkhoff et al., (1993), Wang et al. (1995), Mosandl (1995), Casabianca and Graff (1996) used MDGC technique to determine the enantiomeric distribution of linalool in orange oil. Berneuther and Schreier (1991) used an MS detector. Once again, only values reported by Casabianca and Graff (1996) are sensibly different from the others. They analysed orange oils from Morocco and Spain, respectively obtained by steam distillation and SFE.

Dugo et al. (1994b) and Mondello et al. (1996) determined the enantiomeric ratio of linalool for sweet orange oils of different cultivars by using a fully automated on-line HPLC-HRGC system. They observed that the linalool enantiomeric distribution depends on the cultivar of the fruits. In Italian oils, (−)-linalool/(+)-linalool ratio is generally higher in blond oils, except for the blond oils from cv. Ovale, which shows values similar to the red oils.

BITTER ORANGE OIL

Data on the enantiomeric distribution of some components of bitter orange oil are poor in comparison to the other citrus essential oils. Table 20.5 summarises these results.

Table 20.4 Enantiomeric ratios of some components of cold-pressed sweet orange oils reported in the literature

	1, 2	3	4	5	6	7	8	9	10	11	12	13	14	15	16, 17
α-pinene 1R,5R(+)	100				100	100					96–100	100			
1S,5S(−)	tr				tr	tr					4–tr	tr			
β-pinene 1R,5R(+)	54–73				62–73	61					76	54–73			
1S,5S(−)	46–27				38–27	39					24	46–27			
sabinene 1R,5R(+)															94.6–97.9
1S,5S(−)															5.4–2.1
limonene 4S(−)	tr				tr	0.5	1.1				tr–0.5	tr			0.6
4R(+)	100				100	99.5	98.9				100–99.5	100			99.4
linalool 4R(−)				4.1–5.3			5.6	4–11		3.4		<8	5–9	14–15	
4S(+)				95.9–94.7			94.4	96–89		96.6		>92	95–91	86–85	
α-copaene (+)		16							15						
(−)		84							85						
δ-cadinene (+)									90						
(−)									10						
carvone (+)			59.27												
(−)			40.73												
citronellal (−)							36								
(+)							64								

Notes

1 Mosandl et al. (1990); 2 Hener et al. (1990a); 3 Takeoka et al. (1990); 4 König et al. (1990); 5 Bernreuther and Schreier (1991); 6 Kreis et al. (1991); 7 Rocca et al. (1992); 8 Werkhoff et al. (1993); 9 Dugo et al. (1994b); 10 König et al. (1994); 11 Wang et al. (1995); 12 Casabianca et al. (1995); 13 Mosandl, 1995; 14 Mondello et al. (1996); 15 Casabianca and Graff (1996); 16 Mondello et al. (1997); 17 Dugo et al. (2001).

Table 20.5 Enantiomeric distribution of some components of cold-pressed bitter orange oils reported in the literature

		1, 2, 3	4	5 a	5 b	5 c	5 d	6	7
α-pinene	1R,5R(+)	92	92–95						
	1S,5S(−)	8	8–5						
β-pinene	1R,5R(+)	3	3–6						1.7–3.4
	1S,5S(−)	97	97–94						98.3–96.6
sabinene	1R,5R(+)								44.4–54.5
	1S,5S(−)								55.6–45.5
limonene	4S(−)	1	<1						0.6
	4R(+)	99	>99						99.4
linalool	3R(−)			83	80–82	67–68	66–67	83	80.4–87.5
	3S(+)			17	20–18	33–32	34–33	17	19.6–12.5
terpinen-4-ol	4S(+)								65.3–67.6
	4R(−)								34.7–32.4
α-terpineol	4S(−)								6.8–11.6
	4R(+)								93.2–88.4

Notes

a Italy; b Spain; c Brazil; d Ivory Coast; 1 Mosandl et al. (1990); 2 Hener et al. (1990a); 3 Mosandl (1995); 4 Kreis et al. (1991); 5 Dugo et al. (1994b); 6 Mondello et al. (1996); 7 Mondello et al. (1997).

Mosandl and his co-workers (Mosandl et al., 1990; Hener et al., 1990a; Kreis et al., 1991; Mosandl, 1995) in a research on the enantiomeric distribution of α-pinene, β-pinene and limonene in essential oils, reported data obtained for a commercial sample of bitter orange oil. Dugo et al. (1994b) and Mondello et al. (1996) determined the enantiomeric distribution of linalool in bitter orange oil using an HPLC-HRGC system. In particular, Dugo et al. (1994b) analysed a large number of genuine samples of different geographical origin (13 Italian oils, 2 Spanish oils, 2 Brazilian and 2 from Ivory Coast). Results reported in Table 20.5 show that the linalool enantiomeric ratio is similar for Italian and Spanish oils, and also for Brasilian and Ivory Coast oils, but considerable differences are evident between these two groups of bitter orange oils. Figure 20.10 reports the enantiomeric distribution of linalool for Italian bitter and sweet orange oils, and for mixtures of the two oils. The enantiomeric ratio of linalool in bitter orange oil [(−)/(+) 83:17] resulted very different from that shown in sweet orange oil [(−)/(+) 7:93]. Bitter orange oil is more valuable than sweet orange oil. Sometimes sweet orange oil can be found in bitter orange oil, either deliberately added, or due to the involuntary presence of fruits of sweet orange in bitter oranges industrially processed. The determination of the enantiomeric distribution of linalool can be useful to detect presence of sweet orange in bitter orange oil as low as 5 per cent.

Most recently, again Mondello et al. (1997) and Dugo et al. (2000) determined the enantiomeric distribution of three monoterpene hydrocarbons (β-pinene, sabinene and limonene) and three monoterpene alcohols (linalool, terpinen-4-ol and α-terpineol) in six samples of bitter orange oil. They used a MDGC system developed in their laboratory, based on the use of mechanical valves, to directly analyse during the same GC run the different fractions containing the compounds of interest. Results of the enantiomeric distribution of sabinene, terpinen-4-ol and α-terpineol were reported for

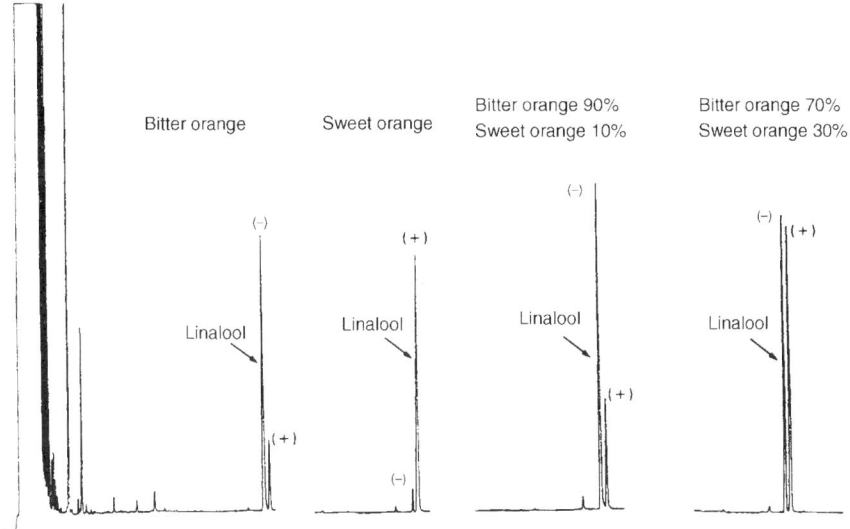

Figure 20.10 Enantiomeric distribution of linalool for Italian bitter and sweet orange oils, and for samples of bitter orange oil with 10% and 30% additions of sweet orange oil. Reproduced with permission from Dugo et al. (1994b).

the first time. Results obtained for linalool were in good agreement with those previously obtained by the same authors (Dugo *et al.*, 1994b; Mondello *et al.*, 1996) by HPLC-HRGC. Values found for β-pinene and limonene are in good agreement with the other results reported in literature, even thought Kreis *et al.* (1991) reported for β-pinene a slightly wider range.

LIME OILS

Lime essential oils industrially produced are of four different types: cold-pressed Key lime oils, type A; cold-pressed Key lime oils type B; cold-pressed Persian lime oils and distilled oils. Data on the enantiomeric distribution of some components of lime oils have been reported only in few papers. Results are listed in Table 20.6. Mosandl *et al.* (1990) and Hener *et al.* (1990a) analysed two samples of lime oils, one of which was self-prepared in laboratory. Moreover, Hener *et al.* (1990a,b) analysed some samples of lime oil of different geographical origin. They determined the enantiomeric distribution of α-pinene, β-pinene and limonene. Mondello *et al.* (1998c) and Dugo *et al.* (2000) analysed lime essential oils to determine the enantiomeric distribution of six components: β-pinene, sabinene, limonene, linalool, terpinen-4-ol and α-terpineol. In this case, they distinguished the type of lime oils analysed: 4 cold-pressed Key lime oils type A; 1 cold-pressed Key lime oil type B; 11 cold-pressed Persian lime oils and 6 distilled lime oils.

Observing data reported in Table 20.6, some conclusions can be made: values reported for α-pinene are in good agreement for all the samples, with the exception of lime oils from Peru and from Florida. As stated by the authors (Hener *et al.*, 1990a,b) the ratio of 57:43 in sample from Peru (probably a sample of known origin) may reflect a natural variation, while the ratio of 24:76 found in one of the samples of unknown origin may reflect a falsification. Unfortunately, no further data on the enantiomeric distribution of α-pinene in lime oil are available. Results obtained for the enantiomeric distribution of limonene determined in all the cold-pressed oils are practically identical. Values of β-pinene in Persian oils show a slightly higher amount of (+) isomer than Key lime oils. If we observe values found by Hener *et al.* (1990a,b) it is possible to see that samples coming from Brazil and from Florida can be Persian lime oils, while the others can be Key oils.

Cold-pressed Key lime oils type A and type B show practically identical values of the enantiomeric distribution of all the components analysed. Cold-pressed Persian lime oils, apart from β-pinene and limonene, show different values for the enantiomeric distribution of the other components: (+)-sabinene, (+)-linalool, (+)-α-terpineol and (−)-terpinene-4-ol in Persian lime oils represent a higher percentage of the total amount of the components than in Key lime oils. As pointed out by the authors (Mondello *et al.*, 1998c; Dugo *et al.*, 2000) the differences between Key and Persian cold-pressed lime oils can be due to the natural characteristics of the two oils, and not to the extraction technology. In fact, Key lime oils type A and type B, show similar values, although they were obtained with two different technologies, while Key lime oils type B and Persian lime oils show different values, although they were obtained using the same extraction technology. Data reported for distilled oils are very different from those shown by cold-pressed oils, with the exception of β-pinene. The other

Table 20.6 Enantiomeric distribution of some components of lime oils reported in the literature

		1, 2	2, 3					4, 5			
			a	b	c	d	e	f	g	b	i
α-pinene	1R,5R(+)	24–30	25	43	30–34	33	23–76				
	1S,5S(−)	76–70	75	57	70–66	67	77–24				
β-pinene	1R,5R(+)	3–10	4	4	8–10	8	2–5	3.4–3.5	3.5	9.1–10.3	3.2–4.0
	1S,5S(−)	97–90	96	96	92–90	92	98–95	96.6–96.5	96.5	90.9–89.7	96.8–96.0
sabinene	1R,5R(+)							15.1–15.2	15.3	18.2–23.4	–
	1S,5S(−)							84.9–84.8	84.7	81.8–76.6	–
limonene	4S(−)	2	6	3	2	2	1–7	2.9–2.6	1.8	0.4–2.7	5.5–8.7
	4R(+)	98	94	97	98	98	99–93	97.1–97.4	98.2	99.6–97.3	94.5–91.3
linalool	3R(−)							71.5–70.2	70.0	54.4–69.3	49.8–50.0
	3S(+)							28.5–29.8	30.0	45.6–30.7	50.2–50.0
terpinen-4-ol	4S(+)							29.2–29.5	29.5	18.6–24.9	42.3–45.0
	4R(−)							70.8–70.5	70.5	81.4–75.1	57.7–55.0
α-terpineol	4S(−)							85.5–84.0	82.8	74.5–80.8	53.3–56.8
	4R(+)							14.5–16.0	17.2	25.5–19.2	46.7–43.2

Notes
All the oils are cold-pressed if not otherwise specified; a Mexico; b Peru; c Brazil; d Florida; e unknown; f Key type A; g Key type B; b Persian; i Distilled; 1 Mosandl et al. (1990); 2 Hener et al. (1990a); 3 Hener et al. (1990b); 4 Mondello et al. (1998); 5 Dugo et al. (2001).

enantiomeric distribution tend to be racemic, and this behaviour is in agreement with the chemistry of the distilled lime oils, described in details by Clark and Chamblee (1992).

BERGAMOT OIL

Bergamot oil is a very valuable oil, widely used by cosmetic and food industries. The composition of the volatile fraction and of the non-volatile residue have been extensively studied. It has been observed that during the production season each component shows large quantitative differences. Therefore, these data are not always useful to determine the authenticity and the quality of a bergamot oil, as it is possible for other valuable oils such as lemon or mandarin. The determination of the enantiomeric distribution of some components of bergamot oil has been proven to provide useful information for authenticity and quality assessment.

Many data have been found in literature on the enantiomeric distribution of monoterpene hydrocarbons, monoterpene alcohols and linalyl acetate. These results are shown in Table 20.7.

Only Mosandl and his co-workers (Mosandl *et al.*, 1990; Hener *et al.*, 1990a; Mosandl, 1995; Juchelka and Mosandl, 1996) determined the enantiomeric distribution of α-pinene in bergamot oils. As can be seen, data obtained for bergamot oils produced with different extraction technologies (cold-pressing, SDE, distillation) are in good agreement. As for α-pinene, data reported for β-pinene do not present significant variations with the kind of bergamot oil analysed, while limonene shows values for genuine cold-pressed oils in accordance with those of distilled oils, but very different from those reported for commercial oils. Only Mondello *et al.* (1997, 1998d) determined the enantiomeric distribution of sabinene. Also in this case, values reported for cold-pressed and distilled oils are in agreement. Mondello *et al.* (1997, 1998d) analysed cold-pressed bergamot oils obtained during a whole productive season, and results obtained showed that values of enantiomeric ratios of β-pinene, limonene and sabinene are stable during the whole season. The results on the enantiomeric distribution of monoterpene hydrocarbons, although they do not vary during the production season, are not able to be used to determine the authenticity of a bergamot oil, since they are very similar to those of other citrus oils.

Most of the papers on the enantiomeric distribution of compounds of bergamot oil refer to linalool and linalyl acetate, that are the most abundant oxygenated compounds of this oil. Since 1996, the sterodifferentiation of these two compounds have been carried out with two separated analysis because a chiral column able to separate the isomers of both linalool and linalyl acetate was not available. In 1996, Juchelka and Mosandl presented for the first time an analysis system that allowed the simultaneous enantioselective analysis of linalool and linalyl acetate as shown in Figure 20.11. The column was a heptakis–(2,3-di-O-acetyl-6-O-tertbutyldimethylsilyl)-β-CD. Dellacassa *et al.* (1997b), Mondello *et al.* (1997, 1998d) and Dugo *et al.* (2000) used 2,3-diethyl-6-O-terbutylsylil-β-CD column to simultaneously analyse linalool and linalyl acetate. In particular, Mondello *et al.* (1997, 1998d) analysed also α-terpineol and terpinen-4-ol, in addition to limonene, β-pinene and sabinene. An example of this analysis is reported in Figure 20.2. Table 20.8 reports the columns used by the different authors for determination of the enantiomeric distribution of linalool and linalyl acetate in bergamot oil.

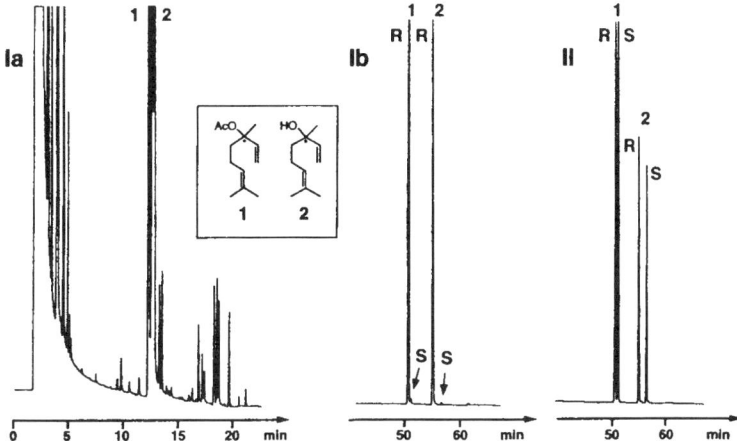

Figure 20.11 Chirospecific analysis of bergamot oil. Linalyl acetate: 1(R)-(−), (S)-(+), linalool: 2(R)-(−), (S)-(+) using MDGC system. Authentic bergamot oil: precolumn (Ia), main column (Ib); commercial bergamot oil: main column (II). Reproduced with permission from Juchelka and Mosandl (1996).

If we consider the results shown in Table 20.7, genuine cold-pressed bergamot oils present a percentage of S(+)-isomer, both for linalool and linalyl acetate, lower than 1 per cent of the total content of the two isomers. Laboratory solvent extracted oils show values for the enantiomeric distribution of linalool and linalyl acetate compatible with cold-pressed genuine oils. In commercial oils and in oils which were obtained by 'drastic' process, such as distillation, this percentage is more than 1 per cent. Therefore, an S-enantiomer content of linalool and linalyl acetate of more than 1 per cent allows the presence of extraneous products in bergamot oils to be detected. In particular, in distilled oils a higher amount of (S)-linalool is formed, but pure (R)-linalyl acetate was detected. Apparently, some linalyl acetate is hydrolised, but the configuration is stable even under such drastic conditions.

During the cold-extraction of bergamot oil, small amount of less valuable bergamot oil can be recovered by processing the residues of the cold-pressed oil extraction. These oils are named depending on the process used: 'Pulizia dischi', 'Torchiati', 'Ricicli', recovered by decantation, centrifugation or by pressing from the liquid and the solid residues and the recycle water at the end of the entire processing day. Distilled oils are recovered by distillation of the residues of the cold-pressing procedures. Previously, the distillation was carried out at atmospheric pressure, while, presently, it is carried out at reduced pressure. Moreover, the phototoxicity of bergapten has been the cause of the demand of cosmetic and perfumery industries of bergapten-free oils, which are treated to obtain an oil with a low content of bergapten. The two methods commonly used are the distillation or a treatment with NaOH.

The determination of the enantiomeric distribution of linalool and linalyl acetate in 'Torchiati', 'Ricicli' and 'Pulizia dischi' bergamot oils, as well as in bergapten-free oils (König *et al.*, 1997; Mondello *et al.*, 1998d) are in the ranges observed for cold-pressed oils. The system used for the recovery of bergamot oil from the residues of the

Table 20.7 Enantiomeric ratios of some components of bergamot oils reported in the literature

		1, 2, 11	3		4		5		6, 14	
		a	a	g	b	a	a	c	d	
α-pinene	R(+)	28								
	S(−)	72								
β-pinene	R(+)	6								
	S(−)	94								
limonene	S(−)	14								
	R(+)	86								
linalool	R(−)				99.5	68.5	55.8–100	74.3	>99.5	
	S(+)				0.5	31.5	44.2–0	25.7	<0.5	
linalyl acetate	S(+)		31–47	50						
	R(−)		69–53	50						

		7			8		9
		a	b	e	a	f	a
linalool	R(−)				0	2	
	S(+)				100	98	
linalyl acetate	R(−)	66	99	78	63	100	
	S(+)	34	1	22	37	0	
	S(+)	60	1	1	25	0–0.5	0
	R(−)	40	99	99	75	100–99.5	100

		10		12			13			15
		b	a	d	a	b	d	a	o	b, d, i, l, m, n, p
α-pinene	R(+)			28.9–32.7	25.4–34.0	30.2				
	S(−)			71.1–67.3	74.6–66.0	69.8				
β-pinene	R(+)			7.4–8.9	4.1–9.0	8.2				
	S(−)			92.6–91.1	95.9–91.0	91.8				
limonene	S(−)			2.0–2.6	1.8–15.0	1.5				
	R(+)			98.0–97.4	98.2–85.0	98.5				
linalool	R(−)			99.0–99.7	54.9–99.3	91.5	99.9–100	63.0–75.6	68.8	100
	S(+)			1.0–0.3	45.1–0.7	8.5	0.1–0	37.0–24.4	31.2	0
linalyl acetate	S(+)	0.5	2–25	<1.0	<1.0–48.1	<1.0	0–0.5	25.0–26.5	0	0
	R(−)	99.5	98–75	>99.0	>99.0–51.9	>99.0	100–99.5	75.0–73.5	100	100

			15	
			o	p
linalool	R(−)		91.3	50
	S(+)		8.7	50

		16	17		18		19, 20			
		d	d	q	d	d	d	b	i	b
α-pinene	R(+)									
	S(−)									
β-pinene	R(+)	<9			6.8–8.9		6.8–9.5	8.2–8.9	8.2–8.7	8.1–9.2
	S(−)	>91			93.2–91.1		93.2–90.5	91.8–91.1	91.8–91.3	91.9–90.8
sabinene	R(−)				14.1–16.0		14.1–18.8	15.2–15.9	15.3–15.7	15.1–16.0
	S(+)				85.9–84.0		85.9–81.2	84.8–84.1	84.7–84.3	84.9–84.0
limonene	S(−)	<3			2.0–2.7		1.9–2.7	2.0–2.3	2.1	2.1–2.2
	R(+)	>97			98.0–97.3		98.1–97.3	98.0–97.7	97.9	97.9–97.8
linalool	R(−)		99.4–100	99.5–99.8	99.5–99.7		99.4–99.7	81.6–98.7	99.6	99.5–99.6
	S(+)		0.6–tr	0.5–0.2	0.5–0.3		0.6–0.3	18.4–1.3	0.4	0.5–0.4
terpinen-4-ol	S(+)				13.4–25.4		9.7–26.3	27.1–31.8	18.4–20.1	19.8–21.5
	R(−)				86.6–74.6		90.3–73.7	72.9–68.2	81.6–79.9	80.2–78.5
α-terpineol	S(+)				49.3–68.1		17.5–69.4	11.2–26.6	31.4–52.1	41.4–50.5
	R(−)				50.7–31.9		82.5–30.6	88.8–73.4	68.6–47.9	58.6–49.5
linalyl acetate	S(+)		tr–0.6	0.2–0.4	0.3–0.2		0.3–0.1	1.1–0.9	0.3–0.2	0.3–0.1
	R(−)	<1	100–99.4	99.8–99.6	99.7–99.8		99.7–99.9	98.9–99.1	99.7–99.8	99.7–99.9
		>99								

Notes

a commercial oils; b laboratory solvent extracted oils; c laboratory steam distilled oils; d genuine cold-pressed oils; e laboratory SDE extracted oils; f oils extracted from pharmaceutical products; g oils extracted from perfumed tea; h distilled oils; i oils treated to remove the fouranocoumarins; l 'Peratoner' (oils obtained by distillation under reduced pressure of residues of cold-extraction); m 'Torchiati' (oils obtained by centrifugation of residues of cold-extraction); n oils obtained from hydro distillation of fresh peels; o 'Feccia' oils (oils obtained by distillation of residues of cold-extraction after isolation of 'Torchiati'); p oils obtained by steam distillation of homogenate of the whole fruits; q laboratory cold-pressed oils; 1 Mosandl et al. (1990); 2 Hener et al. (1990a); 3 Mosandl and Shubert (1990); 4 Schubert and Mosandl (1991); 5 Bernreuther and Schreier (1991); 6 Cotroneo et al. (1992); 7 Weinrich and Nitz (1992); 8 Casabianca and Graff (1994); 9 Ravid et al. (1994); 10 Casabianca et al. (1995); 11 Mosandl (1995); 12 Juchelka and Mosandl (1996); 13 Casabianca and Graff (1996); 14 Verzera et al. (1996); 15 König et al. (1997); 16 Mosandl and Juchelka (1997); 17 Dellacassa et al. (1997b); 18 Mondello et al. (1997); 19 Mondello et al. (1998d); 20 Dugo et al. (2001).

Table 20.8 Chiral columns used for the stereodifferentiation of linalool and linalyl acetate in bergamot oil

Authors	Linalool	Linalyl acetate
Mosandl and Shubert, 1990		Nickel(II)-bis[3-heptafluorobutanoyl-(1R)-camphorate, Ni(HFC)$_2$ in OV-101
Shubert and Mosandl, 1990 Bernreuther and Schreier, 1991	Heptakis-2,3,6-tri-O-ethyl-β-CD in OV-1701 vi Lipodex C (Heptakis-2,3,6-tri-O-pentyl-β-CD)	
Cotroneo et al., 1992 Verzera et al., 1996	2,3,6-tri-O-methyl-β-CD in OV1701	
Weinreich and Nitz, 1992 Casabianca and Graff, 1994 Casabianca et al., 1995	Heptakis-2,3,6-tri-O-metyl-β-CD in polysiloxane	Lipodex E (Octakis-3-O-butyl-2,6-di-O-pentyl-γ-CD) Heptakis-2,3,6-tri-O-methyl-β-CD*
Ravid et al., 1994		Lipodex E (Octakis-3-O-butyl-2,6-di-O-pentyl-γ-CD)
Juchelka and Mosandl, 1996 Dellacassa et al., 1997b		Heptakis-2,3-di-O-acetyl-6-O-(tertbutyldimethylsilyl)-β-CD
Mondello et al., 1997, 1998d Dugo et al., 2001		2,3-di-O-ethyl-6-O-(tert-butyldimethylsilyl)-β-CD in polysiloxane

Note
* Linalyl acetate detected after hydrogenation, as 3,7-dimethyl-3-acetoxyoctane.

cold-pressing process as well as that used for the elimination of bergapten do not seem to influence the enantiomeric distribution of these two components.

Mondello *et al.* (1998d) found both in torchiati and ricicli oils values for S(−)-α-terpineol lower than the lowest values of cold-pressed oils produced in the same period. This behaviour was explained with the hypothesis that R(+)-α-terpineol can be produced by microorganisms being recycling water and solid residues of the cold extraction good culture media, or with the hypothesis of the acid or enzymatic hydrolysis of glycosidically bonded R(+)-α-terpineol.

As previously observed, Mondello *et al.* (1997, 1998d) and Dugo *et al.* (2000) determined the enantiomeric distribution of β-pinene, sabinene, limonene, terpinene-4-ol, α-terpineol, linalool and linalyl acetate in cold-pressed, torchiati, ricicli, pulizia dischi, distilled and bergapten-free bergamot oils using a fully automated, MDGC system. They analysed genuine cold-pressed samples from a whole productive season, and determined the variations of the enantiomeric ratios in function of the harvest period. Figure 20.12 shows the enantiomeric excess of the components analysed during the production season. As can be seen, the enantiomeric distribution of linalool and linalyl acetate seems to be stable during the production period, as well as sabinene. Limonene and β-pinene show slight variations, and these values are in agreement with Juchelka and Mosandl (1996) and Mosandl and Juchelka (1997). The enantiomeric ratio of terpinen-4-ol changes irregularly while α-terpineol changes greatly during the production season.

NEROLI AND PETITGRAIN OILS

Neroli and petitgrain oils are very valuable products, widely used by the perfume and flavour industries. Neroli and petitgrain oils are obtained by steam distillation respectively from fresh blossoms, and from leaves and twigs of bitter orange tree. Some data on the chiral compounds of neroli and petitgrain oils are available. These data may be useful for deduction of characteristic authenticity profiles of these oils. Most of the work on the stereodifferentiation of some compounds of these oils has been carried out by Mosandl and co-workers (Hener *et al.*, 1990a,b; Mosandl *et al.*, 1990; Mosandl, 1995; Juchelka *et al.*, 1996). They determined the enantiomeric distribution of some monoterpene hydrocarbons (α-pinene, β-pinene, limonene), monoterpene alcohols (α-terpineol, linalool, terpinen-4-ol) and linalyl acetate. Moreover, under the experimental conditions used for the chiral separation of monoterpene hydrocarbons, linalool and terpinen-4-ol, also (Z)- and (E)-nerolidol enantiomers may be separated. Only (E)-nerolidol is present in genuine neroli and petitgrain oils, but in this last oil it occurs as a trace compound. So, the enantiomeric distribution of (E)-nerolidol has been determined only in neroli oil. Figure 20.13 reports the enantio MDGC analysis of some components of an authentic neroli oil. Results of the literature are reported in Tables 20.9 and 20.10 for neroli and petitgrain oils respectively. As can be seen, in both oils high enantiomeric purities of R-(−)-linalyl acetate and S-(−)-β-pinene are detected, and may be used as indicator of authenticity. Furthermore, S-(+)-nerolidol of high enantiomeric purity has been detected in authentic neroli oil. This result is in perfect agreement with data reported by König *et al.* (1992). This compound exhibits important olfactory properties, and can be a reliable indicator in the origin control of neroli oil. Juchelka *et al.* (1996) followed the influence of processing and provenance of the plant material on

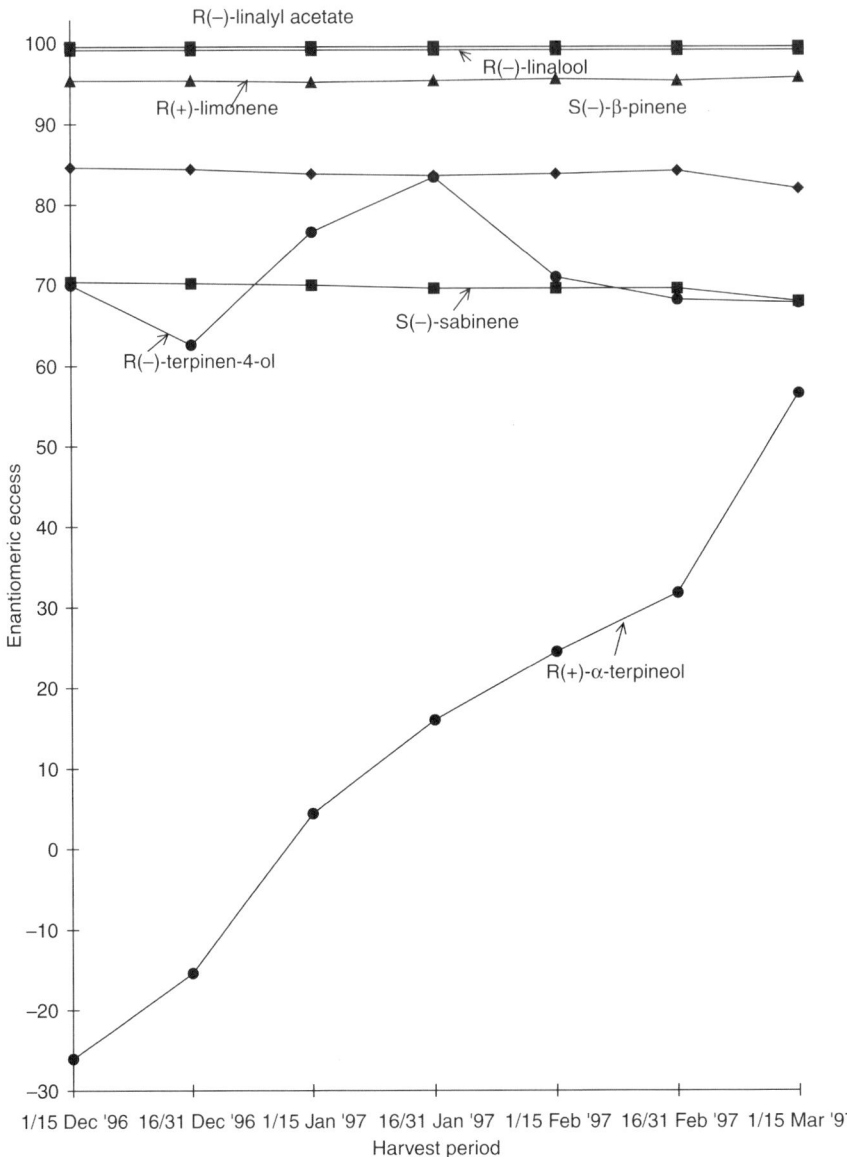

Figure 20.12 Variation of the fortnightly average values of the enantiomeric excess of 4R-(+)-limonene, 1S,5S-(−)-β-pinene, 1S,5S-(−)-sabinene, 4R-(−)-terpinen-4-ol, 4S-(−)-α-terpineol, 3R-(−)-linalool and 3R-(−)-linalyl acetate in cold-pressed bergamot oils during the productive season. Reproduced with permission from Mondello *et al*. (1998d).

the enantiomeric distribution of the compounds under investigation. Results obtained for linalyl acetate, β-pinene and (E)-nerolidol do not vary with provenance and production technology, while the technological conditions used influence the enantiomeric distribution of linalool, that goes toward racemization increasing distillation time.

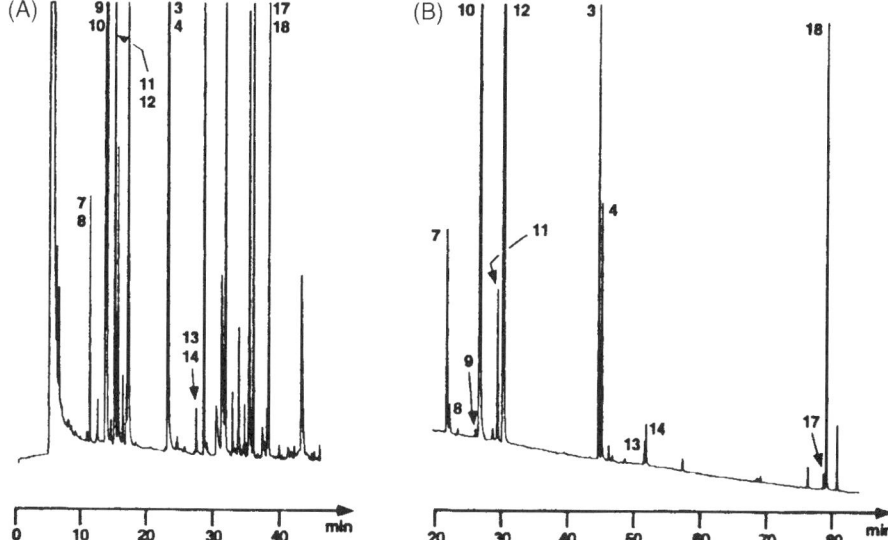

Figure 20.13 Enantio-MDGC analysis of an authentic neroli oil. α-pinene: [7] (S)-(−), [8] (R)-(+); [9] (R)-(+) [10] (S)-(−); limonene: [11] (S)-(−), [12] (R)-(+); linalool: [3] (R)-(−), (4) (S)-(+); terpinen-4-ol: [13] (S)-(+), (14) (R)-(−); (E)-nerolidol: [17] (R)-(−), [18] (S)-(+), using MDGC system. A) precolumn; B) main column. Reproduced with permission from Juchelka *et al.* (1996).

Table 20.9 Enantiomeric ratios of some components of neroli oil reported in the literature

		1, 2, 3, 4a	4b	5b	6	7	
						a	b
α-pinene	1S(−)	77	46–48			86.4–88.2	86.7–89.5
	1R(+)	23	54–52			13.6–11.8	13.3–10.5
β-pinene	1S(−)	96	95–96			99.7–99.5	99.9–96.0
	1R(+)	4	5–4			0.3–0.5	0.1–4.0
limonene	4S(+)	3	8			2.8–3.1	1.7–4.2
	4R(−)	97	92			97.2–96.9	98.3–95.8
α-terpineol	4R(+)			67		69.4–70.1	28.1–71.9
	4S(−)			33		30.6–29.9	39.8–60.2
linalyl acetate	R(−)				95–98.2	96.0–96.2	62.8–97.9
	S(+)				5–1.8	4.0–3.8	37.2–2.1
linalool	R(−)				72–79	71.0–71.4	63.1–78.7
	S(+)				28–21	29.0–28.6	36.9–21.3
terpinen-4-ol	S(+)					36.0–37.6	35.1–58.8
	R(−)					64.0–62.4	64.9–41.2
(E)-nerolidol	R(−)					1.2–1.4	<1.0–33.4
	S(+)					98.8–98.6	>99.0–66.6

Notes
a authentic oils; b commercial oils; *1* Hener *et al.* (1990a); *2* Mosandl *et al.* (1990); *3* Mosandl (1995); *4* Hener *et al.* (1990b); *5* Ravid *et al.* (1995); *6* Casabianca and Graff (1996); *7* Juchelka *et al.* (1996).

Table 20.10 Enantiomeric ratios of some components of petitgrain (bigarade) oil reported in the literature

		1, 2, 3	4a	5a	6		7		
						b	c	a
α-pinene	1S(−)	82					80.1	83.1–88.4
	1R(+)	18					19.9	16.9–11.6
β-pinene	1S(−)	98					99.5	92.0–99.8
	1R(+)	2					0.5	8.0–0.2
limonene	4S(−)	12					3.7	9.2–44.0
	4R(+)	88					96.3	90.8–56.0
linalool	R(−)		53.4–64.9		72.2–73.3	100	70.1	60.7–73.7
	S(+)		46.6–35.1		27.8–26.7	0	29.9	39.3–26.3
α-terpineol	4R(+)			57			73.6	73.0–56.9
	4S(−)			43			26.4	27.0–43.1
linalyl acetate	R(−)				93.4	100	98.4	74.7–99.3
	S(+)				6.6	0	1.6	25.3–0.7
terpinen-4-ol	S(+)						46.0	43.0–49.0
	R(−)						54.0	57.0–51.0

Notes
a commercial oils; b SFE oils; c authentic oils; 1 Hener *et al.* (1990a); 2 Mosandl *et al.* (1990); 3 Mosandl (1995); 4 Berureuther and Schreier (1991); 5 Ravid *et al.* (1995); 6 Casabianca and Graff (1996); 7 Juchelka *et al.* (1996).

REFERENCES

Bernreuther, A. and Schreier, P. (1990) Multidimensional gas chromatography/mass spectrometry: a powerful tool for the direct chiral evaluation of aroma compounds in plant tissues. II. Linalool in essential oils and fruits. *Phytochemical Analysis*, 2, 167–170.

Bicchi, C., D'Amato, A., Manzin, V., Galli, A. and Galli, M. (1994) Cyclodextrin derivatives in the gas chromatographic separation of racemic mixtures of volatile compounds. VII. The use of 2,6-di-O-methyl-3-O-pentyl-β-cyclodextrin diluted in phases with different polarity in the separation of racemates in complex mixtures. *J. Chromatogr. A*, 666, 137–146.

Casabianca, H. and Graff, J.-B. (1996) Chiral analysis of linalool and linalyl acetate in various plants. *EPPOS*, (7), 227–243.

Casabianca, H. and Graff, J.-B. (1994) Separation of linalyl acetate enantiomers: application to the authentication of bergamot food products. *J. High Resolut. Chromatogr.*, 17, 184–186.

Casabianca, H., Graff, J.-B., Jame, P., Perrucchietti, C. and Chastrette, M. (1995) Application of hyphenated techniques to the chromatographic authentication of flavors in food products and perfumes. *J. High Resolut. Chromatogr.*, 18, 279–285.

Clark, B.C. and Chamblee, T.S. (1992) Acid catalysed reactions of citrus oils and other terpene-containing flavours. In *Off-flavors in Food and Beverages*, Elsevier Science Publishers, Amsterdam, pp. 229–275.

Cotroneo, A., Stagno d'Alcontres, I. and Trozzi, A. (1992) On the genuineness of citrus essential oils. Part XXXIV. Detection of added reconstituted bergamot oil in genuine bergamot essential oil by high resolution gas chromatography with chiral capillary columns. *Flav. Fragr. J.*, 7, 15–17.

Dellacassa, E., Lorenzo, D., Moyna, P., Verzera, A., Mondello, L. and Dugo, P. (1997a) Uruguayan essential oils. Part VI. Composition of lemon oil. *Flav. Fragr. J.*, 12, 247–255.

Dellacassa, E., Lorenzo, D., Moyna, P., Verzera, A. and Cavazza, A. (1997b) Uruguayan essential oils. Part V. Composition of bergamot oil. *J. Essent. Oil Res.*, 9, 419–426.

Dugo, G., Lamonica, G., Cotroneo, A., Stagno d'Alcontres, I., Verzera, A., Donato, M.G., Dugo, P. and Licandro, G. (1992a) High resolution gas chromatography for detection of adulterations of citrus cold-pressed essential oils. *Perfum. Flav.*, 17(5), 57–74.

Dugo, G., Stagno d'Alcontres, I., Cotroneo, A. and Dugo, P. (1992b) On the genuineness of citrus essential oils. Part XXXV. Detection of added reconstituted mandarin oil in genuine cold-pressed mandarin essential oil by high resolution gas chromatography with chiral capillary columns. *J. Essent. Oil Res.*, 4, 589–594.

Dugo, G., Stagno d'Alcontres, I., Donato, M.G. and Dugo, P. (1993) On the genuineness of citrus essential oils. Part XXXVI. Detection of added reconstituted lemon oil in genuine cold-pressed lemon essential oil by high resolution gas chromatography with chiral capillary columns. *J. Essent. Oil Res.*, 5, 21–26.

Dugo, G., Verzera, A., Trozzi, A., Cotroneo, A., Mondello, L. and Bartle, K.D. (1994a) Automated HPLC-HRGC: a powerful method for essential oils analysis. Part I. Investigation on enantiomeric distribution of monoterpene alcohols of lemon and mandarin essential oils. *Essenz. Deriv. Agrum.*, 64, 35–44.

Dugo, G., Verzera, A., Cotroneo, A., Stagno d'Alcontres, I., Mondello, L. and Bartle, K.D. (1994b) Automated HPLC-HRGC: a powerful method for essential oil analysis. Part II. Determination of the enantiomeric distribution of linalool in sweet orange, bitter orange and mandarin essential oils. *Flav. Fragr. J.*, 9, 99–104.

Dugo, G., Mondello, L., Cotroneo, A., Bonaccorsi, I. and Lamonica, G. (2001) Enantiomeric distribution of volatile components of citrus essential oils by multidimensional gas chromatography (MDGC) *Perfum. Flav.*, 26(1), 20–35.

Hener, U., Hollnagel, A., Kreis, P., Maas, B., Schmarr, H.-G., Schubert, V., Rettinger, K., Weber, B. and Mosandl, A. (1990a) Direct enantiomer separation of chiral volatiles from complex matrices by multidimensional gaschromatography. In *Flavour Science and Technology*, Wiley, Chichester, West Sussex, England.

Hener, U., Kreis, P. and Mosandl, A. (1990b) Enantiomeric distribution of α-pinene, β-pinene and limonene in essential oils and extracts. Part 2. Oils, perfumes and cosmetics. *Flav. Fragr. J.*, 5, 201–204.

Juchelka, D. and Mosandl, A. (1996) Authenticity profiles of bergamot oil. *Pharmazie*, 51, 417–422.

Juchelka, D., Steil, A., Witt, K. and Mosandl, A. (1996) Chiral compounds of essential oils. XX. Chirality evaluation and authenticity profiles of neroli and petitgrain oils. *J. Essent. Oil Res.*, 8, 487–497.

König, W.A., Krebber, R., Evers, P. and Bruhn, G. (1990) Stereochemical analysis of constituents of essential oils and flavor compounds by enantioselective capillary gas chromatography. *J. High Resolut. Chromatogr.*, 13, 328–332.

König, W.A., Gehrcke, B., Icheln, D., Evers, P., Dönnecke, J. and Wang, W. (1992) New, selectively substituted cyclodextrins as stationary phases for the analysis of chiral constituents of essential oils. *J. High Resolut. Chromatogr.*, 15, 367–372.

König, W.A., Rieck, A., Hardt, I., Gehrcke, B., Kubeczka, K.-H. and Muhle, H. (1994) Enantiomeric composition of the chiral constituents of essential oils. Part 2: sesquiterpene hydrocarbons. *J. High Resolut. Chromatogr.*, 17, 315–320.

König, W.A., Fricke, C., Saritas, Y., Momeni, B. and Hohenfeld, G. (1997) Adulteration or natural variability? Enantioselective gas chromatography in purity control of essential oils. *J. High Resolut. Chromatogr.*, 20, 55–61.

Kreis, P., Hener, U. and Mosandl, A. (1991) Chirale inhaltsstoffe ätherischer öle. 5. Mitteilung: enantiomerenverteilung von α-pinen, β-pinen und limonen in ätherischen ölen und extrakten-citrusfrüchte und citrushaltige getränke. *Dtsch. Lebens. Rundsch.*, 87(1), 8–11.

Mondello, L., Dugo, G., Dugo, P. and Bartle, K.D. (1996) On-line HPLC-HRGC in the analytical chemistry of citrus essential oils. *Perfum. Flav.*, 21(4), 25–49.

Mondello, L., Catalfamo, M., Dugo, P., Proteggente, A.R. and Dugo, G. (1997) Lagascromatografia multidimensionale per l'analisi di miscele complesse. Nota preliminare.

Determinazione della distribuzione enantiomerica di componenti degli olii essenziali agrumari. *Essenz. Deriv. Agrum.*, 67, 62–85.

Mondello, L., Catalfamo, M., Dugo, G. and Dugo, P. (1998a) Multidimensional capillary GC-GC for the analysis of real complex samples. Part I. Development of a fully automated tandem gas chromatography system. *J. Chromatogr. Sci.*, 36, 201–209.

Mondello, L., Catalfamo, M., Proteggente, A.R., Bonaccorsi, I. and Dugo, G. (1998b) Multidimensional capillary GC-GC for the analysis of real complex samples. 3. Enantiomeric distribution of monoterpene hydrocarbons and monoterpene alcohols of mandarin oils. *J. Agric. Food Chem.*, 46, 54–61.

Mondello, L., Catalfamo, M., Dugo, P. and Dugo, G. (1998c) Multidimensional capillary GC-GC for the analysis of real complex samples.Part II. Enantiomeric distribution of monoterpene hydrocarbons and monoterpene alcohols of cold-pressed and distilled lime oils. *J. Microcol. Sep.*, 10, 203–212.

Mondello, L., Verzera, A., Previti, P., Crispo, F. and Dugo, G. (1998d) Multimensional capillary GC-GC for the analysis of complex samples. 5. Enantiomeric distribution of monoterpene hydrocarbons, monoterpene alcohols, and linalyl acetate of bergamot (*Citrus bergamia* Risso et Poiteau) oils. *J. Agric. Food Chem.*, 46, 4275–4282.

Mondello, L., Catalfamo, M., Cotroneo, A., Dugo, G., Dugo, Giacomo, and McNair, H. (1999) Multidimensional capillary GC-GC for the analysis of real complex samples. Part IV. Enantiomeric distribution of monoterpene hydrocarbons and monoterpene alcohols of lemon oil. *J. High Resolut. Chromatogr.*, 22, 350–356.

Mosandl, A., Hener, U., Kreis P. and Schmarr, H.-G. (1990) Enantiomeric distribution of α-pinene, β-pinene and limonene in essential oils and extracts. Part 1. Rutaceae and Gramineae. *Flav. Fragr. J.*, 5, 193–199.

Mosandl, A. and Schubert, V. (1990) Stereoisomeric flavor compounds XXXVII: enantiomer separation of 1-alken-3-yl esters and their chirality evaluation from essential oils using multidimensional gas chromatography (MDGC) *J. Essent. Oil Res.*, 2, 121–132.

Mosandl, A. (1995) Enantioselective capillary gas chromatography and stable isotope ratio mass spectrometry in the authenticity control of flavors and essential oils. *Food. Rev. Int.*, 11, 597–664.

Mosandl, A. and Juchelka, D. (1997) Advances in the authenticity assessment of citrus oils. *J. Essent. Oil Res.*, 9, 5–12.

Ravid, U., Putievsky, E. and Katzir, I. (1994) Chiral GC analysis of enantiomerically pure (R)(−)-linalyl acetate in some Lamiaceae, myrtle and petitgrain essential oils. *Flav. Fragr. J.*, 9, 275–276.

Ravid, U., Putievsky, E. and Katzir, I. (1995) Determination of the enantiomeric composition of α-terpineol in essential oils. *Flav. Fragr. J.*, 10, 281–284.

Rocca, B., Arzouyan, C. and Estienne, J. (1992) Apport de la chromatographie gazeuse en phase chirale dans le controle de l'origine naturelle des aromes. *Ann. Fals. Exp. Chim.*, 85, 327–346.

Schubert, V. and Mosandl, A. (1991) Chiral compounds of essential oils. VIII: stereodifferentiation of linalool using multidimensional gas chromatography. *Phytochemical Analysis*, 2, 171–174.

Starrantino, A., Terranova, G., Dugo, P., Bonaccorsi, I. and Mondello, L. (1997) On the genuineness of citrus oils. Part IL. Chemical characterization of the essential oil of new hybrids of lemon obtained in Sicily. *Flav. Fragr. J.*, 12, 153–161.

Takeoka, G., Flath, R.A, Mon, T.R, Buttery, R.G., Teranishi, R., Güntert, M., Lautamo, R. and Szejtli, J. (1990) Further applications of permethylated β-cyclodextrin capillary gas chromatographic columns. *J. High Resolut. Chromatogr.*, 13, 202–206.

Verzera, A., Lamonica, G., Mondello, L., Trozzi, A. and Dugo, G. (1996) The composition of bergamot oil. *Perfum. Flav.*, 21(6), 19–34.

Wang, X., Jia, C. and Wan, H. (1995) The direct chiral separation of some optically active compounds in essential oils by multidimensional gas chromatography. *J. Chromatogr. Sci.*, 33, 22–25.

Weinrich, B. and Nitz, S. (1992) Influences of processing on the enantiomeric distribution of chiral flavour compounds. Part A: linalyl acetate and terpene alcohols. *Chem. Mikrobiol. Technol. Lebensm.*, 14, 117–124.

Werkhoff, P., Brennecke, S., Bretschneider, W., Güntert, M., Hopp, R. and Surburg, H. (1993) Chirospecific analysis in essential oil, fragrance and flavor research. *Z. Lebensm. Unters Forsch.*, 196, 307–328.

21 Adulteration of citrus oils

David McHale

INTRODUCTION

The peel oils derived from citrus fruits are used extensively as natural flavourings and perfumery ingredients. Their extraction from the fruit is complicated by a number of factors, not the least of which are their low level of occurrence in the fruit and the losses intrinsic to the methods of recovery. From the very earliest days of trading such oils, these difficulties have undoubtedly encouraged unscrupulous producers and traders to seek to improve the profitability of their operations by fraudulent practices. For obvious reasons the perpetrators of such actions have guarded zealously the methods that they have used for extending and cheapening their oils. Thus, the onus has been on the users to recognise doubtful oils and to devise reliable physical and analytical procedures capable of identifying them. Since it was first recognised that commercial oils were frequently adulterated, the users have been under considerable pressure to keep abreast of the more subtle and devious procedures introduced to mask non-authentic oils.

One of the earliest references to the adulteration of citrus oils is attributed to De Domenico (1854) who, in a treatise on the medical virtues of bergamot oil, wrote that it was adulterated in many ways. He reported that there were mixtures with 'so-called' turpentine, with orange and lemon oils and with the oil produced by the distillation of the small unripe rotten or windfall bergamots. There can be no doubt that these and similar methods of adulteration were also used for lemon and orange oils. In fact the situation became so serious for the Sicilian citrus industry that, at the turn of the century, the Italian Government prohibited the export of adulterated citrus oils unless the nature and percentage of any adulterant were declared on the packaging and the bills of lading.

Prior to the early part of the nineteenth century the methods used for the adulteration of citrus oils were relatively crude. They succeeded because at that time there was a lack of knowledge of the composition of such oils and an absence of any reliable analytical procedures for assessing their purity. Methods for determining the physical properties of liquids were not yet fully developed despite the fact that as early as the middle of the seventeenth century Hooke (1665) had determined values for the refractive index of turpentine and orange oil. Furthermore, although Biot (1815) had observed that lemon oil and turpentine were optically active, commercial polarimeters and refractometers did not become available until the last quarter of the nineteenth century. Before then it is likely that the oils were only assessed by appearance, aroma and taste.

The progress made at the end of the nineteenth century on the chemical composition of citrus oils was as valuable to the traders as to the users. For instance, the recognition by

Semmler and Tiemann (1892) that linalyl acetate was a major component of bergamot oil led rapidly to the adulteration of this expensive oil with cheap esters. However, the pioneering work on the analysis of citrus oils undertaken at the turn of the century by Schimmel and Co. laid the foundation for the future recognition of adulterated oils. Much of this work has been reviewed by Gildermeister and Hoffmann (1959) and Guenther (1952).

ADULTERATION OF CITRUS OILS AND DETECTION METHODS

Methods for the detection of the early used adulterants of citrus oils

Because of their relatively low cost and easy availability, petroleum, kerosene, mineral oils and turpentine were frequently used to extend/dilute cold-pressed citrus oils. While it was possible to detect these adulterants by aroma and taste, such methods were subjective and dependent on the skill of the analyst. With the advent of the classical physical methods of analysis such as specific gravity, refractive index and optical rotation, the analyst was provided with the means of recognising adulterated oils but not necessarily the adulterants. Chemical tests were devised to detect petroleum, mineral oil and kerosene in citrus oils. They relied on the fact that these adulterants contained significant amounts of saturated paraffinic hydrocarbons which, unlike the natural terpene hydrocarbons of citrus oils, are inert to fuming sulphuric acid. As a consequence, citrus oils which contained such adulterants showed the presence of unreacted oil after treatment with this acid. However, at best this test was only qualitative.

The addition of turpentine to a citrus oil generally reduced the specific gravity and optical rotation. Because α-pinene, the major component of turpentine, has a lower boiling point than the majority of components of citrus oils and a lower optical rotation than (+)-limonene (the major terpene hydrocarbon), the first ten per cent distillate of a citrus oil adulterated with turpentine will have a lower optical rotation than the original oil. The extent of the reduction in the optical rotation did not correspond directly to the amount of added turpentine because the direction and strength of the optical rotation of α-pinene was dependent on the source of the turpentine.

Undoubtedly, the most difficult adulterants to detect at that time were those derived from the citrus fruits themselves. These comprised two types: citrus oils recovered by steam-distillation of the peel residues from expressed fruit, and the terpenes obtained as by-products in the production of terpeneless oils. It was common practice to extend or dilute genuine oils with these particular materials, either with or without the addition of appropriate amounts of any deficient important ingredient of the resulting 'so-called' genuine oil. Such added ingredients included citral and linalool derived, respectively, from lemon grass oil and Mexican linaloe oil, and linalyl acetate obtained synthetically by the acetylation of linalool. A knowledgeable and ingenious producer could, by careful blending, obtain an adulterated oil with figures within the acceptable values for the physical and chemical constants of a genuine oil.

The role of non-volatile constituents in the detection of adulteration of citrus oils

It was recognised during the early studies into citrus oils that they contained significant amounts of non-volatile components. In fact certain constituents, such as limettin (1)

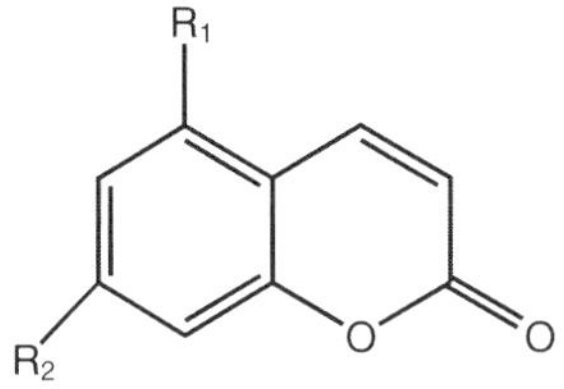

1 $R_1 = R_2 = OCH_3$
2 $R_1 = H$; $R_2 = O(CH_2CH:CCH_3CH_2)_2H$
3 $R_1 = H$; $R_2 = OCH_2CH:CCH_3(CH_2)_2CH\text{-}C(CH_3)_2$ (epoxide)
4 $R_1 = H$; $R_2 = OCH_3$
5 $R_1 = R_2 = O(CH_2)_3H$

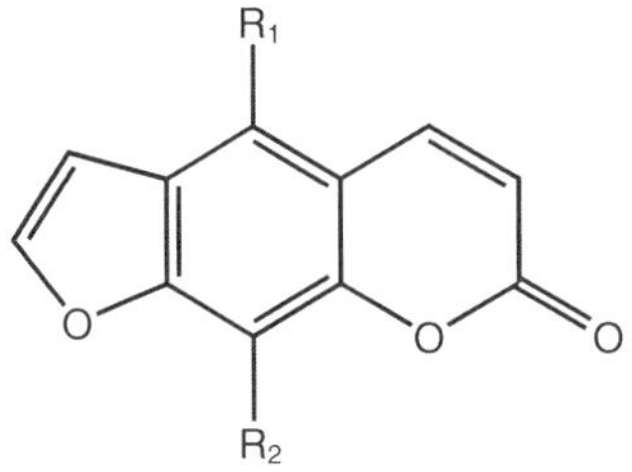

6 $R_1 = OCH_3$; $R_2 = H$
7 $R_1 = O(CH_2CH:CCH_3CH_2)_2H$; $R_2 = OCH_3$
8 $R_1 = R_2 = OCH_3$

Figure 21.1 Chemical structures of coumarins and psoralens present in citrus oils (1–8).

from lime oil (Tilden and Burrows, 1902) and bergaptene (6) from bergamot oil (Pomerantz, 1891), which crystallised directly from the cold-pressed oils or from the evaporation residues, were among the first compounds to be isolated from citrus oils in a pure state (For structures 1–8 see Figure 21.1). To compensate for the reduction in the non-volatile residue resulting from a dilution of a genuine oil with citrus terpenes or other volatile diluents, it was the practice to include also an appropriate amount of a non-volatile adulterant. Materials that were used included cedar wood stearin (Boswigi, 1899), castor oil (Parry, 1909), rosin, and distillation residues from citrus oils.

The direct determination of the non-volatile residue of a cold-pressed citrus oil by evaporation or steam-distillation, while useful in differentiating highly diluted oils, was of limited value where non-volatile adulterants had been added. The identification of such oils remained a problem despite the fact that Morton (1929) had reported that lemon oil absorbed UV light in the region of 311 nm. Surprisingly, it was not until the 1950's that, independently, workers in Sicily and America re-examined the UV absorption data of lemon oil. Sale *et al.* (1953) studied the UV spectrum of various Californian lemon oils over the period 1944–1952 and established that the non-volatile fraction was responsible for the UV spectrum. Neither steam-distilled nor vacuum-distilled lemon oil contained the non-volatile components

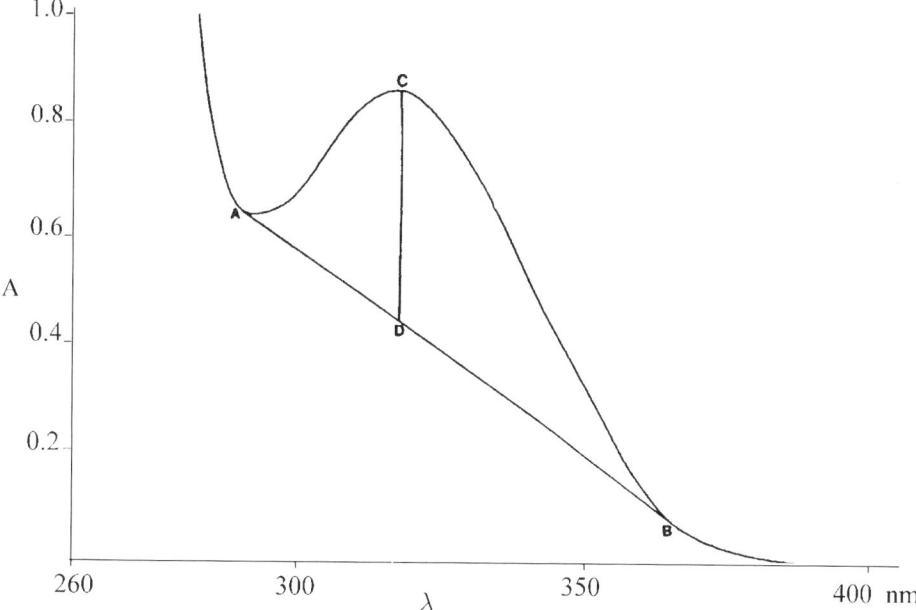

Figure 21.2 A typical CD line plot for lemon oil.

present in cold-pressed oil and this reflected in the UV spectra of the distilled oils. Cold-pressed lemon oil had a characteristic UV spectrum with a maximum at 315 nm but the distilled oils were transparent in this region. Recognising that this difference offered a means of detection of distilled oil in cold-pressed oil, the UV absorbances of a number of genuine lemon oils from various sources were measured in terms of the CD value (Figure 21.2). The study highlighted a wide natural variation in CD values. The minimum CD value observed for an authentic Sicilian oil was 0.49. The minimum for American oils was significantly lower at 0.23. The results obtained by Culterra *et al.* (1952) for Italian lemon oils were in accord with the American results though they expressed their data differently. Known as the Palermo method, the UV transmission data for the authentic oil with the highest values and that with the lowest values were plotted on the same graph. The resulting area between the two plots comprised all the transmission values characteristic of an authentic oil (Figure 21.3). The CD method is more widely used and the current standards for the various expressed citrus oils are given in Table 21.1.

Genuine oils with high CD values tolerate dilution with some distilled oil without detection. It soon became evident to the unscrupulous producers that the extent of dilution could be increased if materials with UV absorption characteristics resembling those of the cold-pressed oil were added to compensate for the lack of absorbance of the distilled oil. Whilst direct measurement of UV absorbance would reveal gross adulteration with a distilled oil, it would not detect unequivocally those oils which also contained UV absorbance enhancers. The detection of such materials generally required a more sophisticated analytical procedure.

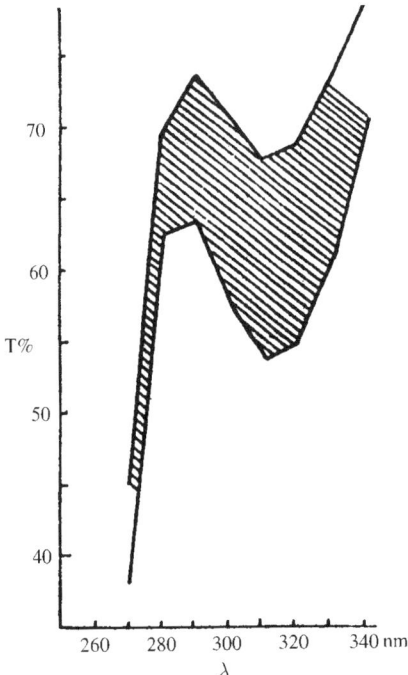

Figure 21.3 The area of 'acceptability' generated by the 'Palermo' method of plotting the maximum and minimum UV transmission data for genuine lemon oils over the range 270–340 nm (c 0.05 per cent v/v in 95 per cent EtOH).

Table 21.1 Minimum acceptable CD values for cold-pressed citrus oils in ethanol

	CD	λ_{max} nm	conc (mg/100 ml)
Italian lemon[a]	0.49	315±3	250
Californian lemon[a]	0.23	315±3	250
Californian orange[a]	0.13	330±3	250
Florida orange[a]	0.24	330±3	250
Lime[b]	0.45	315±3	20
Grapefruit[b]	0.42	319±3	50
Bitter orange[b]	0.32	320±3	50
Mandarin[b]	0.50	323±3	150
Tangerine[b]	0.28	325±3	150

Notes
a USP (1965); b EOA (1970).

The use of chromatographic techniques to evaluate the authenticity of citrus oils

Thin layer chromatography (TLC)

Initially it was thought that limettin (1) was responsible for the UV absorption of lemon oil but Stanley and Vannier (1957a) recognised that the UV maxima were different.

They concluded that there might be similar compounds present in the oil which were shifting the UV maximum. Fractionation of the non-volatile residue of lemon oil by a combination of adsorption column chromatography and TLC established that a series of oxygen heterocyclic compounds were responsible for the characteristic UV absorption spectrum of the oil. The work was subsequently extended to other oils (Stanley, 1963). As shown in Table 21.2, the yields of solid components isolated varied widely across the different *Citrus* species. The figures underestimate the amount initially present because many of the individual components are poorly soluble in the oil and tend to crystallise out on standing, particularly at sub-ambient temperatures. This, coupled with the fact that certain of the isolated compounds show significant differences in the position and intensity of their UV maxima, probably accounts for the wide discrepancy between the data of Tables 21.1 and 21.2. The compounds comprised mainly coumarins, psoralens and polymethoxyflavones and while their structures were well established, information on the levels at which individual compounds were present was not readily obtainable.

Although it was apparent (Chakraborty and Bose, 1956) that partition paper chromatography was useful in the detection and identification of coumarins and psoralens, the application of this procedure for the examination of citrus oils and their residues never found favour. In contrast, the potential value of TLC for the establishment of the authenticity of citrus oils was recognised as soon as the procedure became generally available. Stanley and Vannier (1957b) developed a method for the analysis of the coumarins and psoralens in various citrus oils, based on their chromatographic separation on silicic acid-coated glass strips and subsequent spectrophotometric examination. Observation of the developed strips under long-wave UV light revealed a distinct series of various blue and yellow fluorescent bands. Individual bands were marked and the respective areas collected. The compounds responsible for the fluorescence were recovered from the silicic acid with a suitable solvent and identified by their UV spectrum. The discovery that 7-geranyloxycoumarin (auraptene, 2), a major coumarin of grapefruit oil, did not occur in lemon oil enabled them to identify lemon oils to which grapefruit oil had been added to enhance the CD value. Using a similar procedure, Stanley (1959) detected various synthetic substances such as the so-called 'sun esters' (a mixture of menthyl and *m*-homomenthyl salicylates), methyl anthranilate and 4'-methoxychalcone (Stanley, 1961), which were being used to increase the CD value and mask the addition of distilled lemon oil to cold-pressed oil. The publication by D'Amore and Calapaj (1976) of tables showing the fluorescent TLC patterns obtained when genuine lemon, bergamot, mandarin, and sweet and bitter orange oils were chromatographed using different compositions for the mobile phase, enabled oils with atypical patterns to be recognised. As well as the Rf data, the spectroscopic characteristics for individual

Table 21.2 Yield of solids obtained by chromatography of cold-pressed citrus oils (Stanley, 1963)

	W(%)
Lime, Mexican	6.67
Grapefruit	1.37
Bergamot	0.56
Lemon	0.47
Bitter orange	0.23

bands were also included. However, no estimates of the levels at which individual compounds occurred were given, possibly because of the practical difficulty in achieving complete separation of the bands prior to recovery of the active component. Subsequently, Madsen and Latz (1970) attempted to determine the levels of the individual fluorescent components by direct fluorimetric scanning of the TLC patterns but the results were of limited value.

High performance liquid chromatography (HPLC)

The problems associated with the quantitative determination of the individual non-volatile compounds of citrus oils were overcome with the development of HPLC. With a choice of direct or reversed phase columns and isocratic or gradient elution, conditions could be selected which provided maximum resolution of the non-volatile components of the various citrus oils. The use of a variable wave-length UV detector permitted the response to individual compounds to be optimised. In addition, the operation of UV and fluorescence detectors in series and the application of a stop flow technique enabled those compounds that fluoresced to be readily recognised and their excitation and emission data established (Latz and Ernes, 1978).

With the introduction of photodiode arrays it became possible to check the identity of eluting peaks by scanning their UV absorption spectra. This facilitated the detection and identification of adulterants which migrated with known components of authentic oils. Since sweet orange oil and to a lesser extent grapefruit oil are much more readily obtainable than bitter orange oil, both have been used widely to adulterate the latter. Among the oxygen heterocyclic fraction of grapefruit oil are two coumarins, auraptene (2) and epoxyauraptene (3) which appear to be unique to this *Citrus* species (McHale and Sheridan, 1989). Auraptene is a major component of the fraction and, as can be seen from a comparison of Figures 21.4A and B, migrates faster than any of the oxygen heterocyclic compounds of bitter orange oil. Thus, it offers a means of detecting grapefruit oil in bitter orange oil. In contrast, the detection of sweet orange oil in bitter orange oil by HPLC poses certain problems. These stem primarily from the fact that the heterocyclic fraction of sweet orange oil comprises a mixture of six main polymethoxyflavones but no coumarins or psoralens. Whilst three of these polymethoxyflavones are common to grapefruit oil and bitter orange oil, the best chromatographic conditions for the separation of all six compounds do not correspond with those affording optimum resolution of the coumarins and psoralens of bitter orange oil. Furthermore, detection and quantification are difficult because the polymethoxyflavones show significant differences in the position and intensity of their UV maxima. Nevertheless, the measurement of the UV absorbance of individual coumarins and psoralens relative to the total polymethoxyflavone content can be used to detect sweet orange oil in bitter orange oil.

Gas liquid chromatography (GLC)

In common with other essential oils, knowledge of the volatile components of the citrus peel oils advanced rapidly following the advent of GLC in the 60's. Initially, the resolution using packed columns was far from ideal but careful comparison of the chromatogram of a doubtful oil with that of a genuine oil could highlight differences arising from the addition of volatile adulterants. In an early example (MacLeod *et al.*,

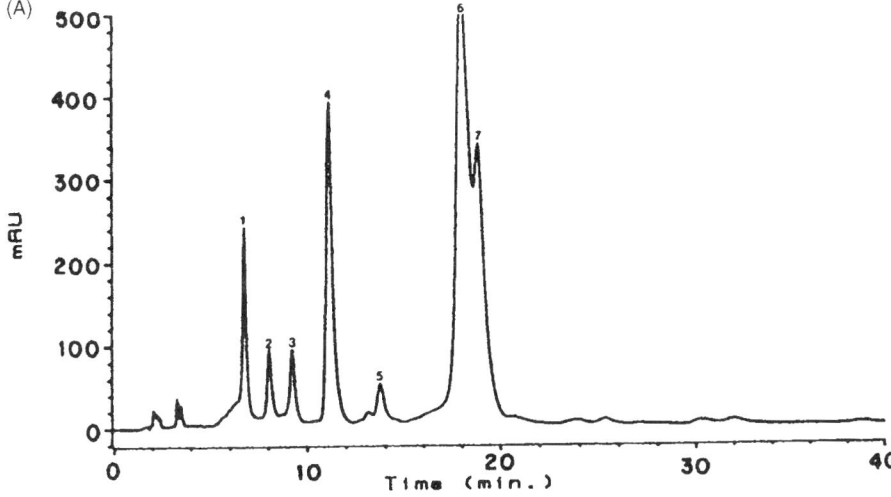

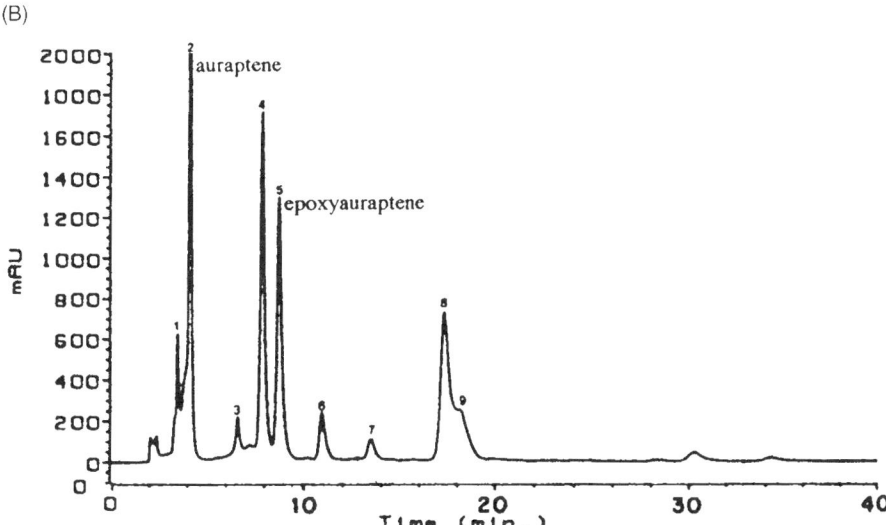

Figure 21.4 Comparison of the HPLC resolution of the oxygen heterocyclic compounds of (A) bitter orange oil and (B) grapefruit oil on a 6 μM Zorbax SIL spherical silica column (25 cm × 4.6 mm i.d.) using isocratic elution with hexane-ethanol (19:1); flow rate 1.5 ml/min; sample volume 20 μl (20 per cent solution of oil in dichloromethane). Detection by UV absorbance at 315 nm.

1964), a commercial sample of cold-pressed lemon oil with minor anomalies in its UV spectrum clearly showed a late running peak on the gas chromatogram. The compound responsible for the peak was isolated by preparative GLC and identified by NMR and IR as benzyl ether. This substance has not been found to occur naturally in lemon oil and was probably added to adjust the density and refractive index of a sub-standard oil.

The improvements in peak resolution afforded by the development of capillary columns and the subsequent coupling of a mass spectrometer at the exit from the GLC, provided the means of obtaining more comprehensive details of the volatile composition of citrus oils. Reproducible retention data and the ability to scan a peak for mass spectral homogeneity meant that volatile contaminants and adulterants could be recognised and identified. Using such a technique Calvarano and Di Giacomo (1970) showed that certain samples of lemon oil contained a minor peak that was not present in genuine oil. The compound responsible for the peak was identified as linalyl acetate and its presence suggested that the samples had been contaminated with traces of bergamot oil during production.

The detailed knowledge of the volatile composition of the various citrus oils has facilitated the detection of one *Citrus* species in another. Despite the fact that there are probably no volatile compounds that are unique to one species, Verzera *et al.* (1987) have developed a procedure which permits one of the commonest forms of adulteration of lemon oil (dilution with sweet orange oil terpenes) to be detected. These workers recognised that orange terpenes contain about 0.10 per cent of δ-3-carene but virtually no α-terpinene or camphene. In contrast, lemon oil is almost free of δ-3-carene but contains about 0.20 per cent of α-terpinene and 0.06 per cent of camphene. Since these compounds are clearly separated from one another and from other components of orange terpenes and lemon oil by GLC, the percentage of δ-3-carene and the ratios of δ-3-carene/α-terpinene and δ-3-carene/camphene can provide a good indication of the addition of orange terpenes to lemon oil. A similar approach based on the value of the ratios δ-3-carene/camphene and δ-3-carene/terpinolene has been suggested for the detection of orange terpenes in bitter orange oil (Dugo *et al.*, 1993).

Apart from the effect on the concentration of the minor hydrocarbons, the addition of orange terpenes to another oil will obviously reduce the level of key flavour components in that oil. Because the aldehyde or citral content of lemon oil is frequently used as a quality parameter, any significant reduction in its concentration in the oil is likely to be detected. Therefore, citral derived from another source is added with the orange terpenes to ensure that the chemical constants are within the acceptable values for a genuine oil. Dugo *et al.* (1984) studied the effect that the addition of citral from different sources had on the GLC pattern of lemon oil. Whilst the results indicated that there were qualitative and/or quantitative differences between the oils with and those without citral added, the differences were not consistent and tended to reflect the concentration of certain impurities in different citrals. In real terms this meant that it was necessary to know the source of the citral in order to confirm its presence in an adulterated lemon oil.

The introduction of chiral stationary phases has provided information on the enantiomeric distribution of the monoterpene hydrocarbons, alcohols, and esters of citrus oils. Using multidimensional gas chromatography (MDGC) involving the on-line coupling of non-chiral with chiral columns (Mosandl and Schubert, 1990), it is possible to determine the enantiomeric distribution of individual volatile constituents of citrus oils. Generally, one enantiomer predominates and in certain cases to the virtual exclusion of the other. Authentic bergamot oil comprises essentially 100 per cent $(R)(-)$-linalool and -linalyl acetate. Weinreich and Nitz (1992) suggest that the presence of any significant amount of the $(S)(+)$-isomers is indicative of adulteration. Care has to be exercised in interpreting such results in the case of linalool since contact of the oil with acidic juice during isolation can bring about some racemisation. However, this would not account for the presence of $(S)(+)$-linalyl acetate.

Stable isotope ratio analysis

Whereas stable isotope analysis has been used successfully for the detection of the adulteration of citrus juices, its use with citrus oils is still under investigation. The details of the theory and the experimental methods involved are beyond the scope of this article but have been well reviewed by Schmidt *et al*. (1998). In brief, the stable isotope pattern of any organic compound will be determined by that of its precursors and the kinetic isotope effects accompanying the reactions leading to its synthesis. The changes involved in the synthetic process are distinct but very small and, therefore, are expressed in the form of a shift (δ-value) of the abundance isotope ratio from that of an international standard.

The most common procedure for determining the carbon isotope ratio of an organic compound involves combustion to CO_2 and direct measurement of the ratio by mass spectrometry. In the case of δ^2H determinations, the water produced in the combustion is reduced to hydrogen prior to measurement of the deuterium/hydrogen (D/H) ratio by mass spectrometry. For obvious reasons it is necessary to ensure the purity of the compound under study and this is best achieved for volatile compounds by the coupling of a gas chromatograph with a combustion interface and an isotope ratio mass spectrometer (GC-C-IRMS). Such a procedure permits the isotope ratio of individual components of citrus oils to be determined.

Braunsdorf *et al*. (1992) report that the natural aliphatic aldehydes of orange oil (octanal, nonanal, decanal and dodecanal) are richer in ^{13}C (i.e. have more positive $\delta^{13}C$ values) than the 'so-called' nature identical compounds. However, Schmidt *et al*. (1998) have indicated that there are insufficient data available in the literature for this to be considered to be a general rule.

Nuclear magnetic resonance (NMR) can also be used to determine stable isotope ratios and affords a means of measuring isotopic patterns within molecules (Martin and Martin, 1990). Although it is possible to determine $^{13}C/^{12}C$ ratios at particular sites in a molecule, the majority of the work to date has been directed at establishing the D/H ratios at specific sites. The method is known as site-specific natural isotope fractionation (SNIF-NMR$^©$). It has the advantage that it leads to much more specific isotopic information about a molecule, thereby permitting natural and synthetic molecules to be clearly differentiated. However, the disadvantages are that the equipment is very expensive, data acquisition times are long, and relatively large quantities of pure compound are needed for analysis.

Current methods used for the adulteration of oils from different *Citrus* species and appropriate means of detection

Sweet orange oil: Citrus sinensis (L) Osbeck

The global cultivation of the sweet orange by far exceeds that of any other citrus fruit and this is reflected in the ready availability of the peel oil and its low cost in relation to that of the other citrus oils. In spite of this, non-authentic or diluted sweet orange oils are still frequently offered as genuine cold-pressed oil. This occurs mainly because large volumes of steam-stripped oil are recovered as a by-product of juice processing, and orange terpenes are available from the concentration of the natural oil. The scale of the operations is such that these materials are very cheap. The terpenes consist mainly of limonene (>94 per cent) and a significant amount of myrcene (ca 2 per cent). If an early cut has not been taken during the distillation they will also contain most of the octanal (ca 0.2 per cent) present in the original oil. The remainder comprises low boiling terpenes. Their main difference from the steam-stripped oil is an absence of the remaining volatile components of orange oil. Dilution of a genuine oil with orange terpenes will obviously reduce the concentration of these volatile components but, unless the dilution has been excessive, their resulting concentrations are likely to remain within the natural variation for the oil. Similarly, the range of CD values observed for natural oils is wide and those with high values will tolerate as much as 50 per cent dilution without falling below the minimum accepted value. Hence, any addition of orange terpenes or steam-stripped oil to a cold-pressed oil may not be detected.

The terpeneless and sesquiterpeneless oils derived from the cold-pressed oil by concentration are much higher priced products. They are important items of commerce because, in comparison to the cold-pressed oil, they have improved solubility, better oxidative stability and greater flavour potential. As the original oil may be concentrated up to a fiftieth of its volume, the composition and yield of the final product can vary significantly. To compensate for such variations and to improve the quality of low grade oils, an unscrupulous producer may add key components from alternative sources to these oils. The detection of such additions has, in the past, not been possible but, with the modern methods now available, it is likely that this will change. For instance, if octanal and decanal are added to enhance the orange character of the oil they will almost certainly be of synthetic origin and the $\delta^{13}C$ values measured for these particular compounds in the oil could be more negative than the values for a genuine oil. However, the number of studies carried out with the synthetic aldehydes is still too limited to give a reliable assessment of the range of their $\delta^{13}C$ values (Schmidt et al., 1998).

Bitter orange oil: Citrus aurantium (L)

In contrast to the sweet orange, the cultivation of the bitter orange is barely sufficient to satisfy commercial demands. These demands include the use of the flower for the preparation of neroli oil and the use of the fresh fruit in marmalade production and for bitter orange oil. Because of its limited availability, the oil is much more expensive than sweet orange oil and is a frequent target for adulteration with cheaper materials. The commonest adulterants for bitter orange oil are sweet orange oil and its terpenes but grapefruit oil and its terpenes are also used, although the volumes of the latter commercially available are much less than those of sweet orange.

As mentioned previously, Verzera et al. (1987) recognised that the level of δ-3-carene in sweet orange oil and its terpenes exceeded that in any of the other citrus oils and, consequently, the value of the ratios δ-3-carene/camphene and δ-3-carene/terpinolene provided a means of detecting the addition of the sweet orange products to bitter orange oil. Since grapefruit oil contains virtually no δ-3-carene, neither it nor its terpenes can be detected in bitter orange oil in this way. Moreover, there are no other components in grapefruit terpenes that can serve as a tracer for its detection in bitter orange oil. However, the addition of grapefruit terpenes may be indicated if the ratio α-pinene/β-pinene is greater than 0.8.

In contrast to its terpenes, the detection of grapefruit oil in bitter orange oil is, as already stated, relatively easily achieved by HPLC. The HPLC pattern of a bitter orange oil that has been adulterated with grapefruit oil will clearly show two additional peaks due to the presence of auraptene (2) and epoxyauraptene (3). For the reasons given previously, HPLC is not appropriate for detecting sweet orange oil in bitter orange oil.

It should perhaps be noted that Dugo *et al.* (1993) pointed out that bitter orange oil might be accidentally contaminated with small amounts of sweet orange and/or lemon oil. This arises because there may be some sweet orange trees within the bitter orange groves and, in addition, the production lines are often those used for lemon oil. Nevertheless, the deliberate adulteration of bitter orange oil with lemon terpenes is known to occur. Unlike pure bitter orange oil, such oils will contain a significant content of α-terpinene derived from the lemon product.

Dugo *et al.* (1994) reported that the linalool content of sweet orange oil is generally greater than that of bitter orange oil and has a different enantiomeric distribution. In sweet orange oil the (S)(+)-isomer predominates, occurring as 89–96 per cent of the total, while in bitter orange oil the linalool consists of 80–87 per cent of the (R)(−)-isomer. Using an automated on-line HPLC-HRGC with a modified β-cyclodextrin coated capillary column to determine the enantiomeric ratio of linalool, they were able to detect the addition of as little as 5 per cent of sweet orange oil to bitter orange oil.

Mandarin oil: Citrus reticulata Blanco

Cold-pressed mandarin oil is a widely used and highly desirable natural flavouring and perfumery ingredient. The genuine oil is produced on a limited scale and, therefore, is significantly more expensive than sweet orange oil. For uses that are less demanding in terms of flavour and odour, there is a cheaper product available that is basically sweet orange terpenes to which γ-terpinene and methyl N-methylanthranilate have been added. This type of product has been used for the adulteration of genuine mandarin oil. However, because the optical rotation of sweet orange oil is approximately 30 per cent higher than that of mandarin oil, it is generally adjusted by the addition of an appropriate amount of (*S*)(−)-limonene, a compound which is produced commercially from the (*S*)(−)-α-pinene of turpentine. Dugo *et al.* (1992) report that chiral analysis using MGDC will detect the atypical enantiomeric distribution of limonene in mandarin oils that have been adulterated in this way.

Adulteration of cold-pressed mandarin oil with the lower quality steam-stripped oil recovered from the screw-pressed residues and the aqueous sludge discharged from the centrifuge, still occurs nowadays. Verzera *et al.* (1992) compared the compositions of the two types of oil and noted that the distilled oil is richer in alcohols such as terpinen-4-ol (9) and α-terpineol (10) (For structures 9–14 see Figure 21.5). The origin of the higher levels of these compounds in the distilled oils is not clear since the likely precursors, sabinene and β-pinene, show no significant difference in content between the two types of oil. Nevertheless, the data provided suggest that the ratios of terpinen-4-ol to *cis*- and *trans*-sabinene hydrate, citronellal, and decanal may be useful in detecting the addition of the steam-stripped oil to cold-pressed oil.

Grapefruit oil: Citrus paradisi Macfadyen

The adulteration of cold-pressed grapefruit oil follows a similar pattern to that used with sweet orange oil. The lower quality steam-stripped product, grapefruit terpenes, and sweet orange oil and its terpenes are all used to extend cold-pressed grapefruit oil. The presence of δ-3-carene in a grapefruit oil is indicative of dilution with orange products. The detection of added steam-stripped oil or grapefruit terpenes still poses

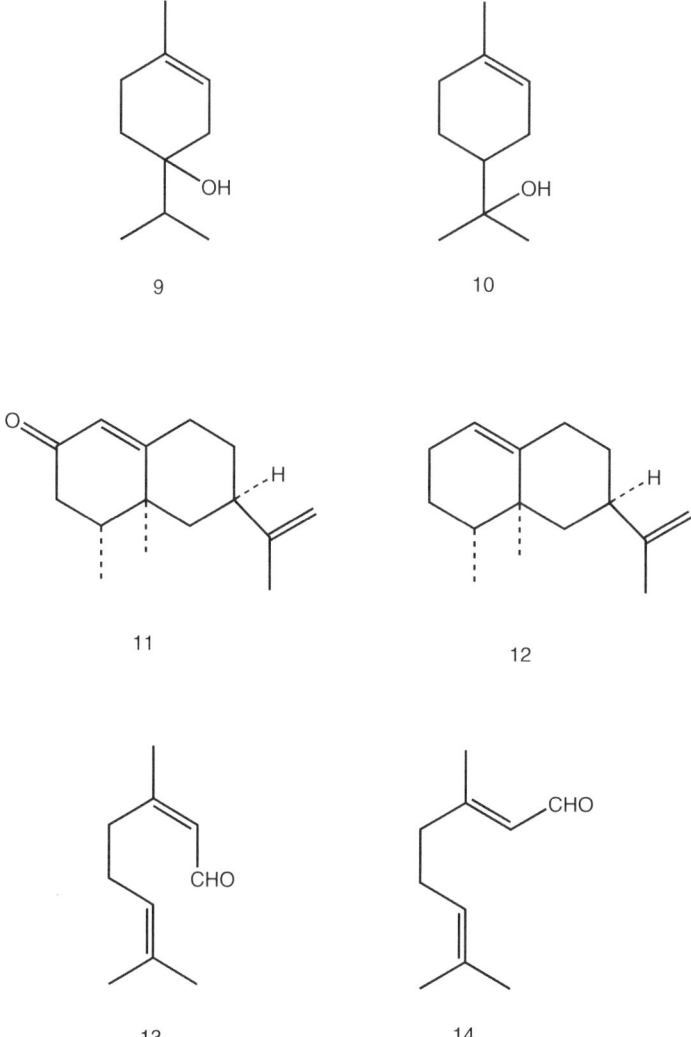

Figure 21.5 Chemical structures of some components of the volatile fraction of citrus oils (9–14).

a problem unless the dilution is so great that the CD value is reduced below the minimum accepted figure. It is possible that UV absorbance enhancers may also be added but oils treated in this way should show an atypical HPLC pattern.

As is the case with sweet orange oil, the higher priced terpeneless and sesquiterpeneless products of grapefruit oil may have their flavour potential enhanced by the addition of certain key components such as nootkatone (11). This compound is produced commercially by the oxidation of valencene (12), a sesquiterpene isolated from sweet orange oil. The detection of this type of adulteration may be possible by stable isotope ratio analysis. Schmidt *et al.* (1998) have pointed out that while the δ^{13}C-value (−30.9‰) of the natural compound is only slightly more positive than a sample (δ^{13}C-value −31.4

to −31.8‰) prepared by the oxidation of valencene, there is a much greater difference in their D/H ratios. In this case, the deuterium content of natural nootkatone is somewhat more negative than the semi-synthetic compound (−160‰ and −112 to −124‰, respectively).

Lemon oil: *Citrus limon* (L) *Burman*

Cold-pressed lemon oil is probably more frequently adulterated than any other citrus oil. The commonest type of adulteration involves dilution with the lower quality steam-stripped oil and the addition of UV absorbance enhancers to mask the reduction in CD value. Although the volatile composition of steam-stripped oil shows minor differences from that of cold-pressed oil, it is difficult to detect its presence in the genuine oil with any degree of certainty. The recognition of lemon oils adulterated with steam-stripped oil relies primarily on the detection of the presence of any UV enhancer. Whilst a few of the commonly used enhancers, such as ethyl *p*-dimethylaminobenzoate, are sufficiently volatile to be detected by GLC, the majority are non-volatile and generally require to be separated by an alternative chromatographic procedure prior to detection/identification by UV-spectroscopy. When suspect oils are compared with genuine oils by HPLC, they often show additional peaks. For instance, in Figure 21.6 the HPLC pattern of a genuine lemon oil is compared with that of a commercial oil suspected of being diluted with steam-stripped oil. There are obviously major differences between the two patterns. The significant early eluting peak (A) present in the commercial oil, but absent from the genuine oil, confirms that ethyl *p*-dimethylaminobenzoate has been added to the former. Furthermore, one of the minor peaks observed in the genuine oil is masked by peak (B) which was shown to be due to the presence of auraptene (2), indicating that grapefruit oil or its residue had also been added to the commercial oil. It would seem very likely that this particular commercial oil was also diluted with steam-stripped oil since the concentrations of the lemon coumarins and psoralens were, in all cases, considerably below those of genuine cold-pressed oil.

It can be seen from Table 21.3 that lime oil has a considerably greater coumarin and psoralen content than any other citrus oil. Because of this, it or the residues that separate when it is stored at sub-ambient temperatures are often used to enhance the CD value of a lemon oil. The presence of 5-geranyloxy-8-methoxypsoralen (7) and

Table 21.3 Oxygen heterocyclic content of citrus peel oils (McHale and Sheridan, 1989)

	W(%)
Lime	6.39
Grapefruit	3.17
Bergamot	2.42
Bitter orange	0.89
Lemon	0.77[a]
Mandarin	0.37
Tangerine	0.25
Sweet orange	0.20

Note
a McHale and Sheridan (1988).

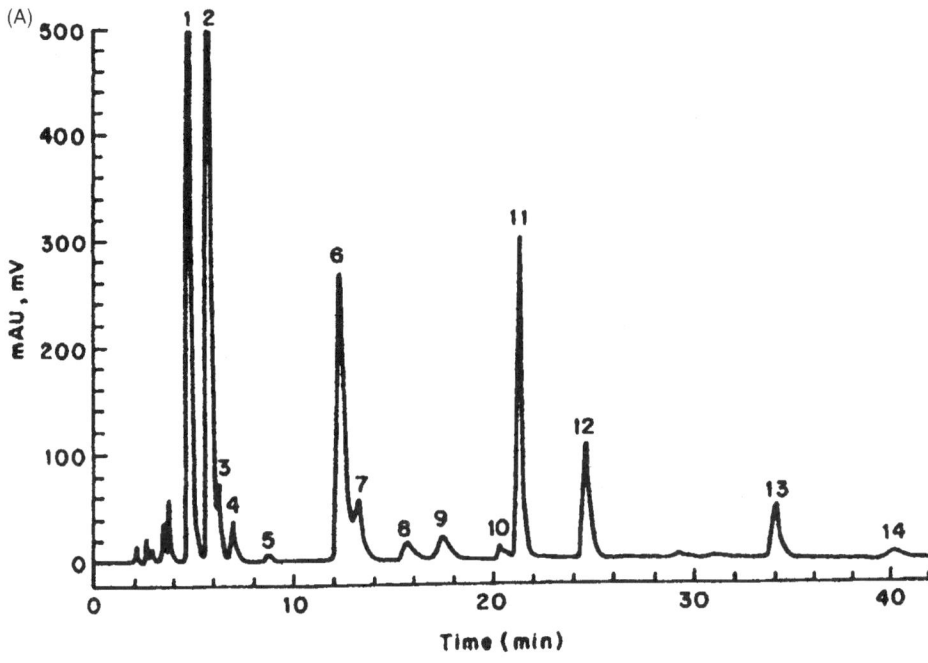
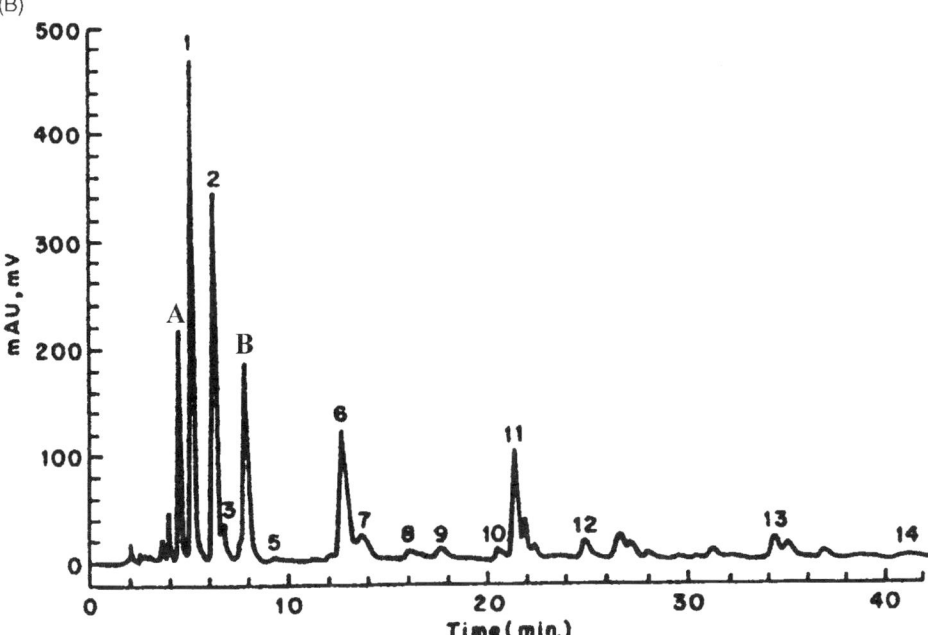

Figure 21.6 Comparison of the HPLC peak patterns of (A) authentic cold-pressed lemon oil and (B) a commercial oil. Column, flow rate, detection and sample volume as in Figure 21.4; gradient elution, solvent A hexane-ethyl acetate (9:1), solvent B hexane-ethanol (9:1), 2–95 per cent B over 25 min. The additional peaks A and B in the commercial oil are due to the presence of ethyl *p*-dimethylaminobenzoate and auraptene.

5,8-dimethoxypsoralen (isopimpinellin 8), which occur at a significant concentration in lime oil but are not found in genuine lemon oil, is indicative of such UV absorption enhancement. Coumarins reported by McHale and Sheridan (1988) as being added to enhance the CD value of lemons oils include 7-methoxycoumarin (herniarin 4), a known trace constituent of genuine lemon oil (Dugo *et al.*, 1998), and 5,7-dipropyloxy-coumarin (5), a compound not known to occur in nature.

Sweet orange oil and its terpenes, and those of lemon oil, are also often used to dilute cold-pressed lemon oil. In such cases, sufficient citral from an alternative source is added to prevent the aldehyde content of the resulting oil falling below the accepted value. Oils which have been diluted with orange products will, in contrast to genuine lemon oil, clearly show the presence of δ-3-carene (Verzera *et al.*, 1987). In addition, they may have relatively high optical rotations since that of sweet orange oil or its terpenes ($[\alpha]_D +98°$ to $+100°$) is significantly greater than that of lemon oil ($[\alpha]_D +57°$ to $+62°$). According to Dugo *et al.* (1993), sufficient (−)-limonene from a cheap source is usually added to counteract the excess of (+)-limonene in the orange product, thereby ensuring that the optical rotation of the diluted oil does not exceed that of a genuine lemon oil. If this practice has been adopted, chiral analysis of the terpene fraction will highlight the elevated level of (−)-limonene present.

Lemon oil terpenes obtained as by-products of the manufacture of terpeneless and sesquiterpeneless oils are also used to extend genuine cold-pressed oil. As was the case with steam-stripped oil, the detection of this type of adulteration is difficult, particularly if the dilution has been carefully carried out and is not excessive. The oils resulting from such treatment generally have low CD values and a low content of oxygenated compounds. The latter can be disguised by the addition of certain key compounds, such as citral, neryl and geranyl acetate, linalool and α-terpineol, which have been derived from a non-lemon source. This may result in atypical $\delta^{13}C$ values for these particular components of the oil.

Terpeneless and sesquiterpeneless lemon oils are, like the corresponding sweet orange products, often brought up to strength by the addition of certain key components. Citral is the major additive and, nowadays, this is generally derived from *Litsea cubeba*. Unfortunately, it is not yet possible to reliably distinguish this from the natural citral of lemon oil. Sheridan *et al.* (1987) compared the $\delta^{13}C$ values of some commercial samples of citral from different sources with that of the natural compound isolated from lemon oil. Apart from a significant difference in the value of the sample from *Cymbopogon citratus* (lemon grass), which is a C4-plant, it can be seen from Table 21.4 that the

Table 21.4 Comparison of $\delta^{13}C$-values for commercial citrals from various sources with those from lemon oil (Sheridan *et al.*, 1987)

Source	GLC Purity%	$\delta^{13}C$ (±0.3‰)
Synthetic ex pinene	98.1	−28.2
Synthetic	99.1	−27.2
Litsea cubeba	99.2	−25.5
Natural	98.1	−23.7
Natural	98.9	−25.9
Commercial ex lemon oil	99.1	−20.6
Isolated ex Greek lemon oil	99.1	−21.8
Isolated from *Cymbopogon citratus*	98.9	−10.9

results were inconclusive, although the two samples from lemon oil had the lowest negative values.

More recent work (Thorpe et al., 1997) has indicated that it may be possible to detect by GLC those lemon oils which contain citral that has been rectified by fractional distillation. During a GC/MS study of the minor constituents of rectified citral from both *Litsea cubeba* and lemon grass oil, it was noticed that among the compounds present there were three that appeared to be isomeric with neral (13) and geranial (14). The similarity of the mass spectra of compounds (15) and (16) suggested that, like neral and geranial, they were *cis*- and *trans*-isomers. As can be seen from Figure 21.7, the third isomer (17) had a significantly different spectrum. The compounds were identified as the isocitrals (15-17) which were reported by Ohloff (1960) to be formed as a result of thermal isomerisation of the 2,3-double bond of citral under distillation conditions. They can also be observed on chromatograms of terpeneless lemon oil but are absent from those of genuine cold-pressed lemon oil. They are well separated from neral and geranial by GLC on a $25\,m \times 0.2\,mm \times 0.33\,\mu m$ HP1 column (Kovats indices: 1220, 1243, 1141, 1159, and 1121 for 13–17, respectively).

Lime oil: Citrus aurantifolia Swingle

Although cold-pressed lime oil is becoming increasingly popular for imparting freshness and stability to lemon-lime flavoured soft drinks and other products, it is still only produced on a minor scale in comparison with the other citrus oils. Furthermore, there are two distinct methods of production in use and these give rise to the different cold-pressed oils known as Type A and Type B (Haro-Guzman, 1980). Most of the oil produced is of the Type A variety and is isolated by centrifugation of the emulsion resulting from the screw-pressing of whole fruit. This oil has an odour strongly resembling that of fresh lime juice. For the production of Type B oil, the outer skin is removed by rasping and washed from the fruit by a stream of water. The oil is then recovered by centrifugation.

As mentioned earlier, cold-pressed lime oil and the non-volatile residue derived from it have been used to enhance the CD value of lemon oils. Nowadays, with the growing market for cold-pressed lime oil, the situation has changed and it is itself sometimes extended with the cheaper and more readily available distilled lemon oil. Oils adulterated in this way are not easily recognisable although a higher than usual optical rotation should be viewed with suspicion.

The production of, and the demand for, distilled lime oil greatly exceed those of the cold-pressed oil. In contrast to other citrus oils, no attempt is made in the case of lime to keep the peel oil and the juice separate during processing. The whole fruit is crushed, either in a roller mill or a screw press, with the result that the oil and the highly acidic juice become intimately mixed. The oil is then recovered by distillation and, because of the acidity of the medium, the bicyclic monoterpenes undergo isomerisation and hydration reactions to form alcohols and ethers (McHale, 1980). Since the quality of the distilled oil is frequently judged by the amount of the oxygen containing fraction, in particular the terpene alcohols, there is a temptation for some suppliers to increase this fraction by the addition of appropriate compounds from alternative sources, thereby also permitting further dilution of the oil with a suitable mixture of terpenes. At present, the recognition of such oils is difficult but is sometimes possible through

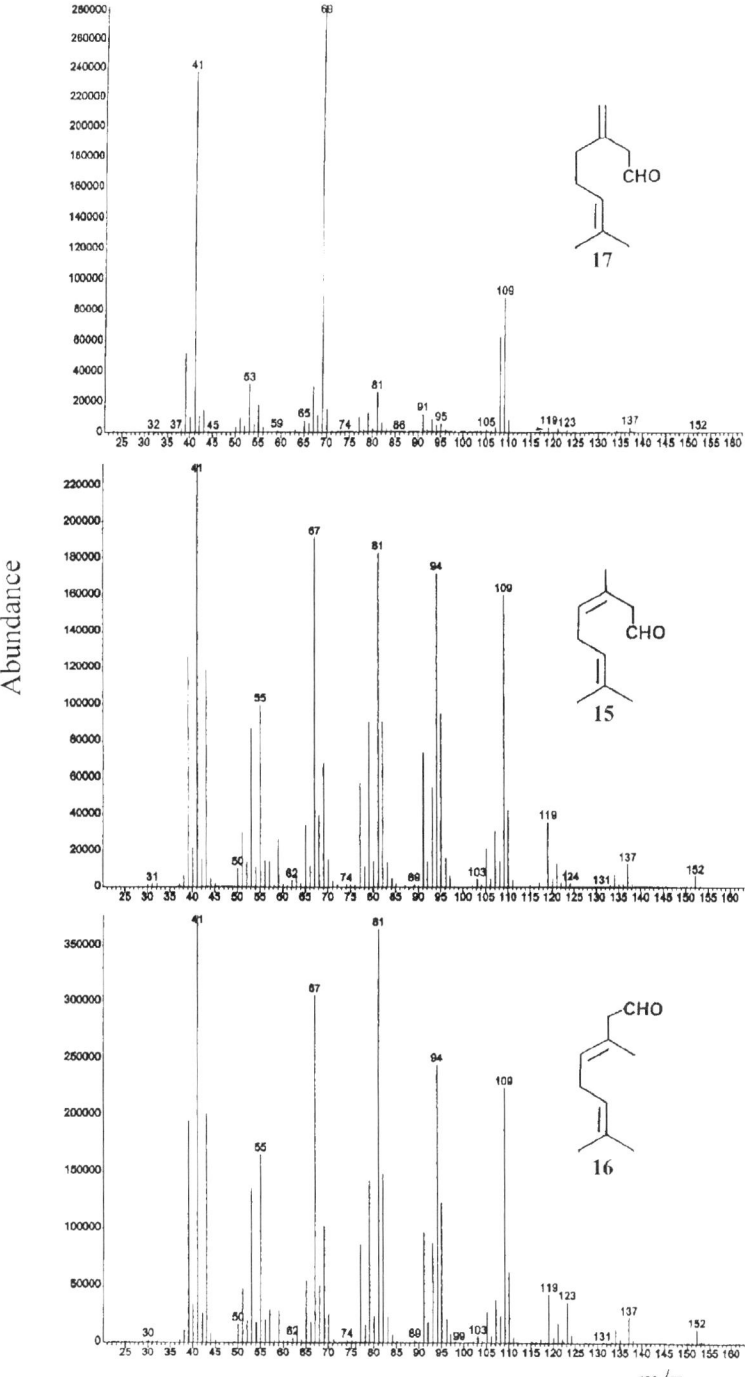

Figure 21.7 GC/MS spectra for the isocitrals (15–17).

the detection of impurities present in the added terpene alcohol fraction. As regards the more recently introduced methods for establishing the authenticity of natural products, doubts exist as to whether these will be effective in solving the problem. For instance, Baigrie et al. (1996) have pointed out that the enantiomeric distibution of the monoterpenes are little changed under the acidic distillation conditions but the terpene alcohols undergo significant racemisation. They conclude that chiral analysis would not be suitable for establishing the authenticity of the distilled oil.

The detection of added α-terpineol in distilled lime oil by stable isotope ratio analysis would also appear to offer little chance of success. It is now generally accepted (Clark and Chamblee, 1992) that the α-terpineol of distilled lime oil is generated mainly from the β-pinene of the cold-pressed oil under the acidic distillation conditions. Since, commercial α-terpineol is also produced from natural α- and/or β-pinene, it would seem unlikely that its $\delta^{13}C$ and $\delta^{2}H$-values will differ significantly from those of the α-terpineol of distilled oil. It is possible that techniques such as SNIF-NMR© will highlight differences in the isotopic ratio at specific molecular sites of commercial α-terpineol and that from lime oil. However, the practicality of using this particular form of analysis to detect the adulteration of commercial samples of lime oil must be questionable at the present time.

Bergamot oil: Citrus bergamia Risso

Bergamot oil is almost exclusively produced in Calabria (Italy) and is the most highly valued of the citrus oils. Its main usage is in the perfumery industry and, because of its limited availability and high cost, it has frequently been subjected to adulteration. Nowadays, the most likely type of adulteration involves the dilution of genuine oil with cheaper substitutes produced by mixing linalool and linalyl acetate with terpenes of different origins. Because of the variability of the quantitative composition of genuine bergamot oil, such additions may be difficult to detect. However, Cotroneo et al. (1992) have pointed out that the linalool of genuine bergamot oil is essentially the pure (R)(−)-enantiomer while that used in the preparation of the cheaper substitutes is likely to be synthetic and, possibly, racemic. They suggest that if the total linalool content of a bergamot oil determined by enantio-MDGC contains more than 0.5 per cent of (S)(+)-linalool then the oil is likely to have been adulterated. Similarly, as the linalyl acetate of genuine oil is almost exclusively the (R)(−)-enantiomer, the presence of any (S)(+)-linalyl acetate will also indicate an adulterated oil (Mosandl and Juchelka, 1997). This is useful since the concentration of linalyl acetate in bergamot oil can exceed that of linalool by two to five-fold, and the latter may be omitted from the mixture used to dilute the genuine oil. Because the dilution of a genuine oil with terpenes will reduce the CD value, an adulterated oil may also contain substances added to compensate for the drop in absorbance. Mondello et al. (1993) reported that certain commercial bergamot oils showed atypical coumarin and psoralen patterns when examined by HPLC. They attributed these to the presence of lime oil or its residues.

Finally, it is important to be aware that if any of the above citrus oils are offered at a price significantly below the accepted market rate, then the quality of the product should be questioned. Deterioration may have occurred, but more likely the oil will have been adulterated.

REFERENCES

Baigrie, B.D., Chisholm, M.G. and Mottram, D.S. (1996) The effects of processing on chiral aroma compounds in fruits and essential oils. In A.J. Taylor and M.S. Mottram (eds), *Flavour Science Recent Developments*, Royal Society of Chemistry, London, pp. 151–157.

Biot, J.B. (1815) *Bull. Soc. Philomath.*, 190 cited by J.R. Partington (1953) in *An Advance Treatise on Physical Chemistry*, **IV**, Longman, London, p. 291.

Boswigi, G. (1899) Oil of lemon: A note on adulteration of the essence with stearin. *Chem. Drug.*, **55**, 710–711.

Braunsdorf, R., Hener, U. and Mosandl, A. (1992) Analytische Differenzierung zwischen naturlich gewachsenen, fermentativ erzeugten und synthetischen (naturidentischen) Aromastoffen. II. Mitt.: GC-C-IRMS Analyse aromarelevanter Aldehyde-Grundlagen und Anwendungsbeispiele. *Z. Lebensm. Unters. Forsch.*, **194**, 426–430.

Calvarano, I. and Di Giacomo, A. (1970) A proposito di una contaminazione dell'essenza di limone. *Essenz. Deriv. Agrum.*, **40**, 135–142.

Chakraborty, D.P. and Bose, P.K. (1956) Paper chromatographic studies of some natural coumarins. *J. Ind. Chem. Soc.*, **23**, 905–910.

Clark Jr., B.C. and Chamblee, T.S. (1992) Acid catalyzed reactions of citrus oils and other terpene-containing flavours. In G. Charalambous (ed.), *Off-Flavours in Foods and Beverages*, Elsevier, Amsterdam, pp. 229–285.

Cotroneo, A., Stagno d'Alcontres, I. and Trozzi, A. (1992) On the genuineness of citrus essential oils. Part XXXIV. Detection of added reconstituted bergamot oil in genuine bergamot essential oil by high resolution gas chromatography with chiral capillary columns. *Flav. Fragr. J.*, **7**, 15–17.

Cultrera, R., Buffa, A. and Trifiro, E. (1952) Spectrophotometric analysis in the evaluation of lemon oils. *Conserve Deriv. Agrum. (Palermo)*, **1**(2), 18–20.

D'Amore, G. and Calapaj, R. (1965) Le sostanze fluorescenti contenute nelle essenze di limone, begamotto, mandarino, arancio dolce, arancio amaro. *Rass. Chim.*, **17**, 264–269.

De Domenico, V. (1854) Sulla virtu medicamentosa dell'essenza di bergamotto. From a reprint held by Stazione Sperimentale per l'Industria delle Essenze, Reggio Calabria.

Dugo, G., Cotroneo, A., Licandro, A., Dugo, G. and Verzera, A. (1984) Ricerche di aggiunte di 'citral' alle essenze di limone mediante gascromatografia ad alta risoluzione. *Riv. Ital. Sostanze Grasse*, **61**, 441–453.

Dugo, G., Stagno d'Alcontres, I., Donato, M.G. and Dugo, P. (1993) On the genuineness of citrus essential oils. Part XXXVI. Detection of added reconstituted lemon oil in genuine cold-pressed lemon essential oil by high resolution gas chromatography with chiral capillary columns. *J. Essent. Oil Res.*, **5**, 21–26.

Dugo, G., Stagno d'Alcontres, I., Cotroneo, A. and Dugo, P. (1992) On the genuineness of citrus essential oils. Part XXXV. Detection of added reconstituted mandarin oil in genuine cold-pressed mandarin essential oil by high resolution gas chromatography with chiral capillary columns. *J. Essent. Oil Res.*, **4**, 589–594.

Dugo, G., Verzera, A., Cotroneo, A., Stagno d'Alcontres, I., Mondello, L. and Bartle, D. (1994) Automated HPLC-HRGC: a powerful method for essential oil analysis. Part II. Determination of the enantiomeric distribution of linalool in sweet orange, bitter orange and mandarin essential oils. *Flav. Fragr. J.*, **9**, 99–104.

Dugo, G., Verzera, A., Stagno d'Alcontres, I., Cotroneo, A. and Ficarra, R. (1993) On the genuineness of citrus essential oils. Part XLI. Italian bitter orange essential oil: composition and detection of contamination and additions of oils and terpenes of sweet orange and of lemon. *Flav. Fragr. J.*, **8**, 25–23.

Dugo, P., Mondello, L., Cogliandro E., Cavazza, A. and Dugo, G. (1998) On the genuineness of citrus essential oils. Part LIII. Determination of the composition of the oxygen heterocyclic fraction of lemon essential oils (*Citrus limon* (L.) Burm. f.) by normal-phase high performance liquid chromatography. *Flav. Fragr. J.*, **13**, 329–334.

Dugo, P., Mondello, L., Cogliandro E., Stagno d'Alcontres, I. and Cotroneo, A. (1994) On the genuineness of citrus essential oils. Part XLVI. Polymethoxylated flavones of the non-volatile residue of Italian sweet orange and mandarin essential oils. *Flav. Fragr. J.*, 9, 105–111.

E.O.A. (1970) *E.O.A. Specification and Standards*, Essential Oil Association of U.S.A. No. 1-1D-4.

Gildermeister, E. and Hoffmann, F. (1959) *Die Aetherischen Oele*, 5, Akademie-Verlag, Berlin.

Guenther, E. (1952) *The Essential Oils*, 3, D. Van Nostrand Book Co., New York.

Haro-Guzman, L. (1980) Les essences de lime mexicaine. Changements dans la composition dus a la technique d'extraction. *Essenz. Deriv. Agrum.*, 50, 332–344.

Hooke, R. (1665) *Micrographia* cited by J.R. Partington (1953) in *An Advance Treatise on Physical Chemistry*, **IV**, Longman, London, p. 23.

Latz, H.W. and Ernes, D.A. (1978) Selective fluorescence detection of citrus oil components separated by high-pressure liquid chromatography. *J. Chromatogr.*, 166, 189–199.

MacLeod Jr., W.D., Lundin, R.E. and Buigues, N.M. (1964) The detection of benzyl ether added to lemon oil. *J. Food Sci.*, 29, 802–803.

Madsen, B.C. and Latz, H.W. (1970) Qualitative and quantitative *in situ* fluorimetry of citrus oil thin-layer chromatograms. *J. Chromatogr.*, 50, 288–303.

Martin, M.L. and Martin, G.J. (1990) Deuterium NMR in the study of site-specific natural isotope fractionation (SNIF-NMR). In P. Dietl, E. Fluck, H. Gunther, R. Kosfeld and J. Seelig (eds), *NMR Basic Principles and Progress*, Vol. 23, Springer Verlag, Berlin, pp. 1–61.

McHale, D. and Sheridan, J.B. (1989) The oxygen heterocyclic compounds of citrus peel oils. *J. Essent. Oil Res.*, 1, 139–149.

McHale, D. (1982) Effect of processing conditions on composition of distilled oil of lime. Paper No. 67, *Proceedings of the VIII International Congress of Essential Oils*, Cannes, October, 1980.

McHale, D. and Sheridan, J.B. (1988) Detection of adulteration of cold-pressed lemon oil. *Flav. Fragr. J.*, 3, 127–133.

Mondello, L., Stagno d'Alcontres, I., Del Duce, R. and Crispo, F. (1993) On the genuineness of citrus essential oils. Part XL. The composition of the coumarins and psoralens of Calabrian bergamot essential oil (*Citrus bergamia* Risso). *Flav. Fragr. J.*, 8, 17–24.

Morton, R.A. (1929) Radiation in connection with essential oils and perfumery chemicals. *Perfumery and Essential Oil Record*, 20, 258–267.

Mosandl, A. and Juchelka, D. (1997) Advances in the authenticity assessment of citrus oils. *J. Essent. Oil Res.*, 9, 5–12.

Mosandl, A. and Schubert, V. (1990) Stereoisomeric flavour compounds XXXVII: enantiomer separation of 1-alken-3-yl esters and their chirality evaluation from essential oils using MDGC. *J. Essent. Oil Res.*, 2, 121–132.

Ohloff, G. (1960) Zur thermischen isomerisation von citral. *Tetrahedron Letters*, 11, 10–14.

Parry, E.J. (1909) Lemon oil. *Chem. Drug.*, 74, 121.

Pomerantz, C. (1891) Uber das bergapten. *Monatsh. f. Chem.*, 12, 379–392.

Sale, J.W., Winkler, W.O., Gnagy, M.J., Hart, F.L., Hess, S.M., Kinney, J.K., Kirsten, G., Marder, J., Miller, D.C. and Wood, G.R. (1953) Analysis of lemon oils. *J. Assoc. Offic. Agr. Chemists*, 36, 112–119.

Schmidt, H.-L, Rossmann, A. and Werner, R.A. (1998) Stable isotope ratio analysis in quality control of flavourings. In E. Ziegler and H. Ziegler (eds), *Flavourings: Production, Composition, Applications and Regulations*, Wiley-VCH, Weinheim, pp. 539–594.

Semmler, F.W. and Tiemann, F. (1892) Ueber sauerstoffhaltige bestandtheile einiger aetherischer oele. *Ber.*, 25, 1180–1188.

Sheridan, J.B., Laurie, W.A. and McHale, D. (1987) Unpublished work.

Stanley, W.L. (1959) Determination of menthyl salicylates in lemon oil. *J. Assoc. Offic. Agr. Chemists*, 42, 643–646.

Stanley, W.L. (1961) A test for chalcones in lemon oil. *J. Assoc. Offic. Agr. Chemists*, 44, 546–548.

Stanley, W.L. (1963) Recent developments in coumarin chemistry. *Aspects of Plant Phenolic Chemistry. Proc. Symp., Plant Phenol Group North America*, Toronto, Ontario, Canada, pp. 79–103.

Stanley, W.L. and Vannier, S.H. (1957a) Chemical composition of lemon oil. I. Isolation of a series of substituted coumarins. *J. Amer. Chem. Soc.*, 79, 3488–3491.

Stanley, W.L. and Vannier, S.H. (1957b) Analysis of coumarin compounds in citrus oils by liquid solid partition. *J. Assoc. Offic. Agr. Chemists*, 40, 582–588.

Thorpe, S.D., Gates, L.M. and McHale, D. (1997) Unpublished work.

Tilden, W.A. and Burrows, H. (1902) The constitution of limettin. *J. Chem. Soc.*, 508–512.

U.S.P. (1965) *United States Pharmacopoeia*, 17th revision, Mack Publishing Co., Easton, Pa., p. 922

Verzera, A., Cotroneo, A., Stagno d'Alcontres, I. and Donato, M.G. (1992) On the genuineness of essential oils. Part XXX. Detection of distilled essential oils added to cold-pressed mandarin essential oils. *J. Essent. Oil Res.*, 4, 273–280.

Verzera, A., Cotroneo, A., Dugo, G. and Salvo, F. (1987) On the genuineness of citrus essential oils. Part XV. Detection of added orange oil terpenes in lemon essential oils. *Flav. Fragr. J.*, 2, 13–16.

Weinreich, B. and Nitz, S. (1992) Influence of processing on the enantiomeric distribution of chiral flavour compounds. Part 1. Linalyl acetate and terpene alcohols. *Chem. Mikrobiol. Technol. Lebensm.*, 14(3/4), 117–124.

22 Contaminants in citrus essential oils

Giacomo Dugo and Giuseppa Di Bella

INTRODUCTION

Recently, enormous importance has been placed on the monitoring of alimentary products for contamination by pesticides and their metabolites, mainly as a result of the pressure from consumers, who are much more concerned today about the quality and integrity of their foodstuffs than they were in the past.

The need to protect crops and crop-yields from attack by parasites is beyond dispute, but unfortunately the massive use of pesticides has significantly altered the ecological balance in many instances and has caused a number of serious health problems.

The organochlorine pesticides most commonly used in the cultivation of citrus fruit, such as dicofol, penetrate the tegumental tissues of the plant. They may also migrate within the plant, but only in very limited quantities. These contaminants are extremely liposoluble. Phosphoric ester (or 'systemic') pesticides, on the other hand, travel through the vascular system of the plant and are absorbed at the cellular level. To avoid potential health problems caused by pesticide residues in the whole fruit, strict regulations have been introduced, imposing threshold limits and quarantine periods, which do not concern us here. However, high organochlorine and organophosphorus residue concentrations are found in the peel, and therefore in the essential oils extracted from it, because of the physico-chemical properties of these pesticides. Although citrus essential oils are not alimentary products as such, producers and dealers often sell them to the food and cosmetic industries, which require products that are pesticide-free or contain very low residue levels.

Unfortunately, contamination of citrus essential oils is not only due to chlorine and phosphorous pesticide residues, but also derives from the presence of phosphorated plasticizers, chloroparaffins, and phthalates. These compounds, present as additives in the plastic components of the machinery used to extract and/or decant the oils, are transferred to the essential oils during the production cycle.

PESTICIDES

Organophosphorus pesticides

Pesticide residues in citrus essential oils have been scrutinised since the 1960s.

Stevens (1967) documented the presence of organophosphorus pesticide residues of 1.5–20 ppm cold-pressed oils obtained from citrus fruit harvested in the crop year 1965–66 in California and Florida.

Table 22.1 Organophosphorus pesticide (ppm) in citrus oils (Leoni and D'Alessandro De Luca, 1978)

	Lemon	Orange	Mandarin	Bergamot
Methyl-parathion	0.52–5.95	n.d.–1.91	2.87–9.30	n.d.–0.76
Parathion	0.22–7.05	0.51–11.52	2.26–9.14	n.d.–1.00
Paraoxon	n.d.	0.73–15.73	n.d.	n.d.
Phentoate	n.d.–3.59	n.d.–0.99	n.d.–4.45	n.d.
Malathion	n.d.–2.87	n.d.–0.26	n.d.–0.69	1.98–8.60
Bromophos	n.d.–0.82	1.81–2.65	n.d.	–
Fenitrothion	n.d.–0.97	n.d.–0.71	n.d.	–

Note
n.d. not detectable.

Günther (1969) found that laboratory-prepared lemon and orange oils from fruit which had been treated with malathion contained residues of 6–10 ppm, whereas commercial oils contained between 12 and 450 ppm of parathion.

Dupuis (1975) detected methidathion levels of 65 ppm in oils obtained from treated orange crops.

The first investigations of essential oils produced in Italy go back to 1978. Leoni *et al.* (1978) were able to separate, by column chromatography on disactivated silica gel, organophosphorus from organochlorine pesticide residues following a rather laborious procedure which involved dissolving the oil in petroleum ether, extracting it with acetonitrile saturated with petroleum ether, and then re-extracting the acetonitrile solution with petroleum ether containing a 2 per cent sodium chloride solution. The residues were analyzed by gas chromatography with an alkaline flame ionization detector (GC-AFID) and an electron capture detector (GC-ECD), using packed columns, and the results obtained are reported in Table 22.1. As can be seen from this table, detectable levels of methylparathion and parathion, which are used in large quantities on citrus crops, were present in almost all the samples analyzed.

In the 1980s, research was undertaken to examine the contamination of lemon, orange, mandarin, and bergamot essential oils produced in Sicily and Calabria (Italy) (Dugo *et al.*, 1987, 1990, 1992, 1994, 1997; Di Bella *et al.*, 1991, 1995). A total of 305 lemon, 146 sweet orange, 84 mandarin and 138 bergamot oil samples were analyzed. The years of production and the analytical conditions used are given in Table 22.2. All the essential oils were of Sicilian or Calabrian origin and were produced between 1983 and 1992. An internal standard was added to the essential oil samples and they were then directly tested for the presence of organophosphorus pesticide residues without a preliminary cleanup procedure. The following summary may be made of the results reported in these papers:

Lemon essential oils

The pesticide residues (expressed in ppm) found in lemon oils of different production years are reported in Table 22.3. The maximum and minimum levels detected for each pesticide are given, along with the percentage of samples contaminated in each case. Between 1983 and 1991, methylparathion, ethylparathion and methidathion were present in the majority

Table 22.2 Analyzed samples and experimental conditions for organophosphorus pesticide analysis

Reference	Production years	Samples analyzed	Injector	Column	Detector
Dugo et al., 1987	1983	33 lemon	Splitter. 250 °C	Capillary fused silica. 25 m × 0.32 mm SE-54. 0.25 µm; 180 °C or 230 °C[a]	NPD 270 °C
Dugo et al., 1990	1984–1986	222 lemon	Splitter. 250 °C	Capillary fused silica. 30 m × 0.25 mm DB-5. 0.25 µm; 170 °C or 230 °C[a]	NPD 270 °C
Di Bella et al., 1991 Dugo et al., 1992	1990–1992	73 sweet orange 50 lemon	PTV. 65–240 °C at 990 °C/min	Capillary fused silica. 30 m × 0.25 mm SPB-5; 75–170 °C (5 min) at 30 °C/min; 170–190 °C at 2 °C/min; 190–265 °C at 30 °C/min	FPD 250 °C
Dugo et al., 1994		73 sweet orange 84 mandarin 9 bergamot			
Di Bella et al., 1995	1992	129 bergamot	PTV. 65–240 °C at 990 °C/min	Capillary fused silica. 5 m × 0.25 mm SE-54. 0.25 µm; 74–140 °C at 30 °C/min; 140–245 °C at 5 °C/min	FPD 250 °C

Notes

[a] 230 °C was used for determination of methyl azinphos and ethyl azinphos; NPD = nitrogen phosphorus detector; FPD = flame photometric detector; PTV = programmed temperature vaporizer.

Table 22.3 Organophosphorus pesticide residues (ppm) in lemon oil (Dugo et al., 1987, 1990, 1992; Di Bella et al., 1995)

	1983[a]	1984[a]	1985[a]	1986[a]	1987[a]	1988[a]	1989[a]	1990–91[a]	All	% of contaminated samples
N° of samples	33	48	39	32	13	57	33	50	305	
Sulphotep	–	–	–	–	–	–	–	0–0.05	0–0.05	0.7
Diazinon	–	0–0.18	0–2.10	0–0.054	0–6.39	0–5.73	0–1.24	0–0.03	0–6.39	29.2
Methyl-parathion	0.65–39.00	0–20.90	0–10.90	0.30–19.30	0.63–18.90	0–35.00	1.09–13.90	0.01–10.27	0–39.00	97.4
Fenitrothion	–	0–14.50	0–0.18	0–2.77	0–0.09	0–1.56	0–0.21	0–0.48	0–14.50	9.2
Methyl-pirimiphos	–	–	–	–	–	–	–	0–0.04	0–0.04	0.7
Malathion	–	0–0.70	0–0.47	0–0.51	–	0–0.31	0–0.32	–	0–0.70	3.6
Ethyl-parathion	tr.–12.10	0.09–17.20	0–12.80	0.62–13.00	0–4.02	0.09–16.00	0.66–10.70	0.03–5.42	0–17.20	99.3
Quinalphos	–	0–14.20	0–10.00	0–7.84	0.07–1.76	0–11.00	0–1.48	0–0.39	0–14.20	53.8
Methidathion	0–54.00	0.06–48.50	0.11–40.10	3.59–40.00	1.14–105.00	0.37–201.00	1.43–33.60	0.01–11.95	0–201.00	98.4
Iodophenphos	–	–	–	–	–	–	–	0–1.04	0–1.04	0.7
Ethyl-bromophos	–	–	–	0–2.79	–	–	–	–	0–2.79	0.3
Ethyl-azinphos	–	–	–	–	–	–	–	0–1.73	0–1.73	4.9

Note
a production season.

Table 22.4 Organophosphorus pesticide residues (ppm) in sweet orange oil (Dugo et al., 1987, 1990, 1992; Di Bella et al., 1991, 1995)

	Sicilian oils					Calabrian oils			
	1989[a]	1991[a]	1992[a]	All	% of contaminated samples	1989[a]	1991[a]	All	% of contaminated samples
N° of samples	42	50	12	104		31	11	42	
Sulphotep	–	–	–	0–0.90	–	–	–	–	–
Dimetoate	0–0.52	0–0.94	–	0–0.94	4.8	0–0.45	–	0–0.45	2.4
Diazinon	0–4.28	0–1.05	0–0.30	0–4.28	53.8	0–3.12	0–0.25	0–3.12	33.3
Methyl-parathion	0–77.20	0.03–14.15	0.34–3.68	0–77.20	98.1	0.04–13.40	0.54–11.55	0.04–13.40	100.0
Fenitrothion	0–0.13	0–0.14	0–0.31	0–0.31	10.6	–	–	–	–
Methyl-pirimiphos	0–0.72	0–0.04	0–0.12	0–0.72	5.8	0–0.12	–	0–0.12	2.4
Malathion	0–1.17	0–0.31	0–0.11	0–1.17	12.5	0–2.19	–	0–2.19	9.5
Ethyl-parathion	0.55–51.40	0.01–6.82	0.52–9.81	0.01–51.40	100.0	1.71–32.10	0.05–3.54	0.05–32.10	100.0
Clorfenvinphos	–	0–0.10	–	0–0.10	1.0	–	–	–	–
Quinalphos	0–11.18	0.01–2.50	0.08–1.02	0–11.18	81.7	–	0.01–2.46	0.01–2.46	26.2
Methidathion	0–97.20	0.06–10.29	0–6.31	0–97.20	97.1	–	0.04–5.83	0.04–5.83	26.2
Ethyl-bromophos	0–12.20	–	0–2.18	0–12.0	4.8	–	–	–	–
Ethion	–	–	0–0.01	0–0.01	1.9	–	–	–	–
Methyl-azinphos	–	0–5.95	0–0.28	0–5.95	8.7	–	0–2.01	0–2.01	2.4
Ethyl-azinphos	–	0–0.03	–	–	1.0	–	–	–	–

Note
a production season.

Table 22.5 Organophosphorus pesticide residues (ppm) in mandarin oil (Dugo et al., 1987, 1990, 1992; Di Bella et al., 1995)

	1990–91[a]	1992[a]	All	% of contaminated samples
N° of samples	10	74	84	
Mevinphos	0–1.28	–	0–1.28	8.3
Sulphotep	0–0.03	–	0–0.03	1.2
Diazinon	0–1.36	0–0.48	0–1.36	10.7
Methyl-parathion	7.10–14.74	0–17.39	0–17.39	88.1
Fenitrothion	–	0–1.13	0–1.13	8.3
Malathion	–	0–0.96	0–0.96	7.1
Ethyl-parathion	1.97–17.74	0–13.35	0–17.74	95.2
Phentoate	–	0–1.26	0–1.26	1.2
Quinalphos	0–1.04	0–1.42	0–1.42	50.0
Mecarbam	–	0–8.92	0–8.92	5.9
Methidathion	0.31–6.03	0–16.02	0–16.02	91.7
Methyl-azinphos	0–1.01	0–0.77	0–1.01	10.7
Ethyl-azinphos	0–0.56	0–0.16	0–0.56	2.4
Coumaphos	–	0–0.56	0–0.56	8.3

Note
a production season.

of samples analyzed from each production year. The oils produced in 1988 showed the highest levels of contamination (with methidathion residues reaching a maximum level of 201 ppm), while those produced in the 1990–1991 season showed the lowest levels.

Sweet orange essential oils

The results obtained for Sicilian and Calabrian sweet orange oils are reported in Table 22.4. Most of the Sicilian oil samples contained methylparathion, ethylparathion, quinalphos and methidathion, and about 50 per cent also contained diazinone. All the Calabrian oil samples contained methylparathion and ethylparathion. As in the case of the lemon oil, the sweet orange oils produced in 1991 and 1992 (both Sicilian and Calabrian oils) were less contaminated than those produced in 1989.

Mandarin essential oils

As can be seen from Table 22.5, the majority of mandarin oil samples contained methylparathion, ethylparathion and methidathion, and 50 per cent also contained quinalphos. A chromatogram of a contaminated mandarin oil is reported in Figure 22.1.

Bergamot essential oils

Compared to the other oils analyzed, bergamot oils showed generally lower levels of contamination and in 50 per cent of the samples the residues were below the detection limits (Table 22.6). The pesticides most frequently detected in the contaminated samples were methylparathion, ethylparathion and methidathion.

Dellacassa et al. (1995, 1999) investigated the levels of organophosphorus pesticide contamination in 83 Uruguayan lemon oils. The 53 oils produced in the south of

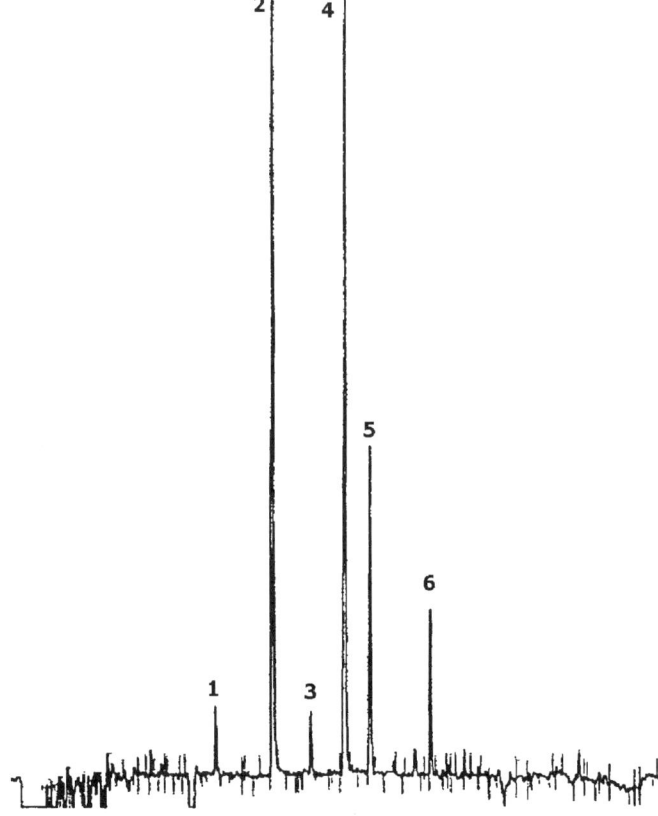

Figure 22.1 FPD chromatogram of a mandarin oil (Dugo *et al.*, 1997): (1) diazinon; (2) methylparathion; (3) fenitrothion; (4) ethylparathion; (5) methylbromophos; (6) methidathion.

Uruguay were all contaminated by fenitrothion and methidathion, with maximum levels of 15.16 and 61.21 ppm respectively, while 45 per cent also contained parathion (maximum level 7.94 ppm). Among the 30 oils produced in the north of the country, 93 per cent were contaminated by fenitrothion (maximum level 2.46 ppm) and 37 per cent contained methidathion (maximum level 7.94 ppm).

Organochlorine pesticides

In addition to organophosphorus pesticides, extensive use is made of organochlorine pesticides in the cultivation of citrus fruit. Since they are extremely liposoluble, these pesticides tend to accumulate in the oil glands of the fruit. Before testing for the presence of organochlorine pesticide residues by gas chromatography, it is necessary first to carry out a cleanup procedure on the samples (Dugo *et al.*, 1997; Saitta *et al.*, 1995). Saitta *et al.* (2000) determined the levels of aldrin, dieldrin, p,p'-DDE, o,p-DDD, endrin, p,p'-DDD, o,p-DDT, dicofol and tetradifon in 148 lemon oils, 123 sweet orange oils, 121 mandarin oils, and 147 bergamot oils produced in Sicily

Table 22.6 Organophosphorus pesticide residues (ppm) in bergamot oil (Dugo et al., 1994; Di Bella et al., 1995)

	1991[a]	1992[a]	All	% of contaminated samples
N° of samples	9	129	138	
Mevinphos	0.01–0.89	–	0.01–0.89	6.5
Sulphotep	0–0.43	–	0–0.43	1.4
Diazinon	0–0.15	0–0.17	0–0.17	1.4
Methyl-parathion	0–19.75	0–17.85	0–19.75	18.1
Fenitrothion	–	0–1.19	0–1.19	0.7
Malathion	0–0.03	–	0–0.03	0.7
Ethyl-parathion	0–6.83	0–2.23	0–6.83	20.3
Phentoate	–	0–1.26	0–1.26	1.2
Quinalphos	0–4.58	0–0.22	0–4.58	9.4
Methidathion	0–0.66	0–19.44	0–19.44	25.4
Ethyl-azinphos	–	0–11.15	0–11.15	0.7

Note
a production season.

and Calabria between 1991 and 1996. The analyses were carried out on a dual channel GC-ECD fitted with two capillary columns and the results obtained for the different oils are summarized below:

Lemon essential oils

The maximum and minimum levels of each pesticide detected are reported in Table 22.7, along with the percentage of contaminated samples and the mean levels of contamination found in every production year. From 1991 to 1996 there was a steady, year-by-year reduction both in the percentage of contaminated samples and in the mean contamination levels. A chromatogram of a contaminated lemon oil is reported in Figure 22.2.

Sweet orange essential oils

The organochlorine residue levels found in the sweet orange oil samples are reported in Table 22.8 and they follow the same trend as those of the lemon oil samples discussed above. The mean levels of dicofol and tetradifon decrease from 1.74 and 0.66 ppm in 1991 to 0.25 and 0.16 ppm in 1996 respectively.

Mandarin essential oils

Table 22.9 shows the organochlorine residues detected in the mandarin oil samples: the mean level of dicofol decreased from 1.96 ppm in 1991 to 0.63 ppm in 1996, while tetradifon levels dropped from 0.95 ppm in 1991 to 0.34 ppm in 1996.

Bergamot essential oils

The bergamot essential oil samples were the least contaminated by organochlorine as well as by organophosphorus pesticide residues (Table 22.10). As early as 1991,

Table 22.7 Organochlorine pesticide residues in lemon oil (ppm) (Saitta *et al.*, 1995, 2000)

	Production season					
	1991	1992	1993	1994	1995	1996
N° of samples	25	24	24	24	26	25
% Contaminated samples	92.0	91.6	83.3	75.0	73.1	64.0
4,4'DCBP						
Range	n.d.–5.16	n.d.–3.18	n.d.–2.95	n.d.–1.95	n.d.–1.24	n.d.–0.94
Mean value	1.72	1.29	1.08	0.66	0.38	0.16
Dicofol						
Range	n.d.–6.94	n.d.–5.24	n.d.–3.24	n.d.–2.37	n.d.–1.24	n.d.–0.88
Mean value	2.26	1.75	1.01	0.76	0.43	0.23
Tetradifon						
Range	n.d.–2.22	n.d.–1.93	n.d.–1.93	n.d.–0.78	n.d.–0.62	n.d.–0.32
Mean value	0.86	0.75	0.45	0.20	0.16	0.08

Note
n.d. not detectable.

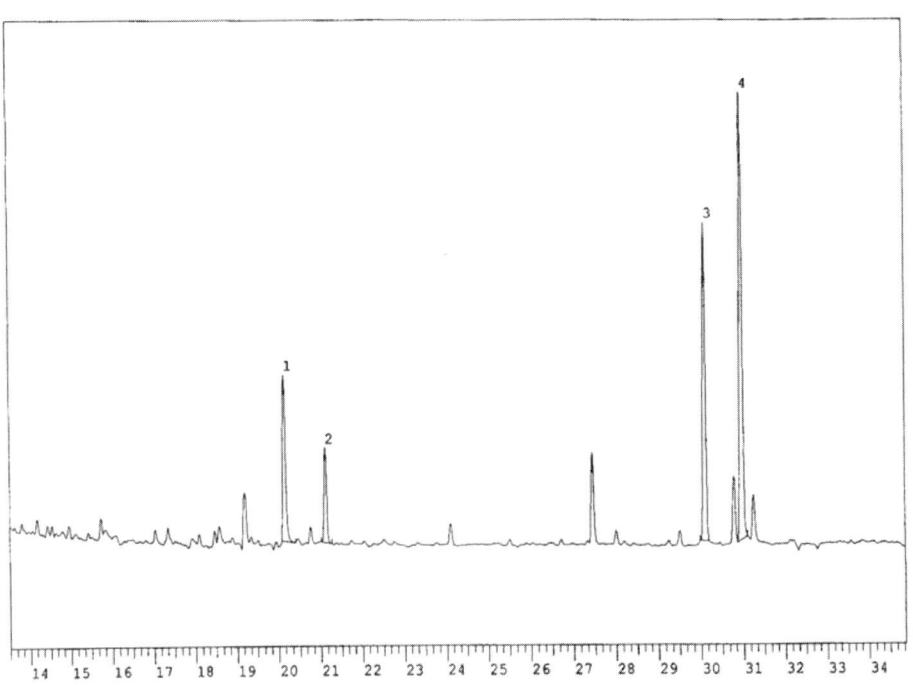

Figure 22.2 ECD chromatogram of a lemon oil (Saitta *et al.*, 2000): (1) 4,4'-dichlorobenzophenone; (2) methylbromophos; (3) dicofol; (4) tetradifon.

Table 22.8 Organochlorine pesticide residues (ppm) in sweet orange oil (Saitta et al., 1995, 2000)

	Production season					
	1991	1992	1993	1994	1995	1996
N° of samples	19	18	23	19	22	22
% Cont. samples	94.7	77.8	78.3	68.4	68.2	54.5
4,4'DCBP						
Range	n.d.–5.48	n.d.–3.97	n.d.–3.94	n.d.–3.12	n.d.–1.22	n.d.–1.49
Mean value	2.41	1.77	1.05	0.85	0.58	0.32
Dicofol						
Range	n.d.–5.02	n.d.–4.11	n.d.–4.01	n.d.–2.16	n.d.–1.12	n.d.–1.05
Mean value	1.74	1.36	0.91	0.76	0.44	0.25
Tetradifon						
Range	n.d.–2.15	n.d.–1.92	n.d.–1.15	n.d.–1.35	n.d.–0.62	n.d.–0.92
Mean value	0.66	0.48	0.41	0.36	0.20	0.16

Note
n.d. not detectable.

Table 22.9 Organochlorine pesticide residues (ppm) in mandarin oil (Saitta et al., 1995, 2000)

	Production season					
	1991	1992	1993	1994	1995	1996
N° of samples	19	20	20	22	20	20
% Cont. samples	94.7	90.0	80.0	77.3	75.0	50.0
4,4'DCBP						
Range	n.d.–4.17	n.d.–2.58	n.d.–2.04	n.d.–4.18	n.d.–2.24	n.d.–1.43
Mean value	1.55	1.10	1.00	1.04	0.56	0.28
Dicofol						
Range	n.d.–5.95	n.d.–2.10	n.d.–2.16	n.d.–3.15	n.d.–3.18	n.d.–1.24
Mean value	1.96	1.11	1.21	0.92	0.63	0.36
Tetradifon						
Range	n.d.–3.12	n.d.–1.24	n.d.–1.54	n.d.–1.01	n.d.–2.24	n.d.–0.92
Mean value	0.95	0.63	0.72	0.43	0.34	0.10

Note
n.d. not detectable.

50 per cent of the samples were uncontaminated and by 1996 the percentage of contaminated samples had dropped to 26.3 per cent.

The Uruguayan lemon oils analyzed by Dellacassa et al. (1995, 1999) did not contain any organochlorine pesticide residues.

PLASTICIZERS

Triarylphosphates, chloroparaffins, phthalates

As mentioned above, a variety of compounds which are generally defined as plasticizers and which are used as technological additives by the plastic materials industry can

Table 22.10 Organochlorine pesticide residues (ppm) in bergamot oil (Saitta et al., 1995; Saitta et al., 2000)

	Production season					
	1991	1992	1993	1994	1995	1996
N° of samples	20	19	19	20	20	19
% Cont. samples	50.0	52.6	31.6	35.0	40.0	26.3
4,4'DCBP						
Range	n.d.–0.92	n.d.–0.72	n.d.–0.52	n.d.–0.15	n.d.–0.92	n.d.–0.32
Mean value	0.17	0.22	0.10	0.02	0.10	0.05
Dicofol						
Range	n.d.–0.81	n.d.–0.96	n.d.–0.56	n.d.–0.24	n.d.–1.18	n.d.–0.42
Mean value	0.20	0.24	0.10	0.04	0.11	0.05
Tetradifon						
Range	n.d.–0.52	n.d.–0.57	n.d.–0.31	n.d.–0.12	n.d.–0.12	n.d.–0.32
Mean value	0.08	0.09	0.04	0.01	0.01	0.03

Note
n.d. not detectable.

contribute to the contamination of citrus essential oils. These substances can be divided into three groups:

1. triarylphosphates (triphenylphosphates, diphenyltolylphosphates, phenylditolylphosphates, tritolylphosphates, ditolylxylylphosphates, tolyldixylylphosphates and trixylylphosphates), which are commonly known as phosphorated plasticizers (Enciclopedia delle Materie Plastiche, 1964; Encyclopedia of Polymer Science and Technology, 1964);
2. chloroparaffins, which are industrial mixtures of chloro-alkanes of different chain length (C10–C20), with a chlorine content between 40 and 70 per cent in weight. They are considered secondary plasticizers or extenders and are always employed together with a primary plasticizer (Encyclopedia of Polymer Science and Technology, 1964; Tedder et al., 1975);

Table 22.11 Experimental conditions for plasticizer analysis in citrus oils (Di Bella, 1998; Saitta et al., 1997)[a] (Di Bella, 1998; Di Bella et al., 2000)[b] (Di Bella, 1998; Di Bella et al., 1999)[c]

Plasticizers analyzed	Injector	Column	Detector
Phosphorated plasticizers[a]	PTV, 65–240 °C at 990 °C/min	Capillary fused silica. 25 m × 0.32 mm MEGA 68. 0.45 µm; 75 (5 min)–100 °C at 7.5 °C/min; 100–170 °C at 2 °C/min; 170 (5 min) –250 °C at 10 °C/min	FPD 250 °C
Chloroparaffins[b]	Splitless (2 min). 250 °C	Capillary fused silica. 30 m × 0.32 mm, Restek RTX-5 e Restek RTX- 1701 0.25 µm; 50 °C (2 min) –150 °C[a] at 25 °C/min; 150–270 °C[a] (20 min) at 4 °C/min	ECD 280 °C
Phthalate esters[c]	Splitless (2 min). 250 °C	Capillary fused silica. 30 m × 0.25 mm DB-5 MS 0.25 µm; 60–275 °C (14 min) at 15 °C/min	MS, SIM

Notes
FPD = flame photometric detector; ECD = electron capture detector; PTV = programmed temperature vaporizer; MS = mass spectrometer; SIM = selected ion monitoring.

Table 22.12 Plasticizer residues (ppm) in citrus oils (Di Bella, 1998; Saitta et al., 1997)[a] (Di Bella, 1998, 2000)[b] (Di Bella, 1998, 1999)[c]

	Sweet orange oils–Italy (98 samples)			Mandarin oils–Italy (96 samples)		
	% of contaminated samples	Range	Mean value	% of contaminated samples	Range	Mean value
Phosphorated plasticizers[a]	25	n.d.–0.62	0.08	49	n.d.–2.95	0.34
Chloroparaffins[b]	33	n.d.–25.6	2.5	38	n.d.–25	5.3
Di-iBuP[c]	94	n.d.–26	1.29	94	n.d.–62	3.99
Di-nBuP[c]	21	n.d.–0.74	0.06	3	n.d.–0.24	–
Bis(2EtHex)P[c]	95	n.d.–29.9	2.98	97	n.d.–9.18	2.1

	Lemon oils–Uruguay (78 samples)			Lemon oils–Italy (102 samples)		
	% of contaminated samples	Range	Mean value	% of contaminated samples	Range	Mean value
Phosphorated plasticizers[a]	100	0.01–5.18	1.06	32	n.d.–0.51	0.08
Chloroparaffins[b]	0	–	–	53	n.d.–60	7.1
Di-iBuP[c]	9	n.d.–0.78	–	47	n.d.–18.9	1.47
Di-nBuP[c]	0	–	–	5	n.d.–0.54	–
Bis(2EtHex)P[c]	95	n.d.–39.4	3.19	63	n.d.–3.18	0.51

Note
n.d. not detectable.

3   phthalic acid esters, which constitute the most important group of plasticizers. In this class of compounds, bis-(2-ethylhexyl)phthalate (bis-(2EtHex)P) is the one most widely used. It is commonly known as dioctylphthalate and it is highly liposoluble. Other commonly used plasticizers of the same family are diisobutyl- and di-n-butylphthalate, (Di-iBuP) and (Di-nBuP) respectively. (Enciclopedia delle Materie Plastiche, 1964; Encyclopedia of Polymer Science and Technology, 1964).

Essential oils which have come into contact with plastic materials during the production cycle may show the presence of plasticizers. A number of studies have been devoted to the detection of these compounds in oils from Italy and South America: phosphorated plasticizers (Saitta *et al.*, 1997; Di Bella, 1998); chloroparaffins (Di Bella, 1998, 2000); phthalate esters (Di Bella, 1998, 1999). These investigations were all carried out by gas chromatography, using capillary columns and selective detectors, under the conditions summarized in Table 22.11. The samples analyzed and the results obtained are collected in Table 22.12. As can be seen from the data shown, all the Uruguayan oil samples contained triarylphosphate residues, but were entirely free of chloroparaffins. In the Italian oils, on the other hand, chloroparaffin residues were present in about 40 per cent of the samples analyzed, while the degree of contamination from triarylphosphates was lower, compared to the Uruguayan oils, in terms both of the percentage of contaminated samples and the mean contamination levels found. A similar overall level of phthalate ester contamination was found in the oil samples from the two countries, although the proportions in which individual phthalate compounds were present varied.

CONCLUSIONS

The studies on organophosphorus and organochlorine residues show that the use of these pesticides in citrus cultivation has decreased in recent years. It is too early to identify a trend in plasticizer contamination, since research in this area is a relatively recent development. However, from the data so far reported it would appear that contamination levels are not determined by the citrus species, but depend only on the history of the oil. Judging from the results reviewed here, the overall levels of essential oil contamination do not give great cause for concern from the strictly toxicological point of view, although it should be pointed out that nothing much is known about the possible combined or cumulative effects of the different substances present.

REFERENCES

Dellacassa, E., Lorenzo, D., Moyna, P., Dugo, G. and Saitta, M. (1995) Contaminazione di oli essenziali agrumari: residui di pesticidi organofosforici e organoclorici negli oli di limone dell'Uruguay. *Proceedings II Congresso Nazionale di Chimica degli Alimenti*, Giardini Naxos, Italy, May 1995, pp. 955–960.

Dellacassa, E., Lorenzo, D., Di Bella, G. and Dugo G. (1999) Pesticide residues in Uruguayan lemon oils. *J. Essent. Oil Res.*, 11, 465–469.

Di Bella, G., Dugo, G., Saitta, M., Salvo, F. and Ziino, M. (1991) Sulla genuinità delle essenze agrumarie. Nota XXXII. Sulla presenza di residui di pesticidi organofosforici negli oli essenziali di arancia dolce prodotti in Sicilia e Calabria. *Exedra*, 5, 5–8.

Di Bella, G., Dugo, G., Salvo, F., Alfa, M. and Saitta, M. (1995) Contaminazione di oli esenziali agrumari: valutazione dei residui di pesticidi organofosforici dal 1983 al 1992. *Proceedings II Congresso Nazionale di Chimica degli Alimenti*, Giardini Naxos, Italy, May 1995, pp. 941–945.

Di Bella, G. (1998) *Contaminazione di Oli Essenziali Agrumari: Plastificanti e Cloroparaffine*. Ph.D. Thesis, University of Messina, Italy.

Di Bella, G., Saitta, M., Pellegrino, M.C., Salvo, F. and Dugo, G. (1999) Contamination of citrus essential oils: the presence of phthalate esters. *J. Agric. Food Chem.*, 47, 1009–1012.

Di Bella, G., Saitta, M., Lo Curto, S., Visco, A. and Dugo, G. (2000) Contamination of citrus essential oils: the presence of chloroparaffin. *J. Agric. Food Chem.*, 48, 4460–4462.

Dugo, G., Salvo, F., Alfa, M. and Dugo, P. (1987) Sulla genuinità delle essenze agrumarie. Nota XVII. Dosaggio rapido di pesticidi organofosforici negli oli essenziali agrumari. *Essenz. Deriv. Agrum.*, 57, 610–619.

Dugo, G., Salvo, F., Saitta, M. and Di Bella, G. (1990) Sulla presenza di residui di pesticidi organofosforici negli oli essenziali agrumari. *Essenz. Deriv. Agrum.*, 60, 428–451.

Dugo, G., Famà, G., Saitta, M. and Stagno d'Alcontres, I. (1992) Sulla genuinità delle essenze agrumarie. Nota XLII. Determinazione di pesticidi organofosforici con rivelatore a fiamma fotometrica. *Essenz. Deriv. Agrum.*, 62, 127–146.

Dugo, G., Di Bella, G., Saitta, M. and Salvo, F. (1994) Sulla genuinità delle essenze agrumarie. Nota XLV. Dosaggio rapido di pesticidi organofosforici negli oli essenziali di bergamotto. *Essenz. Deriv. Agrum.*, 64, 234–247.

Dugo, G., Saitta, M., Di Bella, G. and Dugo, P. (1997) Organophosphorus and organochlorine pesticide residues in Italian citrus oils. *Perfum. Flav.*, 22(4), 33–44.

Dupuis, G. (1975) Pesticide residues in citrus. In *Citrus*, 4th in a series of Ciba Geigy agrochemical technical monographs, pp. 81–88.

Enciclopedia delle Materie Plastiche (1964) L'Industria, Milano, pp. 381–394.

Encyclopedia of Polymer Science and Technology (1964), Vol. 10, John Wiley and Sons, Inc. New York, pp. 228–306.

Guenther, F. (1969) Insecticide residues in California citrus fruits and products. *Res. Rev.*, 1, 28–32.

Leoni, V. and D'Alessandro De Luca, E. (1978) An important aspect of the health problem caused by pesticides: the presence of organophosfate insecticide residues in essential oils. *Essenz. Deriv. Agrum.*, 48, 39–50.

Saitta, M., Dugo, G., Di Bella, G., Salvo, F. and Previti, P. (1995) Contaminazione di oli essenziali agrumari: valutazione dei residui di pesticidi organoclorici. *Proceedings II Congresso Nazionale di Chimica degli Alimenti*, Giardini Naxos, Italy, May 1995, pp. 947–953.

Saitta, M., Di Bella, G., Bonaccorsi, I., Dugo, G. and Dellacassa, E. (1997) Contamination of citrus essential oils: the presence of phosphorated plasticizers. *J. Essent. Oil Res.*, 9, 613–618.

Saitta, M., Di Bella, G., Dugo, G., Salvo, F. and Lo Curto, S. (2000) Organochlorine pesticide residues in Italian citrus essential oils, 1991–1996. *J. Agric. Food Chem.*, 48, 797–801.

Stevens, R.K. (1967) A rapid specific method for gas chromatographic determination of organophosphate pesticides in cold-pressed citrus oils. *J. Assoc. Off. Anal. Chem.*, 50, 1236–1240.

Tedder, J.M., Nechvatal, A. and Jubb, A.H. (1975) Basic Organic Chemistry–part 5 *Industrial Products*, John Wiley and Sons, Inc. New York, pp. 136–163.

23 The market of citrus oils around the world

Angelo Di Giacomo

WORLD PRODUCTION OF CITRUS

Worldwide citrus production evaluated for selected countries during the 1999/2000 season was about 70 million tons, the same level as for 1997/1998. The greatest contributions to the total production were obtained from the US, Italy, China, and Spain (USDA, Foreign Agricultural Service, 2000).

Table 23.1 reports the 1999/2000 production (thousand tons) obtained in the Northern and Southern Hemispheres for each different citrus species. The Northern Hemisphere represents about 70 per cent of the total amount.

Crop year refers to the harvest and marketing period, which usually begin in the autumn and extends to the spring. This corresponds roughly to October–June in the Northern Hemisphere, and April–December at the Southern Hemisphere. For the Southern Hemisphere, the harvest occurs almost entirely during the second year of cultivation.

World citrus production can be reported in percentages of each single citrus species as follows:

Oranges	67.08%
Tangerines	18.98%
Lemons	6.00%
Grapefruits	5.61%
Other citrus	2.33%

The amount of citrus produced in the Mediterranean Basin in the 1999/2000 year was 17,163,000 metric tons, divided as follows:

	tons
Oranges	9,806,000
Tangerines	4,198,000
Lemons	2,216,000
Grapefruits	561,000
Other citrus (Limes, Bergamots, Sour Oranges, etc.)	352,000

In the Mediterranean area the major producer country was Spain (over 5 million tons), followed by Italy, Egypt, Turkey, Morocco and Greece.

Table 23.1 World citrus production (thousand tons) (USDA Foreign Agricultural Service, 2000)

	Northern hemisphere	Southern hemisphere	World
Oranges	28.141	18.834	46.975
Tangerines	12.950	340	13.290
Lemons	3.015	1.190	4.205
Grapefruits	3.561	365	3.926
Other citrus (limes, bergamots, sour oranges, etc.)	1.633	–	1.633
Total	49.300	20.279	70.029

In the Northern Hemisphere however, the United States are the major producer country, with more than 15 millions tons, followed by Mexico and Japan.

In the Southern Hemisphere, Brazil is the lead country for the citrus production with the highest amount of orange production, followed by Argentina.

The world total production of citrus (FAO, Food and Agriculrure Organization, 2000), for the year 1999/2000 is summarised below:

Norhern Hemisphere	63,891.6
Southern Hemisphere	26,995.2
World	90,886.8

WORLD PROCESSING

It has been determined that, of the total world production of 70 millions tons of citrus, about 42 per cent is sent to the transformation industry. In Table 23.2 is reported, for the most common citrus species (oranges, tangerines, grapefruits, lemons), the amount of fruits transformed during the seasons 1997/98–1999/2000 (USDA, Foreign Agricultural Service, 2000).

Of the total citrus fruits processed, the proportionate percentage of each single citrus species for the year 1999–2000 is as follows:

Oranges	82.35%
Tangerines	4.40%
Grapefruits	6.39%
Lemons	5.98%
Other citrus (Limes, Bergamots, Sour Oranges, etc.)	0.88%

The major countries where citrus are industrially processed are (1,000 metric tons):

Northern Hemisphere: China (535), Cuba (560), Greece (324), Israel (418), Italy (1,012), Mexico (636), Spain (1,224), United States (11,589).
Southern Hemisphere: Argentina (936), Brazil (11,179).

Table 23.2 World citrus for processing (thousand tons) (USDA Foreign Agricultural Service, 2000)

	1997/1998	1998/1999	1999/2000
Oranges	25.769	23.404	24.398
Tangerines	1.237	1.146	1.304
Grape-fruits	1.834	1.777	1.892
Lemons	1.752	1.659	1.774
Other citrus	283	267	260
Total	30.875	28.253	29.628

For each single citrus species the following selected countries produced:

Oranges: Argentina (120)*, Australia (280), Brazil (11,179), China (144), Cuba (130), Greece (290), Israel (140), Italy (600), Mexico (360), Morocco (100), South Africa (630), Spain (715), Turkey (105), United States (9,743).
Tangerines/Mandarins: China (391), Italy (90), Japan (157), Spain (280), United States (180).
Grapefruits: Cuba (230), Israel (222), United States (1,302).
Lemons: Argentina (740), Italy (320), Spain (215), United States (740).

* 1000 metric tons.

WORLD PRODUCTION OF ESSENTIAL OILS

During the 1998–1999 season 1,659,000 tons of lemons were transformed by the industries. Given an average yield of about 350 g of essential oil per ton of lemons, we can estimate a total production of about 5,800 tons of essential oil.

However, in our opinion, the data relative to the amount of the different citrus processed by the industries cannot be used to estimate the amount of the relative oil produced. In fact, the yields of essential oils varies within a very large range for different reasons that depend on the raw material and the extraction technology used, and also on the economical advantage estimated prior to the extraction. On the other hand, the productive diagram does not always includes the essential oil extraction.

A complete review of the world production of essential oil is due to Lawrence (1985) who used for this review, data obtained from different sources: Foreign Agricultural Circulars of the USDA, Tropical Products Institute (now called Tropical Development Research Institute) publications, United Nations publications and limited country specific production reviews. The data reported in this review relate to 1984, since from this knowledge it is possible to estimate the growth of the essential oils production occurred during the last 15 years.

In 1984 the world production of the different essential oils was as follows (metric tons): bergamot oil (115), bitter orange oil (32), grapefruit oil (180), lemon oil (2,300), lime oil cold-pressed (160), lime oil distilled (450), mandarin oil (120), orange oil (12,000), tangerine oil (300). The total world production amounted to between 15,000 and 16,000 tons of essential oils.

Table 23.3 World production of essential oils (1998–1999)

Essential oils	Production (tons)	Major producing countries	Other countries
Bergamot oil	65	Italy (50)	Ivory Coast, Guinea, Brazil, Argentina
Bitter orange oil	93	Brazil (20), Ivory Coast (18), Spain (15), Dominican Republic (12), Italy (8)	Argentina, Haiti
Grapefruit oil	468	USA (215), Israel (50), Mexico (65), Cuba (35), Argentina (20), South Africa (18), Cyprus (5), Turkey (10)	Gaza, Jamaica, Lebanon, West Indies
Lemon oil	5,823	Argentina (2,100), USA (1,360), Italy (1,200) Spain (280), Brazil (260), Greece (100), South Africa (90), Ivory Coast (88), Turkey (72)	Cyprus, Israel, Uruguay China, Australia
Lime oil	1,352	Mexico (1,075), Cuba (18), Egypt (5), Spain (4)	Australia, Brazil, India, Ivory Coast, Peru, USA, West Indies
Mandarin oil	330	Italy (250), Spain (30), Argentina (25), Egypt (5)	Brazil, China, Ivory Coast, USA
Orange oil	61,705	Brazil (30,000), USA (28,000), Mexico (1,800), Italy (900), Spain (170), Greece (82), Australia (70), South Africa (65)	Argentina, China, Cyprus, Israel, Morocco, Turkey
Tangerine oil	627	China (175), USA (95), Mexico (85), Spain (80), Japan (45), Argentina (35), Israel (15), Turkey (12), Greece (5)	Brazil, South Africa

The list of the essential oils producing countries is extensive; therefore we consider only the major producing countries: Bergamot oil: Italy (90); Bitter orange oil: Brazil (8), Dominican republic (10), Haiti (10), Italy (2); Grapefruit oil: Brazil (40); Lemon oil: Argentina (480), Brazil (390), Greece (100), Italy (500), Ivory Coast (120), Spain (100), USA (600); Lime cold-pressed oil: Brazil (90), Mexico (25), USA (40); Lime distilled oil: Mexico (180), Haiti (50), Peru (130); Mandarin oil: China (40), Italy (50); Orange oil: Brazil (7,800), USA (2,500); Tangerine oil: Brazil (250).

The reality relative to the year 1984 has since drastically changed. The production of essential oils in the 1998–1999 period exceeds 70,000 tons, more than four times higher than in 1984.

Table 23.3 reports, the amount produced worldwide and the relative major producing countries for each essential oil. For the collection of these data similar sources as those used by Lawrence in the former review were selected, and information obtained from various business, trade institutions and private communications were included.

Compared to the data relative to 1984 all the single productions result increased considerably; a negative trend is noticed only relatively to bergamot oil. This may be explained by the fact that during the 1998–1999, particular climate conditions significantly compromised citrus production.

For lemon oil the major producing countries are Argentina, USA, and Italy. Brazil and USA are the major producers of orange oil; Mexico is the leader country for lime oil production; Italy for bergamot and for mandarin oils; USA for grapefruit oil; China for tangerine oil.

NEGOTIATION AND MARKET PRICES

When purchasing an essential oil the potential consumer will evaluate the sample relative to the stock, in order to determine if this will be compatible with the physico-chemical and chromatographic specifications; at the same time the oil is evaluated based on the aroma quality. It is obvious that the aromatic quality of the oil is the most important parameter for its approval. An essential oil that presents normal specifications but that has poor odour notes it is normally rejected. On the other hand, if an oil presents appropriate aromatic properties but physical-chemical specifications out of range, this may still have a commercial value; in this case the supplier must give the explanations necessary to justify the lack of correspondence of the oil relative to the requested specifications.

According to Lawrence (1993), in the eyes of the user, essential oils are subject to the following sensitivities:

- variability of uncertainty of apply
- origin and seasonal variability
- adulteration
- stability problems
- political influence: tariffs, quotas, regulations
- substitution with reconstituted oils from aroma chemical blends
- price fluctuation
- new scientific knowledge

Until the beginning of the seventies (Di Giacomo, 1973), from a commercial point of view, the authorities of this field were in the area of Messina (Italy). For what concerned, in particular, lemon essential oil the prices were reported in Sicilian pounds (Sicilian pound = 0.318 Kg), which indicated the merchandise free of the packaging, free delivered. The essential oil was commercialised based on a citral content of about 4 per cent; if chemical analysis showed a different content, a price adjustment proportional to the amount to the effective citral content would be applied. Since many years, however, a new trend has been established for the evaluation of citrus essential oils relatively to the official certificate of purity, without considering the amount of citral determined.

During the 1998/99 season, in Italy, the most common citrus essential oils were commercialised according to the following quotations (Di Giacomo, 2000):

Essential oil	Euro/Kg
Winter Lemon	12.91–16.53
Spring/Summer Lemon	12.29–12.91
Distilled Lemon	7.23–7.75
Sweet Orange	0.67–0.72
Bitter Orange	37.18–43.90
Mandarin	30.99–51.65
Bergamot	72.30–82.63
Bergapten-free Bergamot	80.05–90.38
Petitgrain Citronnier	92.96–103.29

Petitgrain Bigarade   77.47–82.63
Petitgrain Bergamottier   77.47–82.63
Petitgrain Mandarinier   113.62

The amounts reported in the table above refer to the prices determined from the marketing between small size producers and the companies who specialised in the final preparation of the stock to be sold to users both on national and international markets. The essential oils in questions are therefore unfiltered, not jet winterised and stocked in temporary drums.

The quotations for lemon oil are slightly higher if produced by 'sfumatrice' compared to the 'pelatrice' quality. Those oils produced by 'organic' fruits have been quoted at higher values; for instance, 'organic' bergamot oil has reached the price of 87.80 Euro/Kg.

Table 23.4 reports the approximate worldwide values of 1 kg of various citrus oils from the market.

Table 23.4 Value of 1 kg of citrus oils of commerce (1999–2000)

Oil	$ value
Bergamot, Italy	57–61
Bergamot bergapten-free, Italy	63–67
Bitter orange, Italy	31–37
Grape-fruit	15
Lemon, USA	14–15
Lemon, Argentina	13–14
Lemon, Italy	17–19
Lemon, Ivory Coast	15–16
Lemon, Spain	13–14
Lime distilled, Mexico	13
Mandarin (red), Italy	35
Mandarin (yellow), Italy	31
Mandarin, China	23
Mandarin, USA	28
Orange, Brazil	1.48
Orange, USA	1.57
Orange, Italy	2.18–2.62

It was Braddock's opinion (1999), that, if some essential oils (e.g. lemon oil and lime oil) presented a higher value, compared to sweet orange essential oil, it was not only because of the more appreciated flavour and aroma of these products, but also, and mainly, for the scarcity of fruits produced and for the low demand for processed juice.

REFERENCES

Braddock, R.J. (1999) *Handbook of Citrus By-products and Processing Technology*, John Wiley & Sons, Inc., New York.
Chemical Market Reporter (1998) Chemical prices, Chem. Market Report, 254(7), 26–35.
Di Giacomo, A. (1973) *Gli Oli Essenziali degli Agrumi*, EPPOS, 55, 672–687.
Di Giacomo, A. (2000) L'industria agrumaria nel 1999. *Essenz. Deriv. Agrum*, 70, 61–66.

FAO–Food and Agriculture Organization of the United Nations (2000) Citrus fruit fresh and processed, CCP: CI/ST/2000.

Lawrence, B.M. (1985) A review of the world production of essential oils (1984). *Perfum. Flav.*, **10**, 66–79.

Lawrence, B.M. (1993) Is the development of an essential oil industry in Malaysia a valuable commercial opportunity? *IFEAT meeting in Kuala Lumpur*, Malaysia, October 31–November 4, 1993. In Lawrence, B.M. (1995) *Essential oils 1992–1994*. Allured Publishing Corporation, Carol Stream, IL 60188–2787 USA, pp. 187–204.

USDA, Foreign Agricultural Service (2000) World Horticultural Trade and U.S. Export Opportunities, *Circular Series FHORT 8-00*, August 2000.

24 Citrus oils in food and beverages: uses and analyses

Enrico Colombo, Claudio Ghizzoni and Dimitri Cagni

USES

Citrus flavours are popular in beverages as well as in other sweet products such as confectionery, cookies and desserts. Lately, they have also become popular in food, particularly in sauces and dressings. Their level of use varies widely, from about 200 ppm in beverages to more than ten times higher in chewing gum. This is related mainly to flavour release because of the particular food system. The main flavour used in soft drinks are cola, lemon-lime and orange, all which are based on citrus oils, and constitute about 80 per cent of all soft drinks in the world. The colas are the main segment in the US, but in international markets orange and other flavours are far more important. One third of the oils is consumed in the US, and the usage is about equally divided over the main flavours: cola, lemon-lime and orange. Soft drinks flavours utilise citrus oils in various ways for technological reasons, mainly based on solubility in water. Cloudy beverages use citrus oils as such because the terpenes provide part of the cloudy appearance. In clear drinks such as lemonades, we typically use extracts which remove most of the terpenes. Concentrated and terpeneless oils are used for the same reasons, as well as for flavour effect or stability. The performances of citrus oils in soft drinks is quite different from most other food applications, because low levels of oil in water at pH 3 or less are employed and this leads to many reactions which affect flavour quality. This types of changes are typical for soft drinks and, in general, are not found in other food or fragrance applications, or they occur at a much lower rate. This means that quality considerations for the soft drink category should be quite different, too. The next large user of citrus flavours is the confectionery industry. Here orange, lemon and lime are the most popular flavours in certain categories, and to a large extent, these flavours are, indeed, based on citrus oils. Twenty per cent of confectionery is citrus flavoured and comprises hardy candy or boiled sweet. There are many other applications besides soft drinks and confectionery, but these are all much smaller in usage, and there are no reliable statistics. Example are ice cream, cookies, desserts, powdered drinks, alcoholic beverages, food, etc. Citrus oils are extremely important for the flavouring of foods and beverages and their consumption is growing steadily at more than three per cent per year. On the other hand, the demands for performance can be expected to become more strict, particularly as far as stability and purity are concerned (Buchel, 1989).

Now we will consider main citrus fruits used for food and beverages flavouring and we will point out their use in food industry.

Bergamot oil

The expressed oil of bergamot is extremely sweet-fruity and fresh with a sharp-like topnote. This is followed by a sweet, rich, oily, herbaceous aroma. Also present are rich floral, balsamic notes. The flavour is somewhat reminiscent of bitter orange. Bergamot is used extensively in perfumery because of its sweetening and freshening qualities. It is an integral part of eau-de-Cologne formulations. It is used more in extract perfumes than in chemistry stressed functional fragrances. The use of bergamot in flavourings is somewhat more limited. It functions as topnote, and can act as a blender or modifier, especially in combinations with other citrus. An interesting use is in tobacco flavourings (Swaine and Swaine, 1988). Another important use of bergamot flavour is in the aromatisation of tea, such as the British speciality *Earl Grey* that owes its aroma to bergamot flavour, other less important uses are in jams and in some alcoholic beverages.

Grapefruit oil

Grapefruit oil has a characteristic aroma and flavour. It has a fresh citrus aroma, one that Arctander characterised as having notes common to both the sweet and bitter orange. This has an interesting woody base note. The primary application for grapefruit in perfumery is as a modifier for other citrus compositions; for example bergamot. However, grapefruit is used more in the flavour industry as a blender and modifier. It adds body to a number of citrus and non-citrus flavourings. Single fold grapefruit oil is used in concentrations of 50–200 ppm in soft drinks. The use of the oil derived from the evaporation of grapefruit juice is becoming more popular in an attempt to recreate fresh and natural grapefruit juice. It is blended in ratio 3:1 with cold-pressed grapefruit and used at 60–120 ppm in juice products. The increasing popularity of this flavourant and the limited supply increases price raise.

Lemon oil

The oil expressed from the lemon rind has a characteristic lemon aroma-sweet, fresh, and sharp. The refreshing, sweet traits of lemon oil make it amenable to many fragrance applications. Lemon adds a refreshing topnote to almost any fragrance. Lemon is used in flavourings as a blender or modifier as well as for its characterising flavour. Beverages and confections consume the greatest quantities of lemon oil. Lemon is also used to mask off-odours.

Distilled lime oil

Distilled lime oil is fresh and sharp. It is characterised as citrusy and terpeny as well as fruity-perfumy. There is a distinct floral (lilac) note. The odour is much more harsh and sharp than expressed lime. Distilled lime is used to flavour soft drinks, especially cola and lemon-lime types, and confections, mainly hard candies. It is also used as modifier in citrus formulations.

Expressed lime oil

The aroma of this oil is considered by many to be the *champagne* of citrus oils. The limited quantity produced makes it highly valued. It has a fresh, rich, sweet, citrus peel odour. Expressed lime oil is much smoother and less harsh than the distilled lime, and in fact, has an odour more akin to lemon peel oil. It is more mellow and full, even perfumery to some. The rather expensive price severely limits the use of lime in flavourings. However, new processing methods are making the oil from the small soft fruit more accessible. It does blend well with and modify other citrus notes especially lemon. It, like lemon, is a good masking agent.

Mandarin oil

In contrast with tangerine, mandarin oil has an amine-type (i.e. fishy) topnote. Although sweet, one would not consider the aroma fresh or refreshing as with other citrus oil. It possesses a rich floral undertone. Its application in flavourings is mainly as a modifier for citrus compositions, especially with orange and bitter orange. It enhances, intensifies and enriches orange character. The major application is in citrus beverage compositions.

Bitter orange oil

This oil possesses fresh citrus topnotes, but is considered less sweet, and even bitter and dry. It has floral and aldehyde characteristics. The tenacity is greater than most other citrus oils. In flavourings, this oil can be used to provide a citrus topnote or act as a modifier in a citrus blend. The major application is in citrus flavourings for beverages, especially liqueurs. It also intensifies the orange character in soft drinks.

Sweet orange oil

The mostly widely used citrus oil is cold-pressed sweet orange oil. It possesses a light sweet, fresh topnote with a fruity and aldehydic characters. Orange oil is widely used in the flavour industry, especially in beverages, and in candies. It can provide the topnote for citrus flavourings as well as the characteristic and most universally accepted flavour. The essence oil is used almost exclusively in the juice industry as it introduces many of the fresh, natural, fruity notes lost during processing. The fundamental orange uses deterpenated oil for character, strength and the clarity needed in finished beverages, gelatine desserts, and hard candies. Blended with mandarin it becomes richer, fuller and sweeter flavour reminiscent of blood orange. This blend is used in cake mixes, as they withstands the rigors of heat.

Tangerine oil

Tangerine should not be used interchangeably with mandarin. Tangerine has a sweeter, fresher quality, especially in the topnote. It has a thinner body and lacks the richness and perfumy character of mandarin. Tangerin is used in commercial flavourings, but primarily as a modifier. It is used in candy and soft drinks, as well as citrus soft drinks (Freeburg *et al.*, 1994).

STABILITY

Recently analysis on degradation mechanisms of main components of citrus flavours were carried out. This is a very important aspect in relation with food industry as there are concrete technological problems concerning citrus flavour stability.

Monoterpenes are important components of many essential oils, particularly citrus oils, which consists of 80–95 per cent monoterpenes, mainly limonene. Limonene is easily oxidised via free radicals, which are converted into a mixture of six unstable hydroperoxides by reaction with oxygen, in a manner similar to the structurally related carvomenthene, α-terpineol and α-pinene. Similar reaction proceeds under intensive ultraviolet irradiation even in absence of photosensibilisers; they are oxidised into hydroperoxides, however, their composition is different. Carvone and carveol and 1,2-limonene oxides belong to important oxidation products that deteriorate the sensory character. During storage of flavoured foods, terpenes are oxidised even in presence of low content of oxygen, intensities of typical citrus flavour notes decrease and off-odours imparted by oxidation products increase, particularly woody, heavy and acid notes. Antioxidants inhibit the flavour deterioration, Rosemary extracts being more active than 1,4-dihydropyridines. The compositions of stored limonene and linalool are effected differently by antioxidants used, which influences are respective sensory profiles. The effect of the particular antioxidant on flavour changes should be taken into account when they are applied to foods and beverages (Pokorni *et al.*, 1998).

Citrus flavours are subjected to deterioration related to medium typology particular in soft drink beverages. In fact beverages typically have low pHs (3.0 or lower), and many of the citrus flavour components undergo acid catalysed degradation reactions. Other components react with dissolved oxygen in the aqueous medium, which is replenished from the package's head space. Some flavour components, such as citral, are so reactive that they completely disappear from a beverage in a matter of weeks at low pH's. As a result of these reactions, it is often necessary to overload a beverage with flavour initially in order to ensure that the consumer will be satisfied when it is consumed many weeks or months later. This is an expensive practice since flavours are often the most costly components of a beverage. Also, as flavours degrade, unpleasant tasting end products are evolved that further limit the shelf-life of a beverage. The conditions under which a beverage is stored play an important role in the persistence of flavour because the rates of all the degradation reactions are strongly influenced by temperature. Excluding physical causes (like scalping) temperature, pH, and the availability of oxygen are the three most important factors effecting the rate of flavour loss in a beverage. An accurate knowledge of the way in which these factors influence flavour loss will allow more informed decisions to be made in the formulation of a product. In Figure 24.1 a scheme of possible pathway of citral degradation is reported (Ikenberry and Saaleb, 1993). These authors used a series of samples stored in particular conditions (the conditions are reported in Figures 24.2A and 24.2B) of pH and temperature. In Figures 24.2A and 24.2B it is also possible to see the influence of pH and temperature in citral degradation. At higher temperature such as 70 °F, there is a further decrease of the amount of citral.

Other studies on the kinetics and mechanisms involved in the light-induced photodegradation of lemon oil have been investigated using singlet molecular oxygen [$O_2(^1\Delta_g)$] time-resolved phosphorescence detection and steady-state irradiation, in organic and micellar aqueous solutions. Limonene and other minor components of lemon oil [especially the terpinenes, efficient $O_2(^1\Delta_g)$ quenchers] are responsible for the photo-degradation.

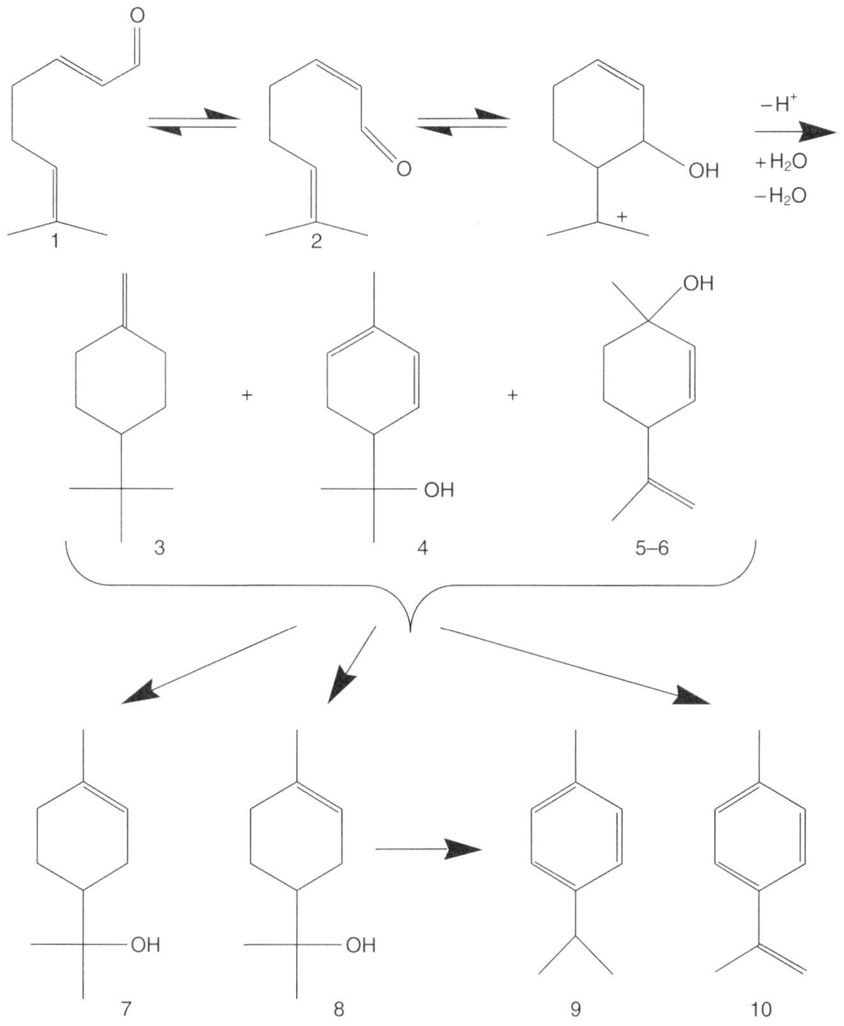

Figure 24.1 Scheme of possible pathway of citral degradation (Ikenberry and Saaleb, 1993).

Upon direct irradiation of lemon oil in the wave-length range 300–500 nm an additional photochemical process operates: the decomposition proceeds from electronically excited states of lemon oil susceptible components, which prevails over the competitive $O_2(^1\Delta_g)$ mechanism. The results should be carefully considered in a practical sense, when lemon oil or lemon oil-containing foods are stored under daylight conditions. Direct and sensitised light exposures cause irreversible transformations of lemon oil (Newmann and Garcia, 1992).

The decomposition of lemon oil induces the formation of many compounds. In order to establish the identity of the odourants formed during peroxidation, a group of researcher isolated by column chromatography some fractions. The fractions isolated were also investigated under the flavour point of view. They founded that the odour of

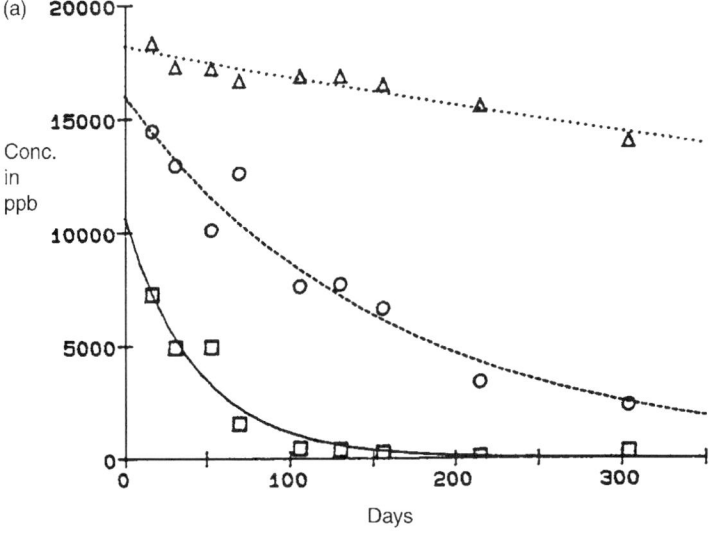

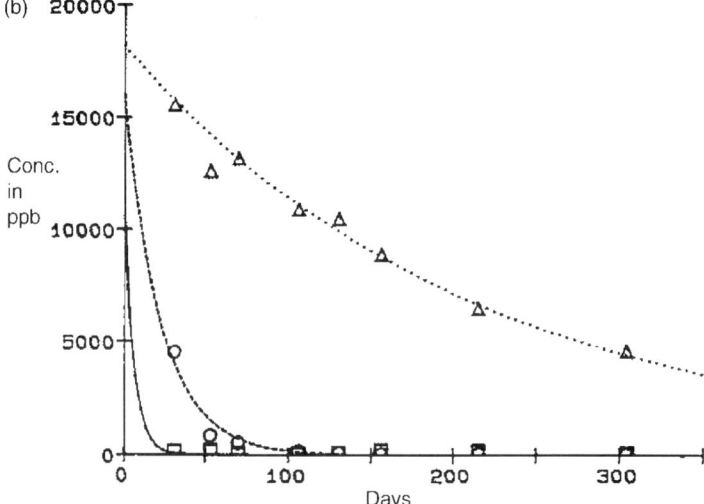

Figure 24.2 Citral degradation effects of temperature and pH (Ikenberry and Saaleb, 1993).

most of these hydroperoxides was very similar and was characterised as *turpentine-like*. The identification experiments together with the aromagram revealed that carvone and *p*-methylacetophenone followed by 4-acetyl-1-methyl-1-cyclohexene and *p*-cresol were the most potent odourants of the peroxidised lemon oil. Some of the identified structures are shown in Figure 24.3, it is possible to see *p*-cymene (a), carvone (g), *p*-cresol (h) and *t*-carveol (f) (Schieberle and Grosch, 1989).

Other comparative studies on beverages containing flavours with or without citral were performed using both sensory and instrumental evaluation analysis and authors show that significant sensory changes exist between the fresh citral containing and the

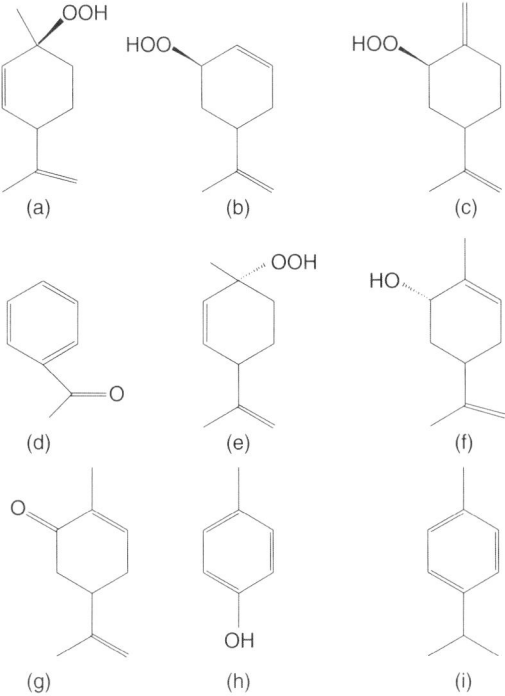

Figure 24.3 Potent odourants from the peroxidation of lemon oil (Schieberle and Grosch, 1989).

citralless beverages. Besides the citralless beverages generally changes less during storage than the citral-containing beverages. They also observed that the citralless products was more stable in terms of throat-burn character (Freeburg *et al.*, 1994).

ANALYSES

Sample preparation

Sample preparation is the initial step in aroma and flavour analysis in order to obtain information that is representative of the material under study. Flavour fraction is often in a very complex matrix, so it's easy to understand the role of sample preparation in flavour analysis. Components have to be isolated from foods and beverages to determine their identity. They must be isolated from water substrate by distillation or extraction, then, detection, identification, and characterisation of individual compounds are possible. It is also possible to know with sensory studies the contribution to biological activity. Each step of all analytical procedure is significant in order to advance to the level of correlating individual chemical compounds with the flavour of food or beverage. However the most important step is the development of the isolation method, the greatest obstacles of the analyst is to define the right way to proceed. For example many food matrix have an abundant quantity of water, so the problem is to concentrate the little amounts of flavour compounds from this solvent. In some foods, the flavour compounds must be extracted from materials that are predominantly carbohydrates. Extraction from carbohydrates provides an environment for leaching of unwanted compounds from the heterogeneous system and creates complications in isolating the desired flavour compounds. The recovery of flavour components depends upon the substrate type which can range from very small to total retrieval of the amounts present in the starting material. Difficulties in isolations process are the artefact formation (given by Maillard reaction for example) and decompositions due to thermal and catalytic processes. The analyst must be certain that the isolated compounds are actually from the starting material rather than from reaction products formed during isolation steps. A good example of a thermally induced alteration is the thermal reaction by which germacrene D transforms into other compounds this is due to the contact with silica gel.

Distillation and distillation-solvent extraction techniques

Probably the ancient isolation technique is the classical distillation process, that is commonly used in commercial laboratories to obtain essential oils and flavouring materials. Distillation has been used effectively for many years to obtain materials for flavour enhancement of products from citrus, apples, and strawberries. Essence recovered from the concentration of orange juice is distilled to obtain sufficient amounts to flavour bakery goods and confectionery products. Other processed foodstuffs could supply similar flavouring materials if systems were installed for recovery of materials presently released into the atmosphere. Probably the most-used method for isolation of volatiles for research is the combination technique of simultaneous steam distillation-extraction (SDE). The SDE concept has been known to organic chemists for a long time. Nickerson and Likens (1966) applied this technique to the study of beer, and Buttery *et al.* (1968) made the system popular for isolating volatile compounds from assorted materials. Flath and Forrey (1977) designed a compact but efficient condensing system, which is now used internationally. This SDE is suited for isolating materials at $100\,°C$ (and atmospheric pressure) or at $50\,°C$ (and $13.3\,kPa$ or $100\,mmHg$) (Teranishi and Kint, 1993). The most recent evolution of distillation-extraction techniques is represented by Solvent Assisted Flavour Evaporation (SAFE), a system which was developed by Engel *et al.* (1999). A compact and a versatile distillation unit (see Figure 24.4) was developed for the fast and careful isolation of volatiles from complex food matrices. In connection with a high vacuum pump ($5 \times 10^{-3}\,Pa$), the new technique, designated solvent assisted flavour evaporation, allows the isolation of volatiles from either solvent extracts, aqueous foods, such as milk or beer, aqueous food suspensions, such as fruit pulps, or even matrices with a high oil content. Application of SAFE to model solutions of selected aroma compounds resulted in higher yields from both solvent extracts or fatty matrices (50 per cent fat) compared to previously used techniques, such as high vacuum transfer. Direct distillation of aqueous fruit pulps in combination with a stable isotope dilution analysis enabled the fast quantification (60 min including MS analysis) of compounds such as he very polar and unstable 4-hydroxy-2,5-dimethil-3(2H)-furanone in strawberries ($3.2\,mg/kg$) and tomatoes ($340\,\mu g/kg$). Furthermore, the direct distillation of aqueous foods, such as beer or orange juice, gave flavourful aqueous distillates free from non-volatile matrix compounds (Engel *et al.*, 1999).

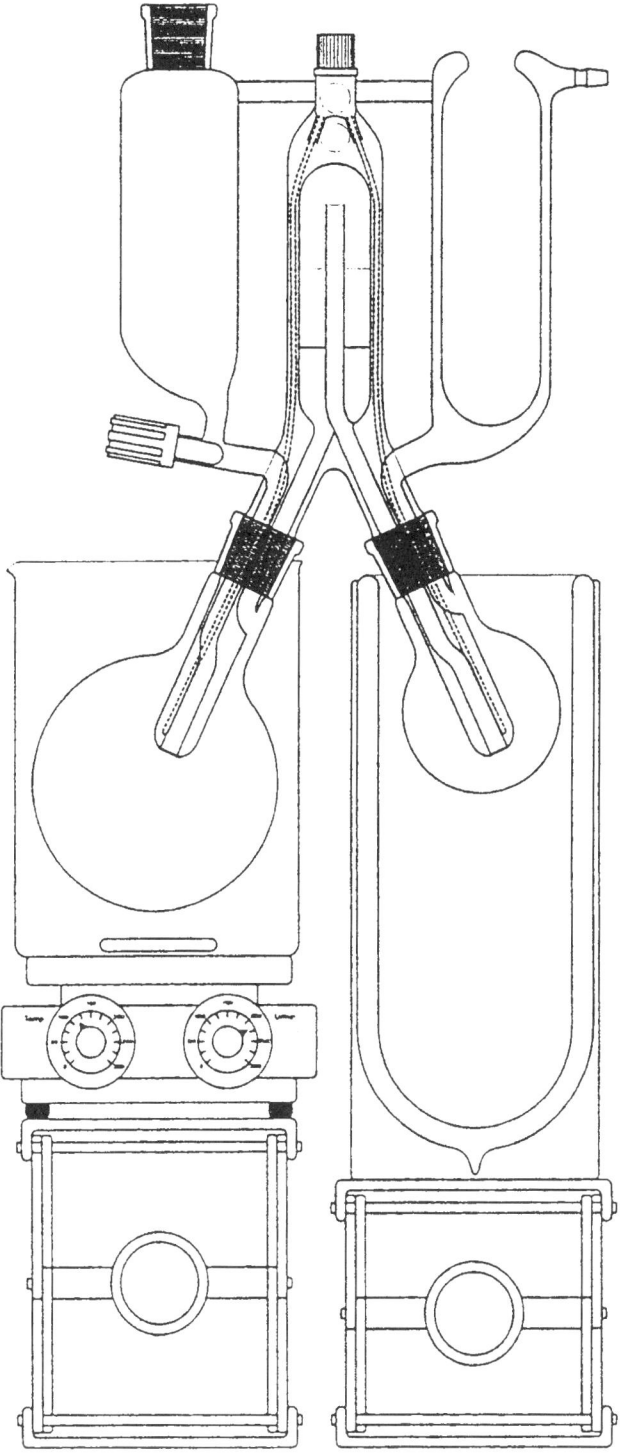

Figure 24.4 View of assembled equipment for SAFE (Engel *et al.*, 1999).

Supercritical fluid extraction (SFE)

After introducing preparation techniques related to distillation, we will deal now with the main extraction techniques. The aim of extraction operations is to avoid the interference of various substances which could affect with analytical operations. Besides classical extraction techniques using a solvent, techniques that are commonly known, there are methods which are based on carbon dioxide as extraction solvent. Supercritical Fluid Extraction (SFE) is a good technique, used in flavour isolation from complex matrix or from fruit pulp, for example from strawberries (Ghizzoni *et al.*, 1997). It is possible to use SFE technique, to isolate citrus flavour from aqueous beverages. The flavour fraction obtained is poor of terpene hydrocarbons and rich in oxygenated components, that are most important for the characteristic flavour of citrus oils (Bezold and Zeltz, 1998).

Adsorption techniques

Now we are going to consider systems based on use of common absorbent materials and which allow extraction, concentration so that chromatographic analysis is possible. The adsorption method of separating volatile compounds using GC goes back as far as 1960 when MacKay (1960) reported that volatile compounds could be concentrated on a cold GC column. Until 1971, the range of compounds that could be isolated was limited, but Schultz *et al.* (1971) showed that compounds with molecular weights up to those of sesquiterpenes could be trapped and analysed by GC. Charcoal is excellent for trapping volatile compounds because it is not deactivated by water and has a large capacity for capturing organic compounds even from dilute aqueous solutions. Subsequently, volatile compounds are desorbed from charcoal with solvents, and the desirable aroma compounds are obtained. With charcoal as an adsorption medium problems arise from thermal decomposition when volatile compounds are thermally desorbed. In recent years, porous polymers have become common as adsorption media (Teranishi and Kint, 1993). Tenax-TA (2,6-diphenyl-*p*-phenylene oxide) is the most widely used porous polymer adsorbent in flavour studies, because it provides the best overall trapping efficiency for organic compounds. Flavour compounds are stripped off the sample matrix by the purge gas and are entrained on a trap packed with adsorbing materials. Purge-and-trap, also referred to as a dynamic headspace sampling (DHS), is a means of enriching headspace gas prior to GC analyses. DHS with thermal desorption has been shown to be highly sensitive and relatively inert in the analyses of fragrances and food flavours. A purge-and-trap system involves a purging gas (He, N_2), a sample holder, and a trap at the exhaust of the system. In developing purge-and-trap methodologies, it is critical to evaluate method parameters. Recently Sucan and Russel (1997), emphasised the importance of experimentally optimising all variable parameters to obtain data that are most reflective of the original composition of foods sample. According to the principle of headspace GC, the sample containing volatile components is placed in a sealed vial and conditioned until the volatile components partition into the vapour space above the sample and reach equilibrium; as a result their concentration in the vapour phase is a function of the concentration in the original mixture. In this method the sample is extracted by inert carrier gas with subsequent GC analysis. It is obviously better to employ a gas as a solvent with its ideal solubility for every volatile component as well as the advantage that a gas is normally available in a higher purity than any liquid solvent, which avoids problems with trace impurity interferences. In addition, a gas

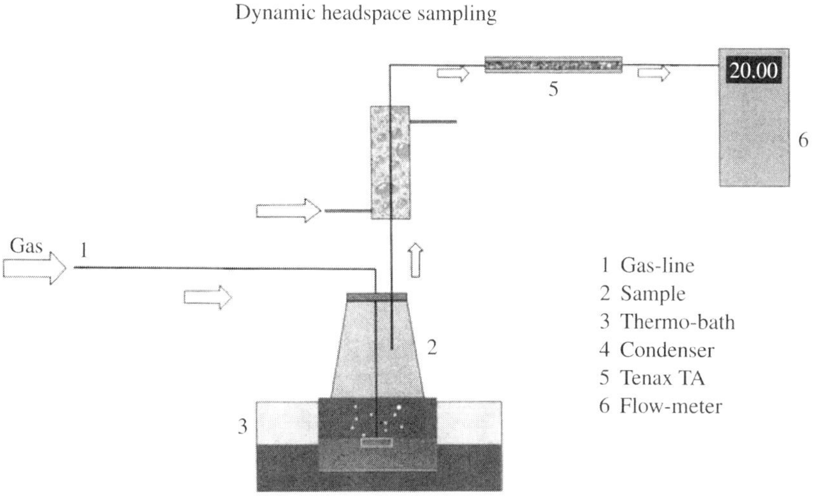

Figure 24.5 Schematic view of dynamic headspace sampling (Cagni and Ghizzoni, 1998).

1. Gas-line
2. Sample
3. Thermo-bath
4. Condenser
5. Tenax TA
6. Flow-meter

does not cause a solvent peak with tailing in the chromatogram. The headspace techniques is a very convenient way of cleaning up a sample before the actual GC analysis. Analytical procedures based on headspace GC are becoming increasingly popular with more and more trace determinations of volatile compounds needed due to the increasing number of ecological problems nowadays. It is preferred if standard GC procedures cause problems with the samples matrix in respect of solubility or thermal stability. A schematic of a DHS system is reported in Figure 24.5 (Cagni and Ghizzoni, 1998). Figure 24.6 shows the GC-MS chromatogram of an orange beverage flavour obtained with DHS as sample preparation method (Cagni and Ghizzoni, 1998).

In order to optimise all the parameters, involved in DHS sampling, Sucan *et al.* (1998) considered the traditional *one variable at a time* approach (i.e. systematically altering one variable at a time until an optimum point is attained) and develop a new way of proceeding in parameters optimisation: using a statistical based approach (Response Surface Methodology, RSM). The use of RSM proved effective in their study: the amounts of volatile flavours recovered could be optimised while maintaining the proportions of flavour components isolated. Two sets of conditions (purging time, sample size, and purge gas flow rate). RSM method allowed the sampling of much greater amount of volatiles than the *one variable at a time* method which increased the GC detector sensitivity by a factor 2.

Solid-phase microextraction (SPME) (Arthur and Pawliszyn, 1990) is an alternative to the above sampling techniques. It is a rapid, inexpensive, solvent-less and easily automated technique for the isolation of organic compounds from gaseous and liquid samples. It is based on the enrichment of components on a polymer- or adsorbent-coated fused-silica fibre by exposing the fibre either directly to the sample or to its headspace. The newly developed solid-phase micro-extraction is increasingly being used for the gas chromatographic determination of a wide variety of volatile and semi-volatile organic compounds in water or aqueous extracts of different substrates. Basically, it

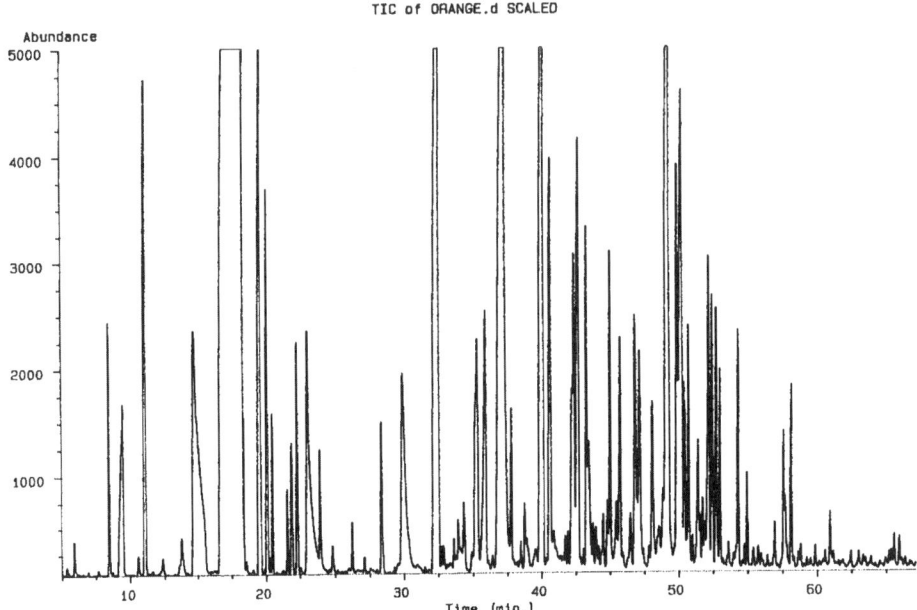

Figure 24.6 Gas chromatogram of orange oil (Cagni and Ghizzoni, 1998).

involves extraction of specific organic analyses directly from aqueous samples, or from the headspace of these samples in closed vials, onto a fused-silica fibre coated with a polymeric liquid phase, polydimethylsiloxane or polyacrylate. After equilibration, the fibre containing the absorbed or adsorbed analyte(s) is ready to be analysed. The analyses are then performed by gas chromatography (GC) using an appropriate column and detector with or without cryo-focusing (Sen *et al.*, 1997). SPME fibre combines sampling and pre-concentration in a single step. After a well-defined adsorption time the fibre is transferred to a standard split/split-less injector, where the organic compounds are thermally desorbed from the polymeric phase. A further advantage of the headspace SPME approach is that samples from virtually any matrix can be analysed since the fibre is not in direct contact with the sample. although care should be taken to release analyses efficiently into the headspace. The detection limits of the headspace SPME technique have been claimed to be at the subpicogram level (Zhang and Pawliszyn, 1995). Yang and Peppard (1994) applied the SPME technique to ground coffee, a fruit juice beverage, and a butter flavour in vegetable oil. They found that the conventional headspace sampling method generally was more sensitive for highly volatile compounds of espresso roast ground coffee, while the SPME headspace method extracted more of the less volatile compounds. In comparison with traditional headspace Tenax adsorption-desorption GC/MS analyses of volatile organic sulfur compounds, SPME technique was less suited for quantitative analyses because the PDMS fibre coating strongly discriminated against more polar and very volatile compounds (Elmore *et al.*, 1997). The chromatograms obtained with three different techniques are shown in Figure 24.7.

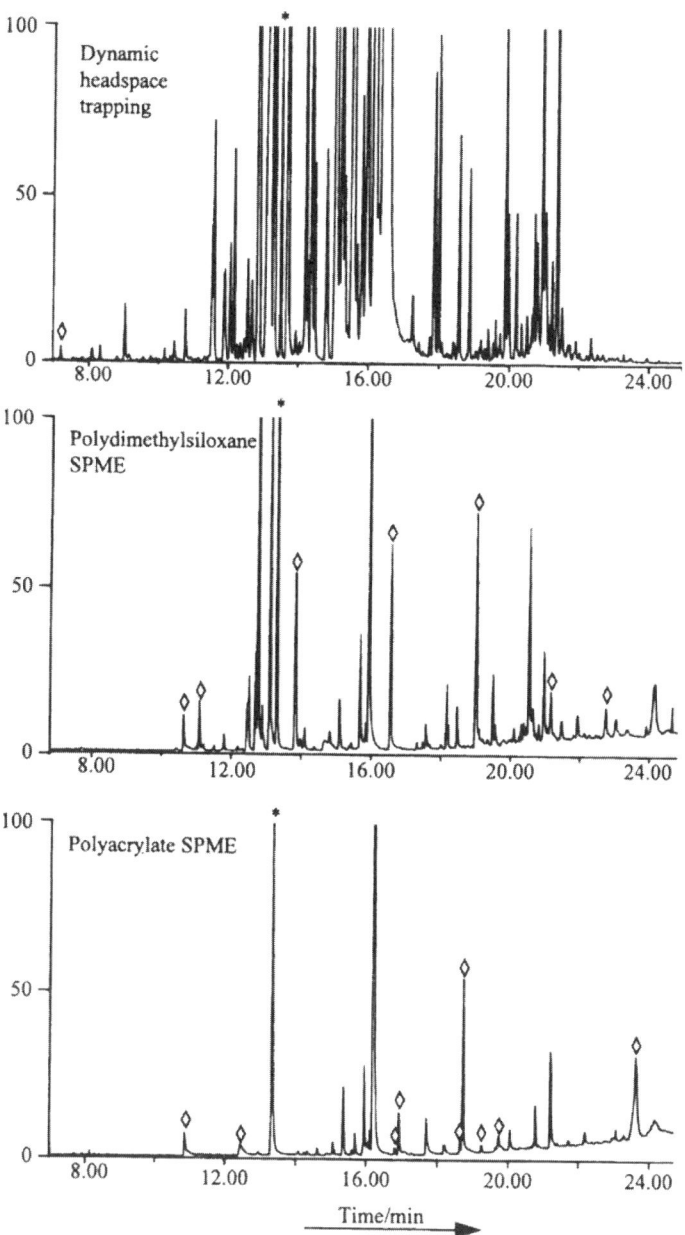

Figure 24.7 Comparison of gas chromatographic cola extracts obtained using three differents techniques (Elmore *et al.*, 1997).

Vacuum headspace method (VHS)

Another way of isolating flavour from complex matrix is the vacuum headspace method (VHS). One objective for the quantitative and qualitative analyses of flavour is to correlate sensory data with instrumental results to determine flavour components of organoleptic importance. The analytical methods used to assay flavours must yield data that are reflective of the original composition of a sample. Unfortunately, the isolation and characterisation of flavour compounds is very challenging because flavour compounds have trace level properties, complex nature, and diverse chemical classes. Flavour analysis on non-thermally treated foods, which are eaten in the row state, needs to be done in a very gentle way. In order to avoid cooked notes and artefacts, fruits have to be worked-up in a way that only the genuine flavour compounds in the right proportions are picked up. Vacuum headspace method (VHS) was first applied by various scientists for the investigation of blossom scents. The first publication was by Daniel Joulain (1968). Since then it has become an established method for analysing blossom scents. More recently, the vacuum headspace technique has been used for flavour analysis of various fruits, such as strawberries, peaches, and cupuaçu. Very recently Tarantilis and Polissiou (1997) reported on the isolation and identification of the flavour compounds of saffron using the vacuum headspace method. If also has to be mentioned, a similar method for isolating flavour volatiles was described in the literature many years ago under the name *high vacuum distillation*. The vacuum headspace method is, in principle, a vacuum steam distillation that takes place with the water of the respective fruit. The volatile flavour compounds distil off and are condensed in the cooled traps. The contents of the traps are then combined and are cold extracted with an organic solvent. The aroma of the flavour concentrates, obtained from fruits by the different isolation method, were compared with the aroma of fresh fruits by the flavourists. The results obtained show the great potentiality of this technique.

Chromatographic analysis

The analytical technique most widely used for flavours and fragrances analyses is surely high resolution gas chromatography (HRGC) with the use of conventional detectors (thermal conductivity, TCD; flame ionisation, FID) and selective detectors (electron-capture, ECD; nitrogen-phosphorous, NPD). The gas chromatography can be coupled to spectroscopic techniques, such as the mass spectroscopy (GC/MS), and Fourier transform infrared spectroscopy (GC/FTIR). These methods will be discussed in detail in a different chapter of this book.

Olfactometric analysis

In the past decades, many detection techniques have been hyphenated to gas chromatography. Less attention has been paid to GC-Olfactometry (GC-O) in which the human nose plays the role of the detector. However, the human nose is often more sensitive than any physical detector, and GC-O exhibits powerful capabilities that can be applied to flavours and perfumes, as well as to any odoriferous products (e.g. pollutants). Olfactometric (or 'sniffing') techniques allow the determination of impact odourants in food. They can be classified into two categories: dilution methods, which are based on successive dilutions of an aroma extract until no odour is perceived at the sniffing port of the chromatograph; and intensity methods, in which the aroma extract is only injected once but the smeller records the odour intensity as a function of time by moving the cursor of a variable resistor. The GC-Olfactometry is a simple to use method, by installing at the end of a chromatographic column

Introduction

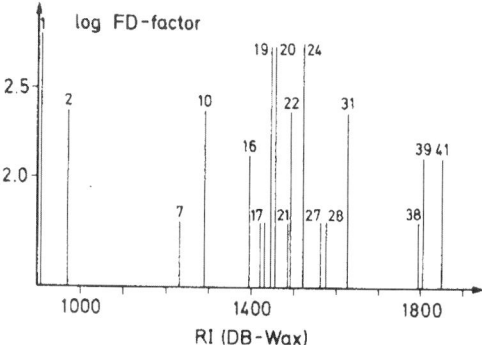

Figure 24.8 Aromagram of rye bread crust obtained using FDF (Maarse, 1991).

a split which allows the sample to be splitted (e.g. 1:50) FID Detector/nose. The peak/odour impression correlation will then be performed by specialised fragrancy chemists. There are different ways to proceed in CG-O technique: one of these is the so-called Charm Analysis. Therefore a major task in flavour chemistry is to distinguish the strongly odour-active compounds from the less odour-active ones. Vital to this flavour characterisation and classification has been the application of sensory technique capable of associating flavour intensity or flavour activity with each chemical constituent. Such a technique is Charm Analysis, a procedure described elsewhere that uses gas chromatography, olfactometry and computer software to quantitate the odour significance of individual volatile constituents. Charm, a bioassay for flavour analysis, combine sniffing of the gas chromatographic effluent with the measurement of ethyl ester standard retention indices. The technique measures the odour intensity of separated volatiles of natural compounds in units of Charm over a range of retention indices. Charm is the ratio of the amount of an odour-active compound to its detection threshold in air. An important feature of Charm is its stability to lead chemical identification into new regions of odour activity. A particular region of new odour activity that underwent further investigation was the odour impact of the unknown compound under the limonene peak (Gaffney *et al.*, 1996). Another way to use GC-O is the FDF (Flavour Dilution Factor) concept. The FDF is obtained diluting stepwise with a solvent an aroma extract until no more odorous compound are detectable in GC-SNIF effluent. The strongest dilution at which a substance is almost detectable is the FDF. This factor is related to odour threshold. In Figure 24.8 is reported an aromagram of rye bread crust obtained using FDF, in logarithmic scale versus column retention index (Maarse, 1991).

There is another way to proceed in the elaboration of quantitative data obtained from a simple GC analysis. Using odour unit (OU), i.e. the ratio between the concentration of a substance in a food and its odour threshold, it is also possible to obtain aromagram. The characterisation of citrus aroma quality was made using the odour unit concept.

There are many examples of the application of GC-O in scientific literature. Besides systems that are based on human nose as a detector, there are machines which uses sensors to determine odours. Electronic/artificial noses are being developed as systems for the automated detection and classification of odours, vapours, and gases. An electronic nose is generally composed of a chemical sensing system (e.g. sensor array or spectrometer) and a pattern recognition system (e.g. artificial neural network). The two main components of an electronic nose are the sensing system and the automated pattern recognition system. The sensing system can be an array of several different sensing elements (e.g. chemical sensors), where each element measures a different property of the sensed chemical, or it can be a single sensing device (e.g. spectrometer) that produces an array of measurements for each chemical, or it can be a combination. Each chemical vapour presented to the sensor array produces a signature or pattern characteristic of the vapour. The quantity and complexity of the data collected by sensors array can make conventional chemical analysis of data in an automated fashion difficult. One approach to chemical vapour identification is to build an array of sensors, where each sensor in the array is designed to respond to a specific chemical. With this approach, the number of unique sensors must be at least as great as the number of chemicals being monitored. It is both expensive and difficult to build highly selective chemical sensors. Artificial neural networks (ANNs), which have been used to analyse complex data and to recognise patterns, are showing promising results in chemical vapour recognition. When an ANN is combined with a sensor array, the number of detectable chemicals is generally greater than the number of sensors. Also, less selective sensors which are generally less expensive can be used with this approach. Once the ANN is trained for chemical vapour recognition, operation consists of propagating the sensor data through the network. Since this is simply a series of vector-matrix multiplications, unknown chemicals can be rapidly identified in the field. An electronic noses is composed of an array of nine tin-oxide vapour sensors, a humidity sensor, and a temperature sensor coupled with an ANN. Although each sensor is designed for a specific chemical, each responds to a wide variety of chemicals. Collectively, these sensors respond with unique signatures (patterns) to different chemicals. During the training process, various chemicals with known mixtures are presented to the system. By training on samples of various chemicals, the ANN learns to recognise the different chemicals. Currently, the biggest field for electronic noses applications is the food industry. Applications of electronic noses in the food industry include quality assessment in food production, inspection of food quality by odour, control of food cooking processes, inspection of fish, monitoring the fermentation process, checking rancidity of mayonnaise, verifying if orange juice is natural, monitoring food and beverage odours, grading whiskey, inspection of beverage containers, checking plastic wrap for containment of onion odour, and automated flavour control. In some instances electronic noses can be used to augment or replace panels of human experts. In other cases, electronic noses can be used to reduce the amount of analytical chemistry that is performed in food production especially when qualitative results will do (Keller *et al.*, 1996). Another kind of electronic nose is the AromaScan system, that is based on an array of 32 conducting polymer sensors, located on a single ceramics substrate which allows them to sense each sample aroma simultaneously. This enables accurate testing of samples with consistent temperature and humidity conditions, giving a truly representative aroma pattern in real time. The tests carried out by AromaScan examined six citrus oils of different geographical origin: Valencia, Grapefruit, Mandarin, SA Tang, Chinese Tang and Florida Tang. After equilibration in pouches for a ten minute period the samples were sprayed onto filter paper for analysis and data selection. The resulting data showed the six samples to have very different aroma characteristics, findings which were confirmed by organoleptic assessments carried out on the same samples by a sensory panel (Tulett, 1996).

Sensorial analysis

We report only one example. The sensorial technique was used in order to verify the conformity of a production lot of a cola based soft drink suspected of deviations from the usual quality standards. For this purpose, the sensorial data were treated with multivariate statistical techniques of validation. An aromatic component, responsible of the citrus fruit taste in the cola based soft drink, was detected and confirmed by gas-chromatography with mass spectrometry detection. In order to evaluate this alteration, comparison is made between some production lot (the standard one and the anomalous one) with two market leader brands. The sensorial analysis permitted to the authors to identify a difference between two production lot, this difference was due to the natural time-variation of the drink (Artoni *et al.*, 1998).

REFERENCES

Arthur, C.L. and Pawliszyn, J. (1990) Solid-phase micro-extraction with thermal desorption using fused-silica optical fibers. *Anal. Chem.*, 62, 2145–2148.

Artoni, A., Gatti, G.C., Zoboli, G.P. and Zoboli, G. (1998) Sensory analysis in the production of soft drinks. *Industria Bevande*, 27, 149–153.

Bezold, A. and Zetzl, C. (1998) Gas extraction of citrus flavour from aqueous and non-aqueous solutions. *Technische Universitat*, Hamburg.

Buchel, J.A. (1989) Flavoring with citrus oils. *Perfum. Flav.*, 14(1), 22–26.

Buttery, R.G., Seifert, R.M., Guadagni, D.G., Black, D.R. and Ling, L.C. (1968) Characterization of some volatile constituents of carrots. *J. Agric. Food Chem.*, 16, 1009–1015.

Cagni, D. and Ghizzoni, C. (1998) Metodi avanzati per l'analisi della componente aromatica delle bevande ipocaloriche e non ipocaloriche. *Symposium: Bevande Innovazione e Protezione*, Milan, November 26, 1998.

Elmore, J.S., Erbahadir, M.A. and Mottram, D.S. (1997) Comparison of dynamic headspace concentration on Tenax with solid phase microextraction for the analysis of aroma volatiles. *J. Agric. Food Chem.*, 45, 2638–2641.

Engel, W., Bahr, W. and Schieberle, P. (1999) Solvent assisted flavour evaporation–a new versatile technique for the careful and direct isolation of aroma compounds from complex food matrices. *Eur. Food Res. Technol.*, 209, 237–241.

Flath, R.A. and Forrey, R.R. (1977) Volatile components of papaya (*Carica papaya* L., Solo variety) *J. Agric. Food Chem.*, 25, 103–109.

Freeburg, E.J., Mistry, B.S. and Reineccius, G.A. (1994) Stability of citral-containing and citralless lemon oils in flavour emulsions and beverages. *Perfum. Flav.*, 19(4), 23–32.

Gaffney, B.M., Havekotte, M., Jacobs, B. and Costa, L. (1996) CharmAnalysis of two *Citrus sinensis* peel oil volatiles. *Perfum. Flav.*, 21(4), 1–5.

Ghizzoni, C., Nova, M., Del Popolo, F. and Porretta, S. (1997) Aroma related compounds commercially strawberry varieties. *International Symposium on Flavour and Sensory Related Aspects*. Cernobbio, Italy. March 1997.

Ikenberry, D.A. and Saaleb, R.K. (1993) The effect of temperature and acidity on the stability of specific lemon flavor components, II. Kinetics. *Dev. Food Sci.*, 32 (*Food Flavours, Ingredients and Composition*), 355–369.

Joulain, D. (1986) Study of the fragrance given off by certain springtime flowers. In E.J. Brunke (ed.), *Progress in essential oil research*. Walter de Gruyter, Berlin, New York, pp. 57–67.

Keller, P.E., Kangas, L.J., Liden, L.H., Hashem, S. and Kouzes, R.T. (1996) Electronic noses and their applications. *World Congress on Neural Networks*. San Diego CA, September 1996.

Maarse, H. (1991) Introduction-characterization. In H. Maarse (ed.), *Volatile Compounds in Food and Beverages*, New York, NY, 1, 25–28.

MacKay, D.A.M. (1960) Determination of trace volatile components contained in gases. In R.P.W. Scott (ed.), *Gas Chromatography* (1960). Butterworths, London, pp. 357–359.

Neumann, M. and Garcia, N.A. (1992) Kinetics and mechanism of the light-induced deterioration of lemon. *J. Agric. Food Chem.*, 40, 957–960.

Nickerson, G.B. and Likens, S.T. (1966) Gas chromatographic evidence for the occurrence of hop oil components in beer. *J. Chromatogr.*, 2, 677–678.

Pokorny, J., Pudil, F., Volfova, J. and Valentova, H. (1998) Changes in the flavour of monoterpenes during their autoxidation under storage conditions. *Dev. Food Sci.*, 40, (*Food Flavors: Formation, Analysis and Packaging Influences*), 667–677.

Schieberle, P. and Grosch, W. (1989) Potent odorants resulting from the peroxidation of lemon oil. *Z. Lebensm. Unters. Forsch.*, 189, 26–31.

Schultz, T.H., Flath, R.A. and Mon, T.R. (1971) Analysis of orange volatiles with vapor sampling. *J. Agric. Food Chem.*, 19, 1060–1065.

Sen, N.P., Seaman, S.W. and Page, B.D. (1997) Rapid semi-quantitative estimation of N-nitrosodibutylamine and N-nitrosodibenzylamine in smoked hams by solid-phase microextraction followed by gas chromatography-thermal energy analysis. *J. Chromatogr.*, 788, 131–140.

Sucan, M.K. and Russel, G.F. (1997) A novel system for purge-and-trap with thermal desorption. Optimization using tomato juice volatile compounds. *J. High Resolut. Chromatogr.*, 20, 310–314.

Sucan, M.K., Fritz-Jung, C. and Ballam, J. (1998) Evaluation of purge-and-trap parameters: optimisation using a statistical design. *Flavor Analysis*, 3, 23–37.

Swaine, R.L. and Swaine, R.L., Jr. (1988) Citrus Oils: processing, technology and applications. *Perfum. Flav.*, 13(6), 1–20.

Tarantilis, P.A. and Polissiou, M.G. (1997) Isolation and identification of the aroma components from saffron (*Crocus sativus*). *J. Agric. Food Chem.*, 45, 459–463.

Teranishi, R. and Kint, S. (1993) Sample Preparation. In T.E. Acree and R. Teranishi (eds), *Flavour Science*. Washington, D.C., 5, 137–143.

Tullett, C., (1996) Citrus oil face aromascan analysis. *Fruit Processing*, 6(2), 66–67.

Yang, X. and Peppard, T. (1994) Solid phase microextraction for flavor analysis. *J. Agric. Food Chem.*, 42, 1925–1930.

Zhang, Z. and Pawliszyn, J. (1995) Quantitative extraction using an internally cooled solid phase microextraction device. *Anal. Chem.*, 67, 34–43.

25 Citrus oils in perfumery and cosmetic products

Felix Buccellato

INTRODUCTION

It does not take an anthropological genius or decades of research to surmise that citrus either as a food or strictly for fragrance value was used long before any written history. Today citrus fruit, juice or peel oils are currently used in everything from food or food preparations, soft drinks, ice cream, candy, pharmaceutical preparations, air fresheners, cleaning products, solvents as well as colognes and fine perfumes. More than 80,000 metric tons are consumed from orange, tangerine, lemon, lime, and grapefruit alone (Nonino, 1997).

MAJOR COMPONENTS OF CITRUS OILS

Listed below in Table 25.1 are the general types of commercial citrus oils available to the fragrance and flavour industry. The columns indicate the percentage of major components commonly shared between the various oils.

If we examine the table we can see that it is organised by groups starting with the highest percentage of limonene. One might think that the products of each group would smell and taste quite similar to one another. This is clearly not the case. Citrus products are a clear example in which the major components are not responsible for the

Table 25.1 Percentage content of major components in citrus oils

Citrus oil	α-pinene	β-pinene	Limonene	γ-terpinene	Linalool	Linalyl acetate	Citral
Orange[a]	0.5	<0.1	95.1	<0.1	0.4		0.1
Grapefruit[b]	0.5	0.2	93.0	0.1	0.1		0.1
Tangerine[c]	0.8	2.0	91.0	3.3	0.5		1.5
Mandarin[a]	2.4	2.0	68.2	20.1	0.1		
Lemon[a]	1.9	13.1	65.2	9.8	0.1		1.9
Lime[d]	2.1	12.9	56.6	13.5	0.2		3.5
Bergamot[a]	1.2	7.0	33.9	7.8	10.4	32.6	0.5
Neroli[e]	1.3	20.2	24.5	3.7	15.6	9.8	1.1
Bitter orange flower absolute[f]	0.1	0.4	5.0		32.0	17.0	
Bitter orange petitgrain[g]	0.1	1.0	1.1	<0.1	27.7	54.0	0.8

Notes
a Dugo *et al.*, 1994; b Boelens, 1991; c Lawrence, 1992; d Dugo *et al.*, 1997; e Mondello *et al.*, 1996; f Buccellato, 1999; g Dugo *et al.*, 1996.

character of the oil. If they were, there would be little difference between orange, grapefruit and tangerine, which are nearly identical at first glance. It is probably best to think of limonene (the major component of most of the citrus oils), as nothing more than a medium in which the important odour characterising components are dissolved.

Although limonene itself, present in citrus oils mainly in its (+)-enantiomeric form, is not the most significant odour-characterising component in these oils, it is still an odour contributor. It has a very fresh pleasing aroma that can be described as citrus and somewhat piney. It is interesting to note that (−)-limonene is different in character and exhibits more of a spearmint character. This enantiomeric difference has been noted by comparing (+)- and (−)-carvone as well (Boelens *et al.*, 1993). Both (+)- and (−)-limonene are fairly low impact aroma contributors and quite easily can be 'coloured' by significant odour contributors like low molecular weight esters, aldehydes and ketones. These are abundant (in number not percentage) in all citrus oils. The other low molecular weight hydrocarbons like α- and β-pinene are in the same performance category as limonene. That is to say, they have low odour impact, mild in character (piney), and do not contribute in a significant way to the citrus aroma of any of the citrus products. A cursory evaluation of commercial materials such as α- or β-pinene or limonene can be quite misleading. They are always contaminated with trace impurities that are carried over from the source material. The most typical case is (+)-limonene that has been isolated from orange peel. The orange character that remains is due mainly to all the trace components from orange peel. How do we know this? When we examine (+)-limonene from a totally synthetic source, e.g. from alpha pinene, we notice that the typical orange character is completely absent. At this point one can begin to appreciate the true nature of (+)-limonene.

The above example and logic can be applied to α- and β-pinene. If we go down the list in Table 25.1 to tangerine and mandarin then to lemon and lime oils we can see that other terpene hydrocarbons start to appear like γ-terpinene. This component, an unsaturated hydrocarbon, starts to introduce additional character to nature's citrus composition. It is somewhat citrus in character but is differentiated by a sharp pungent character that can be described as somewhat spicy similar to black pepper or thyme. This adds a new dimension to the tangerine, mandarin, lemon and lime oils and puts the 'tang' in tangerine and all citrus products.

Upon examination of constituents in Table 25.1, we find bergamot oil and the subsequent products from *Citrus aurantium* L., produce a drastic change in composition. There are large amounts of linalool and linalyl acetate. This marks a change where the citrus character becomes blended and enhanced with a floral element. If one were to describe the character of linalool, it would fall into the category of both lemon-like citrus and lily-like floral. If one were to suppose that nature is the best teacher, as this author does, then bergamot oil is a great example of a marvellously blended and complemented citrus bouquet. One should realise that the combination of linalool, linalyl acetate and limonene does not make a bergamot oil! At the time of this writing Italian bergamot, my personal favourite, contains over 100 components and the closer we look, the more we find (Casabianca, 1994). The elements that comprise bergamot are: hydrocarbons, alcohols, ketones, esters, acids, lactones, ethers and epoxides. This includes saturated and unsaturated versions. The unsaturated versions are the most interesting from a flavour and fragrance point of view. The least explored and perhaps the most important groups of nitrogenous and sulphur containing molecules make a significant contribution to bergamot as well as all other citrus types. The grouping, after bergamot oil, includes neroli, bitter orange flower absolute and bitter orange petitgrain oils. These are all from

Table 25.2 Odour character of some citrus essential oil components

Citrus	Floral	Fruity	Spice	Green/Earthy
α-pinene	phenyl ethyl alcohol	neryl acetate	eugenol	benzyl nitrile
β-pinene	α-terpineol	citronellyl acetate		pyrazines
myrcene	citronellol	geranyl acetate		pyridines
limonene	methylanthranilate[a]	methylanthranilate[a]		
ocimene	geraniol			
linalool[a]	linalool[a]			
linalyl acetate	nerol			
	farnesol			
	nerolidol			

Note
a Methylanthranilate and linalool display an important duality in character.

C. aurantium and are somewhat related. Much of the same rhetoric described in the paragraph pertaining to bergamot applies to all three of these oils. The literature is full of compositional information and it is not necessary to examine them here.

Some of the components present in citrus oils are reported in Table 25.2, where they are grouped by their primary odour character.

CHARACTERS OF CITRUS OILS

Orange is the most abundant citrus utilised in the world. The real character donating components in orange oil are trace components that are various low to medium molecular weight saturated and unsaturated aldehydes, ketones, as well as esters and alcohols. For as much as we have learned about the constituents of orange oil (over 100 components), the fragrance and flavour industry cannot make a good orange flavour without using the natural product.

Grapefruit oil is characterised by specific sulphur containing components as well as the woody, sesquiterpenic section that provides a very rich long lasting character.

Tangerine and mandarin oils are characterised by different unsaturated trace aldehydes as well as amino esters like methyl *N*-methyl anthranilate.

Lemon oil is characterised by citral, (neral and geranial) and other aldehydes both saturated and unsaturated. These, along with trace sulphurous notes contribute significantly to the character that defines good fresh lemon. However, it is these very important materials that are unstable and short-lived. For this reason, there are no excellent lemon flavoured soft drinks that can approach the flavour of fresh lemonade. There are many good products, some achieved through stable chemistry research, which we will discuss later.

Lime oils have a very distinctive aroma and flavour, which is imparted by terpinolene, α-fenchyl alcohol and various cyclohexadienes. Unfortunately at this point in industry we have not adequately identified the real characterising components of lime oils. There are many important sulphur compounds waiting to be discovered in lime oils that will enhance our ability to create a more nature-like lime aroma. Lime oils suffer the same fate as lemon oil in that they are unstable, easily oxidised and quality in soft drinks is only a shadow of nature.

Bergamot oil is one of the most beautiful oils in the citrus family. However, as can be seen in Table 25.1, it is a hybrid of citrus notes as well as floral notes. The main components, which are terpene hydrocarbons, terpene alcohols and esters, are all enhanced by traces of aldehydes, lactones, as well as trace sulphur components that are so wide ranging that bergamot oil is practically a fragrance by itself.

Neroli oil is the steam-distilled flowers of *C. aurantium*. This is an increasingly rare and expensive oil. While it is related to bitter orange flower absolute and bitter orange petitgrain oil, there is nothing like it. It is a hybrid character with unique citrus, floral, green, and nearly bread-like notes that has a performance all of its own. This is due to the combination of unsaturated terpene hydrocarbons, linalool, linalyl acetate, aldehydes both unsaturated and saturated. Furthermore, pyrazine, pyridine and sulphur compounds present at trace levels play an extremely important role in the overall impression and uniqueness of this oil.

Bitter orange flower absolute is the solvent/alcohol extract of the *C. aurantium* flower. This is characterised by linalool, linalyl acetate and various esters blended with methyl N-methyl anthranilate (an important note in mandarin and tangerine oils), along with indole, and lactones. This is a decidedly floral but complex note that adds dimension and longevity to any citrus composition. This is one of the most complex oils used in our industry. It contains approximately 20 components over 0.1 per cent, and over 100 trace ingredients (Buccellato, 1999).

Bitter orange flower water absolute is obtained from bitter orange flower by solvent extraction of the water-phase produced by steam distillation. This unique product is used, as neroli oil and bitter orange flower absolute, in a wide variety of applications in both flavour and fragrance.

Bitter orange petitgrain oil (*bitter orange leaf oil*) is obtained by steam distillation of the leaves and twigs of *C. aurantium*. It is characterised by terpene hydrocarbons, linalool and linalyl acetate, various other acetates, aldehydes and most importantly, as in neroli oil, by pyrazines and sulphur components. Bitter orange petitgrain oil is a very beautiful and inexpensive material that is used in a wide variety of applications ranging from men's and women's colognes as well as cosmetic and functional products. The major components limonene, linalool and linalyl acetate can be described as the supporting medium only. The star performing components are the trace sulphur ingredients and various substituted pyrazines, pyridines and thiazoles. The pyrazines in particular are extraordinarily potent and stable in a wide variety of applications. In addition, they are the most important characterising components in bitter orange petitgrain oil. It is a wonder that more leaf oils from other citrus species are not used in the same quantity.

Bitter orange leaf water absolute is obtained by solvent extraction of the water-phase produced by steam distillation of leaves and twigs of the bitter orange tree. The main characterising components are pyrazines and sulfur components which create the earthy, leafy and potent aroma which is common to all leaf oils.

Bitter orange petitgrain absolute, is obtained by alcohol extraction of the leaves and twigs of the bitter orange tree. This product is not very common but it is commercially available. However it is expensive which may explain why its use is limited.

Other Petitgrain oils do not enjoy the same status of bitter orange petitgrain oil (bitter orange leaf oil). It should be noted that each citrus crop could be producing an oil called petitgrain. They are more often referred to as leaf oils. There are potentially four leaf oils that could be produced in the United States. These are sweet orange, lemon, lime and probably most interesting grapefruit leaf oils. Pricing as well as manufacturing, high labour costs and pruning processes might be the reason that these are not viable commercial products.

USAGE OF CITRUS OILS

How long have citrus products been used for either fragrance or flavour by man? Is this question even important to answer? We are always looking for the 'first recorded' use of an item to pinpoint a date.

The real answers to questions like these are elusive. It can be assumed that these items have enjoyed flavour and fragrance use long before recorded history. It is known that citron seeds with fruit tissue attached to them were excavated on the island of Cyprus dating back to the thirteenth century BC (Feliks, 1994). One can easily assume that uses do not spring into being overnight. It could be safely assumed that long before any civilisation or written history, man has been using these fruits not for thousands, but for millions of years.

Some of the earliest world famous colognes have always used citrus products of some sort. Imperial 1850 and Eau Imperiale 1861 by Guerlain included the use of bergamot, lemon, lime, mandarin, orange and orange blossom oils (Bedoukian, 1993). The earliest European fragrances were based on citrus oils like lemon, lime, bergamot, orange, mandarin and petitgrain oils. The classic accords like Eau de Cologne (created in 1750) and 4711 have been the foundations for many fragrances in history and have evolved into today's modern citrus blends including CK One by Calvin Klein.

If we take a look at the history of the fragrance industry we can, nearly at random, pick the popular fragrances of each decade and take a cursory look at the construction.

Women's fragrances

1921–Chanel 5 utilises lemon and bergamot oils along with synthetic aldehydes used possibly for the first time in amounts larger than usual and larger than permitted by nature. What this means is that whenever natural products such as citrus are used, a portion of their performance and beauty is a result of the naturally occurring aldehydes that are part of their complex. This is true for other natural products as well but it is particularly true for citrus products. The use of synthetic aldehydes, saturated and unsaturated, impart and enhance a citrus character as well as providing a fresh character to the overall composition. The additional effect is to introduce unparalleled power and performance. This was especially unusual in 1921.

1927–Arpege by Lanvin uses bergamot and neroli oils on top of a heavier oriental chypre and amber complex. The citrus section is also supported by the use of the now very popular synthetic aldehydes, which enhance the citrus section.

1936–Muguet de Bois by Coty uses bergamot and orange oils to enhance the natural floralcy of the muguet character.

1945–Vent Vert by Balmain uses lemon, lime, bergamot orange and mandarin oils enhance the vetivert and natural green notes.

1947–L'air du Temps by Nina Ricci is a light floral muguet that is extended and enhanced by bergamot, lemon and orange oils. This light 'springtime' fragrance accord is still very popular today. Many fragrances both old and new have followed this formula of light floral notes blended with judicious amounts of citrus. This blend creates universally accepted wonderful fragrances, which are still popular today.

1956–Diorissimo by Dior is a wonderful floral muguet blend that is enhanced by bergamot and lemon oils much the same way as L'air du Temps.

1969–O de Lancôme by Lancôme utilises bergamot, lemon and petitgrain oils.

1973–Charlie by Revlon contains bergamot oil and touches of other citrus notes to enhance the underlying floralcy of jasmine, tuberose and gardenia.

1974–Cristalle by Chanel owes its top note to lemon and bergamot oils providing a fresh clean fragrance.

1978–Lauren by Cosmair is a most beautiful fruity muguet, which is heightened by bergamot and orange blossom oils.

1981–Giorgio by Giorgio Beverly Hills has a powerful tuberose floral foundation, which is combined with bergamot and orange blossom oils.

1996–Allure by Chanel is a modern floral using the time-tested combination of lemon and bergamot oils.

1997–Contradiction by Calvin Klein is a complex woody amber blend which becomes fresh and clean with the use of bergamot, lemon, mandarin and orange oils.

Men's fragrances

1949–English Leather by Mem Corporation contains bergamot, lemon, orange and petitgrain oils to provide a bright and fresh top note to an oriental oakmoss blend.

1966–Eau Sauvage by Dior contains bergamot, lemon, lime, orange, mandarin and petitgrain oils.

1974–Eau de Guerlain contains bergamot, lemon, lime, orange, mandarin and petitgrain oils.

1982–Eau de Cologne by Hermes contains bergamot, lemon, lime, orange, mandarin and petitgrain oils.

The 3 preceding fragrances can be thought of as modern versions of the nearly antique 4711 which is based on citrus products like bergamot, lemon, lime, orange, mandarin and petitgrain oils. These three really feature citrus products almost exclusively.

1995–Polo Sport by Ralph Lauren is a modern sophisticated fragrance, which employs lemon and bergamot oils very skilfully.

1998–Cool Water by Davidoff uses bergamot and lemon oils among some new synthetic materials, which we will cover a bit later in the following text.

By now it can be seen that citrus products are widespread throughout the fragrance industry. It is difficult to say how important these materials are. One should not be deceived by thinking that all the above fragrances are the same or even similar because of

Table 25.3 Top selling fragrances in 1999 organised into odour categories

Citrus blends	Woody blends	Floral blends	Oriental blends
CK One	Polo	Pleasures	Obsession
Eternity men	Obsession men	Eternity	White Diamonds
Polo Sport	Aromatics Elixir	Tresor	
Tommy	Safari men	Allure	
Polo Sport women		Jessica McLintock	
Cool Water		White Linen	
Hugo		Chanel 5	
Drakkar Noir		Tommy Girl	
Escape men		Contradiction	
		Beautiful	

the widespread repeated use of the same materials. It is the very nature of citrus to absorb, enhance and modify the identity of any fragrance in which they are used. It is also due to the artistic skill of the creative perfumers who use these wonderful items.

The top selling fragrances in 1999 have a story to tell. If we examine them, organise them into odour categories (Table 25.3) and examine them philosophically we may see a pattern. The categories include unisex, women's and men. As can be seen, the two most popular categories are floral blends and citrus blends.

The citrus blends are of course, the topic of this book and chapter. As we focus on a few of these fragrances we will see, not only what is being used in fragrances today but additionally, one may realise why citrus items are so universally popular and ubiquitous.

When we look at Table 25.3 it is difficult to identify any fragrance that does not contain some citrus. Almost every fragrance created or designed for almost any product utilises at least one citrus ingredient. More often than not, several citrus oils are combined. What is the reason for this? Citrus products have the ability to lift the character of the fragrance and extend and enhance those notes that are present and create a fresh and lively aroma profile. As previously noted, limonene, along with linalool and linalyl acetate themselves do not impart the character of the fruits. It is the trace ingredients that are doing all the work. It seems that when any citrus product is judiciously used in fragrances, they easily assume the character of the components that are already present. A great example of this is the Shalimar fragrance. Upon examination by the newly initiated, this seems like a balsamic, vanilla and powder like fragrance. The fact that Shalimar has a high percentage bergamot and other citrus oils may not seem obvious. This situation is typical in perfumery.

CK One, introduced in 1994 is an extremely popular and top selling fragrance worldwide. This universal appeal is in part attributable to the prodigious use of bergamot, lemon and mandarin oils. This is a fresh citrus cologne that is artfully blended with other ingredients to create a new entity with its own unique identity. This composition uses notes of pineapple and papaya for the fruity character, cardamom and amber for the spice section, jasmine, lily, rose and freesia for floralcy, cedar, orris, sandalwood and oakmoss for the woody section. These are all supported by amber and musk. This is great example of how commonly available ingredients can be utilised in a unique way to create a successful fragrance with universal appeal. This can be said for all successful fragrances and applies to all the examples above and many not cited here.

Eternity for Men, introduced in 1989 utilises mandarin, lemon oils and notes of orange blossom and petitgrain oils for an Old World twist. This citrus blend is supported by basil, tarragon, sage and coriander for a spice section, geranium and lavender for a herbal section, cedar, vetivert, oakmoss, rosewood with a hint of jasmine amber and oakmoss for a classic blend which is still popular and well received today.

Polo Sport, introduced in 1993 uses lemon and bergamot oils, along with jasmine, rose, cyclamen and tagette for the fruity floral aspects. Geranium, lavender and artemesia form the herbal section and oakmoss, cedar sandal and guaiacwood rest on a base of musk and amber. These combined with a hint of nutmeg form a classic blend of forest and deep rich woody notes that are lifted as a result of the citrus section.

Tommy Hilfiger, introduced in 1995 employs the use of bergamot, lemon, lime, mandarin and grapefruit oils. In addition to wild berry notes there is an herbal section using lavender, and mint along with spices including nutmeg and cinnamon. These are blended with an amber and musk foundation rounded out with sandalwood and cedar. While this is a sophisticated and complex fragrance, it is the citrus section that makes the whole composition fresh and lively.

Cool Water, introduced in 1988 blends bergamot, lemon and orange blossom oils with dry woody and amber notes, which is fresh, unique and long lasting. This amber and musk based composition combines the fresh aroma of herbs like rosemary, lavender, mint, sage and coriander blended with geranium, jasmine, honeysuckle and lily. The foundation is composed of woody components like orris, sandalwood and cedar. It also uses dihydromyrcenol, which is an important synthetic fresh citrus character. This enhances the lemon note and provides a unique twist to the citrus core.

Hugo for Men, introduced in 1995 combines lemon, lime, orange and bergamot oils with oakmoss and other deep rich woody forest notes that are very rich and Old World like in a modern era. These notes are blended with lavender, geranium, mint, sage and other spices. These are further blended with touches of jasmine and muguet. Cedar, sandalwood and vetivert support these notes, which magnify and enhance the grapefruit character.

Drakkar Noir, introduced in 1982 utilises mandarin, lemon and bergamot oils and like Cool Water, utilises a generous portion of dihydromyrcenol. This is bolstered by a spice blend of basil, cinnamon and cardamom. This blend is linked to a foundation of patchouli, pine, cedar, sandalwood and oakmoss. We will examine the importance of dihydromyrcenol shortly.

Escape for Men, introduced in 1993 is a classic type blending bergamot, lemon and grapefruit oils with new and synthetic citrus like ozone and fresh notes that extend and enhance the citrus character. This modern fragrance is blended with birch, juniper, eucalyptus and pine which all complement the citrus section. The foundation is based on artemesia with touches of rose and cyclamen, which has an element of citrus character, by itself.

It has often been said, that the prestige market often leads the way for popular trends. This is due in large part to the massive advertising and marketing campaigns that are launched for each fragrance or product. It is true that a good fragrance will have a tremendous effect for second purchase and continued long-term success, but a mediocre product with great advertising, packaging and distribution will win every time over a great product that no one knows about.

After a new type (or a similar type) of fragrance has been introduced and market demand created, a second tier of related functional products such as soap, hair care, creams, lotions, candles, and air freshener products soon can be expected to follow.

Citrus fragrances are intrinsically prone to two types of problems, rapid oxidation and insolubility in hydrophilic systems such as shower gels. The fragrances which are high in citrus are often perceived as clean and fresh and are precisely the type desired for these products but their short shelf life, stability and incompatibility presents a technical challenge that is often overcome by the use of surrogate materials that impart citrus notes which do not oxidise quickly and are more soluble in hydrophilic systems.

Synthetic citrus notes

As mentioned throughout the previous text, there are many synthetic products available for use in perfumery. They include, but are not limited to, synthetic aldehydes, alcohols, acetals and to a limited extent, ketones. There are also some nitrogen and sulphur components that are available for enhancement and stability purposes. When one examines the information available, it is possible to find approximately 200 synthetic components that are primarily citrus in character.

As referenced above in Cool Water and Drakkar Noir dihydromyrcenol (2,6-dimethyl-7-octen-2-ol) is a synthetic ingredient that possesses a fresh lemon-lime like aroma. Dihydromyrcenol, like linalool (3,7-dimethyl-1,6-octadien-3-ol) is a tertiary alcohol, but the odour and stability of these two compounds is quite different. While linalool, a major component of bergamot oil has a somewhat citrus floral character dihydromyrcenol is decidedly more lemon lime like and extraordinarily stable. Aside from the intrinsic quality of the odour character the real virtue is the extraordinary stability of this molecule in a variety of media. It does not easily oxidise or deteriorate in any media. This problem of oxidation and deterioration is common to all citrus products so to have a material that does not deteriorate is a tremendous tool in the perfumer's stability arsenal.

Another important synthetic citrus note is dimetol (2,6-dimethyl-heptan-2-ol), a stable lemon like aroma with a bit more floralcy than dihydromyrcenol. This is used in Paco Rabanne along with bergamot, petitgrain oils and other citrus aldehydes, which enhance the citrus top note.

Since aldehydes themselves, exhibit a citrus character and will enhance any citrus performance and they are almost always used in addition to citrus products. They are however, not as stable as one might need under a variety of circumstances or in functional products like soap, detergents or other products where oxidation or reactivity presents a problem. To replace the use of unstable aldehydes the perfume industry has resorted to the use of nitriles. These nitriles are less prone to oxidation and are less reactive under a variety of circumstances. Some examples are geranyl and neryl nitriles, citronellyl nitrile and decyl and dodecyl nitriles to name a few. At this point in time we are aware of at least sixty synthetic nitriles many of which possess a citrus character. Another choice in this arena is the use of acetals, one might say these are aldehydes where the reactivity has been deactivated by reacting the aldehydes in advance. Many of these acetals have a citrus character albeit somewhat different and unique from their aldehyde progenitors. Examples would be neral and geranial, both dimethyl and diethyl acetals, citronellal dimethyl acetal, octanal dimethyl and diethyl acetals etc. We are currently using over seventy various acetals at this point in time.

Until this point we have addressed the issue of oxidation and reactivity of citrus products for which we look to the above examples for solutions or alternatives. We have not yet addressed the issue of longevity or lasting ability for a citrus character. It is possible, but often difficult, to obtain a satisfactory citrus character that would be sufficiently long lasting. The intrinsic character of citrus is to be light fresh and as a consequence, is typically short lived. In order to accomplish this, we often turn to Schiff's bases for the solution. Many Schiff's bases of aldehydes, both straight chain and unsaturated along with citral and other branched or cyclic aldehydes are reacted with amines like methyl anthranilate. This provides an unusual character peculiar to Schiff's bases but related to citrus. These Schiff's bases possess a particular citrus-like character that has tremendous longevity and enhanced stability in paper air freshener products as well as in functional products. These materials do not provide the citrus fresh top notes that we appreciate, but they do impart an undertone that is quite long lasting and in effect, mimics the peel like character.

Can any pattern be seen from the above examples? It should be clear that citrus notes are ubiquitous in the fragrance industry. The few isolated examples that have been extracted for illustration are only the tip of a very large iceberg. It should be remembered that the very few examples shown in this text are only a one-person view. This is a very limited view and for that reason I must apologise for any omissions of important products or inaccuracies that may be presented. However the overall philosophical view of citrus use is fairly representative of what and how citrus products in general are used.

FUTURE USES OF CITRUS PRODUCTS

What will be the future of citrus use in fragrance? It is inevitable that change will occur. The fragrance industry has seen many changes, some good, some less so. Many products have been eliminated or drastically reduced for toxicological reasons. The best example is natural cold-pressed bergamot oil, that, as illustrated by the preceding examples, is used in many fragrances. This beautiful material is restricted from skin use due to the presence of bergaptene (5-methoxy psoralene), which exhibits phototoxic effects. This seems a shame because the effects and consequences of using this material may be questionable. When one thinks of the centuries of use without any real incident, one has to question if the use of this material should have been restricted at all. Whenever the bergaptene free or distilled version of bergamot oil is used, there is enormous lack of depth and beauty of which one normally experiences from use of the natural cold-pressed bergamot oil. One should also be reminded of the many people who are making lemonade and hand squeezing fresh lemon and lime (which are also restricted), into tea and other food preparations throughout the world and are not complaining of photosensitisation. The amount that is deposited directly onto the skin from this activity is often many hundreds of times more than the exposure that anyone would receive from a cologne and thousands of times more than anyone would be exposed to from citrus used in a functional product. Our industry, through science has chosen to err on the side of safety, not on the side of practicality or artistic sensibility.

Many chemists around the world are doing great research on natural products. This is an area where many newly identified products will enhance our understanding of nature. As indicated throughout this text, the area of greatest importance is the sulphur and nitrogen components that abound in nature. These are critically important in citrus products. New products are waiting to be identified, synthesised and brought into commercial use. This is the least understood and probably the most important area in the fragrance and flavour industry at this point in time. They are part of the reason that the natural products smell and taste so great. Synthetic fragrances and flavours do not smell or taste bad (different) because they are synthetic, they taste bad (different) because they are much simpler and less complete than nature.

There is always new information and materials that are being introduced to the fragrance industry as well as new challenges in new products. New products always require new ideas. I feel very certain that the continued use of citrus products both synthetic and natural (hopefully), will be assured for many centuries to come.

REFERENCES

Bedoukian, P. (1993) Eau Imperiale by Guerlain. *Perfum. Flav.*, 18(4), 33.

Boelens, M.H. (1991) A critical review on the chemical composition of citrus oils. *Perfum. Flav.*, 16(2), 17–34.

Boelens, M.H., Boelens, H. and van Gemert, L.J. (1993) Sensory properties of optical isomers. *Perfum. Flav.*, 18(6), 1–16.

Buccellato, F. (1999) Unpublished work.

Casabianca, H. (1994) Bergamot oil. *J. High Resolut. Chromatogr.*, 17, 184–186.

Dugo, G., Cotroneo, A., Del Duce, R., Donato, M.G., Dugo, Giacomo, Dugo, P., Lamonica, G., Licandro, G., Mondello, L., Stagno d'Alcontres, I., Trozzi, A. and Verzera, A. (1994) The composition of the volatile fraction of the Italian citrus essential oils. *Perfum. Flav.*, 19(6), 29–51.

Dugo, G., Mondello, L., Cotroneo, A., Stagno d'Alcontres, I., Basile, A., Previti, P., Dugo, P. and Bartle, K.D. (1996) Characterization of Italian citrus petitgrain oils. *Perfum. Flav.*, 21(3), 17–28.

Dugo, P., Mondello, L., Lamonica, G. and Dugo, G. (1997) Characterization of cold-pressed Key and Persian lime oils by GC, GC/MS, HPLC and physicochemical indices. *J. Agric. Food. Chem.*, 45, 3608–3616.

Feliks, Y. (1994) *Fruit Trees in the Biblical and Talmudic Literature*, Rubin Mass Press, Jerusalem.

Fleisher, A. (1996) The first citrus of the western world. *Perfum. Flav.*, 21(6), 11–16.

Lawrence, B.M. (1992) Progress in essential oils. *Perfum. Flav.*, 17(4), 39–44.

Mondello, L., Dugo, G., Dugo, P. and Bartle, K.D. (1996) On-line HPLC-HRGC in the analytical chemistry of citrus essential oils. *Perfum. Flav.*, 21(5), 25–49.

Nonino, E. (1997) Where is the citrus industry going? *Perfum. Flav.*, 22(2), 53–58.

26 Legislation of citrus oils

Friedrich Grundschober

INTRODUCTION

Citrus oils are used as ingredients of flavourings and fragrances as well as chemicals. Their use is regulated by legislations for foods, cosmetics and chemical substances. The composition of consumer products is regulated with regard to safety and labelling aspects. In addition to official regulations issued by governments, there also exist Codes of Practice issued by industry organizations.

Citrus oils are covered by regulations for flavourings, fragrances and dangerous substances. A survey of legislation has to cover the main industrialized areas, Europe and USA, as well as the regional and international harmonization of legislation by the European Union, the FAO/WHO Foods Standard Programme, the Codex Alimentarius as well as IFRA (International Fragrance Association) and IOFI (International Organization of the Flavour Industry).

FLAVOURINGS

European Union

The European Flavourings Directive 88/388/EEC (1988) covers citrus oils by the definition of *flavouring preparation* which is obtained by appropriate physical processes including distillation and solvent extraction from material of vegetable origin (Art. 1,2,c). Citrus oils or individual substances isolated from citrus oils can be used in *natural* flavourings (Art. 9,2,1st paragraph).

If the sales description of the flavouring contains a reference to a foodstuff or a flavouring source, the word *natural* or any other word having substantially the same meaning, may not be used unless the flavouring component has been isolated by appropriate physical processes, enzymatic or microbiological processes or traditional food-preparation processes solely or almost solely from the foodstuff or the flavouring source concerned (Art. 9,2,2nd paragraph of the above cited directive).

This wording allows the conclusion that all known methods for preparing citrus oils as well as the methods for deterpenization will lead to natural flavourings. Whether the term *natural* can be combined with the name of the fruit depends on an interpretation of *solely or almost solely*. There is no doubt that *solely* means 100 per cent, the *almost solely* may be between 90 per cent and 95 per cent. It should also be taken into account that the legislator has no intention to abolish the traditional practice of blending e.g. orange oil with small amounts of mandarin oil etc.

There exist no detailed legislation for essential oils. The harmonization of the EU flavouring legislation is starting with the listing and evaluation of defined flavouring substances. Natural flavourings are regulated by the restrictions for active principles in Annex II of the Flavouring Directive 88/388/EEC. The active principles of Annex II are not occurring in citrus oils and have therefore not to be controlled. The only restrictions of interest applicable to all flavourings are the limits for heavy metals: 3 mg/kg As, 10 mg/kg Pb, 1 mg/kg Cd and 1 mg/kg Hg (Art. 4,a).

USA

Essential oils are part of the definition of *natural flavourings*. In the list of essential oils that are generally recognized as safe (GRAS) the following oils are mentioned (US Code of Federal Regulation):

Bergamot	*Citrus aurantium* (L.) ssp. *bergamia* Wright et Arn.
Citrus	*Citrus* ssp.
Curaçao orange	*Citrus aurantium* (L.)
Grapefruit	*Citrus paradisi* Macf.
Lemon	*Citrus limon* (L.) Burm.f.
Lime	*Citrus aurantifolia* Swingle
Mandarin	*Citrus reticulata* Blanco
Neroli, bigarade	*Citrus aurantium* (L.)
Orange bitter	*Citrus aurantium* (L.)
Orange sweet	*Citrus sinensis*
Petitgrain	*Citrus aurantium* (L.)
Petitgrain lemon	*Citrus limon* (L.) Burm.f.
Petitgrain mandarin or tangerine	*Citrus reticulata* Blanco
Tangerine	*Citrus reticulata* Blanco

These oils can be used as ingredients of food flavourings provided that purity requirements are respected. The Tolerance Residue Levels of certain pesticides in the oil are:

40 CFR		ppm
180.449	Avermectin B_1 (time-limited)	0.1
185.1000	Chlorpyrifos *O,O*-diethyl *O*-(3,5,6-trichloro-2-pyridyl) phosphorothioate and its metabolite 3,5,6-trichloro-2-pyridinol	25.0
185.3650	Imazalil 1-[2-(2,4-dichlorophenyl)-2-(2-propenyloxy) ethyl]-1H-imidazole and its metabolite 1-(2,4-dichlorophenyl)-2-(1H-imidazole)-1yl)-1-ethanol	25.0
185.4000	Metalaxyl *N*-(2,6-dimethylphenyl)-*N*-(methoxyacetyl) alanine methyl ester and its metabolites containing the 2,6-dimethylaniline moiety and *N*-(2-hydroxy methyl-6-methyl)-*N*-methoxyacetyl)-(alanine methylester)	7.0

185.2950	Ethyl 3-methyl-4-(methylthio)phenyl (1-methyl-ethyl) phosphoramidate and its cholinesterase-inhibiting metabolites ethyl 3-methyl-4(methylsulfinyl) phenyl (1-methylethyl)-phophoramidate and ethyl 3-methyl-4-(methyl sulfonyl)-phenyl (1-methylethyl) phosphoramidate	25.0

Other countries

Citrus oils are not specially restricted for use in foods.

In case that pesticide residue levels exist for certain foods, essential oils should not contribute higher pesticide levels to these foods than authorized. The pesticide residues in essential oils are not specially regulated.

Council of Europe

The Committee of Experts on Flavouring Substances of the Council of Europe, Partial Agreement has issued a report with lists of evaluated source materials for natural flavourings indicating also the part of the plant normally used and evaluated for safety (Council of Europe, 1981). The following citrus fruits are listed as source materials:

Bitter orange, Neroli	*Citrus aurantium* (L.) ssp. *amara* L.
Bergamot	*Citrus aurantium* (L.) ssp. *bergamia* (Riss. & Poiteau) Engl.
Lemon	*Citrus limonum* (L.) N.L. Burman
Grapefruit	*Citrus maxima* (J. Burman) Merr.
Lime	*Citrus aurantifolia*
Mandarin	*Citrus reticulata* Blanco
Sweet orange	*Citrus sinensis* (L.) Osbeck

The flesh of these fruits is classified as food, the rind and the flowers are commonly added to foodstuffs in small quantities and the use is considered acceptable. The leaves are temporarily acceptable and insufficient information is available for an adequate assessment of their potential long-term toxicity.

Codex Alimentarius

The Joint FAO/WHO Food Standards Programme has issued General Requirements for Natural Flavourings (1988). It is recognized that natural flavourings may contain food additives as far as these are necessary for the production, storage and application of the flavourings and as far as these are present in amounts which would not perform a technological function in the finished food. This means for example that synthetic antioxidants can be used to extend the shelf life of an essential oil used in natural flavourings.

FRAGRANCES

The use of fragrance ingredients is not regulated in detail by specific regulations as in the case of flavourings. Fragrances are covered by cosmetics regulations and the manufacturers have to respect the general provision of all product legislations not to harm consumers.

Many national authorities recognize the Code of Practice and the Guidelines of the IFRA as an expert opinion on the safe use of fragrance ingredients and they control compliance with the IFRA recommendations (1974).

IFRA Code and guidelines

The fragrance industry has set up the Research Institute for Fragrance Materials RIFM to investigate toxicity, phototoxicity, irritating and sensitizing potential, etc. of fragrance ingredients. Based on the results of research sponsored by RIFM or found in the scientific literature, IFRA has issued recommendations restricting the use of certain fragrance ingredients (IFRA Code of Practice).

Citrus oils contain variable amounts of phototoxic furanocoumarins or psoralenes. The most widely found ingredient is bergapten or 5-methoxypsoralene (5-MOP). Its occurrence in bergamot oil has to be reduced before it can be used in applications exposed to sunlight. Other citrus oils have lower contents of psoralenes and the use of these oils has to be limited.

IFRA is recommending for all citrus oils applied on areas of skin exposed to sunshine, excluding bath preparations, soaps and other products, which are washed-off the skin, that the total level of bergapten in the consumer product should not exceed 0.0015 per cent or 15 ppm.

Where the level of bergapten has not been determined, the limits specified for the individual oils have to be used. The recommended restrictions for furanocoumarin containing citrus essential oils are:

	A	B
Bergamot oil expressed	2	0.4
Bitter orange oil expressed	7	1.4
Grapefruit oil expressed	20	4
Lemon oil cold pressed	10	2
Lime oil expressed	3.5	0.7

A: Percentage in the compound used at 20% in consumer product
B: Percentage in the consumer product.

Since phototoxic effects are additional, it is necessary to reduce the use levels accordingly if combinations of phototoxic oils are used. The sum of the concentrations of all phototoxic fragrance ingredients, expressed in per cent of their recommended maximum level in the consumer product, shall not exceed 100.

Petitgrain mandarin oil, tangerine oil cold pressed and mandarin oil cold pressed contain only small amounts of phototoxic furanocoumarins. These levels are not high enough to require special restrictions if used alone, but if used in combination with one or the other phototoxic essential oil, attention should be paid that the total level of bergapten (5-MOP) in the consumer product does not exceed 15 ppm.

Limonene and essential oils containing substantial amounts of it should only be used when the level of peroxides is kept to the lowest practical level for instance by adding antioxidants at the time of production. Such products should have a peroxide value of less than 20 millimoles peroxides per liter. This recommendation is applicable for all citrus oils.

European Union

A list of substances which must not form part of the composition of cosmetic products is given in Annex II of the European Cosmetics Directive 76/768/EEC. No. 358 (1995) of this list concerns 7H-Furo (3,2-g)[1]-benzopyran-7-one and its alkyl substituted derivatives (e.g. trioxalen and 8-methoxypsoralen), except for normal content in natural essences. This interdiction has therefore no practical consequences for citrus oils containing psoralens.

The eighteenth Commission Directive 95/34/EC regulated a very special use of bergamot oil. Furanocoumarines were recognized to be photomutagenic and photocarcinogenic. They shall be below 1 mg/kg in sun protection and bronzing products. The consequence of this low limit is that bergamot oil is no longer used in bronzing products.

An inventory of ingredients employed in cosmetic products was published by Commission Decision of May 8, 1996. It contains in section II a representative, but not exhaustive list of perfume and aromatic raw materials.

The following citrus oils and materials derived from citrus oils are listed:

EINECS No.	CAS No.	Name
227-813-5	5989-27-5	(R)-*p*-mentha-1,8-diene (*d*-limonene)
232-433-8	8028-48-6	Orange, sweet (*Citrus sinensis*)
266-034-5	65996-98-7	Terpenes and terpenoids, limonene fraction. Derived from citrus oils. Contains at least 80% limonene
277-143-2	72968-50-4	Orange, sour (*Citrus aurantium*)
284-515-8	84929-31-7	Lemon (*Citrus limonum*)
284-521-0	84929-38-4	Mandarin orange (*Citrus nobilis*)
289-612-9	89957-91-5	Bergamot (*Citrus bergamia* Risso)
289-904-6	90045-43-5	Grapefruit (*Citrus paradisi* M.)
290-010-3	90063-52-8	Lime (*Citrus aurantifolia*)
296-429-8	92704-01-3	Bergamot (*Citrus bergamia melarosa*)
297-608-3	93685-55-3	*Citrus medica acida*
297-916-8	93763-95-2	Tangelo (*Citrus paradisi* × *C. reticulata*)
304-454-3	94266-47-4	Citrus
307-891-8	97766-30-8	Orange, sweet, Valencia (*Citrus sinensis*)

This inventory does not have the legal meaning of a positive list.

USA

The US cosmetics legislation has no detailed dispositions concerning fragrances. The US FDA has announced that compliance with the IFRA recommendations is controlled by the analysis of finished products from the market.

Other countries

Fragrances are not regulated in detail. Compliance with the IFRA Guidelines is required or expected in many countries.

CHEMICAL SUBSTANCES LEGISLATION

The workers involved in manufacturing, transport, handling and transformation of essential oils have to be informed about potential risks and should be instructed how to avoid any possible harm. The most obvious risk of citrus oils is their flammability. Oral toxicity or toxicity after skin contact are not important. The known phototoxicity is not relevant, since exposure to sunlight is not encountered in the industrial environment.

A serious, potentially life threatening property of certain essential oils is their aspiration hazard. Serious pulmonary effects due to aspiration of the liquid into the lung may occur in case of accidental ingestion.

The aspiration hazard is associated with liquid hydrocarbons having a relatively low viscosity and low surface tension. Essential oils and fragrance compounds containing more than 10 per cent of these hydrocarbons may show aspiration hazard, unless determined otherwise based on their physico-chemical properties.

European Union

The Dangerous Substances Directive 67/548/EEC (1993) requires the classification and labelling of chemical substances and natural extracts, essential oils, etc. Symbols, risk phrases (R-phrases), safety phrases (S-phrases) and classification requirements are presented in the annexes to the Commission Directive 93/21/EEC (1993), the eighteenth amendment of the Dangerous Substances Directive.

The symbols Xn Harmful (Andreas cross), Xi Irritant (Andreas cross) and N Dangerous for the environment (dead tree and dead fish) and the following risk phrases are of special interest:

- R10 Flammable. Used for liquid substances and preparations having a flash point between 21 °C and 55 °C
- R38 Irritating to skin
- R43 May cause sensitization by skin contact
- R50/53 Very toxic to aquatic organisms, may cause long term adverse effects in the aquatic environment.

The conditions for classifying substances and preparations as R65 Harmful: may cause lung damage if swallowed, has been published in the O.J. of the EC, L355/618 (30.12.98). Liquid substances and preparations present an aspiration hazard because of their low viscosity under the following conditions: contain at least 10 per cent of aliphatic, alicyclic and aromatic hydrocarbons and have either a flow time of less than 30 sec in a 3 mm ISO cup (ISO 2431) or a kinetic viscosity measured by a glass capillary viscometer (ISO 3104/3105) or by rotational viscometry (ISO 3129) of less than 7×10^{-6} m^2/sec at 40 °C.

Substances and preparations do not need to be classified if they have a surface tension greater than 33 mN/m at 25 °C.

The following safety advice is also relevant:

- S2 Keep out of the reach of children
- S24 Avoid contact with skin
- S37 Wear suitable gloves
- S60 This material and its container must be disposed of as hazardous waste
- S61 Avoid release to the environment
- S62 If swallowed do not induce vomiting: seek medical advice immediately and show this container or label.

The EU Commission has officially classified the substance 601-029-00-7

EC Number	CAS Number	Name
205-341-0	138-86-3	Dipentene, limonene
227-813-5	5989-27-5	(R)-*p*-Mentha-1,8-diene, *d*-limonene
227-815-6	5989-54-8	(S)-*p*-Mentha-1,8-diene, *l*-limonene
229-977-3	6876-12-6	*trans*-1-Methyl-4-(1-methylvinyl) cyclohexene
231-732-0	7705-14-8	(±)-1-Methyl-4-(1-methylvinyl) cyclohexene

Xi; N; R10-38-43-50/53; S(2)24-37-60-61 (O.J. of the EC L 305/1-2, 12-13; 16.11.1998). The classification sensitizing is justified by the peroxide content of technical limonene.

The qualities used by the fragrance industry have however a low peroxide content and are not sensitizing. A working group of experts from the industry, organized by IFRA and other organizations, is discussing and recommending classifications and labels for materials not yet classified officially. This group is recommending the following classification for 266-034-5; 65996-98-7 Terpenes limonene fraction Xn; R(10)-65; S62.

The same classification is also recommended for bergamot, grapefruit, lemon, mandarin, neroli, orange and tangerine oils. The label declaration as flammable may be required for some qualities, but not for others. The decision to use R10 or not should be made after the determination of the flashpoint of each quality.

USA

The Toxic Substances Control Act–TSCA Chemical Substances inventory issued by the US Environmental Protection Agency (1979) lists the following citrus derived products:

CAS Number	Name
8003-31-9	Mandarin oils
8007-75-8	Bergamot oils

8008-26-2	Lime oils, *Citrus aurantifolia*
8008-56-8	Lemon oils, *Citrus limonia*
8008-57-9	Orange oils, *Citrus sinensis*
8014-17-3	Petitgrain Mandarin, *Citrus reticulata*
8016-20-4	Grapefruit oils
8016-38-4	Neroli oils, Oils of orange blossom, orange flowers
8016-85-1	Tangerine oils, expressed
8030-28-2	Orange flower water, Absolute petitgrain
8048-51-9	Petitgrain citronnier oil, Petitgrain lemon oil
68606-94-0	Orange oils, terpene-free
68606-99-5	Petitgrain oils, sapon., rectified
68607-01-2	Tangerine oils, terpene-free
68648-33-9	Oil of Bergamote, furanocoumarin free, psoralen-free
68648-39-5	Oil of petitgrain lemon
68648-39-5	Lemon oils, terpene free
68915-85-5	Petitgrain oils, terpene-free
68916-02-9	Bitter orange oils, terpene-free
68916-03-0	Bitter orange oils, acetylated
68916-46-1	Grapefruit oils, terpene-free
68916-83-6	Lime oils, psoralen-free
68916-84-7	Lime oils, terpene-free
68916-88-1	Lemon oils, pectin-free
68916-89-2	Lemons oils, psoralene-free
68917-06-6	Orange oils, de-oiled
68917-07-7	Orange oils, psoralene free
68917-17-9	Mixed grapefruit and shaddock oils, terpene-free
68917-20-4	Mandarin oils, terpene-free
68991-22-0	Bergamot oils, acetylated
	Terpenes and terpenoids of
65996-98-7	Limonene fraction
68917-80-6	Bergamot oil
68917-81-7	Bitter orange oil
68608-34-4	Citrus oil
68917-32-8	Grapefruit oil
68917-33-9	Lemon oil
68917-71-5	Lime oil
68953-04-8	Mandarin oil
69917-57-7	Mixed bitter and sweet orange oil
68917-58-8	Mixed grapefruit and shaddock oil
68647-72-3	Orange oil
68917-61-3	Petitgrain oil
68608-38-8	Tangerine oil

Other countries

The regulations of the EU and the USA have influenced regulations in other countries. The use of citrus oils is not restricted in other countries.

REFERENCES

Commission Decision of 8 May 1996 establishing an inventory and a common nomenclature of ingredients employed in cosmetic products. *O.J. of the EC, L132/1-684* (1.6.96).

Commission Directive 93/21/EEC of 27 April 1993 adapting to technical progress for the 18th time Council Directive 67/548/EEC on the approximation of the laws, regulations and administrative provisions relating to the classification, packaging and labelling of dangerous substances. *O.J. of the EC, L110 A/1-86* (4.5.93).

18th Commission Directive 95/34/EEC of 10 July 1995 on adapting to technical progress the Cosmetics Directive 76/768/EEC. *O.J. of the EC No L167/19-21* (18.7.95).

Commission Directive on the adaptation to technical progress of the Council Directive 67/548 EEC on dangerous substances, *O.J. L 305/1-2, 12-13* (16.11.1998).

Commission Directive on the adaptation to technical progress of the Council Directive 67/548 EEC on dangerous substances, *O.J. L 355/618* (30.12.1998).

Council of Europe 1981, *Flavouring Substances and Natural Sources of Flavourings*, 3rd Edition, Maisonneuve, Strasbourg.

Flavouring Directive 88/388/EEC. *O.J. of the EC, L 184/61* (15.7.88).

IFRA Code of Practice (1974, regularly updated), Annex 2 Industry Guidelines to Restrict Ingredient Usage, as Amended. International Fragrance Association–IFRA, Brussels.

Joint FAO/WHO Food Standards Programme, General requirements for natural flavourings, *Codex Alimentarius* Vol. **XIV**, Suppl. 1 (1988).

Toxic Substances Control Act Chemical Substances Inventory. US Environmental Protection Agency, Office of Toxic Substances Washington DC 20460, May 1979.

US Code of Federal Regulations, Title 21, paragraphs 101.22 and 182.20.

27 Citrus species and their essential oils in traditional medicine

Antonio Imbesi and Anna De Pasquale

INTRODUCTION

The genus *Citrus* L. (Fam. *Rutaceae*) contains a large number of species (more than 400) (*INDEX Kewensis*, 1997) along with innumerable varieties, cultivars, etc. All cultivated species probably derive from plants native to tropical and subtropical zones of Southeast Asia (Tutin *et al.*, 1968).

India would appear to be the original cradle of the *Citrus* genus. We find references to their usage in ancient Hindu medicine as *Amara-Koscha* (Royle, 1937) under the names *Jambira* (*Citrus acida*) and *Narānga* (*Citrus aurantium*). The lemon is one of the remedies found in numerous treatises on Vedic-Brahminic medicine, the most important of which is the *Susruta* (1300 BC) (De Pasquale, 1984).

According to Bretschneider (1871), the *Pent' ts'ao Kang Mu*, a book of Materia medica that draws together knowledge dating back thousands of years BC and is considered a true Pharmacopoeia, includes the fruits of *Citrus digitata* and *Citrus japonica* in section IV/2 (Mountain fruits).

Of the hundreds of species belonging to the *Citrus* genus, only a small number were extensively cultivated and acclimatised, initially in neighbouring countries and later, at the time of the conquests of Alexander the Great (330 BC), also in Greece and Palestine.

There is reliable documentation of the use of a very small number of *Citrus* species in the traditional medicine of the countries of origin and of the countries where the various species have naturalised.

Opinion is fairly divided (Imbesi, 1962) about the date when the most important species, including those used for medicinal purposes, were introduced to Europe and were first acclimatised and cultivated. The fact that synonymity between species and their numerous varieties is not always clear has also caused confusion.

The oldest existing record of the *Citrus* genus is the discovery in excavations of the city of Nippur, south of Babylon, of seeds identified as those of the citron, *Citrus medica* Risso. A biblical quotation (Leviticus, XXIII, 40) is believed to refer to this species, or one of its varieties *Citrus medica* var. *lageriformis* Roem. A citron fruit is also depicted on a coin minted by Simon Maccabeus in the fourth year of the *Redemption of Zion* (136 BC) (Tolkowsky, 1938).

Various authors have identified the fruits of the 'tree of knowledge of good and evil' (Genesis II, 9,17; III, 6) in the Scriptures as the fruits of *Citrus*, and some Renaissance artists conformed to this interpretation in their *representations* of the Garden of Eden (Moldenke and Moldenke, 1952). Van Eyck's famous polypty *Adoration of the Mystic*

Lamb, kept in the church of St. Bavon in Gand, portrays Eve holding an indeterminate fruit that bears a strong resemblance to a lemon (Moldenke and Moldenke, 1952).

Apicius Caelius (third century AD) recommended preserving the *citria* in plaster and Alexander Trallianus (sixth century) prescribed the fruit flesh and juice of the χίτρον του εντός (lemon) (Tschirch, 1910, p. 592).

Theophrastus Heresius, a philosopher of the school of Aristotle, who was born at Eresos in Lesbos in 371 BC, drawing from the protocols of Alexander the Great's Asian expeditions, in Book IV (Chap. IV, 2) of *Historia plantarum* provides the oldest botanical description of the citron, calling it μηλον μεδιχὸν χαι περσιχὸν (Tschirch, 1910, p. 547), from its regions of origin, Media and Persia. He adds that placing the inedible fruit and the leaves among clothes protects them against worms (clothes moths?) and when administered in wine in cases of poisoning they trigger evacuation of the alvus and hence eliminate the poison; they also freshen bad breath; these reports were picked up again by later writers (e.g. Pliny, 1831; Dioscorides, 1968).

The citron was also the first species of the *Citrus* genus known in Europe. The first attempts to acclimatise it in Italy date from the time of Pliny. Wehrmahn (1912) found a representation of it in a fresco in Herculaneum.

Most authors believe that the lemon was unknown to the ancient Greeks and Romans and was introduced to the Mediterranean area and the rest of Europe from India, via Persia and Palestine, much later than the citron. Fluckiger and Hanbury (1878) maintain that the tree was introduced to Europe by the Arabs, but are uncertain about the exact date. This opinion is generally shared among even the most recent specialist publications and treatises on Pharmacognosy (e.g. Braverman, 1949; Dezani 1953; Fenaroli, 1963; Guareschi, 1897; Romeo, 1930; Trease, 1961, etc.), which set the date of its introduction as between the tenth and twelfth centuries. Tschirch (1917), who cites information from the ancient authors, suggests that these may sometimes refer to the citron. He observes that the lemon does not appear in the wall paintings of Pompeii and is not mentioned by most of the agricultural publications of the Roman writers, and concludes generically that it reached Europe and Italy much later than the citron.

However, the recent discovery of the paintings in the orchard house (*casa del frutteto*) in Pompeii in which lemon trees are accurately depicted, clearly proves that this species, along with others of the *Citrus* genus such as the citron and perhaps also the orange and lime (Casella, 1950; Tolkowsky, 1938; Wehrmann, 1912), was well-known and widely cultivated in Italy many centuries prior to the Arab expansion.

In *Naturalis Historia* (Plinii, 1831), book XIII/16, Pliny (Caius Plinius Secundus, 23–79 AD), known as Pliny the Elder, talks of a tree that bears fragrant fruit with a bitter taste and is grown as an ornamental. In book XXIII/56 he specifies that the fruits (*citrea*) and seeds, when placed in wine, are used to counteract venom, and that their decoction is used to perfume the mouth; the seeds, if eaten, are harmful to pregnant women but are used for treating stomach ailments.

In the same century the Greek Pedacius Dioscorides from Anazarbus in Cilicia (born 50 AD), a Roman army physician under Nero, travelled with the troops through Egypt, Africa, Spain and Italy and brought together all known information on medicaments in use in a book entitled '*della materia*' (on the raw materials provided by nature), a title to which copyists later added the adjective medicinal, since the therapeutic uses of each drug were given along with their description. In this book we find the 'Persica mela' (*Citrus medica*), the fruit of which was held to be useful for the stomach and the belly.

Dioscorides' book was translated into Latin under the title '*De Materia Medica*' and was highly influential not only in the Roman and Arab worlds but also well into the Middle Ages. Dioscorides' fame was such that he was referred to by Dante Alighieri in the Divine Comedy

> *e vidi il buon accoglitor del quale*
> *Dioscoride dico*
>
> (Inf. IV, 139–140)
>
> *I saw the good collector of medicinals,*
> *I mean Dioscorides*
>
> (Inf. IV, 139–140)

The many reprints of Dioscorides' book, in Latin and Vulgar, frequently expanded with annotations by the various authors, became the Bible of doctors and apothecaries.

One of the most famous re-editions (1557) is that of Pietro Andrea Mattioli (1504–1577) of Siena, who commented and illustrated Dioscorides' descriptions with figures of the plants. Up until the eighteenth century this re-edition was the textbook for medical schools, used to train generations of physicians. In the extensive commentary to the chapter *mele di Media, ovvero Cedromele, che dà latini si chiamano Citria*, Mattioli distinguishes citrons from oranges and lemons and recounts legends telling of the potency of citron fruits against asp venom. He also describes the efficacy of the juice against plague and cholera and enumerates a series of generic actions performed by the fruit's peel and flesh, described principally as *amara* (bitter). He maintains that the effects of the citron are not substantially different from those of the lemon, the '*pomi d'Adamo o Lomia*' or the orange, adding that in Italy there exist three species of oranges, '*acetosi, mezani, e dolci*' (bitter, slightly bitter, and sweet).

The *Theatrum Sanitatis* (1940), codex No. 4184 of the Casenatense Library in Rome, from the late fourteenth century, contains descriptions of the activity and toxicity of *Cetrona et Citra* together with miniatures of the plants.

Castore Durante (1529–1590), a doctor and citizen of Rome, illustrated a *Herbario* (Durante, 1585) with splendid engravings of plants from European countries and from the East and West Indies and described their medicinal virtues with verses in Latin. He also provided an extensive commentary on the methods of preparation of suitable pharmaceutical forms and on the ways they were to be used in the most serious infirmities. Of the *Citrus* genus he described the properties of the citrons (in Latin *Citria mala*), the lemons (in Latin *Mala limonia*) and the oranges (in Latin *Aurancia*), interspersing their real (today confirmed) properties with a series of presumed therapeutic effects based on groundless popular belief. For example, orange flower water was held to be invaluable in combating pestilential fevers in the presence of petechiae, for strengthening the heart, easing childbirth, quickening the spirits, and so on, while the seeds were believed to be effective against scorpion stings (!!). The first edition of the *Herbario* was published in 1585 and was followed by many others, all of which enjoyed enormous popularity.

In the Italian translation of the fourth edition (1766) of the '*Dizionario ovvero Trattato Universale delle Droghe Semplici*' (Dictionary or Universal Treatise of Simples) by Nicolas Lemery, the heading *Citreum* includes *Cedrum* or *Citro* or *Malum Citreum*, of which the leaves and the flowers were used on rare occasions as *cordials and fortifiers*, more frequently the fruit peel to *fortify the heart, the stomach and the brain and to withstand poison; the juice is cordial, refreshing, good for calming heat of the blood, for precipitating bile, for quenching thirst, and for counteracting venom; the seed is always cordial, and effective for resisting corruptions.* The fruit, picked with cloves, should be kept in the pocket during epidemics and sniffed every so often to provide protection from contagion.

The same heading includes another species of citron, called the sweet citron, which is not used in medicine, and *Citron essential oil*, obtained from a species of Italian citron or bergamot (bergamot pear) which is *cordial, stomachic, cephalic, and effective for warding off malign humours.*

The lemons (*limones, sive limonia mala*) are included under a separate heading, and the virtues of the peel, the juice and the seeds were reputed to be similar to those of the citrons. The seeds were also believed to be useful in helminthiasis.

THE *CITRUS* IN THE OLD PHARMACOPOEIAS

The importance of some species of *Citrus* (orange, lemon, citron) in therapy and pharmacy received official recognition with the appearance of the first pharmacopoeias.

In the 1550 edition of the *El Ricettario del l'Arte et Universita de Medici, et Spetiali della Citta di Firenze* we find the recipe for a *Sciroppo di Acetosità di Limoni*. Later editions (Ricettario Fiorentino, 1802) included preparations using the leaves, fruit peel, fresh orange flowers, fresh citron fruit juice (*Citrus limonia* off., *C. medica* Linn.) and the peel of the fruit of lemon, *Mela Rosa*, bergamot etc. These were considered varieties of citron and were used for preparing *Acqua Carminativa Comune*. Orange and lemon peel was used for preparing *Acqua di Fior d'Aranci* (*Vulgo Acqua Lanfa*). The following are also described: Waters of whole citron or orange, lemon and bergamot peel; troches of orange or citron or lime, from the peel of the fruit; orange, bergamot, citron, lemon or *Mela Rosa* peel oil; Lemon juice syrup (*Sciroppo d'Acetosità di Limoni*) and Orange or Citron Peel Syrup.

The *Antidotarium* of Carolus Clusius, published in Antwerp in 1561, describes how to prepare conserves of *citriorum, malorum medicorum* and *limonum*, and *Syrupus acetositatis citri Mesuae*, which amongst other things was reputed to mitigate burning fever and pestilence and to be effective against drunkenness and dizziness. The *Syrupus acetositatis limonum* or the *Syrupus e corticibus Citri* was held to fortify the brain, the stomach and the heart and freshen the breath (*oris gratiam commendat*).

The *Officina Medicamentorum et Methodus*, published in Valencia in 1601, describes *Syrupus De Corticibus, Succus ex acetositate Limonum* and *Acqua florii citranguli, id est Aqua Naphae*.

In the *Pharmacopea Dogmaticorum Restituta* by Iosepho Quercetano (1608), counsellor and royal physician, orange and lemon peel are used as part of the complex preparation of *Aqua imperialis communis, et paratu facilis* and of *Elixir vitae maius, admirandum remedium ad inueteratos morbos, peneque desperatos propulsandos, sanitatem conseruandam, vitamque ipsam prorogandam*; of *Aliud elixi paratu facillimum*, of *Aqua theriacalis, cordialis, et of Bezoardica, omnibus cordis pathematis, pestiferisque affectibus, miro progressu conferens, sudoresque mouens*. Citron seeds were used in *Aqua antepileptica maior Quercet*, in *Alia aqua ad epilepsiam, apoplexiam et paralysim Quercet*, citron peel in *Aqua dysenterica* and lemon juice in *Potio ad Gonorrhoeam Virulentam seu foetidam*. Citron and lemon juice (*acetositat citrij o de limonio*) were used in the preparation of *Syrupi refrigerantes* and *Syrupi aromatum* or as

Bilem flauam conquentes and *Syrupus helleboratus minor o maior Quercetani*, the latter two of which were reputed useful for myriad ailments affecting various organs, including head, chest, stomach, liver, uterus, spleen, etc.

In the 1641 edition of the *Pharmacopoeia Bruxellensis*, orange fruit is one of the components used for producing *Aqua catharralis*, while citron peel and seeds were used for preparing *Acetum Theriacale*, lemon or citron peel for preparing *Liquores per deliquium*, and lemon juice or syrup was widely used in various preparations. Orange, citron and lemon peel were used for preparing candied fruit, and citron fruit flesh (*Medullae Citriorum, vel Limonum magnorum*) for obtaining the *Conserva Acredinis Citri*. We find these drugs in various preparations, such as powder to combat the plague, aromatic species, and antidote formulations.

The *Pharmacopoea Ultrajectina* of 1656 includes a recipe for lemon syrup. In the *Pharmacopoea Hagiensis* (1659) orange peel is used for preparing *Aqua Hysterica* or is candied (*condita*), while citron peel is one of the many ingredients of *Aqua Vitae Composita*.

Also in the *Pharmacia Antverpiensis* of 1661 in the formulation of *Species Aromaticae, Antidota* and *Opiata* we find the peel or seeds of citron and lemon. However, it is interesting to note that in the chapter on cosmetic formulations (*De Cosmeticis*), appearing for the first time in a Pharmacopoeia, we find the citron (*Citra incisa et contusa*) in the preparation of *Aqua cosmetica*; lemon juice with egg white was used to for preparing *Pomatum de limonibus* or was included in preparations of *Pasta Manualis*. In the *Aquae Chymicae* section, citron and orange peel and citron juice and bruised seeds were among the many ingredients of *Aqua Theriacalis*, while lemon and citron featured in equal proportions amongst the ingredients of *Aqua cardiaca frigida Saxoniae*.

The *Antidotarium Gandavense* of 1663 lists orange and citron flowers, citron seeds, and citron and lemon juice among the simples used in pharmacy, and citron peel as one of the ingredients of *Cerevisia laxativa*; it also refers to syrups of citron juice, citrus peel (*per se purgati*: having a laxative action?), lemon juice, candied orange, citron and lemon. Furthermore, citron seeds were included in the *Confectio de hyacintho* and in the *Pulvis ad vermes*, and the peel in the *Species laetitiae Galeni*.

In 1629 Ioannes Baptista Cortesius, Bononiense, doctor and philosopher and professor of Practical medicine at the Gymnasio Messinese, compiled a *Pharmacopea seu Antidotarium Messanense*, for which he consulted not only the works of Dioscorides, Galen, Valerius Cordus, etc., but also more recent works and the Pharmacopoeias and Antidotaria of the times. Here too we find syrups of citron or lemon juice and citron seeds and peel in the *Syrupus de Succo Endiuiae compositus Gentilis* with '*refrigeranti et exsiccandi*' properties, citron or lemon juice in the *Syrupus de Pomis cum sena Quercetani* with a purgative effect. Citron flowers are one of the ingredients in a syrup to counteract malignant fever. Citron and lemon juices are classified as *Aque frigide*, and *Aqua thoracica* which contained them was used as an expectorant in the treatment of pneumonia, pleurisy, etc. An *Electuarium de Citro solutiuum* was held to be very useful for continuous fevers. The *Elixir Thadei Florentini*, one of the ingredients of which was citron peel, was considered highly effective against tremors, paralysis, neck pains, for delaying balding and reducing kidney stones. Various theriacal waters, electuaries and antidotes contain citron peel and lemon juice in their compositions. We also find candied orange, lemon and citron peel which fortify (*roborant*) the stomach and heart and freshen the breath.

The Roman Pietro Castelli, a philosopher and physician and disciple of Andrea Cisalpino, and professor of practical medicine at the *Mamertinorum Gymnasio* in Messina, as lecturer in simples (*Simplicium lectoris*) added practical demonstrations (*Ostensio simplicium*) to theory lessons. In 1638 in Messina he created the first *Garden of Simples* in Sicily, which he illustrated in detail in a booklet of which only a few copies survive (Castelli, 1640). In 1678 the University of Messina was closed down and the *Hortus Messanensis* destroyed by the Spanish, against whom the inhabitants of Messina had risen up four years earlier. Thanks to a well-organised network of correspondents, Castelli had grown Sicilian species in the garden together with exotic species introduced from India, Arabia, America, Egypt, China, Ponto, Persia, Spain, Belgium, Peru and Turkey, as Castelli, himself wrote in a letter sent to Domenico Panaroli (Università di Messina, 1989) '*Maxima cum voluptate in meo horto messanensis ubi admiror......tot plantas e dissitis regionibus, ex India nimirum, Arabia, America, Aegypto, China, Ponto, Perside, Hispania, Belgio, Hollandia, Perù et Turchia huc allatas proficere et germinare......*'. Herbal plants, *che non sono altro che semplici* (which were none other than simples), were grown in the garden. Herbalists had to know their forms, names, smell, colour and taste and the manner and times for gathering, drying and preserving them, while Physicians had to know *le particolari nature, le proprietà, l'uso e la quantità da oprarsi* (their special natures, properties and uses and the quantities in which they should be used). Among the simples that were cultivated and whose properties physicians and apothecaries had to be familiar with were: *Malus Citria (Malus Medica Matt, Citrum Trago), Malus Citria monstruosa cornuta* and *Malus Limonia Citrata, Limones magni (Malus limonia acida crassa pulpa), Limones parui acida (tenui cortice), Limones parui, piriformes (Peretti Messanensium), Lima (cortice laeuia), Aurantia malus (fructu acido), Aurantia malus subdulcis, Aurantia malus fructu dulci, Aurantia malus chinensis*. The fact that these plants were grown in the Hortus testifies to their use in medicine in Messina.

In the *Catanense Dispensatorium* of Nicolao Catanuto (1666), the various parts of the lemon or citron are included in complex preparations such as *Confectio Hyacinthi apud Poterium*, useful even against plagues, *Electuarium Pliris arcoticon Nicolai* which cured melancholy, weakness of the stomach and heart and syncope, was useful for epileptics, beneficial for the memory and countered debilitation of the brain, and *Aquam ad febrem malignam Minadoi*.

In the *Antidotario Romano* (1678) translated by Ippolito Ceccarelli, the chapter entitled '*Delli canditi e conserve*' describes how candied citron fruits and peel should be prepared, while the chapter '*Delli Composti Aromatici*' provides the recipe for *Confectio de Hyacyntho Petri Castelli*, explaining the role of each ingredient. The citron seed was held to be a *theriac against venomous bites, comforting the heart and the substance of vital spirits*. The syrups described include the *Syrupus e corticibus citri*, and finally there is a recipe for making *oglio di scorze di cedro* (citron peel oil) and *oglio di scorze di cedro composto* (compound citron peel oil).

In the *Pharmacopea Brugensis* of 1697, citron seeds feature among the various components of *Aqua antiscorbutica*. Syrups of orange and lemon juice and peel were among the ingredients of a *Decoctum cordiale et salutivum*, which also included citron seeds; along with many bizarre ingredients such as sapphires, emeralds and topazes, the *Confectio Hyacynthorum* also contained shelled citron seeds.

In a list of simples used in pharmaceutical preparations, the Pharmacopea Almeriana of 1726 includes the peel and fruit of *Aurantium vulgare* and the peel of *Aurantium ex Curaçao* and the citron, which were used for preparing *Aqua aurantiorum, Aqua antihysterica, Aqua Vitae composita, Spiritus salis volatilis aromaticus, Spiritus carminativus, Spiritus aurantiorum* and *Spiritus citrorum*, as well as the *Elixir, Tinctura Alexipharmaca, Syrupus purgans, Syrupus stomachalis, Confectio Hamech* which also contained several purgative drugs, and *Pulvis antinephriticus*.

The *Codex Medicamentaius seu Pharmacopea Parisensis* of 1748 describes a large number of preparations based on Citrus. The heading *Arantia malus* lists *Aurantium acri medulla vulgare* (bitter orange), *Aurantium sylvestre medulla acri* (wild orange) and *Aurantium dulci medulla* (sweet orange), from which are prepared: waters or essential oil distilled from the flowers (*Aqua Napha*), antiscorbutic syrup and stomachic tinctures from the peel; *Aqua Theriacale, Imperiale, Isterica*, stomachic and cordial elixir from the flavedo. Cosmetic products were also obtained, such as *Aquam divinam, Aquam imperialem, Odoratam seu milleflorum, Aquam ad gengivas per infusionem, Pastillos odoratos profumo, Spiritum lavandulae compositum*. The peel of *Limon vulgaris* was used as an ingredient in *Aquam ad Gingivas, Opiata Salomonis* and *Confectionem de Hyacyntho*.

In the *Pharmacopea Collegii Regalis Medicorum Londinensis* of 1767, the juice of *Aurantiorum hispalensium* was an ingredient in the recipe for *Succi scorbutici (Scorbutic juice)*, while the flavedo of the fruit was used for preparing an *Aqua distillata* (Distilled water) and a '*spirituosa*' (aromatic) distilled water. A *Vinum Ipecacoanhae* included orange flavedo, which is also used as an ingredient in a bitter tincture. There is also a recipe for a syrup produced with lemon juice.

In the *Farmacopea Cerusica* of Dr. Teobaldo Rebaudengo (1772), orange syrup is added as a flavouring agent to the formula for *Opiata antivenerea*. Leaves, flowers and peel of the fruit of *Citrus aurantium* and lemon peel are used in the *Pharmacopea Austriaca Provincialis* of 1798 for obtaining distilled waters, essential oils, syrups and tinctures.

The *Pharmacopea Saxonica* of 1820 includes flowers, leaves, mature and immature fruits and fruit peel of *Aurantia mala*, and flavedo, juice and essential oil of lemon fruits. With reference to bergamot essential oil it is stated that the tree is a native of Barbados, but that in Gaul and Italy it is obviously freshly distilled (*expressed*). Orange peel extract, together with other extracts, is an ingredient of *Elixir temperatium balsamicum sive viscerale Frid. Hoffmann*, and of *Elixir viscerale (sive corticum aurantiorum compositum Ph. Boruss.)*.

The 1821 edition of the *Codex Medicamentarium sive Pharmacopea Danica* mentions not only the fruit, flavedo, flowers and leaves of *Citrus aurantium*, the peel of *Aurantiorum curassaviensium cortices* and the juice and flavedo of *Citrus medica*, but also bergamot essential oil, stating that it originates from Italy. Various preparations of the drugs obtained from the above-mentioned species are found in this pharmacopoeia, such as the *Species fumales optimae*, which also contain citron essential oil.

The *Codex Medicamentarius seu Pharmacopea Lusitanica*, which dates from about the same period (1822), describes preparations of *Malus limonia acida* (*Citrus medica β*) and of *Aurantium Hispalense* (*Citrus aurantium* L.).

In the first part of the *Farmacopea* (1825) of Antonio Ferrarini, pharmacist, member of the Health Commission of the city and province of Bologna and formerly lecturer (*ripetitore*) in Pharmacy at the University, describing '*Materie ricavate dalle sostanze vegetabili animali e vegetali*', we find *ARANCIO off., Foglie, Fiori, frutto e corteccia*. As the author writes poetically, '*This plant originated in ASIA, and for its beauty and utility spurred many nations to possess it; thus it was taken on a pilgrimage to Portugal, Provence, America, China, and to many other southern provinces of Europe*'. He mentions both bitter and sweet orange and describes the uses of the fruit peel, fresh flowers, and in the case of sweet orange also the juice. He draws together and intermingles various *Citrus* species, varieties and hybrids under the heading *CEDRO-Scorza, sugo e seme Sin. Citrus medica Cedrato*, and provides the so-called *distinctive characteristics* of the fruits, which he divides into sections: the first includes the *Citrons* (C. di Firenze, C, coronato, C. mela rosa, C. a pera, C. dolce di Spagna) and the second the *Lemons* (Limoncello di Napoli, Peretta di Napoli, Spongino, Lumie – Lumia cedrata, Bergamotta, Popona etc.). Various preparations are obtained, such as *Acqua aromatica di cedro* (*Citron aromatic water*), *Sciroppo con acido citrico* (*Citron juice syrup*). The chapter entitled '*Preparazioni e composizioni Chimiche e Farmaceutiche*' (Preparation and Chemical and Pharmaceutical Compositions) describes the preparation of *Crystallised citric acid* and *Purified fluid citric acid* starting out from lemon juice. The fresh flowers are used to prepare *Acqua di fiori di arancio* or *Aqua nanfa*.

The aromatic tincture with sulphuric acid or *Elisir di vetriuolo di Mynsicht officinale*, or the *Alcool con China composto dell'Huxam* contain fresh orange peel together with other aromas. Citron peel and lemon juice are among the components of Theriacal Water. Dry orange peel is one of the ingredients of '*Alcool allungato con genziana composto*' or '*Infusione amara Edimburghese officinale*'. Fresh citron and orange peels are candied or are used to prepare conserves. Fresh peels of orange, bergamot, citron or lemon are used to prepare aromatic essential oils. The composition of *Alcoolato di Cedro composto* (*Sin. Acqua di Colonia officinale*) includes essential oils of bergamot, citron, cedrate, orange blossom, and those of rosemary, lavender, cinnamon and lemon balm. Orange leaves are included, together with extract of cinchona bark and valerian root, in the *Elettuario Antiepilettico*. Fresh lemon juice is an ingredient in the recipe for *gelatina di pane*(?). Jujube, liquorice and rubber tablets are aromatised with citron water. The Pharmacopoeia also includes Syrup of citron juice, Aromatic syrup of citron, Syrup of orange peel and Vino scillitico with dry orange peel.

The *Codice Farmaceutico Romano–Teorico e Pratico*, compiled and published by order of His Holiness Pope Pius IX in 1868, mentions the orange, specifying that *Col succo della sua frutta si compone una graditissima e refrigerante bevanda, conosciuta sotto il nome di aranciata, le cortecce di questi assoggettate alla distillazione emettono un olio essenziale delicatissimo* (*Olio di Neroli*) (the juice from its fruit can be used to produce a very pleasant and refreshing drink, known by the name of orangeade; when distilled, fruit peel produces a very delicate essential oil (neroli oil)). The peels are effective as stimulants and are used in various tonic and purifying syrups and elixirs. Orange blossom aromatic water is very effective against verminous afflictions. *Aromatic Waters*, troches and other preparations of citron are used as stimulants, antiseptics and sedatives; the essential oil is used for aromatising many medicinal drugs and for forming troches of citron. The leaves yield decoctions and astringent and antinervous tonic infusions.

Tilli, in the *Catalogus Plantarum Horti Pisani* (1723), lists the species of *Citrus* grown in the garden of simples: *Citreum vulgare Tournef.–Malum Citreum; Citreum dulci medulla Tournef. Malum Citreum, dulce medulla; Citreum magno fructu Tournef.; Citria Malus, cum fructu magno; Citrum Corcyraeum Ferr.; Citreum Cretense Ferr.; Citroides vulgo Citratum, fructu magno, plerunque turbinato; Citroides seu Citratum Florentinum, fructu mucrunato; Citroides seu Citratum Florentinum fructu minori; Citroides seu Citratum Florentinum odoratissimum*. He also lists 25 species of *Limon* and 19 species of *Aurantium*.

THE *CITRUS* IN THE CHEMIST'S SHOP IN THE SEVENTEENTH AND EIGHTEENTH CENTURIES

Robert Talbor, nicknamed Talbot, first assistant in an apothecary shop, then apothecary (*apothicaire*) in Cambridge and '*Pyrétiatre*' (one who cures fevers) in London and later in

France at the court of Louis XIV, realised that the posology and method of use of extract of cinchona bark in treating malaria had to be adapted to each patient. In 1682 he published *'Le Remède Anglais pour la Guérison des Fièvres'*, which contains a detailed description of his remedy for fevers based on extract of cinchona bark, including the note that in order to keep the drugs used in the preparation secret he added lemon together with the other aromas (Delaveau, 1982).

In the seventeenth century the deeds of sale of apothecary shops included lists of the entire contents of the shop. In one such list drawn up for the sale of an apothecary shop in Carpentras (Provence) early in the seventeenth century we find distilled orange water (Delaveau, 1982). In Savoy, to ward off the plague, lemon was boiled together with other medicinal herbs in white wine and the potion was drunk before journeying to unwholesome places. Large quantities of lemons were taken on long sea journeys to allay scurvy.

Napoleon Bonaparte had a liking for an eau de cologne with a formula containing essences of various *Citrus* species (bergamot, lemon, orange, neroli) as well as essence of rosemary.

The recipe for lemon balm water, also known as *Carmelite water*, developed by the Barefooted Carmelites of Paris in 1611, contains fresh lemon peel. The recipe for this preparation is reported as *Aqua Melissae composita seu Vulgo Aqua Carmelitana seu Spiritus Melissa compositus* in the *Pharmacopea Manualis reformata*, written by Benedicto Mojon in 1784.

According to Lagriffe (1979) the formula for aperitifs and bitters not only includes cardamom, juniper and rhubarb but also the peel of bigarade (bitter orange) in a sequence that a devotee described as follows:

> Pour bien tonifier, du genièvre en grains
> De la rhubarbe pour libérer ventre et reins
> Pour faire digérer, un peu de gentiane
> Joindre pour empêcher l'estomac d'éructer
> Quelches graines d'amome
> Ajouter quelches zestes de bigarade
> Pour éclairer de célestes rêves phosphorescents
> La crâne et l'exciter.

> Use grains of juniper as an effective tonic,
> Rhubarb to free the belly and the kidneys,
> A little gentian to aid digestion,
> Add a few cardamom seeds
> to prevent the stomach from eructing,
> Add bigarade peel
> to lighten and stimulate the head
> with blue phosphorescent dreams.

THE *CITRUS* IN PHARMACOLOGY TREATISES AND IN THERAPY FROM THE EIGHTEENTH TO TWENTIETH CENTURIES

In the eighteenth and nineteenth centuries, all Materia medica and Pharmacology treatises reported drugs obtained from *Citrus* species, already present in the above-mentioned

Pharmacopoeias (Boehraave, 1772; De Rochefort, 1789; Edwards and Vavasseur, 1829; Chevallier and Richard, 1830; Ferrarini, 1825; Semmola, 1836; Cassola, 1838; Targioni-Tozzetti, 1847; Bouchardat, 1855; Orosi, 1856–57; Cantani, 1887).

Boerhaave (1772) attributes to *Citrus* fruits the property of curing various illnesses (*morbes*), and lists citron oil among remedies for fevers in general, heart disease (*Pulvis cardiacus, calidus, narcoticus*), or to be used together with other medicinals against burning fevers (*In siti febbrili, Decoctum in valida siti et debilitati*); as an antiemetic (*Haustus anti-emeticus*), antiscorbutic (*Antiscorbutica frigidiuscula*), colluttorium (*Colluttoria oris. In Calidis*), in treating dropsy (*Mistura aromatica, cardiaca, acida, sitim sedans, vires vitales excitans, lymphae fluorem concilians*), infirmities in pregnant women (*ad gravidarum morbos*), as an aromatic cardiac medicated wine (*Vinum medicatum, aromaticum, cardiacum*) or in an acid aromatic cardiac mixture, and also in *Lue Venerea* as *Mistura anodina e diaforetica*.

An interesting reference is to be found in the *Cours Elementaire di Matière Médicale* by De Rochefort (1789). He reports that the antispasmodic action of orange leaves was little known in the past and that this action, comparable to that of valerian, had been discovered no more than 30–40 years earlier (i.e. in the first half of the eighteenth century), and furthermore that he had seen it used to remedy even severe cases of 'nervous' epilepsy in children. And since the decoction's very bitter taste made it unpleasant to drink, especially for children, he recommended using the pulverised leaf incorporated in bolus or in jam. He added that this antispasmodic was highly recommended by the school of Vienna and by Tissot and that orange blossom water was used to correct the flavour of purgative potions and other unpleasant tasting medicaments.

In the last century, lemon juice, the juice and peel of bitter orange, and the peel of citron were used as tonic bitters, as stomachics and carminatives, and for treating scurvy. Lemon juice was an ingredient in stomachic potions effective against vomiting and hiccups, and was also recommended for jaundice and feverish illnesses, especially in the case of strong thirst (Edwards and Vavasseur, 1829). These authors mention the so-called '*Arancelli*' (*Aurantia curassaventia, in other words the small oranges that fall from the tree before ripening, which have properties similar to those of orange peel*), which in France were used to make *cautery balls* and in England were used in the same way as orange peel and were included in the composition of numerous medicinals. Chevalier and Richard (1830) also report the use of these *small oranges*, gathered when they are the size of a cherry, for preparing a highly bitter, aromatic and stomachic tincture.

Semmola (1836) includes lemons and oranges among *Aciduli vegetali*, while essential oil of Cedrate (*Citrus malus*), lemon, orange, bergamot, neroli and orange leaves are included in the chapter on *Aromatic simples*. He adds that Lebreton had extracted a bitter substance that he called *hesperidine* from immature oranges.

Cassola (1838) mentions a method for preserving juice of *cedri* (citron?) so that it can be taken on long journeys, as it is effective in the prevention of scurvy.

Targioni Tozzetti (1847), professor of Botany and Materia Medica at the Arcispedale di S. Maria Nuova in Florence, attributed numerous virtues to lemon (effective against dysentery, and in treating febrile and inflammatory illnesses, putrid, adynamic or typhoid fevers the juice was used in drinks and in enemas) and referred to its use as an antiscorbutic, vermifuge and antiputrid agent and for treating heart diseases (Devees in Targioni, 1847), syphilitic affections (Rollo in Targioni, 1847) and dropsy (Cahen in Targioni, 1847). Iced lemonade was used for internal haemorrhages, for which reason Evratt recommended it for local use in uterine haemorrhages, and it was also used as an antidote for poisoning by alcohol or narcotics, etc. Cooked orange flesh, again according to Targioni Tozzetti (1847), was applied by Wright as a poultice to foetid ulcers; orange peel was an ingredient in vermouths and antiscorbutic medicinal tinctures; the pulverised peel, in doses from half a dram to two drams, was used in various illnesses as a stomachic, stimulant, cordial, emmenagogue and vermifuge. He added that Heistero, Riverio, Verloff and others used it for intermittent fevers, especially quartan agues. Orange flower water, or *Aqua Nanfa*, was widely used as an antihysteric, cordial, cephalic, carminative, and anodyne agent. An infusion of candied flowers was used as a diaphoretic, as a sedative for coughs, and in rheumatic ailments. An infusion of the leaves was used as a diaphoretic, anodyne and fortifier against certain nervous and hysterical afflictions, and against weakness of stomach. De Haen and Lochner (Targioni Tozzetti, 1847) confirm its good effect in the treatment of epilepsy. The decoction of orange leaves was used in enemas as a stimulant, carminative, etc. The dried and pulverised leaves were used in the treatment of numerous neuroses, painful twitching, convulsions and Saint Vitus dance. Targioni Tozzetti adds that the effects of the juice and peel of the fruit of sweet orange are similar to those of bitter orange.

We find many of these uses in the second and subsequent third edition, which was translated into Italian, of the *Manuale di Materia Medica e di Terapeutica Comparata e di Farmacia* by Bouchardat (1855a, 1856b), who states in the preface that in reporting the physiological properties and therapeutic uses of the drugs he referred to the works of various authors, including many of those mentioned above.

Cantani (1887), like the other authors, reports a whole series of therapeutic uses of the leaves, flowers, unripe fruits, and the peel (flavedo) of ripe fruits of orange or lemon, but acknowledges in the foreword that *detailed* research had not been conducted into the action of the various drugs obtained from the *Citrus* genus.

The third edition (1856–57) of Orosi's *Farmacologia Teorica e Pratica* lists the various drugs extracted from sweet orange, strong bitter orange (*Aurantium acre*) and citron for their anodyne, stomachic, fortifying, carminative and cordial effects. In the second part of the book he gives the composition and methods of preparation of orange flower and whole citron aromatic water (*Aqua Nanfa o Lanfa*), citron syrup, orange peel spirits, *Acqua Senza Pari* or *Sanspairelle* – an alcoholic solution containing essential oil of bergamot, lemon, citron and rosemary, *Spirito gengivale* which contained fresh citron peel among its ingredients, and the above-mentioned Carmelite Water.

Dujardin-Beaumetz (1889), in his *Dizionario Terapeutico di Materia Medica, di Farmacologia, di Tossicologia e delle acque minerali* (Therapeutic Dictionary of Materia Medica, Pharmacology, Toxicology and mineral waters), published towards the end of the nineteenth century, reports under the heading *Orangers* the therapeutic virtues already attributed by previous authors to the infusion of leaves, wine and syrup of orange peel.

In the first half of the twentieth century, the industrialised nations had not yet seen the proliferation of synthetic drugs that would have such a major impact on the prevention and cure of numerous illnesses and which, together with improved conditions of hygiene and diet, would contribute to lengthening human life expectancy. In these countries, particularly in the Mediterranean area where citrus species grow luxuriantly, the fruits of some species (lemon in particular, but also orange, bergamot and citron) were cheap enough for all pockets. They were considered a divine gift and were widely used in therapy and hygiene.

According to an old popular Sicilian saying, orange is golden in the morning, silver at midday, but leaden in the evening, meaning that it would be harmful in the evening.

Lemon juice was used for washing greasy hair every day for a month to make it soft and shiny. After finishing the housework, women would wipe their hands with half a lemon; many home-made creams based on glycerine and lemon were used to soften the skin and to get rid of stains, wrinkles and chapping. Lemon juice diluted with water was considered the best eyewash and many actresses of the times gave written accounts of this on their photographs.

In the *Annali di Odontojatria*, Angelo Chiavaro writes of his conviction, 'from clinical experience, of the effective disinfectant, astringent, acidulous and stimulating action that the juice of citrus fruits has on tooth caries and on gums if prescribed in aqueous solution for prolonged washes of the tissues of the mouth' (Chiavaro, 1934).

The widespread use of citrus fruits attracted the attention of the medical profession, which began to note their effects in various clinical cases. Clinicians of the times prescribed the use of lemon or orange juice against scurvy due to its high vitamin C content, and paediatricians recommended administering a few drops of orange juice to children fed on sterilised milk. Lemon juice was used as an adjuvant for the gastric juices in order to improve digestion, and tonic tisanes for the stomach were prepared with lemon or orange peel.

As late as 1931, in an inquiry conducted by the Camera Agrumaria of Reggio Calabria (Citrus, 1931) and reported by national newspapers, we find that one of the doctors at the Ospedale Maggiore in Milan considered that lemon was useful for its antiscorbutic and antirachitic properties and that its juice had a beneficial digestive and thirst-quenching effect. It was also reported that some people considered it useful in treating uricaemia, gout and gravel and effective against arteriosclerosis.

According to other clinicians Pende (1926) maintained that the lemon countered obesity and allayed arteriosclerosis and hypertension, while orange was rich in vitamins and had a strong antiscorbutic, antihaemorrhagic and antiarthritic action. Even today orange peel is considered a mild sedative.

In a Congress of the American Dentistry Association held in Denver in the 1930s, Dr. Hanke, of the Otho Sprague Foundation of the University of Chicago, revealed the results of two years of study and research. He declared that adding lemon and orange juice to the diet improved children's general growth and that administering a litre of orange juice every day for ten days produced a marked improvement to even the most severe inflammations of the oral cavity (Editorial, Citrus, 1930a).

According to a London newspaper, orange juice or the juice of seeds chewed but not swallowed offered the most effective protection then known to science against grippe and pulmonary illnesses (Editorial, Citrus, 1930b).

In an issue of the *Giornale Italiano di malattie veneree* (Italian journal of venereal diseases) of 1884, it was reported that 3–4 injections of fresh lemon decoction could be used successfully in blennorrhagia without the need to resort to other therapeutic measures. This remedy also had the advantage that it could be used without the need to wait for all the inflammatory effects to disappear (Editorial, Citrus, 1959). Taking up from these studies, P. Stella (1929) wrote that he had used lemon juice in the prophylaxis of gonorrheal conjunctivitis in new-born babies with infected mothers and that in none of the cases was the disease found to develop.

CITRUS IN TRADITIONAL MEDICINE

Citrus in traditional Asiatic medicine

In a comparative study of the use of herbal drugs in the traditional medicines of India and Europe, Puri (1971) found a marked similarity between the drugs used in the two continents. He attributed this not only to the similarity of the vegetation in the two areas, but also to the influence that traditional Indian medicine, in particular the *Atherveda*, one of the most ancient repositories of human knowledge, had on Egypt, Greece and Rome. He listed the principal uses of a small number of these drugs, including bitter orange peel, which in India is used as an aromatic, stomachic, tonic, astringent and carminative agent, and lemon, which is used as a flavouring and for its carminative and stomachic effects.

In the *Valmiki-Ramayana*, written after the Vedas and one of the most sacred of all religious books which enumerates the virtues of the medicinal plants that Lord Rama (Vishnu) met during his fourteen-year journey around different parts of India, Karnick and Hocking (1975) identified and listed fifty of these drugs with their use as described in the Ayurvedica (or native Indian) system of medicine. The immature fruit of *Citrus aurantifolia* (Christm) Swingle was used as an fortifier, cardiotonic, laxative, antihelminthic and for combating fatigue; the mature fruit as a sweetener, laxative and aphrodisiac; a decoction of the flowers for allaying fevers; the juice of the fruit as a tonic and for treating cases of swelling of the spleen; and the fruit peel as an anthelminthic.

In the *Ayurveda* the fruit of *Limonia acidissima* L. was used as a sour, sweet, acrid, refrigerant, aphrodisiac, alexipharmic; it cured dysentery, removed biliousness, *vata, tridosha*, asthma, tumours, and leucorrhoea. Juice was placed in the ear to cure earache. The unripe fruit was alexipharmic; the seed cured heart diseases; the essential oil was acrid, astringent, alexiteric and destroyed biliousness; the flowers were an antidote to poisons; the leaves were useful during vomiting, hiccoughing and dysentery.

A description of the use of *Citrus* species in traditional European medicine clearly shows how the ancient medical knowledge of India travelled to Europe and became an integral part of the medical culture.

In a medical investigation carried out from 1989 to 1992 in the Varanasi District of Uttar Pradesh, Bajpai *et al.* (1995) report information on traditional species of Ayurvedic medicine used by the population living in rural areas to alleviate ailments. Lemon juice (*Citrus aurantium* L.) mixed with ground root bark of *Strychnos nux-vomica* L. is taken for loose motions and to stimulate the appetite; lemon juice, mixed with fruit juice of *Emblica officinalis* Gaertn (Euphorbiaceae), is taken to stop dysentery. In India a fruit poultice of *Citrus sinensis* Osbeck is applied in some skin conditions such as psoriasis.

A number of *Citrus* species have been used in traditional Chinese medicine.

The dry pericarp of the fruits of pomelo, *C. reticulata* Blanco (mandarin), ripe or unripe orange and *C. medica* L. is used mainly as an expectorant or stomachic; the pericarp of the unripe fruit of *C. reticulata* is used amongst other things in the treatment of hernia; the dry unripe fruit of orange and its cultivated variants and sweet orange are used in the treatment of diarrhoea, anal prolapse and frank prolapse; the seeds of *C. reticulata* Blanco (mandarin; tangerine) are used principally in the treatment of hernia (Tang and Eisenbrand, 1992).

The dried fruit of orange is included among the antishock herbs used in the traditional Chinese medicine Tai-fu, which generally treated shock symptomatically. In Chinese folk medicine, this drug was commonly used to treat indigestion and to relieve abdominal distension and ptosis of the anus or uterus (Huang, 1993).

While numerous effects of *Citrus* fruits are described in *Medicinal Plants of China* by Duke and Ayensu (1985), the orange fruit is antiemetic, antitussive, diaphoretic, digestive, carminative and expectorant, and can be used to treat abdominal pain, diarrhoea, chest congestion, cancer, hematachezia, rectal prolapse, rectocele, splenitis, stomach ailments, tenesmus, uterine prolapse. The seed and the pericarp are used for anorexia, chest pain, colds, coughs, hernia, orchitis and nausea, and can be crushed and applied to freckles and pimples. The pomelo leaf is boiled and placed on painful places and swellings and ulcers and is used as a depurative; the peel is antivinous, aromatic, bitter and stomachic and is used for boils, cholera, colds, dyspepsia, itching and nausea; the seeds cooked with pork or *Illicium* are used for bladder pain, hernia and swollen genitals. The sweet orange died peel is used for anorexia, cold coughs, malignant breast sores and phlegm.

According to Chinese tradition (Chang and But, 1987), the not completely mature fruits ('Zhiqiao') of *Citrus aurantium* L., *C. aurantium* L. var. *amara* Engl. or *C. Wilsonii* Tanaka *disperse stagnant vital energy, have an expectorant effect and act as a digestant. They are mainly used to treat excessive accumulation of phlegm, thoracic distress, distention of the costal region, dyspepsia, eructation, vomiting, dysentery with tenesmus, and prolapse of the uterus or rectum.*

A short summary of the uses of *Citrus* species in East and Southeast Asia (India, China, Indonesia, Philippines) is given by Perry (1980). The rind or juice of fruit of various species are employed in the traditional medicine as a fragrant, stomachic, carminative, antiscorbutic, antiemetic, antivinous, diuretic and expectorant. A fruit peel decoction is used in coughs, colds, dyspepsia, as an antidote for fish poison, and in ointments on acne and eczema. Some juices (probably sour) are employed as a gargle for the treatment of sore, inflamed or suppurated throats and to cleanse or dress wounds. The leaves are considered to be sedative and antispasmodic, while a decoction of the leaves is applied hot to treat aches and swellings.

The sour (bitter) orange is used in Vietnam as a diaphoretic and purgative, in the Philippines as a food for the sick, particularly in febrile, inflammatory and scorbutic conditions (Watt and Breyer-Brandwijk, 1962).

Essential lemon oil is held to have some value as an antibiotic '*to treat typhus, meningococcus and other bacteria bacilli*, but care must be exercised in its use, as too large a dose is stupefying' (Pételot, 1952).

Citrus in traditional medicine of Africa

In Saudi Arabia the fruit of *Citrus medica* L. var. *turung* Bonavia is used for stomach-ache and fevers. A concentrated infusion of the fruit mixed with salt and honey is used against coughs (Qedan, 1974).

In Arabian medicine the juice, dried and fresh fruit, peel and bark of lime are used as remedies for cataracts, colds and fevers, chest pains and earache and as a stomachic.

A variety of sweet orange (*safargal*) cultivated in Oman and its fruit is considered a general tonic to improve the stomach, and eaten 2–3 times a day is used in treating jaundice. Smelling the plant or fruit of safargal is held to be good for the heart. In Yemen orange peel is eaten at breakfast mixed with milk to make the stomach strong and stimulate the appetite; fruit pulp is believed to strengthen the heart (Ghazanfar, 1994).

The bitter orange is cultivated in North Africa. An infusion of the bark (fruit peel ??) is used for stomach pains and dysentery, the bark (fruit peel ??) as a digestive, antispasmodic, diaphoretic and mild sedative, and an infusion of the dried flowers as a nerve sedative and antispasmodic (hysteria, spasms, hiccups). Water distilled from the flowers is given to babies to calm them and help them sleep. An infusion of the flowers is stimulant and antidiarrhoeic (Boulos, 1983).

In East Africa (Kokwaro, 1976) the root of lime is used as a snake bite treatment, to relieve abdominal pains, to treat gonorrhoea (as a decoction) and as an aphrodisiac (cooked with meat). The juice obtained from the leaves is squeezed into the ear for ear-ache.

In South Africa lemon juice is used with salt as a ringworm remedy, in the Transvaal with honey as a cough remedy and with ginger for a cold. Swahili women apply the juice to the vagina to produce contraction after childbirth (Watt and Breyer-Brandwijk, 1962).

Lime leaves are a Malay remedy for headache, and the infusion is used for fevers with slight jaundice and bilious fever. The fruit juice is a cough reliever, a tonic and a remedy for stomach-ache and is used externally as a cleanser of and stimulant to wound surfaces. Roasted slices of lime are applied to chronic sores and yaws, but when applied to the skin followed by exposure to sunlight they produce a photodynamic reaction (Watt and Breyer-Brandwijk, 1962). In West Africa (Ayensu, 1978) lime leaf preparations are used as a gargle and for treating pneumonia (mixed with a large number of other plants), malarial fever, fever and eye diseases (boiled with orange leaves), fever with jaundice, gonorrhoea (as a vapour bath), headache, bilious fever (with *Ocimum* leaves) and stomach-ache (with *Areca catechu*), and as a mouth wash. The rind of *C. medica* or *C. acida* is used against helminthiasis and the root bark of *C. nobilis* as an antirheumatic (Oliver-Bever, 1984).

In the traditional Senegalese Pharmacopoeia (Kerharo and Adam, 1974) a decoction of lime leaves is used as a drink for treating urinary retention and affections of the respiratory tract; the bark of the root, in the form of a decoction with twigs including leaves and seeds, is highly considered for its diuretic and antiblennorrhagic action. The fruit juice is used not only for aromatising, but also as a vehicle for numerous febrifuge and antidysenteric preparations, and is recommended for aphthae in children, etc.; the juice, after adding the essence, is used for massaging as an invigorating, febrifuge and sudorific agent and also an antiophidic remedy.

Citrus in traditional medicine of the New World

The above description of the uses of *Citrus* in traditional and popular medicine refers mainly to Mediterranean countries. After the discovery of America, cultivation of the most widely used Citrus species spread to the New World. There is documentary evidence of the presence of bitter orange plantations in Vera Cruz and in Mexico in the second half of the sixteenth century (Soto, 1935), whereas a few years later (1568) in America there were plantations of lemons and sweet oranges (Cavendish, 1935). Cultivation gradually extended to Florida, Louisiana, California, etc. and today the original naturalised species and the numerous varieties derived from them are so widespread throughout the United States that *Citrus* fruits have a place not only in the diet but also in traditional medicine.

In America the fruit, peel and oils of bitter orange and sweet orange are used as a carminative, aromatic and stomachic. Orange peel and oil are used as a flavouring in food, drinks and medicine and the juice as a source of Vitamin C. Lime fruit (*C. Limetta* Risso) is used more for flavouring than medicinally, although the juice is used as an antiscorbutic (Wren, 1989).

In Mexico (Martinez, 1959) the leaves of bitter orange (*naranjo agrio*) are used for their antispasmodic, tonic and febrifuge properties and in treating nervous excitation, palpitations and epilepsy. P. Font Quer (1962), in describing the virtues (*virtudes*) of the citron, lemon, bitter orange and sweet orange, re-examines the work of Dioscorides, and outlines the current use of the drugs obtained from these plants, the lemon in particular, stating that it 'has countless esteemers and constitutes a panacea believed to cure an infinite number of illnesses'.

Morton (1975a,b) includes the mandarin (*mandarin orange; C. reticulata* Blanco) among folk remedy plants identified in Northern Venezuela. The decoction of the fruit peel is believed to have a hypoglycaemic effect while the peel is considered an antibiotic.

The lime (*C. aurantifolia* Swingle) has a multitude of uses in folk medicine throughout the tropics: the juice is taken to halt diarrhoea, it is considered an effective disinfectant in the eyes of newborn infants and for wounds, diluted in water, hot or cold, it is gargled for throat ailments and taken internally to relieve fevers, liver complaints and oedema; it also serves as a diuretic. In Yucatan the juice is used in diabetes, in rheumatism and in atherosclerosis and the leaf infusion serves as an antispasmodic, sudorific and sedative; the root decoction is given as a treatment for gonorrhoea. In Curaçao lime juice is added to 'tea' of *Cymbopogon citratus* taken to relieve bronchitis. In Trinidad lime juice is used against thrush and toothache and also as a rub on erysipelas and is applied to the chest in the case of pneumonia; a 'tea' of flower buds is taken as a soporific and a root decoction is applied on scorpion stings. In Jamaica, lime juice mixed with the crushed fruits of *Solanum aculeatum* is said to cure ringworm. In the Bahamas the juice is used to repel mosquitoes and if one is bitten by these insects the juice will relieve the itching; the leaf decoction is used as a tonic and as a remedy for blood pressure (Morton, 1975b).

In Central America bitter orange juice is used as diuretic and purgative and as a remedy for coughs and fevers; a decoction of the peel is administered to relieve flatulence. The leaves are commonly utilised in folk medicine in the form of an infusion or decoction as a diuretic, cardiac stimulant and sedative and for reinforcing other remedies for fever (Yucatan), for treating gall bladder problems, hypertension and when the stomach is upset (Curaçao), and in a decoction with *Annona muricata* leaves as a remedy for asthma, as a sedative and for promoting good sleep (Curaçao). The decoction or essential oil of the leaves and flowers are considered antispasmodic; the fresh and dried flowers are used in a 'tea' as a stomachic and sedative; an ointment prepared from the flowers is applied on skin disease; the peel decoction is used to relieve stomach-ache and as a bitter tonic; the fruit juice is used as a remedy for colds and sore throat. The sweet orange peel decoction is a remedy for dysentery and intestinal worms and for flatulence. In Curaçao a decoction of the peel is taken when one lacks appetite; the leaf infusion as an antispasmodic or in treating hysteria, palpitations, convulsions and epileptic attacks, in gastralgia and bronchitis. The Mayas in Yucatan use the leaves 'tea' as a sudorific. The 'tea' of the leaves is taken in Trinidad for pneumonia and as a depurative after child birth; the flower decoction is a popular sedative. In Ecuador an extract of orange seed is used in the treatment of malaria (Morton, 1975b).

In the Caicos Islands (Morton, 1977) at the southern end of the Bahamas chain, acid juice of the lime fruit (*C. aurantifolia* Swingle) or of the bitter orange is a remedy for typhoid, as they say 'give plenty of sours'.

In the *Amazonian Ethnobotanical Dictionary* by Duke and Vasquez (1994) it is reported that the decoction of lime root, taken a cup-a-day during the menses, is said to act as a contraceptive; the juice of one lime, taken by both partners just before intercourse, will prevent conception; lime peel is used as an antidandruff, decongestant and sedative; the flowers as tonic are used for cramps and enteritis. The lemon leaves infusion is a digestive.

On the strength of their continuing widespread use in medicine and pharmacy, the drugs obtained from the various species of *Citrus* are included in modern Pharmacopoeias (Imbesi, 1964):

Citrus aurantium L. var. *amara* L.: Exocarp (flavedo), exocarp and mesocarp (bitter orange peel), flowers, leaves, fruit, unripe fruit, fruit essential oil, flower essential oil (neroli).

Citrus aurantium L. var. *bergamia* Risso: Exocarp (flavedo), fruit, exocarp essential oil.

Citrus aurantium L. var. *dulcis* L.: Exocarp (flavedo), exocarp and mesocarp (sweet orange peel), fruit, exocarp essential oil, fruit juice.

Citrus aurantium L. spp. *natsudaidai* Hayata: Exocarp and mesocarp.

Citrus medica L. var. *limetta* Risso. Fruit, exocarp essential oil.

Citrus medica L. spp. *limonum* (Risso) Hook f.: Exocarp (flavedo), exocarp and mesocarp (peel), fruit, seed, flavedo essential oil, fruit juice.

Citrus medica L. var. *vulgaris* Risso: Exocarp and mesocarp, fruit, seed, flavedo essential oil.

Bergamot in traditional medicine

A short chapter is devoted here to a *Citrus* species that was introduced to Europe towards the end of the seventeenth century and has a habitat limited to a small area on the Strait of Messina: the bergamot.

The bergamot (*Citrus bergamia* Risso et Poiteau) is one of the cultivated forms of the *Citrus* genus along with the many other varieties of orange, lemon and citron. Its origin began to be a subject of discussion at the end of the eighteenth century. In his *'Traité du Citrus'* of 1811, Gallesio notes that the plant has characteristics in common both with the orange and the lemon and therefore should be considered a hybrid of these two.

The bergamot, like other citrus species, was unknown to the ancient Greeks and Romans and was introduced to Europe many years after the citron, lime and orange. Like the other *Citrus* species already mentioned, the bergamot is a native of India, where the fruit was called *Limbu* in Punjabi, *niboo* in Bengali and *neamboo* in Hindi.

In these regions it is still used against scurvy, dyspepsia and as an antiemetic, antipyretic and antiseptic.

There exist various theories regarding the origin of the name, the most likely of which appears to be from bergamot pear (deriving from Beg-ármûdi, which in Turkish means 'the prince's pear'), which the bergamot resembles.

It is not known with certainty when the plant first appeared in Europe. It appears to have been introduced as an ornamental at the end of the seventeenth century, but was first cultivated for the fragrance of its fruits towards 1750 around Reggio Calabria, later spreading to the Ionic and Tyrrhenian coasts.

The first reports of the presence of the bergamot in Europe date from 1693: in *'Le Parfumeur François'*, printed in Lyon, it is indicated as *Essence de Cedra ou Bergamotte*, which according to the author, a certain *'le sieur Barbe parfumeur'*, is obtained from the fruits of a *Citrus* species grafted onto the trunk of a bergamot pear.

In 1713 Volkammer of Nuremberg published a splendid book on the *Citrus* tribes entitled *'Hesperides Norimbergenses'*, dedicating a chapter to the *Limon bergamotta*, which he called *gloria limonum et fructus inter omnes nobilissimus*, and wrote that the Italians obtained one of the finest of essences from this fruit.

Ever since the plant was introduced to Calabria, the essential oil obtained from the fruit was renowned in perfumery for its fresh, delicate aroma. However it was also widely used in the popular medicine of this region as a cicatrisant in treating burns or varicose veins, as a microbicide, analgesic, antipyretic and antiphlogistic, and in treating furunculosis, pediculosis and dandruff.

In countries where bergamot is processed, the workers extracting the essential oil of the fruits never exhibit suppuration of lesions even if the lesions are relatively serious or complicated by abrasion or considerable loss of blood.

In the second half of the eighteenth century, the volatile oil was added in drops to lime tea or coffee as an antimalarial agent and had long been used as an anthelminthic and in treating scabies. For internal use, it was used as a sedative in a dose of 2–5 drops. It was also administered in drops on a sugar lump against insomnia.

Country people still use it to alleviate toothache and to disinfect and cicatrise lesions inflicted while working the bergamot. A few drops in water of the essence itself or of a well-known brand of soap solution of the essence is used by women for washing in the first few days following childbirth.

Massaging with bergamot oil is an infallible remedy for chapped, sore hands and feet and chilblains in cold winter weather. One of the preparations for use against chilblains, to be rubbed in 3 times a day, consists of 6 g of bergamot oil, 6 g of mint oil, 2.50 g of camphor and 100 g of castor oil. This oil is considered so useful and precious that even today houses are seldom without a bottle.

The use of the bergamot oil in popular medicine was immediately extended to the official medicine of the times, right from when the plant was first introduced to Europe. According to Fluckiger and Hambury (1878), bergamot oil was already included in a list of medicaments in an apothecary shop in the small German city of Giessen, printed in 1688.

In Pharmacology treatises from the last century we find an entire series of uses of bergamot oil. It was used in pharmacy to improve the smell of medicinal ointments and unguents, to prepare tooth-powder, hair oils and cosmetic preparations, and as an ingredient in the so-called 'smelling mixtures'. In medicine it was used on its own or in an alcohol solution or in other preparations for massaging in cases of chronic rheumatism, paralysis and weakness of the limbs; for internal use it served as a taeniacide. In 1805 a work, published in Messina by Calabrò and Ansalone, described the balsamic virtue of bergamot essence in treating lesions.

Some of the popular uses of bergamot essence have been re-examined and their utility confirmed. In 1932 A. Spinelli (1932), provided an extensive *in vitro* and *in vivo* demonstration that it could constitute a new antiseptic for use in surgery. It could substitute a 10 per cent iodine tincture since two successive brush applications of pure essential oil were sufficient to sterilise a given portion of skin. Along with perfect asepsis, there was a complete absence of local irritation or symptoms attributable to absorption of the essential oil.

Corassini (1934), reporting the clinical results of a more than ten month long experimentation project, in sterile and non-sterile surgery departments under his direction, using a product based on natural bergamot essential oil, observed that the product not only had an excellent disinfectant action on wounds and sores, but also promoted keratinisation of ulcers, while it had no or very little harmful effect on healthy or inflamed tissues. He also observed that the oil prepared artificially, a procedure used in that period, did not have the same effects.

After further concordant observations, a number of surgeons began to use it in a 15 per cent alcohol solution, as a substitute for iodine tincture, for preparing the operating field and for medicating infected lesions. Although alcohol solutions are unstable, a stable aqueous soapy solution was successfully prepared. The antiseptic action of the essential oil is a result firstly of the presence of phenols, alcohols and aldehydes which act directly on germs, and secondly of the hyperemia and positive leukocytosis induced in the wound, which increase the organism's defences. In 1935 G. De Nava confirmed that bergamot oil can usefully be applied in the field of odontostomatology since it has a marked specificity of action in the pathology of the tooth socket and an antiputrid power on roots with pulp gangrene or which are highly infected.

According to Maimone, who in 1940 tried it out on soldiers affected with scabies, the soapy solution cured all the patients treated, generally after 3–4 days of treatment.

In 1979 Ricci and Rovesti confirmed the efficacy of bergamot oil in treating furunculosis, acne and trichoses, since it reduces dandruff and hair loss.

Due to the presence of bergaptene, which has harmful biological effects such as photosensitisation, mutagenicity and carcinogenicity, good results were nonetheless obtained in the treatment of psoriasis as this psoralene blocks proliferation of skin cells and prevents them from whitening, which is a characteristic manifestation of the disease.

This fruit, whose bitter taste that reflects the harshness of the land for the people who inhabit it, encloses within its golden skin a strong but delicate fragrance, rich in beneficial virtues and offering hope of a cure.

Citrus as a current folk remedy in Mediterranean countries

In Mediterranean countries, we find that citrus species have an important place not only in the diet but also in self-medication. The lemon, in particular, appears to constitute a kind of panacea for all ills and many of its uses are based on popular beliefs, e.g. to treat ophthalmias slices of lemon are applied to the temples for a sufficient length of time to induce burning, in typhoid fevers roast lemon slices are applied to the forehead (Pitrè, 1870). Other uses are found in traditional medicine and have a therapeutic purpose: 10 per cent bergamot essence in 70° alcohol as a local antiseptic;

distilled water of *C. aurantium* var. *dulcis* L. and syrup of *Capsella bursa pastoris* as an antihaemorrhagic in menorrhagia during puberty and menopause (Penso, 1980).

In cosmetics (Penso, 1983), orange flower water, neroli oil and bergamot oil are ingredients in the formulation of colognes, perfumes and creams for dry skin (orange flower water). Massaging the face with lemon juice reduces hyperactivity of the sebaceous glands of greasy skin. Ten per cent bergamot essential oil in 70° alcohol, followed by an emollient mask based on starch and oil to reduce burning or cutaneous tension is useful in the treatment of comedones. Lemon masks serve to soften the skin. Our grandmothers used to rub slices of lemon onto the skin of their hands to keep the skin smooth and soft. A dry skin cream contains orange flower water, almond oil, cocoa butter, beeswax and an emulsifier.

Diced lemon is used in recipes for antiseborrheic baths. A mixture of lemon juice and avocado oil can be rubbed onto fragile finger nails.

In aromatherapy citron oil is used as an internal antiseptic; lemon essential oil incorporated with creams or milks can perform an antiseptic function in skin infections; it is a good antiseborrheic, helps prevent wrinkling and is soothing for the hands; orange flower water, the Medieval *Aqua Nanfa*, is well-known for its soothing, antiseptic, refreshing and deodorant action; sweet orange oil has a sedative action (Falchi Delitala and Drago, 1984).

In addition to preparations of Galenic forms, the above-mentioned *Citrus* species are included in numerous medicinal specialities from all over the world, not just as flavour and smell enhancers but also due to their vitamin C content, their bitter-tonic action and their antispasmodic effect.

The importance of the use of *Citrus* species in diet and traditional medicine is reflected in the widespread use of their images on postage stamps: the orange features on stamps from all of 30 countries (Albania, Algeria, Formosa, Cyprus, Gabon, Guinea, Honduras, Iraq, Israel, Italy, Jamaica, Lebanon, Montserrat, Mozambique, Pakistan, Paraguay, Rhodesia, Russia, Spanish, Marocco, Turkey, etc.). Fruits depicted on other country's stamps are the orange and lemon (Algeria), the lemon (Gabon, Guinea, Lebanon), the lime (Dominica, Montserrat), the grapefruit (Honduras, Somalia) and the citron (North Vietnam) (Griffenhagen, 1967).

In a critical reappraisal based on the most recent pharmacological research, several of the numerous therapeutic properties attributed to some species of the *Citrus* genus in the traditional medicine of the countries, where they have spread and acclimatised, have been shown to be well-founded and the active ingredients responsible for these activities have been identified. The anti-oxidant activity of anthocyanin, present particularly in pigmented oranges, has led to a revaluation of this generous gift of nature.

The golden fruits of the mythological garden of the Hesperides, the oranges of a bright red colour that reflects the fire of Etna, the inebriating perfume of the *zagara* (orange blossom), the white flowers that adorn brides, are the manifestation of the gifts brought to us by *Citrus* species: nutriment, beauty, health and poetry.

Kennst du das Land? Wo die Citronen blühn
Im dunkeln Laub die Gold-Orangen glühn...

Do you know the land where lemon trees bloom,
where golden oranges glisten among the dark leaves?

(J.W. Goethe–*Mignon*, verses 1–2)

REFERENCES

Antidotario Romano, translated by I. Ceccarelli (1678) In Venetia, presso Gio. Francesco Valuasense, pp. 14–15, 78–80, 135, 232–233.

Antidotarium Gandavense (1663), facsimile reprint 1974, De Backer, Gent, pp. 12–15, 29, 35, 36, 44–45, 50, 57, 60, 71, 74, 88, 124, 131.

Ayensu, E.-S. (1978) *Medicinal Plants of West Africa*, Reference Publications, Algonac Michigan, pp. 230–231.

Bajpai, A., Ojha, J.-K. and Sant, H.-R. (1995) Medicobotany of the Varanasi District, Uttar Pradesh, India. *Int. J. Pharmacognosy*, 33, 172–176.

Boerhaave, H. (1772) *Libellus de Materia Medica et Remediorum Formulis quae Serviunt. Aphorismis de Cognoscendis et Curandis Morbis*, Editio sexta, Nicaeae, apud Gabrielem Floteront, pp. 9, 25, 26, 28, 31, 33, 38, 45, 97, 115, 123, 125, 169–170, 179–180, 192–195, 202–203, 230.

Bouchardat (1855a; 1856b) *Manuale di Materia Medica e di Terapeutica Comparata e di Farmacia*, Tradotto in italiano ed annotato da B. Salemi, II (ed.), Stamperia di Giambattista Lorsnaider, Palermo (a) II (ed.) pp. 203–204; (b) III (ed.) pp. 231–232.

Boulos, L. (1983) *Medicinal Plants of North Africa*, Reference Publications Inc., Algonac Michigan, pp. 155, 200.

Braverman, J.B.S. (1949) *Citrus Products*, Interscience Publishers, New York, NY, p. 4.

Bretschneider (1871) *On the Knowledge Possessed by the Ancient Chinese of the Arabs and Arabian Colonies*, etc. London, in Tschirch A., Band I (1910) p. 519.

Camera Agrumaria di Reggio Calabria, (1931) La pubblica salute ed i prodotti ortofrutticoli ed agrumari. *Citrus*, 17, 35–38.

Cantani, A. (1887) *Manuale di Farmacologia Clinica* (Materia Medica e Terapeutica), II (ed.), Vallardi, Milano, Vol. II, pp. 133–137, Vol. III, pp. 53–57.

Casella, D. (1950) La frutta nelle pitture pompeiane. *Pompeiana*, pp. 355–386.

Cassola, F. (1838) *Trattato di Chimica Elementare*, Vol. IV, Stamperia e Cartiera del Fibreno, Napoli, p. 202.

Castelli, P. (1640) *Hortus Messanensis*, Messanae, Typis Viduae Ioannis Francisci Bianco, p. 13, 25, 29.

Catanuto, N. (1666) *Catanense Dispensatorium sive Antidotarium*, Ex Typographia Iosephi Bisagni, Catanae, pp. 35–36, 46–48, 186–187.

Cavendish, ref. by Tyozaburo Tanaka (1935).

Chang, H.-M. and But, P.-P.-H. (1987) *Pharmacology and Applications of Chinese Materia Medica*, Vol. 2, Word Scientific, Singapore, pp. 848–851.

Chevallier, A. and Richard, A. (1830) *Dizionario delle Droghe Semplici e Composte*, Tomo I, Tasso, Venezia, pp. 270–272, 330.

Chiavaro, A. (1934) Il succo di arance e limoni in odontoiatria. *Annali di Odontoiatria*, 17, 136, ref. by *Fitoterapia*, 1934, 10, 58–59.

Clusius, C. (1561) *Antidotarium*, Antwerp, facsimile reprint 1973, De Backer, Ghent, pp. 41–42, 56–58.

Codex Medicamentarius seu Pharmacopoea Danica (1821) Lipsiae, apud Fr. Fleischer, pp. 19–20, 24, 39–41, 129, 146, 149, 176–177, 184–185.

Codex Medicamentarius seu Pharmacopoea Lusitanica (1822) Lipsiae et Soraviae, apud Fr. Fleischer, pp. 14–15, 65, 192.

Codex Medicamentarius seu Pharmacopoea Parisiensis ex Mandato Facultatis Medicinae (1748) Parisiis, Apud Guillelmum Cavelier, pp. XII–XIII, XXXI, LXX.

Codice Farmaceutico Romano (1868) Roma, tipi della Civiltà Cattolica, pp. 5–6.

Corassini, G. (1934) Un nuovo antisettico in chirurgia: il 'Salbeol'. *Citrus*, 20, 257–259.

Cortesi, I.-B. (1629) *Pharmacopoeia seu Antidotarium Messanense*, Messanae, Ex Typis Pietri Brea, pp. 65, 193–194, 196, 201, 308.

De Pasquale, A. (1984) Pharmacognosy: the oldest modern Science. *J. Ethnopharmacol.*, 11, 1–16.
De Rochefort, D. (1789) *Cours Élémentaire de Matière Médicale suivi d'un Prècis de l'Art de Formuler*, Tome II, Méquignon, Paris, pp. 84–87, 386.
Delaveau, P. (1982) *Histoire et Renouveau des Plantes Médicinales*, Paris, Ed. Albin Michel, pp. 187–190, 239.
Delaveau, P. (1987) *Les Épices. Histoire, Description et Usage des Différents Epices, Aromates et Condiments*, Paris, Ed. Albin Michel, p. 139.
Dezani, S. and Guidetti, E. (1953) *Trattato di Farmacognosia*, II (ed.), UTET, Torino, p. 350.
Dioscoride, *The Greek Herbal of Dioscorides*, illustrated by Byzantine A.D. 512, Englished by John Goodyer, A.D., 1655, Edited and first Printed, A.D. 1933 by Gunther, R.T. Hafner Pb. Co., London and New York, 1968, pp. 84–85.
Dujardin-Beaumetz, M. (1889) *Dictionaire de Thérapeutique, de Matière Médicale, de Pharmacologie, de Toxicologie et des Eaux Minérales*, Tome IV, Octave Doin Ed., Paris, pp. 77–79.
Duke, J.-A. and Ayensu, E.-S. (1985) *Medicinal Plants of China*, Vol. 2, Reference Publications Inc., Algonac-Michigan, pp. 571–573.
Duke, J.-A. and Vasquez, R. (1994) *Amazonian Ethnobotanical Dictionary*, CRC Press, Boca Raton, p. 52.
Durante, C. (1585) *Herbario Nuovo*, Roma, appresso Bartholomeo Bonfandino e Tito Diani, p. 40, 101, 259.
Editorial staff (1930a) Le Vitamine 'C' dei succhi d'arancio e di limone. *Citrus*, 16, 403.
Editorial staff (1930b) Virtù medicinali dei semi d'arancio. *Citrus*, 16, 200.
Editorial staff (1959) Un decotto di limone secondo Mannino sarebbe un rimedio efficace contro i gonococchi. *Giornale Italiano Malattie Veneree* (1884) 19, 361, ref. by *Documentazione CIBA*, n. 5, 155.
Edwards, H.-M. and Vavasseur, P. (1829) *Manuale di Materia Medica ossia Breve Descrizione de' Medicamenti*, Tomo I, Minerva, Napoli, pp. 159–161, 185–186.
El Ricettario dell'Arte, et Universita de Medici, et Spetiali della Citta di Firenze, Florence 1550, facsimile reprint 1973, De Backer, Gent, p. 94.
Falchi Delitala, L. and Drago, G. (1984) *Gli Olii Essenziali. Aromaterapia. Aromacosmesi*, Arti grafiche Russo, Caserta, pp. 64–74.
Fenaroli, G. (1963) *Sostanze Aromatiche Naturali*, Vol. I, Hoepli, Milano, p. 696.
Ferrarini, A. (1825) *Farmacopea*, I ed., Bologna, per le stampe del Sassi, pp. 16, 41–42, 225–226, 245, 251, 257–260, 269, 285, 335, 371, 427, 482–483.
Flückiger, F.-A. and Hanbury, D. (1878) *Histoire des Drogues d'Origine Végétale*, Tome I, Doin, Paris, pp. 212–233.
Font Quer, P. (1962) *Plantas Medicinales. El Dioscórides Renovado*. Editorial Labor, Barcelona, pp. 434–438.
Gallesio, (1811) *Traité du Citrus*, Paris, ref. by Flückiger and Hanbury, 1878, Tome 1, p. 213.
Ghazanfar, S.-A. (1994) *Handbook of Arabian Medicinal Plants*, CRC Press, Boca Raton, pp. 187–188.
Griffenhagen, G. (1967) *Drug and Pharmacy on Stamps*, Pb. American Topical Ass., Milwaukee, Wisconsin, pp. 48–49.
Guareschi, I. (1897) *Commentario della Farmacopea Italiana*, UTET, Torino, I/1, p. 405; I/2, pp. 272, 366.
Huang, K.-C. (1993) *The Pharmacology of Chinese Herbs*, CRC Press, Boca Raton, pp. 107–109, 173, 177.
Imbesi, A. (1962–63) Le piante medicinali nei dipinti pompeiani e l'introduzione del limone in Italia. *Lavori Istituto Farmacognosia Università di Messina*, 2, 421–431.
Imbesi, A. (1964) *Index plantarum quae in Omnium Populorum Pharmacopoeis Sunt adhuc Receptae*, Messina, pp. 274–281.
Index Kewensis (1997) Oxford University Press, on compact disc version 2.0.

Karnick, C.-R. and Hocking G.-M. (1975) Ethnobotanical records of drug plants described in Valmiki Ramayana and their uses in the Ayurvedic System of medicine. *Quart. J. Crude Res.*, 13, 143–154.

Kerharo, J. and Adam, J.-G. (1974) *La Pharmacopée Sénégalaise Traditionnelle. Plantes Medicinales et Toxique*, Vigot Frères, Paris, pp. 708–710.

Kokwaro, J.-O. (1976) *Medicinal Plants of East Africa*, East African Literature Bureau, Kampala, p. 195.

Lagriffe, L. (1979) *Le Livre des Epices, Condiments et Aromates*, Paris, Tchou et Morel, ref. by Delaveau, 1987.

Lemery, N. (1766) *Dizionario ovvero Trattato Universale delle Droghe Semplici*, IV (ed.), Butella e Perlin, Venezia, p. 37, 81, 185.

Maimone, D. (1940) L'essenza di bergamotto nella cura della scabbia. *Il Dermosifilografo*, 15(7).

Martinez, M. (1959) *Les Plantas Medicinales de México*, Cuarta Ed., Ed. Botas, Mexico, p. 460.

Mattioli, M.P.A. (1557) *I Discorsi di M. Pietro Mattioli Medico Sanese, ne i Sei Libri della Materia Medicinale di Pedacio Dioscoride Anazarbeo*, Valgrisi e Costantini, Vinegia, pp. 142–146.

Mojon, B. (1784) *Pharmacopoea Manualis reformatam*, Genuae, apud Repettum, pp. 13–14.

Moldenke, H.N. and Moldenke, A.L. (1952) *Plants of the Bible*, The Ronald Press Co., New York, NY, pp. 185–286.

Morton, J.-F. (1975a) Current folk remedies of northern Venezuela. *Quart. J. Crude Drug Res.*, 13, 97–121.

Morton, J.-F. (1975b) *Atlas of Medicinal Plants of Middle America*, Thomas Pb., Springfield, Illinois, pp. 373–377.

Morton, J.-F. (1977) Medicinal and other plants used by people on North Caicos (Turks and Caicos Island, West Indies). *Quart. J. Crude Drug Res.*, 15, 1–24.

Officina Medicamentorum et Methodus Recte eodem Componendi, cum Variis Scholiis, et Aliis quam Plurimis, Ipsi Operi Necessaris, ex Sententia Valentinorum Pharmacopolarum, Auctore Eorundem Collegio (1601), Apud Iohannem Chrystomum Garriz, Valencia, p. 33, 46, 57.

Oliver-Bever, B. (1986) *Medicinal Plants of Tropical West Africa*, Cambridge University Press, Cambridge, pp. 47, 174, 212.

Orosi, G. (1856–1857) *Farmacologia Teorica e Pratica o Farmacopea Italiana*, III (ed.) Vincenzo Mansi Ed., Livorno, parte I, pp. 178–179, parte II, pp. 624–626, 685, 774–777, 781–783.

Pende, N. (1926) Il succo del limone quale preventivo e curativo dell'arteriosclerosi. *Citrus*, 12, 120.

Penso, G. (1980) *Piante Medicinali nella Terapia Medica*, OEMF, Milano, p. 195, 248.

Penso, G. (1983) *Piante Medicinali nella Cosmetica*, OEMF, Milano, p. 156, 181, 205, 231, 247, 252.

Perry, L.-M. (1980) *Medicinal Plants of East and Southeast Asia. Attributed Properties and Uses*, The Mitt Press, Cambridge, Massachussetts and London, pp. 361–363.

Pételot, A. (1952) Les plantes médicinales du Cambodge, du Laos et du Viet-nam. *Arch. Recherches Agron. Camboge, Laos, Vietnam*, 14, 165; ref. by Perry (1980).

Pharmacia Antverpiensis (1661), facsimile reprint 1974, De Backer, Gent, pp. 40, 44–45, 66, 68, 108, 128, 220, 222, 231–232, 235.

Pharmacopoea Almeriana (1726–1795), facsimile reprint 1973, De Backer, Gent, pp. 7–8, 23, 39, 42, 44, 49, 50, 54, 58, 62, 80–81, 85, 88, 98–99, 101, 137.

Pharmacopoea Austriaco-Provincialis (1798), Venetiis, typis Sebastiani Valle, pp. 5, 12, 51–52, 88, 114, 116, 127.

Pharmacopoea Hagiensis (1659), facsimile reprint 1973, De Backer, Gent, pp. 2–3, 21, 36.

Pharmacopoea Saxonica (1820), Dresdae, Sumtibus G.M. Waltheri, pp. 18–19, 23, 48, 166, 200, 310–311, 329.

Pharmacopoea Ultrajectina (1656), facsimile reprint 1974, De Backer, Gent, p. 27.

Pharmacopoeia Brugensis (1697), facsimile reprint 1973, De Backer, Gent, p. 8, 24, 32, 37, 57, 73, 78, 90, 113, 136.

Pharmacopoeia Bruxellensis (1641), facsimile reprint 1973, De Backer, Gent, pp. 22, 25–29, 39, 59, 62, 70, 79–80, 85–88, 97, 102–103, 108–109, 126–127, 142.

Pharmacopoeia Collegii Regalis Medicorum Londinensis una cum Meadiana (1767), Venetiis, apud Laurentium Basilium, pp. 21–22, 44, 47, 55, 57, 65–66.

Pitrè, G. (1870–1913) *Medicina Popolare Siciliana*, Ristampa anastatica 1978, Il Vespro Siciliano, Palermo, pp. 275–331.

Plinii Caii Secundi (1831) *Historia Naturalis, ex Recensione I. Harduini et Recentiorum Adnotationibus*, Pomba, Torino, l. XIII/16.

Puri, H.-S. (1971) A comparative study of folklore vegetable drugs of Europe and India. *Acta Phytotherapeutica*, **18**, 21–33.

Qédan, S. (1974) Heimische Arzneipflanzen der Arabischen Volksmedizin. *Planta med.*, **26**, 65–74.

Quercetano, I. (1608) *Pharmacopoea Dogmaticorum Restituta*, Venetiis, Apud Iohannem Antonium et Iacohum de Franciscis, pp. 22–29, 41, 65, 96–97, 100, 109.

Rebaudengo, T. (1772) *Farmacopea Cerusica*, Vercelli, presso Giuseppe Panialis, p. 142.

Ricci, G.-H. and Rovesti, P. (1979) Impieghi cosmetici dell'olio essenziale di bergamotto. *EPPOS*, **61**, 18–19.

Ricettario Fiorentino Nuovamente Compilato e Ridotto all'uso Moderno (1802), Pietro Sola, Venezia, pp. 13, 17, 60, 63–64, 76–77, 104, 124, 127, 138.

Romeo, G. (1930) *Le Essenze degli Agrumi (Esperidee)*, Tip. La Sicilia, Messina.

Royle (1837) *An Essay of the Antiquity of Hindoo Medicine*, London, ref. by Tschirch, A. (1910) Band I, p. 505.

Semmola, G. (1836) *Saggio Chimico-medico su la Preparazione, Facoltà ed Uso dei Medicamenti*, Tip. Giuseppe Severino, Napoli, pp. 44–45, 145–146.

Soto (1557) ref. by Tyozaburo Tanaka (1935).

Spinelli, A. (1932) L'essenza di bergamotto nuovo antisettico nella pratica chirurgica. *Policlinico Sez. Chirurgica*, **39**, 5.

Stella, P. (1929) L'azione del succo di limone sul gonococco. *Citrus*, **15**, 169.

Tang, W. and Eisenbrand, G. (1992) *Chinese Drugs of Plant Origin. Chemistry, Pharmacology and Use in Traditional and Moderne Medicine*, Springer-Verlag, Berlin, pp. 337–349.

Targioni Tozzetti, A. (1847) *Corso di botanica Medico-Farmaceutica*, Vincenzo Batelli e Compagni, Firenze, pp. 307–311.

Theatrum Sanitatis, Codice 4182 della R. Biblioteca Casanatense (1940) Libreria dello Stato, Roma, p. 56.

Tilli, M.-A. (1723) *Catalogus Plantarum Horti Pisani*, Florentiae, apud Taitinium et Franchium, p. 42, 100.

Tolkowsky, S. (1938) *Hesperides, a History of the Culture and the Use of Citrus Fruits*, Londres.

Trease, G.E. (1961) *A textbook of Pharmacognosy*, VIII (ed.), Baillière, Tindall a. Cox, London, pp. 341–350.

Tschirch, A. (1910–1925) *Handbuch der Pharmacognosie*, Tauchnitz Leipzig, Band I (1910) pp. 547, 592; Band II (1917) p. 851.

Tutin, T.G., Heywood, V.H., Burges N.A., Moore, D.M., Valentine, D.H., Walters, S.M. and Webb, D.A. (1968) *Flora Europea*, University Press, Cambridge, Vol. II, pp. 229–230.

Tyozaburo Tanaka (1935) L'acclimatazione degli agrumi fuori dei paesi d'origine, *Citrus* **21**, 73–78.

Università degli Studi di Messina, Facoltà di Scienze, Matematiche, Fisiche e Naturali (1989) *L'Orto Botanico*, Messina, Samperi, p. 4.

Vanden Zande, I. (1697) *Pharmacopoeia Brugensis*, facsimile reprint 1973, De Backer, Gent, pp. 8–9, 24, 32, 37, 57–58, 78, 90, 112.

Volkammer, (1713) *Hesperides Norimbergenses*, l. 3, cap. 26, 156, ref. by Flückiger, F.-A. and Hanbury, D. 1878, Tome 1, p. 223.

Watt, J.-M. and Breyer-Brandwijk, M.-G. (1962) *The Medicinal and Poisonous Plants of Southern and Eastern Africa*, II (ed.), E.a.S. Livingstone LTD., Edimburg and London, pp. 912–917.

Wehrmann, (1912) Ber. d. Bot. Ges., p. 99, ref. by Imbesi, (1962–63).

Wren, R.-C. (1989) *Potter's New Cyclopaedia of Botanical Drugs and Preparations*, C.W. Daniel Co., Essex, p. 172, 205.

28 The biological activity of citrus oils

Giuseppe Bisignano and Antonella Saija

INTRODUCTION

Citrus oils are largely employed as aromatizers in the food and pharmaceutical industries. In particular, lemon and bergamot essential oils are used in the cosmetic industry, for the production of perfumes, detergents and body-care products, mostly thanks to their fragrance and solvent properties. Besides, they are present in various pharmaceutical preparations for gynaecological, ophthalmic and surgical use, as also in dentistry, because of their well-known antiseptic properties. Apart from being used as disinfectants, thanks to just these antimicrobial properties, recent studies point out the possibility of employing citrus essential oils and/or their active principles in clinical fields (also due to their wide range of activity and their reduced resistance phenomena), but also as preservatives in the food industry and as alternative pesticides in integrated programmes.

Citrus essential oils are rarely utilized, as such, in the pharmacotherapeutic field. However, much has been achieved as regards knowledge of the biological properties of the active principles isolated from these essential oils. As a result, some of these active principles (or compounds derived from them) are being successfully employed in therapy (for example, 5-methoxypsoralen in the treatment of psoriasis and vitiligo); the possibility is also being closely examined of utilizing them in the prevention of certain pathological conditions (for example, *d*-limonene in the chemoprevention of tumoral disease). Furthermore, extensive toxicological studies have proved to be fundamental in regulating, on the basis of scientific criteria, the use of citrus essential oils and/or their active principles not only in the pharmacotherapeutic field, but also in quite different fields (such as that of the cosmetic industry and in de-fatting products).

The following paragraphs thus represent an attempt to exhaustively delineate, by means of numerous papers reported in the international scientific literature, the variegated profile of the biological and pharmacological activities of citrus essential oils and especially, of their active principles, as also to punctuate their actual and/or potential applications in the pharmacotherapeutic field.

ANTIMICROBIAL ACTIVITY

Some essential oils are traditionally used in the phytotherapy of infective diseases, both for their established *in vitro* efficacy against bacteria and fungi frequently isolated in man, and for the reduced resistance phenomena and the broad range of activity, for which the oil components are, either singly or synergically, responsible. The mechanism of action could be led back both to a damage of the bacterial or fungal cell, and to an indirect, immunomodulating and prevalently chemotactic action (Camporese, 1997).

As far back as 1948, Piacentini had already referred to the antiseptic potency of orange, bergamot and lemon oils, greater than that of phenol. Murdock and Allen (1960) showed that the addition of orange oil or d-limonene increased the preservative properties of sodium benzoate.

The essential oils can be a highly interesting source of molecules able to inhibit the growth of microorganisms, fungi and insects. Because of their volatile nature and their low toxicity in mammals, some groups of compounds, such as monoterpenes, are being studied for their possible use as preservatives in food products. Interest is being focussed not only on the volatile oil components of citrus oils, but also on the components of the non-volatile fraction: coumarins, psoralens and flavonoids. These compounds are considered phytoalexins, because they are produced by the plant as a response to a challenge by other organisms; they can be toxic against bacteria, fungi, protozoa and insects.

Dabbah *et al.* (1970) assayed the activity of some citrus essential oils, of their deterpenated fractions and of some of their components (+)-limonene, geraniol and terpineol on food-contaminating bacteria. Limonene, the principal component of the volatile fraction of citrus essential oils, proved, as such, to possess antibacterial, antifungal and insecticidal activity. Moreover, its two enantiomers have been observed to have a different activity. In a study carried out on 25 strains, (+)-limonene resulted active against 22 strains, whereas (−)-limonene against 16 strains (Lis-Balchin *et al.*, 1996). The growth of *Staphylococcus aureus* was inhibited by all the citrus oils and their derivatives assayed, with a percentual reduction of the initial inoculum ranging between 100 per cent for the terpene-less citrus oils and 67 per cent for the lemon essential oil. Inhibition of *Pseudomonas* spp. (18 strains) ranged between 87 per cent for lemon essential oil and 100 per cent for terpineol. The orange, lemon and lime terpene-less oils considerably reduced (from 86 per cent to 100 per cent) the initial inoculus of *Salmonella senftenberg*, *Escherichia coli* and *S. aureus*. d-Limonene stored for over four years showed a considerable activity against *S. senftenberg*; this confirms the statement by Zukerman (1951), that oxidized d-limonene is more inhibitory than the freshly-distilled product.

Orange essential oil acted on *Salmonella* spp. by inhibiting its growth in a range between 84 per cent for *S. enteriditis* and 99 per cent for *S. gallinarum* and *S. norwich*. Terpineol, instead, completely inhibited the growth of all 24 serotypes of *Salmonella* tested. Orange essential oil acted as bacteriostatic against the other strains belonging to the *Enterobacteriaceae* group; growth inhibition oscillated between 88 per cent for *Aerobacter aerogenes* and 100 per cent for *Alcaligens faecalis* and for two of the 12 strains of *E. coli* assayed. Growth inhibition of the *Proteus* group was approximately 95 per cent. In the same way as for the *Salmonella* group, terpineol completely inhibited the growth of all the bacteria tested, with the exception of *Serratia marcescens*, for which strain a growth inhibition of 89 per cent was registered. Bitter orange essential oil inhibited the growth of *Pseudomonas* and of other Gram-negative bacteria (10–100 per cent), as also terpineol (94–100 per cent), that, however, proved to be less efficacious against *Pseudomonas* spp. and *Achromobacter* spp., when its action was compared with that against *Enterobacteriaceae*.

Chaibi *et al.* (1997) tested the activity of essential oils extracted from various plants, among which orange and grapefruit oils, on vegetative cells and spores of *Clostridium botulinum* and *Bacillus cereus*. They inhibited both the germination of the spores and the multiplication of the vegetative cells, acting as sporostatics, on the *Clostridium botulinum* strain, at concentrations between 300 and 500 ppm; conversely, on the *Bacillus cereus* strain, grapefruit oil acted as a sporostatic (300–500 ppm) and orange oil as a sporicidal (>300 ppm). Moreover, it was possible to establish that the two essential oils acted during different phases of the spore cycle; in particular, grapefruit oil appears to act in the spore commitment-to-germinate stage, whereas orange oil acts in a successive stage. In any case, Gram-positive bacteria were more sensitive than Gram-negative bacteria to the inhibitory action of the citrus essential oils and their derivatives assayed.

Alderman *et al.* (1976) demonstrated that citrus essential oils suppress growth of *Aspergillus parasiticus* and aflatoxin production by this bacterial strain in a range between 500 and 7,000 ppm. Orange and lemon essential oils were more potent inhibitors of growth and aflatoxin production than their principal component d-limonene. Subba *et al.* (1967) observed that 2,000 ppm of orange oil, but not of lemon oil, inhibited the growth of *Aspergillus flavus*, an aflatoxin-producing strain.

Bergamot essential oil from various sources, produced in Italy and in the Ivory Coast, and reconstituted in laboratory, was analysed as regards its antibacterial property against saprophytic and pathogenic organisms, among which food contaminants. Gram-positive bacteria proved to be the most sensitive to the essential oils, although, of the *Listeria* genus, only *L. seeligeri* resulted sensitive to the oils tested, while *L. monocytogenes* and *L. ivanovii* were weakly inhibited only by the non-diluted oils. The different activity shown against Gram-positive and Gram-negative bacteria can lead to the hypothesis that it might be the result of an action on cell wall components (Ferrini *et al.* 1998). Lemon essential oil has proved to be active against *L. monocytogenes* and efficacious against all the strains isolated from various kinds of food (Lis-Balchin *et al.*, 1997).

Penicillium digitatum and *Penicillium italicum* are the pathogens responsible for the worst damage to fruits of citrus spp. after harvesting. They infect the fruit through microinjuries produced in the flavedo during harvesting and processing. These microinjuries can also involve the essential oil-containing glands, causing the oil to overflow; Norman *et al.* (1967) and Mc Calley *and* Torrees-Grifol (1992) have shown that the damaged fruit releases a greater quantity of oil constituents than the non-damaged fruit. It can be hypothesized that conidia of *Penicillium* spp. come into contact with the essential oil, that can thus play a role in the pathogenic process. Caccioni *et al.* (1998) evaluated the effects that the essential oils of *C. sinensis, C. aurantium, C. deliciosa, C. paradisii* and of the hybrid of *C. sinensis* × *Poncirus trifoliata* have on the growth of these pathogens. 'Citrange' essential oil proved to be the most active inhibitor, followed by the lemon oil produced from fruits picked in February. The activity of grapefruit, tangerine, bitter orange and lemon (fruits picked in November) oils was less efficacious, but, at any rate, good. The activity of sweet orange essential oil proved to be weak, especially against *P. digitatum*.

Bitter orange essential oil was assayed on clinical isolates of various bacterial and fungal species. It has shown antimicrobial activity against clinical isolates, endowed with multiple resistance, of *S. aureus* (46 strains), *E. coli* (37 strains) and *Pseudomonas aeruginosa* (36 strains). The action mechanism of citrus essential oils against microorganisms is unknown. Lipid solubility and effects on the bacterial surface might be involved, as also the terpene-oxidized derivatives present in citrus essential oils. The bitter orange essential oil has manifested fungistatic and fungicidal properties on clinical isolates of *Candida albicans*, as also on four clinically isolated dermatophytes and two isolates of *Tinea rubrum* (Sonbol *et al.*, 1995). Bitter orange essential oil was assayed as an antifungal agent on patients affected by *Tinea* (Ramadan *et al.*, 1996). The *in vitro* antifungal activity of the oil proves to be superior to its antibacterial activity.

Citrus essential oils are endowed with an acaricidal activity. The essential oils of *C. maxima*, *C. reticulata*, *C. suncris*, *C. sinensis* and *C. hystrix* were tested as to this activity against *Boophilus microplus*. At a 1:10 dilution, the essential oils of *C. reticulata* and *C. maxima* showed the highest acaridal effect, twice as active as (+)-limonene, and also the highest larvicidal activity, while the essential oils of *C. hystrix* and *C. suncris* (unripe fruits) exhibited a poor larvicidal activity (1.5 times higher than that of (+)-limonene). The essential oils of other *C.* species had a weak acaricidal effect (Chungsamarnyart *et al.*, 1996).

Citrus essential oils have proved to be toxic to insects. Lemon (*C. limon*), grapefruit (*C. paradisi*) and orange (*C. sinensis*) peel oils were assayed for their insecticidal activity against larvae and adults of *Culex pipiens* and *Musca domestica*. Lemon oil resulted in being the most efficacious against the larvae and adults of *C. pipiens*. Grapefruit oil was the most toxic against adults of *M. domestica*, while lemon oil was the most toxic against *Musca* larvae. Orange oil was, in any case, the least efficacious against larvae and adults of both species (Shalaby *et al.*, 1998). The peel essential oils of Balady oranges and mandarins disrupted growth and development of *Spodoptera littoralis* larvae (LD_{50} values of 0.26 and 0.547/100 ml water for orange and mandarin essential oils respectively) (Omer *et al.*, 1997). Experiments carried out on *Callosobruchus maculatus*, *Sitophilus zeamais* and *Dermestes maculatus* proved that they cause adult death, but have little or no activity on eggs and larvae originating from surviving insects. Topical toxicity tests effected on *D. maculatus* have shown that contact toxicity of citrus essential oils is of little or no importance, while an elevated mortality rate was reported when the treated insects were confined in air-tight glass chambers, thus confirming the volatile nature of the toxic components. Tests were moreover carried out on *C. maculatus*, in order to establish which oil components are responsible for the insecticidal activity. Of the 30 components assayed, 22 (including (+)-limonene) were shown to have strong fumigant insecticidal activity, but none of them can, singly, account for the activity found in a typical citrus essential oil, such as lime essential oil. They, thus, act synergically amongst themselves (Don-Pedro, 1996a,b,c).

Yajima *and* Munakata (1979) were the first authors to report that some non-volatile essential oil components, such as coumarins and furanocoumarins, inhibit several phytophages and that furanocoumarins are toxic to *Spodoptere eridania* larvae. As regards insecticidal activity, four major components (5,7-dimethoxycoumarin, 8-geranoxypsoralen, 5-geranoxypsoralen, 5-geranoxy-7-methoxycoumarin) were isolated and identified from the non-volatile fraction of lemon essential oil. These four compounds were assayed by means of topical application on *Sitophilus oryzae* and *Callosobruchus maculatus*. The results obtained indicated that the compounds containing a geranoxy substituent are active against these insects; moreover it is possible that some of these compounds might act synergically amongst themselves or with other oil minor components (Su *et al.*, 1997). There are various hypotheses as to which components, that might act synergically amongst themselves, should be considered responsible for the inhibitory activity of citrus essential oils. Psoralen, 5-methoxypsoralen (5-MOP) and 8-methoxypsoralen (8-MOP), activated by UV light at 365 nm for 60 min, were assayed as antimicrobial agents on *Micrococcus luteus*, *E. coli* and *L. monocytogenes*. *L. monocytogenes* was inhibited by the three furanocoumarins at concentrations ranging between

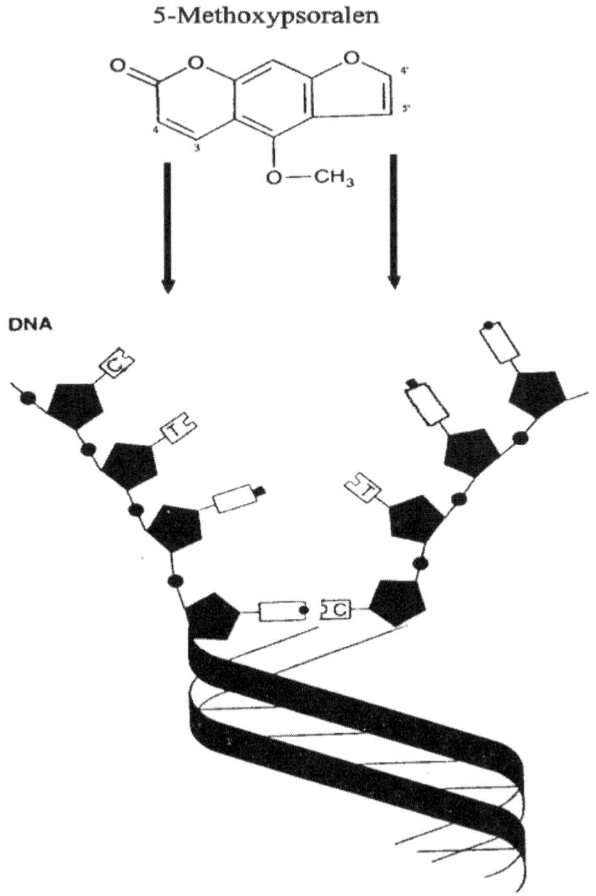

5-Methoxypsoralen

Figure 28.1 Possible interaction of 5-methoxypsoralen double bonds with DNA pyrimidine bases cytosine (C) and thymine (T).

2 and 10 µg/ml, whereas only psoralen and 8-MOP inhibited the growth of *E. coli* and *M. luteus* strains. The difference in activity can be ascribed to the presence of methoxy substituents on the furanocoumarinic ring; the substitution in the basic structure, proper to psoralen, reduces or eliminates the photoreactivity and, as a consequence, reduces its antimicrobial efficacy. Lime peel extract and cold-pressed oil and 5-MOP inhibited the growth of *L. monocytogenes*, but not of *E. coli* O157:H7, while the growth of *M. luteus* was inhibited only by the cold-pressed lime oil. This suggests the presence, in the oil, of other antimicrobial agents, such as citral (Ulate-Rodriguez *et al.*, 1997a,b). Moreover, Lin *et al.* (1989) treated bacteria (*E. coli* and *S. aureus*), RNA viruses and DNA viruses with UVA radiation (wavelength 320–400 nm) in the presence of 8-MOP, a component of lemon essential oil, and showed that such a system is able to inhibit all the microorganisms tested. It is hypothesised that, in the presence of UV light, furanocoumarins act on DNA replication and transcription processes (Figure 28.1);

another mechanism proposed involves furanocoumarin-photooxidation products, that are aldehydes, hydrogen peroxide and carboxylic acids.

Vargas *et al.* (1999) isolated three permethoxylated flavones, dehydroabietic acid and linoleil monoglyceride from the non-volatile residue of orange essential oil. All these compounds proved to be active against phytopathogenic species and food contaminants. Hexa- and eptamethoxyflavone showed fungicidal activity against *Geotrichum candidum*, that is not inhibited by the broad-spectrum fungicide Benomyl; moreover, these compounds proved to have an efficacious antioxidant activity, similar to that of tert-butyl-4-hydroxyanisol, a widely-used synthetic antioxidant.

Also citral (3,7-dimethyl-2,6-octadienale), a mixture of the two geometric isomeric aldehydes neral and geranial, is, by several authors (Ben-Yehoshua *et al.*, 1992; Caccioni *et al.*, 1995; Rodov *et al.*, 1995), believed to be responsible for the inhibitory activity. Asthana *et al.* (1992) were the first authors to report that citral ehibits UV-A (315–400 nm) light enhanced oxygen-dependent toxicity against a series of *E. coli* strains differing in DNA repair and catalase proficiency. In fact, *E. coli* strains carrying a gene leading to catalase deficiency are particularly sensitized to inactivation by citral and UV-A treatment. Even on *Fusarium oxysporum* and *F. solani*, pathogens of citrus roots, citral toxicity is potentiated in the presence of UV-A rays, while it is negligible in the dark. When the plasmid was treated with citral in the presence of UV-A rays, its covalently-closed circular conformation was observed to change to an open circular conformation and, lastly, to a linear form. The conformational change of the plasmid corresponds to a reduction in the transforming activity.

Citral is one of the compounds having antifungal activity isolated from lemon flavedo; these compounds make up the first line of fruit resistance against the attack of *Penicillium digitatum*, a pathogen chiefly responsible for the postharvest fruit deterioration. Rodov *et al.* (1995) established the existence of a relationship between fruit age, flavedo citral concentration and fruit resistance to deterioration. Citral is present in a 1.5–2 fold higher concentration in the flavedo of young mature lemons than in that of older yellow fruits. Together with a reduction of citral concentration, yellow lemon flavedo extract reveals an increased concentration of another compound, the monoterpene ester neryl acetate. This latter compound has a poor inhibitory activity and, besides, at concentrations lower than 500 ppm promotes the development of the pathogen organism. Together with reduction in citral concentration, the authors observed a reduction in flavedo antifungal properties, which leads to the hypothesis that a relationship exists between a high citral concentration in oil glands of young fruits and their resistance to postharvest decay. In inoculation experiments with *P. digitatum*, the compounds located in young lemon oil glands were able to inhibit the pathogen *in situ*, while the gland content of old fruits proved to be inactive, and, possibly, even to stimulate disease development. It has been demonstrated that esogenic application of citral on ripe green lemons inoculated with *P. digitatum* inhibits the development of the pathogenic agent; this suggests that it can possibly be employed to check the decay process, thus reducing the use of fungicides and consequently toxic residues on the fruit. However, phytotoxic damage caused by citral, visible as dark spots on fruit peel, limits practical applications (Ben-Yehoushua *et al.*, 1992).

The differences found between young and ripe fruits cannot be attributed merely to the difference in citral content. Yellow and green lemons differ also in their ability to produce induced antifungal materials, such as the phytoalexin scoparone. In fact, the preformed antifungal material present in lemon flavedo, containing citral, limettine

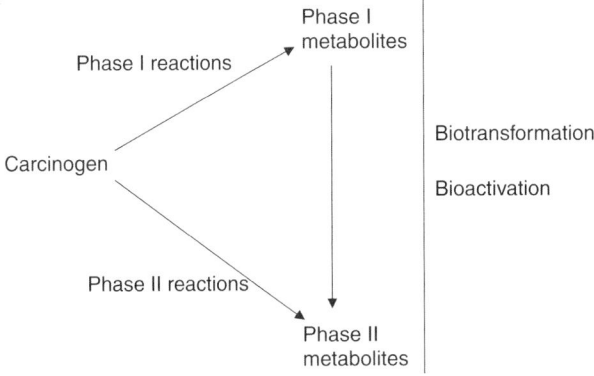

Figure 28.2 Biotransformation/bioactivation pathways of carcinogens.

(5,7-dimethoxycoumarin) and 5-geranoxy-7-methoxycoumarin, represents the first defence line against pathogen attack. The second defence line is made up of scoparone (6,7-dimethoxycoumarin), the production of which is induced by a pathogen challenge (Arimoto *et al.*, 1986; Kim *et al.*, 1991). In fact, the biosynthesis of scoparone was associated with resistance of mature fruits of the hybrid citrus tangelo Nova and of the two species *C. paradisi* and *C. aurantium* to infection by *Phytophthora parasitica* (Ortuño *et al.*, 1997).

BERGAMOT OIL AND 5-METHOXYPSORALEN

Obtained from the peel of *C. bergamia*, bergamot oil has been used for centuries as a fundamental ingredient of many perfumes. This widespread use of bergamot oil depends on its particular fragrance that however rapidly disappears when the essential oil is exposed to solar radiations.

Application of perfumes containing bergamot oil on skin areas exposed to solar radiation can cause undesired phototoxic effects such as oedema, long-lasting erythema and successively a pronounced cutaneous pigmentation; these symptoms make up the classic clinical picture of Berlock dermatitis. This pathological condition is ascribable to 5-methoxypsoralen (5-MOP, also known as bergapten; Figure 28.1), present in pure bergamot essential oil in concentrations up to 3,000–3,600 ppm.

In the last decade, various studies have been carried out to evaluate the potential toxicity of the bergamot oil present in cosmetic preparations used as tanning lotions or sunscreens; the melanogenic agent present in these preparations is, in fact, represented by 5-MOP.

The phototoxic, photomutagenic and photocarcinogenic properties of bergamot oil lead essentially back to 5-MOP and other components, such as bergamottine (5-geranoxy-psoralen) being contained in it. These properties, exhaustively studied even in experimental conditions simulating solar exposure by various research groups (Averbeck *et al.*, 1990; Dubertret *et al.*, 1990a; Morlière *et al.*, 1990; Young *et al.*, 1990), can be attributed to the ability of 5-MOP to form stable adducts with DNA in the presence of UV radiation (see following pages for further details). On the other hand, bergamottine, which is the major UVB plus UVA radiation absorber in the bergamot oil, is rapidly degraded following UV radiation exposure. This photodegradation and its consequent phototoxicity, can be inhibited by the addition, in cosmetic preparations, of opportune concentrations of sunscreens (such as Parsol MCX®, a UVB sunscreen of the cinnamate family) (Dubertret et al., 1990b; Morlière et al., 1990; Moysan et al., 1993; Treffel et al., 1991). However, contrary to many furanocoumarins, bergamottine does not strongly interact with DNA (Aubin et al., 1991; Morlière et al., 1991).

In apparent contradiction to the above, in a study on healthy volunteers, Potten et al. (1993) have demonstrated that 5-MOP added to a lotion containing Parsol MCX® has a greater protective effect against DNA damage induced in cutaneous cells by exposure to radiation from a solar simulator (UVA+UVB). So far, no regulation has been established by the European Union relative to the use of 5-MOP, although in the United Kingdom the inclusion of 5-MOP in cosmetic sun-tanning preparations is already, though not officially, banned. The IFRA (International Fragrance Association) recommends the use of bergamot oil in concentrations not higher than 0.4 per cent in preparations to be applied on the skin exposed to solar radiation (except for preparations that can be washed off the skin, such as soaps). In accordance with IFRA guidelines, most perfume producers limit the 5-MOP content to 75 ppm and use distilled bergamot oil from which 5-MOP has been removed. Also 8-methoxypsoralen (8-MOP) is banned in cosmetic products unless as a natural component of the essential oils. A recent paper by Clark and Wilkinson (1998) points out the possibility of insurgence of phototoxic contact dermatitis caused by 5-MOP contained in bergamot oil used in aromatherapy.

The peculiar photosensibilizing properties of 5-MOP also provide this psoralen with a phototherapeutic activity, which is being fully exploited in the so-called PUVA therapy (psoralen plus UVA). The administration (topic or systemic) of a psoralen (5-MOP, 8-MOP or 4,5',8-trimethoxypsoralen, TMP) together with UVA irradiation has been clearly demonstrated to be therapeutically efficacious in the treatment of two radically different pathological conditions, psoriasis and vitiligo. Detailed papers as to the aspects relative to pharmacodynamic and pharmacokinetic properties, tolerability and dosage schedules of the psoralens in PUVA therapy are widely present in international scientific literature (McNeely and Goa, 1998). Vitiligo is characterized by the spontaneous appearance, often progressive and persistent, of depigmented cutaneous patches; this pathology is due to an alteration in melanocyte functionality and thus in melanin production; besides it seems that the T-lymphocytes might be involved in melanocyte destruction characterizing this pathology, by a sort of auto-immune mechanism. Psoriasis, this too a chronic pathological disease condition, is characterized by raised, scaly, reddened cutaneous plaques, resulting from hyperproliferation of the epidermis and inflammation of both epidermal and dermal layers. PUVA therapy has not only given good results in the topical treatment of vitiligo, but, when orally administered, it is considered the systemic treatment of choice for both vitiligo and psoriasis. The rationale for the use of PUVA therapy in the treatment of vitiligo and psoriasis is rather complex. The photoactivated molecules of the psoralen form mono- or bi-functional adducts (cross-links) between their 4',5' and/or 3,4 double bonds and the pyrimidine bases (thymine and cytosine) of DNA strands (Figure 28.1). These cross-links are thought to prevent DNA replication and thus the hyperproliferative cutaneous state, a condition peculiar to psoriasis. Furthermore the photoactivation of the psoralen causes the generation of reactive oxygen species, such as superoxide; this stimulates melanocyte proliferation and increases pigmentation, certainly fundamental factors for the beneficial effect of PUVA therapy in the treatment of vitiligo. Besides, the effects of UV exposure on cutaneous cytokine release may be enhanced by 5-MOP. It is evident that the efficacy of this kind of combined therapy stems from a well-studied risk to benefit ratio. In particular, the efficacy of PUVA therapy is based on the capacity of psoralen to increase skin sensitivity to the beneficial effects of the UV light (pigmentation and hypoproliferation, due to photoexcited psoralen molecules), and to provoke simultaneously a photoadaptation (pigmentation), thus protecting the skin from the toxic effects of UV radiation (erythema, edema, exfoliation). The effects of PUVA therapy on the enzymes regulating melanogenesis are not completely clear, even though PUVA therapy-induced melanogenesis seems to be consequent to an increased expression of tyrosinase (Mengeaud and Ortonne, 1996). Besides, it seems that the repigmentation of vitiligo lesions following PUVA therapy results from the migration of melanocytes from hair follicles, a migration stimulated by cytokines released in the skin following UV irradiation. Moreover, PUVA therapy brings about a destruction of T-lymphocytes, resulting in a reduction of melanocyte destruction and thus in an inhibition of the further development of new vitiligo lesions. On the other hand, photoprotection (consisting in an increase in skin thickness and melanin production) can contribute to the protection of the skin, especially of the pigmented areas, from the toxic effects of UV radiation during exposure. As regards psoriasis treatment, the hypoproliferative effect on the epidermal cells, due to the photoexcitation of psoralen molecules, is certainly fundamental, but does not cause erythema, edema or exfoliation. Also in this case, increase in melanine production can contribute to the protection of the skin from the effects of prolonged exposure to UV radiation. However, care must be taken not to induce an excessive production of melanine, that might interfere with UV penetration through the skin and thus reduce the efficacy of the therapeutic treatment.

In oral PUVA therapy, the standard dose of 5-MOP is 1.2 mg/kg, to be taken 2 hours before exposure to UV radiation. It is generally necessary to complete 20 sessions at a rate of 2–4 sessions a week. Duration of UV irradiation should be carefully controlled. Exposure to solar radiation and administration of photosensitising drugs, such as sulphonamides, phenothiazines and tetracyclines, should be carefully avoided during the treatment. Contraindications for the use of 5-MOP include skin diseases aggravated by sunlight (lupus erythematosus, epithelioma and melanoma), the presence of precancerous skin lesions (because of the presumed carcinogenic potential of PUVA therapy) and previous treatments with arsenic compounds. Side-effects noticeable during short-term PUVA therapy may be gastrointestinal (due to the psoralen) or cutaneous (because of UVA radiation). In this kind of combined therapy, oral administration of 5-MOP is usually preferred to that of 8-MOP because of a marked lesser incidence and severity of side-effects. However one must keep in mind that 5-MOP-treated patients generally need a higher total exposure to UVA radiation than those treated with 8-MOP. In fact, the minimum mean phototoxic dose was significantly higher (30–50 per cent within individuals) for 5-MOP (1.2 mg/kg, per os) compared to that of 8-MOP (0.6 mg/kg). Moreover, nothing is yet known as regards 5-MOP long-term tolerability. In particular, skin cancer risk following PUVA treatment has been demonstrated for 8-MOP and, thus, given the similar mechanism of action, it is possible (also if not ascertained) for 5-MOP; this fact imposes great caution in the application of these therapies. However the risk can also be reduced by a careful selection of the patients and by taking precautions (for instance, use of sunscreens or proper clothing during exposure to sunlight).

Besides other factors, the physico-chemical properties of each psoralen determine its choice. For example, TMP shows a weaker skin penetration (consequent to its lesser lipophilicity) with respect to 5-MOP and 8-MOP (Said et al., 1997). Moreover, significant concentrations of 5-MOP and 8-MOP, but not of TMP, have been found in the skin of guinea pigs after oral or intraperitoneal administration (Park et al., 1993). As a consequence, TMP seems to be the best-suited for topical application; on the other hand, because of the relationship between lipophilia and skin permeation, in generalized skin disorders only 5-MOP and 8-MOP can be administered either orally or topically.

Finally, it is interesting to note that the use of bergamot essential oil is recommended to alleviate stress and depression and to induce sleep. Oral administration of 5-MOP at a dose of 40 mg/day for 7 days (without UV radiation) seems to cause an improvement in the condition of patients with depressive disorders. If confirmed, this would be of further benefit in the treatment of patients affected by psoriasis or vitiligo, that, often involving an emotional aspect, potentially develop stress and insomnia.

ANTITUMOR ACTIVITY OF THE ACTIVE PRINCIPLES OF CITRUS ESSENTIAL OILS

The ever-increasing incidence of the tumoral disease in the world population indicates the pressing urgency to develop not only more advanced diagnostic and therapeutic strategies, but also new efficacious approaches for the prevention of cancer.

It is well-known that various environmental factors can play a role in the development of tumors; but it is unimaginable that it could be possible to eliminate or remove all carcinogenic agents from the environment, whether they be mutagenic agents or promoters. Chemoprevention is, in fact, one of the new strategies developed for cancer prevention. Chemoprevention means the prevention of cancer by means of a chemical approach, that is by the administration of a chemical non-toxic compound during the carcinogenic process. It is interesting to note that many chemopreventive agents, known or being studied today, are of natural origin and often found in vegetable foods widely present in the human diet.

Monoterpenes are non-nutritive components of the diet largely present in the essential oils from *Citrus* and other plants. A certain number of monoterpenes have a chemopreventive or even chemotherapeutic activity, in that they prevent carcinogenesis both during the initial stage and the promotion/progress stage and can be efficacious even in the treatment of both early and advanced cancers. *d*-Limonene, the main component of orange and lemon essential oils, and other monoterpenes, such as perillyl alcohol, are well-known as having a significant antitumor activity (Crowell, 1997, 1999; Gould, 1997; Hohl, 1996).

The present paragraph reviews not only the antitumor activity of *d*-limonene, but also that of perillyl alcohol and perillic acid, compounds derived from *d*-limonene metabolism. However, though not a component of citrus essential oils, perillyl alcohol can be found in nature in the essential oils of lavender, mint, cherry and celery.

In vivo studies have shown that *d*-limonene and correlated compounds (perillyl alcohol and perillic acid) are able to prevent and treat various types of cancer, such as breast,

Table 28.1 Antitumor activity of d-limonene in some *in vivo* experimental models reported in papers published in the last ten years

Carcinogen	Species	Organ	References
NMU	Rat	Mammary	Chander *et al.*, 1994
DBA	Rat	Mammary	Haag and Gould, 1994
DEN; DEN/PB	Mouse	Liver	Giri *et al.*, 1999
DEN+GLU-P-1; DEN+DMN	Rat	Liver	Hirose *et al.*, 1995
AZM	Rat	Colon	Kawamori *et al.*, 1996
DEN+NMU+BBN+DMH+DHPN	Rat	Multiorgan	Kimura *et al.*, 1996
NNK	Mouse	Lung, liver	Morse and Toburen, 1996
BOP	Hamster	Pancreas	Nakaizumi *et al.*, 1997
MNNG	Rat	Stomach	Uedo *et al.*, 1999
MNNG+NaCl	Rat	Stomach	Yano *et al.*, 1999
NNK	Mouse	Lung, forestomach	Wattenberg and Coccia, 1991

Notes
AZM azoxymethane; BBN N-butyl-N-(4-hydroxybutyl)nitrosamine; BOP N-nitrosobis(2-oxopropyl)amine; DBA 7,12-dimethylbenz[a]anthracene; DEN N-diethylnitrosamine; DHPN dihydroxy-di-N-propylnitrosamine; DMH 1,2-dimethylhydrazine; DMN N-dimethylnitrosamine; GLU-P-1 2-amino-6-methyldipyridol[1,2-*a*:3',2'-*d*]imidazole; MNNG N-methyl-N'-nitro-N-nitrosoguanidine; NaCl sodium chloride; NMU nitrosomethylurea; NNK 4-(methylnitrosamino)-1-(3-pyridyl)-1-butanone; PB phenobarbital.

liver, lung and pancreas cancer (Table 28.1); *in vitro* data suggest that these compounds can be efficacious even in the treatment of neuroblastoma and leukemia.

The antitumor efficacy of d-limonene and perillyl alcohol are ascribable to various cellular and molecular mechanisms. First of all, the chemopreventive effects of d-limonene and perillyl alcohol during the initiation phase of carcinogenesis seem due to the inhibition of Phase I enzymes (that activate carcinogenic agents) or to the induction of Phase II enzymes (that metabolize the carcinogens, increasing their detoxification) (Figure 28.2). Heterocyclic amines formed during combustion processes represent one of the most important groups of environmental carcinogens. These amines generally need metabolic activation by the hepatic microsomal enzymes and successive O-acetylation or O-sulfation before being able to form adducts in their target organs. d-Limonene has proved to strongly inhibit the formation of breast tumours induced by 7,12-dimethylbenz[a]anthracene in the rat (Haag *et al.*, 1992), and of skin cancer induced by benzo[a]pyrene in the mouse (Van Durren and Goldschmitd, 1976). Besides, lemon and orange essential oils and d-limonene inhibit, in female A/J mice, the development of lung or forestomach neoplasia induced by 4-(methylnitrosamino)-1-(3-pyridyl)-1-butanone (NNK), a tobacco-specific carcinogen (Wattenberg and Coccia, 1991); d-limonene has proved to be efficacious when administered, orally or intraperitoneally, at a short time interval before carcinogen challenge, thus suggesting a direct inhibitory effect of d-limonene on NNK activation. Successively, by means of *in vitro* and *in vivo* experiments, Morse and Toburen (1996) have shown that d-limonene and, more markedly, perillyl alcohol inhibits cytochrome P-450-mediated metabolic activation of NNK, and, in particular, the formation, by α-hydroxylation, of two metabolites, 4-hydroxy-1-(3-pyridyl)-1-butanone and 4-oxo-4-(3-pyridyl)butyric acid, which give rise to formation of methylating and pyridyloxobutylating species that can alkylate DNA and thus result in neoplastic initiation. In accordance with these data, d-limonene has manifested a good protective effect against hepatocarcinogenesis induced by dimethylnitrosamine or 2-amino-6-methyldipyrido[1,2-*a*:3',2'-d]imidazole in F344 male rats initiated with *N*-diethylnitrosamine (Hirose *et al.*, 1995).

An inverse correlation seems to exist in man between colon cancer risk and content, in the gastrointestinal tissue, of glutathione-S-transferase (GST), a family of detoxifying enzymes of which several isoforms are known (α, μ, π and θ). *d*-Limonene supplemented in the diet of male Wistar rats increased hepatic GSTα and colonic GSTπ levels (van Lieshout *et al.*, 1998a). Besides, *d*-limonene also increased the activity of both selenium-dependent and non selenium-dependent glutathione peroxidase (GPx) in the oesophagus and of nonselenium-dependent GPx in the colon (van Lieshout *et al.*, 1998b). This increase in GST and GPx levels can result in a greater detoxification, thus contributing to the explanation of *d*-limonene antitumor activity. In agreement with these results, a *d*-limonene-supplemented diet in the mouse prevented liver glutathione depletion following a prolonged paracetamol challenge (Reicks *and* Crankshaw, 1993).

Besides acting during the initiation phase (by modifying the metabolism of the cancerogens), monoterpenes also possess chemopreventive and chemotherapeutic effects in the tumoral promotion/progression phase. These effects are consequent to multiple and as yet not-well-understood molecular mechanisms (induction of cellular apoptosis, tumor redifferentiation, inhibition of post-translational isoprenylation of proteins regulating cellular growth).

d-Limonene prevents the development of azoxymethane-induced aberrant crypt foci in the colon of F344 male rats (Kawamori *et al.*, 1996). Reddy *et al.* (1997) similarly demonstrated also for perillyl alcohol a chemoprotective effect against azoxymethane-induced colon cancer in F344 rats. Azoxymethane is metabolized in the liver to methylazoxymethanol, which is then further metabolized in the liver and colon to give a highly reactive electrophile compound, methyldiazonium ion, that can methylate cellular nucleophiles, including DNA. It can thus be supposed that the inhibition of azoxymethane-related carcinogenesis may be due to a modulation of the carcinogen metabolism. At any rate, in these studies the protective effect of perillyl alcohol seemed to be mediated by the loss of tumoral cells by means of apoptosis, while that of *d*-limonene seemed to be consequent to a diminished cellular proliferation. Instead, the protective effect of *d*-limonene against gastric carcinogenesis induced, in Wistar rats, by *N*-methyl-*N*'-nitro-*N*-nitrosoguanidine and enhanced by sodium chloride, is mediated by an increase of apoptosis and a decrease in DNA synthesis in cancer cells (Uedo *et al.*, 1999; Yano *et al.*, 1999).

Also as regards the protective effect of *d*-limonene on *N*-nitrosobis(2-oxopropyl)amine-induced pancreatic cancerogenesis in the hamster, Nakaizumi *et al.* (1997) evidenziated that *d*-limonene inhibits the development of pancreatic carcinomas not by enhancing tumoral cell loss through apoptosis, but rather by inhibiting cellular proliferation through inhibition of *ras*-p21 oncoprotein isoprenylation. In fact, *d*-limonene and the correlated monoterpenes can selectively inhibit the isoprenylation of certain proteins, a post-translational modification in which an isoprenic group is covalently attached to the carboxylic termination. This inhibiting effect of *d*-limonene regards mostly proteins having a molecular weight of 20,000–26,000, which are small G-proteins like those belonging to the *ras*-21 family. This can bring about alterations in the transduction of the signal and result in an altered genic expression. For example, treatment with *d*-limonene or perillyl alcohol is accompanied by an increased expression of the mannose-6-phosphate/insulin-like growth factor II receptor, together with an increase in the production and activation of the powerful cytostatic factor TGF-β1 (transforming growth factor β1) (Belanger, 1998; Gould, 1997). Stayrook et al. (1998) have shown that, in pancreatic tumor cells, treatment with perillyl alcohol inhibits the prenylation of H-*ras* but not K-*ras* proteins. Besides, the antitumor effect of *d*-limonene on N-nitrosodiethylamine- or N-nitrosodiethylamine/phenobarbital-induced hepatocarcinogenesis in AKR rats is correlated to the inhibition of the over-expression of the oncoproteins *c-jun* and *c-myc* (Giri et al., 1999). Finally, Wistar–Furth female rats chronically fed with perillyl alcohol showed inhibition, in *in situ* mammary gland epithelial cells, not only of coenzyme Q synthesis, but also of prenylation of some substrates (having transforming properties) of type I geranyl-geranyl protein transferase, such as RhoA (Ren and Gould, 1998).

However the molecular mechanisms by which the monoterpenes repress protein prenylation are not as yet clarified, and can also be considered as completely distinct from the inhibition of HMG-CoA (3-hydroxy-3-methylglutaryl coenzyme A) reductase or of Ftasi (farnesyl-transferase) (Hohl, 1996). For example, perillic acid selectively inhibits the production of interleukin-2 and interleukin-10 in human mitogen-activated T lymphocytes by depleting cellular G-protein content (Schulz et al., 1997). Moreover, treatment with perillyl alcohol protects against the carcinogenic effects of UVB light as demonstrated both *in vivo* in transgenic mice and *in vitro* in cultured human keratinocytes; this protective effect is consequent to the inhibition of UVB light-induced AP-1 transactivation (Barthelman et al., 1998). Besides, in the colonic adenocarcinoma SW480 PA cell line, perillyl alcohol has proved to alter early phases of cholesterol synthesis with a mechanism independent of an effect on HMG-CoA reductase (Cerda et al., 1999). Finally, Bardon et al. (1998) have recently demonstrated that *d*-limonene, perillyl alcohol and perillic acid (order of potency: perillyl alcohol > perillic acid > *d*-limonene) inhibit the proliferation of some mammary tumoral cell lines, the hormone-dependent T-47D and MCF7 cells and the hormone-independent MDA-MB-231 cells. This antiproliferative activity is correlated to an accumulation of cells in the G_1 phase of the cell cycle and a fall in the proportion of cells in the S phase. This G_1-S arrest is preceeded by a decrease in cyclin D1 mRNA levels. Further studies are necessary to clarify whether the sensibility of these cell lines is correlated to the arrangement of their hormonal receptors or to other genetic characteristics. Moreover, the greater efficacy (reported in many papers present in literature) of perillyl alcohol and perillic acid compared to that of *d*-limonene leads to the hypothesis that the *d*-limonene metabolites significantly contribute to its *in vivo* activity (Hardcastle et al., 1999). On the other hand, the greater *in vivo* antitumor activity of perillyl alcohol with respect to *d*-limonene can, at least partly, lead back to differences in the pharmacokinetic characteristics of these two monoterpenes (Haag and Gould, 1994). Finally, a recent study by Kimura et al. (1996) did not establish any chemopreventive effect of *d*-limonene in a multiorgan carcinogenesis model in the rat; this shows that monoterpene mechanisms of action and/or effects can differ depending on the animal species, the cancerogen used and the organ/tissue observed.

At the moment, *d*-limonene and perillyl alcohol are being studied in phase I clinical trials in patients with advanced cancer. Unfortunately, the results obtained with the therapeutic protocols used so far have not been not particularly gratifying, at least as regards perillyl alcohol (Belanger, 1998; Ripple et al., 1998), also because a significative incidence of side-effects, in particular gastrointestinal ones, has been observed.

As regards d-limonene, apart from the phase I clinical trial, a limited phase II evaluation in breast cancer has also been carried out (Vigushin et al., 1998); the results seem quite encouraging, also because d-limonene appears to have been well-tolerated at doses that can have a clinical effect. In fact, chemotherapeutic activity and toxicity of d-limonene have been evaluated in patients with advanced cancer, at preclinical level in a phase I study (to determine the maximum tolerated dose, MDT) and in a limited phase II study (in breast cancer patients). A group of 32 patients with solid refractory tumors was treated with d-limonene, administered orally in doses between 0.5 and $12\,g/m^2$ a day ($8\,g/m^2$ a day in the case of patients included in the phase II study) in 21-day cycles. d-Limonene toxicity seemed limited to gastrointestinal side-effects, which, however, were never particularly severe, neither were serious signs of toxicity observed in other organs or systems. The MDT was of $8.8\,g/m^2$ a day. In 3 patients with colorectal carcinoma treated with $0.5\,g/m^2$ a day, the disease remained stable for more than 6 months (between 7 and 12 months). In one subject with breast cancer, a partial positive result was observed for about 11 months. Unfortunately no positive result was observed in the successive phase II study.

Some studies are still in progress to evaluate whether the antitumor efficacy of d-limonene can be potentiated (and/or its toxicity lessened when used chronically at high doses) by association with other drugs (e.g. 4-hydroxyandrostenedione, flavonoids, tocotrienols, mevinolin, etc.) having different mechanisms of action (Broitman et al., 1996; Carroll et al., 1997; Chander et al., 1994).

Naturally, once the mechanisms of action of citrus monoterpenes are known, interesting perspectives may open up as regards their use, also in other than chemotherapeutic field. For instance, the addition of d-limonene (or of extracts of C. sudachi and C. junos flavedo) decreases the mutagenicity of 1-methyl-1,2,3,4-tetrahydro-β-carboline-3-carboxylic acid, a compound present in soy sauce, whose mutagenic activity is strongly enhanced in the presence of nitrite and ethanol (Higashimoto et al., 1998). Moreover, the modulation of G-proteins and the consequent antiproliferative activity of perillyl alcohol can be useful in controlling the development of intima hyperplasia in vein grafts, as observed after oral administration of perillyl alcohol in New Zealand rabbits (Fulton et al., 1997).

Researches focused on the individualization of new plant compounds endowed with chemopreventive activities have pointed out the antitumor efficacy of auraptene, a coumarin compound, and of tangeretin, a polymethoxyflavone.

Auraptene shows spasmolytic (Yamada et al., 1997) and anti-platelet aggregation action (Chen et al., 1995; Teng et al., 1992); it has recently been proved to be a chemopreventive agent, active against cancerogenesis in various organs (particularly colon, oral cavity and skin) both in the initiation and post-initiation phases. Auraptene has prevented, in male F344 rats, the development of azoxymethane-induced aberrant crypt foci (lesions considered biomarkers of colon cancer in rodents and in man) (Tanaka et al., 1997, 1998a); this inhibitory effect seems to be due to an increase, in the liver and colon, in the activity of enzymes involved in phase II metabolic reactions (in particular, glutathione S-transferase and quinone reductase), and the inhibition of cellular proliferation and lipid peroxidation in the colon mucosa. Similar mechanisms (activation of enzymes involved in the detoxification of cancerogens and inhibition of increased cellular proliferation) have also been demonstrated as regards prevention of 4-nitroquinoline 1-oxide-induced oral carcinogenesis in F344 rats (Tanaka et al., 1998b). Furthermore, auraptene showed a good chemopreventive effect in a two-stage carcinogenesis model with topical application of 7,12-dimethylbenz[a]anthracene and of 12-O-tetradecanoylphorbol-13-acetate on the skin of ICR mice (Murakami et al., 1997); the mechanism by which auraptene seems to carry out its chemopreventive activity in this experimental model is the suppression of leucocyte activation, since auraptene suppresses 12-O-tetradecanoylphorbol-13-acetate-induced superoxide generation in differentiated human promyelocytic HL-60 cells.

Tangeretin (5,6,7,8,4'-pentamethoxyflavone) is present in numerous citrus. Tangerine peel is widely used in Japan in Kampo medicine in the treatment of tumor patients. Tangeretin inhibits *in vitro* cell growth of promyelocytic human HL-60 leukaemia and efficaciously induces their apoptosis; this effect is significantly attenuated in the presence of Zn^{2+} ions, which are known inhibitors of Ca^{2+}-dependent endonuclease activity (Hirano et al. 1995). Moreover cytotoxicity of tangeretin against HL-60 cells and normal human lymphocytes is extremely low. Furthermore, tangeretin inhibits *in vitro* growth of JCS cells, induces their differentiation into macrophages and polymorphonucleates and decreases their tumorigenicity *in vivo* (Mak et al., 1996). These data indicate that tangeretin could be a potential candidate for the treatment of various forms of myeloide leukaemia, without, moreover, having serious side effects on immune system cells.

According to Kandaswami et al. (1991), the greater antiproliferative efficacy of tangeretin and of other methoxyflavonoids, compared with the corresponding non-methoxylated flavonoids, observed *in vitro* on a human squamous cell carcinoma cell line, is due to a relatively higher cellular uptake consequent to the lesser hydrophilicity of the methoxylated compounds. Besides being correlated to modifications in the expression of some cytochrome P450 isoforms with a transcriptional or/and post-transcriptional mechanism (Canivenc-Lavier et al., 1996), tangeretin activity also seems related to a modulation of intercellular communication. For example, in rat liver epithelial cells, tangeretin antagonizes the tumor promoter 12-O-tetradecanoylphorbol-13-acetate induced inhibition of gap junctional intercellular communication (Chaumontet et al., 1994, 1997); there is, however, a certain discrepancy in the data obtained *in vivo* in a rat liver short-term carcinogenesis (Chaumontet et al., 1996). Moreover, tangeretin, as also tamoxiphen, have proved to be able to induce an upregulation of the functions of the E-cadherin/catenin invasion-suppressor complex (Vermeulen et al., 1996); furthermore, tangeretin and tamoxiphen together have shown *in vitro* additive effects as regards inhibition of both growth and invasive properties of human mammary cancer cells. Unfortunately, *in vivo* experiments on nude mice in which human MCF-7/6 mammary adenocarcinoma cells were inoculated, showed that tangeretin treatment not only does not inhibit tumoral growth, but also completely neutralizes the inhibitory effects of tamoxiphen (Bracke et al., 1999); in fact, *in vitro* tangeretin inhibits the cytolytic effects of murin natural killer cells on MCF-7/6 cells.

d-LIMONENE AS ENHANCER OF PERCUTANEOUS ABSORPTION

Transdermal administration and transdermal drug delivery systems are becoming more popular, and various drugs are administered successfully in this way for both systemic and local effects. The main advantage of this route is that it avoids the hepatic first pass effect; it is also to be recommended in some drugs in order to avoid problematic side

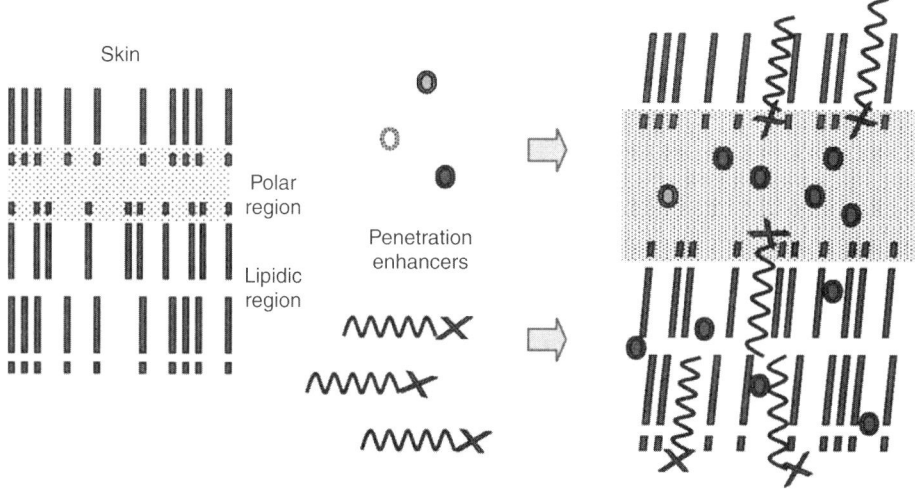

Figure 28.3 Postulated mechanism of action for hydrophilic (small circles) and hydrophobic (undulate lines) penetration enhancers.

effects such as in the case of aspirin and similar drugs (McAdam *et al.*, 1996). However transcutaneous delivery is heavily limited by the permeation characteristics of the stratum corneum (SC) and for many drugs it is insufficient to obtain and maintain efficacious systemic levels.

Several methods are being studied to increase drug penetration through biological membranes. For a certain number of drugs this goal can be achieved by the so-called *penetration enhancers*. An ideal *enhancer* should be pharmacologically inactive, non-irritant and should not provoke irreversible alterations in the SC (Figure 28.3).

Much interest is at the moment focussed on the use, as *penetration enhancers*, of molecules of natural origin, such as d-limonene and, in general, terpenes. d-Limonene, in fact, responds to the above-mentioned requisites, that is, low potential in producing skin irritation, induction of reversible SC alterations and efficacy as enhancer even in low concentrations.

d-Limonene is capable of increasing transdermal permeation of 5-fluorouracil (Yamane *et al.*, 1995), tamoxiphen (Zhao *and* Singh, 1998), aspirin (McAdam *et al.*, 1996), indomethacin (Okabe *et al.*, 1989), diclofenac sodium (Arellano *et al.*, 1996), ziduvudine and haloperidol (Almirall *et al.*, 1996). It has proved to be very efficacious in increasing transport of lipophilic molecules such as indomethacin and diclofenac sodium, whereas it is less effective in the transport of propanolol (a prototype of hydrophilic molecules). According to Kunta *et al.* (1997), this is due to the fact that d-limonene is a hydrophobic terpene lacking hydroxylic groups and thus unable to form hydrogen bonds. Other authors, however, give discordant results. For example, Almirall *et al.* (1996) have observed that the percutaneous transport of haloperidol, but not that of chlorpromazine, is increased by d-limonene, although chlorpromazine is more lipophilic than haloperidol. Moreover, in combination with propylene glycol, d-limonene increases permeation, through mouse skin, of caffeine (a hydrophilic model drug) and hydrocortisone (a polar steroid), but not of triamcinolone acetonide (Godwin *and* Mickniak, 1999).

The mechanism of penetration enhancers is generally due to an increased drug diffusivity through the skin, consequent to an alteration of intercellular lipids and/or intracellular proteins and to an increased partitioning of the drug in the SC (Barry, 1988, 1991). Most probably, the enhancer effect of *d*-limonene is mostly due to a reduction of skin resistance to drug permeation and a consequent increase in the drug diffusion coefficient (Cornwell *et al.*, 1996; Yamane *et al.*, 1995; Zhao and Singh, 1998). Lately, Moghimi *et al.* (1996a,b, 1997) have investigated the mechanism of *d*-limonene by using a lamellar matrix that mimes the structural and barrier properties of the SC lipids. Increasing concentrations of *d*-limonene initially cause a decrease in the lamellar structure, followed by a lamellar-to-cubic phase transition and, lastly, by a complete change of the matrix to a dispersed system of crystalline and mesoform phases in a continuous liquid phase. The same experimental model has helped to clarify that the capability of *d*-limonene in increasing 5-fluorouracil skin permeation is, at least partly, due to interaction with the SC proteins (Moghimi *et al.*, 1996b, 1997). It has recently been observed that a 5 per cent *d*-limonene/ethanol solution plus iontophoresis enhances the permeation of the luteinizing hormone releasing hormone (LHRH) through human skin, because it causes a disruption of the keratin pattern and a swelling of SC cell layers (Bhatia and Singh, 1999).

MISCELLANEA

Other biological activities of bergamottine

The non-volatile total residue (NVTR) of bergamot oil (*C. bergamia* Risso) possesses significant cardiovascular properties, demonstrated both in *in vivo* (pitressin-induced coronary spasm, arrythmias and pressure alterations, oubaine-induced ventricular arrythmias in the guinea-pig), and *in vitro* (isolated and perfused rat heart) models with a mechanism similar to that of verapamil (Occhiuto and Circosta, 1996a). These effects of the bergamot oil NVTR seem ascribable to bergamottine (5-geranyloxypsoralen) which, in the same above-mentioned models, manifested significant antiangina and antiarrythmic properties (Occhiuto and Circosta, 1996b). It is probable that the mechanism involved in the cardiovascular properties of this furanocoumarin might be an alteration in calcium metabolism. Its calcium-antagonist properties indicate that bergamottine has the characteristics of anti-arrythmic drug of class IV (inhibitor of calcium transport) (Occhiuto and Circosta, 1997).

Occhiuto *et al.* (1995) have also studied, in the mouse, the muscle relaxant and analgesic activity of the bergamot oil NVTR and its effects on gross behaviour, pentobarbital-induced sleep time and pentylentetrazol-induced seizures. The NVTR has proved to possess important depressive effects at the neurosomatic level of the CNS, because it presents sedative, analgesic and anticonvulsant properties together with low systemic toxicity (LD_{50} in the mouse after intraperitoneal administration: >1.5 g/Kg). These depressive effects on the CNS make up an interesting scientific background that justifies the employment of bergamot essential oil as a sedative in traditional medicine.

In vivo studies in the mouse and *in vitro* studies in mouse keratinocytes have shown that bergamottine blocks the formation of DNA adducts and tumor initiation by polycyclic aromatic hydrocarbons, such as benz[a]pyrene and 7,12-dimethylbenz[a]anthracene, probably through the inhibition of specific cytochromes P450 involved in the metabolic activation of these hydrocarbons (Cai *et al.*, 1997a,b).

In recent years, dietary uptake of grapefruit juice has been proved to enhance oral bioavailability of various drugs metabolized by the cytochrome P450 3A4, both by inhibiting the activity and by diminishing the intestinal content of this cytochrome. Bergamottine, 6',7'-dihydroxybergamottine and a closely correlated dimer seem to be responsible for this activity; however bergamottine needs to be metabolized in order to carry out its inactivating effect (He *et al.*, 1998; Schmiedlin-Ren *et al.*, 1997). It is probable that bergamottine might be a mechanism-based inactivator of the cytochrome P450 3A4, involving a modification of apoP450 in the active site of the enzyme rather than heme adduct formation or heme fragmentation. Similarly, other furanocoumarins, such as 5-methoxypsoralen and 8-methoxypsoralen, are mechanism-based inhibitors of some cytochrome P450 isoforms (Koenigs *and* Trager, 1998a,b).

Antiplatelet properties of auraptene

The antiplatelet actions of auraptene have recently been demonstrated by Teng *et al.* (1992) and Chen *et al.* (1995). These studies seem to show that, *in vitro*, this coumarin induces an antiaggregating effect by inhibiting thromboxane A2 formation and phosphoinositide breakdown.

TOXICOLOGY

Contact dermatitis and citrus essential oils

Citrus essential oils are known to have been responsible for the appearance of various kinds of contact dermatitis (irritation, delayed allergic eczematous reactions, immediate urticaria and vesicular eruptions, phototoxicity) in man. Exposure to citrus fruits is certainly greater in certain professional categories (workers responsible for cultivation, transformation or sale) even though the problem of possible cutaneous reaction is surely made more complex by the contemporaneous presence of other substances that can act as allergens or irritants (for example, pesticide residues and preservatives). Moreover, because of its fragrance, *d*-limonene is added in low concentrations to some household products and, as it acts as a solvent, it is used as an additive in cleaning products; particularly, because of its much less harmful environmental impact, *d*-limonene is being put to new uses to substitute other certainly more toxic chemical compounds (such as chlorinated hydrocarbons, chlorofluorocarbons and various organic solvents) for defatting and industrial cleaning. Dipentene, the racemic mixture of *d*-(+)-limonene and of *l*-(-)-limonene, is used as a solvent in the painting industry. Finally, *d*-limonene is employed for topical application as *penetration enhancer* in pharmaceutical and cosmetic products.

d-Limonene itself does not seem to cause cutaneous sensibilization. However some of its oxidation products, formed when *d*-limonene is exposed to air and daylight, have been shown to be strong allergens. These oxidation products are *cis*- and *trans*-limonene-1,2-oxide, *cis*- and *trans*-carveol and some limonene hydroperoxides (Nilsson *et al.*, 1996). Except for carveol, these compounds cause evident skin sensitization reactions in guinea-pigs (Kalberg et al., 1992). Results of studies on man, too, seem to show that d-limonene should be classified as allergenic because of the formation of oxidation products with allergenic properties (Kalberg and Dooms-Goossens, 1997; Kalberg et al., 1999). In any case, the allergenicity rate of d-limonene depends on the time of exposure to air, seeing that this also regulates the relationship between the various oxidation products formed. Moreover, as the formation of these oxidation products brings about a change in the technical properties of the solvent, antioxidants, such as 2,6-di-t-butyl-4-methylphenol, are often added to commercial products.

Finally it must be pointed out that some cases of allergic contact dermatitis have been recognized as due not to d-limonene, but to other minor components of citrus essential oils, such as citral and geraniol (Cardullo et al., 1989). It has lately been demonstrated that citral itself is one of the most unstable components of lemon essential oil to UV light exposure (Iwanami et al., 1997); nothing, however, is known as to the toxicity of the products derived from citral photolysis.

The nephrotoxicity of d-limonene

The observation that d-limonene may be responsible for the appearance of renal adenomas and adenocarcinomas in male rats goes back to the beginning of the 90s. Although it is now certain that d-limonene does not cause cancer risk in man, these studies are cited here because they are an interesting example of a compound whose carcinogenic activity is strictly species- and sex-specific (Hard, 1998; Jonker et al., 1996; Rasonyi et al., 1999; Whysner and Williams, 1996). d-Limonene belongs to a group of chemical compounds denominated CIGA (*Chemicals Inducing* α_{2u}-Globulin Accumulation). α_{2u}-Globulin is a low molecular weight protein, found in significant quantities in male rat urine, but in much lower quantities (about 100 times less) in female rat urine and is not present in human urine. Exposure of male rats to d-limonene results in the production of hyaline-droplet formation, restricted to the P_2 segment of the proximal convoluted tubules. This nephropathy is characterized by necrosis and reactive proliferation of the tubular cells with tubule dilatation; moderate quantities of α_{2u}-globulin released by exfoliated tubular cells are present in urine. d-Limonene and one of its metabolites, d-limonene-1,2-oxide, bind to this protein, inhibiting its degradation by lysosomal enzymes in the epithelial cells of the P_2 segment of the proximal convoluted tubules. This inhibition is responsible for the development of hyaline-droplet nephropathy.

Toxicity of citral

Citral (3,7-dimethyl-2,6-octadien-1-al) is a mixture of two geometric isomers, geranial and neral. Citral is widely used in food, cosmetic and detergent industries for its aroma and lemon fragrance. However few data are available as regards its acute and chronic toxicity.

The α,β-unsaturated carbonyl group of citral is perhaps responsible for some morphological and consequently functional membrane alterations. The citral-induced hemolysis of rat erythrocytes has been ascribed to the reactive aldehyde group that might cause peroxidation of the polyunsaturated fatty acids in the cell membrane (Segal and Milo-Goldzweig, 1985; Tamir et al., 1984). Finally, citral can inhibit retinol oxidation to retinoic acid in mouse epidermis after topical application (Connor, 1988). Its α,β unsaturated aldehydic chemical structure has also suggested a potential carcinogenicity and teratogenicity of citral. However urinary metabolites of citral, identified in a study carried out on male F344 rats, seem to arise from metabolic pathways different from nucleophilic addition to the double-bond (Diliberto et al., 1990). Moreover, the fact that it is rapidly eliminated from the organism suggests that, in any case, even repeated exposure to citral cannot bring about its systemic accumulation.

Bergapten and ovarian function

It has recently been observed that dietary administration of bergapten can have negative effects on body growth and fecundity in Wistar rats (Diawara et al., 1997, 1999). In fact, these compounds reduce circulating oestrogen levels and the number of implantation sites and corpora lutea in female rats (coupled with non-treated males). The reduction of the ovarian follicular function and ovulation observed in animals treated with bergapten can be partially explained by an increase in oxidative metabolism and by conjugation of oestrogens.

BIOAVAILABILITY

As regards 5-MOP, a considerable interindividual variation is evident for all the pharmacokinetic parameters studied. Apart from the kind of galenic preparation (micronized or non-micronized formulations), the absorption of 5-MOP administered orally depends upon the time of day when the drug is taken; in fact, higher values of plasmatic C_{max} and of AUC have been determined when the drug is administered in the evening rather than in the morning. C_{max} values are higher also when 5-MOP is given to patients after a standard low-fat meal, rather than to patients fasted for at least 8 hours (Ehrsson et al., 1994). In literature t_{max} values are reported ranging between 0.8 and 5 hours. It is important to know t_{max} value since exposure to UVA radiation should ideally coincide with the time of maximum skin concentration (that temporarily coincides with plasmatic C_{max}) of 5-MOP. Exposure to UVA light should thus take place 2 or 3 hours after having taken the drug. Metabolites of 5-MOP (especially as glucuronic acid conjugates) are already present in the urine 2 hours after drug administration; a small fraction of the drug is eliminated in the urine in an unmodified form. These metabolites are probably formed in the liver or also in the intestinal wall. The existence of a saturable first pass effect has been demonstrated for 8-MOP; given the stuctural likeness, this effect is likely to exist also for 5-MOP. Albumin is the principal carrier of 5-MOP in the serum (Muret et al., 1993). Half-life during elimination ranges between 2.2–3 hours. Nothing is known as regards possible interactions with other drugs or chemical compounds. However, 5-MOP inhibits the metabolism, dependent on CYP1A2, of caffeine (Bendriss et al., 1996).

The pharmacokinetic parameters of d-limonene have been evaluated in cancer patients (Vigushin et al., 1998). d-Limonene is well absorbed after oral administration. The principal circulating metabolites are perillic acid, dihydroperillic acid, limonene-1,2-diol, limonene-8,9-diol, uroterpenol and a perillic acid isomer. The two isomers of perillic acid, dihydroperillic acid and limonene-8,9-diol, are to be found in the urine as glucuronides (Poon et al., 1996). The intratumoral levels of d-limonene and uroterpenol are higher than the corresponding plasma levels. Friedman et al. (1994) report that significantly higher concentrations of limonene are present in the air expired from lungs of patients suffering from hepatic disorders (especially non-cholestatic) compared to normal subjects. However, the analysis of diet habits and the biochemical investigations carried out on these subjects indicates a dietetic origin for these high levels of limonene rather than to alterations in the absorption, metabolism and elimination of limonene. The pharmacokinetic properties of d-limonene have been studied also in Sprague-Dawley rats after oral and intravenous administration of 200 mg/Kg (Chen et al., 1998). After intravenous administration, the blood concentration/time curve presents a biphasic decline (initial $t_{1/2} = 12.4$ min; terminal $t_{1/2} = 280$ min). The plasma/erythrocyte partition is 0.84 and plasma protein binding of d-limonene is about 50 per cent. After oral administration the bioavailability of d-limonene is >40 per cent; also in this case the blood concentration/time curve undergoes a biphasic decline (initial $t_{1/2} = 34$ min; terminal $t_{1/2} = 337$ min).

The pharmacokinetic properties of perillyl alcohol were examined in cancer patients (Ripple et al., 1998). As for d-limonene, the principal circulating metabolites are perillic acid and cis- and trans-dihydroperillic acids, whose plasmatic C_{max} is reached 2–3 hours and 3–5 hours respectively after administration of perillyl alcohol. The half-life of these two metabolites is about 2 hours. Besides perillyl alcohol, both perillic and dihydroperillic acids were identified also in the urine. More recently, perillyl alcohol has been found in the plasma in a non-modified form after intravenous administration in the rat and after oral administration in man (Zhang et al., 1999). Finally, the pharmacokinetic characteristics of perillic acid have been studied in CD2F1 mice after intravenous administration (Ikechukwu Ezennia et al., 1997). Plasma concentrations of perillic acid decline in a monoexponential way. The rather short half-life time (about 16 min) suggests that frequent administration or continued infusion are necessary to maintain efficacious blood concentrations over a prolonged period of time. From the calculated values of apparent distribution volume, one can deduce that perillic acid is concentrated mostly in extravascular tissues.

ACKNOWLEDGMENT

The authors are in debt to Dr. Maria Grazia Laganà for her skilful assistance in the drawing up of this review.

REFERENCES

Alderman, G.G. and Marth, E.H. (1976) Inhibition of growth and aflatoxin production of *Aspergillus parasiticus* by citrus oil. *Z. Lebensm. Unters.-Forsch.*, **160**, 353–358.

Almirall, M., Montana, J., Escribano, E., Obach, R. and Berrozpe, J.D. (1996) Effect of d-limonene, α-pinene and cineole on in vitro transdermal human skin penetration of chlorpromazine and haloperidol. *Arzneim.-Forsch./Drug Res.*, **46**, 676–680.

Arellano, A., Santoyo, S., Martin, C. and Ygartua, P. (1996) Enhancing effect of terpenes on the in-vitro percutaneous-absorption of diclofenac sodium. *Int. J. Pharm.*, **130**, 141–145.

Arimoto, Y., Homma, Y. and Misato, T. (1986a) Studies on citrus melanose and citrus stem-end rot by *Diaporthe citri* (Faw) Wolf. Part 4. Antifungal substance in melanose spot. *Ann. Phytopathol. Soc. Jpn.*, **52**, 39–46.

Arimoto, Y., Homma, Y. and Misato, T. (1986b) Studies on citrus melanose and citrus stem-end rot by *Diaporthe citri* (Faw) Wolf. Part 5. Identification of phytoalexin in melanose spot. *Ann. Phytopathol. Soc. Jpn.*, 52, 620–625.

Asthana, A., Larson, R.A., Markey, K.A. and Tuveson, R.W. (1992) Mechanisms of citral phototoxicity. *Photochem. Photobiol.*, 56, 211–222.

Aubin, F., Humbert, P. and Agache, P. (1994) Effects of a new psoralen, 5-geranoxypsoralen, plus UVA radiation on murine ATPase positive Langerhans cells. *J. Dermatol. Sci.*, 7, 176–184.

Averbeck, D., Dubertret, L., Young, A.R. and Morlière, P. (1990) Genotoxicity of bergapten and bergamot oil in *Saccharomyces cerevisiae*. *J. Photochem. Photobiol. B: Biol.*, 7, 209–229.

Bardon, S., Picard, K. and Martel, P. (1998) Monoterpenes inhibit cell growth, cell cycle progression and cyclin D1 gene expression in human breast cancer cell lines. *Nutr. Cancer*, 32, 1–7.

Barry, B.W. (1988) Action of skin penetration enhancers–the lipid protein partioning theory. *Int. J. Cosmet. Sci.*, 10, 281–293.

Barry, B.W. (1991) Lipid protein-partioning theory of skin penetration enhancement. *J. Control. Release*, 15, 237–248.

Barthelman, M., Chen, W., Gensler, H.L., Huang, C., Dong, Z. and Bowden, G.T. (1998) Inhibitory effects of perillyl alcohol on UVB-induced murine skin cancer and AP-1 transactivation. *Cancer Res.*, 58, 7111–7116.

Belanger, J.T. (1998) Perillyl alcohol: applications in oncology. *Altern. Med. Rev.*, 3, 448–457.

Bendriss, E.K., Bechtel, Y., Humbert, P.H., Paintaud, G., Magnette, J., Agache, P. and Bechtel, P.R. (1996) Inhibition of caffeine metabolism by 5-methoxypsoralen in patients with psoriasis. *Brit. J. Clin. Pharmacol.*, 41, 421–424.

Ben-Yehoshua, S., Rodov, V., Kim, J.J. and Carmeli, S. (1992) Preformed and induced antifungal materials of citrus fruits in relation to the enhancement of decay resistance by heat and ultraviolet treatments. *J. Agric. Food Chem.*, 40, 1217–1221.

Bhatia, K.S. and Singh, J. (1999) Effect of linolenic acid/ethanol or limonene/ethanol and iontophoresis on the in vitro percutaneous absorption of LHRH and ultrastructure of human epidermis. *Int. J. Pharm.*, 180, 235–250.

Bracke, M.E., Depypere, H.T., Van Marck, V.L., Vennekens, K.M., Vanluchene, E., Nuytinck, M., Serreyn, R. and Marrel, M.M. (1999) Influence of tangeretin on tamoxifen's therapeutic benefit in mammary cancer. *J. Natl. Cancer. Inst.*, 91, 354–359.

Broitman, S.A., Wilkinson, J., Cerda, S. and Branch, S.K. (1996) Effects of monoterpenes and mevinolin on murine colon tumor CT-26 in vitro and its hepatic 'metastases' in vivo. *Adv. Exp. Med. Biol.*, 401, 111–130.

Caccioni, D.R., Deans, S.G. and Ruberto, G. (1995) Inhibitory effect of citrus fruit essential oil components on *Penicillium italicum* and *P. digitatum*. *Petria*, 5, 177–182.

Caccioni, D.R., Guizzardi, M., Biondi, D.M., Renda, A. and Ruberto, G. (1998) Relationship between volatile components of citrus fruit essential oils and antimicrobial action on *Penicillium digitatum* and *P. italicum*. *Int. J. Food Microbiol.*, 43, 73–79.

Cai, Y., Baer-Dubowska, W., Ashwood-Smith, M. and Di Giovanni, J. (1997a) Inhibitory effects of naturally occuring coumarins on the metabolic activation of benz(a)anthracene in cultured mouse keratinocytes. *Carcinogenesis*, 18, 215–222.

Cai, Y., Kleiner, H., Johnston, D., Dubowski, A., Bostic, S., Ivie, W. and Di Giovanni, J. (1997b) Effect of naturally occuring coumarins on the formation of epidermal DNA adducts and skin tumors induced by benzo(a)pyrene and 7,12-dimethylbenz(a)anthracene in SENCAR mice. *Carcinogenesis*, 18, 1521–1527.

Camporese, A. (1997) Rapporto tra composizione chimica ed attività antimicrobica degli oli essenziali. *Erboristeria Domani*, 20, 64–73.

Canivenc-Lavier, M.C., Vernevaut, M.F., Totis, M., Siess, M.H., Magdalou, J. and Suschetet, M. (1996) Comparative effects of flavonoids and model inducers on drug-metabolizing enzymes in rat liver. *Toxicology*, 114, 19–27.

Cardullo, A.C., Ruzkowski, A.M. and De Leo, V. (1989) Allergic contact dermatitis resulting from sensitivity to citrus peel, geraniol and citral. *J. Am. Acad. Dermatol.*, 21, 395–397.

Carroll, K.K., Guthrie, N., Gapor, A. and Chambers, A.F. (1997) Effect of limonene on the proliferation of MDA-MB-435 human breast cancer cells, alone or in combination with flavonoids and/or tocotrienols. *Faseb J.*, 11, 2146.

Cerda, S.R., Wilkinson, J. 4th, Branch, S.K. and Broitman, S.A. (1999) Enhancement of sterol synthesis by the monoterpene perillyl alcohol is unaffected by competitive 3-hydroxy-3-methylglutaryl-CoA reductase inhibition. *Lipids*, 34, 605–615.

Chaibi, A., Ababouch, L.H., Belasri, K., Boucetta, S. and Busta, F.F. (1997) Inhibition of germination and vegetative growth of *Bacillus cereus* T and *Clostridium botulinum* 62A spores by essentials oils. *Food Microbiol.*, 14, 161–174.

Chander, S.K., Lansdown, A.G.B., Luqmani, Y.A., Gomm, J.J., Coope, R.C., Gould, N. and Coombes, R.C. (1994) Effectiveness of combined limonene and 4-hydroxyandrostenedione in the treatment of NMU-induced rat mammary tumors. *Br. J. Cancer*, 69, 879–882.

Chaumontet, C., Bex, V., Gaillard-Sanchez, I., Seillan-Heberden, C., Suschetet, M. and Martel, P. (1994) Apigenin and tangeretin enhance gap junctional intercellular communication in rat liver epithelial cells. *Carcinogenesis*, 15, 2325–2330.

Chaumontet, C., Suschetet, M., Honikman-Leban, E., Krutovskikh, V.A., Berges, R., LeBon, A.M., Heberden, C., Shahin, M.M., Yamasaki, H. and Martel, P. (1996) Lack of tumor-promoting effects of flavonoids: on rat liver preneoplastic foci and on in vivo and in vitro gap junctional intercellular communication. *Nutr. Cancer*, 26, 251–263.

Chaumontet, C., Droumaguet, C., Bex, V., Heberden, C., Gaillard-Sanchez, I. and Martel, P. (1997) Flavonoids (apigenin, tangeretin) counteract tumor promoter-induced inhibition of intercellular communication of rat liver epithelial cells. *Cancer Lett.*, 114, 207–210.

Chen, I.S., Lin, Y.C., Tsai, I.L., Teng, C.M., Ko, F.N., Ishikawa, T. and Ishii, H. (1995) Coumarins and anti-platet aggregation constituents from *Zanthoxylum schinifolium*. *Phytochem.*, 39, 1091–1097.

Chen, J., Montanari, A.M. and Widmer, W.W. (1997) Two new polymethoxylated flavones, a class of compounds with potential anticancer activity, isolated from cold-pressed dancy tangerine peel oil solids. *J. Agr. Food Chem.*, 45, 364–368.

Chen, H., Chan, K.K. and Budd, T. (1998) Pharmacokinetics of d-limonene in the rat by GC-MS assay. *J. Pharmaceut. Biomed. Anal.*, 17, 631–640.

Chungsamarnyart, N. and Janswan, W. (1996) Acaricidal activity of peel oil of *Citrus* spp. on *Boophilus microplus*. *Kasertat J. Nat. Sci.*, 30, 112–117.

Clark, S.M. and Wilkinson, S.M. (1998) Phototoxic contact dermatitis from 5-methoxypsoralen in aromatherapy oil. *Contact Dermatitis*, 38, 289–290.

Connor, M.J. (1988) Oxidation of retinol to retinoic acid as a requirement for biological activity in mouse epidermis. *Cancer Res.*, 48, 7038–7041.

Cornwell, P.A., Barry, B.W., Bouwstra, J.A. and Gooris, G.S. (1996) Modes of action of terpene penetration enhancers in human skin differential scanning calorimetry, small-angle X-ray-diffraction and enhancer uptake studies. *Int. J. Pharm.*, 127, 9–26.

Crowell, P.L. (1997) Monoterpenes in breast cancer chemoprevention. *Breast Cancer Res. Treat.*, 46, 191–197.

Crowell, P.L. (1999) Prevention and therapy of cancer by dietary monoterpenes. *J. Nutr.*, 129, 775S-778S.

Dabbah, R., Edwards, V.M. and Moats, W.A. (1970) Antimicrobial action of some citrus fruits oils on selected food-borne bacteria. *Appl. Microbiol.*, 19, 27–31.

Diawara, M.M., Allison, T., Kulkosky, P. and Williams, D.E. (1997) Psoralen-induced growth inhibition in Wistar rats. *Cancer Lett.*, 114, 159–160.

Diawara, M.M., Chavez, K.J., Hoyer, P.B., Williams, D.E., Dorsch, J., Kulkosky, P. and Franklin, M.R. (1999) A novel group of ovarian toxicants: the psoralens. *J. Biochem. Mol. Toxicol.*, 13, 195–203.

Diliberto, J.J., Srinivas, P., Overstreet, D., Usha, G., Burka, L.T. and Birnbaum, L.S. (1990) Metabolism of citral, an α,β-unsaturated aldehyde, in male F344 rats. *Drug Metab. Dispos.*, 18, 866–875.

Don-Pedro, K.N. (1996a) Fumigant toxicity is the major route of insecticidal activity of citrus peel essential oils. *Pest. Sci.*, 46, 71–78.

Don-Pedro, K.N. (1996b) Investigation of single and joint fumigant insecticidal activity of citrus peel oil components. *Pest. Sci.*, 46, 79–84.

Don-Pedro, K.N. (1996c) Fumigant toxicity of citrus peel oils against adult and immature stages of storage insect pests. *Pest. Sci.*, 47, 213–223.

Dubertret, L., Morlière, P., Averbeck, D. and Young, A.R. (1990a) The photochemistry and photobiology of bergamot oil as a perfume ingredient: an overview. *J. Photochem. Photobiol. B:Biol.*, 7, 362–365.

Dubertret, L., Serrat-Tricazes, D., Jeanmougin, M., Morlière, P., Averbeck, D. and Young, A.R. (1990b) Phototoxic properties of perfumes containing bergamot oil on human skin: photoprotective effect of UVA and UVB sunscreens. *J. Photochem. Photobiol. B:Biol.*, 7, 251–259.

Ehrsson, H., Wallin, I., Ros, A.M., Eksborg, S. and Berg, M. (1994) Food-induced increase in bioavailability of 5-methoxypsoralen. *Eur. J. Clin. Pharmacol.*, 46, 375–377.

Ferrini, A.M., Mannoni, V., Hodzic, S., Salvatore, G. and Aureli, P. (1998) Antimicrobial activity of bergamot oil in relation to chemical composition and different origin. *EPPOS*, (Spec. Num.), 139–153.

Friedman, M.I., Preti, G., Deems, R.O., Friedman, L.S., Munoz, S.J. and Maddrey, W.C. (1994) Limonene in expired long air of patients with liver disease. *Digest. Dis. Sci.*, 39, 1672–1676.

Fulton, G.J., Barber, L., Svendsen, E., Hagen, P. and Davies, M.G. (1997) Oral monoterpene therapy (perillyl alcohol) reduces vein graft intimal hyperplasia. *J. Surg. Res.*, 69, 128–134.

Giri, R.K., Parija, T. and Das, B.R. (1999) d-Limonene chemoprevention of hepatocarcinogenesis in AKR mice: inhibition of c-jun and c-myc. *Oncol. Rep.*, 6, 1123–1127.

Godwin, D.A. and Michnik, B.B. (1999) Influence of drug lipophilicity on terpenes as transdermal penetration enhancers. *Drug Dev. Ind. Pharm.*, 25, 905–915.

Gould, M.N. (1997) Cancer chemoprevention and therapy by monoterpenes. *Environ. Health Perspect.*, 105, 977–979.

Haag, J.D. and Gould, M.N. (1994) Mammary carcinoma regression induced by perillyl alcohol, a hydroxylated analog of limonene. *Cancer Chemother. Pharmacol.*, 34, 477–483.

Hard, G.C. (1998) Mechanism of chemically induced renal carcinogenesis in the laboratory rodent. *Toxicol. Pathol.*, 26, 104–112.

Hardcastle, I.R., Rowlands, M.G., Barber, A.M., Grimshaw, R.M., Mohan, M.K., Nutley, B.P. and Jarman, M. (1999) Inhibition of protein prenylation by metabolites of limonene. *Biochem. Pharmacol.*, 57, 801–809.

He, K., Iyer, K.R., Hayes, R.N., Sinz, M.W., Woolf, T.F. and Hollenberg, P.F. (1998) Inactivation of cytochrome P450 3A4 by bergamottin, a component of grapefruit juice. *Chem. Res. Toxicol.*, 11, 252–259.

Higashimoto, M., Yamato, H., Kinouchi, T. and Ohnishi, Y. (1998) Inhibitory effects of citrus fruits on the mutagenicity of 1-methyl-1,2,3,4,-tetrahydro-beta-carboline-3-carboxylic acid treated with nitrite in the presence of ethanol. *Mutat. Res.*, 415, 219–226.

Hirano, T., Abe, K., Gotoh, M. and Oka, K. (1995) Citrus flavone tangeretin inhibits leukaemic HL-60 cell growth partially through induction of apoptosis with less cytotoxicity on normal lymphocytes. *Brit. J. Cancer*, 72, 1380–1388.

Hirose, M., Hasegawa, R., Kimura, J., Akagi, K., Yoshida, Y., Tanaka, H., Miki, T., Satoh, T., Wakabayashi, K., Ito, N. and Shirai, T. (1995) Inhibitory effects of 1-O-hexyl-2,3,5-trimethylhydroquinone (HTHQ), green tea catechins and other antioxidant on 2-amino-6-methyldipyrido[1,2-a:3',2'd]imidazole (GLU-P-1)-induced rat hepatocarcinogenesis and dose-dependent inhibition by HTHQ of lesion indunction by Glu-P-1 or 2-amino-3,8-dimethylimidazo[4,5-f]quinoxaline (MeIQx). *Carcinogenesis*, 16, 3049–3055.

Hohl, R.J. (1996) Monoterpenes as regulators of malignant cell proliferation. *Adv. Exp. Med. Biol.*, 401, 137–146.

Ikechukwu Ezennia, E., Phillips, L.R., Wolfe, T.L. and Tabibi, S.E. (1997) Analysis of perillic acid in plasma by reversed-phase high-performance liquid chromatography with ultraviolet detection. *J. Chromatogr. B*, 688, 354–358.

Iwanami, Y., Tateba, H., Kodama, N. and Kishino, K. (1997) Changes of lemon flavor components in an aqueous solution during UV irradiation. *J. Agric. Food Chem.*, 45, 463–466.

Jonker, D., Woutersen, R.A. and Feron, V.J. (1996) Toxicity of mixtures of nephrotoxicants with similar or dissimilar mode of action. *Food Chem. Toxicol.*, 34, 1075–1082.

Kandaswami, C., Perkins, E., Soloniuk, D.S., Drzewiecki, G. and Middleton, E. Jr (1991) Antiproliferative effects of citrus flavonoids on a human squamous cell carcinoma in vitro. *Cancer Lett.*, 56, 147–152.

Karlberg, A.-T., Magnusson, K. and Nilsson, U. (1992) Air oxidation of *d*-limonene (the citrus solvent) creates potent allergens. *Contact Dermatitis*, 26, 332–340.

Karlberg, A.-T. and Dooms-Goossens, A.D. (1997) Contact allergy to oxidized *d*-limonene among dermatitis patients. *Contact Dermatitis*, 36, 201–206.

Karlberg, A.-T., Basketter, D., Goossens, A. and Lepoittevin, J.P. (1999) Regulatory classification of substances oxidized to skin sensitizers on exposure to air. *Contact Dermatitis*, 40, 183–188.

Kawamori, T., Tanaka, T., Hirose, Y., Ohnishi, M. and Mori, H. (1996) Inhibitory effects of d-limonene on the development of colonic aberrant crypt foci induced by azoxymethane in F344 rats. *Carcinogenesis*, 17, 369–372.

Kim, J.J., Ben-Yehoshua, S., Shapiro, B., Hensis, Y. and Carmeli, S. (1991) Accumulation of scoparone in heat-treated lemon fruit inoculated with *Penicillium digitatum* Sacc.. *Plant Physiol.*, 97, 880–885.

Kimura, J., Takahashi, S., Ogiso, T., Yoshida, Y., Akagi, K., Hasegawa, R., Kurata, M., Hirose, M. and Shirai, T. (1996) Lack of chemoprevention effects of the monoterpene d-limonene in a rat multiorgan carcinogenesis model. *Jpn. J. Cancer Res.*, 87, 589–594.

Koenigs, L.L. and Trager, W.F. (1998a) Mechanism-based inactivation of cytochrome P450 2B1 by 8-methoxypsoralen and several other furanocoumarins. *Biochemistry*, 37, 13184–13193.

Koenigs, L.L. and Trager, W.F. (1998b) Mechanism-based inactivation of cytochrome P450 2A6 by furanocoumarins. *Biochemistry*, 37, 10047–10061.

Kunta, J.R., Goskonda, V.R., Brotherton, H.O., Khan, M.A. and Reddy, I.K. (1997) Effect of menthol and related terpenes on the percutaneous absorption of propanolol across excised hairless mouse skin. *J. Pharm. Sci.*, 86, 1369–1373.

Lin, L., Wiesehahn, G.P., Morel, P.A. and Corash, L. (1989) Use of 8-methoxypsoralen and long-wavelength ultraviolet radiation for decontamination of platelet concentrates. *Blood*, 74, 517–525.

Lis-Balchin, M., Ochoka, R.J., Deans, S.G., Asztemborska, M. and Haret, S. (1996) Bioactivity of the enantiomers of limonene. *Med. Sci. Res.*, 24, 309–310.

Lis-Balchin, M. and Deans, S.G. (1997) Bioactivity of selected plant essential oils against *Listeria monocytogenes*. *J. Appl. Microbiol.*, 82, 759–762.

Mak, N.K., Wong-Leung, Y.L., Chan, S.C., Wen, J., Leung, K.N. and Fung, M.C. (1996) Isolation of anti-leukemia compounds from *Citrus reticulata*. *Life Sci.*, 58, 1269–1276.

McAdam, B., Keimowitz, R.M., Maher, M. and Fitzgerald, D.J. (1996) Transdermal modification of platet function: an aspirin patch system results in marked suppression of platet cyclooxygenase. *J. Pharmacol. Toxicol. Methods*, 277, 559–564.

McCalley, D. and Torres-Grifol, J.F. (1992) Analysis from volatiles in good and bad conditions by gas cromatography-mass spectrometry. *Analyst*, 117, 721–725.

McNeely, W. and Goa, K.L. (1998) 5-Methoxypsoralen. A review of its effects in psoriasis and vitiligo. *Drugs*, 56, 667–690.

Mengeaud, V. and Ortonne, J.-P. (1996) PUVA (5-methoxypsoralen plus UVA) enhances melanogesis and modulates expression of melanogenic proteins in cultured melanocytes. *J. Invest. Dermatol.*, **107**, 57–62.
Moghimi, H.R., Williams, A.C. and Barry, B.W. (1996a) A lamellar matrix model for stratum corneum intercellular lipids. 3. Effects of terpene penetration enhancers on the release of 5-fluorouracil and estradiol from the matrix. *Int. J. Pharm.*, **145**, 37–47.
Moghimi, H.R., Williams, A.C. and Barry, B.W. (1996b) A lamellar matrix model for stratum corneum intercellular lipids. 4. Effects of terpene penetration enhancers on the permeation of 5-fluorouracil and estradiol through the matrix. *Int. J. Pharm.*, **145**, 49–59.
Moghimi, H.R., Williams, A.C. and Barry, B.W. (1997) A lamellar matrix model for stratum corneum intercellular lipids. 5. Effects of terpene penetration enhancers on the structure and thermal behavior of the matrix. *Int. J. Pharm.*, **146**, 41–54.
Morlière, P., Huppe, G., Averbeck, D., Young, A.R., Santus, R. and Dubertret, L. (1990) In vitro photostability and photosensitizing properties of bergamot oil. Effects of a cinnamate sunscreen. *J. Photochem. Photobiol. B:Biol.*, **7**, 199–208.
Morlière, P., Bazin, M., Dubertret, L., Santus, R., Sa E. Melo, T., Huppe, G., Gaigle, J., Forlt, P. and Bernard, A. (1991) Photoreactivity of 5-geranoxypsoralen and lack of photoreaction with DNA. *Photochem. Photobiol.*, **53**, 13–19.
Morse, M.A. and Toburen, A.L. (1996) Inhibition of metabolic activation of 4-(methylnitrosamino)-1-(3-pyridyl)-1-butanone by limonene. *Cancer Lett.*, **104**, 211–217.
Moysan, A., Morlière, P., Averbeck, D. and Dubertret, L. (1993) Evaluation of phototoxic and photogenotoxic risk associated with the use of photosensitizers in suntan preparations: application to tanning preparations containing bergamot oil. *Skin Pharmacol.*, **6**, 282–291.
Murakami, A., Kuki, W., Takahashi, Y., Yonei, H., Nakamura, Y. Ohto, Y., Ohigashi, H. and Koshimizu, K. (1997) Auraptene, a citrus coumarin, inhibits 12-O-tetradecanoylphorbol-13-acetate-induced tumor promotion in ICR mouse skin, possibly through suppression of superoxide generation in leucocytes. *Jpn. J. Cancer. Res.*, **88**, 443–452.
Murdock, D.L. and Allen, W.E. (1960) Germicidal effect of orange peel oil and d-limonene in water and orange juice. Fungicidal properties against yeasts. *Food Technol.*, **14**, 441–445.
Muret, P., Humbert, P., Makki, S., Bechtel, P. Urien, S. and Tillement, J.P. (1993) Serum free 5-methoxypsoralen fraction in health and psoriasis: relationship with human serum albumin concentration. *Arch. Dermatol. Res.*, **285**, 287–290.
Nakaizumi, A., Baba, M., Uehara, H., Iishi, H. and Tatsuta, M. (1997) d-Limonene inhibits N-nitrosobis(2-oxopropyl)amine induced hamster pancreatic carcinogenesis. *Cancer Lett.*, **117**, 99–103.
Nilsson, U., Bergh, M., Shao, L.P. and Karlbergf, A-T. (1996) Analysis of contact allergenic compounds in oxidized d-limonene. *Chromatographia*, **42**, 199–205.
Norman, S., Craft, C.C. and Davis, P.L. (1967) Volatiles from injured and uninjured Valencia oranges at different temperatures. *J. Food Sci.*, **32**, 656–659.
Occhiuto, F., Limardi, F. and Circosta, C. (1995) Effects of the non-volatile residue from the essential oil of *Citrus bergamia* on the central nervous system. *Int. J. Pharmacognosy*, **33**, 198–203.
Occhiuto, F. and Circosta, C. (1996a) Cardiovascular properties of the non-volatile total residue from the essential oil of *Citrus bergamia*. *Int. J. Pharmacognosy*, **34**, 128–133.
Occhiuto, F. and Circosta, C. (1996b) Antianginal and antiarrhythmic effects of bergamottine, a furocoumarin isolated from bergamot oil. *Phytother. Res.*, **10**, 491–496.
Occhiuto, F. and Circosta, C. (1997) Investigations to characterize the antiarrhythmic action of bergamottine, a furocoumarin isolated from bergamot oil. *Phytother. Res.*, **11**, 450–453.
Okabe, H., Takayama, K., Ogura, A. and Nagai, T. (1989) Effect of limonene and related compounds on the percutaneous absorption of indomethacin. *Drug Design and Delivery*, **4**, 313–321.
Omer, E.A., Youssef, A.A., Abo-Zeid, E.N. and Sharaby, A. (1997) Biochemical studies on the essential oils of Balady orange and mandarine. *Egypt. J. Horticolture*, **24**, 207–218.

Ortuño, A., Botia, J.M., Fuster, M.D., Porras, I., Garcia-Lidon, A. and Del Rio, J.A. (1997) Effect of scoparone (6,7-dimethoxycoumarin) biosynthesis on the resistance of tangelo nova, *Citrus paradisi* and *Citrus aurantium* fruits against *Phytophtora parasitica. J. Agric. Food Chem.*, 45, 2740–2743.

Park, Y.K., Hann, S.K., Im, S. and Kim, S. (1993) Skin concentration of 8-methoxypsoralen, 5-methoxypsoralen and 4,5,8-trimethoxypsoralen in guinea pigs. *J. Dermatol. Sci.*, 5, 19–24.

Piacentini, G. (1948) Antiseptic and disinfectant property of essence of bergamot, orange and lemon in aqueous solution, against spore formers. *Ann. Igiene*, 58, 1.

Poon, G.K., Vigushin, D., Griggs, L.J., Rowlands, M.G., Coombes, R.C. and Jarman, M. (1996) Identification and characterization of limonene metabolites in patients with advanced cancer by liquid-chromatography mass-spectrometry. *Drug Metab. Disposition*, 24, 565–571.

Potten, C.S., Chadwick, C.A., Cohen, A.J., Nikaidos, O., Matsunaga, T., Schipper, N.W. and Young, A.R. (1993) DNA damage in UV-irradiated human skin in vivo: automated direct measurement by image analysis (thymine dimers) compared with indirect measurement (unscheduled DNA synthesis) and protection by 5-methoxypsoralen. *Int. J. Radiat. Biol.*, 63, 313–324.

Ramadan, W., Mourad, B., Ibrahim, S. and Sonbol, F. (1996) Oil of a bitter orange: new topical antifungal agent. *Int. J. Dermatol.*, 35, 448–449.

Rasonyi, T., Schlatter, J. and Dietrich, D.R. (1999) The role of alpha 2u-globulin in ochratoxin A induced renal toxicity and tumors in F344 rats. *Toxicol. Lett.*, 104, 83–92.

Reddy, B.S., Wang, C.-X., Samaha, H., Lubet, R., Steele, V.E. and Kelloff, G.J. (1997) Chemoprevention of colon carcinogenesis by dietary perillyl alcohol. *Cancer Res.*, 57, 420–425.

Reicks, M.M. and Crankshaw, D. (1993) Effects of d-limonene on hepatic microsomal monooxygenase activity and paracetamol-induced glutathione depletion in mouse. *Xenobiotica*, 23, 809–819.

Ren, Z. and Gould, M.N. (1998) Modulation of small G protein isoprenylation by anticancer monoterpenes in *in situ* mammary gland epithelial cells. *Carcinogenesis*, 19, 827–832.

Ripple, G.H., Goud, M.N., Stewart, J.A., Tutsch, K.D., Arzoomanian, R.Z., Alberti, D., Feierabend, C., Pomplun, M., Wilding, G. and Bailey, H.H. (1998) Phase I clinical trial of perillyl alcohol administered daily. *Clin. Cancer Res.*, 4, 1159–1164.

Rodov, V., Ben-Yehoshua, S., Fang, D.Q., Kim, J.J. and Ashkenazi, R. (1995) Preformed antifungal compounds of lemon fruit: citral and its relation to disease resistance. *J. Agric. Food Chem.*, 43, 1057–1061.

Said, A., Makki, S., Muret, P., Toubin, G., Humbert, P. and Millet, J. (1997) Spectrofluorimetric determination of 5-methoxypsoralen pharmacokinetic in patients' serum. *Exp. Dermatol.*, 6, 57–63.

Schmiedlin-Ren, P., Edwards, D.J., Fitzsimmons, M.E., He, K., Lown, K.S., Woster, P.M., Rahman, A., Thummel, K.E., Fisher, J.M., Hollenberg, P.F. and Watkins, P.B. (1997) Mechanism of enhanced oral availability of CYP3A4 substrates by grapefruit constituents. Decreesad enterocyte CYP3A4 concentration and mechanism-based inactivation by furanocoumarins. *Drug Metab. Dispos.*, 25, 1228–1233.

Schulz, S., Rinhold, D., Schmidt, H., Ansorge, S. and Hollt, V. (1997) Perillyl acid inhibits Ras/ MAPkinase-driven IL-2 production in humane T lymphocytes. *Biochem. Biophys. Res. Comm.*, 241, 720–725.

Segal, R. and Milo-Goldzweig, I. (1985) The hemolytic activity of citral-II. Glutathione depletion in citral treated erythrocytes. *Biochem. Pharmacol.*, 34, 4117–4119.

Shalaby, A.A., Allam, K.A.M., Mostafa, A.A. and Fahmy, S.M.E. (1998) Insecticidal properties of citrus oils against *Culex pipiens* and *Musca domestica. J. Egypt. Soc. Parasitol.*, 28, 595–606.

Sonbol, F.I., Ibrahim, S.M. and Mohamed, B.M. (1995) Antimicrobial activity of oil of bitter orange. *Alexandria J. Pharm. Sci.*, 9, 107–109.

Stayrook, K.R., McKinzie, J.H., Barbhaiya, L.H. and Crowell, P.L. (1998) Effects of the antitumor agent perillyl alcohol on H-Ras vs. K-Ras farnesylation and signal transduction in pancreatic cells. *Anticancer Res.*, **18**, 823–828.

Su, H.C.F. and Horvat, R. (1997) Isolation and characterization of four major components from insecticidally active lemon peel extract. *J. Agric. Food Chem.*, **35**, 509–511.

Subba, M.S., Soumithri, T.C. and Suryanarayana Rao, R. (1967) Antimicrobial action of citrus oils. *J. Food Sci.*, **32**, 225–227.

Tamir, I., Abramovici, A., Milo-Goldzweig, I. and Segal, R. (1984) The hemolytic activity of citral: evidence for free radical partecipation. *Biochem. Pharmacol.*, **33**, 2945–2950.

Tanaka, T., Kawabata, K., Kakumoto, M., Makita, H., Hara, A., Mori, H., Satoh, K., Hara, H., Murakami, A., Kuki, W., Takahashi, Y., Yonei, H., Koshimizu, K. and Ohigashi, H. (1997) *Citrus* auraptene inhibits chemically induced colonic aberrant crypt foci in male F344 rats. *Carcinogenesis*, **18**, 2155–2161.

Tanaka, T., Kawabata, K., Kakumoto, M., Matsunaga, K., Mori, H., Murakami, A., Kuki, W., Takahasi, Y., Yonei, H., Satoh, K., Hara, A., Maeda, M., Ota, T., Odashima, S., Koshimizu, K. and Ohigashi, H. (1998a) Chemoprevention of 4-nitroquinoline 1-oxide-induced oral carcinogenesis by citrus auraptene in rats. *Carcinogenesis*, **19**, 425–431.

Tanaka,, T., Kawabata, K., Kakumoto, M., Hara, A., Murakami, A., Kuki, W., Takahashi, Y., Yonei, H., Maeda, M., Ota, T., Odashima, S., Yamane, T., Koshimizu, K. and Ohigashi, H. (1998b) *Citrus* auraptene exerts dose-dependent chemopreventive activity in rat large bowel tumorigenesis: the inhibition correlates with suppression of cell proliferation and lipid peroxidation and with induction of phase II drug metabolizing enzymes. *Cancer Res.*, **58**, 2550–2556.

Teng, C.M., Li, H.L., Wu, T.S., Huang, S.C. and Huang, T.F. (1992) Antiplatet action of some coumarin compounds isolated from plant sources. *Tromb. Res.*, **66**, 549–557.

Treffel, P., Makki, S., Faivre, B., Humbert, P., Blanc, D. and Agache, P. (1991) Citropten and bergapten suction blister fluid concentrations after solar product application in man. *Skin Pharmacol.*, **4**, 100–108.

Uedo, N., Tatsuta, M., Iishi, H., Baba, M., Sakai, N., Yano, H. and Otani, T. (1999) Inhibition by d-limonene of gastric carcinogenesis induced by N-methyl-N'-nitrosoguanidine in Wistar rats. *Cancer Lett.*, **137**, 131–136.

Ulate-Rodriguez, J., Schafer, H.W., Zottola, E.A. and Davidson, P.M. (1997a) Inhibition of *Listeria monocytogenes, Escherichia coli* O157:H7 and *Micrococcus luteus* by linear furanocoumarins in culture media. *J. Food Protect.*, **60**, 1046–1049.

Ulate-Rodriguez, J., Schafer, H.W., Zottola, E.A. and Davidson, P.M. (1997b) Inhibition of *Listeria monocytogenes, Escherichia coli* O157:H7 and *Micrococcus luteus* by linear furanocoumarins in a model food system. *J. Food Protect.*, **60**, 1050–1054.

VanDurren, B.L. and Goldschmidt, B.M. (1976) Carcinogenic and tumor promoting agents in tobacco carcinogenesis. *J. Natl. Cancer Inst.*, **56**, 1237–1242.

van Lieshout, E.M., Ekkel, M.P., Bedaf, M.M., Nijoff, W.A. and Peters, W.H. (1998a) Effects of dietary anticarcinogens on rat gastrointestinal glutathione peroxidase activity. *Oncol. Rep.*, **5**, 959–963.

van Lieshout, E.M., Posner, G.H., Woodard, B.T. and Peters, W.H. (1998b) Effects of the sulphorosane analog compound 30, indole-3-carbinol, d-limonene or relafen on glutathione S-transferases and glutathione peroxidase of the rat digestive tract. *Biochim. Biophys. Acta*, **1379**, 325–336.

Vargas, I., Sanz, I., Moya, P. and Prima-Yufera, E. (1999) Antimicrobial and antioxidant compounds in the non-volatile fraction of expressed orange essential oil. *J. Food Protect.*, **62**, 929–932.

Vermeulen, S., van Mark, V., van Hoorde, L., van Roy, F., Bracke, M. and Mareel, M. (1996) Regulation of the invasion suppressor function of the cadherin/catenin complex. *Pathol. Res. Pract.*, **192**, 694–707.

Vigushin, D.M., Poon, G.K., Boddy, A., English, J., Halbert, G.W., Pagonis, C., Jarman, M. and Coombes, R.C. (1998) Phase I and pharmacokinetic study of d-limonene in patients with advanced cancer. *Cancer Chemother. Pharmacol.*, 42, 111–117.

Yamada, Y., Okamoto, M., Kikuzaki, H. and Nakatani, N. (1997) Spasmolytic activity of auraptene analogs. *Biosci. Biotechnol. Biochem.*, 61, 740–742.

Yamane, M.A., Williams, A.C. and Barry, B.W. (1995) Terpene penetration enhancers in propyleneglycol/water co-solvent system-effectiveness and mechanism of action. *J. Pharm. Pharmacol.*, 47, 978–989.

Yano, H., Tatsuta, M., Iishi, H., Baba, M., Sakai, N. and Uedo, N. (1999) Attenuation by d-limonene of sodium chloride-enhanced gastric carcinogenesis induced by N-methyl-N'-nitro-N-nitrosoguanidine in Wistar rats. *Int. J. Cancer*, 82, 665–668.

Young, A.R., Walker, S.L., Kinley, J.S., Plastow, S.R., Averbeck, D., Morlière, P. and Dubertret, L. (1990) Phototumorigenesis studies of 5-methoxypsoralen in bergamot oil: evaluation and modification of risk of human use in an albino mouse skin model. *J. Photochem. Photobiol. B:Biol.*, 7, 231–250.

Wattenberg, L.W. and Coccia, J.B. (1991) Inhibition of 4-(methylnitrosamino)-1-(3-pyridyl)-1-butanone carcinogenesis in mice by d-limonene and citrus fruit oils. *Carcinogenesis*, 12, 115–117.

Whysner, J. and Williams, G.M. (1996) d-Limonene mechanist data and risk assessment: absolute species-specific cytotoxicity, enhanced cell proliferation and tumor promotion. *Pharmacol. Ther.*, 71, 127–136.

Zhang, Z., Chen, H., Chan, K.K., Budd, T. and Ganapathi, R. (1999) Gas chromatographic-mass spectrometric analysis of perillyl alcohol and metabolites in plasma. *J. Chromatogr. B Biomed. Sci. Appl.*, 728, 85–95.

Zhao, K. and Singh, J. (1998) Mechanism of percutaneous absorption of tamoxifen by terpenes: eugenol, d-limonene and menthone. *J. Control. Rel.*, 55, 253–260.

Zukerman, I. (1951) Effect of oxidized d-limonene on microorganisms. *Nature*, 168, 517.

Author index

Abd Ar-Ramân 7
Abd el Latif ben Yusuf, Muvaffaq ed-Din 7
Accree, T.E. 294
Acosta, Jose de 12
Adeishvili, N. 439
Aertsen, Pieter 10
Agarwal, S.G. 332
Agricultura de Jardines (Los Rios) 10
Agriculturae Opus (Palladius) 5
Agricultural Research Service 71
Akin, D. 107
Al-Istakhart 7
Al-Masudi 7
Alberti, Leandro 9
Albrigo, L.G. 49
Alderman, G.G. 604
Alessandro, R.T. 237
Alexander the Great 4, 577
Allen, W.E. 603
Almirall, M. 617
Alstom ACB 103
Altenburger, R. 339
Amaha, M. 328, 339, 431, 432, 448, 455
Amarakosha (Sanskrit dictionary) 3
Annali di Odontojatria (Chiavaro) 588
Antidotario Romano (Ceccarelli) 582
Antidotarium (Clusius) 580
Antidotarium Gandavense (1663) 581
Antiphanes 5
Apicius Caelius 5, 578
Araki, C. 347
Arcimboldo, Giuseppe 10
Arctander, S. 411
Armentano, L. 68
Asano, K.-I. 327
Ashoor, S.H.M. 213, 218, 220, 228–9, 243, 252, 285, 286
Atheneus of Alexandria 4
Attaway, J.A. 251, 257, 443, 446, 448, 450, 453, 455
Attaway, S. 77
Augustin, Miguel 10
Ault, W.C. 136
Ayedoun, A.M. 303, 435, 439
Ayensu, E.-S. 590
Ayurveda 589
Azzouz, M.A. 230

Baaliouamer, A. 265, 303, 319, 431, 432, 438, 448
Baber, Zeher-ed-din Muhammed 3
Baigrie, B.D. 514
Bajpai, A. 589
Baker, R.A. 93
Balaban, M.O. 103
Balansa 63
Balducci-Pergolotti, Francesco 8
Banerjee, B.C. 453

Bardon, S. 614
Barillà, Nicola 64
Barkley, P. 59
Barth, D. 297
Barukadze, N.S. 265
Baschenis, Evaristo 11
Baser, K.H.C. 284
Beasley, L.R.M. 78, 88
Beck, C.R. 357
ben Yusuf *see* Abd el Latif ben Yusuf
Benedetti, Andrea 10
Bennet of Messina 66
Beotian, The (Antiphanes) 5
Beradze, L.V. 220, 258, 286, 301
Berger, R.G. 248
Bernhard, R.A. 202, 218, 220, 228–9, 243, 252, 285, 286
Bernreuther, A. 478, 479, 487, 492
Berry, R.E. 370, 373
Besson, Jacques 63
Beyeren, Abraham Van 10
Bianchini, J.P. 373
Bicchi, C. 466, 469, 473, 474
Bigi, B. 414
Bignon, J. 104
Bimbi 11
Biot, J.B. 496
Blanco Tirado, C. 256, 271, 446, 451
Blondus Flavius 8
Bodenheimer, F.S. 53
Boelema, Maerten 10
Boelens, M.H. 211, 214, 247, 291, 417, 418, 428
Boerhaave, H. 586
Bogert, M.T. 376
Böhme, H. 369, 373
Bolti, Rinaldo 11
Bomare, Valmont de 8
Bononiense, Ionnes Baptista Cortesius 581
Bordoloi, A.K. 327
Born, R. 373
Boselli, Felice 11
Bott, T.R. 399
Botticelli, Sandro 10
Bouchardat 587
Braconnot, H. 68
Braddock, R.J. 68, 76, 163, 251, 537
Braunsdorf, R. 505
Braverman, J.B.S. 64, 67, 71, 395
Bretschneider 577
Brogden, W.B. 207, 219, 229, 244, 253, 286
Bronzino, Angiolo 10
Brown, W.O. 67
Buccellato, F. 417, 418, 419
Buiarelli, F. 373
Buigues, N.M. 206
Buttery, R.G. 546

Caccioni, D.R.L. 215, 216, 226, 250, 270, 299, 300
Calabrò, G. 242, 243, 273, 357
Calapaj, R. 355, 365, 373, 501
Caldwell, A.G. 355
Calvarano, I. 207, 319, 428, 429, 434, 435, 440–1, 452, 453, 504
Calvarano, M. 202, 207, 248, 249, 256, 273, 274, 284, 376, 414, 456, 457
Cameron, J.W. 328
Campi, Vincenzo 10
Cano, M.P. 104
Cantani, A. 587
Capannelli, G. 99
Caporale, G. 377
Capparella, M. 54
Cappello, C. 222, 256, 290, 338
Carmelites 585
Cartoni, G.P. 255, 295, 432
Casabianca, H. 463, 466, 470, 471, 474, 478, 479, 487, 491, 492
Cassola, F. 586
Castelli, Giovanni Paolo 11
Castelli, Pietro 581–2
Castle, W.S. 57
Catalogus Plantarum Horti Pisani (Tilli) 584
Catanense Dispensatorium (Catanuto) 582
Catanuto, Nicolao 582
Cavazzòlo, Paolo (Morando) 10
Ceccarelli, Ippolito 582
Cemeroglu, B. 84
Cézanne, Paul 15
Chace, E.M. 69
Chaibi, A. 603
Chamblee, T.S. 204, 223, 233, 234, 291, 296, 406, 409, 411
Chapot, H. 49
Charaka samhita (medical book) 3
Chardin, Jean-Simeon 10
Chen, C.S. 92
Chen, I.S. 619
Cheng, Y.S. 328, 448, 453, 455
Chevalier, A. 586
Chi Han 2
Chialva, F. 319
Chiavaro, Angelo 588
Chirico, Giorgio de 15
Chouchi, D. 280
Cianciolo, 66
Cieri, U.R. 357, 361
Cirio, U. 54
Cirus (Cyrus) the Great 3, 4
Cisalpino, Andrea 581
Cittadini, Pier Francesco 11
Claesz, Pieter 10
Clark, B.C. 233, 234, 409, 411
Clark, J.R. 202
Clark, S.M. 609
Cloanzius Verus 5
Clusius *see* Lécluse, Charles de
Codex Medicamentarius seu Pharmacopea Parisensis (1748) 583
Codex Medicamentarium sive Pharmacopea Danica (1821) 583
Codex Medicamentarius seu Pharmacopea Lusitanica (1822) 583

Codice Farmaceutico Romano-Teorico e Pratico (1868) 584
Cohn, E. 51
Coleman, R.L. 258
Conte, G. 202, 286
Corassini, G. 595
Coreggio, Antonio Allegri da 10
Corrêa, Gaspar 8
Corsali, Andrea 8
Cotroneo, A. 204, 293, 338, 487
Couchi, D. 212, 213, 400
Couilloud, R. 49
Council of Europe 570
Cours Elementaire di Matière Médicale (de Rochefort) 586
Crandal, P.G. 92
Crescenzi, Piero de' 8
Crescimanno, F.G. 439
Crespi, Daniel 10
Crismer, M.L. 376
Criss, B. 67
Crusaders 8
Cultrera, R. 124, 499
Currò, P. 273, 357
Cutuli, G. 56, 57
Cyrus the Great *see* Cirus the Great

Dabbah, R. 603
Dalmasso, A. 51
D'Amore, G. 242, 243, 355, 365, 373, 501
Dancy, G.C. 15
Dante Alighieri 579
Darvas, B. 55
Davis, F.S. 49
De agricultura opusculum (Venuto) 9
De Cicco, V. 57
De Domenico, V. 496
De La Torre, P.C. 208
De re coquinaria (Apicius Caelius) 5
De Rochefort, D. 586
De Vottero, L.R. 433, 440
Deipnosophistae (*wise men in conversation*) (Atheneus of Alexandria) 4
Della Porta, G. 250
Della Robbia, Luca 10
Dellacassa, E. 280, 291, 302, 466, 487, 523, 527
Delucchi, V.L. 49
Demarne, F. 432, 451, 455
Deshpande, S.S. 96
Di Giacomo, A. 63, 66, 72, 103, 105, 202, 207, 220, 243, 258, 273, 286, 287, 376, 380, 434, 435, 504
Di Martino, E. 49
Diocletian 5
Dioscorides, Pedanius 4, 578–9
Dixon, C.W. 273
Dizionario ovvero Trattato Universale delle Droghe Semplici (Lemery) 579
Dizionario Terapeutico di Materia Medica (Dujardin-Beaumetz) 587
Dodge, F.D. 369
Doglia, G. 319
don Felipe (Spanish noble) 13
Drescher, R.W. 284
Dreyer, D.L. 370
Du Pont 108

Dugo, Giacomo 519, 521, 522, 523, 524, 525
Dugo, Giovanni 83, 204, 211, 222, 225, 226, 233, 237, 246, 247, 265, 280, 286, 287, 291, 293, 319, 338, 406, 428, 463, 466, 467, 468, 469, 470, 471, 473, 474, 478, 479, 480, 481, 484, 487, 504, 507, 510
Dugo, P. 236, 359, 362, 365, 366, 367, 373, 374, 376, 377, 380
Dujardin-Beaumetz, M. 587
Duke, J.-A. 590, 593
Duncan, L.W. 51, 224
Dung, N.X. 329, 342
DuPuis, C. 273, 274
Dupuis, G. 519
Dupuis, Pier 10
Durante, Castore 579
Dykstra, K. 365
Dzhanikashvili, M.I. 334, 344

Ebeling, W. 49
Edwards, D.J. 240
Ekundayo, O. 449, 450, 455, 457
El Ricettario dell'Arte et Universita de Medici et Spetiali della Città di Firenze (1550) 580
El Samay, S.K. 213, 237, 266
El-Kebeer, M.E. 213
Empoli, Jacopo da 10
Engel, W. 546
Escher, S. 334
Espinosa, Juan Bautista de 10
Eyck, Jan van 577

Faas, W.E. 233, 236, 406
Fabriano, Gentile da 10
Falcando, Hugo 8, 9
Fanghänel, E. 207
Fantain-Latour, Henri J.T. 15
Farmacologia Teorica e Pratica (Orosi) 587
Farmacopea Cerusica (Rebaudengo) 583
Farmacopea (Ferrarini) 583
Faulhabert, S. 189
Fellers, P.J. 85
Ferrarini, Antonio 583
Ferrarius, Baptista 11
Filomena, M. 109
Fisher, J.F. 357, 365, 370, 373, 381
Fisher-Ayloff-Cook, K.P. 98
Flath, R.A. 546
Flegel, Georg 10
Fleisher, A. 338, 339, 340–2, 396, 442, 443, 444, 446, 449
Fleisher, Z. 338, 339, 340–2, 442, 443, 444, 446, 449
Florentinus 5
Fluckinger 63
Font Quer, P. 592
Foppens Van Es, Jacob 10
Formàcek, V. 250, 290, 428
Forrey, R.R. 546
Forte, Luca 11
Fox, K. 102, 104
Fra Angelico 10
Friedman, M.I. 621

Galenus 4
Galizia, Fede 10
Gallesio, G. 8
Gallo, Augustin 10

Gama, Vasco da 8
Gao, S.Y. 327
Garbo, Raffaellino del 10
Gargilius Martial 5
Garnsey, S.M. 59, 60
Garzoni, Giovanna 11
Gaubius 63
Gaydou, E.M. 373, 374
Gazea, F. 299, 319
Geoffroy Cl.J. 63
Geoponica (Florentinus) 5
Germanà, M.A. 298, 417, 431, 432, 439, 442, 450, 455
Gesner, Conrad 63
Gessi, Francesco 10
Gherardi, S. 88
Ghiberti, Lorenzo 10
Ghirlandaio, Demenico 10
Giambologna, Fleming 10
Gildmeister, E. 414
Gionfriddo, F. 377
Glandian, R. 361
Goettsch, H.B.G. 108
Goldsworthy, L.J. 374
Goldthwaithe, N.E. 68
Goodner, J.K. 103
Goretti, G. 434
Gostoli, C. 108
Gozzoli, Benozzo 10
Graff, J.-B. 466, 474, 478, 479, 487, 491, 492
Grant, P. 92
Grohmann, K. 106
Guatelli, G. 92
Guenther, E. 63, 65, 122, 123, 135, 201, 205, 377, 413, 421, 423, 497, 519
Guilbert, H.R. 66
Gumpf, D.J. 60
Günther, H. 285
Gupta, G.N. 337
Gupta, J.C. 337
Gurib-Fakim, A. 431, 432, 451, 455

Haensel, Heinrich 66
Haggag, E.G. 304, 439, 457
Hamen y Leon, Juan Van Der 10
Hanke, Dr 588
Haro-Guzmán, L. 230, 233, 236, 362, 406, 409
Hata, K. 380
Hattori, S. 347
Haubruge, E. 213, 222, 265, 297, 428
Hauser, A. 429
Hawkins, Sir Richard 10
Hayakawa, I. 103
Heda, Willem Claesz 10
Heem, Jan Davidsz de 10
Heintz, Joseph the younger 10
Hener, U. 466, 473, 479, 482, 487, 491, 492
Hérisset, A. 273
Hernandez, E. 99
Herrera, Alonso de 10
Hesperides, sive de Malorum aureorum Cultura et Usu Libri Quatuor (Ferrarius) 11
Historia naturalis (Pliny the Elder) 5
Hocking, G.-M. 589
Hodgson, R.W. 16
Hoffmann, F. 414, 497

Hofsommer, H.J. 105
Hooke, R. 496
Hopper, Edward 15
Hua, Y.-F. 327, 446, 450
Huang, M.-B. 216
Huang, Y.-Z. 282, 284, 319, 328, 332, 333, 334, 335, 337, 339, 342–3, 345, 453
Huet, R. 258, 273, 274, 279, 286, 330, 338, 409
Huffman, J.W. 219
Hughes, Griffiths 12
Hunter, G.L.K. 207, 219, 229, 230, 244, 253, 286
Hussein, M.M. 244, 253
Hyfnawy, M.S. 441, 443

Ibn al-Awwâm 7
Ibn Botlal, Abdul Hassan al Muchtar 7
Ibn el-Beithârvis 7
Ibn Jamiya 7
Ibn Sina, Abn Ali al-Hussayn Ibn Abdallah 7
Ibn Suleyman 7
Ibn-el-Wahshya 7
Ifuku, Y. 344
Ikeda, R.M. 258, 273
Imagawa, K. 347
Imperial Conservatory 2
Inoma, S. 212, 222, 250, 255, 259, 280, 344
Inserra, R.N. 51

Jansen, E. 99
Jantan, I. 238, 239, 327, 328, 330, 457
Jennings, E.G. 186
Jeppson, L.R. 49
Jimenez, R. 211, 214, 247, 291
Jones, E.R.H. 355
Joseph(us) Flavius 4
Joulain, D. 552
Juchelka, D. 423, 484, 485, 487, 488, 491, 492
Jurd, L. 356, 370

Kainrath, P. 355
Kaiser, R. 419
Kalf, Willem 10
Kamiyama, S. 328, 334, 339, 345, 431, 432, 438, 448, 455, 456
Kandaswami, C. 616
Karawya, M.S. 431, 432, 441, 443, 453
Kariyone, T. 369
Karnick, C.-R. 589
Kato, Y. 344
Kekelidze, N.A. 220, 224, 258, 286, 300, 301, 334, 335, 344, 345
Kesterson, J.W. 136, 163, 251
Kharebava, L.G. 345, 439
Khurdiya, D.S. 238
Kim, H. 347
Kimball, D.A. 109
Kimura, J. 614
King, M.B. 399
Kirbaslar, S.I. 282, 284
Kirk, W.A. 69
Kitahara, T. 334
Knorr, D. 101, 103, 104
Kojiki (Report on ancient history) 3
Koketsu, M. 211, 222, 236, 246, 254, 255, 259, 265, 279, 290
Kolb, E. 90

Komatsu, S. 369
König, W.A. 478, 479, 487, 488
Kovats, E. 202, 242, 243, 406, 409
Kreis, P. 466, 473, 480, 482
Kubeczka, K.H. 250, 290, 428
Kugler, E. 202, 242, 243
Kumamoto, J. 383
Kumar, S. 453
Kumar, U. 301, 439
Kunta, J.R. 617
Kuriyama, T. 327
Kusunose, H. 216, 217

Ladwig, H. 102
Lagriffe, L. 585
Lakszner, K. 282, 296
Lamparsky, D. 279, 419
Lancas, F. 236, 254, 255, 290
Lander, D. 104
Lanuzza, F. 211, 282, 296
Latz, H.W. 357, 361, 362, 502
Lawrence, B.M. 205, 330, 357, 376, 426, 428, 535, 536
Lécluse, Charles de (Carolus Clusius) 580
Lee, C.-S. 328, 448, 453, 455
Lee, N.A. 419
Lemery, Nicolas 579
Lenggenhager, T. 98, 106
Leonardo da Vinci 10
Liberti, A. 202, 286
Libri, Girolamo dai 10
Licandro, G. 292, 293
Likens, S.T. 546
Lin, L. 606
Lin, Z.-K. 216, 327, 417, 431, 432, 446, 450
Linard, Jacques 10
Linares, F. 54
Lippo, Fra Filippo 10
Lo Castro 64
Lo Giudice, V. 51
Lodge, N. 292
Longo, S. 52
Los Rios, Greorio de 10
Lota, M.L. 338, 339, 340, 342, 343–4
Lotto, Lorenzo 10
Luini, Bernardino 10
Lund, E.D. 332, 333, 456
Lung-Yang His 2
Luyckx, Cerstiaen 10
Lyndon, R. 107

Ma, L. 417
McCalley, D. 604
MacDonald, H.M. 84
McDuff, O.R. 136
McHale, D. 233, 234, 357, 362, 363, 365, 366, 370, 371, 373, 374, 381, 383, 409, 414, 509
MacKay, D.A.M. 548
MacLeod, A.J. 273
MacLeod, W.D. 206, 219, 230, 286
Macrobius 5
Madsen, B.C. 357, 361, 362, 502
Maekawa, K. 206
Maheshwari, M.L. 238
Maier, V.P. 105
Maimone, D. 595

Majláth, P. 286
Mandanpal nighunt (medical text) 3
Manet, Edouard 15
Mantegna, Andrea 10
Manuale di Materia Medica e di Terapeutica Comparata e di Farmacia (Bouchardat) 587
Manyoshu (poetic Anthology) (Otomo Yakamoci) 3
Marais, L.J. 60
Marr, I.M. 240
Matile, P. 339
Matsuno, T. 369, 370
Mattioli, Pietro 9
Mattioli, Pietro Andrea 579
Maupetit 281
Maurer, B. 429
Mazza, G. 204, 246, 247, 279, 295
Mead, S.W. 66
Medical Materials (Theophrastus) 4
Medici, Lorenzo the Magnificent 9
Mehlitz, A. 207
Meklati, B.Y. 303, 431
Melendez, Luis 10
Melendreras, F.A. 435, 438
Mertens, Wouter 10
Micali, G. 211, 249, 282, 296
Minas, Th. 207
Mincione, B. 72
Miskiewicz, M.A. 297, 417, 418, 419, 422, 428
Miyake, M. 335
Moghimi, H.R. 618
Mojon, Benedicto 585
Mojonnier Bros 95
Monardes, Nicolas 10
Mondello, L. 195, 211, 223, 249, 259, 282, 319, 417, 426, 428, 434, 440, 441, 442, 446, 447, 450, 464, 466, 469, 470, 471, 474, 475, 478, 480, 481, 482, 484, 487, 488, 490, 514
Morando *see* Cavazzòlo, Paolo
Moretto da Brescia 10
Moreuil, C. 330
Morse, M.A. 612
Morton, J.-F. 592
Morton, R.A. 498
Mosandl, A. 423, 461, 466, 473, 476, 478, 479, 480, 481, 482, 484, 485, 487, 488, 491, 492
Moshonas, M.G. 219, 230, 234, 251, 255, 259, 286, 336, 451
Mossman, D.D. 376
Moyler, D.A. 398
Mulder, G.J. 376
Muller, H. 101
Muller, J.G. 96
Munakata 605
Munari, Cristoforo 11
Murail, M.C. 258
Murdock, D.L. 603
Murray, R.D.H. 355, 382
Musaraki (writer) 3
Myers, F.S. 222, 223

Nakaizumi, A. 613
Namba, T. 344
Napoleon Bonaparte 585
Natural history of Indies (d'Oviedo) 11–12
Natural and moral history of Indies (Acosta) 12
Naturalis Historia (Pliny the Elder) 578

Naves, 421
Nelson, E.K. 374
Nemec, S. 332, 333
Nicandrus of Kalophone 4
Nickerson, G.B. 546
Nigam, M.C. 336, 337
Nigg, H.N. 362
Nihon Shoki (Report on Japan) 3
Nikdel, S. 102
Nippur 577
Nishida, R. 294
Nishimura, K. 333–4
Nisperos-Carriedo, M.O. 338
Nitz, S. 487, 504
Njoroge, S.M. 216, 217, 240, 302, 334
Noirmoutier, Anna Maria de la Tremoille de 63, 148
Nomura, D. 369, 370
Nordby, H.E. 370, 381
Norman, S. 604
Nuvolone, Panfilo 10

Obra de Agriculture (Herrera) 10
Occhiuto, F. 618
Officina Medicamentorum et Methodus (1601) 580
Ogihara, K. 345, 451
Ohloff, G. 512
Ohme, C. 376
Ohta, H. 335
Oporto, A. 211, 417, 428
Oppius 5
Orosi, G. 587
Orsini, Flavio, Duke of Bracciano 63, 148
Ortiz, J.M. 25, 208, 274, 319, 431, 432, 453
Ortiz Marcide, J.M. 319
Osajima, Y. 335
Oviedo, Gonzalo d' 11–12
Owusu-Yaw, J. 394

Paladin, Perroni 65
Palladius, Rutilius Taurus Aemilianus 5
Paolo, Giovanni di 10
Papanikolau, X. 215, 431
Papyri Oxyrrhichus 5
Parish, M. 78
Parmigianino 10
Patnayak, K.C. 365
Pelletier, B. 104
Pende, N. 588
Pennisi, L. 287
Pent' ts'ao Kang Mu 577
Peppard, T. 550
Peratoner 66
Peres Zayas, J. 328
Perry, L.-M. 590
Petracek, P.D. 227
Petronius 5
Peyron, L. 414, 432, 433, 440, 446, 452, 453, 455, 457
Pharmacia Antiverpiensis (1661) 581
Pharmacopea Almeriana (1726) 582
Pharmacopea Brugensis (1697) 582
Pharmacopea Collegii Regalis Medicorum Londinensis (1767) 583
Pharmacopea Dogmaticorum Restitua (Quercetano) 580
Pharmacopea Manualis reformata (Mojon) 585

Pharmacopea Saxonica (1820) 583
Pharmacopea seu Antidotarium Messanense (Bononiense) 581
Pharmacopoea Hagiensis (1659) 581
Pharmacopoea Ultrajectina (1656) 581
Pharmacopoeia Bruxellensis (1641) 581
Piacentini, G. 603
Pidel, A.R. 244, 253
Pieranzoni, Nicolò 10
Pieribattesti, J.C. 330
Pietsch, G. 369, 373
Pliny the Elder 5, 578
Plutarc(h) 4
Poiana, M. 338
Polissou, M.G. 552
Pomerantz, C. 376
Pontanus, Jovianus 10
Porcaro, P.J. 380
Porta, Giovanni Battista della 10, 63
Potten, C.S. 609
Prager, M.J. 297, 417, 418, 419, 422, 428
Protopapadakis, E. 215, 431
Pudil, F. 330
Puri, H.-S. 589

Quayle, H.J. 54
Quercetano, Iosepho 580

Ramteke, R.S. 95
Ravid, U. 466, 469, 491, 492
Realfonso, Tommaso 11
Rebaudengo, Teobaldo 583
Recco, Giuseppe 11
Reddy, B.S. 613
Redon, Odilon 15
Reece, P.C. 25
Reignier, R. 330
Reineccius, G.A. 230
Remy, M. 418, 419
Renoir, Pierre 15
Reuther, W. 331, 338, 339
Ricci, G.-H. 595
Ricciardi, A.I.A. 279
Richard, A. 586
Riemenschneider, R.W. 136
Riganesis, M. 374
Rispoli, G. 202, 286
Robinson, R. 374
Rocca, B. 463, 466, 473, 479
Rodanò, C. 116
Rodier, Clément 15
Rodighero, G. 377
Rodov, V. 607
Roessler, Y. 55
Rojahn, W. 376
Romeo, G. 157
Rosen, D. 52, 55
Rossellino, Antonio 10
Rouse, A.H. 84
Rousseau, Henri Julien 15
Rovesti, P. 595
Ruberto, G. 225, 226, 319
Ruoppolo, Gian Battista 10
Russel, G.F. 548
Rusznyák, St 68
Rysstad, G. 83

Saito, M. 297, 524
Sakakibara, H. 347
Sakamoto, K. 344
Sale, J.W. 498
Salerno, M. 56, 57
Salini, Tommaso 10
Samish, Z. 64
Sanchez, J.L. 290
Sardi, T.C. 208
Sato, A. 330
Sattar, A. 301
Satyricon (Petronius) 5
Sawada, M. 292
Sawamura, M. 216, 217, 225, 227, 239, 240, 284, 296, 302–3, 327, 334, 344
Scacciati 11
Scarlata, 66
Schenk, H.P. 279
Schmidt, E. 357
Schmidt, H.-L. 508
Schofield, T.F. 83
Schreier, P. 478, 479, 487, 492
Schubert, V. 487
Schultz, T.H. 548
Schwob, R. 65, 330
Scora, R.W. 1, 228, 229, 328, 332, 383, 438
Scribonius Largus 5
Sebastian, Stoskopff 10
Sebastiani, E. 203
Segovia, C.B. 67
Semmler, F.W. 497
Semmola, G. 586
Serres, Olivier de 10
Sestini, abbot Domenico 63–4, 69
Seymour, T.H. 84
Shaw, P.E. 205, 206, 209, 217, 219, 222, 223, 228, 234, 241, 248, 251, 253, 255, 256, 257, 258, 259, 272, 284, 285, 286, 331, 336, 451
Sheridan, J.B. 357, 362, 363, 365, 366, 370, 371, 373, 374, 381, 509, 511
Shih Ching (*Book of odes*) 2
Shinoda, N. 334
Shiota, H. 342
Shomer, R. 89
Shu, C.K. 370
Shubiak, P. 380
Shulz, H. 224
Signorelli, Luca 10
Silvaticus, Matteus 8
Simon Maccabeus 4, 577
Sinclair, W.B. 163
Sindreu, R.J. 211, 417, 428
Slater, C.A. 202, 229, 409, 411
Snyders, Frans 10
Soffer, T. 89
Son, Joris Van 10
Soulari, M. 207
Späth, E. 355, 376
Speciale, Nicolò 8
Spinelli, A. 595
Spitellcr, G. 358, 373, 381
Srinivas, S.R. 417
Ssu Hsiang-Ju 2
Stanley, W.L. 355, 356, 357, 361, 362, 370, 381, 500, 501
Staroscik, J.A. 290, 292

Starrantino, A. 466
Stella, P. 588
Sterneicht, M. 54
Stevens, R.K. 518
Straus, D.A. 286
Strozzi, Bernardo 10
Sucan, M.K. 548, 549
Sugisawa, H. 266, 267
Sugiyama, K. 297
Sulser, H. 219
Sun, D. 227
Sun, H.-D. 339
Sung Yu 2
Surburg, H. 420
Swift, L.J. 373
Swingle, W.T. 16, 25
Szent-Györgyi, A. 68
Szepesy, L. 282, 296

Tajima, K. 334
Takács, J. 286
Takeoka, G. 478, 479
Talbot, Robert 584–5
Talhouk, A.S. 49
Tanaka, T. 19, 318, 327, 330, 331, 335, 338, 343
Tarantilis, P.A. 552
Tateo, F. 395
Tatum, J.H. 370, 373, 383
Teng, C.M. 619
Théâtre d'Agriculture et ménage des champs (Serres) 10
Theatrum Sanitatis (14th century) 579
Theophrastus of Eresus/Heresius 4, 5, 578
Tiemann, F. 497
Tilden, W.A. 357
Till, M.-A. 584
Timmons, D.E. 66
Tiziano 10
Toburen, A.L. 612
Tolkowski, S. 9
Torres-Grifol, J.F. 604
Toyoda, T. 330, 345, 414, 420
Tozzetti, Targioni 586–7
Trallianus, Alexander 578
Trama, L.A. 357, 365, 370, 373
Trozzi, A. 270, 293
TungFeng-So 2
Tuzcu, O. 213
Tyissen, H.A.C. 96
Tzamtzis, N.E. 395

Uchida, K. 266, 267
Udine, Giovanni da 10
Urlaub, R. 98
Usai, M. 270, 299

Valenzuela, R. 54
Valmiki-Ramayana 589
Van de Velde, Jan Jansz 10
Van Gogh, Vincent 15
Van Nisterooij, M. 97
Van Os, D. 365
Van Pelt, W.H.J.M. 96

Van Soden, 376
Van Weelden, G. 97
Vannier, S.H. 355, 357, 361, 362, 370, 500, 501
Vargas, I. 607
Varjas, L. 55
Vasquez, R. 593
Vassallo, Anton Maria 11
Vauquelin, M. 68
Veldhuis, M.K. 207, 219, 244
Venturella, P. 374
Venuto, Antonio 9
Vernin, G. 380, 429, 433, 440
Verrocchio, Andrea 10
Versteig, C. 84
Verzera, A. 83, 270, 293, 319, 407, 487, 506
Viggiani, G. 55
Virgil 5
Vitry, Jacques de 8
Volkammer of Nuremberg 593
Von Soden, H. 376
Vovlas, N. 51

Wajasaney Samhita (Brahma texts) 2
Wang, D.-J. 328
Wang, X. 478, 479
Watanabe, I. 334
Watkins, W.T. 409, 411
Wehrmahn 578
Weinreich, B. 487, 504
Weissman, S. 78
Wen, M. 297, 302, 439
Wenninger, J.A. 219, 273
Werkhoff, P. 478, 479
Wicker, L. 84
Wiegand, B. 93
Will, R.T. 66, 69
Wilson 609
Wilson, A.A. 290, 292
Wilson, C.P. 68
Wilson, C.W. 222, 223, 224, 228, 248, 255, 331
Wölcker 373
Wolstromer, R.J. 286

Yajima, I. 344, 347, 605
Yakamochi, Otomo 3
Yamada, T. 292
Yang, H.J. 419
Yang, R.H. 238, 334
Yang, S. 216, 302
Yang, X. 550
Yeh, P.-H. 327
Yokoyama, F. 218
Yoshida, J. 273
Yukawa, C. 334

Zalkow, L.H. 219
Zani, F. 275
Zheng, H.-J. 327
Zhu, L.-F. 216, 327, 339, 342
Ziegler, E. 207, 219, 230, 244, 273
Ziegler, H. 358, 373, 381
Zukerman, I. 603
Zurbarán, Francisco de 10

Subject index

acidless orange; varieties 19; *see also* blood orange; sour orange; sweet orange; trifoliate orange; yuzo
adulteration; background 496–7; bergamot oil 514; bitter orange oil 506–7; chromatographic techniques 500–5; detection methods 497–514; early use 497; grapefruit oil 508–9; lemon oil 509–12; lime oil 512–14; mandarin oil 507–8; non-volatile constituents 497–500; sweet orange oil 505–6
Alfa Laval 94
American Machinery Corporation (AMC) 124
American Society of Mechanical Engineers 66
analysis techniques 179–97; absorption 548–51; chromatographic 552; distillation/distillation-solvent extraction 546; fast GC/fast GC-MS 191–5; interactive use of MS data/linear retention indices 195–7; multidimensional gas chromatography (MDGC) 185–8; olfactometric 552–5; on-line HPLC-HRGC and HPLC-HRGC/MS 180–5; sample preparation 545–52; sensorial 555; stable isotope ratio mass spectrometry 188–91, 505; supercritical fluid extraction (SFE) 548; vacuum headspace method (VHS) 551–2; *see also* chromatography
APV Baker Ltd 94, 95
Argentina 58
Australia 58, 60
Automatic Machinery and Electronic Inc.; Birillatrici extractors 80; Brown extractors 79–80
Avena extractor 122–3

Bearss lime *see* Persian lime
bergamot (*C. bergamia*); names for 593–4; petitgrain oil 452–3; traditional medicine 593–5
bergamot cold-pressed peel oil; adulteration/detection methods 514; Calabrian machine extractor 64; characters 560; chiral compounds 484–9; furanocoumarins free 144; oxygen heterocyclic compounds 376–80; pesticides 523–4, 525–6; uses 540; volatile fraction composition 272–84
biogas 175
biological activity; antimicrobial 602–8; antitumor 611–16; auraptene antiplatelet properties of 619; background 602; bergamot and 5-methoxypsoralen (5-MOP) 608–11; bergamottine 618–19; bioavailability 621–2; d-limonene as enhancer of percutaneous absorption 616–18; toxicology 619–21
Bireley's Inc. 67
bitter orange (*C. aurantium*) 148–9; extracts from flowers 413–23, flowers distillation 150, flowers extraction by volatile solvents 150; flowers harvest, transport, storage 149; leaves harvest 151; leaves steam distillation 151; names for 7; petitgrain oil 426–32

bitter orange cold-pressed peel oil; adulteration/detection methods 506–7; characters 560; chiral compounds 478–82; oxygen heterocyclic compounds 365–9; uses 541; volatile fraction composition 206–17
blood orange; varieties 19, 20; *see also* acidless orange; sour orange; sweet orange; trifoliate orange; yuzo
Brazil 12, 19, 58, 60
brined citrus 69
brined peel 171–2
Brown International Corporation 65, 118–19, 127
by-products 159; brined/candied peel 171–2; citrus sections 174; comminuted 172–3; d-limonene 159–66; dry peel 159–66; fermentation products 174–5; flavonoids 169–70; juice cells/standard pulp 173–4; molasses 159–66; naringin 170–1; pectin 166–9

C. aurantifolia (Christm.) Swing. (key lime) 1, 19, 21
C. aurantium L. (bitter orange) 1, 2, 19, 22
C. aurantium var. *myrtifolia* (myrtle-leafed orange/chinotto) 22
C. bergamia (bergamot) 22
C. celebica 19
C. clementina (clementine) 21
C. combara 19
C. deliciosa Ten. (Mediterranean mandarin) 21
C. excelsa 19
C. grandis/maxima (L.) Osb. (pummelo) 1, 2, 19, 22
C. halimii 1
C. hystrix (papeda) 19
C. ichangensis 19
C. indica 19
C. jambhiri (rough lemon) 3
C. junos (yuzu) 2, 19
C. kerrii 19
C. latifolia Tan. (Persian/Bearss lime) 21
C. latipes 19
C. limon (L.) Burm. (lemon) 1, 19, 21
C. limonimedica (citron/orange hybrid) 3
C. macrophylla 19
C. macroptera 19
C. medica L. (citron) 23
C. medica var. *sarcodactylis* (fingered citron) 23, 25
C. micrantha 19
C. nobilis Lour. (mandarin) 21
C. paradisi Macf. (grapefruit) 1, 19, 21
C. reticulata Blanco (mandarin) 1, 19, 21
C. sinensis Osb. (sweet/common orange) 1, 2, 3, 19
C. tachibana 19
C. tangerine Hort. ex Tan. (tangerine) 21
C. temple Hort. (mandarin) 21
C. unshiu (Mark.) Marc. (satsuma) 20
C. wilsonii 19
calamansi *see* calamondin/calamansi
calamondin/calamansi (*C. madurensis*) 335–6; juice oil 338; leaf oil 337; peel oil 336–7

Subject index 639

California Fruit Growers' Exchange 68
Camera Agrumaria 588
candied peel 171–2
Chin (*Tsin*) dynasty 2
China 1–2
chinotto *see* myrtle-leafed orange
chiral compounds; background 461–3; bergamot oil 484–9; bitter orange oil 478–82; clementine oil 477; lemon oil 463–70; lime oils 482–4; mandarin oil 470–6; neroli oil 489–92; petitgrain oil 489–92; sweet orange oil 476–8
Chou dynasty 2
chromatography 552; gas liquid (GLC) 502–5; high performance liquid (HPLC) 502; multidimensional gas (MDGC) 185–8; terpeneless/sesquiterpeneless oils 395–6; thin layer (TLC) 500–2; *see also* analysis techniques
Ch'u (citrus tree) 3
citrange 15
citrangequat 15
citrangor 15
citremon 15
citric acid 67–8
Citromat 67
citron (*C. medica*) 1, 2, 10, 23; cultivation of 8; flower oil 339; leaf oil 339; names for 3; peel oil 338–9
citron etrog (*C. medica* var. *ethrog*) 339; leaf oil 341–2; peel oil 340
citron/orange hybrid (*C. limonimedica*) 3
citrumelos 15
citrus; artistic depictions of 5, 10–11; by-products 69; culinary uses 5, 8; cultivation of 8, 10, 13–14; as folk remedy in Mediterranean countries 595–6; from antiquity to first millenium 1–5; from middle ages to nineteenth century 6–15; fruits 19–26; genus/subgenus 17–19; industry 63–9; morphology 26–31; names for 2–8; new species 14–15; origins 1; physiology 31–4; processing flowsheet 71–6; sections 174; as source of vitamin C 10; taxonomy 16–19; trees 3
clementine (*C. clementina*) 15, 20–1, 319; 'Clemenules' 21; 'Fina' 21; leaf oil 319; 'Monreal' 21; peel oil 319
clouding extracts *see* soluble solids/clouding extracts
combava *see* papeda
comminuted 172–3
common orange *see* sweet orange
contaminants; background 518; pesticides 518–27; plasticizers 527–30
cosmetics *see* perfume/cosmetics

Daubron system 96
d-limonene 163–4
diseases 55–6; blast and black pit 57; graft-transmissible pathogens 57, 58–60; Mal secco 56; Phytophthora 56–7; postharvest fungal attacks 57, 58t
distilled oil composition; chemical transformations 409–11; key/Persian limes 402–6; lemon, mandarin, grapefruit 402–3, 407–8
distilled peel oil; historical background 153–4; lime 154; peratoner 154; technology 154–8
Dominica 64
dry peel 159–63

enzyme treatment 141–2
essential oil 63–6; adulteration of 496–514; alcohol soluble citrus flavour 140–1; analyses 179–97, 545–55; chiral compounds of 461–92; cold-press extraction 116–32; coloured orange 143; contaminants 418–30; determination of content 115–16; distribution/location 114–15; dried flavourings 144–5; extraction 72–6, 79–81; factors which influence quality 136–8; furanocoumarins free bergamot 144; methods of separation 132–4; perfumery/cosmetics 557–67; preservation/storage 134–6; recovery by distillation 138–40; special technologies 141–4; stability of 542–5; treatment of de-pulped peel 124–30; uses of 539–41
etrog citron (*C. medica* var. *ethrog*) *see* citron etrog
European Community (EC) 99
European Union (EU) 165, 568–9, 572, 573–4
evaporation 91–2; cryogenic evaporators with heat pump 95; other types 95–6; plates evaporators 94–5; tube bundle evaporators with falling film 92–3; tube bundle evaporators with thermal recompression 93–4
extraction 72–6, 78; cold-press 116–32; essential oil before juice 80–1; finishing operations 81–2; juice before essential oil 79–80; press method 129–30; separating oil/juice simultaneously 79; supercritical solvents 142–3; volatile solvents 150

feed mill 160; drying 162–3; lime treatment 160–1; milling 160; pelletizing 163; pressing 161–2; superheated steam 163; waste heat evaporator 162
fermentation products 174–5
fertilization *see* nutrition/fertilization
fingered citron (*C. medica* var. *sarcodactylis*) 2, 23, 25; leaf oil 343–4; peel oil 342–3
flavonoids 68–9, 169–70
Florida Department of Citrus 102
flower oils 328–30, 339, 345–7 *see also* neroli oils
Food and Drug Administration (FDA) 165
Food Machinery Corporation (FMC) 65, 67, 79, 82, 88, 92, 104, 108, 130–2, 165, 173
forbidden fruit (grapefruit) 12
Fortunella spp (kumquat) 1, 2, 16, *17*
Fraser-Brace Engineering Company 123–4
Fratelli Indelicato 65, 67
frozen concentrated orange juice (FCOJ) 165
fruit; harvesting 71; preliminary treatment 71–2; processing 72–6 and tables; sizing 71–2
Fukien (China) 1

Gasquet 96
Gea Wiegand 93
Genoa 9
graft-transmissible pathogens 57, 58t; blight 60; exocortis 59; ringspot-psorosis complex 59; stubborn 60; Tristeza 58–9
grapefruit (*C. paradisi*) 1, 12, 21; distilled oil composition 402–3, 408; hybrids 25; petitgrain oil 453; varieties 21, *23*
grapefruit cold-pressed peel oil; adulteration/detection methods 508–9; characters 559; oxygen heterocyclic compounds 369–71; uses 540; volatile fraction composition 217–28

Greece 4–5, 15
Grenco Process Technology B.V. 96
Gulf Machinery 95

Han dynasty 2
Han Wu Ti kingdom 2
hand of Buddha (*Fingered citron*) 2
hazard analysis critical control point (HACCP) 87
herbicides 37–40
hesperidin 170
Hesperidium (citrus fruit) 4
hybrids 3, 15, 23, 24; intergeneric 25–6; mandarins 24; other 25; tangelos 25
Hyland-Stanford Corporation 65

innovative technologies; debittering 105–7; high electric field pulses technology 104; high hydrostatic pressure (pascalization) 103–4; membrane concentration 107–8; non-thermal processes 102–3; non-traditional pasteurization 101; non-traditional thermal processes 101–2; ultra high pressure technology 104
integrated pest management (IPM) 54, 55
intergeneric hybrids 25; back crosses 26; bigeneric 26; trigeneric 26
irrigation 43–4
Israel 14, 58, 64, 65
Italy 8, 9–11, 13–14, 19, 64–5, 67

Japan 3, 103
jelly grade (JG) 166
juice; cells/standard pulp 173–4; contamination of 78; extraction of 66–7, 72–6; oils 330–1, 335, 338, 347–8; production of 66–67
juice concentration 90–1; evaporation 91–6; freeze-concentration 96–7
juice technology; background 77–8; clear 97–100; concentration 90–7; de-aeration/de-oiling 82–3; dehydrated 100–1; extraction 78–82; innovating processes 101–8; not-from-concentrate juices 86–90; pasteurization process 83–6; single strength from concentrate 90; storage/shipment 108–10

key lime (*C. aurantifolia*) *see* lime;
Krupp 104
kumquat (*Fortunella spp*) 1, 2, 16, 17
Kuo Han 2

leaf oils 319, 324–6, 328, 330, 332, 334–5, 337, 339, 341–2, 343–4, 345; *see also* petitgrain oils
legislation; background 568; chemical substances 573–5; codex alimentarius 570; flavourings 568–70; fragrances 570–3; IFRA code/guidelines 571–2
lemon (*C. limon*) 1, 2–3, 10, 21; artistic depictions of 5; cultivation of 8; distilled oil composition 402–3, 407; names for 6–7; petitgrain oil 432–9; varieties 21
lemon cold pressed peel oil; adulteration/detection methods 509–12; characters 559; chiral compounds 463–70; oxygen heterocyclic compounds 357–61; pesticides 519, 523, 525; uses 540; volatile fraction composition 284–304
lime (*C. aurantifolia* and *C. latifolia*) 1, 2, 3, 5, 21; artistic depictions of 5; cultivation of 8; distillation of 154–8; distilled oil composition 402, 404–6; names for 7; petitgrain oil 453–7; uses 540
lime cold-pressed peel oils; adulteration/detection methods 512–14; characters 559; chiral compounds 482–4; oxygen heterocyclic compounds 361–5; uses 541; volatile fraction composition 228–41
mandarin (*C. reticulata*) 1; confusion over 2; cultivation of 15; distilled oil composition 402–3, 407–8; hybrids 24, 25; Mediterranean 21; petitgrain oil 440–3; Satsuma group 20; tangerine group 20–1; varieties 2–3, 15, 20–1
mandarin cold-pressed peel oil; adulteration/detection methods 507–8; chiral compounds 470–6; oxygen heterocyclic compounds 371–6; pesticides 523, 525; uses 541; volatile fraction composition 241–51, 256
marmalade 2, 12, 14
Mazzoni SpA 95
Media 3–4
medicine; African traditions 590–1; Asiatic traditions 589–90; background 577–80; bergamot 593–5; folk remedies in Mediterranean countries 595–6; New World traditions 591–3; old pharmacopeoias 580–4; pharmacology treatises/therapy from eighteenth to twentieth centuries 585–8; seventeenth/eighteenth century chemist's shops 584–5
Mediterranean mandarin 21
Mexico 12, 14, 54, 99
mikan *see* satsuma
Minneola (hybrid) 15
Mitsubishi Heavy Industry 103
molasses 159–63
morphology; flowers 28–9; fruit 29–31, *30*, 34t; leaves *18*, 27–8; root system 26–7; seeds 31; trunk/branches 27
myrtle-leafed orange (*C. aurantium* var. *myrtifolia*) 318–19; leaf oil 319; peel oil 319

Nara period 3
naringin 170–1
navel orange 19, *20*
neroli oil 148–51; adulteration of 420–3; characters 560; chiral compounds 489–92; composition 413–18
not-from-concentrate (NFC) juices 77, 86–90, 165
nutrition/fertilization; calcium 42; fertirrigation 43; leaf analysis 41; manure 42; mineral intake 40–3; phosphorous 42; potassium 42; soil analysis 41; timing 42

oils *see* essential oil
orange *see* acidless orange; bitter orange; blood orange; sweet orange; trifoliate orange; yuzo
oxygen heterocyclic compounds 384–6; background 355–7; bergamot 376–80; biogenesis 381–3; bitter orange 365–9; chemotaxonomic considerations 383–4; distribution in peel oils 357–80; grapefruit 369–71; isolation artefacts 380–1; lemon 357–61; lime 361–5; mandarin 371–6; sweet orange 371–6

Palestine 4, 8, 67
papeda 17, **18**–19
papeda (*C. hystrix*) 17; characteristics 18–19; juice oil 330–1; leaf oil 330; peel oil 330; subsections 19
pasteurization process 83–5; traditional thermal 85–6
pectin 68, 166; fibre 169; production 166–9
peel oils 319, 320–3, 327, 330, 331, 333–4, 336–7, 338–9, 340, 342–3, 344–5; *see also* volatile fraction of cold-pressed citrus peel oil
Pelatrice Moscato 123
Pelatrice speciale 120–2
Peratoner centrifugation process 66
Peratoner distillation process 158
perfume/cosmetics; *Allure* by Chanel 562; *Arpege* by Lanvin 561; bergamot 560; bitter orange flower absolute 560; bitter orange flower water absolute 560; bitter orange leaf water absolute 560; bitter orange petitgrain absolute 560; bitter orange petitgrain oil 560; *Chanel 5* 561; *Charlie* by Revlon 562; *CK One* by Calvin Klein 563; *Contradiction* by Calvin Klein 562; *Cool Water* by Davidoff 562, 564; *Cristalle* by Chanel 562; *Diorissimo* by Dior 562; *Drakkar Noir* 564; *Eau de Cologne* by Hermes 562; *Eau de Guerlain* 562; *Eau Sauvage* by Dior 562; *English Leather* by Mem Corporation 562; *Escape for Men* by Calvin Klein 564; *Eternity for Men* by Calvin Klein 563–4; female 561–2; future products 566–7; *Giorgio* by Giorgio Beverly Hills 562; grapefruit oil 559; *Hugo for Men* by Hugo Boss 564; *L'air du Temps* by Nin Ricci 562; *Lauren* by Cosmair 562; lemon oil 559; lime oil 559; major citrus oil components 557–9; male 562–5; *Muguet de Bois* by Coty 561; neroli 560; *O de Lancôme* 562; orange 559; other petitgrain oils 560–1; Paco Rabanne 565; *Polo Sport* by Ralph Lauren 562, 564; synthetic citrus notes 565–6; tangerine/mandarin oil 559; *Tommy Hillfiger* 564; use of citrus in 596; *Vent Vert* by Balmain 561
Persia 3–4
Persian lime (*C. latifolia*); *see* lime
pesticides; organochlorine 524–7; organophosphorus 518–24
pests 49; ants 51; aphids 50t, 51; armored scales 50, 51; autocidal control 54; beetles 50, 51; biological control 52–3; bugs 50; chemical control 54–5; cultural control 53; fruitflies 50, 51; integrated pest management (IPM) 55; management techniques 52; margarodid scale 50; mealybugs 50; mechanical/physical control 53–4; mites 49, 50; moths 51; nematodes 51–2; soft scales 50, 51; thrips 50; whiteflies 50, 51
petitgrain oil 148–51; background 425–6; bergamot 452–3; bitter orange 426–32; characters 560; chiral compounds 489–92; grapefruit 453; key lime 453–7; lemon 432–9; mandarin 440–3; Persian lime 457; sweet orange 446–52; tangerine 443–6
physiology; flowering 31–2; fruit growth 32; fruit maturation 33; germination 31; growth 31; pollination, fertilisation, fruit sets 32; senescence 33

Pipkin Peel Oil Press 65
plasticizers 527–30
Polycitrus 67
Polycitrus extractor 119–20
Pompei 5, 6
Poncirus (trifoliate orange) 17
production *see* world production
pruning 45–6, 47, 48
pummelo (*C. grandis/maxima*) 1, 8, 10, 14, 19, 22; flower oil 328–30; leaf oil 328; names for 7; peel oil 327; varieties of 2
Pure Pulse Technologies Inc 104

Renaissance 8, 9–12
reverse osmosis (RO) 143
Rome 5
Rotary Juice Press 67

rough lemon (*C. jambhiri*) 3; leaf oil 332–3; peel oil 331

satsuma (*C. unshiu*) 3; flower oil 345–7; juice oil 347–8; leaf oil 345; peel oil 344–5; varieties 20
Scheele process 67
Schmidt Bretten GmbH 94
scurvy 10
SeparaSystem 108
sesquiterpeneless oils *see* terpeneless/sesquiterpeneless oils; Shaddock *see* pummelo
Sicily 4, 5, 9, 14, 65, 66, 67–8
slow-folding machines 127
soil; care of 37–40; choice of 36–7
soluble solids/clouding extracts; core-wash 165; pulp wash-wesos 164–5; second pressure extracts 165–6
sour orange (*C. aurantium*) 1, 2, 3, 5, 7, 8, 318; Bergamot 22; Bouquet de Fleurs 22; Chinotto 22; cultivation of 8; Granito/Abers 22; names for 7; standard 22; substitutes 15; *see also* acidless orange; bitter orange; blood orange; sweet orange; trifoliate orange; yuzu
Spain 13–14, 58, 65
sterile insect technique (SIT) 54
Struthers 96
sweet orange (*C. sinensis*) 1, 2, 3, 5, 10, 19; introduction into Europe 8–9; names for 9; petitgrain oil 446–52; varieties 19; *see also* acidless orange; blood orange; sour orange; trifoliate orange; yuzo
sweet orange cold-pressed peel oil; adulteration/detection methods 505–6; chiral compounds 476–78, 479; oxygen heterocyclic compounds 371–6; pesticides 523, 525; uses 541
volatile fraction composition 256–71
Szechaw (China) 1

tangelo (grapefruit/mandarin hybrid) 15; varieties 25
tangerine (*C. tangerine*) 15, 20–1; petitgrain oil 443–6; prototype 15; variety 21
tangerine cold-pressed peel oil; characters 559; oxygen heterocyclic compounds 374; uses 541; volatile fraction composition 241, 251–6

taxonomy 16; Aurantioideae 17;
  Balsamocitrinae 16; Citreae 16; Clausenae 16;
  genus 16–19; Rutaceae 16; Triphasilinae 16
terpeneless/sesquiterpeneless oils 400–1;
  background 391; carbon dioxide 399–400;
  chromatography 395–6; counter current 396,
  398; folding 394–5; processes 391–6; PTFE
  packed column 396–400; washing 392–4
Thailand 1
toxicology; bergapten and ovarian function 621;
  citral 620–1; contact dermatitis 619–20;
  nephrotoxicity of d-limonene 620
trifoliate orange (*Poncirus*) 17; *see also* acidless
  orange; blood orange; sour orange;
  sweet orange; yuzo
Tunisia 5
Tusculum 5

ultrafiltration (UF) 143
Union Carbide 96
United States of America 12–13, 19, 58, 65–6,
  68, 569–70, 572, 574–5
Unshiu Mikan (*C. unshiu*) *see* satsuma

Villa del Casale (Sicily) 4
volatile fraction of cold-pressed citrus peel oil;
  background 201–6; bergamot 272–84; bitter
  orange o 206–17; grapefruit 217–28; lemon
  284–304; lime 228–41; mandarin 241–51, 256;
  sweet orange 256–71; tangerine 241, 251–6
Votator systems 96

Wakayama Nokyo Food Industry 103
Wakayama Prefectural Agricultural Processing 103
weed control 37–40
West Indies 12, 64, 67
white orange *see* sweet orange
world production 532–3; essential oils 534–5;
  negotiation/market prices 536–7;
  processing 533–4

Yellow Kan (Hung Kan) 2
yuzu (*C. junos*) 2, 19; juice oil 335; leaf oil 334–5;
  peel oil 333–4; *see also* acidless orange; sour orange;
  sweet orange; trifoliate orange

zamboa (pummelo) 7